# Student Solutions Manual

# Calculus of a Single Variable
## Early Transcendental Functions

### SIXTH EDITION

**Ron Larson**
The Pennsylvania State University,
The Behrend College

**Bruce Edwards**
University of Florida

Prepared by

**Bruce Edwards**
University of Florida

CENGAGE
Learning·

Australia • Brazil • Mexico • Singapore • United Kingdom • United States

For product information and technology assistance, contact us at **Cengage Learning Customer & Sales Support, 1-800-354-9706**.

For permission to use material from this text or product, submit all requests online at **www.cengage.com/permissions** Further permissions questions can be emailed to **permissionrequest@cengage.com**.

ISBN-13: 978-1-285-77480-0
ISBN-10: 1-285-77480-9

**Cengage Learning**
200 First Stamford Place, 4th Floor
Stamford, CT 06902
USA

Cengage Learning is a leading provider of customized learning solutions with office locations around the globe, including Singapore, the United Kingdom, Australia, Mexico, Brazil, and Japan. Locate your local office at: **www.cengage.com/global**.

Cengage Learning products are represented in Canada by Nelson Education, Ltd.

To learn more about Cengage Learning Solutions, visit **www.cengage.com**.

Purchase any of our products at your local college store or at our preferred online store **www.cengagebrain.com**.

Printed in the United States of America
2 3 4 5 6 7 17 16 15

# CONTENTS

# C H A P T E R  1
# Preparation for Calculus

# C H A P T E R  1
# Preparation for Calculus

## Section 1.1   Graphs and Models

**1.** $y = -\frac{3}{2}x + 3$

$x$-intercept: $(2, 0)$

$y$-intercept: $(0, 3)$

Matches graph (b).

**3.** $y = 3 - x^2$

$x$-intercepts: $\left(\sqrt{3}, 0\right), \left(-\sqrt{3}, 0\right)$

$y$-intercept: $(0, 3)$

Matches graph (a).

**5.** $y = \frac{1}{2}x + 2$

| $x$ | $-4$ | $-2$ | 0 | 2 | 4 |
|-----|------|------|---|---|---|
| $y$ | 0 | 1 | 2 | 3 | 4 |

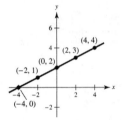

**7.** $y = 4 - x^2$

| $x$ | $-3$ | $-2$ | 0 | 2 | 3 |
|-----|------|------|---|---|---|
| $y$ | $-5$ | 0 | 4 | 0 | $-5$ |

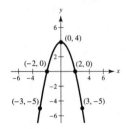

**9.** $y = |x + 2|$

| $x$ | $-5$ | $-4$ | $-3$ | $-2$ | $-1$ | 0 | 1 |
|-----|------|------|------|------|------|---|---|
| $y$ | 3 | 2 | 1 | 0 | 1 | 2 | 3 |

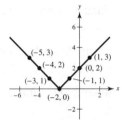

**11.** $y = \sqrt{x} - 6$

| $x$ | 0 | 1 | 4 | 9 | 16 |
|-----|---|---|---|---|----|
| $y$ | $-6$ | $-5$ | $-4$ | $-3$ | $-2$ |

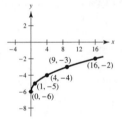

**13.** $y = \dfrac{3}{x}$

| $x$ | $-3$ | $-2$ | $-1$ | 0 | 1 | 2 | 3 |
|-----|------|------|------|---|---|---|---|
| $y$ | $-1$ | $-\frac{3}{2}$ | $-3$ | Undef. | 3 | $\frac{3}{2}$ | 1 |

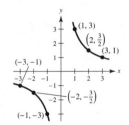

**15.** $y = \sqrt{5 - x}$

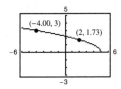

(a) $(2, y) = (2, 1.73)$   $\left(y = \sqrt{5 - 2} = \sqrt{3} \approx 1.73\right)$

(b) $(x, 3) = (-4, 3)$   $\left(3 = \sqrt{5 - (-4)}\right)$

**17.** $y = 2x - 5$

$y$-intercept: $y = 2(0) - 5 = -5;\ (0, -5)$

$x$-intercept: $0 = 2x - 5$

$\qquad\qquad 5 = 2x$

$\qquad\qquad x = \frac{5}{2};\ \left(\frac{5}{2}, 0\right)$

**19.** $y = x^2 + x - 2$

$y$-intercept: $y = 0^2 + 0 - 2$

$\qquad\qquad y = -2;\ (0, -2)$

$x$-intercepts: $0 = x^2 + x - 2$

$\qquad\qquad 0 = (x + 2)(x - 1)$

$\qquad\qquad x = -2, 1;\ (-2, 0), (1, 0)$

**21.** $y = x\sqrt{16 - x^2}$

$y$-intercept: $y = 0\sqrt{16 - 0^2} = 0;\ (0, 0)$

$x$-intercepts: $0 = x\sqrt{16 - x^2}$

$\qquad\qquad 0 = x\sqrt{(4 - x)(4 + x)}$

$\qquad\qquad x = 0, 4, -4;\ (0, 0), (4, 0), (-4, 0)$

**23.** $y = \dfrac{2 - \sqrt{x}}{5x + 1}$

$y$-intercept: $y = \dfrac{2 - \sqrt{0}}{5(0) + 1} = 2\ ;\ (0, 2)$

$x$-intercept: $0 = \dfrac{2 - \sqrt{x}}{5x + 1}$

$\qquad\qquad 0 = 2 - \sqrt{x}$

$\qquad\qquad x = 4\ ;\ (4, 0)$

**25.** $x^2 y - x^2 + 4y = 0$

$y$-intercept: $0^2(y) - 0^2 + 4y = 0$

$\qquad\qquad y = 0;\ (0, 0)$

$x$-intercept: $x^2(0) - x^2 + 4(0) = 0$

$\qquad\qquad x = 0;\ (0, 0)$

**27.** Symmetric with respect to the $y$-axis because

$$y = (-x)^2 - 6 = x^2 - 6.$$

**29.** Symmetric with respect to the $x$-axis because

$$(-y)^2 = y^2 = x^3 - 8x.$$

**31.** Symmetric with respect to the origin because

$$(-x)(-y) = xy = 4.$$

**33.** $y = 4 - \sqrt{x + 3}$

No symmetry with respect to either axis or the origin.

**35.** Symmetric with respect to the origin because

$$-y = \frac{-x}{(-x)^2 + 1}$$

$$y = \frac{x}{x^2 + 1}.$$

**37.** $y = \left|x^3 + x\right|$ is symmetric with respect to the $y$-axis

because $y = \left|(-x)^3 + (-x)\right| = \left|-(x^3 + x)\right| = \left|x^3 + x\right|.$

**39.** $y = 2 - 3x$

$y = 2 - 3(0) = 2,\ y$-intercept

$0 = 2 - 3(x) \Rightarrow 3x = 2 \Rightarrow x = \frac{2}{3},\ x$-intercept

Intercepts: $(0, 2), \left(\frac{2}{3}, 0\right)$

Symmetry: none

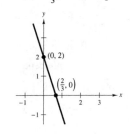

**41.** $y = 9 - x^2$

$y = 9 - (0)^2 = 9,\ y$-intercept

$0 = 9 - x^2 \Rightarrow x^2 = 9 \Rightarrow x = \pm 3,\ x$-intercepts

Intercepts: $(0, 9), (3, 0), (-3, 0)$

$y = 9 - (-x)^2 = 9 - x^2$

Symmetry: $y$-axis

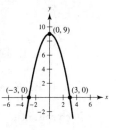

**43.** $y = x^3 + 2$

$y = 0^3 + 2 = 2$, $y$-intercept

$0 = x^3 + 2 \Rightarrow x^3 = -2 \Rightarrow x = -\sqrt[3]{2}$, $x$-intercept

Intercepts: $\left(-\sqrt[3]{2},\ 0\right),\ (0,\ 2)$

Symmetry: none

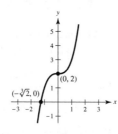

**45.** $y = x\sqrt{x + 5}$

$y = 0\sqrt{0 + 5} = 0$, $y$-intercept

$x\sqrt{x + 5} = 0 \Rightarrow x = 0, -5$, $x$-intercepts

Intercepts: $(0,\ 0),\ (-5,\ 0)$

Symmetry: none

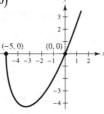

**47.** $x = y^3$

$y^3 = 0 \Rightarrow y = 0$, $y$-intercept

$x = 0$, $x$-intercept

Intercept: $(0,\ 0)$

$-x = \left(-y\right)^3 \Rightarrow -x = -y^3$

Symmetry: origin

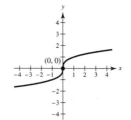

**49.** $y = \dfrac{8}{x}$

$y = \dfrac{8}{0} \Rightarrow$ Undefined $\Rightarrow$ no $y$-intercept

$\dfrac{8}{x} = 0 \Rightarrow$ No solution $\Rightarrow$ no $x$-intercept

Intercepts: none

$-y = \dfrac{8}{-x} \Rightarrow y = \dfrac{8}{x}$

Symmetry: origin

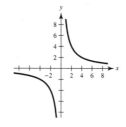

**51.** $y = 6 - |x|$

$y = 6 - |0| = 6$, $y$-intercept

$6 - |x| = 0$

$6 = |x|$

$x = \pm 6$, $x$-intercepts

Intercepts: $(0,\ 6),\ (-6,\ 0),\ (6,\ 0)$

$y = 6 - |-x| = 6 - |x|$

Symmetry: $y$-axis

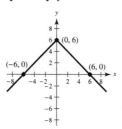

**53.** $y^2 - x = 9$

$y^2 = x + 9$

$y = \pm\sqrt{x + 9}$

$y = \pm\sqrt{0 + 9} = \pm\sqrt{9} = \pm 3$, $y$-intercepts

$\pm\sqrt{x + 9} = 0$

$x + 9 = 0$

$x = -9$, $x$-intercept

Intercepts: $(0,\ 3),\ (0,\ -3),\ (-9,\ 0)$

$\left(-y\right)^2 - x = 9 \Rightarrow y^2 - x = 9$

Symmetry: $x$-axis

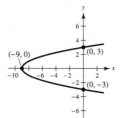

**55.** $x + 3y^2 = 6$

$$3y^2 = 6 - x$$

$$y = \pm\sqrt{\frac{6 - x}{3}}$$

$$y = \pm\sqrt{\frac{6 - 0}{3}} = \pm\sqrt{2}, \ y\text{-intercepts}$$

$x + 3(0)^2 = 6$

$$x = 6, \ x\text{-intercept}$$

Intercepts: $(6, 0), \left(0, \sqrt{2}\right), \left(0, -\sqrt{2}\right)$

$x + 3(-y)^2 = 6 \Rightarrow x + 3y^2 = 6$

Symmetry: $x$-axis

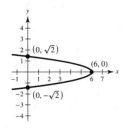

**57.** $x + y = 8 \Rightarrow y = 8 - x$

$4x - y = 7 \Rightarrow y = 4x - 7$

$8 - x = 4x - 7$

$$15 = 5x$$

$$3 = x$$

The corresponding $y$-value is $y = 5$.

Point of intersection: $(3, 5)$

**59.** $x^2 + y = 6 \Rightarrow y = 6 - x^2$

$x + y = 4 \Rightarrow y = 4 - x$

$6 - x^2 = 4 - x$

$$0 = x^2 - x - 2$$

$$0 = (x - 2)(x + 1)$$

$$x = 2, -1$$

The corresponding y-values are $y = 2$ (for $x = 2$) and $y = 5$ (for $x = -1$).

Points of intersection: $(2, 2), (-1, 5)$

**61.** $x^2 + y^2 = 5 \Rightarrow y^2 = 5 - x^2$

$x - y = 1 \Rightarrow y = x - 1$

$5 - x^2 = (x - 1)^2$

$5 - x^2 = x^2 - 2x + 1$

$0 = 2x^2 - 2x - 4 = 2(x + 1)(x - 2)$

$$x = -1 \text{ or } x = 2$$

The corresponding $y$-values are $y = -2$ (for $x = -1$) and $y = 1$ (for $x = 2$).

Points of intersection: $(-1, -2), (2, 1)$

**63.** $y = x^3 - 2x^2 + x - 1$

$y = -x^2 + 3x - 1$

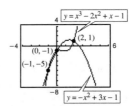

Points of intersection: $(-1, -5), (0, -1), (2, 1)$

Analytically, $x^3 - 2x^2 + x - 1 = -x^2 + 3x - 1$

$$x^3 - x^2 - 2x = 0$$

$$x(x - 2)(x + 1) = 0$$

$$x = -1, 0, 2.$$

**65.** $y = \sqrt{x + 6}$

$y = \sqrt{-x^2 - 4x}$

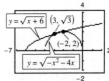

Points of intersection: $(-2, 2), \left(-3, \sqrt{3}\right) \approx (-3, 1.732)$

Analytically,   $\sqrt{x + 6} = \sqrt{-x^2 - 4x}$

$$x + 6 = -x^2 - 4x$$

$$x^2 + 5x + 6 = 0$$

$$(x + 3)(x + 2) = 0$$

$$x = -3, -2.$$

**67.** (a) Using a graphing utility, you obtain
$y = 0.005t^2 + 0.27t + 2.7.$

(b)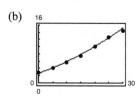

(c) For 2020, $t = 40$.

$$y = 0.005(40)^2 + 0.27(40) + 2.7$$
$$= 21.5$$

The GDP in 2020 will be \$21.5 trillion.

**69.**
$$C = R$$
$$2.04x + 5600 = 3.29x$$
$$5600 = 3.29x - 2.04x$$
$$5600 = 1.25x$$
$$x = \frac{5600}{1.25} = 4480$$

To break even, 4480 units must be sold.

**71.** $y = kx^3$

(a) $(1, 4)$:    $4 = k(1)^3 \Rightarrow k = 4$

(b) $(-2, 1)$:    $1 = k(-2)^3 = -8k \Rightarrow k = -\frac{1}{8}$

(c) $(0, 0)$:    $0 = k(0)^3 \Rightarrow k$ can be any real number.

(d) $(-1, -1)$:  $-1 = k(-1)^3 = -k \Rightarrow k = 1$

**73.** Answers may vary. *Sample answer*:

$y = (x + 4)(x - 3)(x - 8)$ has intercepts at
$x = -4$, $x = 3$, and $x = 8$.

**75.** (a) If $(x, y)$ is on the graph, then so is $(-x, y)$ by $y$-axis symmetry. Because $(-x, y)$ is on the graph, then so is $(-x, -y)$ by $x$-axis symmetry. So, the graph is symmetric with respect to the origin. The converse is not true. For example, $y = x^3$ has origin symmetry but is not symmetric with respect to either the $x$-axis or the $y$-axis.

(b) Assume that the graph has $x$-axis and origin symmetry. If $(x, y)$ is on the graph, so is $(x, -y)$ by $x$-axis symmetry. Because $(x, -y)$ is on the graph, then so is $(-x, -(-y)) = (-x, y)$ by origin symmetry. Therefore, the graph is symmetric with respect to the $y$-axis. The argument is similar for $y$-axis and origin symmetry.

**77.** False. $x$-axis symmetry means that if $(-4, -5)$ is on the graph, then $(-4, 5)$ is also on the graph. For example, $(4, -5)$ is not on the graph of $x = y^2 - 29$, whereas $(-4, -5)$ is on the graph.

**79.** True. The $x$-intercepts are $\left( \dfrac{-b \pm \sqrt{b^2 - 4ac}}{2a}, 0 \right)$.

# Section 1.2    Linear Models and Rates of Change

**1.** $m = 2$

**3.** $m = -1$

**5.** $m = \dfrac{2 - (-4)}{5 - 3} = \dfrac{6}{2} = 3$

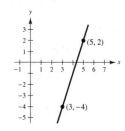

**7.** $m = \dfrac{1 - 6}{4 - 4} = \dfrac{-5}{0}$, undefined.

The line is vertical.

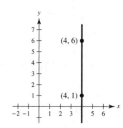

**9.** $m = \dfrac{\dfrac{2}{3} - \dfrac{1}{6}}{-\dfrac{1}{2} - \left(-\dfrac{3}{4}\right)} = \dfrac{\dfrac{1}{2}}{\dfrac{1}{4}} = 2$

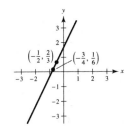

**11.**

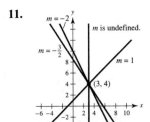

**13.** Because the slope is 0, the line is horizontal and its equation is $y = 2$. Therefore, three additional points are $(0, 2)$, $(1, 2)$, $(5, 2)$.

**15.** The equation of this line is

$$y - 7 = -3(x - 1)$$
$$y = -3x + 10.$$

Therefore, three additional points are $(0, 10)$, $(2, 4)$, and $(3, 1)$.

**17.** $y = \frac{3}{4}x + 3$

$$4y = 3x + 12$$
$$0 = 3x - 4y + 12$$

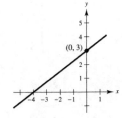

**19.** $y = \frac{2}{3}x$

$$3y = 2x$$
$$0 = 2x - 3y$$

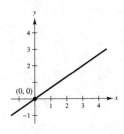

**21.** $y + 2 = 3(x - 3)$

$$y + 2 = 3x - 9$$
$$y = 3x - 11$$
$$0 = 3x - y - 11$$

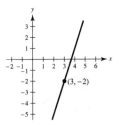

**23.** (a)  Slope $= \dfrac{\Delta y}{\Delta x} = \dfrac{1}{3}$

(b)

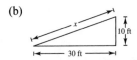

By the Pythagorean Theorem,

$$x^2 = 30^2 + 10^2 = 1000$$
$$x = 10\sqrt{10} \approx 31.623 \text{ feet.}$$

**25.** $y = 4x - 3$

The slope is $m = 4$ and the $y$-intercept is $(0, -3)$.

**27.** $x + 5y = 20$

$$y = -\tfrac{1}{5}x + 4$$

Therefore, the slope is $m = -\frac{1}{5}$ and the $y$-intercept is $(0, 4)$.

**29.** $x = 4$

The line is vertical. Therefore, the slope is undefined and there is no $y$-intercept.

**31.** $y = -3$

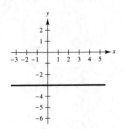

**33.** $y = -2x + 1$

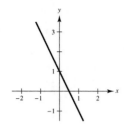

**35.** $y - 2 = \frac{3}{2}(x - 1)$

$y = \frac{3}{2}x + \frac{1}{2}$

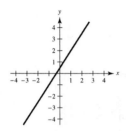

**37.** $2x - y - 3 = 0$

$y = 2x - 3$

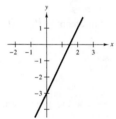

**39.** $m = \dfrac{8 - 0}{4 - 0} = 2$

$y - 0 = 2(x - 0)$

$y = 2x$

$0 = 2x - y$

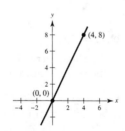

**41.** $m = \dfrac{8 - 0}{2 - 5} = -\dfrac{8}{3}$

$y - 0 = -\dfrac{8}{3}(x - 5)$

$y = -\dfrac{8}{3}x + \dfrac{40}{3}$

$8x + 3y - 40 = 0$

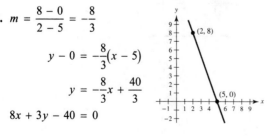

**43.** $m = \dfrac{8 - 3}{6 - 6} = \dfrac{5}{0}$, undefined

The line is horizontal.

$x = 6$

$x - 6 = 0$

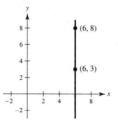

**45.** $m = \dfrac{\dfrac{7}{2} - \dfrac{3}{4}}{\dfrac{1}{2} - 0} = \dfrac{\dfrac{11}{4}}{\dfrac{1}{2}} = \dfrac{11}{2}$

$y - \dfrac{3}{4} = \dfrac{11}{2}(x - 0)$

$y = \dfrac{11}{2}x + \dfrac{3}{4}$

$0 = 22x - 4y + 3$

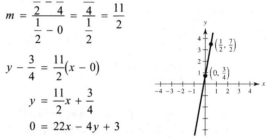

**47.**    $x = 3$

$x - 3 = 0$

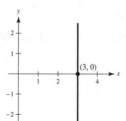

**49.**    $\dfrac{x}{2} + \dfrac{y}{3} = 1$

$3x + 2y - 6 = 0$

**51.** $\dfrac{x}{a} + \dfrac{y}{a} = 1$

$\dfrac{1}{a} + \dfrac{2}{a} = 1$

$\dfrac{3}{a} = 1$

$a = 3 \Rightarrow x + y = 3$

$x + y - 3 = 0$

**53.**  $\dfrac{x}{2a} + \dfrac{y}{a} = 1$

$\dfrac{9}{2a} + \dfrac{-2}{a} = 1$

$\dfrac{9 - 4}{2a} = 1$

$5 = 2a$

$a = \dfrac{5}{2}$

$\dfrac{x}{2\left(\frac{5}{2}\right)} + \dfrac{y}{\left(\frac{5}{2}\right)} = 1$

$\dfrac{x}{5} + \dfrac{2y}{5} = 1$

$x + 2y = 5$

$x + 2y - 5 = 0$

**55.**  The given line is vertical.

(a)  $x = -7$, or $x + 7 = 0$

(b)  $y = -2$, or $y + 2 = 0$

**57.**  $x - y = -2$

$y = x + 2$

$m = 1$

(a)   $y - 5 = 1(x - 2)$

$y - 5 = x - 2$

$x - y + 3 = 0$

(b)   $y - 5 = -1(x - 2)$

$y - 5 = -x + 2$

$x + y - 7 = 0$

**59.**  $4x - 2y = 3$

$y = 2x - \dfrac{3}{2}$

$m = 2$

(a)  $y - 1 = 2(x - 2)$

$y - 1 = 2x - 4$

$0 = 2x - y - 3$

(b)   $y - 1 = -\dfrac{1}{2}(x - 2)$

$2y - 2 = -x + 2$

$x + 2y - 4 = 0$

**61.**  $5x - 3y = 0$

$y = \dfrac{5}{3}x$

$m = \dfrac{5}{3}$

(a)   $y - \dfrac{7}{8} = \dfrac{5}{3}\left(x - \dfrac{3}{4}\right)$

$24y - 21 = 40x - 30$

$0 = 40x - 24y - 9$

(b)   $y - \dfrac{7}{8} = -\dfrac{3}{5}\left(x - \dfrac{3}{4}\right)$

$40y - 35 = -24x + 18$

$24x + 40y - 53 = 0$

**63.**  The slope is 250.

$V = 1850$ when $t = 2$.

$V = 250(t - 2) + 1850$

$\quad = 250t + 1350$

**65.**  The slope is $-1600$.

$V = 17,200$ when $t = 2$.

$V = -1600(t - 2) + 17,200$

$\quad = -1600t + 20,400$

**67.**  $m_1 = \dfrac{1 - 0}{-2 - (-1)} = -1$

$m_2 = \dfrac{-2 - 0}{2 - (-1)} = -\dfrac{2}{3}$

$m_1 \neq m_2$

The points are not collinear.

**69.**  Equations of perpendicular bisectors:

$y - \dfrac{c}{2} = \dfrac{a - b}{c}\left(x - \dfrac{a + b}{2}\right)$

$y - \dfrac{c}{2} = \dfrac{a + b}{-c}\left(x - \dfrac{b - a}{2}\right)$

Setting the right-hand sides of the two equations equal and solving for $x$ yields $x = 0$.

Letting $x = 0$ in either equation gives the point of intersection:

$\left(0, \dfrac{-a^2 + b^2 + c^2}{2c}\right).$

This point lies on the third perpendicular bisector, $x = 0$.

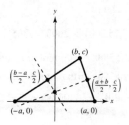

**71.** Equations of altitudes:

$$y = \frac{a - b}{c}(x + a)$$

$$x = b$$

$$y = -\frac{a + b}{c}(x - a)$$

Solving simultaneously, the point of intersection is

$$\left(b, \frac{a^2 - b^2}{c}\right).$$

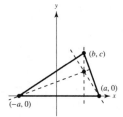

**73.** $ax + by = 4$

(a) The line is parallel to the $x$-axis if $a = 0$ and $b \neq 0$.

(b) The line is parallel to the $y$-axis if $b = 0$ and $a \neq 0$.

(c) Answers will vary. *Sample answer*: $a = -5$ and $b = 8$.

$$-5x + 8y = 4$$

$$y = \tfrac{1}{8}(5x + 4) = \tfrac{5}{8}x + \tfrac{1}{2}$$

(d) The slope must be $-\tfrac{5}{2}$.

Answers will vary. *Sample answer*: $a = 5$ and $b = 2$.

$$5x + 2y = 4$$

$$y = \tfrac{1}{2}(-5x + 4) = -\tfrac{5}{2}x + 2$$

(e) $a = \tfrac{5}{2}$ and $b = 3$.

$$\tfrac{5}{2}x + 3y = 4$$

$$5x + 6y = 8$$

**75.** Find the equation of the line through the points $(0, 32)$ and $(100, 212)$.

$$m = \tfrac{180}{100} = \tfrac{9}{5}$$

$$F - 32 = \tfrac{9}{5}(C - 0)$$

$$F = \tfrac{9}{5}C + 32$$

or

$$C = \tfrac{1}{9}(5F - 160)$$

$$5F - 9C - 160 = 0$$

For $F = 72°$, $C \approx 22.2°$.

**77.** (a) Current job:    $W_1 = 0.07s + 2000$

New job offer: $W_2 = 0.05s + 2300$

(b)

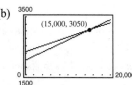

Using a graphing utility, the point of intersection is $(15{,}000, 3050)$.

Analytically, $W_1 = W_2$

$$0.07s + 2000 = 0.05s + 2300$$

$$0.02s = 300$$

$$s = 15{,}000$$

So, $W_1 = W_2 = 0.07(15{,}000) + 2000 = 3050$.

When sales exceed $15{,}000$, the current job pays more.

(b) No, if you can sell $20{,}000$ worth of goods, then $W_1 > W_2$.

(**Note:** $W_1 = 3400$ and $W_2 = 3300$ when $s = 20{,}000$.)

**79. (a)** Two points are (50, 780) and (47, 825).

The slope is

$$m = \frac{825 - 780}{47 - 50} = \frac{45}{-3} = -15.$$

$$p - 780 = -15(x - 50)$$

$$p = -15x + 750 + 780 = -15x + 1530$$

or

$$x = \frac{1}{15}(1530 - p)$$

**(b)**

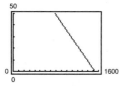

If $p = 855$, then $x = 45$ units.

**(c)** If $p = 795$, then $x = \frac{1}{15}(1530 - 795) = 49$ units

**81.** The tangent line is perpendicular to the line joining the point (5, 12) and the center (0, 0).

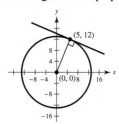

Slope of the line joining (5, 12) and (0, 0) is $\frac{12}{5}$.

The equation of the tangent line is

$$y - 12 = \frac{-5}{12}(x - 5)$$

$$y = \frac{-5}{12}x + \frac{169}{12}$$

$$5x + 12y - 169 = 0.$$

**83.** $x - y - 2 = 0 \Rightarrow d = \dfrac{\left|1(-2) + (-1)(1) - 2\right|}{\sqrt{1^2 + 1^2}}$

$$= \frac{5}{\sqrt{2}} = \frac{5\sqrt{2}}{2}$$

**85.** A point on the line $x + y = 1$ is (0, 1). The distance from the point (0, 1) to $x + y - 5 = 0$ is

$$d = \frac{\left|1(0) + 1(1) - 5\right|}{\sqrt{1^2 + 1^2}} = \frac{\left|1 - 5\right|}{\sqrt{2}} = \frac{4}{\sqrt{2}} = 2\sqrt{2}.$$

**87.** If $A = 0$, then $By + C = 0$ is the horizontal line $y = -C/B$. The distance to $(x_1, y_1)$ is

$$d = \left| y_1 - \left( \frac{-C}{B} \right) \right| = \frac{|By_1 + C|}{|B|} = \frac{|Ax_1 + By_1 + C|}{\sqrt{A^2 + B^2}}.$$

If $B = 0$, then $Ax + C = 0$ is the vertical line $x = -C/A$. The distance to $(x_1, y_1)$ is

$$d = \left| x_1 - \left( \frac{-C}{A} \right) \right| = \frac{|Ax_1 + C|}{|A|} = \frac{|Ax_1 + By_1 + C|}{\sqrt{A^2 + B^2}}.$$

(Note that $A$ and $B$ cannot both be zero.) The slope of the line $Ax + By + C = 0$ is $-A/B$.

The equation of the line through $(x_1, y_1)$ perpendicular to $Ax + By + C = 0$ is:

$$y - y_1 = \frac{B}{A}(x - x_1)$$
$$Ay - Ay_1 = Bx - Bx_1$$
$$Bx_1 - Ay_1 = Bx - Ay$$

The point of intersection of these two lines is:

$$Ax + By = -C \quad\quad \Rightarrow A^2x + ABy = -AC \quad\quad\quad (1)$$
$$Bx - Ay = Bx_1 - Ay_1 \Rightarrow \underline{B^2x - ABy = B^2x_1 - ABy_1} \quad\quad (2)$$
$$(A^2 + B^2)x = -AC + B^2x_1 - ABy_1 \;\; \text{(By adding equations (1) and (2))}$$

$$x = \frac{-AC + B^2x_1 - ABy_1}{A^2 + B^2}$$

$$Ax + By = -C \quad\quad \Rightarrow ABx + B^2y = -BC \quad\quad\quad (3)$$
$$Bx - Ay = Bx_1 - Ay_1 \Rightarrow \underline{-ABx + A^2y = -ABx_1 + A^2y_1} \quad (4)$$
$$(A^2 + B^2)y = -BC - ABx_1 + A^2y_1 \;\; \text{(By adding equations (3) and (4))}$$

$$y = \frac{-BC - ABx_1 + A^2y_1}{A^2 + B^2}$$

$$\left( \frac{-AC + B^2x_1 - ABy_1}{A^2 + B^2}, \frac{-BC - ABx_1 + A^2y_1}{A^2 + B^2} \right) \text{point of intersection}$$

The distance between $(x_1, y_1)$ and this point gives you the distance between $(x_1, y_1)$ and the line $Ax + By + C = 0$.

$$d = \sqrt{\left[ \frac{-AC + B^2x_1 - ABy_1}{A^2 + B^2} - x_1 \right]^2 + \left[ \frac{-BC - ABx_1 + A^2y_1}{A^2 + B^2} - y_1 \right]^2}$$

$$= \sqrt{\left[ \frac{-AC - ABy_1 - A^2x_1}{A^2 + B^2} \right]^2 + \left[ \frac{-BC - ABx_1 - B^2y_1}{A^2 + B^2} \right]^2}$$

$$= \sqrt{\left[ \frac{-A(C + By_1 + Ax_1)}{A^2 + B^2} \right]^2 + \left[ \frac{-B(C + Ax_1 + By_1)}{A^2 + B^2} \right]^2} = \sqrt{\frac{(A^2 + B^2)(C + Ax_1 + By_1)^2}{(A^2 + B^2)^2}} = \frac{|Ax_1 + By_1 + C|}{\sqrt{A^2 + B^2}}$$

**89.** For simplicity, let the vertices of the rhombus be $(0, 0)$, $(a, 0)$, $(b, c)$, and $(a + b, c)$, as shown in the figure.

The slopes of the diagonals are then $m_1 = \dfrac{c}{a + b}$ and

$m_2 = \dfrac{c}{b - a}$. Because the sides of the rhombus are

equal, $a^2 = b^2 + c^2$, and you have

$$m_1 m_2 = \frac{c}{a + b} \cdot \frac{c}{b - a} = \frac{c^2}{b^2 - a^2} = \frac{c^2}{-c^2} = -1.$$

Therefore, the diagonals are perpendicular.

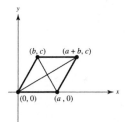

**91.** Consider the figure below in which the four points are collinear. Because the triangles are similar, the result immediately follows.

$$\frac{y_2^* - y_1^*}{x_2^* - x_1^*} = \frac{y_2 - y_1}{x_2 - x_1}$$

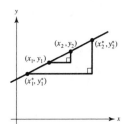

**93.** True.

$$ax + by = c_1 \Rightarrow y = -\frac{a}{b}x + \frac{c_1}{b} \Rightarrow m_1 = -\frac{a}{b}$$

$$bx - ay = c_2 \Rightarrow y = \frac{b}{a}x - \frac{c_2}{a} \Rightarrow m_2 = \frac{b}{a}$$

$$m_2 = -\frac{1}{m_1}$$

**95.** True. The slope must be positive.

# Section 1.3   Functions and Their Graphs

**1.** (a) $f(0) = 7(0) - 4 = -4$

(b) $f(-3) = 7(-3) - 4 = -25$

(c) $f(b) = 7(b) - 4 = 7b - 4$

(d) $f(x - 1) = 7(x - 1) - 4 = 7x - 11$

**3.** (a) $g(0) = 5 - 0^2 = 5$

(b) $g(\sqrt{5}) = 5 - (\sqrt{5})^2 = 5 - 5 = 0$

(c) $g(-2) = 5 - (-2)^2 = 5 - 4 = 1$

(d) $g(t - 1) = 5 - (t - 1)^2 = 5 - (t^2 - 2t + 1)$
$\qquad\qquad = 4 + 2t - t^2$

**5.** (a) $f(0) = \cos(2(0)) = \cos 0 = 1$

(b) $f\left(-\dfrac{\pi}{4}\right) = \cos\left(2\left(-\dfrac{\pi}{4}\right)\right) = \cos\left(-\dfrac{\pi}{2}\right) = 0$

(c) $f\left(\dfrac{\pi}{3}\right) = \cos\left(2\left(\dfrac{\pi}{3}\right)\right) = \cos\dfrac{2\pi}{3} = -\dfrac{1}{2}$

(d) $f(\pi) = \cos(2(\pi)) = 1$

**7.** $\dfrac{f(x + \Delta x) - f(x)}{\Delta x} = \dfrac{(x + \Delta x)^3 - x^3}{\Delta x} = \dfrac{x^3 + 3x^2\Delta x + 3x^2(\Delta x)^2 + (\Delta x)^3 - x^3}{\Delta x} = 3x^2 + 3x\Delta x + (\Delta x)^2,\ \Delta x \neq 0$

**9.** $\dfrac{f(x) - f(2)}{x - 2} = \dfrac{\left(1/\sqrt{x - 1} - 1\right)}{x - 2}$

$\qquad = \dfrac{1 - \sqrt{x - 1}}{(x - 2)\sqrt{x - 1}} \cdot \dfrac{1 + \sqrt{x - 1}}{1 + \sqrt{x - 1}} = \dfrac{2 - x}{(x - 2)\sqrt{x - 1}\left(1 + \sqrt{x - 1}\right)} = \dfrac{-1}{\sqrt{x - 1}\left(1 + \sqrt{x - 1}\right)},\ x \neq 2$

**11.** $f(x) = 4x^2$

Domain: $(-\infty, \infty)$

Range: $[0, \infty)$

**13.** $f(x) = x^3$

Domain: $(-\infty, \infty)$

Range: $(-\infty, \infty)$

**15.** $g(x) = \sqrt{6x}$

Domain: $6x \geq 0$

$x \geq 0 \Rightarrow [0, \infty)$

Range: $[0, \infty)$

**17.** $f(x) = \sqrt{16 - x^2}$

$16 - x^2 \geq 0 \Rightarrow x^2 \leq 16$

Domain: $[-4, 4]$

Range: $[0, 4]$

**Note:** $y = \sqrt{16 - x^2}$ is a semicircle of radius 4.

**19.** $f(t) = \sec \dfrac{\pi t}{4}$

$\dfrac{\pi t}{4} \neq \dfrac{(2n + 1)\pi}{2} \Rightarrow t \neq 4n + 2$

Domain: all $t \neq 4n + 2$, $n$ an integer

Range: $(-\infty, -1] \cup [1, \infty)$

**21.** $f(x) = \dfrac{3}{x}$

Domain: all $x \neq 0 \Rightarrow (-\infty, 0) \cup (0, \infty)$

Range: $(-\infty, 0) \cup (0, \infty)$

**23.** $f(x) = \sqrt{x} + \sqrt{1 - x}$

$x \geq 0$  and  $1 - x \geq 0$

$x \geq 0$  and  $x \leq 1$

Domain: $0 \leq x \leq 1 \Rightarrow [0, 1]$

**25.** $g(x) = \dfrac{2}{1 - \cos x}$

$1 - \cos x \neq 0$

$\cos x \neq 1$

Domain: all $x \neq 2n\pi$, $n$ an integer

**27.** $f(x) = \dfrac{1}{|x + 3|}$

$|x + 3| \neq 0$

$x + 3 \neq 0$

Domain: all $x \neq -3$

Domain: $(-\infty, -3) \cup (-3, \infty)$

**29.** $f(x) = \begin{cases} 2x + 1, & x < 0 \\ 2x + 2, & x \geq 0 \end{cases}$

(a) $f(-1) = 2(-1) + 1 = -1$

(b) $f(0) = 2(0) + 2 = 2$

(c) $f(2) = 2(2) + 2 = 6$

(d) $f(t^2 + 1) = 2(t^2 + 1) + 2 = 2t^2 + 4$

(**Note:** $t^2 + 1 \geq 0$ for all $t$.)

Domain: $(-\infty, \infty)$

Range: $(-\infty, 1) \cup [2, \infty)$

**31.** $f(x) = \begin{cases} |x| + 1, & x < 1 \\ -x + 1, & x \geq 1 \end{cases}$

(a) $f(-3) = |-3| + 1 = 4$

(b) $f(1) = -1 + 1 = 0$

(c) $f(3) = -3 + 1 = -2$

(d) $f(b^2 + 1) = -(b^2 + 1) + 1 = -b^2$

Domain: $(-\infty, \infty)$

Range: $(-\infty, 0] \cup [1, \infty)$

**33.** $f(x) = 4 - x$

Domain: $(-\infty, \infty)$

Range: $(-\infty, \infty)$

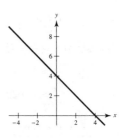

**35.** $h(x) = \sqrt{x - 6}$

Domain:

$x - 6 \geq 0$

$x \geq 6 \Rightarrow [6, \infty)$

Range: $[0, \infty)$

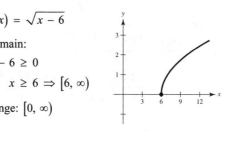

**37.** $f(x) = \sqrt{9 - x^2}$

Domain: $[-3, 3]$

Range: $[0, 3]$

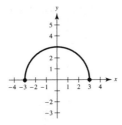

**39.** $g(t) = 3 \sin \pi t$

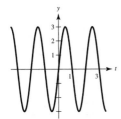

Domain: $(-\infty, \infty)$

Range: $[-3, 3]$

**41.** The student travels $\dfrac{2 - 0}{4 - 0} = \dfrac{1}{2}$ mi/min during the first

4 minutes. The student is stationary for the next
2 minutes. Finally, the student travels

$\dfrac{6 - 2}{10 - 6} = 1$ mi/min during the final 4 minutes.

**43.** $x - y^2 = 0 \Rightarrow y = \pm\sqrt{x}$

$y$ is not a function of $x$. Some vertical lines intersect the graph twice.

**45.** $y$ is a function of $x$. Vertical lines intersect the graph at most once.

**47.** $x^2 + y^2 = 16 \Rightarrow y = \pm\sqrt{16 - x^2}$

$y$ is not a function of $x$ because there are two values of $y$ for some $x$.

**49.** $y^2 = x^2 - 1 \Rightarrow y = \pm\sqrt{x^2 - 1}$

$y$ is not a function of $x$ because there are two values of $y$ for some $x$.

**51.** The transformation is a horizontal shift two units to the right.

Shifted function: $y = \sqrt{x - 2}$

**53.** The transformation is a horizontal shift 2 units to the right and a vertical shift 1 unit downward.

Shifted function: $y = (x - 2)^2 - 1$

**55.** $y = f(x + 5)$ is a horizontal shift 5 units to the left. Matches d.

**57.** $y = -f(-x) - 2$ is a reflection in the $y$-axis, a reflection in the $x$-axis, and a vertical shift downward 2 units. Matches c.

**59.** $y = f(x + 6) + 2$ is a horizontal shift to the left 6 units, and a vertical shift upward 2 units. Matches e.

**61.** (a) The graph is shifted 3 units to the left.

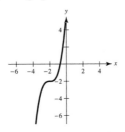

(b) The graph is shifted 1 unit to the right.

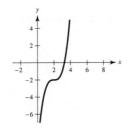

(c) The graph is shifted 2 units upward.

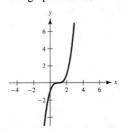

(d) The graph is shifted 4 units downward.

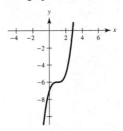

(e) The graph is stretched vertically by a factor of 3.

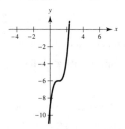

(g) The graph is a reflection in the *x*-axis.

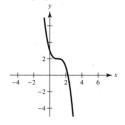

(f) The graph is stretched vertically by a factor of $\frac{1}{4}$.

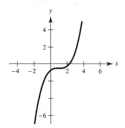

(h) The graph is a reflection about the origin.

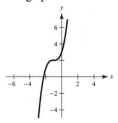

**63.** $f(x) = 3x - 4$,   $g(x) = 4$

(a) $f(x) + g(x) = (3x - 4) + 4 = 3x$

(b) $f(x) - g(x) = (3x - 4) - 4 = 3x - 8$

(c) $f(x) \cdot g(x) = (3x - 4)(4) = 12x - 16$

(d) $f(x)/g(x) = \dfrac{3x - 4}{4} = \dfrac{3}{4}x - 1$

**65.** (a) $f\big(g(1)\big) = f(0) = 0$

(b) $g\big(f(1)\big) = g(1) = 0$

(c) $g\big(f(0)\big) = g(0) = -1$

(d) $f\big(g(-4)\big) = f(15) = \sqrt{15}$

(e) $f\big(g(x)\big) = f\big(x^2 - 1\big) = \sqrt{x^2 - 1}$

(f) $g\big(f(x)\big) = g\big(\sqrt{x}\big) = \big(\sqrt{x}\big)^2 - 1 = x - 1,\ (x \geq 0)$

**67.** $f(x) = x^2$, $g(x) = \sqrt{x}$

$(f \circ g)(x) = f\big(g(x)\big)$

$\qquad = f\big(\sqrt{x}\big) = \big(\sqrt{x}\big)^2 = x,\ x \geq 0$

Domain: $[0, \infty)$

$(g \circ f)(x) = g\big(f(x)\big) = g\big(x^2\big) = \sqrt{x^2} = |x|$

Domain: $(-\infty, \infty)$

No. Their domains are different. $(f \circ g) = (g \circ f)$ for $x \geq 0$.

**69.** $f(x) = \dfrac{3}{x}$, $g(x) = x^2 - 1$

$(f \circ g)(x) = f\big(g(x)\big) = f\big(x^2 - 1\big) = \dfrac{3}{x^2 - 1}$

Domain: all $x \neq \pm 1 \Rightarrow (-\infty, -1) \cup (-1, 1) \cup (1, \infty)$

$(g \circ f)(x) = g\big(f(x)\big)$

$\qquad = g\left(\dfrac{3}{x}\right) = \left(\dfrac{3}{x}\right)^2 - 1 = \dfrac{9}{x^2} - 1 = \dfrac{9 - x^2}{x^2}$

Domain: all $x \neq 0 \Rightarrow (-\infty, 0) \cup (0, \infty)$

No, $f \circ g \neq g \circ f$.

**71.** (a) $(f \circ g)(3) = f\big(g(3)\big) = f(-1) = 4$

(b) $g\big(f(2)\big) = g(1) = -2$

(c) $g\big(f(5)\big) = g(-5)$, which is undefined

(d) $(f \circ g)(-3) = f\big(g(-3)\big) = f(-2) = 3$

(e) $(g \circ f)(-1) = g\big(f(-1)\big) = g(4) = 2$

(f) $f\big(g(-1)\big) = f(-4)$, which is undefined

**73.** $F(x) = \sqrt{2x - 2}$

Let $h(x) = 2x$, $g(x) = x - 2$ and $f(x) = \sqrt{x}$.

Then, $(f \circ g \circ h)(x) = f(g(2x)) = f((2x) - 2) = \sqrt{(2x) - 2} = \sqrt{2x - 2} = F(x)$.

[Other answers possible]

**75.** (a) If $f$ is even, then $\left(\frac{3}{2}, 4\right)$ is on the graph.

(b) If $f$ is odd, then $\left(\frac{3}{2}, -4\right)$ is on the graph.

**77.** $f$ is even because the graph is symmetric about the $y$-axis. $g$ is neither even nor odd. $h$ is odd because the graph is symmetric about the origin.

**79.** $f(x) = x^2(4 - x^2)$

$f(-x) = (-x)^2(4 - (-x)^2) = x^2(4 - x^2) = f(x)$

$f$ is even.

$f(x) = x^2(4 - x^2) = 0$

$x^2(2 - x)(2 + x) = 0$

Zeros: $x = 0, -2, 2$

**81.** $f(x) = x \cos x$

$f(-x) = (-x) \cos(-x) = -x \cos x = -f(x)$

$f$ is odd.

$f(x) = x \cos x = 0$

Zeros: $x = 0$, $\dfrac{\pi}{2} + n\pi$, where $n$ is an integer

**83.** Slope $= \dfrac{4 - (-6)}{-2 - 0} = \dfrac{10}{-2} = -5$

$y - 4 = -5(x - (-2))$

$y - 4 = -5x - 10$

$y = -5x - 6$

For the line segment, you must restrict the domain.

$f(x) = -5x - 6$, $-2 \leq x \leq 0$

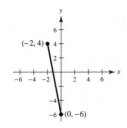

**85.** $x + y^2 = 0$

$y^2 = -x$

$y = -\sqrt{-x}$

$f(x) = -\sqrt{-x}$, $x \leq 0$

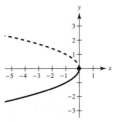

**87.** Answers will vary. *Sample answer*: Speed begins and ends at 0. The speed might be constant in the middle:

**89.** Answers will vary. *Sample answer*: In general, as the price decreases, the store will sell more.

**91.** $y = \sqrt{c - x^2}$

$y^2 = c - x^2$

$x^2 + y^2 = c$, a circle.

For the domain to be $[-5, 5]$, $c = 25$.

**93.** (a) $T(4) = 16°$, $T(15) \approx 23°$

(b) If $H(t) = T(t - 1)$, then the changes in temperature will occur 1 hour later.

(c) If $H(t) = T(t) - 1$, then the overall temperature would be 1 degree lower.

**95.** (a)

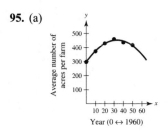

(b)   $A(25) \approx 445$  (Answers will vary.)

**97.** $f(x) = |x| + |x - 2|$

If $x < 0$, then $f(x) = -x - (x - 2) = -2x + 2$.

If $0 \le x < 2$, then $f(x) = x - (x - 2) = 2$.

If $x \ge 2$, then $f(x) = x + (x - 2) = 2x - 2$.

So,

$$f(x) = \begin{cases} -2x + 2, & x \le 0 \\ 2, & 0 < x < 2. \\ 2x - 2, & x \ge 2 \end{cases}$$

**99.** $f(-x) = a_{2n+1}(-x)^{2n+1} + \cdots + a_3(-x)^3 + a_1(-x)$

$\phantom{f(-x)} = -\left[ a_{2n+1}x^{2n+1} + \cdots + a_3x^3 + a_1x \right]$

$\phantom{f(-x)} = -f(x)$

Odd

**101.** Let $F(x) = f(x)g(x)$ where $f$ and $g$ are even. Then $F(-x) = f(-x)g(-x) = f(x)g(x) = F(x)$.

So, $F(x)$ is even. Let $F(x) = f(x)g(x)$ where $f$ and $g$ are odd. Then

$F(-x) = f(-x)g(-x) = \left[ -f(x) \right]\left[ -g(x) \right] = f(x)g(x) = F(x)$.

So, $F(x)$ is even.

**103.** By equating slopes, $\dfrac{y - 2}{0 - 3} = \dfrac{0 - 2}{x - 3}$

$$y - 2 = \frac{6}{x - 3}$$

$$y = \frac{6}{x - 3} + 2 = \frac{2x}{x - 3},$$

$$L = \sqrt{x^2 + y^2} = \sqrt{x^2 + \left( \frac{2x}{x - 3} \right)^2}.$$

**105.** False. If $f(x) = x^2$, then $f(-3) = f(3) = 9$, but $-3 \ne 3$.

**107.** True. The function is even.

**109.** False. The constant function $f(x) = 0$ has symmetry with respect to the $x$-axis.

**111.** First consider the portion of $R$ in the first quadrant: $x \ge 0$, $0 \le y \le 1$ and $x - y \le 1$; shown below.

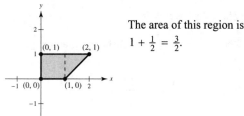

The area of this region is $1 + \frac{1}{2} = \frac{3}{2}$.

By symmetry, you obtain the entire region R:

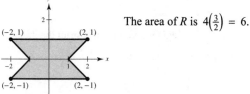

The area of $R$ is $4\left(\frac{3}{2}\right) = 6$.

## Section 1.4  Fitting Models to Data

**1.** (a) and (b)

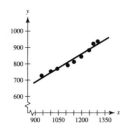

Yes, the data appear to be approximately linear.

The data can be modeled by equation
$y = 0.6x + 150$. (Answers will vary).

(c) When $x = 1075$, $y = 0.6(1075) + 150 = 795$.

**3.** (a) $d = 0.066F$

(b)

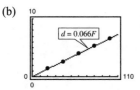

The model fits the data well.

(c) If $F = 55$, then $d \approx 0.066(55) = 3.63$ cm.

**5.** (a) Using a graphing utility, $y = 0.122x + 2.07$

The correlation coefficient is $r \approx 0.87$.

(b)

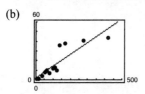

(c) Greater per capita energy consumption by a country
tends to correspond to greater per capita gross
national income. The three countries that most
differ from the linear model are Canada, Japan, and
Italy.

(d) Using a graphing utility, the new model
is $y = 0.142x - 1.66$.

The correlation coefficient is $r \approx 0.97$.

**7.** (a) Using graphing utility,
$$S = 180.89x^2 - 205.79x + 272.$$

(b)

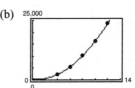

(c) When $x = 2$, $S \approx 583.98$ pounds.

(d) $\dfrac{2370}{584} \approx 4.06$

The breaking strength is approximately 4 times
greater.

(e) $\dfrac{23,860}{5460} \approx 4.37$

When the height is doubled, the breaking strength
increases approximately by a factor of 4.

**9.** (a) $y = -1.806x^3 + 14.58x^2 + 16.4x + 10$

(b)

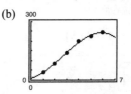

(c) If $x = 4.5$, $y \approx 214$ horsepower.

**11. (a)** $y_1 = -0.0172t^3 + 0.305t^2 - 0.87t + 7.3$

$y_2 = -0.038t^2 + 0.45t + 3.5$

$y_3 = 0.0063t^3 + -0.072t^2 + 0.02t + 1.8$

**(b)**
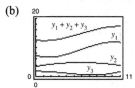

$y_1 + y_2 + y_3 = -0.0109t^3 + 0.195t^2 - 0.40t + 12.6$

For 2014, $t = 14$. So,

$y_1 + y_2 + y_3 = -0.0109(14)^3 + 0.195(14)^2 - 0.40(14) + 12.6$

$\approx 15.31$ cents/mile

**13. (a)** Yes, $y$ is a function of $t$. At each time $t$, there is one and only one displacement $y$.

**(b)** The amplitude is approximately

$(2.35 - 1.65)/2 = 0.35$.

The period is approximately

$2(0.375 - 0.125) = 0.5$.

**(c)** One model is $y = 0.35 \sin(4\pi t) + 2$.

**(d)**
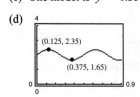

The model appears to fit the data.

**15.** Answers will vary.

**17.** Yes, $A_1 \le A_2$. To see this, consider the two triangles of areas $A_1$ and $A_2$:

For $i = 1, 2$, the angles satisfy $\alpha i + \beta i + \gamma i = \pi$. At least one of $\alpha_1 \le \alpha_2$, $\beta_1 \le \beta_2$, $\gamma_1 \le \gamma_2$ must hold.

Assume $\alpha_1 \le \alpha_2$. Because $\alpha_2 \le \pi/2$ (acute triangle), and the sine function increases on $[0, \pi/2]$, you have

$A_1 = \frac{1}{2}b_1c_1 \sin \alpha_1 \le \frac{1}{2} b_2c_2 \sin \alpha_1$

$\le \frac{1}{2}b_2c_2 \sin \alpha_2 = A_2$

# Section 1.5   Inverse Functions

**1.** (a)   $f(x) = 5x + 1$

$g(x) = \dfrac{x-1}{5}$

$f(g(x)) = f\left(\dfrac{x-1}{5}\right) = 5\left(\dfrac{x-1}{5}\right) + 1 = x$

$g(f(x)) = g(5x+1) = \dfrac{(5x+1)-1}{5} = x$

(b)

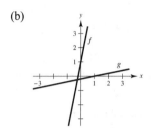

**3.** (a)   $f(x) = x^3$

$g(x) = \sqrt[3]{x}$

$f(g(x)) = f\left(\sqrt[3]{x}\right) = \left(\sqrt[3]{x}\right)^3 = x$

$g(f(x)) = g(x^3) = \sqrt[3]{x^3} = x$

(b)

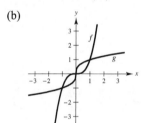

**5.** (a)   $f(x) = \sqrt{x-4}$

$g(x) = x^2 + 4, \quad x \geq 0$

$f(g(x)) = f(x^2 + 4)$

$\qquad = \sqrt{(x^2+4)-4} = \sqrt{x^2} = x$

$g(f(x)) = g\left(\sqrt{x-4}\right)$

$\qquad = \left(\sqrt{x-4}\right)^2 + 4 = x - 4 + 4 = x$

(b)

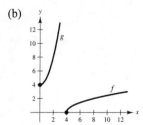

**7.** (a)   $f(x) = \dfrac{1}{x}$

$g(x) = \dfrac{1}{x}$

$f(g(x)) = \dfrac{1}{1/x} = x$

$g(f(x)) = \dfrac{1}{1/x} = x$

(b)

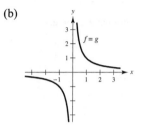

**9.** Matches (c)

**11.** Matches (a)

**13.** $f(x) = \frac{3}{4}x + 6$

One-to-one; has an inverse

**15.** $f(\theta) = \sin\theta$

Not one-to-one; does not have an inverse

**17.** $h(s) = \dfrac{1}{s-2} - 3$

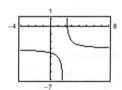

One-to-one; has an inverse

**19.** $g(t) = \dfrac{1}{\sqrt{t^2+1}}$

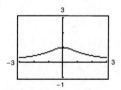

Not one-to-one; does not have an inverse

**21.** $g(x) = (x + 5)^3$

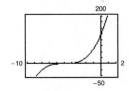

One-to-one; has an inverse

**23.** $f(x) = \dfrac{x^4}{4} - 2x^2$

Not one-to-one; $f$ does not have an inverse.

**25.** $f(x) = 2 - x - x^3$

One-to-one; has an inverse

**27.** (a) $f(x) = 2x - 3 = y$

$$x = \dfrac{y + 3}{2}$$

$$y = \dfrac{x + 3}{2}$$

$$f^{-1}(x) = \dfrac{x + 3}{2}$$

(b)

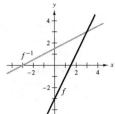

(c) The graphs of $f$ and $f^{-1}$ are reflections of each other in the line $y = x$.

(d) Domain of $f$:  all real numbers
Range of $f$:  all real numbers
Domain of $f^{-1}$:  all real numbers
Range of $f^{-1}$:  all real numbers

**29.** (a) $f(x) = x^5 = y$

$$x = \sqrt[5]{y}$$

$$y = \sqrt[5]{x}$$

$$f^{-1}(x) = \sqrt[5]{x} = x^{1/5}$$

(b)

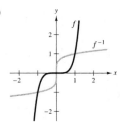

(c) The graphs of $f$ and $f^{-1}$ are reflections of each other in the line $y = x$.

(d) Domain of $f$:  all real numbers
Range of $f$:  all real numbers
Domain of $f^{-1}$:  all real numbers
Range of $f^{-1}$:  all real numbers

**31.** (a) $f(x) = \sqrt{x} = y$

$$x = y^2$$

$$y = x^2$$

$$f^{-1}(x) = x^2, \quad x \geq 0$$

(b)

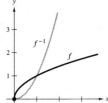

(c) The graphs of $f$ and $f^{-1}$ are reflections of each other in the line $y = x$.

(d) Domain of $f$:  $x \geq 0$
Range of $f$:  $y \geq 0$
Domain of $f^{-1}$:  $x \geq 0$
Range of $f^{-1}$:  $y \geq 0$

**33.** (a) $f(x) = \sqrt{4 - x^2} = y, \quad 0 \le x \le 2$

$$4 - x^2 = y^2$$
$$x^2 = 4 - y^2$$
$$x = \sqrt{4 - y^2}$$
$$y = \sqrt{4 - x^2}$$
$$f^{-1}(x) = \sqrt{4 - x^2}, \quad 0 \le x \le 2$$

(b)

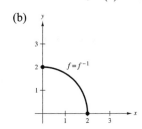

(c) The graphs of $f$ and $f^{-1}$ are reflections of each other in the line $y = x$. In fact, the graphs are identical.

(d) Domain of $f$:    $0 \le x \le 2$

Range of $f$:    $0 \le y \le 2$

Domain of $f^{-1}$:   $0 \le x \le 2$

Range of $f^{-1}$:    $0 \le y \le 2$

**35.** (a) $f(x) = \sqrt[3]{x - 1} = y$

$$x - 1 = y^3$$
$$x = y^3 + 1$$
$$y = x^3 + 1$$
$$f^{-1}(x) = x^3 + 1$$

(b)

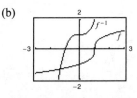

(c) The graphs of $f$ and $f^{-1}$ are reflections of each other in the line $y = x$.

(d) Domain of $f$:    all real numbers

Range of $f$:    all real numbers

Domain of $f^{-1}$:   all real numbers

Range of $f^{-1}$:    all real numbers

**37.** (a) $f(x) = x^{2/3} = y, \quad x \ge 0$

$$x = y^{3/2}$$
$$y = x^{3/2}$$
$$f^{-1}(x) = x^{3/2}, \quad x \ge 0$$

(b)

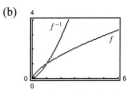

(c) The graphs of $f$ and $f^{-1}$ are reflections of each other in the line $y = x$.

(d) Domain of $f$:    $x \ge 0$

Range of $f$:    $y \ge 0$

Domain of $f^{-1}$:   $x \ge 0$

Range of $f^{-1}$:    $y \ge 0$

**39.** (a) $f(x) = \dfrac{x}{\sqrt{x^2 + 7}} = y$

$$x = y\sqrt{x^2 + 7}$$
$$x^2 = y^2(x^2 + 7) = y^2 x^2 + 7y^2$$
$$x^2(1 - y^2) = 7y^2$$
$$x = \frac{\sqrt{7}y}{\sqrt{1 - y^2}}$$
$$y = \frac{\sqrt{7}x}{\sqrt{1 - x^2}}$$
$$f^{-1}(x) = \frac{\sqrt{7}x}{\sqrt{1 - x^2}}, \quad -1 < x < 1$$

(b)

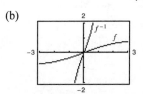

(c) The graphs of $f$ and $f^{-1}$ are reflections of each other in the line $y = x$.

(d) Domain of $f$:    all real numbers

Range of $f$:    $-1 < y < 1$

Domain of $f^{-1}$:   $-1 < x < 1$

Range of $f^{-1}$:    all real numbers

**41.**

| $x$ | 0 | 1 | 2 | 3 |
|-----|---|---|---|---|
| $f(x)$ | 1 | 2 | 3 | 4 |

| $x$ | 1 | 2 | 3 | 4 |
|-----|---|---|---|---|
| $f^{-1}(x)$ | 0 | 1 | 2 | 4 |

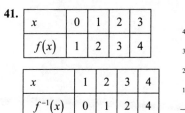

**43.** (a) Let $x$ be the number of pounds of the commodity costing 1.25 per pound. Because there are 50 pounds total, the amount of the second commodity is $50 - x$. The total cost is

$$y = 1.25x + 1.60(50 - x)$$
$$= -0.35x + 80, \quad 0 \le x \le 50.$$

(b) Find the inverse of the original function.

$$y = -0.35x + 80$$
$$0.35x = 80 - y$$
$$x = \tfrac{100}{35}(80 - y)$$

Inverse: $y = \tfrac{100}{35}(80 - x) = \tfrac{20}{7}(80 - x)$

$x$ represents cost and $y$ represents pounds.

(c) Domain of inverse is $62.5 \le x \le 80$.

The total cost will be between \$62.50 and \$80.00.

(d) If $x = 73$ in the inverse function,

$$y = \tfrac{100}{35}(80 - 73) = \tfrac{100}{5} = 20 \text{ pounds.}$$

**45.** $f(x) = \sqrt{x - 2}, \ x \ge 2$

$f$ is one-to-one; has an inverse.

$$y = \sqrt{x - 2}, \ x \ge 2, \ y \ge 0$$
$$y^2 = x - 2$$
$$x = y^2 + 2$$
$$f^{-1}(x) = x^2 + 2, \ x \ge 0$$

**47.** $f(x) = -3$

Not one-to-one; does not have an inverse.

**49.** $f(x) = ax + b$

$f$ is one-to-one; has an inverse.

$$ax + b = y$$
$$x = \frac{y - b}{a}$$
$$y = \frac{x - b}{a}$$
$$f^{-1}(x) = \frac{x - b}{a}, \ a \ne 0$$

**51.** $f(x) = (x - 4)^2$ on $[4, \infty)$

$f$ passes the Horizontal Line Test on $[4, \infty)$, so it is one-to-one.

**53.** $f(x) = \dfrac{4}{x^2}$ on $(0, \infty)$

$f$ passes the Horizontal Line Test on $(0, \infty)$, so it is one-to-one.

**55.** $f(x) = \cos x$ on $[0, \pi]$

$f$ passes the Horizontal Line Test on $[0, \pi]$, so it is one-to-one.

**57.** $f(x) = (x - 3)^2$ is one-to-one for $x \ge 3$.

$$(x - 3)^2 = y$$
$$x - 3 = \sqrt{y}$$
$$x = \sqrt{y} + 3$$
$$y = \sqrt{x} + 3$$
$$f^{-1}(x) = \sqrt{x} + 3, \ x \ge 0$$

(Answer is not unique.)

**59.** (a) $f(x) = (x + 5)^2$

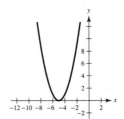

(b) $f$ is one-to-one on $[-5, \infty)$. (Note that $f$ is also one-to-one on $(-\infty, -5]$.)

(c) $f(x) = (x + 5)^2 = y, \qquad x \ge -5$
$$x + 5 = \sqrt{y}$$
$$x = \sqrt{y} - 5$$
$$y = \sqrt{x} - 5$$
$$f^{-1}(x) = \sqrt{x} - 5$$

(d) Domain of $f^{-1}$: $x \ge 0$

**61.** (a)  $f(x) = \sqrt{x^2 - 4x}$

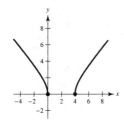

(b) $f$ is one-to-one on $[4, \infty)$. (Note that $f$ is also

one-to-one on $(-\infty, 0]$.)

(c) $f(x) = \sqrt{x^2 - 4x} = y,\ x \geq 4$

$$x^2 - 4x = y^2$$

$$x^2 - 4x + 4 = y^2 + 4$$

$$(x - 2)^2 = y^2 + 4$$

$$x - 2 = \sqrt{y^2 + 4}$$

$$x = 2 + \sqrt{y^2 + 4}$$

$$y = 2 + \sqrt{x^2 + 4}$$

$$f^{-1}(x) = 2 + \sqrt{x^2 + 4}$$

(d) Domain of $f^{-1}$: $x \geq 0$

**63.** (a)  $f(x) = 3 \cos x$

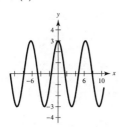

(b) $f$ is one-to-one on $[0, \pi]$. (other answers possible)

(c) $f(x) = 3 \cos x = y$

$$\cos x = \frac{y}{3}$$

$$x = \arccos\left(\frac{y}{3}\right)$$

$$y = \arccos\left(\frac{x}{3}\right)$$

$$f^{-1}(x) = \arccos\left(\frac{x}{3}\right)$$

(d) Domain of $f^{-1}$: $-3 \leq x \leq 3$

**65.**  $f(x) = x^3 + 2x - 1$

$$f(1) = 2 = a \Rightarrow f^{-1}(2) = 1$$

**67.**  $f(x) = \sin x$

$$f\left(\frac{\pi}{6}\right) = \frac{1}{2} = a \Rightarrow f^{-1}\left(\frac{1}{2}\right) = \frac{\pi}{6}$$

**69.**  $f(x) = x^3 - \dfrac{4}{x}$

$$f(2) = 6 = a \Rightarrow f^{-1}(6) = 2$$

**In Exercises 71–73, use the following.**

$f(x) = \frac{1}{8}x - 3$ and $g(x) = x^3$

$f^{-1}(x) = 8(x + 3)$ and $g^{-1}(x) = \sqrt[3]{x}$

**71.** $\left(f^{-1} \circ g^{-1}\right)(1) = f^{-1}\left(g^{-1}(1)\right) = f^{-1}(1) = 32$

**73.** $\left(f^{-1} \circ f^{-1}\right)(6) = f^{-1}\left(f^{-1}(6)\right) = f^{-1}(72) = 600$

**In Exercises 75–77, use the following.**

$f(x) = x + 4$ and $g(x) = 2x - 5$

$f^{-1}(x) = x - 4$ and $g^{-1}(x) = \dfrac{x + 5}{2}$

**75.** $\left(g^{-1} \circ f^{-1}\right)(x) = g^{-1}\left(f^{-1}(x)\right)$

$$= g^{-1}(x - 4)$$

$$= \frac{(x - 4) + 5}{2}$$

$$= \frac{x + 1}{2}$$

**77.** $(f \circ g)(x) = f\left(g(x)\right)$

$$= f(2x - 5)$$

$$= (2x - 5) + 4$$

$$= 2x - 1$$

So, $(f \circ g)^{-1}(x) = \dfrac{x + 1}{2}$.

**Note:** $(f \circ g)^{-1} = g^{-1} \circ f^{-1}$

**79.** (a) $f$ is one-to-one because it passes the Horizontal Line Test.

(b) The domain of $f^{-1}$ is the range of $f$: $[-2, 2]$.

(c) $f^{-1}(2) = -4$ because $f(-4) = 2$.

**81.**

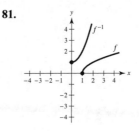

**83.** $y = \arcsin x$

(a)

| $x$ | $-1$ | $-0.8$ | $-0.6$ | $-0.4$ | $-0.2$ | 0 | 0.2 | 0.4 | 0.6 | 0.8 | 1 |
|---|---|---|---|---|---|---|---|---|---|---|---|
| $y$ | $-1.571$ | $-0.927$ | $-0.644$ | $-0.412$ | $-0.201$ | 0 | 0.201 | 0.412 | 0.644 | 0.927 | 1.571 |

(b)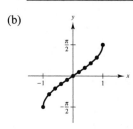

(c)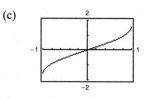

(d) Symmetric about origin: $\arcsin(-x) = -\arcsin x$

Intercept: $(0, 0)$

**85.** $y = \arccos x$

$\left(-\dfrac{\sqrt{2}}{2}, \dfrac{3\pi}{4}\right)$ because $\cos\left(\dfrac{3\pi}{4}\right) = -\dfrac{\sqrt{2}}{2}.$

$\left(\dfrac{1}{2}, \dfrac{\pi}{3}\right)$ because $\cos\left(\dfrac{\pi}{4}\right) = \dfrac{1}{2}.$

$\left(\dfrac{\sqrt{3}}{2}, \dfrac{\pi}{6}\right)$ because $\cos\left(\dfrac{\pi}{6}\right) = \dfrac{\sqrt{3}}{2}.$

**87.** $\arcsin \dfrac{1}{2} = \dfrac{\pi}{6}$

**89.** $\arccos \dfrac{1}{2} = \dfrac{\pi}{3}$

**91.** $\arctan \dfrac{\sqrt{3}}{3} = \dfrac{\pi}{6}$

**93.** $\text{arccsc}\left(-\sqrt{2}\right) = -\dfrac{\pi}{4}$

**95.** $\arccos (0.8) \approx 2.50$

**97.** $\text{arcsec} (1.269) = \arccos\left(\dfrac{1}{1.269}\right) \approx 0.66$

**99.** $\cos\left[\arccos(-0.1)\right] = -0.1$

**In Exercises 101–105, use the triangle.**

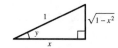

**101.**     $y = \arccos x$

$\cos y = x$

**103.** $\tan y = \dfrac{\sqrt{1 - x^2}}{x}$

**105.** $\sec y = \dfrac{1}{x}$

**107.** (a) $\sin\left(\arctan\dfrac{3}{4}\right) = \dfrac{3}{5}$

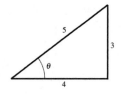

(b) $\sec\left(\arcsin\dfrac{4}{5}\right) = \dfrac{5}{3}$

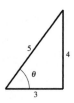

**109.** (a) $\cot\left[\arcsin\left(-\dfrac{1}{2}\right)\right] = \cot\left(-\dfrac{\pi}{6}\right) = -\sqrt{3}$

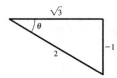

(b) $\csc\left[\arctan\left(-\dfrac{5}{12}\right)\right] = -\dfrac{13}{5}$

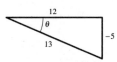

**111.** $y = \cos\left(\arcsin 2x\right)$

$\theta = \arcsin 2x$

$y = \cos \theta = \sqrt{1 - 4x^2}$

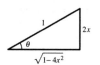

**113.** $y = \sin(\text{arcsec } x)$

$\theta = \text{arcsec } x, \; 0 \le \theta \le \pi, \; \theta \ne \dfrac{\pi}{2}$

$y = \sin \theta = \dfrac{\sqrt{x^2 - 1}}{|x|}$

The absolute value bars on $x$ are necessary because of the restriction $0 \le \theta \le \pi$, $\theta \ne \pi/2$, and $\sin \theta$ for this domain must always be nonnegative.

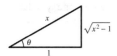

**115.** $y = \tan\!\left(\text{arcsec } \dfrac{x}{3}\right)$

$\theta = \text{arcsec } \dfrac{x}{3}$

$y = \tan \theta = \dfrac{x^2 - 9}{3}$

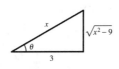

**121.** $y = \arccos x$

$y = \arctan x$

The point of intersection is given by

$f(x) = \arccos x - \arctan x = 0, \; \cos(\arccos x) = \cos(\arctan x).$

$x = \dfrac{1}{\sqrt{1 + x^2}}$

$x^2\!\left(1 + x^2\right) = 1$

$x^4 + x^2 - 1 = 0$ when $x^2 = \dfrac{-1 + \sqrt{5}}{2}.$

So, $x = \pm\sqrt{\dfrac{-1 + \sqrt{5}}{2}} \approx \pm 0.7862.$

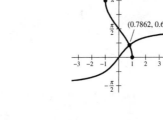

Point of intersection: $(0.7862, 0.6662)$ $\left[\text{Because } f(-0.7862) = \pi \ne 0.\right]$

**123.** Let $y = f(x)$ be one-to-one. Solve for $x$ as a function of $y$. Interchange $x$ and $y$ to get $y = f^{-1}(x)$. Let the domain of $f^{-1}$ be the range of $f$. Verify that $f\!\left(f^{-1}(x)\right) = x$ and $f^{-1}\!\left(f(x)\right) = x.$

Example:

$f(x) = x^3$

$y = x^3$

$x = \sqrt[3]{y}$

$y = \sqrt[3]{x}$

$f^{-1}(x) = \sqrt[3]{x}$

**117.** $\arcsin(3x - \pi) = \tfrac{1}{2}$

$3x - \pi = \sin\!\left(\tfrac{1}{2}\right)$

$x = \tfrac{1}{3}\!\left[\sin\!\left(\tfrac{1}{2}\right) + \pi\right] \approx 1.207$

**119.** $\arcsin \sqrt{2x} = \arccos \sqrt{x}$

$\sqrt{2x} = \sin\!\left(\arccos \sqrt{x}\right)$

$\sqrt{2x} = \sqrt{1 - x}, \; 0 \le x \le 1$

$2x = 1 - x$

$3x = 1$

$x = \tfrac{1}{3}$

**125.** The trigonometric functions are not one-to-one. So, their domains must be restricted to define the inverse trigonometric functions.

**127.** $\arctan \dfrac{9}{x} = \arcsin \dfrac{9}{\sqrt{x^2 + 81}}$

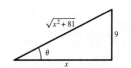

**129.** (a) $\operatorname{arccsc} x = \arcsin \dfrac{1}{x}, \ |x| \geq 1$

Let $y = \operatorname{arccsc} x$.

Then for $-\dfrac{\pi}{2} \leq y < 0$ and $0 < y \leq \dfrac{\pi}{2}$,

$\csc y = x \Rightarrow \sin y = \dfrac{1}{x}$.

So, $y = \arcsin\left(\dfrac{1}{x}\right)$. Therefore,

$\operatorname{arccsc} x = \arcsin\left(\dfrac{1}{x}\right)$.

(b) $\arctan x + \arctan\dfrac{1}{x} = \dfrac{\pi}{2}, \ x > 0$

Let $y = \arctan x + \arctan(1/x)$.

Then $\tan y = \dfrac{\tan(\arctan x) + \tan\left[\arctan(1/x)\right]}{1 - \tan(\arctan x)\,\tan\left[\arctan(1/x)\right]}$

$= \dfrac{x + (1/x)}{1 - x(1/x)}$

$= \dfrac{x + (1/x)}{0}$ (which is undefined).

So, $y = \pi/2$. Therefore,

$\arctan x + \arctan(1/x) = \pi/2$.

**131.** $f(x) = \arcsin(x - 1)$

$x - 1 = \sin y$

$x = 1 + \sin y$

Domain: $[0, 2]$

Range: $[-\pi/2, \pi/2]$

$f(x)$ is the graph of arcsin $x$ shifted right one unit.

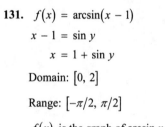

**133.** $f(x) = \operatorname{arcsec} 2x$

$2x = \sec y$

$x = \dfrac{1}{2}\sec y$

Domain: $(-\infty, -1/2], [1/2, \infty)$

Range: $[0, \pi/2), (\pi/2, \pi]$

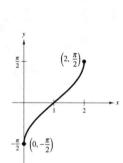

**135.** Because $f(-3) = 8$ and $f$ is one-to-one, you have

$f^{-1}(8) = -3$.

**137.** Let $f$ and $g$ be one-to-one functions.

Let $(f \circ g)(x) = y$, then $x = (f \circ g)^{-1}(y)$. Also:

$(f \circ g)(x) = y$

$f(g(x)) = y$

$g(x) = f^{-1}(y)$

$x = g^{-1}(f^{-1}(y))$

$x = (g^{-1} \circ f^{-1}(y))$

So, $(f \circ g)^{-1}(y) = (g^{-1} \circ f^{-1})(y)$ and

$(f \circ g)^{-1} = g^{-1} \circ f^{-1}$.

**139.** Let $y = \sin^{-1}x$. Then $\sin y = x$ and

$\cos(\sin^{-1}x) = \cos(y) = \sqrt{1 - x^2}$, as indicated in the figure.

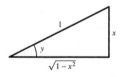

**141.** False. Let $f(x) = x^2$.

**143.** False

$\arcsin^2 0 + \arccos^2 0 = 0 + \dfrac{\pi^2}{2} \neq 1$

**145.** True

**147.** (a)  arccot $x = y$ if and only if cot $y = x$,

$0 < y < \pi$.

For $x > 0$, cot $y > 0$ and $0 < y < \dfrac{\pi}{2}$.

So, tan $y = \dfrac{1}{x} > 0$ and $y = \arctan\left(\dfrac{1}{x}\right)$.

For $x = 0$, $\operatorname{arccot}(0) = \dfrac{\pi}{2}$.

For $x < 0$, cot $y < 0$ and $\dfrac{\pi}{2} < y < \pi$.

So, tan $y = \dfrac{1}{x} < 0$ and $\arctan\left(\dfrac{1}{x}\right) < 0$.

Therefore, you need to add $\pi$ to get $y = \pi + \arctan\left(\dfrac{1}{x}\right)$.

(b)  $y = \operatorname{arcsec} x$ if and only if sec $y = x$, $|x| \geq 1$, $0 \leq y \leq \pi$, $y \neq \dfrac{\pi}{2}$.

So, cos $y = \dfrac{1}{x}$ and $y = \arccos\left(\dfrac{1}{x}\right)$.

(c)  $y = \operatorname{arccsc} x$ if and only if csc $y = x$, $|x| \geq 1$, $-\dfrac{\pi}{2} \leq y \leq \dfrac{\pi}{2}$, $y \neq 0$.

So, sin $y = \dfrac{1}{x}$ and $y = \arcsin\left(\dfrac{1}{x}\right)$.

**149.**  $\tan(\arctan x + \arctan y) = \dfrac{\tan(\arctan x + \arctan y)}{1 - \tan(\arctan x)\tan(\arctan y)} = \dfrac{x + y}{1 - xy}, \; xy \neq 1$

So,

$\arctan x + \arctan y = \arctan\left(\dfrac{x + y}{1 - xy}\right), \; xy \neq 1$.

Let $x = \dfrac{1}{2}$ and $y = \dfrac{1}{3}$.

$\arctan\left(\dfrac{1}{2}\right) + \arctan\left(\dfrac{1}{3}\right) = \arctan\dfrac{\frac{1}{2} + \frac{1}{3}}{1 - \left(\frac{1}{2} \cdot \frac{1}{3}\right)} = \arctan\dfrac{\frac{5}{6}}{1 - \frac{1}{6}} = \arctan\dfrac{\frac{5}{6}}{\frac{5}{6}} = \arctan 1 = \dfrac{\pi}{4}$

**151.**  $y = ax^2 + bx + c$. Interchange $x$ and $y$, and solve for $y$ using the quadratic formula.

$ay^2 + by + c - x = 0$

$$y = \frac{-b \pm \sqrt{b^2 - 4a(c - x)}}{2a}$$

Because $x \leq \dfrac{-b}{2a}$, use the negative sign.

$$f^{-1}(x) = \frac{-b - \sqrt{b^2 - 4ac + 4ax}}{2a}$$

**153.** $f$ is one-to-one if $f(x_1) = f(x_2)$ implies $x_1 = x_2$. So assume

$$f(x_1) = f(x_2)$$

$$\frac{ax_1 + b}{cx_1 + d} = \frac{ax_2 + b}{cx_2 + d}$$

$$acx_1x_2 + adx_1 + bcx_2 + bd = acx_1x_2 + adx_2 + bcx_1 + bd$$

$$adx_1 + bcx_2 = adx_2 + bcx_1$$

$$(ad - bc)x_1 = (ad - bc)x_2.$$

So, $x_1 = x_2$ if $ad - bc \neq 0$. To find $f^{-1}$, solve for $x$ as follows.

$$y = \frac{ax + b}{cx + d}$$

$$ycx + yd = ax + b$$

$$(yc - a)x = b - yd$$

$$x = \frac{b - yd}{yc - a}$$

$$f^{-1}(x) = \frac{b - dx}{cx - a}$$

## Section 1.6   Exponential and Logarithmic Functions

**1.** (a) $25^{3/2} = 5^3 = 125$

(b) $81^{1/2} = 9$

(c) $3^{-2} = \frac{1}{3^2} = \frac{1}{9}$

(d) $27^{-1/3} = \frac{1}{27^{1/3}} = \frac{1}{3}$

**3.** (a) $(5^2)(5^3) = 5^{2+3} = 5^5 = 3125$

(b) $(5^2)(5^{-3}) = 5^{2-3} = 5^{-1} = \frac{1}{5}$

(c) $\frac{5^3}{25^2} = \frac{5^3}{5^4} = \frac{1}{5}$

(d) $\left(\frac{1}{4}\right)^2 2^6 - \frac{2^6}{2^4} = 2^2 = 4$

**5.** (a) $e^2(e^4) = e^6$

(b) $(e^3)^4 = e^{12}$

(c) $(e^3)^{-2} = e^{-6} = \frac{1}{e^6}$

(d) $\frac{e^5}{e^3} = e^2$

**7.** $3^x = 81 \Rightarrow x = 4$

**9.** $6^{x-2} = 36 \Rightarrow x - 2 = 2 \Rightarrow x = 4$

**11.** $\left(\frac{1}{2}\right)^x = 32 \Rightarrow 2^{-x} = 32 \Rightarrow -x = 5 \Rightarrow x = -5$

**13.** $\left(\frac{1}{3}\right)^{x-1} = 27 \Rightarrow 3^{1-x} = 27 \Rightarrow 1 - x = 3 \Rightarrow x = -2$

**15.** $4^3 = (x + 2)^3 \Rightarrow 4 = x + 2 \Rightarrow x = 2$

**17.** $x^{3/4} = 8 \Rightarrow x = 8^{4/3} = 2^4 = 16$

**19.** $e^x = 5 \Rightarrow x = \ln 5 \approx 1.609$

**21.** $e^{-2x} = e^5 \Rightarrow -2x = 5 \Rightarrow x = -\frac{5}{2}$

**23.** $\left(1 + \frac{1}{1,000,000}\right)^{1,000,000} \approx 2.718280469$

$$e \approx 2.718281828$$

$$e > \left(1 + \frac{1}{1,000,000}\right)^{1,000,000}$$

**25.** $y = 3^x$

| $x$ | $-2$ | $-1$ | $0$ | $1$ | $2$ |
|-----|------|------|-----|-----|-----|
| $y$ | $\frac{1}{9}$ | $\frac{1}{3}$ | $1$ | $3$ | $9$ |

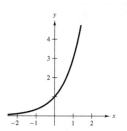

**27.** $y = \left(\frac{1}{3}\right)^x = 3^{-x}$

| x | -2 | -1 | 0 | 1 | 2 |
|---|---|---|---|---|---|
| y | 9 | 3 | 1 | $\frac{1}{3}$ | $\frac{1}{9}$ |

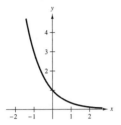

**29.** $f(x) = 3^{-x^2}$

| x | 0 | ±1 | ±2 |
|---|---|---|---|
| y | 1 | $\frac{1}{3}$ | 0.0123 |

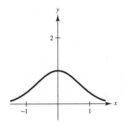

**31.** $y = e^{-x}$

| x | -1 | 0 | 1 |
|---|---|---|---|
| y | e | 1 | $\frac{1}{e}$ |

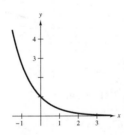

**33.** $y = e^x + 2$

| x | -2 | -1 | 0 | 1 | 2 |
|---|---|---|---|---|---|
| y | $\frac{1}{e^2} + 2$ | $\frac{1}{e} + 2$ | 3 | $e + 2$ | $e^2 + 2$ |

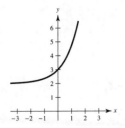

**35.** $h(x) = e^{x-2}$

| x | 0 | 1 | 2 | 3 | 4 |
|---|---|---|---|---|---|
| y | $e^{-2}$ | $e^{-1}$ | 1 | $e$ | $e^2$ |

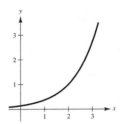

**37.** $y = e^{-x^2}$

Symmetric with respect to the *y*-axis

Horizontal asymptote

$y = 0$

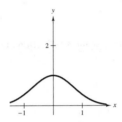

**39.** $f(x) = \dfrac{1}{3 + e^x}$

Because $e^x > 0$, $3 + e^x > 0$.

Domain: all real numbers

**41.** $f(x) = \sqrt{1 - 4^x}$

$1 - 4^x \geq 0 \Rightarrow 4^x \leq 1 \Rightarrow x \ln 4 \leq \ln 1 = 0$

Domain: $x \leq 0$

**43.** $f(x) = \sin e^{-x}$

Domain: all real numbers

**45.** (a)

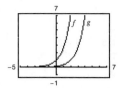

Horizontal shift 2 units to the right.

(b)

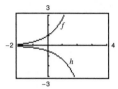

A reflection in the $x$-axis and a vertical shrink.

(c)

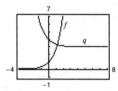

Vertical shift 3 units upward and a reflection in the $y$-axis.

**47.** $y = Ce^{ax}$

Horizontal asymptote: $y = 0$

Matches (c)

**49.** $y = C\left(1 - e^{-ax}\right)$

Vertical shift $C$ units

Reflection in both the $x$- and $y$-axes

Matches (a)

**51.** $y = Ca^x$

$(0, 2)$: $2 = Ca^0 = C$

$(3, 54)$: $54 = 2a^3$

$\quad 27 = a^3$

$\quad 3 = a$

$\quad y = 2\left(3^x\right)$

**53.** $f(x) = \ln x + 1$

Vertical shift 1 unit upward

Matches (b)

**55.** $f(x) = \ln(x - 1)$

Horizontal shift 1 unit to the right

Matches (a)

**57.** $e^0 = 1$

$\ln 1 = 0$

**59.** $\ln 2 = 0.6931...$

$e^{0.6931...} = 2$

**61.** $f(x) = 3 \ln x$

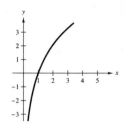

Domain: $x > 0$

**63.** $f(x) = \ln 2x$

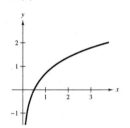

Domain: $x > 0$

**65.** $f(x) = \ln(x - 3)$

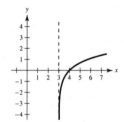

Domain: $x > 3$

**67.** $h(x) = \ln(x + 2)$

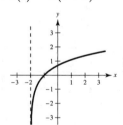

Domain: $x > -2$

**69.** 8 units upward: $e^x + 8$

Reflected in $x$-axis: $-\left(e^x + 8\right)$

$\quad y = -\left(e^x + 8\right) = -e^x - 8$

**71.** 5 units to the right: $\ln(x - 5)$

1 unit downward: $\ln(x - 5) - 1$

$$y = \ln(x - 5) - 1$$

**73.** $f(x) = e^{2x}$

$$g(x) = \ln\sqrt{x} = \frac{1}{2}\ln x$$

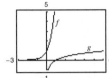

**75.** $f(x) = e^x - 1$

$$g(x) = \ln(x + 1)$$

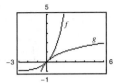

**77.** (a)     $y = e^{4x-1}$

$$\ln y = 4x - 1$$

$$\ln y + 1 = 4x$$

$$x = \tfrac{1}{4}(\ln y + 1)$$

$$f^{-1}(x) = \tfrac{1}{4}(\ln x + 1)$$

(b)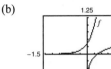

(c) $f^{-1}(f(x)) = f^{-1}(e^{4x-1}) = \tfrac{1}{4}(\ln e^{4x-1} + 1) = \tfrac{1}{4}(4x - 1 + 1) = x$

$f(f^{-1}(x)) = f\left(\tfrac{1}{4}(\ln x + 1)\right) = e^{(\ln x + 1) - 1} = e^{\ln x} = x$

**79.** (a)     $y = 2\ln(x - 1)$

$$\frac{y}{2} = \ln(x - 1)$$

$$e^{y/2} = x - 1$$

$$x = 1 + e^{y/2}$$

$$f^{-1}(x) = 1 + e^{x/2}$$

(b)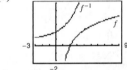

(c) $f^{-1}(f(x)) = f^{-1}(2\ln(x - 1)) = 1 + e^{\ln(x-1)} = 1 + x - 1 = x$

$f(f^{-1}(x)) = f(1 + x^{x/2}) = 2\ln\left[(1 + e^{x/2}) - 1\right] = 2\left(\dfrac{x}{2}\right) = x$

**81.** $\ln e^{x^2} = x^2$

**83.** $e^{\ln(5x+2)} = 5x + 2$

**85.** $-1 + \ln e^{2x} = -1 + 2x$

**87.** (a) $\ln 6 = \ln 2 + \ln 3 \approx 1.7917$

(b) $\ln\frac{2}{3} = \ln 2 - \ln 3 \approx -0.4055$

(c) $\ln 81 = 4\ln 3 \approx 4.3944$

(d) $\ln\sqrt{3} = \frac{1}{2}\ln 3 \approx 0.5493$

**89.** $\ln\dfrac{x}{4} = \ln x - \ln 4$

**91.** $\ln\dfrac{xy}{z} = \ln x + \ln y - \ln z$

**93.** $\ln\left(x\sqrt{x^2 + 5}\right) = \ln x + \ln(x^2 + 5)^{1/2}$

$$= \ln x + \tfrac{1}{2}\ln(x^2 + 5)$$

**95.** $\ln\sqrt{\dfrac{x-1}{x}} = \ln\left(\dfrac{x-1}{x}\right)^{1/2} = \dfrac{1}{2}\ln\left(\dfrac{x-1}{x}\right)$

$$= \frac{1}{2}\big[\ln(x - 1) - \ln x\big]$$

$$= \frac{1}{2}\ln(x - 1) - \frac{1}{2}\ln x$$

**97.** $\ln 3e^2 = \ln 3 + 2\ln e = 2 + \ln 3$

**99.** $\ln x + \ln 7 = \ln(x \cdot 7) = \ln(7x)$

**101.** $\ln(x - 2) - \ln(x + 2) = \ln \dfrac{x - 2}{x + 2}$

**103.** $\dfrac{1}{3}\Big[2\ln(x + 3) + \ln x - \ln(x^2 - 1)\Big] = \dfrac{1}{3}\ln\dfrac{x(x + 3)^2}{x^2 - 1}$

$$= \ln \sqrt[3]{\dfrac{x(x + 3)^2}{x^2 - 1}}$$

**105.** $2\ln 3 - \dfrac{1}{2}\ln(x^2 + 1) = \ln 9 - \ln\sqrt{x^2 + 1} = \ln\dfrac{9}{\sqrt{x^2 + 1}}$

**107.** (a) $e^{\ln x} = 4$

$\quad\quad x = 4$

(b) $\ln e^{2x} = 3$

$\quad\quad 2x = 3$

$\quad\quad x = \dfrac{3}{2}$

**109.** (a) $\ln x = 2$

$\quad\quad x = e^2 \approx 7.389$

(b) $e^x = 4$

$\quad\quad x = \ln 4 \approx 1.386$

**111.** $\quad e^x > 5$

$\quad\quad \ln e^x > \ln 5$

$\quad\quad\quad x > \ln 5$

**113.** $-2 < \ln x < 0$

$\quad e^{-2} < x < e^0 = 1$

$\quad \dfrac{1}{e^2} < x < 1$

**115.**

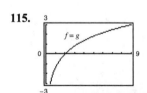

**117.** The domain of the natural logarithmic function is $(0, \infty)$ and the range is $(-\infty, \infty)$. The function is continuous, increasing, and one-to-one, and its graph is concave downward. In addition, if $a$ and $b$ are positive numbers and $n$ is rational, then $\ln(1) = 0$, $\ln(a \cdot b) = \ln a + \ln b$, $\ln(a^n) = n \ln a$, and $\ln(a/b) = \ln a - \ln b$.

**119.** $f(x) = e^x$. Domain is $(-\infty, \infty)$ and range is $(0, \infty)$. $f$ is continuous, increasing, one-to-one, and concave upwards on its entire domain.

$$\lim_{x \to -\infty} e^x = 0 \text{ and } \lim_{x \to \infty} e^x = \infty$$

**121.**

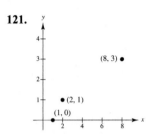

| $x$ | 1 | 2 | 8 |
|---|---|---|---|
| $y$ | 0 | 1 | 3 |

(a) $y$ is an exponential function of $x$:  False

(b) $y$ is a logarithmic function of $x$:  True; $y = \log_2 x$

(c) $x$ is an exponential function of $y$:  True; $2^y = x$

(d) $y$ is a linear function of $x$:  False

**123.** (a) $\beta = \dfrac{10}{\ln 10} \ln\left(\dfrac{I}{10^{-16}}\right)$

$$= \dfrac{10}{\ln 10}\Big[\ln I - \ln 10^{-16}\Big]$$

$$= \dfrac{10}{\ln 10}\Big[\ln I + 16\ln 10\Big]$$

$$= \dfrac{10}{\ln 10}\ln I + 160$$

$$= 10 \log_{10} I + 160$$

**125.** False

$\quad \ln x + \ln 25 = \ln(25x) \neq \ln(x + 25)$

**127.**

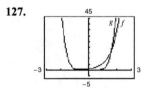

The graphs intersect three times: $(-0.7899, 0.2429)$, $(1.6242, 18.3615)$ and $(6, 46{,}656)$.

The function $f(x) = 6^x$ grows more rapidly.

**129.** $f(x) = \ln\left(x + \sqrt{x^2 + 1}\right)$

(a)

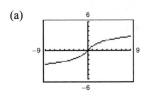

Domain: $-\infty < x < \infty$

(b) $f(-x) = \ln\left(-x + \sqrt{x^2 + 1}\right)$

$$= \ln\left[\frac{\left(-x + \sqrt{x^2 + 1}\right)\left(-x - \sqrt{x^2 + 1}\right)}{\left(-x - \sqrt{x^2 + 1}\right)}\right]$$

$$= \ln\left[\frac{\left(x^2\sqrt{x^2 + 1}\right)}{\left(-x - \sqrt{x^2 + 1}\right)}\right]$$

$$= \ln\left[\frac{1}{\left(x + \sqrt{x^2 + 1}\right)}\right]$$

$$= -\ln\left(x + \sqrt{x^2 + 1}\right) = -f(x)$$

(c) $\qquad y = \ln\left(x + \sqrt{x^2 + 1}\right)$

$$e^y = x + \sqrt{x^2 + 1}$$

$$\left(e^y - x\right)^2 = x^2 + 1$$

$$2xe^y = e^{2y} - 1$$

$$x = \frac{e^{2y} - 1}{2e^x}$$

**131.** $n = 12$

$12! = 12 \cdot 11 \cdot 10 \cdots 3 \cdot 2 \cdot 1 = 479{,}001{,}600$

Stirlings Formula:

$$12! \approx \left(\frac{12}{e}\right)^{12} \sqrt{2\pi(12)} \approx 475{,}687{,}487$$

# Review Exercises for Chapter 1

**1.** $y = 5x - 8$

$x = 0$: $y = 5(0) - 8 = -8 \Rightarrow (0, -8)$, $y$-intercept

$y = 0$: $0 = 5x - 8 \Rightarrow x = \frac{8}{5} \Rightarrow \left(\frac{8}{5}, 0\right)$, $x$-intercept

**3.** $y = \dfrac{x - 3}{x - 4}$

$x = 0$: $y = \dfrac{0 - 3}{0 - 4} = \dfrac{3}{4} \Rightarrow \left(0, \dfrac{3}{4}\right)$, $y$-intercept

$y = 0$: $0 = \dfrac{x - 3}{x - 4} \Rightarrow x = 3 \Rightarrow (3, 0)$, $x$-intercept

**5.** $y = x^2 + 4x$ does not have symmetry with respect to either axis or the origin.

**7.** Symmetric with respect to both axes and the origin because:

$$y^2 = \left(-x^2\right) - 5 \qquad (-y)^2 = x^2 - 5 \qquad (-y)^2 = (-x)^2 - 5$$

$$y^2 = x^2 - 5 \qquad\quad y^2 = x^2 - 5 \qquad\quad y^2 = x^2 - 5$$

**9.** $y = -\dfrac{1}{2}x + 3$

$y$-intercept: $y = -\dfrac{1}{2}(0) + 3 = 3$

$(0, 3)$

$x$-intercept: $-\dfrac{1}{2}x + 3 = 0$

$-\dfrac{1}{2}x = -3$

$x = 6$

$(6, 0)$

Symmetry: none

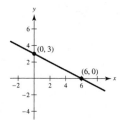

**11.** $y = x^3 - 4x$

$y$-intercept: $y = 0^3 - 4(0) = 0$

$(0, 0)$

$x$-intercepts: 

$x^3 - 4x = 0$

$x(x^2 - 4) = 0$

$x(x - 2)(x + 2) = 0$

$x = 0, 2, -2$

$(0, 0), (2, 0), (-2, 0)$

Symmetric with respect to the origin because

$(-x)^3 - 4(-x) = -x^3 + 4x = -(x^3 - 4x).$

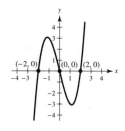

**13.** $y = 2\sqrt{4 - x}$

$y$-intercept: $y = 2\sqrt{4 - 0} = 2\sqrt{4} = 4$

$(0, 4)$

$x$-intercept: $2\sqrt{4 - x} = 0$

$\sqrt{4 - x} = 0$

$4 - x = 0$

$x = 4$

$(4, 0)$

Symmetry: none

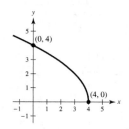

**15.** $5x + 3y = -1 \Rightarrow y = \frac{1}{3}(-5x - 1)$

$x - y = -5 \Rightarrow y = x + 5$

$\frac{1}{3}(-5x - 1) = x + 5$

$-5x - 1 = 3x + 15$

$-16 = 8x$

$-2 = x$

For $x = -2$, $y = x + 5 = -2 + 5 = 3$.

Point of intersection is: $(-2, 3)$

**17.** $x - y = -5 \Rightarrow y = x + 5$

$x^2 - y = 1 \Rightarrow y = x^2 - 1$

$x + 5 = x^2 - 1$

$0 = x^2 - x - 6$

$0 = (x - 3)(x + 2)$

$x = 3$ or $x = -2$

For $x = 3$, $y = 3 + 5 = 8$.

For $x = -2$, $y = -2 + 5 = 3$.

Points of intersection: $(3, 8), (-2, 3)$

**19.**

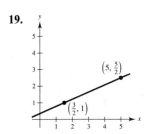

$$\text{Slope} = \frac{\left(\dfrac{5}{2}\right) - 1}{5 - \left(\dfrac{3}{2}\right)} = \frac{\dfrac{3}{2}}{\dfrac{7}{2}} = \frac{3}{7}$$

**21.** $y - (-5) = \frac{7}{4}(x - 3)$

$$y + 5 = \frac{7}{4}x - \frac{21}{4}$$

$$4y + 20 = 7x - 21$$

$$0 = 7x - 4y - 41$$

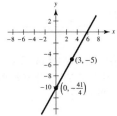

**23.**     $y - 0 = -\frac{2}{3}\left(x - (-3)\right)$

$$y = -\frac{2}{3}x - 2$$

$$2x + 3y + 6 = 0$$

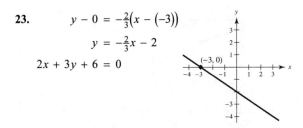

**25.** $y = 6$

Slope: 0

$y$-intercept: $(0, 6)$

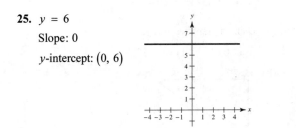

**27.** $y = 4x - 2$

Slope: 4

$y$-intercept: $(0, -2)$

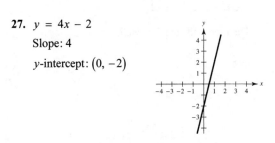

**29.**     $m = \dfrac{2 - 0}{8 - 0} = \dfrac{1}{4}$

$$y - 0 = \frac{1}{4}(x - 0)$$

$$y = \frac{1}{4}x$$

$$4y - x = 0$$

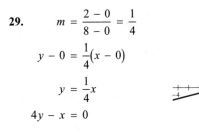
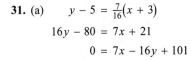

**31. (a)**     $y - 5 = \frac{7}{16}(x + 3)$

$$16y - 80 = 7x + 21$$

$$0 = 7x - 16y + 101$$

**(b)** $5x - 3y = 3$ has slope $\frac{5}{3}$.

$$y - 5 = \frac{5}{3}(x + 3)$$

$$3y - 15 = 5x + 15$$

$$0 = 5x - 3y + 30$$

**(c)** $3x + 4y = 8$

$$4y = -3x + 8$$

$$y = \frac{-3}{4}x + 2$$

Perpendicular line has slope $\dfrac{4}{3}$.

$$y - 5 = \frac{4}{3}\left(x - (-3)\right)$$

$$3y - 15 = 4x + 12$$

$$4x - 3y + 27 = 0 \ \text{ or } \ y = \frac{4}{3}x + 9$$

**(d)** Slope is undefined so the line is vertical.

$$x = -3$$

$$x + 3 = 0$$

**33.** The slope is $-850$.

$V = -850t + 12{,}500$.

$V(3) = -850(3) + 12{,}500 = \$9950$

**35.** $f(x) = 5x + 4$

**(a)** $f(0) = 5(0) + 4 = 4$

**(b)** $f(5) = 5(5) + 4 = 29$

**(c)** $f(-3) = 5(-3) + 4 = -11$

**(d)** $f(t + 1) = 5(t + 1) + 4 = 5t + 9$

**37.** $f(x) = 4x^2$

$$\frac{f(x + \Delta x) - f(x)}{\Delta x} = \frac{4(x + \Delta x)^2 - 4x^2}{\Delta x}$$

$$= \frac{4\left(x^2 + 2x\Delta x + (\Delta x)^2\right) - 4x^2}{\Delta x}$$

$$= \frac{4x^2 + 8x\Delta x + 4(\Delta x)^2 - 4x^2}{\Delta x}$$

$$= \frac{8x\Delta x + 4(\Delta x)^2}{\Delta x}$$

$$= 8x + 4\Delta x, \quad \Delta x \neq 0$$

**39.** $f(x) = x^2 + 3$

Domain: $(-\infty, \infty)$

Range: $[3, \infty)$

**41.** $f(x) = -|x + 1|$

Domain: $(-\infty, \infty)$

Range: $(-\infty, 0]$

**43.** $x - y^2 = 6$

$$y = \pm\sqrt{x - 6}$$

Not a function because there are two values of $y$ for some $x$.

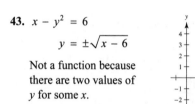

**45.** $y = \dfrac{|x - 2|}{x - 2}$

$y$ is a function of $x$ because there is one value of $y$ for each $x$.

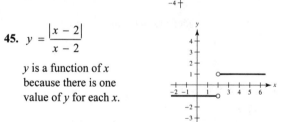

**47.** $f(x) = x^3 - 3x^2$

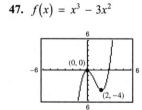

(a) The graph of $g$ is obtained from $f$ by a vertical shift down 1 unit, followed by a reflection in the $x$-axis:

$$g(x) = -\left[f(x) - 1\right] = -x^3 + 3x^2 + 1$$

(b) The graph of $g$ is obtained from $f$ by a vertical shift upwards of 1 and a horizontal shift of 2 to the right.

$$g(x) = f(x - 2) + 1 = (x - 2)^3 - 3(x - 2)^2 + 1$$

**49.** (a) $f(x) = x^2(x - 6)^2$

The leading coefficient is positive and the degree is even so the graph will rise to the left and to the right.

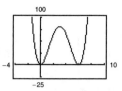

(b) $g(x) = x^3(x - 6)^2$

The leading coefficient is positive and the degree is odd so the graph will rise to the right and fall to the left.

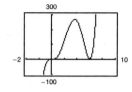

(c) $h(x) = x^3(x - 6)^3$

The leading coefficient is positive and the degree is even so the graph will rise to the left and to the right.

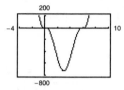

**51.** For company (a) the profit rose rapidly for the first year, and then leveled off. For the second company (b), the profit dropped, and then rose again later.

**53.** (a) $y = -1.204x + 64.2667$

(b)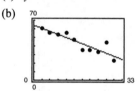

(c) The data point $(27, 44)$ is probably an error. Without this point, the new model is $y = -1.4344x + 66.4387$.

**55.** (a) Using a graphing utility,

$$y = 0.61t^2 + 11.0t + 172$$

(b)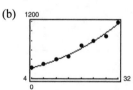

The model fits the data well.

**57.** (a)    $f(x) = \frac{1}{2}x - 3$

$$y = \frac{1}{2}x - 3$$

$$2(y + 3) = x$$

$$2(x + 3) = y$$

$$f^{-1}(x) = 2x + 6$$

(b)

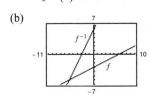

(c)   $f^{-1}(f(x)) = f^{-1}\left(\frac{1}{2}x - 3\right) = 2\left(\frac{1}{2}x - 3\right) + 6 = x$

$$f(f^{-1}(x)) = f(2x + 6) = \frac{1}{2}(2x + 6) - 3 = x$$

**59.** (a)    $f(x) = \sqrt{x + 1}$

$$y = \sqrt{x + 1}$$

$$y^2 - 1 = x$$

$$x^2 - 1 = y$$

$$f^{-1}(x) = x^2 - 1, \quad x \geq 0$$

(b)

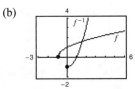

(c)   $f^{-1}(f(x)) = f^{-1}\left(\sqrt{x + 1}\right) = \sqrt{\left(x^2 - 1\right)^2} - 1 = x$

$$f(f^{-1}(x)) = f(x^2 - 1) = \sqrt{(x^2 - 1) + 1}$$

$$= \sqrt{x^2} = x \text{ for } x \geq 0$$

**61.** (a)    $f(x) = \sqrt[3]{x + 1}$

$$y = \sqrt[3]{x + 1}$$

$$y^3 - 1 = x$$

$$x^3 - 1 = y$$

$$f^{-1}(x) = x^3 - 1$$

(b)

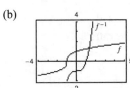

(c)   $f^{-1}(f(x)) = f^{-1}\left(\sqrt[3]{x + 1}\right)$

$$= \left(\sqrt[3]{x + 1}\right)^3 - 1 = x$$

$$f(f^{-1}(x)) = f(x^3 - 1) = \sqrt[3]{(x^3 - 1) + 1} = x$$

**63.** $f(x) = 2 \arctan(x + 3)$

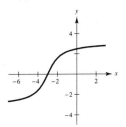

**65.** Let $\theta = \arcsin \frac{1}{2}$.

$$\sin \theta = \frac{1}{2}$$

$$\sin\left(\arcsin\frac{1}{2}\right) = \sin \theta = \frac{1}{2}$$

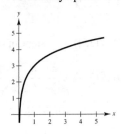

**67.** $f(x) = e^x$ matches (d).

The graph is increasing and the domain is all real $x$.

**69.** $f(x) = \ln(x + 1) + 1$ matches (c).

The graph is increasing and the domain is $x > -1$.

**71.** $f(x) = \ln x + 3$

Vertical shift three units upward

Vertical asymptote: $x = 0$

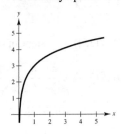

**73.** $\ln \sqrt[5]{\dfrac{4x^2 - 1}{4x^2 + 1}} = \dfrac{1}{5} \ln \dfrac{(2x - 1)(2x + 1)}{4x^2 + 1}$

$$= \frac{1}{5}\left[\ln(2x - 1) + \ln(2x + 1) - \ln(4x^2 + 1)\right]$$

**75.** $\ln 3 + \dfrac{1}{3}\ln(4 - x^2) - \ln x = \ln 3 + \ln \sqrt[3]{4 - x^2} - \ln x$

$$= \ln\left(\frac{3\sqrt[3]{4 - x^2}}{x}\right)$$

**77.** $\ln\sqrt{x + 1} = 2$

$$\sqrt{x + 1} = e^2$$

$$x + 1 = e^4$$

$$x = e^4 - 1 \approx 53.598$$

**79. (a)**    $f(x) = \ln \sqrt{x}$

$y = \ln \sqrt{x}$

$e^y = \sqrt{x}$

$e^{2y} = x$

$e^{2x} = y$

$f^{-1}(x) = e^{2x}$

**(b)**

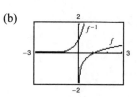

**(c)** $f^{-1}(f(x)) = f^{-1}(\ln \sqrt{x}) = e^{2\ln \sqrt{x}} = e^{\ln x} = x$

$f(f^{-1}(x)) = f(e^{2x}) = \ln \sqrt{e^{2x}} = \ln e^x = x$

**81.** $f = e^{-x/2}$

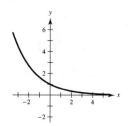

# Problem Solving for Chapter 1

**1. (a)**    $x^2 - 6x + y^2 - 8y = 0$

$(x^2 - 6x + 9) + (y^2 - 8y + 16) = 9 + 16$

$(x - 3)^2 + (y - 4)^2 = 25$

Center: $(3, 4)$; Radius: 5

**(b)** Slope of line from $(0, 0)$ to $(3, 4)$ is $\dfrac{4}{3}$. Slope of tangent line is $-\dfrac{3}{4}$. So, $y - 0 = -\dfrac{3}{4}(x - 0) \Rightarrow y = -\dfrac{3}{4}x$    Tangent line

**(c)** Slope of line from $(6, 0)$ to $(3, 4)$ is $\dfrac{4 - 0}{3 - 6} = -\dfrac{4}{3}$.

Slope of tangent line is $\dfrac{3}{4}$. So, $y - 0 = \dfrac{3}{4}(x - 6) \Rightarrow y = \dfrac{3}{4}x - \dfrac{9}{2}$    Tangent line

**(d)** $-\dfrac{3}{4}x = \dfrac{3}{4}x - \dfrac{9}{2}$

$\dfrac{3}{2}x = \dfrac{9}{2}$

$x = 3$

Intersection: $\left(3, -\dfrac{9}{4}\right)$

**3.** $H(x) = \begin{cases} 1, & x \geq 0 \\ 0, & x < 0 \end{cases}$

**(a)** $H(x) - 2 = \begin{cases} -1, & x \geq 0 \\ -2, & x < 0 \end{cases}$

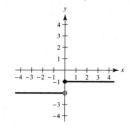

(b) $H(x - 2) = \begin{cases} 1, & x \geq 2 \\ 0, & x < 2 \end{cases}$

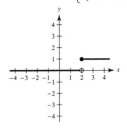

(c) $-H(x) = \begin{cases} -1, & x \geq 0 \\ 0, & x < 0 \end{cases}$

(d) $H(-x) = \begin{cases} 1, & x \leq 0 \\ 0, & x > 0 \end{cases}$

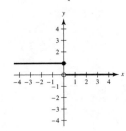

(e) $\frac{1}{2}H(x) = \begin{cases} \frac{1}{2}, & x \geq 0 \\ 0, & x < 0 \end{cases}$

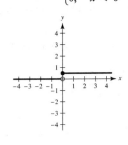

(f) $-H(x - 2) + 2 = \begin{cases} 1, & x \geq 2 \\ 2, & x < 2 \end{cases}$

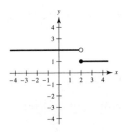

**5.** (a) $x + 2y = 100 \Rightarrow y = \dfrac{100 - x}{2}$

$$A(x) = xy = x\left(\frac{100 - x}{2}\right) = -\frac{x^2}{2} + 50x$$

Domain: $0 < x < 100$ or $(0, 100)$

(b)

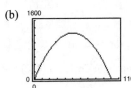

Maximum of 1250 m² at $x = 50$ m, $y = 25$ m.

(c) $A(x) = -\frac{1}{2}(x^2 - 100x)$

$\qquad = -\frac{1}{2}(x^2 - 100x + 2500) + 1250$

$\qquad = -\frac{1}{2}(x - 50)^2 + 1250$

$A(50) = 1250$ m² is the maximum.

$x = 50$ m, $y = 25$ m

**7.** The length of the trip in the water is $\sqrt{2^2 + x^2}$, and the length of the trip over land is $\sqrt{1 + (3 - x)^2}$.

So, the total time is $T = \dfrac{\sqrt{4 + x^2}}{2} + \dfrac{\sqrt{1 + (3 - x)^2}}{4}$ hours.

**9.** (a) Slope $= \dfrac{9 - 4}{3 - 2} = 5$. Slope of tangent line is less than 5.

(b) Slope $= \dfrac{4 - 1}{2 - 1} = 3$. Slope of tangent line is greater than 3.

(c) Slope $= \dfrac{4.41 - 4}{2.1 - 2} = 4.1$. Slope of tangent line is less than 4.1.

(d) Slope $= \dfrac{f(2 + h) - f(2)}{(2 + h) - 2}$

$= \dfrac{(2 + h)^2 - 4}{h}$

$= \dfrac{4h + h^2}{h}$

$= 4 + h, \; h \neq 0$

(e) Letting $h$ get closer and closer to 0, the slope approaches 4. So, the slope at $(2, 4)$ is 4.

**11.** $f(x) = y = \dfrac{1}{1 - x}$

(a) Domain: all $x \neq 1$ or $(-\infty, 1) \cup (1, \infty)$

Range: all $y \neq 0$ or $(-\infty, 0) \cup (0, \infty)$

(b) $f(f(x)) = f\left(\dfrac{1}{1 - x}\right) = \dfrac{1}{1 - \left(\dfrac{1}{1 - x}\right)} = \dfrac{1}{\dfrac{1 - x - 1}{1 - x}} = \dfrac{1 - x}{-x} = \dfrac{x - 1}{x}$

Domain: all $x \neq 0, 1$ or $(-\infty, 0) \cup (0, 1) \cup (1, \infty)$

(c) $f(f(f(x))) = f\left(\dfrac{x - 1}{x}\right) = \dfrac{1}{1 - \left(\dfrac{x - 1}{x}\right)} = \dfrac{1}{\dfrac{1}{x}} = x$

Domain: all $x \neq 0, 1$ or $(-\infty, 0) \cup (0, 1) \cup (1, \infty)$

(d) The graph is not a line. It has holes at $(0, 0)$ and $(1, 1)$.

**13.** (a) $\dfrac{I}{x^2} = \dfrac{2I}{(x - 3)^2}$

$x^2 - 6x + 9 = 2x^2$

$x^2 + 6x - 9 = 0$

$x = \dfrac{-6 \pm \sqrt{36 + 36}}{2}$

$= -3 \pm \sqrt{18}$

$\approx 1.2426, \, -7.2426$

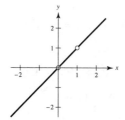

(b) $\dfrac{I}{x^2 + y^2} = \dfrac{2I}{(x - 3)^2 + y^2}$

$(x - 3)^2 + y^2 = 2(x^2 + y^2)$

$x^2 - 6x + 9 + y^2 = 2x^2 + 2y^2$

$x^2 + y^2 + 6x - 9 = 0$

$(x + 3)^2 + y^2 = 18$

Circle of radius $\sqrt{18}$ and center $(-3, 0)$.

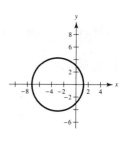

**15.**

$$d_1 d_2 = 1$$

$$\left[(x+1)^2 + y^2\right]\left[(x-1)^2 + y^2\right] = 1$$

$$(x+1)^2(x-1)^2 + y^2\left[(x+1)^2 + (x-1)^2\right] + y^4 = 1$$

$$\left(x^2 - 1\right)^2 + y^2\left[2x^2 + 2\right] + y^4 = 1$$

$$x^4 - 2x^2 + 1 + 2x^2y^2 + 2y^2 + y^4 = 1$$

$$\left(x^4 + 2x^2y^2 + y^4\right) - 2x^2 + 2y^2 = 0$$

$$\left(x^2 + y^2\right)^2 = 2\left(x^2 - y^2\right)$$

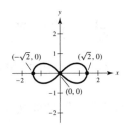

Let $y = 0$. Then $x^4 = 2x^2 \Rightarrow x = 0$ or $x^2 = 2$.

So, $(0, 0)$, $\left(\sqrt{2},\, 0\right)$ and $\left(-\sqrt{2},\, 0\right)$ are on the curve.

# CHAPTER 2
# Limits and Their Properties

# CHAPTER 2
# Limits and Their Properties

## Section 2.1   A Preview of Calculus

1. Precalculus: $(20 \text{ ft/sec})(15 \text{ sec}) = 300 \text{ ft}$

3. Calculus required: Slope of the tangent line at $x = 2$ is the rate of change, and equals about 0.16.

5. (a)  Precalculus: Area $= \frac{1}{2}bh = \frac{1}{2}(5)(4) = 10$ sq. units

   (b)  Calculus required:  Area $= bh$
   $$\approx 2(2.5)$$
   $$= 5 \text{ sq. units}$$

7. $f(x) = 6x - x^2$

(a)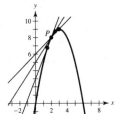

(b)  slope $= m = \dfrac{(6x - x^2) - 8}{x - 2} = \dfrac{(x - 2)(4 - x)}{x - 2}$

   $$= (4 - x), x \neq 2$$

   For $x = 3, m = 4 - 3 = 1$

   For $x = 2.5, m = 4 - 2.5 = 1.5 = \dfrac{3}{2}$

   For $x = 1.5, m = 4 - 1.5 = 2.5 = \dfrac{5}{2}$

(c)  At $P(2, 8)$, the slope is 2. You can improve your approximation by considering values of $x$ close to 2.

9. (a)  Area $\approx 5 + \frac{5}{2} + \frac{5}{3} + \frac{5}{4} \approx 10.417$

   Area $\approx \frac{1}{2}\left(5 + \frac{5}{1.5} + \frac{5}{2} + \frac{5}{2.5} + \frac{5}{3} + \frac{5}{3.5} + \frac{5}{4} + \frac{5}{4.5}\right) \approx 9.145$

   (b)  You could improve the approximation by using more rectangles.

## Section 2.2   Finding Limits Graphically and Numerically

1.

| $x$ | 3.9 | 3.99 | 3.999 | 4.001 | 4.01 | 4.1 |
|-----|-----|------|-------|-------|------|-----|
| $f(x)$ | 0.2041 | 0.2004 | 0.2000 | 0.2000 | 0.1996 | 0.1961 |

$\displaystyle\lim_{x \to 4}\dfrac{x - 4}{x^2 - 3x - 4} \approx 0.2000$   $\left(\text{Actual limit is } \dfrac{1}{5}.\right)$

3.

| $x$ | −0.1 | −0.01 | −0.001 | 0.001 | 0.01 | 0.1 |
|-----|------|-------|--------|-------|------|-----|
| $f(x)$ | 0.9983 | 0.99998 | 1.0000 | 1.0000 | 0.99998 | 0.9983 |

$\displaystyle\lim_{x \to 0}\dfrac{\sin x}{x} \approx 1.0000$   $(\text{Actual limit is } 1.)$ $(\text{Make sure you use radian mode.})$

**5.**

| x | −0.1 | −0.01 | −0.001 | 0.001 | 0.01 | 0.1 |
|---|---|---|---|---|---|---|
| f(x) | 0.9516 | 0.9950 | 0.9995 | 1.0005 | 1.0050 | 1.0517 |

$$\lim_{x \to 0} \frac{e^x - 1}{x} \approx 1.0000 \quad \text{(Actual limit is 1.)}$$

**7.**

| x | 0.9 | 0.99 | 0.999 | 1.001 | 1.01 | 1.1 |
|---|---|---|---|---|---|---|
| f(x) | 0.2564 | 0.2506 | 0.2501 | 0.2499 | 0.2494 | 0.2439 |

$$\lim_{x \to 1} \frac{x - 2}{x^2 + x - 6} \approx 0.2500 \quad \left(\text{Actual limit is } \frac{1}{4}.\right)$$

**9.**

| x | 0.9 | 0.99 | 0.999 | 1.001 | 1.01 | 1.1 |
|---|---|---|---|---|---|---|
| f(x) | 0.7340 | 0.6733 | 0.6673 | 0.6660 | 0.6600 | 0.6015 |

$$\lim_{x \to 1} \frac{x^4 - 1}{x^6 - 1} \approx 0.6666 \quad \left(\text{Actual limit is } \frac{2}{3}.\right)$$

**11.**

| x | −6.1 | −6.01 | −6.001 | −6 | −5.999 | −5.99 | −5.9 |
|---|---|---|---|---|---|---|---|
| f(x) | −0.1248 | −0.1250 | −0.1250 | ? | −0.1250 | −0.1250 | −0.1252 |

$$\lim_{x \to -6} \frac{\sqrt{10 - x} - 4}{x + 6} \approx -0.1250 \quad \left(\text{Actual limit is } -\frac{1}{8}.\right)$$

**13.**

| x | −0.1 | −0.01 | −0.001 | 0.001 | 0.01 | 0.1 |
|---|---|---|---|---|---|---|
| f(x) | 1.9867 | 1.9999 | 2.0000 | 2.0000 | 1.9999 | 1.9867 |

$$\lim_{x \to 0} \frac{\sin 2x}{x} \approx 2.0000 \quad \text{(Actual limit is 2.)} \text{(Make sure you use radian mode.)}$$

**15.**

| x | 1.9 | 1.99 | 1.999 | 2.001 | 2.01 | 2.1 |
|---|---|---|---|---|---|---|
| f(x) | 0.5129 | 0.5013 | 0.5001 | 0.4999 | 0.4988 | 0.4879 |

$$\lim_{x \to 2} \frac{\ln x - \ln 2}{x - 2} \approx 0.5000 \quad \left(\text{Actual limit is } \frac{1}{2}.\right)$$

**17.** $\lim\limits_{x \to 3} (4 - x) = 1$

**19.** $\lim\limits_{x \to 2} f(x) = \lim\limits_{x \to 2} (4 - x) = 2$

**21.** $\lim\limits_{x \to 2} \dfrac{|x - 2|}{x - 2}$ does not exist.

For values of $x$ to the left of 2, $\dfrac{|x - 2|}{(x - 2)} = -1$, whereas

for values of $x$ to the right of 2, $\dfrac{|x - 2|}{(x - 2)} = 1$.

**23.** $\lim\limits_{x \to 0} \cos(1/x)$ does not exist because the function oscillates between $-1$ and $1$ as x approaches 0.

**25.** (a) $f(1)$ exists. The black dot at $(1, 2)$ indicates that
$$f(1) = 2.$$

(b) $\lim\limits_{x \to 1} f(x)$ does not exist. As x approaches 1 from the left, $f(x)$ approaches 3.5, whereas as $x$ approaches 1 from the right, $f(x)$ approaches 1.

(c) $f(4)$ does not exist. The hollow circle at $(4, 2)$ indicates that $f$ is not defined at 4.

(d) $\lim\limits_{x \to 4} f(x)$ exists. As x approaches 4, $f(x)$ approaches 2: $\lim\limits_{x \to 4} f(x) = 2.$

**27.**

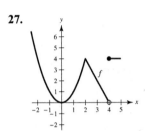

$\lim\limits_{x \to c} f(x)$ exists for all values of $c \neq 4.$

**29.** One possible answer is

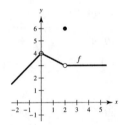

**31.** You need $\left| f(x) - 3 \right| = \left| (x + 1) - 3 \right| = \left| x - 2 \right| < 0.4.$
So, take $\delta = 0.4.$ If $0 < \left| x - 2 \right| < 0.4,$ then
$$\left| x - 2 \right| = \left| (x + 1) - 3 \right| = \left| f(x) - 3 \right| < 0.4, \text{ as desired.}$$

**33.** You need to find $\delta$ such that $0 < \left| x - 1 \right| < \delta$ implies
$$\left| f(x) - 1 \right| = \left| \frac{1}{x} - 1 \right| < 0.1. \text{ That is,}$$

$$-0.1 < \frac{1}{x} - 1 < 0.1$$

$$1 - 0.1 < \frac{1}{x} < 1 + 0.1$$

$$\frac{9}{10} < \frac{1}{x} < \frac{11}{10}$$

$$\frac{10}{9} > x > \frac{10}{11}$$

$$\frac{10}{9} - 1 > x - 1 > \frac{10}{11} - 1$$

$$\frac{1}{9} > x - 1 > -\frac{1}{11}.$$

So take $\delta = \frac{1}{11}.$ Then $0 < \left| x - 1 \right| < \delta$ implies

$$-\frac{1}{11} < x - 1 < \frac{1}{11}$$

$$-\frac{1}{11} < x - 1 < \frac{1}{9}.$$

Using the first series of equivalent inequalities, you obtain

$$\left| f(x) - 1 \right| = \left| \frac{1}{x} - 1 \right| < 0.1.$$

**35.** $\lim\limits_{x \to 2}(3x + 2) = 3(2) + 2 = 8 = L$

$$\left| (3x + 2) - 8 \right| < 0.01$$

$$\left| 3x - 6 \right| < 0.01$$

$$3\left| x - 2 \right| < 0.01$$

$$0 < \left| x - 2 \right| < \frac{0.01}{3} \approx 0.0033 = \delta$$

So, if $0 < \left| x - 2 \right| < \delta = \frac{0.01}{3},$ you have

$$3\left| x - 2 \right| < 0.01$$

$$\left| 3x - 6 \right| < 0.01$$

$$\left| (3x + 2) - 8 \right| < 0.01$$

$$\left| f(x) - L \right| < 0.01.$$

**37.** $\lim\limits_{x \to 2}(x^2 - 3) = 2^2 - 3 = 1 = L$

$$\left|(x^2 - 3) - 1\right| < 0.01$$

$$\left|x^2 - 4\right| < 0.01$$

$$\left|(x + 2)(x - 2)\right| < 0.01$$

$$\left|x + 2\right|\left|x - 2\right| < 0.01$$

$$\left|x - 2\right| < \frac{0.01}{\left|x + 2\right|}$$

If you assume $1 < x < 3$, then $\delta \approx 0.01/5 = 0.002$.

So, if $0 < \left|x - 2\right| < \delta \approx 0.002$, you have

$$\left|x - 2\right| < 0.002 = \frac{1}{5}(0.01) < \frac{1}{\left|x + 2\right|}(0.01)$$

$$\left|x + 2\right|\left|x - 2\right| < 0.01$$

$$\left|x^2 - 4\right| < 0.01$$

$$\left|(x^2 - 3) - 1\right| < 0.01$$

$$\left|f(x) - L\right| < 0.01.$$

**39.** $\lim\limits_{x \to 4}(x + 2) = 4 + 2 = 6$

Given $\varepsilon > 0$:

$$\left|(x + 2) - 6\right| < \varepsilon$$

$$\left|x - 4\right| < \varepsilon = \delta$$

So, let $\delta = \varepsilon$. So, if $0 < \left|x - 4\right| < \delta = \varepsilon$, you have

$$\left|x - 4\right| < \varepsilon$$

$$\left|(x + 2) - 6\right| < \varepsilon$$

$$\left|f(x) - L\right| < \varepsilon.$$

**41.** $\lim\limits_{x \to -4}\left(\frac{1}{2}x - 1\right) = \frac{1}{2}(-4) - 1 = -3$

Given $\varepsilon > 0$:

$$\left|\left(\frac{1}{2}x - 1\right) - (-3)\right| < \varepsilon$$

$$\left|\frac{1}{2}x + 2\right| < \varepsilon$$

$$\frac{1}{2}\left|x - (-4)\right| < \varepsilon$$

$$\left|x - (-4)\right| < 2\varepsilon$$

So, let $\delta = 2\varepsilon$.

So, if $0 < \left|x - (-4)\right| < \delta = 2\varepsilon$, you have

$$\left|x - (-4)\right| < 2\varepsilon$$

$$\left|\frac{1}{2}x + 2\right| < \varepsilon$$

$$\left|\left(\frac{1}{2}x - 1\right) + 3\right| < \varepsilon$$

$$\left|f(x) - L\right| < \varepsilon.$$

**43.** $\lim\limits_{x \to 6} 3 = 3$

Given $\varepsilon > 0$:

$$\left|3 - 3\right| < \varepsilon$$

$$0 < \varepsilon$$

So, any $\delta > 0$ will work.

So, for any $\delta > 0$, you have

$$\left|3 - 3\right| < \varepsilon$$

$$\left|f(x) - L\right| < \varepsilon.$$

**45.** $\lim\limits_{x \to 0} \sqrt[3]{x} = 0$

Given $\varepsilon > 0$: $\left|\sqrt[3]{x} - 0\right| < \varepsilon$

$$\left|\sqrt[3]{x}\right| < \varepsilon$$

$$\left|x\right| < \varepsilon^3 = \delta$$

So, let $\delta = \varepsilon^3$.

So, for $0\left|x - 0\right|\delta = \varepsilon^3$, you have

$$\left|x\right| < \varepsilon^3$$

$$\left|\sqrt[3]{x}\right| < \varepsilon$$

$$\left|\sqrt[3]{x} - 0\right| < \varepsilon$$

$$\left|f(x) - L\right| < \varepsilon.$$

**47.** $\lim\limits_{x \to -5}\left|x - 5\right| = \left|(-5) - 5\right| = \left|-10\right| = 10$

Given $\varepsilon > 0$: $\left|\left|x - 5\right| - 10\right| < \varepsilon$

$$\left|-(x - 5) - 10\right| < \varepsilon \quad (x - 5 < 0)$$

$$\left|-x - 5\right| < \varepsilon$$

$$\left|x - (-5)\right| < \varepsilon$$

So, let $\delta = \varepsilon$.

So for $\left|x - (-5)\right| < \delta = \varepsilon$, you have

$$\left|-(x + 5)\right| < \varepsilon$$

$$\left|-(x - 5) - 10\right| < \varepsilon$$

$$\left|\left|x - 5\right| - 10\right| < \varepsilon \quad (\text{because } x - 5 < 0)$$

$$\left|f(x) - L\right| < \varepsilon.$$

**49.** $\lim\limits_{x \to 1} \left(x^2 + 1\right) = 1^2 + 1 = 2$

Given $\varepsilon > 0$:   $\left|\left(x^2 + 1\right) - 2\right| < \varepsilon$

$$\left|x^2 - 1\right| < \varepsilon$$

$$\left|(x + 1)(x - 1)\right| < \varepsilon$$

$$\left|x - 1\right| < \frac{\varepsilon}{|x + 1|}$$

If you assume $0 < x < 2$, then $\delta = \varepsilon/3$.

So for $0 < |x - 1| < \delta = \dfrac{\varepsilon}{3}$, you have

$$\left|x - 1\right| < \frac{1}{3}\varepsilon < \frac{1}{|x + 1|}\varepsilon$$

$$\left|x^2 - 1\right| < \varepsilon$$

$$\left|\left(x^2 + 1\right) - 2\right| < \varepsilon$$

$$\left|f(x) - 2\right| < \varepsilon.$$

**51.** $\lim\limits_{x \to \pi} f(x) = \lim\limits_{x \to \pi} 4 = 4$

**53.** $f(x) = \dfrac{\sqrt{x + 5} - 3}{x - 4}$

$\lim\limits_{x \to 4} f(x) = \dfrac{1}{6}$

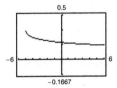

The domain is $[-5, 4) \cup (4, \infty)$.

The graphing utility does not show the hole at $\left(4, \dfrac{1}{6}\right)$.

**55.** $f(x) = \dfrac{x - 9}{\sqrt{x} - 3}$

$\lim\limits_{x \to 9} f(x) = 6$

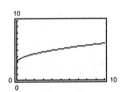

The domain is all $x \geq 0$ except $x = 9$. The graphing utility does not show the hole at $(9, 6)$.

**57.** $C(t) = 9.99 - 0.79\left[\!\left[-(t - 1)\right]\!\right]$

(a)

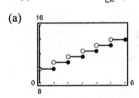

(b)

| $t$ | 3 | 3.3 | 3.4 | 3.5 | 3.6 | 3.7 | 4 |
|---|---|---|---|---|---|---|---|
| $C$ | 11.57 | 12.36 | 12.36 | 12.36 | 12.36 | 12.36 | 12.36 |

$\lim\limits_{t \to 3.5} C(t) = 12.36$

(c)

| $t$ | 2 | 2.5 | 2.9 | 3 | 3.1 | 3.5 | 4 |
|---|---|---|---|---|---|---|---|
| $C$ | 10.78 | 11.57 | 11.57 | 11.57 | 12.36 | 12.36 | 12.36 |

The $\lim\limits_{t \to 3} C(t)$ does not exist because the values of $C$ approach different values as $t$ approaches 3 from both sides.

**59.** $\lim\limits_{x \to 8} f(x) = 25$ means that the values of $f$ approach 25 as $x$ gets closer and closer to 8.

**61.** (i) The values of $f$ approach different numbers as $x$ approaches $c$ from different sides of $c$:

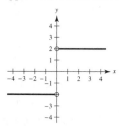

(ii) The values of $f$ increase without bound as $x$ approaches $c$:

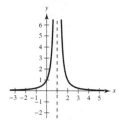

(iii) The values of $f$ oscillate between two fixed numbers as $x$ approaches $c$:

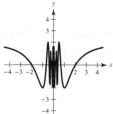

**63.** (a) $C = 2\pi r$

$$r = \frac{C}{2\pi} = \frac{6}{2\pi} = \frac{3}{\pi} \approx 0.9549 \text{ cm}$$

(b) When $C = 5.5: r = \dfrac{5.5}{2\pi} \approx 0.87535 \text{ cm}$

When $C = 6.5: r = \dfrac{6.5}{2\pi} \approx 1.03451 \text{ cm}$

So $0.87535 < r < 1.03451$.

(c) $\displaystyle\lim_{x \to 3/\pi} (2\pi r) = 6; \varepsilon = 0.5; \delta \approx 0.0796$

**65.** $f(x) = (1 + x)^{1/x}$

$$\lim_{x \to 0}(1 + x)^{1/x} = e \approx 2.71828$$

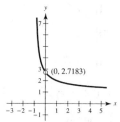

| $x$ | $f(x)$ | $x$ | $f(x)$ |
|---|---|---|---|
| $-0.1$ | 2.867972 | 0.1 | 2.593742 |
| $-0.01$ | 2.731999 | 0.01 | 2.704814 |
| $-0.001$ | 2.719642 | 0.001 | 2.716942 |
| $-0.0001$ | 2.718418 | 0.0001 | 2.718146 |
| $-0.00001$ | 2.718295 | 0.00001 | 2.718268 |
| $-0.000001$ | 2.718283 | 0.000001 | 2.718280 |

**67.**

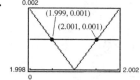

Using the zoom and trace feature, $\delta = 0.001$. So $(2 - \delta, 2 + \delta) = (1.999, 2.001)$.

**Note:** $\dfrac{x^2 - 4}{x - 2} = x + 2$ for $x \neq 2$.

**69.** False. The existence or nonexistence of $f(x)$ at $x = c$ has no bearing on the existence of the limit of $f(x)$ as $x \to c$.

**71.** False. Let

$$f(x) = \begin{cases} x - 4, & x \neq 2 \\ 0, & x = 2 \end{cases}$$

$$f(2) = 0$$

$$\lim_{x \to 2} f(x) = \lim_{x \to 2}(x - 4) = 2 \neq 0$$

**73.** $f(x) = \sqrt{x}$

$$\lim_{x \to 0.25} \sqrt{x} = 0.5 \text{ is true.}$$

As $x$ approaches $0.25 = \frac{1}{4}$ from either side, $f(x) = \sqrt{x}$ approaches $\frac{1}{2} = 0.5$.

**75.** Using a graphing utility, you see that

$$\lim_{x \to 0} \frac{\sin x}{x} = 1$$

$$\lim_{x \to 0} \frac{\sin 2x}{x} = 2, \text{ etc.}$$

So, $\lim_{x \to 0} \dfrac{\sin nx}{x} = n$.

**77.** If $\lim_{x \to c} f(x) = L_1$ and $\lim_{x \to c} f(x) = L_2$, then for every $\varepsilon > 0$, there exists $\delta_1 > 0$ and $\delta_2 > 0$ such that

$|x - c| < \delta_1 \Rightarrow |f(x) - L_1| < \varepsilon$ and $|x - c| < \delta_2 \Rightarrow |f(x) - L_2| < \varepsilon$. Let $\delta$ equal the smaller of $\delta_1$ and $\delta_2$.

Then for $|x - c| < \delta$, you have $|L_1 - L_2| = |L_1 - f(x) + f(x) - L_2| \le |L_1 - f(x)| + |f(x) - L_2| < \varepsilon + \varepsilon$.

Therefore, $|L_1 - L_2| < 2\varepsilon$. Since $\varepsilon > 0$ is arbitrary, it follows that $L_1 = L_2$.

**79.** $\lim_{x \to c} [f(x) - L] = 0$ means that for every $\varepsilon > 0$ there

exists $\delta > 0$ such that if

$0 < |x - c| < \delta$,

then

$|(f(x) - L) - 0| < \varepsilon$.

This means the same as $|f(x) - L| < \varepsilon$ when

$0 < |x - c| < \delta$.

So, $\lim_{x \to c} f(x) = L$.

**81.** The radius $OP$ has a length equal to the altitude $z$ of the

triangle plus $\dfrac{h}{2}$. So, $z = 1 - \dfrac{h}{2}$.

Area triangle $= \dfrac{1}{2}b\left(1 - \dfrac{h}{2}\right)$

Area rectangle $= bh$

Because these are equal, $\dfrac{1}{2}b\left(1 - \dfrac{h}{2}\right) = bh$

$$1 - \frac{h}{2} = 2h$$

$$\frac{5}{2}h = 1$$

$$h = \frac{2}{5}.$$

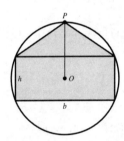

# Section 2.3   Evaluating Limits Analytically

**1.**

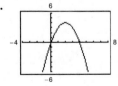

(a) $\lim_{x \to 4} h(x) = 0$

(b) $\lim_{x \to -1} h(x) = -5$

**3.**

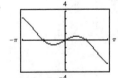

$f(x) = x \cos x$

(a) $\lim_{x \to 0} f(x) = 0$

(b) $\lim_{x \to \pi/3} f(x) \approx 0.524$

$\left(= \dfrac{\pi}{6}\right)$

**5.** $\lim\limits_{x \to 2} x^3 = 2^3 = 8$

**7.** $\lim\limits_{x \to 0} (2x - 1) = 2(0) - 1 = -1$

**9.** $\lim\limits_{x \to -3} (x^2 + 3x) = (-3)^2 + 3(-3) = 9 - 9 = 0$

**11.** $\lim\limits_{x \to -3} (2x^2 + 4x + 1) = 2(-3)^2 + 4(-3) + 1$
$$= 18 - 12 + 1 = 7$$

**13.** $\lim\limits_{x \to 3} \sqrt{x + 1} = \sqrt{3 + 1} = 2$

**15.** $\lim\limits_{x \to -4} (x + 3)^2 = (-4 + 3)^2 = 1$

**17.** $\lim\limits_{x \to 2} \dfrac{1}{x} = \dfrac{1}{2}$

**19.** $\lim\limits_{x \to 1} \dfrac{x}{x^2 + 4} = \dfrac{1}{1^2 + 4} = \dfrac{1}{5}$

**21.** $\lim\limits_{x \to 7} \dfrac{3x}{\sqrt{x + 2}} = \dfrac{3(7)}{\sqrt{7 + 2}} = \dfrac{21}{3} = 7$

**23.** $\lim\limits_{x \to \pi/2} \sin x = \sin \dfrac{\pi}{2} = 1$

**25.** $\lim\limits_{x \to 1} \cos \dfrac{\pi x}{3} = \cos \dfrac{\pi}{3} = \dfrac{1}{2}$

**27.** $\lim\limits_{x \to 0} \sec 2x = \sec 0 = 1$

**29.** $\lim\limits_{x \to 5\pi/6} \sin x = \sin \dfrac{5\pi}{6} = \dfrac{1}{2}$

**31.** $\lim\limits_{x \to 3} \tan\left(\dfrac{\pi x}{4}\right) = \tan \dfrac{3\pi}{4} = -1$

**33.** $\lim\limits_{x \to 0} e^x \cos 2x = e^0 \cos 0 = 1$

**35.** $\lim\limits_{x \to 1} \left( \ln 3x + e^x \right) = \ln 3 + e$

**37.** (a) $\lim\limits_{x \to 1} f(x) = 5 - 1 = 4$

   (b) $\lim\limits_{x \to 4} g(x) = 4^3 = 64$

   (c) $\lim\limits_{x \to 1} g(f(x)) = g(f(1)) = g(4) = 64$

**39.** (a) $\lim\limits_{x \to 1} f(x) = 4 - 1 = 3$

   (b) $\lim\limits_{x \to 3} g(x) = \sqrt{3 + 1} = 2$

   (c) $\lim\limits_{x \to 1} g(f(x)) = g(3) = 2$

**41.** (a) $\lim\limits_{x \to c}\left[5g(x)\right] = 5 \lim\limits_{x \to c} g(x) = 5(2) = 10$

   (b) $\lim\limits_{x \to c}\left[f(x) + g(x)\right] = \lim\limits_{x \to c} f(x) + \lim\limits_{x \to c} g(x) = 3 + 2 = 5$

   (c) $\lim\limits_{x \to c}\left[f(x)g(x)\right] = \left[\lim\limits_{x \to c} f(x)\right]\left[\lim\limits_{x \to c} g(x)\right] = (3)(2) = 6$

   (d) $\lim\limits_{x \to c} \dfrac{f(x)}{g(x)} = \dfrac{\lim\limits_{x \to c} f(x)}{\lim\limits_{x \to c} g(x)} = \dfrac{3}{2}$

**43.** (a) $\lim\limits_{x \to c}\left[f(x)\right]^3 = \left[\lim\limits_{x \to c} f(x)\right]^3 = (4)^3 = 64$

   (b) $\lim\limits_{x \to c}\sqrt{f(x)} = \sqrt{\lim\limits_{x \to c} f(x)} = \sqrt{4} = 2$

   (c) $\lim\limits_{x \to c}\left[3f(x)\right] = 3 \lim\limits_{x \to c} f(x) = 3(4) = 12$

   (d) $\lim\limits_{x \to c}\left[f(x)\right]^{3/2} = \left[\lim\limits_{x \to c} f(x)\right]^{3/2} = (4)^{3/2} = 8$

**45.** $f(x) = \dfrac{x^2 - 1}{x + 1} = \dfrac{(x + 1)(x - 1)}{x + 1}$ and

   $g(x) = x - 1$ agree except at $x = -1$.

   $\lim\limits_{x \to -1} f(x) = \lim\limits_{x \to -1} g(x) = \lim\limits_{x \to -1} (x - 1) = -1 - 1 = -2$

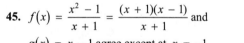

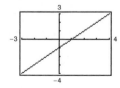

**47.** $f(x) = \dfrac{x^3 - 8}{x - 2}$ and $g(x) = x^2 + 2x + 4$ agree except at $x = 2$.

   $\lim\limits_{x \to 2} f(x) = \lim\limits_{x \to 2} g(x) = \lim\limits_{x \to 2}(x^2 + 2x + 4)$
   $$= 2^2 + 2(2) + 4 = 12$$

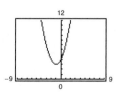

**49.** $f(x) = \dfrac{(x+4)\ln(x+6)}{x^2-16}$ and $g(x) = \dfrac{\ln(x+6)}{x-4}$

agree except at $x = -4$.

$$\lim_{x\to-4} f(x) = \lim_{x\to-4} g(x) = \frac{\ln 2}{-8} \approx -0.0866$$

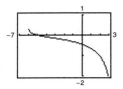

**51.** $\displaystyle\lim_{x\to0}\frac{x}{x^2-x} = \lim_{x\to0}\frac{x}{x(x-1)} = \lim_{x\to0}\frac{1}{x-1} = \frac{1}{0-1} = -1$

**53.** $\displaystyle\lim_{x\to4}\frac{x-4}{x^2-16} = \lim_{x\to4}\frac{x-4}{(x+4)(x-4)}$

$$= \lim_{x\to4}\frac{1}{x+4} = \frac{1}{4+4} = \frac{1}{8}$$

**55.** $\displaystyle\lim_{x\to-3}\frac{x^2+x-6}{x^2-9} = \lim_{x\to-3}\frac{(x+3)(x-2)}{(x+3)(x-3)}$

$$= \lim_{x\to-3}\frac{x-2}{x-3} = \frac{-3-2}{-3-3} = \frac{-5}{-6} = \frac{5}{6}$$

**57.** $\displaystyle\lim_{x\to4}\frac{\sqrt{x+5}-3}{x-4} = \lim_{x\to4}\frac{\sqrt{x+5}-3}{x-4}\cdot\frac{\sqrt{x+5}+3}{\sqrt{x+5}+3}$

$$= \lim_{x\to4}\frac{(x+5)-9}{(x-4)(\sqrt{x+5}+3)} = \lim_{x\to4}\frac{1}{\sqrt{x+5}+3} = \frac{1}{\sqrt{9}+3} = \frac{1}{6}$$

**59.** $\displaystyle\lim_{x\to0}\frac{\sqrt{x+5}-\sqrt{5}}{x} = \lim_{x\to0}\frac{\sqrt{x+5}-\sqrt{5}}{x}\cdot\frac{\sqrt{x+5}+\sqrt{5}}{\sqrt{x+5}+\sqrt{5}}$

$$= \lim_{x\to0}\frac{(x+5)-5}{x(\sqrt{x+5}+\sqrt{5})} = \lim_{x\to0}\frac{1}{\sqrt{x+5}+\sqrt{5}} = \frac{1}{\sqrt{5}+\sqrt{5}} = \frac{1}{2\sqrt{5}} = \frac{\sqrt{5}}{10}$$

**61.** $\displaystyle\lim_{x\to0}\frac{\dfrac{1}{3+x}-\dfrac{1}{3}}{x} = \lim_{x\to0}\frac{3-(3+x)}{(3+x)3(x)} = \lim_{x\to0}\frac{-x}{(3+x)(3)(x)} = \lim_{x\to0}\frac{-1}{(3+x)3} = \frac{-1}{(3)3} = -\frac{1}{9}$

**63.** $\displaystyle\lim_{\Delta x\to0}\frac{2(x+\Delta x)-2x}{\Delta x} = \lim_{\Delta x\to0}\frac{2x+2\Delta x-2x}{\Delta x} = \lim_{\Delta x\to0}\frac{2\Delta x}{\Delta x} = \lim_{\Delta x\to0}2 = 2$

**65.** $\displaystyle\lim_{\Delta x\to0}\frac{(x+\Delta x)^2-2(x+\Delta x)+1-(x^2-2x+1)}{\Delta x} = \lim_{\Delta x\to0}\frac{x^2+2x\Delta x+(\Delta x)^2-2x-2\Delta x+1-x^2+2x-1}{\Delta x}$

$$= \lim_{\Delta x\to0}(2x+\Delta x-2) = 2x-2$$

**67.** $\displaystyle\lim_{x\to0}\frac{\sin x}{5x} = \lim_{x\to0}\left[\left(\frac{\sin x}{x}\right)\left(\frac{1}{5}\right)\right] = (1)\left(\frac{1}{5}\right) = \frac{1}{5}$

**75.** $\displaystyle\lim_{x\to\pi/2}\frac{\cos x}{\cot x} = \lim_{x\to\pi/2}\sin x = 1$

**69.** $\displaystyle\lim_{x\to0}\frac{\sin x(1-\cos x)}{x^2} = \lim_{x\to0}\left[\frac{\sin x}{x}\cdot\frac{1-\cos x}{x}\right]$

$$= (1)(0) = 0$$

**77.** $\displaystyle\lim_{x\to0}\frac{1-e^{-x}}{e^x-1} = \lim_{x\to0}\frac{1-e^{-x}}{e^x-1}\cdot\frac{e^{-x}}{e^{-x}} = \lim_{x\to0}\frac{(1-e^{-x})e^{-x}}{1-e^{-x}}$

$$= \lim_{x\to0}e^{-x} = 1$$

**71.** $\displaystyle\lim_{x\to0}\frac{\sin^2 x}{x} = \lim_{x\to0}\left[\frac{\sin x}{x}\sin x\right] = (1)\sin 0 = 0$

**79.** $\displaystyle\lim_{t\to0}\frac{\sin 3t}{2t} = \lim_{t\to0}\left(\frac{\sin 3t}{3t}\right)\left(\frac{3}{2}\right) = (1)\left(\frac{3}{2}\right) = \frac{3}{2}$

**73.** $\displaystyle\lim_{h\to0}\frac{(1-\cos h)^2}{h} = \lim_{h\to0}\left[\frac{1-\cos h}{h}(1-\cos h)\right]$

$$= (0)(0) = 0$$

**81.** $f(x) = \dfrac{\sqrt{x+2} - \sqrt{2}}{x}$

| $x$ | −0.1 | −0.01 | −0.001 | 0 | 0.001 | 0.01 | 0.1 |
|---|---|---|---|---|---|---|---|
| $f(x)$ | 0.358 | 0.354 | 0.354 | ? | 0.354 | 0.353 | 0.349 |

It appears that the limit is 0.354.

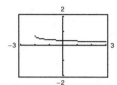

The graph has a hole at $x = 0$.

Analytically, $\displaystyle\lim_{x\to0}\dfrac{\sqrt{x+2}-\sqrt{2}}{x} = \lim_{x\to0}\dfrac{\sqrt{x+2}-\sqrt{2}}{x}\cdot\dfrac{\sqrt{x+2}+\sqrt{2}}{\sqrt{x+2}+\sqrt{2}}$

$$= \lim_{x\to0}\dfrac{x+2-2}{x\left(\sqrt{x+2}+\sqrt{2}\right)} = \lim_{x\to0}\dfrac{1}{\sqrt{x+2}+\sqrt{2}} = \dfrac{1}{2\sqrt{2}} = \dfrac{\sqrt{2}}{4} \approx 0.354.$$

**83.** $f(x) = \dfrac{\dfrac{1}{2+x} - \dfrac{1}{2}}{x}$

| $x$ | −0.1 | −0.01 | −0.001 | 0 | 0.001 | 0.01 | 0.1 |
|---|---|---|---|---|---|---|---|
| $f(x)$ | −0.263 | −0.251 | −0.250 | ? | −0.250 | −0.249 | −0.238 |

It appears that the limit is −0.250.

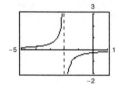

The graph has a hole at $x = 0$.

Analytically, $\displaystyle\lim_{x\to0}\dfrac{\dfrac{1}{2+x}-\dfrac{1}{2}}{x} = \lim_{x\to0}\dfrac{2-(2+x)}{2(2+x)}\cdot\dfrac{1}{x} = \lim_{x\to0}\dfrac{-x}{2(2+x)}\cdot\dfrac{1}{x} = \lim_{x\to0}\dfrac{-1}{2(2+x)} = -\dfrac{1}{4}.$

**85.** $f(t) = \dfrac{\sin 3t}{t}$

| $t$ | −0.1 | −0.01 | −0.001 | 0 | 0.001 | 0.01 | 0.1 |
|---|---|---|---|---|---|---|---|
| $f(t)$ | 2.96 | 2.9996 | 3 | ? | 3 | 2.9996 | 2.96 |

It appears that the limit is 3.

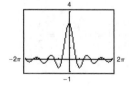

The graph has a hole at $t = 0$.

Analytically, $\displaystyle\lim_{t\to0}\dfrac{\sin 3t}{t} = \lim_{t\to0}3\left(\dfrac{\sin 3t}{3t}\right) = 3(1) = 3.$

**87.** $f(x) = \dfrac{\sin x^2}{x}$

| $x$ | $-0.1$ | $-0.01$ | $-0.001$ | $0$ | $0.001$ | $0.01$ | $0.1$ |
|---|---|---|---|---|---|---|---|
| $f(x)$ | $-0.099998$ | $-0.01$ | $-0.001$ | ? | $0.001$ | $0.01$ | $0.099998$ |

It appears that the limit is 0.

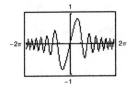

The graph has a hole at $x = 0$.

Analytically, $\displaystyle\lim_{x \to 0}\frac{\sin x^2}{x} = \lim_{x \to 0} x\left(\frac{\sin x^2}{x}\right) = 0(1) = 0.$

**89.** $f(x) = \dfrac{\ln x}{x - 1}$

| $x$ | $0.5$ | $0.9$ | $0.99$ | $1.01$ | $1.1$ | $1.5$ |
|---|---|---|---|---|---|---|
| $f(x)$ | $1.3863$ | $1.0536$ | $1.0050$ | $0.9950$ | $0.9531$ | $0.8109$ |

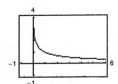

It appears that the limit is 1.

Analytically, $\displaystyle\lim_{x \to 1}\frac{\ln x}{x - 1} = 1.$

**91.** $\displaystyle\lim_{\Delta x \to 0}\frac{f(x + \Delta x) - f(x)}{\Delta x} = \lim_{\Delta x \to 0}\frac{3(x + \Delta x) - 2 - (3x - 2)}{\Delta x} = \lim_{\Delta x \to 0}\frac{3x + 3\Delta x - 2 - 3x + 2}{\Delta x} = \lim_{\Delta x \to 0}\frac{3\Delta x}{\Delta x} = 3$

**93.** $\displaystyle\lim_{\Delta x \to 0}\frac{f(x + \Delta x) - f(x)}{\Delta x} = \lim_{\Delta x \to 0}\frac{\dfrac{1}{x + \Delta x + 3} - \dfrac{1}{x + 3}}{\Delta x}$

$\qquad\qquad = \displaystyle\lim_{\Delta x \to 0}\frac{x + 3 - (x + \Delta x + 3)}{(x + \Delta x + 3)(x + 3)} \cdot \frac{1}{\Delta x}$

$\qquad\qquad = \displaystyle\lim_{\Delta x \to 0}\frac{-\Delta x}{(x + \Delta x + 3)(x + 3)\Delta x}$

$\qquad\qquad = \displaystyle\lim_{\Delta x \to 0}\frac{-1}{(x + \Delta x + 3)(x + 3)} = \frac{-1}{(x + 3)^2}$

**95.** $\displaystyle\lim_{x \to 0}(4 - x^2) \le \lim_{x \to 0} f(x) \le \lim_{x \to 0}(4 + x^2)$

$\qquad\qquad 4 \le \displaystyle\lim_{x \to 0} f(x) \le 4$

Therefore, $\displaystyle\lim_{x \to 0} f(x) = 4.$

**97.** $f(x) = |x|\sin x$

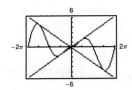

$\displaystyle\lim_{x \to 0}|x|\sin x = 0$

**99.** $f(x) = x\sin\dfrac{1}{x}$

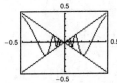

$\displaystyle\lim_{x \to 0}\left(x\sin\frac{1}{x}\right) = 0$

**101.** (a) Two functions $f$ and $g$ agree at all but one point (on an open interval) if $f(x) = g(x)$ for all $x$ in the interval except for $x = c$, where $c$ is in the interval.

(b) $f(x) = \dfrac{x^2 - 1}{x - 1} = \dfrac{(x + 1)(x - 1)}{x - 1}$ and

$g(x) = x + 1$ agree at all points except $x = 1$.

(Other answers possible.)

**103.** If a function $f$ is squeezed between two functions $h$ and $g$, $h(x) \le f(x) \le g(x)$, and $h$ and $g$ have the same limit $L$ as $x \to c$, then $\lim\limits_{x \to c} f(x)$ exists and equals $L$.

**105.** $f(x) = x$, $g(x) = \sin x$, $h(x) = \dfrac{\sin x}{x}$

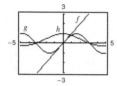

When the $x$-values are "close to" 0 the magnitude of $f$ is approximately equal to the magnitude of $g$. So, $|g|/|f| \approx 1$ when $x$ is "close to" 0.

**107.** $s(t) = -16t^2 + 500$

$$
\begin{aligned}
\lim_{t \to 2} \frac{s(2) - s(t)}{2 - t} &= \lim_{t \to 2} \frac{-16(2)^2 + 500 - \left(-16t^2 + 500\right)}{2 - t} \\
&= \lim_{t \to 2} \frac{436 + 16t^2 - 500}{2 - t} \\
&= \lim_{t \to 2} \frac{16\left(t^2 - 4\right)}{2 - t} \\
&= \lim_{t \to 2} \frac{16(t - 2)(t + 2)}{2 - t} \\
&= \lim_{t \to 2} -16(t + 2) = -64 \text{ ft/sec}
\end{aligned}
$$

The paint can is falling at about 64 feet/second.

**109.** $s(t) = -4.9t^2 + 200$

$$
\begin{aligned}
\lim_{t \to 3} \frac{s(3) - s(t)}{3 - t} &= \lim_{t \to 3} \frac{-4.9(3)^2 + 200 - \left(-4.9t^2 + 200\right)}{3 - t} \\
&= \lim_{t \to 3} \frac{4.9\left(t^2 - 9\right)}{3 - t} \\
&= \lim_{t \to 3} \frac{4.9(t - 3)(t + 3)}{3 - t} \\
&= \lim_{t \to 3} \left[-4.9(t + 3)\right] \\
&= -29.4 \text{ m/sec}
\end{aligned}
$$

The object is falling about 29.4 m/sec.

**111.** Let $f(x) = 1/x$ and $g(x) = -1/x$. $\lim\limits_{x \to 0} f(x)$ and $\lim\limits_{x \to 0} g(x)$ do not exist. However,

$$
\lim_{x \to 0} \left[f(x) + g(x)\right] = \lim_{x \to 0} \left[\frac{1}{x} + \left(-\frac{1}{x}\right)\right] = \lim_{x \to 0}[0] = 0
$$

and therefore does not exist.

**113.** Given $f(x) = b$, show that for every $\varepsilon > 0$ there exists a $\delta > 0$ such that $\left|f(x) - b\right| < \varepsilon$ whenever $\left|x - c\right| < \delta$. Because $\left|f(x) - b\right| = |b - b| = 0 < \varepsilon$ for every $\varepsilon > 0$, any value of $\delta > 0$ will work.

**115.** If $b = 0$, the property is true because both sides are equal to 0. If $b \ne 0$, let $\varepsilon > 0$ be given. Because $\lim\limits_{x \to c} f(x) = L$, there exists $\delta > 0$ such that

$$\left|f(x) - L\right| < \varepsilon/|b| \text{ whenever } 0 < |x - c| < \delta.$$ So, whenever $0 < |x - c| < \delta$, we have

$$|b|\left|f(x) - L\right| < \varepsilon \quad \text{or} \quad \left|bf(x) - bL\right| < \varepsilon$$

which implies that $\lim\limits_{x \to c}\left[bf(x)\right] = bL$.

**117.**

$$
\begin{aligned}
-M\left|f(x)\right| &\le f(x)g(x) \le M\left|f(x)\right| \\
\lim_{x \to c}\left(-M\left|f(x)\right|\right) &\le \lim_{x \to c} f(x)g(x) \le \lim_{x \to c}\left(M\left|f(x)\right|\right) \\
-M(0) &\le \lim_{x \to c} f(x)g(x) \le M(0) \\
0 &\le \lim_{x \to c} f(x)g(x) \le 0
\end{aligned}
$$

Therefore, $\lim\limits_{x \to c} f(x)g(x) = 0$.

**119.** Let

$$
f(x) = \begin{cases} 4, & \text{if } x \ge 0 \\ -4, & \text{if } x < 0 \end{cases}
$$

$\lim\limits_{x \to 0}\left|f(x)\right| = \lim\limits_{x \to 0} 4 = 4.$

$\lim\limits_{x \to 0} f(x)$ does not exist because for $x < 0$, $f(x) = -4$ and for $x \ge 0$, $f(x) = 4$.

**121.** The limit does not exist because the function approaches 1 from the right side of 0 and approaches $-1$ from the left side of 0.

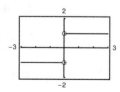

**123.** True.

**125.** False. The limit does not exist because $f(x)$ approaches 3 from the left side of 2 and approaches 0 from the right side of 2.

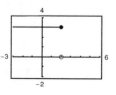

**129.** $f(x) = \dfrac{\sec x - 1}{x^2}$

(a) The domain of $f$ is all $x \neq 0,\ \pi/2 + n\pi$.

(b)

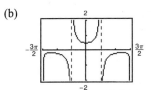

The domain is not obvious. The hole at $x = 0$ is not apparent.

(c) $\displaystyle\lim_{x \to 0} f(x) = \dfrac{1}{2}$

(d) $\dfrac{\sec x - 1}{x^2} = \dfrac{\sec x - 1}{x^2} \cdot \dfrac{\sec x + 1}{\sec x + 1} = \dfrac{\sec^2 x - 1}{x^2(\sec x + 1)}$

$\qquad = \dfrac{\tan^2 x}{x^2(\sec x + 1)} = \dfrac{1}{\cos^2 x}\left(\dfrac{\sin^2 x}{x^2}\right)\dfrac{1}{\sec x + 1}$

So, $\displaystyle\lim_{x \to 0}\dfrac{\sec x - 1}{x^2} = \lim_{x \to 0}\dfrac{1}{\cos^2 x}\left(\dfrac{\sin^2 x}{x^2}\right)\dfrac{1}{\sec x + 1}$

$\qquad\qquad\qquad = 1(1)\left(\dfrac{1}{2}\right) = \dfrac{1}{2}.$

# Section 2.4   Continuity and One-Sided Limits

**1.** (a) $\displaystyle\lim_{x \to 4^+} f(x) = 3$

    (b) $\displaystyle\lim_{x \to 4^-} f(x) = 3$

    (c) $\displaystyle\lim_{x \to 4} f(x) = 3$

The function is continuous at $x = 4$ and is continuous on $(-\infty, \infty)$.

**3.** (a) $\displaystyle\lim_{x \to 3^+} f(x) = 0$

    (b) $\displaystyle\lim_{x \to 3^-} f(x) = 0$

    (c) $\displaystyle\lim_{x \to 3} f(x) = 0$

The function is NOT continuous at $x = 3$.

**5.** (a) $\displaystyle\lim_{x \to 2^+} f(x) = -3$

    (b) $\displaystyle\lim_{x \to 2^-} f(x) = 3$

    (c) $\displaystyle\lim_{x \to 2} f(x)$ does not exist

The function is NOT continuous at $x = 2$.

**7.** $\displaystyle\lim_{x \to 8^+}\dfrac{1}{x + 8} = \dfrac{1}{8 + 8} = \dfrac{1}{16}$

**9.** $\displaystyle\lim_{x \to 5^+}\dfrac{x - 5}{x^2 - 25} = \lim_{x \to 5^+}\dfrac{x - 5}{(x + 5)(x - 5)}$

$\qquad\qquad = \displaystyle\lim_{x \to 5^+}\dfrac{1}{x + 5} = \dfrac{1}{10}$

**11.** $\displaystyle\lim_{x \to -3^-}\dfrac{x}{\sqrt{x^2 - 9}}$ does not exist because $\dfrac{x}{\sqrt{x^2 - 9}}$

decreases without bound as $x \to -3^-$.

**13.** $\displaystyle\lim_{x \to 0^-}\dfrac{|x|}{x} = \lim_{x \to 0^-}\dfrac{-x}{x} = -1$

**15.** $\displaystyle\lim_{\Delta x \to 0^-}\dfrac{\dfrac{1}{x + \Delta x} - \dfrac{1}{x}}{\Delta x} = \lim_{\Delta x \to 0^-}\dfrac{x - (x + \Delta x)}{x(x + \Delta x)} \cdot \dfrac{1}{\Delta x} = \lim_{\Delta x \to 0^-}\dfrac{-\Delta x}{x(x + \Delta x)} \cdot \dfrac{1}{\Delta x}$

$\qquad\qquad\qquad\qquad\qquad = \displaystyle\lim_{\Delta x \to 0^-}\dfrac{-1}{x(x + \Delta x)}$

$\qquad\qquad\qquad\qquad\qquad = \dfrac{-1}{x(x + 0)} = -\dfrac{1}{x^2}$

**17.** $\displaystyle\lim_{x \to 3^-} f(x) = \lim_{x \to 3^-}\dfrac{x + 2}{2} = \dfrac{5}{2}$

**19.** $\displaystyle\lim_{x \to \pi}\cot x$ does not exist because

$\displaystyle\lim_{x \to \pi^+}\cot x$ and $\displaystyle\lim_{x \to \pi^-}\cot x$ do not exist.

**21.** $\displaystyle\lim_{x \to 4^-}\left(5[\![x]\!] - 7\right) = 5(3) - 7 = 8$

$\left([\![x]\!] = 3 \text{ for } 3 \le x < 4\right)$

**23.** $\lim\limits_{x \to 3}\left(2 - [\![-x]\!]\right)$ does not exist because

$$\lim_{x \to 3^-}\left(2 - [\![-x]\!]\right) = 2 - (-3) = 5$$

and

$$\lim_{x \to 3^+}\left(2 - [\![-x]\!]\right) = 2 - (-4) = 6.$$

**25.** $\lim\limits_{x \to 3^+} \ln(x - 3) = \ln 0$

does not exist.

**27.** $\lim\limits_{x \to 2^-} \ln\left[x^2(3 - x)\right] = \ln\left[4(1)\right] = \ln 4$

**29.** $f(x) = \dfrac{1}{x^2 - 4}$

has discontinuities at $x = -2$ and $x = 2$ because $f(-2)$ and $f(2)$ are not defined.

**31.** $f(x) = \dfrac{[\![x]\!]}{2} + x$

has discontinuities at each integer $k$ because

$$\lim_{x \to k^-} f(x) \neq \lim_{x \to k^+} f(x).$$

**33.** $g(x) = \sqrt{49 - x^2}$ is continuous on $[-7, 7]$.

**35.** $\lim\limits_{x \to 0^-} f(x) = 3 = \lim\limits_{x \to 0^+} f(x). f$ is continuous on $[-1, 4]$.

**37.** $f(x) = \dfrac{6}{x}$ has a nonremovable discontinuity at $x = 0$

because $\lim\limits_{x \to 0} f(x)$ does not exist.

**39.** $f(x) = 3x - \cos x$ is continuous for all real $x$.

**41.** $f(x) = \dfrac{1}{4 - x^2} = \dfrac{1}{(2 - x)(2 + x)}$ has nonremovable

discontinuities at $x = \pm 2$ because $\lim\limits_{x \to 2} f(x)$ and

$\lim\limits_{x \to -2} f(x)$ do not exist.

**43.** $f(x) = \dfrac{x}{x^2 - x}$ is not continuous at $x = 0, 1$.

Because $\dfrac{x}{x^2 - x} = \dfrac{1}{x - 1}$ for $x \neq 0$, $x = 0$ is

a removable discontinuity, whereas $x = 1$ is a nonremovable discontinuity.

**45.** $f(x) = \dfrac{x}{x^2 + 1}$ is continuous for all real $x$.

**47.** $f(x) = \dfrac{x + 2}{x^2 - 3x - 10} = \dfrac{x + 2}{(x + 2)(x - 5)}$

has a nonremovable discontinuity at $x = 5$ because $\lim\limits_{x \to 5} f(x)$ does not exist, and has a removable discontinuity at $x = -2$ because

$$\lim_{x \to -2} f(x) = \lim_{x \to -2} \frac{1}{x - 5} = -\frac{1}{7}.$$

**49.** $f(x) = \dfrac{|x + 7|}{x + 7}$

has a nonremovable discontinuity at $x = -7$ because $\lim\limits_{x \to -7} f(x)$ does not exist.

**51.** $f(x) = \begin{cases} x, & x \leq 1 \\ x^2, & x > 1 \end{cases}$

has a **possible** discontinuity at $x = 1$.

**1.** $f(1) = 1$

**2.** $\left.\begin{array}{l} \lim\limits_{x \to 1^-} f(x) = \lim\limits_{x \to 1^-} x = 1 \\ \lim\limits_{x \to 1^+} f(x) = \lim\limits_{x \to 1^+} x^2 = 1 \end{array}\right\} \lim\limits_{x \to 1} f(x) = 1$

**3.** $f(-1) = \lim\limits_{x \to 1} f(x)$

$f$ is continuous at $x = 1$, therefore, $f$ is continuous for all real $x$.

**53.** $f(x) = \begin{cases} \dfrac{x}{2} + 1, & x \leq 2 \\ 3 - x, & x > 2 \end{cases}$

has a **possible** discontinuity at $x = 2$.

**1.** $f(2) = \dfrac{2}{2} + 1 = 2$

**2.** $\lim\limits_{x \to 2^-} f(x) = \lim\limits_{x \to 2^-} \left(\dfrac{x}{2} + 1\right) = 2$

$\lim\limits_{x \to 2^+} f(x) = \lim\limits_{x \to 2^+} (3 - x) = 1$ $\Big\}$ $\lim\limits_{x \to 2} f(x)$ does not exist.

Therefore, $f$ has a nonremovable discontinuity at $x = 2$.

**55.** $f(x) = \begin{cases} \tan\dfrac{\pi x}{4}, & |x| < 1 \\ x, & |x| \ge 1 \end{cases}$

$= \begin{cases} \tan\dfrac{\pi x}{4}, & -1 < x < 1 \\ x, & x \le -1 \text{ or } x \ge 1 \end{cases}$

has **possible** discontinuities at $x = -1$, $x = 1$.

**1.** $f(-1) = -1$ $\qquad$ $f(1) = 1$

**2.** $\lim\limits_{x \to -1} f(x) = -1$ $\qquad$ $\lim\limits_{x \to 1} f(x) = 1$

**3.** $f(-1) = \lim\limits_{x \to -1} f(x)$ $\qquad$ $f(1) = \lim\limits_{x \to 1} f(x)$

$f$ is continuous at $x = \pm 1$, therefore, $f$ is continuous for all real $x$.

**57.** $f(x) = \begin{cases} \ln(x + 1), & x \ge 0 \\ 1 - x^2, & x < 0 \end{cases}$

has a **possible** discontinuity at $x = 0$.

**1.** $f(0) = \ln(0 + 1) = \ln 1 = 0$

**2.** $\lim\limits_{x \to 0^-} f(x) = 1 - 0 = 1$

$\lim\limits_{x \to 0^+} f(x) = 0$ $\Big\}$ $\lim\limits_{x \to 0} f(x)$ does not exist.

So, $f$ has a nonremovable discontinuity at $x = 0$.

**59.** $f(x) = \csc 2x$ has nonremovable discontinuities at integer multiples of $\pi/2$.

**61.** $f(x) = [\![x - 8]\!]$ has nonremovable discontinuities at each integer $k$.

**63.** $f(1) = 3$

Find $a$ so that $\lim\limits_{x \to 1^-} (ax - 4) = 3$

$a(1) - 4 = 3$

$a = 7.$

**65.** Find $a$ and $b$ such that $\lim\limits_{x \to -1^+} (ax + b) = -a + b = 2$ and $\lim\limits_{x \to 3^-} (ax + b) = 3a + b = -2$.

$\begin{aligned} a - b &= -2 \\ (+)\,3a + b &= -2 \\ \hline 4a &= -4 \\ a &= -1 \\ b &= 2 + (-1) = 1 \end{aligned}$ $\qquad$ $f(x) = \begin{cases} 2, & x \le -1 \\ -x + 1, & -1 < x < 3 \\ -2, & x \ge 3 \end{cases}$

**67.** $f(1) = \arctan(1 - 1) + 2 = 2$

Find $a$ such that $\lim\limits_{x \to 1^-} (ae^{x-1} + 3) = 2$

$ae^{1-1} + 3 = 2$

$a + 3 = 2$

$a = -1.$

**69.** $f(g(x)) = (x - 1)^2$

Continuous for all real $x$

**71.** $f(g(x)) = \dfrac{1}{(x^2 + 5) - 6} = \dfrac{1}{x^2 - 1}$

Nonremovable discontinuities at $x = \pm 1$

**73.** $y = [\![x]\!] - x$

Nonremovable discontinuity at each integer

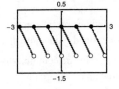

**75.** $g(x) = \begin{cases} x^2 - 3x, & x > 4 \\ 2x - 5, & x \leq 4 \end{cases}$

Nonremovable discontinuity at $x = 4$

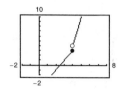

**77.** $f(x) = \dfrac{x}{x^2 + x + 2}$

Continuous on $(-\infty, \infty)$

**79.** $f(x) = 3 - \sqrt{x}$

Continuous on $[0, \infty)$

**81.** $f(x) = \sec\dfrac{\pi x}{4}$

Continuous on:

$\ldots, (-6, -2), (-2, 2), (2, 6), (6, 10), \ldots$

**83.** $f(x) = \begin{cases} \dfrac{x^2 - 1}{x - 1}, & x \neq 1 \\ 2, & x = 1 \end{cases}$

Since $\displaystyle\lim_{x \to 1} f(x) = \lim_{x \to 1} \dfrac{x^2 - 1}{x - 1} = \lim_{x \to 1} \dfrac{(x - 1)(x + 1)}{x - 1}$

$= \displaystyle\lim_{x \to 1}(x + 1) = 2,$

$f$ is continuous on $(-\infty, \infty)$.

**85.** $f(x) = \dfrac{\sin x}{x}$

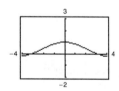

The graph **appears** to be continuous on the interval $[-4, 4]$. Because $f(0)$ is not defined, you know that $f$ has a discontinuity at $x = 0$. This discontinuity is removable so it does not show up on the graph.

**87.** $f(x) = \dfrac{\ln(x^2 + 1)}{x}$

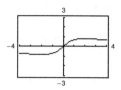

The graph **appears** to be continuous on the interval $[-4, 4]$. Because $f(0)$ is not defined, you know that $f$ has a discontinuity at $x = 0$. This discontinuity is removable so it does not show up on the graph.

**89.** $f(x) = \frac{1}{12}x^4 - x^3 + 4$ is continuous on the interval $[1, 2]$. $f(1) = \frac{37}{12}$ and $f(2) = -\frac{8}{3}$. By the Intermediate Value Theorem, there exists a number $c$ in $[1, 2]$ such that $f(c) = 0$.

**91.** $h$ is continuous on the interval $\left[0, \dfrac{\pi}{2}\right]$.

$h(0) = -2 < 0$ and $h\left(\dfrac{\pi}{2}\right) \approx 0.91 > 0$. By the Intermediate Value Theorem, there exists a number c in $\left[0, \dfrac{\pi}{2}\right]$ such that $h(c) = 0$.

**93.** $f(x) = x^3 + x - 1$

$f(x)$ is continuous on $[0, 1]$.

$f(0) = -1$ and $f(1) = 1$

By the Intermediate Value Theorem, $f(c) = 0$ for at least one value of $c$ between 0 and 1. Using a graphing utility to zoom in on the graph of $f(x)$, you find that $x \approx 0.68$. Using the *root* feature, you find that $x \approx 0.6823$.

**95.** $g(t) = 2\cos t - 3t$

$g$ is continuous on $[0, 1]$.

$g(0) = 2 > 0$ and $g(1) \approx -1.9 < 0$.

By the Intermediate Value Theorem, $g(c) = 0$ for at least one value of $c$ between 0 and 1. Using a graphing utility to zoom in on the graph of $g(t)$, you find that $t \approx 0.56$. Using the *root* feature, you find that $t \approx 0.5636$.

**97.** $f(x) = x + e^x - 3$

$f$ is continuous on $[0, 1]$.

$f(0) = e^0 - 3 = -2 < 0$ and $f(1) = 1 + e - 3 = e - 2 > 0$.

By the Intermediate Value Theorem, $f(c) = 0$ for at least one value of $c$ between 0 and 1. Using a graphing utility to zoom in on the graph of $f(x)$, you find that $x \approx 0.79$. Using the *root* feature, you find that $x \approx 0.7921$.

**99.** $f(x) = x^2 + x - 1$

$f$ is continuous on $[0, 5]$.

$f(0) = -1$ and $f(5) = 29$

$-1 < 11 < 29$

The Intermediate Value Theorem applies.

$x^2 + x - 1 = 11$

$x^2 + x - 12 = 0$

$(x + 4)(x - 3) = 0$

$x = -4$ or $x = 3$

$c = 3 (x = -4$ is not in the interval.)

So, $f(3) = 11$.

**101.** $f(x) = x^3 - x^2 + x - 2$

$f$ is continuous on $[0, 3]$.

$f(0) = -2$ and $f(3) = 19$

$-2 < 4 < 19$

The Intermediate Value Theorem applies.

$x^3 - x^2 + x - 2 = 4$

$x^3 - x^2 + x - 6 = 0$

$(x - 2)(x^2 + x + 3) = 0$

$x = 2$

$(x^2 + x + 3$ has no real solution.)

$c = 2$

So, $f(2) = 4$.

**103.** (a) The limit does not exist at $x = c$.

(b) The function is not defined at $x = c$.

(c) The limit exists at $x = c$, but it is not equal to the value of the function at $x = c$.

(d) The limit does not exist at $x = c$.

**105.** If $f$ and $g$ are continuous for all real $x$, then so is $f + g$ (Theorem 2.11, part 2). However, $f/g$ might not be continuous if $g(x) = 0$. For example, let $f(x) = x$ and $g(x) = x^2 - 1$. Then $f$ and $g$ are continuous for all real $x$, but $f/g$ is not continuous at $x = \pm 1$.

**107.** True

1. $f(c) = L$ is defined.

2. $\lim_{x \to c} f(x) = L$ exists.

3. $f(c) = \lim_{x \to c} f(x)$

All of the conditions for continuity are met.

**109.** False. A rational function can be written as $P(x)/Q(x)$ where $P$ and $Q$ are polynomials of degree $m$ and $n$, respectively. It can have, at most, $n$ discontinuities.

**111.** The functions agree for integer values of $x$:

$g(x) = 3 - [\![-x]\!] = 3 - (-x) = 3 + x$

$f(x) = 3 + [\![x]\!] = 3 + x$    for $x$ an integer

However, for non-integer values of $x$, the functions differ by 1.

$f(x) = 3 + [\![x]\!] = g(x) - 1 = 2 - [\![-x]\!]$.

For example,

$f\left(\frac{1}{2}\right) = 3 + 0 = 3,\ g\left(\frac{1}{2}\right) = 3 - (-1) = 4$.

**113.** $C(t) = \begin{cases} 0.40, & 0 < t \le 10 \\ 0.40 + 0.05 [\![t - 9]\!], & t > 10,\ t \text{ not an integer} \\ 0.40 + 0.05(t - 10), & t > 10,\ t \text{ an integer} \end{cases}$

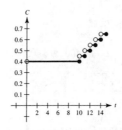

There is a nonremovable discontinuity at each integer greater than or equal to 10.

Note: You could also express $C$ as

$C(t) = \begin{cases} 0.40, & 0 < t \le 10 \\ 0.40 - 0.05 [\![10 - t]\!], & t > 10 \end{cases}$

**115.** Let $s(t)$ be the position function for the run up to the campsite. $s(0) = 0$ ($t = 0$ corresponds to 8:00 A.M., $s(20) = k$ (distance to campsite)). Let $r(t)$ be the position function for the run back down the mountain: $r(0) = k$, $r(10) = 0$. Let $f(t) = s(t) - r(t)$.

When $t = 0$ (8:00 A.M.),

$f(0) = s(0) - r(0) = 0 - k < 0$.

When $t = 10$ (8:00 A.M.), $f(10) = s(10) - r(10) > 0$.

Because $f(0) < 0$ and $f(10) > 0$, then there must be a value $t$ in the interval $[0, 10]$ such that $f(t) = 0$. If $f(t) = 0$, then $s(t) - r(t) = 0$, which gives us $s(t) = r(t)$. Therefore, at some time $t$, where $0 \le t \le 10$, the position functions for the run up and the run down are equal.

**117.** Suppose there exists $x_1$ in $[a, b]$ such that $f(x_1) > 0$ and there exists $x_2$ in $[a, b]$ such that $f(x_2) < 0$. Then by the Intermediate Value Theorem, $f(x)$ must equal zero for some value of $x$ in $[x_1, x_2]$ (or $[x_2, x_1]$ if $x_2 < x_1$). So, $f$ would have a zero in $[a, b]$, which is a contradiction. Therefore, $f(x) > 0$ for all $x$ in $[a, b]$ or $f(x) < 0$ for all $x$ in $[a, b]$.

**119.** If $x = 0$, then $f(0) = 0$ and $\lim_{x \to 0} f(x) = 0$. So, $f$ is continuous at $x = 0$.

If $x \ne 0$, then $\lim_{t \to x} f(t) = 0$ for $x$ rational, whereas $\lim_{t \to x} f(t) = \lim_{t \to x} kt = kx \ne 0$ for $x$ irrational. So, $f$ is not continuous for all $x \ne 0$.

**121.** (a)

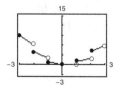

(b) There appears to be a limiting speed and a possible cause is air resistance.

**123.** $f(x) = \begin{cases} 1 - x^2, & x \le c \\ x, & x > c \end{cases}$

$f$ is continuous for $x < c$ and for $x > c$. At $x = c$, you need $1 - c^2 = c$. Solving $c^2 + c - 1$, you obtain

$c = \dfrac{-1 \pm \sqrt{1 + 4}}{2} = \dfrac{-1 \pm \sqrt{5}}{2}$.

**125.** $f(x) = \dfrac{\sqrt{x + c^2} - c}{x}, c > 0$

Domain: $x + c^2 \ge 0 \Rightarrow x \ge -c^2$ and $x \ne 0, [-c^2, 0) \cup (0, \infty)$

$\lim_{x \to 0} \dfrac{\sqrt{x + c^2} - c}{x} = \lim_{x \to 0} \dfrac{\sqrt{x + c^2} - c}{x} \cdot \dfrac{\sqrt{x + c^2} + c}{\sqrt{x + c^2} + c} = \lim_{x \to 0} \dfrac{(x + c^2) - c^2}{x[\sqrt{x + c^2} + c]} = \lim_{x \to 0} \dfrac{1}{\sqrt{x + c^2} + c} = \dfrac{1}{2c}$

Define $f(0) = 1/(2c)$ to make $f$ continuous at $x = 0$.

**127.** $h(x) = x[\![x]\!]$

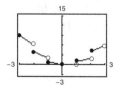

$h$ has nonremovable discontinuities at

$x = \pm 1, \pm 2, \pm 3, \dots.$

**129.** The statement is true.

If $y \geq 0$ and $y \leq 1$, then $y(y-1) \leq 0 \leq x^2$, as desired. So assume $y > 1$. There are now two cases.

Case 1:  If $x \leq y - \frac{1}{2}$, then $2x + 1 \leq 2y$ and

$$y(y-1) = y(y+1) - 2y$$
$$\leq (x+1)^2 - 2y$$
$$= x^2 + 2x + 1 - 2y$$
$$\leq x^2 + 2y - 2y$$
$$= x^2$$

Case 2: If $x \geq y - \frac{1}{2}$

$$x^2 \geq \left(y - \frac{1}{2}\right)^2$$
$$= y^2 - y + \frac{1}{4}$$
$$> y^2 - y$$
$$= y(y-1)$$

In both cases, $y(y-1) \leq x^2$.

# Section 2.5   Infinite Limits

**1.** $\displaystyle\lim_{x \to -2^+} 2 \left| \frac{x}{x^2 - 4} \right| = \infty$

$\displaystyle\lim_{x \to -2^-} 2 \left| \frac{x}{x^2 - 4} \right| = \infty$

**3.** $\displaystyle\lim_{x \to -2^+} \tan \frac{\pi x}{4} = -\infty$

$\displaystyle\lim_{x \to -2^-} \tan \frac{\pi x}{4} = \infty$

**5.** $f(x) = \dfrac{1}{x - 4}$

As $x$ approaches 4 from the left, $x - 4$ is a small negative number. So,

$$\lim_{x \to 4^-} f(x) = -\infty$$

As $x$ approaches 4 from the right, $x - 4$ is a small positive number. So,

$$\lim_{x \to 4^+} f(x) = \infty$$

**7.** $f(x) = \dfrac{1}{(x - 4)^2}$

As $x$ approaches 4 from the left or right, $(x - 4)^2$ is a small positive number. So,

$$\lim_{x \to 4^+} f(x) = \lim_{x \to 4^-} f(x) = \infty.$$

**9.** $f(x) = \dfrac{1}{x^2 - 9}$

| $x$ | −3.5 | −3.1 | −3.01 | −3.001 | −2.999 | −2.99 | −2.9 | −2.5 |
|------|------|------|-------|--------|--------|-------|-------|--------|
| $f(x)$ | 0.308 | 1.639 | 16.64 | 166.6 | −166.7 | −16.69 | −1.695 | −0.364 |

$\displaystyle\lim_{x \to -3^-} f(x) = \infty$

$\displaystyle\lim_{x \to -3^+} f(x) = -\infty$

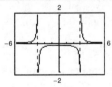

**11.** $f(x) = \dfrac{x^2}{x^2 - 9}$

| $x$ | $-3.5$ | $-3.1$ | $-3.01$ | $-3.001$ | $-2.999$ | $-2.99$ | $-2.9$ | $-2.5$ |
|---|---|---|---|---|---|---|---|---|
| $f(x)$ | 3.769 | 15.75 | 150.8 | 1501 | $-1499$ | $-149.3$ | $-14.25$ | $-2.273$ |

$\lim\limits_{x \to -3^-} f(x) = \infty$

$\lim\limits_{x \to -3^+} f(x) = -\infty$

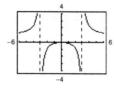

**13.** $f(x) = \dfrac{1}{x^2}$

$\lim\limits_{x \to 0^+} \dfrac{1}{x^2} = \infty = \lim\limits_{x \to 0^-} \dfrac{1}{x^2}$

Therefore, $x = 0$ is a vertical asymptote.

**15.** $f(x) = \dfrac{x^2}{x^2 - 4} = \dfrac{x^2}{(x + 2)(x - 2)}$

$\lim\limits_{x \to -2^-} \dfrac{x^2}{x^2 - 4} = \infty$ and $\lim\limits_{x \to -2^+} \dfrac{x^2}{x^2 - 4} = -\infty$

Therefore, $x = -2$ is a vertical asymptote.

$\lim\limits_{x \to 2^-} \dfrac{x^2}{x^2 - 4} = -\infty$ and $\lim\limits_{x \to 2^+} \dfrac{x^2}{x^2 - 4} = \infty$

Therefore, $x = 2$ is a vertical asymptote.

**17.** $g(t) = \dfrac{t - 1}{t^2 + 1}$

No vertical asymptotes because the denominator is never zero.

**19.** $f(x) = \dfrac{3}{x^2 + x - 2} = \dfrac{3}{(x + 2)(x - 1)}$

$\lim\limits_{x \to -2^-} \dfrac{3}{x^2 + x - 2} = \infty$ and $\lim\limits_{x \to -2^+} \dfrac{3}{x^2 + x - 2} = -\infty$

Therefore, $x = -2$ is a vertical asymptote.

$\lim\limits_{x \to 1^-} \dfrac{3}{x^2 + x - 2} = -\infty$ and $\lim\limits_{x \to 1^+} \dfrac{3}{x^2 + x - 2} = \infty$

Therefore, $x = 1$ is a vertical asymptote.

**21.** $f(x) = \dfrac{x^2 - 2x - 15}{x^3 - 5x^2 + x - 5}$

$= \dfrac{(x - 5)(x + 3)}{(x - 5)(x^2 + 1)}$

$= \dfrac{x + 3}{x^2 + 1}, \; x \ne 5$

$\lim\limits_{x \to 5} f(x) = \dfrac{5 + 3}{5^2 + 1} = \dfrac{15}{26}$

There are no vertical asymptotes. The graph has a hole at $x = 5$.

**23.** $f(x) = \dfrac{e^{-2x}}{x - 1}$

$\lim\limits_{x \to 1^-} f(x) = -\infty$ and $\lim\limits_{x \to 1^+} = \infty$

Therefore, $x = 1$ is a vertical asymptote.

**25.** $h(t) = \dfrac{\ln(t^2 + 1)}{t + 2}$

$\lim\limits_{t \to -2^-} h(t) = -\infty$ and $\lim\limits_{t \to -2^+} = \infty$

Therefore, $t = -2$ is a vertical asymptote.

**27.** $f(x) = \dfrac{1}{e^x - 1}$

$\lim\limits_{x \to 0^-} f(x) = -\infty$ and $\lim\limits_{x \to 0^+} f(x) = \infty$

Therefore, $x = 0$ is a vertical asymptote.

**29.** $f(x) = \csc \pi x = \dfrac{1}{\sin \pi x}$

Let $n$ be any integer.

$\lim\limits_{x \to n} f(x) = -\infty$ or $\infty$

Therefore, the graph has vertical asymptotes at $x = n$.

**31.** $s(t) = \dfrac{t}{\sin t}$

$\sin t = 0$ for $t = n\pi$, where $n$ is an integer.

$\lim\limits_{t \to n\pi} s(t) = \infty$ or $-\infty$ (for $n \ne 0$)

Therefore, the graph has vertical asymptotes at $t = n\pi$, for $n \ne 0$.

$\lim\limits_{t \to 0} s(t) = 1$

Therefore, the graph has a hole at $t = 0$.

**33.** $\lim\limits_{x\to-1}\dfrac{x^2-1}{x+1}=\lim\limits_{x\to-1}(x-1)=-2$

Removable discontinuity at $x=-1$

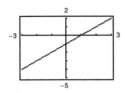

**35.** $\lim\limits_{x\to-1^+}\dfrac{x^2+1}{x+1}=\infty$

$\lim\limits_{x\to-1^-}\dfrac{x^2+1}{x+1}=-\infty$

Vertical asymptote at $x=-1$

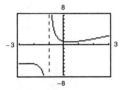

**37.** $\lim\limits_{x\to-1^+}\dfrac{1}{x+1}=\infty$

**39.** $\lim\limits_{x\to2^+}\dfrac{x}{x-2}=\infty$

**41.** $\lim\limits_{x\to-3^-}\dfrac{x+3}{\left(x^2+x-6\right)}=\lim\limits_{x\to-3^-}\dfrac{x+3}{(x+3)(x-2)}$

$=\lim\limits_{x\to-3^-}\dfrac{1}{x-2}=-\dfrac{1}{5}$

**43.** $\lim\limits_{x\to0^-}\left(1+\dfrac{1}{x}\right)=-\infty$

**45.** $\lim\limits_{x\to-4^-}\left(x^2+\dfrac{2}{x+4}\right)=-\infty$

**47.** $\lim\limits_{x\to0^+}\dfrac{2}{\sin x}=\infty$

**49.** $\lim\limits_{x\to8^-}\dfrac{e^x}{(x-8)^3}=-\infty$

**51.** $\lim\limits_{x\to(\pi/2)^-}\ln|\cos x|=\ln\left|\cos\dfrac{\pi}{2}\right|=\ln 0=-\infty$

**53.** $\lim\limits_{x\to(1/2)^-}x\sec\pi x=\lim\limits_{x\to(1/2)^-}\dfrac{x}{\cos\pi x}=\infty$

**55.** $f(x)=\dfrac{x^2+x+1}{x^3-1}=\dfrac{x^2+x+1}{(x-1)(x^2+x+1)}$

$\lim\limits_{x\to1^+}f(x)=\lim\limits_{x\to1^+}\dfrac{1}{x-1}=\infty$

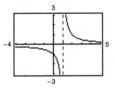

**57.** $f(x)=\dfrac{1}{x^2-25}$

$\lim\limits_{x\to5^-}f(x)=-\infty$

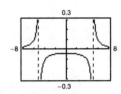

**59.** A limit in which $f(x)$ increases or decreases without bound as $x$ approaches $c$ is called an infinite limit. $\infty$ is not a number. Rather, the symbol

$\lim\limits_{x\to c}f(x)=\infty$

says how the limit fails to exist.

**61.** One answer is

$f(x)=\dfrac{x-3}{(x-6)(x+2)}=\dfrac{x-3}{x^2-4x-12}.$

**63.**

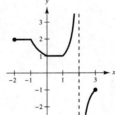

**65.** (a)

| $x$ | 1 | 0.5 | 0.2 | 0.1 | 0.01 | 0.001 | 0.0001 |
|---|---|---|---|---|---|---|---|
| $f(x)$ | 0.1585 | 0.0411 | 0.0067 | 0.0017 | $\approx 0$ | $\approx 0$ | $\approx 0$ |

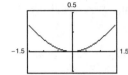

$$\lim_{x \to 0^+} \frac{x - \sin x}{x} = 0$$

(b)

| $x$ | 1 | 0.5 | 0.2 | 0.1 | 0.01 | 0.001 | 0.0001 |
|---|---|---|---|---|---|---|---|
| $f(x)$ | 0.1585 | 0.0823 | 0.0333 | 0.0167 | 0.0017 | $\approx 0$ | $\approx 0$ |

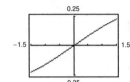

$$\lim_{x \to 0^+} \frac{x - \sin x}{x^2} = 0$$

(c)

| $x$ | 1 | 0.5 | 0.2 | 0.1 | 0.01 | 0.001 | 0.0001 |
|---|---|---|---|---|---|---|---|
| $f(x)$ | 0.1585 | 0.1646 | 0.1663 | 0.1666 | 0.1667 | 0.1667 | 0.1667 |

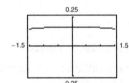

$$\lim_{x \to 0^+} \frac{x - \sin x}{x^3} = 0.1667 \, (1/6)$$

(d)

| $x$ | 1 | 0.5 | 0.2 | 0.1 | 0.01 | 0.001 | 0.0001 |
|---|---|---|---|---|---|---|---|
| $f(x)$ | 0.1585 | 0.3292 | 0.8317 | 1.6658 | 16.67 | 166.7 | 1667.0 |

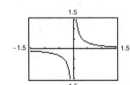

$$\lim_{x \to 0^+} \frac{x - \sin x}{x^4} = \infty \ \text{ or } n > 3, \ \lim_{x \to 0^+} \frac{x - \sin x}{x^n} = \infty.$$

**67.** (a)  $r = \dfrac{2(7)}{\sqrt{625 - 49}} = \dfrac{7}{12}$ ft/sec

(b)  $r = \dfrac{2(15)}{\sqrt{625 - 225}} = \dfrac{3}{2}$ ft/sec

(c)  $\displaystyle \lim_{x \to 25^-} \dfrac{2x}{\sqrt{625 - x^2}} = \infty$

**69. (a)** $A = \frac{1}{2}bh - \frac{1}{2}r^2\theta = \frac{1}{2}(10)(10\tan\theta) - \frac{1}{2}(10)^2\theta = 50\tan\theta - 50\theta$

Domain: $\left(0, \frac{\pi}{2}\right)$

**(b)**

| $\theta$ | 0.3 | 0.6 | 0.9 | 1.2 | 1.5 |
|---|---|---|---|---|---|
| $f(\theta)$ | 0.47 | 4.21 | 18.0 | 68.6 | 630.1 |

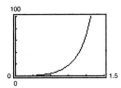

**(c)** $\lim\limits_{\theta \to \pi/2^-} A = \infty$

**71.** False. For instance, let

$f(x) = \dfrac{x^2 - 1}{x - 1}$ or

$g(x) = \dfrac{x}{x^2 + 1}$.

**73.** False. The graphs of $y = \tan x$, $y = \cot x$, $y = \sec x$ and $y = \csc x$ have vertical asymptotes.

**75.** Let $f(x) = \dfrac{1}{x^2}$ and $g(x) = \dfrac{1}{x^4}$, and $c = 0$.

$\lim\limits_{x\to0}\dfrac{1}{x^2} = \infty$ and $\lim\limits_{x\to0}\dfrac{1}{x^4} = \infty$, but $\lim\limits_{x\to0}\left(\dfrac{1}{x^2} - \dfrac{1}{x^4}\right) = \lim\limits_{x\to0}\left(\dfrac{x^2 - 1}{x^4}\right) = -\infty \neq 0$.

**77.** Given $\lim\limits_{x\to c} f(x) = \infty$, let $g(x) = 1$. Then

$\lim\limits_{x\to c}\dfrac{g(x)}{f(x)} = 0$ by Theorem 1.15.

**79.** $f(x) = \dfrac{1}{x - 3}$ is defined for all $x > 3$.

Let $M > 0$ be given. You need $\delta > 0$ such that

$f(x) = \dfrac{1}{x - 3} > M$ whenever $3 < x < 3 + \delta$.

Equivalently, $x - 3 < \dfrac{1}{M}$ whenever

$|x - 3| < \delta$, $x > 3$.

So take $\delta = \dfrac{1}{M}$. Then for $x > 3$ and

$|x - 3| < \delta$, $\dfrac{1}{x - 3} > \dfrac{1}{8} = M$ and so $f(x) > M$.

# Review Exercises for Chapter 2

**1.** Calculus required. Using a graphing utility, you can estimate the length to be 8.3. Or, the length is slightly longer than the distance between the two points, approximately 8.25.

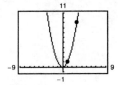

**3.** $f(x) = \dfrac{x - 3}{x^2 - 7x + 12}$

| $x$ | 2.9 | 2.99 | 2.999 | 3 | 3.001 | 3.01 | 3.1 |
|---|---|---|---|---|---|---|---|
| $f(x)$ | −0.9091 | −0.9901 | −0.9990 | ? | −1.0010 | −1.0101 | −1.1111 |

$\lim\limits_{x \to 3} f(x) \approx -1.0000$ (Actual limit is −1.)

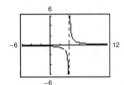

**5.** $h(x) = \dfrac{4x - x^2}{x} = \dfrac{x(4 - x)}{x} = 4 - x, \; x \neq 0$

(a) $\lim\limits_{x \to 0} h(x) = 4 - 0 = 4$

(b) $\lim\limits_{x \to -1} h(x) = 4 - (-1) = 5$

**7.** $\lim\limits_{x \to 1}(x + 4) = 1 + 4 = 5$

Let $\varepsilon > 0$ be given. Choose $\delta = \varepsilon$. Then for $0 < |x - 1| < \delta = \varepsilon$, you have

$$|x - 1| < \varepsilon$$
$$|(x + 4) - 5| < \varepsilon$$
$$|f(x) - L| < \varepsilon.$$

**9.** $\lim\limits_{x \to 2}\left(1 - x^2\right) = 1 - 2^2 = -3$

Let $\varepsilon > 0$ be given. You need

$$\left|1 - x^2 - (-3)\right| < \varepsilon \Rightarrow \left|x^2 - 4\right| = |x - 2||x + 2| < \varepsilon \Rightarrow |x - 2| < \frac{1}{|x + 2|}\varepsilon$$

Assuming $1 < x < 3$, you can choose $\delta = \dfrac{\varepsilon}{5}$.

So, for $0 < |x - 2| < \delta = \dfrac{\varepsilon}{5}$, you have

$$|x - 2| < \frac{\varepsilon}{5} < \frac{\varepsilon}{|x + 2|}$$
$$|x - 2||x + 2| < \varepsilon$$
$$\left|x^2 - 4\right| < \varepsilon$$
$$\left|4 - x^2\right| < \varepsilon$$
$$\left|\left(1 - x^2\right) - (-3)\right| < \varepsilon$$
$$|f(x) - L| < \varepsilon.$$

**11.** $\lim\limits_{x \to -6} x^2 = (-6)^2 = 36$

**13.** $\lim\limits_{x \to 6}(x - 2)^2 = (6 - 2)^2 = 16$

**15.** $\displaystyle\lim_{x\to4}\frac{4}{x-1}=\frac{4}{4-1}=\frac{4}{3}$

**17.** $\displaystyle\lim_{t\to-2}\frac{t+2}{t^2-4}=\lim_{t\to-2}\frac{1}{t-2}=-\frac{1}{4}$

**19.** $\displaystyle\lim_{x\to4}\frac{\sqrt{x-3}-1}{x-4}=\lim_{x\to4}\frac{\sqrt{x-3}-1}{x-4}\cdot\frac{\sqrt{x-3}+1}{\sqrt{x-3}+1}$

$$=\lim_{x\to4}\frac{(x-3)-1}{(x-4)\left(\sqrt{x-3}+1\right)}$$

$$=\lim_{x\to4}\frac{1}{\sqrt{x-3}+1}=\frac{1}{2}$$

**21.** $\displaystyle\lim_{x\to0}\frac{\left[1/(x+1)\right]-1}{x}=\lim_{x\to0}\frac{1-(x+1)}{x(x+1)}=\lim_{x\to0}\frac{-1}{x+1}=-1$

**23.** $\displaystyle\lim_{x\to0}\frac{1-\cos x}{\sin x}=\lim_{x\to0}\left(\frac{x}{\sin x}\right)\left(\frac{1-\cos x}{x}\right)=(1)(0)=0$

**25.** $\displaystyle\lim_{x\to1}e^{x-1}\sin\frac{\pi x}{2}=e^0\sin\frac{\pi}{2}=1$

**27.** $\displaystyle\lim_{\Delta x\to0}\frac{\sin\left[(\pi/6)+\Delta x\right]-(1/2)}{\Delta x}=\lim_{\Delta x\to0}\frac{\sin(\pi/6)\cos\Delta x+\cos(\pi/6)\sin\Delta x-(1/2)}{\Delta x}$

$$=\lim_{\Delta x\to0}\frac{1}{2}\cdot\frac{(\cos\Delta x-1)}{\Delta x}+\lim_{\Delta x\to0}\frac{\sqrt{3}}{2}\cdot\frac{\sin\Delta x}{\Delta x}=0+\frac{\sqrt{3}}{2}(1)=\frac{\sqrt{3}}{2}$$

**29.** $\displaystyle\lim_{x\to c}\left[f(x)g(x)\right]=\left[\lim_{x\to c}f(x)\right]\left[\lim_{x\to c}g(x)\right]$

$$=(-6)\left(\tfrac{1}{2}\right)=-3$$

**31.** $\displaystyle\lim_{x\to c}\left[f(x)+2g(x)\right]=\lim_{x\to c}f(x)+2\lim_{x\to c}g(x)$

$$=-6+2\left(\tfrac{1}{2}\right)=-5$$

**33.** $\displaystyle f(x)=\frac{\sqrt{2x+9}-3}{x}$

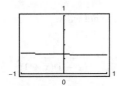

The limit appears to be $\dfrac{1}{3}$.

| $x$ | $-0.01$ | $-0.001$ | $0$ | $0.001$ | $0.01$ |
|---|---|---|---|---|---|
| $f(x)$ | $0.3335$ | $0.3333$ | ? | $0.3333$ | $0.331$ |

$\displaystyle\lim_{x\to0}f(x)\approx0.3333$

$$\lim_{x\to0}\frac{\sqrt{2x+9}-3}{x}\cdot\frac{\sqrt{2x+9}+3}{\sqrt{2x+9}+3}=\lim_{x\to0}\frac{(2x+9)-9}{x\left[\sqrt{2x+9}+3\right]}=\lim_{x\to0}\frac{2}{\sqrt{2x+9}+3}=\frac{2}{\sqrt{9}+3}=\frac{1}{3}$$

**35.** $f(x) = \lim\limits_{x \to 0} \dfrac{20\left(e^{x/2} - 1\right)}{x - 1}$

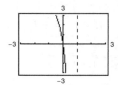

The limit appears to be 0.

| $x$ | −0.1 | −0.01 | −0.001 | 0.001 | 0.01 | 0.1 |
|---|---|---|---|---|---|---|
| $f(x)$ | 0.8867 | 0.0988 | 0.0100 | −0.0100 | −0.1013 | −1.1394 |

$\lim\limits_{x \to 0} f(x) \approx 0.0000$

$\lim\limits_{x \to 0} \dfrac{20\left(e^{x/2} - 1\right)}{x - 1} = \dfrac{20\left(e^0 - 1\right)}{0 - 1} = \dfrac{0}{-1} = 0$

**37.** $v = \lim\limits_{t \to 4} \dfrac{s(4) - s(t)}{4 - t}$

$\quad = \lim\limits_{t \to 4} \dfrac{\left[-4.9(16) + 250\right] - \left[-4.9t^2 + 250\right]}{4 - t}$

$\quad = \lim\limits_{t \to 4} \dfrac{4.9\left(t^2 - 16\right)}{4 - t}$

$\quad = \lim\limits_{t \to 4} \dfrac{4.9(t - 4)(t + 4)}{4 - t}$

$\quad = \lim\limits_{t \to 4} \left[-4.9(t + 4)\right] = -39.2 \text{ m/sec}$

The object is falling at about 39.2 m/sec.

**39.** $\lim\limits_{x \to 3^+} \dfrac{1}{x + 3} = \dfrac{1}{3 + 3} = \dfrac{1}{6}$

**41.** $\lim\limits_{x \to 4^-} \dfrac{\sqrt{x} - 2}{x - 4} = \lim\limits_{x \to 4^-} \dfrac{\sqrt{x} - 2}{x - 4} \cdot \dfrac{\sqrt{x} + 2}{\sqrt{x} + 2}$

$\quad = \lim\limits_{x \to 4^-} \dfrac{x - 4}{(x - 4)\left(\sqrt{x} + 2\right)}$

$\quad = \lim\limits_{x \to 4^-} \dfrac{1}{\sqrt{x} + 2}$

$\quad = \dfrac{1}{4}$

**43.** $\lim\limits_{x \to 2^-} \left(2[\![x]\!] + 1\right) = 2(1) + 1 = 3$

**45.** $\lim\limits_{x \to 2} f(x) = 0$

**47.** $\lim\limits_{t \to 1} h(t)$ does not exist because $\lim\limits_{t \to 1^-} h(t) = 1 + 1 = 2$
and $\lim\limits_{t \to 1^+} h(t) = \frac{1}{2}(1 + 1) = 1.$

**49.** $f(x) = x^2 - 4$ is continuous for all real $x$.

**51.** $f(x) = \dfrac{4}{x - 5}$ has a nonremovable discontinuity at
$x = 5$ because $\lim\limits_{x \to 5} f(x)$ does not exist.

**53.** $f(x) = \dfrac{x}{x^3 - x} = \dfrac{x}{x\left(x^2 - 1\right)} = \dfrac{1}{(x - 1)(x + 1)}, \; x \neq 0$

has nonremovable discontinuities at $x = \pm 1$
because $\lim\limits_{x \to -1} f(x)$ and $\lim\limits_{x \to 1} f(x)$ do not exist,
and has a removable discontinuity at $x = 0$ because

$\lim\limits_{x \to 0} f(x) = \lim\limits_{x \to 0} \dfrac{1}{(x - 1)(x + 1)} = -1.$

**55.** $f(2) = 5$

Find $c$ so that $\lim\limits_{x \to 2^+} (cx + 6) = 5.$

$c(2) + 6 = 5$

$2c = -1$

$c = -\dfrac{1}{2}$

**57.** $f(x) = -3x^2 + 7$

Continuous on $(-\infty, \infty)$

**59.** $f(x) = \sqrt{x - 4}$

Continuous on $[4, \infty)$

**61.** $g(x) = 2e^{[\![x]\!]/4}$ is continuous on all intervals $(n, n + 1)$, where $n$ is an integer. $g$ has nonremovable discontinuities at each $n$.

**63.** $f(x) = \dfrac{3x^2 - x - 2}{x - 1} = \dfrac{(3x + 2)(x - 1)}{x - 1}$

$\lim\limits_{x\to1} f(x) = \lim\limits_{x\to1} (3x + 2) = 5$

Removable discontinuity at $x = 1$

Continuous on $(-\infty, 1) \cup (1, \infty)$

**65.** $f$ is continuous on $[1, 2]$. $f(1) = -1 < 0$ and $f(2) = 13 > 0$. Therefore by the Intermediate Value Theorem, there is at least one value $c$ in $(1, 2)$ such that $2c^3 - 3 = 0$.

**67.** $f(x) = \dfrac{x^2 - 4}{|x - 2|} = (x + 2)\left[\dfrac{x - 2}{|x - 2|}\right]$

(a) $\lim\limits_{x\to2^-} f(x) = -4$

(b) $\lim\limits_{x\to2^+} f(x) = 4$

(c) $\lim\limits_{x\to2} f(x)$ does not exist.

**69.** $f(x) = \dfrac{3}{x}$

$\lim\limits_{x\to0^-} \dfrac{3}{x} = -\infty$

$\lim\limits_{x\to0^+} \dfrac{3}{x} = \infty$

Therefore, $x = 0$ is a vertical asymptote.

**71.** $f(x) = \dfrac{x^3}{x^2 - 9} = \dfrac{x^3}{(x + 3)(x - 3)}$

$\lim\limits_{x\to-3^-} \dfrac{x^3}{x^2 - 9} = -\infty$ and $\lim\limits_{x\to-3^+} \dfrac{x^3}{x^2 - 9} = \infty$

Therefore, $x = -3$ is a vertical asymptote.

$\lim\limits_{x\to3^-} \dfrac{x^3}{x^2 - 9} = -\infty$ and $\lim\limits_{x\to3^+} \dfrac{x^3}{x^2 - 9} = \infty$

Therefore, $x = 3$ is a vertical asymptote.

**73.** $g(x) = \dfrac{2x + 1}{x^2 - 64} = \dfrac{2x + 1}{(x + 8)(x - 8)}$

$\lim\limits_{x\to-8^-} \dfrac{2x + 1}{x^2 - 64} = -\infty$ and $\lim\limits_{x\to-8^+} \dfrac{2x + 1}{x^2 - 64} = \infty$

Therefore, $x = -8$ is a vertical asymptote.

$\lim\limits_{x\to8^-} \dfrac{2x + 1}{x^2 - 64} = -\infty$ and $\lim\limits_{x\to8^+} \dfrac{2x + 1}{x^2 - 64} = \infty$

Therefore, $x = 8$ is a vertical asymptote.

**75.** $g(x) = \ln(25 - x^2) = \ln[(5 + x)(5 - x)]$

$\lim\limits_{x\to5} \ln(25 - x^2) = 0$

$\lim\limits_{x\to-5} \ln(25 - x^2) = 0$

Therefore, the graph has holes at $x = \pm5$. The graph does not have any vertical asymptotes.

**77.** $\lim\limits_{x\to1^-} \dfrac{x^2 + 2x + 1}{x - 1} = -\infty$

**79.** $\lim\limits_{x\to-1^+} \dfrac{x + 1}{x^3 + 1} = \lim\limits_{x\to-1^+} \dfrac{1}{x^2 - x + 1} = \dfrac{1}{3}$

**81.** $\lim\limits_{x\to0^+} \left(x - \dfrac{1}{x^3}\right) = -\infty$

**83.** $\lim\limits_{x\to0^+} \dfrac{\sin 4x}{5x} = \lim\limits_{x\to0^+} \left[\dfrac{4}{5}\left(\dfrac{\sin 4x}{4x}\right)\right] = \dfrac{4}{5}$

**85.** $\lim\limits_{x\to0^+} \dfrac{\csc 2x}{x} = \lim\limits_{x\to0^+} \dfrac{1}{x \sin 2x} = \infty$

**87.** $\lim\limits_{x\to0^+} \ln(\sin x) = -\infty$

**89.** $C = \dfrac{80{,}000p}{100 - p}, 0 \le p < 100$

(a) $C(15) \approx \$14{,}117.65$

(b) $C(50) = \$80.000$

(c) $C(90) = \$720{,}000$

(d) $\lim\limits_{p\to100^-} \dfrac{80{,}000p}{100 - p} = \infty$

## Problem Solving for Chapter 2

**1.** (a)  Perimeter $\triangle PAO = \sqrt{x^2 + (y-1)^2} + \sqrt{x^2 + y^2} + 1$

$$= \sqrt{x^2 + (x^2-1)^2} + \sqrt{x^2 + x^4} + 1$$

Perimeter $\triangle PBO = \sqrt{(x-1)^2 + y^2} + \sqrt{x^2 + y^2} + 1$

$$= \sqrt{(x-1)^2 + x^4} + \sqrt{x^2 + x^4} + 1$$

(b)  $r(x) = \dfrac{\sqrt{x^2 + (x^2-1)^2} + \sqrt{x^2 + x^4} + 1}{\sqrt{(x-1)^2 + x^4} + \sqrt{x^2 + x^4} + 1}$

| $x$ | 4 | 2 | 1 | 0.1 | 0.01 |
|---|---|---|---|---|---|
| Perimeter $\triangle PAO$ | 33.02 | 9.08 | 3.41 | 2.10 | 2.01 |
| Perimeter $\triangle PBO$ | 33.77 | 9.60 | 3.41 | 2.00 | 2.00 |
| $r(x)$ | | 0.98 | 0.95 | 1 | 1.05 | 1.005 |

(c)  $\displaystyle\lim_{x \to 0^+} r(x) = \dfrac{1 + 0 + 1}{1 + 0 + 1} = \dfrac{2}{2} = 1$

**3.** (a)  There are 6 triangles, each with a central angle of $60° = \pi/3$. So,

$$\text{Area hexagon} = 6\left[\frac{1}{2}bh\right] = 6\left[\frac{1}{2}(1)\sin\frac{\pi}{3}\right] = \frac{3\sqrt{3}}{2} \approx 2.598.$$

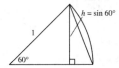
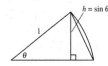

$$\text{Error} = \text{Area (Circle)} - \text{Area (Hexagon)} = \pi - \frac{3\sqrt{3}}{2} \approx 0.5435$$

(b)  There are $n$ triangles, each with central angle of $\theta = 2\pi/n$. So,

$$A_n = n\left[\frac{1}{2}bh\right] = n\left[\frac{1}{2}(1)\sin\frac{2\pi}{n}\right] = \frac{n\sin(2\pi/n)}{2}.$$

(c)

| $n$ | 6 | 12 | 24 | 48 | 96 |
|---|---|---|---|---|---|
| $A_n$ | 2.598 | 3 | 3.106 | 3.133 | 3.139 |

(d)  As $n$ gets larger and larger, $2\pi/n$ approaches 0. Letting $x = 2\pi/n$, $A_n = \dfrac{\sin(2\pi/n)}{2/n} = \dfrac{\sin(2\pi/n)}{(2\pi/n)}\pi = \dfrac{\sin x}{x}\pi$

which approaches $(1)\pi = \pi$.

**5.** (a) Slope $= -\dfrac{12}{5}$

(b) Slope of tangent line is $\dfrac{5}{12}$.

$$y + 12 = \frac{5}{12}(x - 5)$$

$$y = \frac{5}{12}x - \frac{169}{12} \text{ Tangent line}$$

(c) $Q = (x, y) = \left(x, -\sqrt{169 - x^2}\right)$

$$m_x = \frac{-\sqrt{169 - x^2} + 12}{x - 5}$$

(d) $\displaystyle\lim_{x \to 5} m_x = \lim_{x \to 5} \frac{12 - \sqrt{169 - x^2}}{x - 5} \cdot \frac{12 + \sqrt{169 - x^2}}{12 + \sqrt{169 - x^2}}$

$$= \lim_{x \to 5} \frac{144 - \left(169 - x^2\right)}{(x - 5)\left(12 + \sqrt{169 - x^2}\right)}$$

$$= \lim_{x \to 5} \frac{x^2 - 25}{(x - 5)\left(12 + \sqrt{169 - x^2}\right)}$$

$$= \lim_{x \to 5} \frac{(x + 5)}{12 + \sqrt{169 - x^2}} = \frac{10}{12 + 12} = \frac{5}{12}$$

This is the same slope as part (b).

**7.** (a) $3 + x^{1/3} \geq 0$

$$x^{1/3} \geq -3$$

$$x \geq -27$$

Domain: $x \geq -27,\ x \neq 1$ or $[-27, 1) \cup (1, \infty)$

(b)
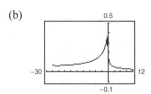

(c) $\displaystyle\lim_{x \to -27^+} f(x) = \frac{\sqrt{3 + (-27)^{1/3}} - 2}{-27 - 1} = \frac{-2}{-28} = \frac{1}{14}$

$$\approx 0.0714$$

(d) $\displaystyle\lim_{x \to 1} f(x) = \lim_{x \to 1} \frac{\sqrt{3 + x^{1/3}} - 2}{x - 1} \cdot \frac{\sqrt{3 + x^{1/3}} + 2}{\sqrt{3 + x^{1/3}} + 2}$

$$= \lim_{x \to 1} \frac{3 + x^{1/3} - 4}{(x - 1)\left(\sqrt{3 + x^{1/3}} + 2\right)}$$

$$= \lim_{x \to 1} \frac{x^{1/3} - 1}{\left(x^{1/3} - 1\right)\left(x^{2/3} + x^{1/3} + 1\right)\left(\sqrt{3 + x^{1/3}} + 2\right)}$$

$$= \lim_{x \to 1} \frac{1}{\left(x^{2/3} + x^{1/3} + 1\right)\left(\sqrt{3 + x^{1/3}} + 2\right)}$$

$$= \frac{1}{(1 + 1 + 1)(2 + 2)} = \frac{1}{12}$$

**9.** (a) $\displaystyle\lim_{x \to 2} f(x) = 3$: $g_1, g_4$

(b) $f$ continuous at 2: $g_1$

(c) $\displaystyle\lim_{x \to 2^-} f(x) = 3$: $g_1, g_3, g_4$

**11.**

(a) $f(1) = [\![1]\!] + [\![-1]\!] = 1 + (-1) = 0$

$f(0) = 0$

$f\left(\tfrac{1}{2}\right) = 0 + (-1) = -1$

$f(-2.7) = -3 + 2 = -1$

(b)   $\lim\limits_{x \to 1^-} f(x) = -1$

$\lim\limits_{x \to 1^+} f(x) = -1$

$\lim\limits_{x \to 1/2} f(x) = -1$

(c)   $f$ is continuous for all real numbers except $x = 0, \pm 1, \pm 2, \pm 3, \ldots$

**13.** (a)

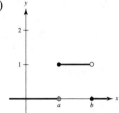

(b)  (i)   $\lim\limits_{x \to a^+} P_{a,b}(x) = 1$

(ii)   $\lim\limits_{x \to a^-} P_{a,b}(x) = 0$

(iii)   $\lim\limits_{x \to b^+} P_{a,b}(x) = 0$

(iv)   $\lim\limits_{x \to b^-} P_{a,b}(x) = 1$

(c)   $P_{a,b}$ is continuous for all positive real numbers except $x = a, b$.

(d)   The area under the graph of $U$, and above the $x$-axis, is 1.

# CHAPTER 3
## Differentiation

# CHAPTER 3
# Differentiation

## Section 3.1   The Derivative and the Tangent Line Problem

**1.** At $(x_1, y_1)$, slope $= 0$.

At $(x_2, y_2)$, slope $= \frac{5}{2}$.

**3.** (a), (b)

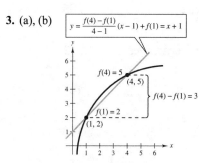

$$y = \frac{f(4) - f(1)}{4 - 1}(x - 1) + f(1) = x + 1$$

(c) $y = \dfrac{f(4) - f(1)}{4 - 1}(x - 1) + f(1)$

$\qquad = \dfrac{3}{3}(x - 1) + 2$

$\qquad = 1(x - 1) + 2$

$\qquad = x + 1$

**5.** $f(x) = 3 - 5x$ is a line. Slope $= -5$

**7.** Slope at $(2, -5) = \lim\limits_{\Delta x \to 0} \dfrac{g(2 + \Delta x) - g(2)}{\Delta x}$

$\qquad = \lim\limits_{\Delta x \to 0} \dfrac{(2 + \Delta x)^2 - 9 - (-5)}{\Delta x}$

$\qquad = \lim\limits_{\Delta x \to 0} \dfrac{4 + 4(\Delta x) + (\Delta x)^2 - 4}{\Delta x}$

$\qquad = \lim\limits_{\Delta x \to 0} (4 + \Delta x) = 4$

**9.** Slope at $(0, 0) = \lim\limits_{\Delta t \to 0} \dfrac{f(0 + \Delta t) - f(0)}{\Delta t}$

$\qquad = \lim\limits_{\Delta t \to 0} \dfrac{3(\Delta t) - (\Delta t)^2 - 0}{\Delta t}$

$\qquad = \lim\limits_{\Delta t \to 0} (3 - \Delta t) = 3$

**11.** $f(x) = 7$

$f'(x) = \lim\limits_{\Delta x \to 0} \dfrac{f(x + \Delta x) - f(x)}{\Delta x}$

$\qquad = \lim\limits_{\Delta x \to 0} \dfrac{7 - 7}{\Delta x}$

$\qquad = \lim\limits_{\Delta x \to 0} 0 = 0$

**13.** $f(x) = -10x$

$f'(x) = \lim\limits_{\Delta x \to 0} \dfrac{f(x + \Delta x) - f(x)}{\Delta x}$

$\qquad = \lim\limits_{\Delta x \to 0} \dfrac{-10(x + \Delta x) - (-10x)}{\Delta x}$

$\qquad = \lim\limits_{\Delta x \to 0} \dfrac{-10x - 10\Delta x + 10x}{\Delta x}$

$\qquad = \lim\limits_{\Delta x \to 0} \dfrac{-10\Delta x}{\Delta x}$

$\qquad = \lim\limits_{\Delta x \to 0} (-10) = -10$

**15.** $h(s) = 3 + \dfrac{2}{3}s$

$h'(s) = \lim\limits_{\Delta s \to 0} \dfrac{h(s + \Delta s) - h(s)}{\Delta s}$

$\qquad = \lim\limits_{\Delta s \to 0} \dfrac{3 + \dfrac{2}{3}(s + \Delta s) - \left(3 + \dfrac{2}{3}s\right)}{\Delta s}$

$\qquad = \lim\limits_{\Delta s \to 0} \dfrac{3 + \dfrac{2}{3}s + \dfrac{2}{3}\Delta s - 3 - \dfrac{2}{3}s}{\Delta s}$

$\qquad = \lim\limits_{\Delta s \to 0} \dfrac{\dfrac{2}{3}\Delta s}{\Delta s} = \dfrac{2}{3}$

**17.** $f(x) = x^2 + x - 3$

$$f'(x) = \lim_{\Delta x \to 0} \frac{f(x + \Delta x) - f(x)}{\Delta x}$$

$$= \lim_{\Delta x \to 0} \frac{(x + \Delta x)^2 + (x + \Delta x) - 3 - (x^2 + x - 3)}{\Delta x}$$

$$= \lim_{\Delta x \to 0} \frac{x^2 + 2x(\Delta x) + (\Delta x)^2 + x + \Delta x - 3 - x^2 - x + 3}{\Delta x}$$

$$= \lim_{\Delta x \to 0} \frac{2x(\Delta x) + (\Delta x)^2 + \Delta x}{\Delta x}$$

$$= \lim_{\Delta x \to 0} (2x + \Delta x + 1) = 2x + 1$$

**19.** $f(x) = x^3 - 12x$

$$f'(x) = \lim_{\Delta x \to 0} \frac{f(x + \Delta x) - f(x)}{\Delta x}$$

$$= \lim_{\Delta x \to 0} \frac{\left[(x + \Delta x)^3 - 12(x + \Delta x)\right] - \left[x^3 - 12x\right]}{\Delta x}$$

$$= \lim_{\Delta x \to 0} \frac{x^3 + 3x^2 \Delta x + 3x(\Delta x)^2 + (\Delta x)^3 - 12x - 12\,\Delta x - x^3 + 12x}{\Delta x}$$

$$= \lim_{\Delta x \to 0} \frac{3x^2 \Delta x + 3x(\Delta x)^2 + (\Delta x)^3 - 12\,\Delta x}{\Delta x}$$

$$= \lim_{\Delta x \to 0} \left(3x^2 + 3x\,\Delta x + (\Delta x)^2 - 12\right) = 3x^2 - 12$$

**21.** $f(x) = \dfrac{1}{x - 1}$

$$f'(x) = \lim_{\Delta x \to 0} \frac{f(x + \Delta x) - f(x)}{\Delta x}$$

$$= \lim_{\Delta x \to 0} \frac{\dfrac{1}{x + \Delta x - 1} - \dfrac{1}{x - 1}}{\Delta x}$$

$$= \lim_{\Delta x \to 0} \frac{(x - 1) - (x + \Delta x - 1)}{\Delta x(x + \Delta x - 1)(x - 1)}$$

$$= \lim_{\Delta x \to 0} \frac{-\Delta x}{\Delta x(x + \Delta x - 1)(x - 1)}$$

$$= \lim_{\Delta x \to 0} \frac{-1}{(x + \Delta x - 1)(x - 1)}$$

$$= -\frac{1}{(x - 1)^2}$$

**23.** $f(x) = \sqrt{x + 4}$

$$f'(x) = \lim_{\Delta x \to 0} \frac{f(x + \Delta x) - f(x)}{\Delta x}$$

$$= \lim_{\Delta x \to 0} \frac{\sqrt{x + \Delta x + 4} - \sqrt{x + 4}}{\Delta x} \cdot \left(\frac{\sqrt{x + \Delta x + 4} + \sqrt{x + 4}}{\sqrt{x + \Delta x + 4} + \sqrt{x + 4}}\right)$$

$$= \lim_{\Delta x \to 0} \frac{(x + \Delta x + 4) - (x + 4)}{\Delta x\left[\sqrt{x + \Delta x + 4} + \sqrt{x + 4}\right]}$$

$$= \lim_{\Delta x \to 0} \frac{1}{\sqrt{x + \Delta x + 4} + \sqrt{x + 4}} = \frac{1}{\sqrt{x + 4} + \sqrt{x + 4}} = \frac{1}{2\sqrt{x + 4}}$$

**25. (a)** $f(x) = x^2 + 3$

$$f'(x) = \lim_{\Delta x \to 0} \frac{f(x + \Delta x) - f(x)}{\Delta x}$$

$$= \lim_{\Delta x \to 0} \frac{\left[(x + \Delta x)^2 + 3\right] - (x^2 + 3)}{\Delta x}$$

$$= \lim_{\Delta x \to 0} \frac{x^2 + 2x\Delta x + (\Delta x)^2 + 3 - x^2 - 3}{\Delta x}$$

$$= \lim_{\Delta x \to 0} \frac{2x\Delta x + (\Delta x)^2}{\Delta x}$$

$$= \lim_{\Delta x \to 0} (2x + \Delta x) = 2x$$

At $(-1, 4)$, the slope of the tangent line is

$$m = 2(-1) = -2.$$

The equation of the tangent line is

$$y - 4 = -2(x + 1)$$
$$y - 4 = -2x - 2$$
$$y = -2x + 2$$

**(b)**

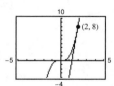

**(c)** Graphing utility confirms $\dfrac{dy}{dx} = -2$ at $(-1, 4)$.

**27. (a)** $f(x) = x^3$

$$f'(x) = \lim_{\Delta x \to 0} \frac{f(x + \Delta x) - f(x)}{\Delta x}$$

$$= \lim_{\Delta x \to 0} \frac{(x + \Delta x)^3 - x^3}{\Delta x}$$

$$= \lim_{\Delta x \to 0} \frac{x^3 + 3x^2\Delta x + 3x(\Delta x)^2 + (\Delta x)^3 - x^3}{\Delta x}$$

$$= \lim_{\Delta x \to 0} \frac{3x^2\Delta x + 3x(\Delta x)^2 + (\Delta x)^3}{\Delta x}$$

$$= \lim_{\Delta x \to 0} \left(3x^2 + 3x\,\Delta x + (\Delta x)^2\right) = 3x^2$$

At $(2, 8)$, the slope of the tangent is $m = 3(2)^2 = 12$. The equation of the tangent line is

$$y - 8 = 12(x - 2)$$
$$y - 8 = 12x - 24$$
$$y = 12x - 16.$$

**(b)**

**(c)** Graphing utility confirms $\dfrac{dy}{dx} = 12$ at $(2, 8)$.

**29.** (a) $f(x) = \sqrt{x}$

$$f'(x) = \lim_{\Delta x \to 0} \frac{f(x + \Delta x) - f(x)}{\Delta x}$$

$$= \lim_{\Delta x \to 0} \frac{\sqrt{x + \Delta x} - \sqrt{x}}{\Delta x} \cdot \frac{\sqrt{x + \Delta x} + \sqrt{x}}{\sqrt{x + \Delta x} + \sqrt{x}}$$

$$= \lim_{\Delta x \to 0} \frac{(x + \Delta x) - x}{\Delta x\left(\sqrt{x + \Delta x} + \sqrt{x}\right)}$$

$$= \lim_{\Delta x \to 0} \frac{1}{\sqrt{x + \Delta x} + \sqrt{x}} = \frac{1}{2\sqrt{x}}$$

At $(1, 1)$, the slope of the tangent line is $m = \dfrac{1}{2\sqrt{1}} = \dfrac{1}{2}$.

The equation of the tangent line is

$$y - 1 = \frac{1}{2}(x - 1)$$

$$y - 1 = \frac{1}{2}x - \frac{1}{2}$$

$$y = \frac{1}{2}x + \frac{1}{2}.$$

(b)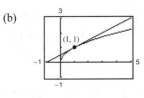

(c) Graphing utility confirms
$\dfrac{dy}{dx} = \dfrac{1}{2}$ at $(1, 1)$.

**31.** (a) $f(x) = x + \dfrac{4}{x}$

$$f'(x) = \lim_{\Delta x \to 0} \frac{f(x + \Delta x) - f(x)}{\Delta x}$$

$$= \lim_{\Delta x \to 0} \frac{(x + \Delta x) + \dfrac{4}{x + \Delta x} - \left(x + \dfrac{4}{x}\right)}{\Delta x}$$

$$= \lim_{\Delta x \to 0} \frac{x(x + \Delta x)(x + \Delta x) + 4x - x^2(x + \Delta x) - 4(x + \Delta x)}{x(\Delta x)(x + \Delta x)}$$

$$= \lim_{\Delta x \to 0} \frac{x^3 + 2x^2(\Delta x) + x(\Delta x)^2 - x^3 - x^2(\Delta x) - 4(\Delta x)}{x(\Delta x)(x + \Delta x)}$$

$$= \lim_{\Delta x \to 0} \frac{x^2(\Delta x) + x(\Delta x)^2 - 4(\Delta x)}{x(\Delta x)(x + \Delta x)}$$

$$= \lim_{\Delta x \to 0} \frac{x^2 + x(\Delta x) - 4}{x(x + \Delta x)}$$

$$= \frac{x^2 - 4}{x^2} = 1 - \frac{4}{x^2}$$

At $(-4, -5)$, the slope of the tangent line is $m = 1 - \dfrac{4}{(-4)^2} = \dfrac{3}{4}$.

The equation of the tangent line is

$$y + 5 = \frac{3}{4}(x + 4)$$

$$y + 5 = \frac{3}{4}x + 3$$

$$y = \frac{3}{4}x - 2.$$

(b)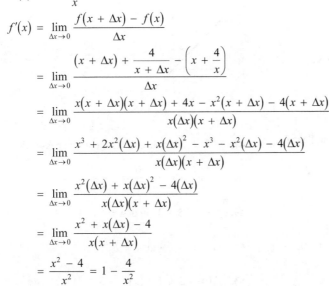

(c) Graphing utility confirms
$\dfrac{dy}{dx} = \dfrac{3}{4}$ at $(-4, -5)$.

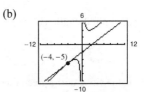

**33.** Using the limit definition of derivative, $f'(x) = 2x$.

Because the slope of the given line is 2, you have

$$2x = 2$$

$$x = 1$$

At the point $(1, 1)$ the tangent line is parallel to

$2x - y + 1 = 0$. The equation of this line is

$$y - 1 = 2(x - 1)$$

$$y = 2x - 1.$$

**35.** From Exercise 27 we know that $f'(x) = 3x^2$.

Because the slope of the given line is 3, you have

$$3x^2 = 3$$

$$x = \pm 1.$$

Therefore, at the points $(1, 1)$ and $(-1, -1)$ the tangent

lines are parallel to $3x - y + 1 = 0$.

These lines have equations

$$y - 1 = 3(x - 1) \quad \text{and} \quad y + 1 = 3(x + 1)$$

$$y = 3x - 2 \qquad\qquad y = 3x + 2.$$

**37.** Using the limit definition of derivative,

$$f'(x) = \frac{-1}{2x\sqrt{x}}.$$

Because the slope of the given line is $-\dfrac{1}{2}$, you have

$$-\frac{1}{2x\sqrt{x}} = -\frac{1}{2}$$

$$x = 1.$$

Therefore, at the point $(1, 1)$ the tangent line is parallel to

$x + 2y - 6 = 0$. The equation of this line is

$$y - 1 = -\frac{1}{2}(x - 1)$$

$$y - 1 = -\frac{1}{2}x + \frac{1}{2}$$

$$y = -\frac{1}{2}x + \frac{3}{2}.$$

**39.** The slope of the graph of $f$ is 1 for all $x$-values.

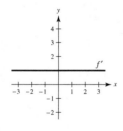

**41.** The slope of the graph of $f$ is negative for
$x < 4$, positive for $x > 4$, and 0 at $x = 4$.

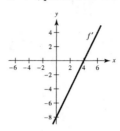

**43.** The slope of the graph of $f$ is negative for $x < 0$ and
positive for $x > 0$. The slope is undefined at $x = 0$.

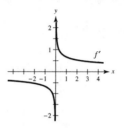

**45.** Answers will vary.

*Sample answer:* $y = -x$

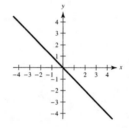

**47.** $g(4) = 5$ because the tangent line passes through $(4, 5)$.

$$g'(4) = \frac{5 - 0}{4 - 7} = -\frac{5}{3}$$

**49.** $f(x) = 5 - 3x$ and $c = 1$

**51.** $f(x) = -x^2$ and $c = 6$

**53.** $f(0) = 2$ and $f'(x) = -3, -\infty < x < \infty$

$$f(x) = -3x + 2$$

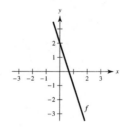

**55.** Let $(x_0, y_0)$ be a point of tangency on the graph of $f$.

By the limit definition for the derivative,

$f'(x) = 4 - 2x$. The slope of the line through $(2, 5)$ and

$(x_0, y_0)$ equals the derivative of $f$ at $x_0$:

$$\frac{5 - y_0}{2 - x_0} = 4 - 2x_0$$

$$5 - y_0 = (2 - x_0)(4 - 2x_0)$$

$$5 - (4x_0 - x_0^2) = 8 - 8x_0 + 2x_0^2$$

$$0 = x_0^2 - 4x_0 + 3$$

$$0 = (x_0 - 1)(x_0 - 3) \Rightarrow x_0 = 1, 3$$

Therefore, the points of tangency are $(1, 3)$ and

$(3, 3)$, and the corresponding slopes are 2 and –2. The

equations of the tangent lines are:

$$y - 5 = 2(x - 2) \qquad y - 5 = -2(x - 2)$$
$$y = 2x + 1 \qquad\qquad y = -2x + 9$$

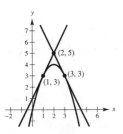

**57. (a)**   $f(x) = x^2$

$$f'(x) = \lim_{\Delta x \to 0} \frac{f(x + \Delta x) - f(x)}{\Delta x}$$

$$= \lim_{\Delta x \to 0} \frac{(x + \Delta x)^2 - x^2}{\Delta x}$$

$$= \lim_{\Delta x \to 0} \frac{x^2 + 2x(\Delta x) + (\Delta x)^2 - x^2}{\Delta x}$$

$$= \lim_{\Delta x \to 0} \frac{\Delta x(2x + \Delta x)}{\Delta x}$$

$$= \lim_{\Delta x \to 0} (2x + \Delta x) = 2x$$

At $x = -1$, $f'(-1) = -2$ and the tangent line is

$$y - 1 = -2(x + 1) \quad \text{or} \quad y = -2x - 1.$$

At $x = 0$, $f'(0) = 0$ and the tangent line is $y = 0$.

At $x = 1$, $f'(1) = 2$ and the tangent line is

$$y = 2x - 1.$$

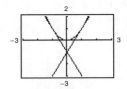

For this function, the slopes of the tangent lines are
always distinct for different values of $x$.

**(b)**   $g'(x) = \lim_{\Delta x \to 0} \dfrac{g(x + \Delta x) - g(x)}{\Delta x}$

$$= \lim_{\Delta x \to 0} \frac{(x + \Delta x)^3 - x^3}{\Delta x}$$

$$= \lim_{\Delta x \to 0} \frac{x^3 + 3x^2(\Delta x) + 3x(\Delta x)^2 + (\Delta x)^3 - x^3}{\Delta x}$$

$$= \lim_{\Delta x \to 0} \frac{\Delta x\left(3x^2 + 3x(\Delta x) + (\Delta x)^2\right)}{\Delta x}$$

$$= \lim_{\Delta x \to 0} \left(3x^2 + 3x(\Delta x) + (\Delta x)^2\right) = 3x^2$$

At $x = -1$, $g'(-1) = 3$ and the tangent line is

$$y + 1 = 3(x + 1) \quad \text{or} \quad y = 3x + 2.$$

At $x = 0$, $g'(0) = 0$ and the tangent line is $y = 0$.

At $x = 1$, $g'(1) = 3$ and the tangent line is

$$y - 1 = 3(x - 1) \quad \text{or} \quad y = 3x - 2.$$

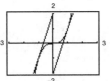

For this function, the slopes of the tangent lines are
sometimes the same.

**59.** $f(x) = \dfrac{1}{2}x^2$

(a)

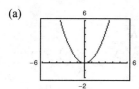

$f'(0) = 0, f'(1/2) = 1/2, f'(1) = 1, f'(2) = 2$

(b) By symmetry: $f'(-1/2) = -1/2, f'(-1) = -1, f'(-2) = -2$

(c)

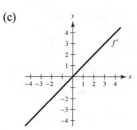

(d) $f'(x) = \lim\limits_{\Delta x \to 0} \dfrac{f(x + \Delta x) - f(x)}{\Delta x} = \lim\limits_{\Delta x \to 0} \dfrac{\frac{1}{2}(x + \Delta x)^2 - \frac{1}{2}x^2}{\Delta x} = \lim\limits_{\Delta x \to 0} \dfrac{\frac{1}{2}\left(x^2 + 2x(\Delta x) + (\Delta x)^2\right) - \frac{1}{2}x^2}{\Delta x} = \lim\limits_{\Delta x \to 0}\left(x + \dfrac{\Delta x}{2}\right) = x$

**61.** $g(x) = \dfrac{f(x + 0.01) - f(x)}{0.01}$

$\quad = \left[2(x + 0.01) - (x + 0.01)^2 - 2x + x^2\right]100$

$\quad = 2 - 2x - 0.01$

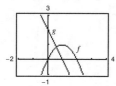

The graph of $g(x)$ is approximately the graph of $f'(x) = 2 - 2x$.

**63.** $f(2) = 2(4 - 2) = 4, f(2.1) = 2.1(4 - 2.1) = 3.99$

$f'(2) \approx \dfrac{3.99 - 4}{2.1 - 2} = -0.1 \quad \left[\text{Exact: } f'(2) = 0\right]$

**65.** $f(x) = x^2 - 5, c = 3$

$\quad f'(3) = \lim\limits_{x \to 3} \dfrac{f(x) - f(3)}{x - 3}$

$\qquad = \lim\limits_{x \to 3} \dfrac{x^2 - 5 - (9 - 5)}{x - 3}$

$\qquad = \lim\limits_{x \to 3} \dfrac{(x - 3)(x + 3)}{x - 3}$

$\qquad = \lim\limits_{x \to 3}(x + 3) = 6$

**67.** $f(x) = x^3 + 2x^2 + 1, c = -2$

$\quad f'(-2) = \lim\limits_{x \to -2} \dfrac{f(x) - f(-2)}{x + 2}$

$\qquad = \lim\limits_{x \to -2} \dfrac{(x^3 + 2x^2 + 1) - 1}{x + 2}$

$\qquad = \lim\limits_{x \to -2} \dfrac{x^2(x + 2)}{x + 2} = \lim\limits_{x \to -2} x^2 = 4$

**69.** $g(x) = \sqrt{|x|}, c = 0$

$$g'(0) = \lim_{x \to 0} \frac{g(x) - g(0)}{x - 0} = \lim_{x \to 0} \frac{\sqrt{|x|}}{x}. \text{ Does not exist.}$$

As $x \to 0^-, \dfrac{\sqrt{|x|}}{x} = \dfrac{-1}{\sqrt{|x|}} \to -\infty.$

As $x \to 0^+, \dfrac{\sqrt{|x|}}{x} = \dfrac{1}{\sqrt{x}} \to \infty.$

Therefore $g(x)$ is not differentiable at $x = 0$.

**71.** $f(x) = (x - 6)^{2/3}, c = 6$

$$f'(6) = \lim_{x \to 6} \frac{f(x) - f(6)}{x - 6}$$

$$= \lim_{x \to 6} \frac{(x - 6)^{2/3} - 0}{x - 6} = \lim_{x \to 6} \frac{1}{(x - 6)^{1/3}}.$$

Does not exist.

Therefore $f(x)$ is not differentiable at $x = 6$.

**73.** $h(x) = |x + 7|, c = -7$

$$h'(-7) = \lim_{x \to -7} \frac{h(x) - h(-7)}{x - (-7)}$$

$$= \lim_{x \to -7} \frac{|x + 7| - 0}{x + 7} = \lim_{x \to -7} \frac{|x + 7|}{x + 7}.$$

Does not exist.

Therefore $h(x)$ is not differentiable at $x = -7$.

**75.** $f(x)$ is differentiable everywhere except at $x = 3.$ (Discontinuity)

**77.** $f(x)$ is differentiable everywhere except at $x = -4.$ (Sharp turn in the graph)

**79.** $f(x)$ is differentiable on the interval $(1, \infty).$ (At $x = 1$ the tangent line is vertical.)

**81.** $f(x) = |x - 5|$ is differentiable everywhere except at $x = -5.$ There is a sharp corner at $x = 5$.

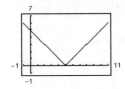

**83.** $f(x) = x^{2/5}$ is differentiable for all $x \ne 0$. There is a sharp corner at $x = 0$.

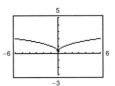

**85.** $f(x) = |x - 1|$

The derivative from the left is

$$\lim_{x \to 1^-} \frac{f(x) - f(1)}{x - 1} = \lim_{x \to 1^-} \frac{|x - 1| - 0}{x - 1} = -1.$$

The derivative from the right is

$$\lim_{x \to 1^+} \frac{f(x) - f(1)}{x - 1} = \lim_{x \to 1^+} \frac{|x - 1| - 0}{x - 1} = 1.$$

The one-sided limits are not equal. Therefore, $f$ is not differentiable at $x = 1$.

**87.** $f(x) = \begin{cases} (x - 1)^3, & x \le 1 \\ (x - 1)^2, & x > 1 \end{cases}$

The derivative from the left is

$$\lim_{x \to 1^-} \frac{f(x) - f(1)}{x - 1} = \lim_{x \to 1^-} \frac{(x - 1)^3 - 0}{x - 1}$$

$$= \lim_{x \to 1^-} (x - 1)^2 = 0.$$

The derivative from the right is

$$\lim_{x \to 1^+} \frac{f(x) - f(1)}{x - 1} = \lim_{x \to 1^+} \frac{(x - 1)^2 - 0}{x - 1}$$

$$= \lim_{x \to 1^+} (x - 1) = 0.$$

The one-sided limits are equal. Therefore, $f$ is differentiable at $x = 1.$ $(f'(1) = 0)$

**89.** Note that $f$ is continuous at $x = 2$.

$$f(x) = \begin{cases} x^2 + 1, & x \le 2 \\ 4x - 3, & x > 2 \end{cases}$$

The derivative from the left is

$$\lim_{x \to 2^-} \frac{f(x) - f(2)}{x - 2} = \lim_{x \to 2^-} \frac{(x^2 + 1) - 5}{x - 2}$$

$$= \lim_{x \to 2^-} (x + 2) = 4.$$

The derivative from the right is

$$\lim_{x \to 2^+} \frac{f(x) - f(2)}{x - 2} = \lim_{x \to 2^+} \frac{(4x - 3) - 5}{x - 2} = \lim_{x \to 2^+} 4 = 4.$$

The one-sided limits are equal. Therefore, $f$ is differentiable at $x = 2.$ $(f'(2) = 4)$

**91.**

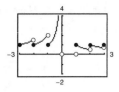

Let $g(x) = \dfrac{[\![x]\!]}{x}$.

For $f(x) = [\![x]\!]$,

$$\lim_{x \to 0^-} \frac{f(x) - f(0)}{x - 0} = \lim_{x \to 0^-} \frac{[\![x]\!] - 0}{x} = \lim_{x \to 0^-} \frac{[\![x]\!]}{x} = \lim_{x \to 0^-} [\![x]\!] \cdot \lim_{x \to 0^-} \frac{1}{x} = -1 \cdot \lim_{x \to 0^-} \frac{1}{x} = \lim_{x \to 0^-} \frac{-1}{x} = \infty.$$

On the other hand,

$$\lim_{x \to 0^+} \frac{f(x) - f(0)}{x - 0} = \lim_{x \to 0^+} \frac{[\![x]\!] - 0}{x} = \lim_{x \to 0^+} \frac{[\![x]\!]}{x} = \lim_{x \to 0^+} [\![x]\!] \cdot \lim_{x \to 0^+} \frac{1}{x} = 0 \cdot \lim_{x \to 0^+} \frac{1}{x} = 0.$$

So, $f$ is not differentiable at $x = 0$ because $\lim\limits_{x \to 0} \dfrac{f(x) - f(0)}{x - 0}$ does not exist. $f$ is differentiable for all $x \neq n$, $n$ an integer.

**93.** False. The slope is $\lim\limits_{\Delta x \to 0} \dfrac{f(2 + \Delta x) - f(2)}{\Delta x}$.

**95.** False. If the derivative from the left of a point does not equal the derivative from the right of a point, then the derivative does not exist at that point. For example, if $f(x) = |x|$, then the derivative from the left at $x = 0$ is $-1$ and the derivative from the right at $x = 0$ is $1$. At $x = 0$, the derivative does not exist.

**97.** $f(x) = \begin{cases} x \sin(1/x), & x \neq 0 \\ 0, & x = 0 \end{cases}$

Using the Squeeze Theorem, you have
$-|x| \leq x \sin(1/x) \leq |x|, x \neq 0.$

So, $\lim\limits_{x \to 0} x \sin(1/x) = 0 = f(0)$ and $f$ is continuous at $x = 0$.

Using the alternative form of the derivative, you have

$$\lim_{x \to 0} \frac{f(x) - f(0)}{x - 0} = \lim_{x \to 0} \frac{x \sin(1/x) - 0}{x - 0} = \lim_{x \to 0}\left(\sin \frac{1}{x}\right).$$

Because this limit does not exist ($\sin(1/x)$ oscillates between $-1$ and $1$), the function is not differentiable at $x = 0$.

$$g(x) = \begin{cases} x^2 \sin(1/x), & x \neq 0 \\ 0, & x = 0 \end{cases}$$

Using the Squeeze Theorem again, you have
$-x^2 \leq x^2 \sin(1/x) \leq x^2, x \neq 0.$

So, $\lim\limits_{x \to 0} x^2 \sin(1/x) = 0 = g(0)$
and $g$ is continuous at $x = 0$. Using the alternative form of the derivative again, you have

$$\lim_{x \to 0} \frac{g(x) - g(0)}{x - 0} = \lim_{x \to 0} \frac{x^2 \sin(1/x) - 0}{x - 0}$$
$$= \lim_{x \to 0} x \sin \frac{1}{x} = 0.$$

Therefore, $g$ is differentiable at $x = 0$, $g'(0) = 0$.

# Section 3.2   Basic Differentiation Rules and Rates of Change

**1.** (a)   $y = x^{1/2}$

$\quad\quad y' = \frac{1}{2}x^{-1/2}$

$\quad\quad y'(1) = \frac{1}{2}$

   (b)   $y = x^3$

$\quad\quad y' = 3x^2$

$\quad\quad y'(1) = 3$

**3.** $y = 12$

$\quad y' = 0$

**5.** $y = x^7$

$\quad y' = 7x^6$

**7.** $y = \dfrac{1}{x^5} = x^{-5}$

$\quad y' = -5x^{-6} = -\dfrac{5}{x^6}$

**9.** $y = \sqrt[5]{x} = x^{1/5}$

$\quad y' = \dfrac{1}{5}x^{-4/5} = \dfrac{1}{5x^{4/5}}$

**11.** $f(x) = x + 11$

$\quad f'(x) = 1$

**13.** $f(t) = -2t^2 + 3t - 6$

$\quad f'(t) = -4t + 3$

**15.** $g(x) = x^2 + 4x^3$

$\quad g'(x) = 2x + 12x^2$

**17.** $s(t) = t^3 + 5t^2 - 3t + 8$

$\quad s'(t) = 3t^2 + 10t - 3$

**19.** $y = \dfrac{\pi}{2}\sin\theta - \cos\theta$

$\quad y' = \dfrac{\pi}{2}\cos\theta + \sin\theta$

**21.** $y = x^2 - \frac{1}{2}\cos x$

$\quad y' = 2x + \frac{1}{2}\sin x$

**23.** $y = \frac{1}{2}e^x - 3\sin x$

$\quad y' = \frac{1}{2}e^x - 3\cos x$

| | *Function* | *Rewrite* | *Differentiate* | *Simplify* |
|---|---|---|---|---|
| **25.** | $y = \dfrac{5}{2x^2}$ | $y = \dfrac{5}{2}x^{-2}$ | $y' = -5x^{-3}$ | $y' = -\dfrac{5}{x^3}$ |
| **27.** | $y = \dfrac{6}{(5x)^3}$ | $y = \dfrac{6}{125}x^{-3}$ | $y' = -\dfrac{18}{125}x^{-4}$ | $y' = -\dfrac{18}{125x^4}$ |
| **29.** | $y = \dfrac{\sqrt{x}}{x}$ | $y = x^{-1/2}$ | $y' = -\dfrac{1}{2}x^{-3/2}$ | $y' = -\dfrac{1}{2x^{3/2}}$ |

**31.** $f(x) = \dfrac{8}{x^2} = 8x^{-2}, (2, 2)$

$\quad f'(x) = -16x^{-3} = -\dfrac{16}{x^3}$

$\quad f'(2) = -2$

**33.** $\quad y = 2x^4 - 3, (1, -1)$

$\quad\quad y' = 8x^3$

$\quad\quad y'(1) = 8$

**35.** $f(\theta) = 4\sin\theta - \theta, (0, 0)$

$\quad f'(\theta) = 4\cos\theta - 1$

$\quad f'(0) = 4(1) - 1 = 3$

**37.** $f(t) = \frac{3}{4}e^t, \left(0, \frac{3}{4}\right)$

$\quad f'(t) = \frac{3}{4}e^t$

$\quad f(0) = \frac{3}{4}e^0 = \frac{3}{4}$

**39.** $g(t) = t^2 - \dfrac{4}{t^3} = t^2 - 4t^{-3}$

$\quad g'(t) = 2t + 12t^{-4} = 2t + \dfrac{12}{t^4}$

**41.** $f(x) = \dfrac{4x^3 + 3x^2}{x} = 4x^2 + 3x$

$\quad f'(x) = 8x + 3$

**43.** $f(x) = \dfrac{x^3 - 3x^2 + 4}{x^2} = x - 3 + 4x^{-2}$

$f'(x) = 1 - \dfrac{8}{x^3} = \dfrac{x^3 - 8}{x^3}$

**45.** $y = x(x^2 + 1) = x^3 + x$

$y' = 3x^2 + 1$

**47.** $f(x) = \sqrt{x} - 6\sqrt[3]{x} = x^{1/2} - 6x^{1/3}$

$f'(x) = \dfrac{1}{2}x^{-1/2} - 2x^{-2/3} = \dfrac{1}{2\sqrt{x}} - \dfrac{2}{x^{2/3}}$

**49.** $f(x) = 6\sqrt{x} + 5\cos x = 6x^{1/2} + 5\cos x$

$f'(x) = 3x^{-1/2} - 5\sin x = \dfrac{3}{\sqrt{x}} - 5\sin x$

**51.** $f(x) = x^{-2} - 2e^x$

$f'(x) = -2x^{-3} - 2e^x = \dfrac{-2}{x^3} - 2e^x$

**53.** (a) $y = x^4 - 3x^2 + 2$

$y' = 4x^3 - 6x$

At $(1, 0)$: $y' = 4(1)^3 - 6(1) = -2$

Tangent line:  $y - 0 = -2(x - 1)$

$y = -2x + 2$

$2x + y - 2 = 0$

(b)

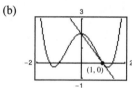

**55.** (a) $g(x) = x + e^x$

$g'(x) = 1 + e^x$

At $(0, 1)$: $g'(0) = 1 + 1 = 2$

Tangent line: $y - 1 = 2(x - 0)$

$y = 2x + 1$

(b)

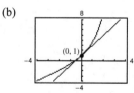

**57.** $y = x^4 - 2x^2 + 3$

$y' = 4x^3 - 4x$

$= 4x(x^2 - 1)$

$= 4x(x - 1)(x + 1)$

$y' = 0 \Rightarrow x = 0, \pm 1$

Horizontal tangents: $(0, 3), (1, 2), (-1, 2)$

**59.** $y = \dfrac{1}{x^2} = x^{-2}$

$y' = -2x^{-3} = -\dfrac{2}{x^3}$ cannot equal zero.

Therefore, there are no horizontal tangents.

**61.** $y = -4x + e^x$

$y' = -4 + e^x = 0$

$e^x = 4$

$x = \ln 4$

Horizontal tangent: $(\ln 4, -4\ln 4 + 4)$

**63.** $y = x + \sin x, 0 \le x < 2\pi$

$y' = 1 + \cos x = 0$

$\cos x = -1 \Rightarrow x = \pi$

At $x = \pi$: $y = \pi$

Horizontal tangent: $(\pi, \pi)$

**65.** $k - x^2 = -6x + 1$  Equate functions.

$-2x = -6$  Equate derivatives.

So, $x = 3$ and $k - 9 = -18 + 1 \Rightarrow k = -8$.

**67.** $\dfrac{k}{x} = -\dfrac{3}{4}x + 3$  Equate functions.

$-\dfrac{k}{x^2} = -\dfrac{3}{4}$  Equate derivatives.

So, $k = \dfrac{3}{4}x^2$ and $\dfrac{\frac{3}{4}x^2}{x} = -\dfrac{3}{4}x + 3 \Rightarrow \dfrac{3}{4}x = -\dfrac{3}{4}x + 3$

$\Rightarrow \dfrac{3}{2}x = 3$

$\Rightarrow x = 2$

$\Rightarrow k = 3$.

**69.**  $kx^3 = x + 1$   Equate equations.

$3kx^2 = 1$     Equate derivatives.

So, $k = \dfrac{1}{3x^2}$ and

$\left(\dfrac{1}{3x^2}\right)x^3 = x + 1$

$\dfrac{1}{3}x = x + 1$

$x = -\dfrac{3}{2}, k = \dfrac{4}{27}.$

**71.** The graph of a function $f$ such that $f' > 0$ for all $x$ and the rate of change of the function is decreasing (i.e., $f'' < 0$) would, in general, look like the graph below.

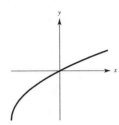

**73.** $g(x) = f(x) + 6 \Rightarrow g'(x) = f'(x)$

**75.**

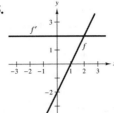

If $f$ is linear then its derivative is a constant function.

$f(x) = ax + b$

$f'(x) = a$

**77.** Let $(x_1, y_1)$ and $(x_2, y_2)$ be the points of tangency on $y = x^2$ and $y = -x^2 + 6x - 5$, respectively.

The derivatives of these functions are:

$y' = 2x \Rightarrow m = 2x_1$ and $y' = -2x + 6 \Rightarrow m = -2x_2 + 6$

$m = 2x_1 = -2x_2 + 6$

$x_1 = -x_2 + 3$

Because $y_1 = x_1^2$ and $y_2 = -x_2^2 + 6x_2 - 5$:

$m = \dfrac{y_2 - y_1}{x_2 - x_1} = \dfrac{\left(-x_2^2 + 6x_2 - 5\right) - \left(x_1^2\right)}{x_2 - x_1} = -2x_2 + 6$

$\dfrac{\left(-x_2^2 + 6x_2 - 5\right) - \left(-x_2 + 3\right)^2}{x_2 - \left(-x_2 + 3\right)} = -2x_2 + 6$

$\left(-x_2^2 + 6x_2 - 5\right) - \left(x_2^2 - 6x_2 + 9\right) = \left(-2x_2 + 6\right)\left(2x_2 - 3\right)$

$-2x_2^2 + 12x_2 - 14 = -4x_2^2 + 18x_2 - 18$

$2x_2^2 - 6x_2 + 4 = 0$

$2\left(x_2 - 2\right)\left(x_2 - 1\right) = 0$

$x_2 = 1$ or $2$

$x_2 = 1 \Rightarrow y_2 = 0, x_1 = 2$ and $y_1 = 4$

So, the tangent line through $(1, 0)$ and $(2, 4)$ is

$y - 0 = \left(\dfrac{4 - 0}{2 - 1}\right)(x - 1) \Rightarrow y = 4x - 4.$

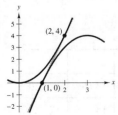

$x_2 = 2 \Rightarrow y_2 = 3, x_1 = 1$ and $y_1 = 1$

So, the tangent line through $(2, 3)$ and $(1, 1)$ is

$y - 1 = \left(\dfrac{3 - 1}{2 - 1}\right)(x - 1) \Rightarrow y = 2x - 1.$

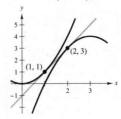

**79.** $f(x) = 3x + \sin x + 2$

$f'(x) = 3 + \cos x$

Because $|\cos x| \leq 1$, $f'(x) \neq 0$ for all $x$ and $f$ does not have a horizontal tangent line.

**81.** $f(x) = \sqrt{x}, (-4, 0)$

$f'(x) = \dfrac{1}{2}x^{-1/2} = \dfrac{1}{2\sqrt{x}}$

$\dfrac{1}{2\sqrt{x}} = \dfrac{0 - y}{-4 - x}$

$4 + x = 2\sqrt{x}\,y$

$4 + x = 2\sqrt{x}\sqrt{x}$

$4 + x = 2x$

$x = 4, \ y = 2$

The point $(4, 2)$ is on the graph of $f$.

Tangent line: $y - 2 = \dfrac{0 - 2}{-4 - 4}(x - 4)$

$\qquad 4y - 8 = x - 4$

$\qquad\quad 0 = x - 4y + 4$

**83.** $f'(1)$ appears to be close to $-1$.

$f'(1) = -1$

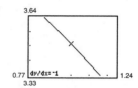

**85.** (a)  One possible secant is between $(3.9, 7.7019)$ and $(4, 8)$:

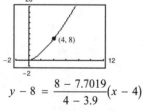

$y - 8 = \dfrac{8 - 7.7019}{4 - 3.9}(x - 4)$

$y - 8 = 2.981(x - 4)$

$y = S(x) = 2.981x - 3.924$

(b)  $f'(x) = \dfrac{3}{2}x^{1/2} \Rightarrow f'(4) = \dfrac{3}{2}(2) = 3$

$T(x) = 3(x - 4) + 8 = 3x - 4$

The slope (and equation) of the secant line approaches that of the tangent line at $(4, 8)$ as you choose points closer and closer to $(4, 8)$.

(c)  As you move further away from $(4, 8)$, the accuracy of the approximation $T$ gets worse.

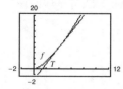

(d)

| $\Delta x$ | $-3$ | $-2$ | $-1$ | $-0.5$ | $-0.1$ | $0$ | $0.1$ | $0.5$ | $1$ | $2$ | $3$ |
|---|---|---|---|---|---|---|---|---|---|---|---|
| $f(4 + \Delta x)$ | 1 | 2.828 | 5.196 | 6.548 | 7.702 | 8 | 8.302 | 9.546 | 11.180 | 14.697 | 18.520 |
| $T(4 + \Delta x)$ | $-1$ | 2 | 5 | 6.5 | 7.7 | 8 | 8.3 | 9.5 | 11 | 14 | 17 |

**87.** False. Let $f(x) = x$ and $g(x) = x + 1$. Then

$f'(x) = g'(x) = x$, but $f(x) \neq g(x)$.

**89.** False. If $y = \pi^2$, then $dy/dx = 0$. ($\pi^2$ is a constant.)

**91.** True. If $g(x) = 3f(x)$, then $g'(x) = 3f'(x)$.

**93.** $f(t) = 4t + 5$,    $[1, 2]$

$f'(t) = 4$. So, $f'(1) = f'(2) = 4$.

Instantaneous rate of change is the constant 4. Average rate of change:

$$\frac{f(2) - f(1)}{2 - 1} = \frac{13 - 9}{1} = 4$$

(These are the same because $f$ is a line of slope 4.)

**95.** $f(x) = -\dfrac{1}{x}$,    $[1, 2]$

$f'(x) = \dfrac{1}{x^2}$

Instantaneous rate of change:

$(1, -1) \Rightarrow f'(1) = 1$

$\left(2, -\dfrac{1}{2}\right) \Rightarrow f'(2) = \dfrac{1}{4}$

Average rate of change:

$$\frac{f(2) - f(1)}{2 - 1} = \frac{(-1/2) - (-1)}{2 - 1} = \frac{1}{2}$$

**97.** $g(x) = x^2 + e^x$,    $[0, 1]$

$g'(x) = 2x + e^x$

Instantaneous rate of change:

$(0, 1)\colon g'(0) = 1$

$(1, 1 + e)\colon g'(1) = 2 + e \approx 4.718$

Average rate of change:

$$\frac{g(1) - g(0)}{1 - 0} = \frac{(1 + e) - (1)}{1} = e \approx 2.718$$

**99.** (a) $s(t) = -16t^2 + 1362$

$v(t) = -32t$

(b) $\dfrac{s(2) - s(1)}{2 - 1} = 1298 - 1346 = -48$ ft/sec

(c) $v(t) = s'(t) = -32t$

When $t = 1$: $v(1) = -32$ ft/sec

When $t = 2$: $v(2) = -64$ ft/sec

(d) $-16t^2 + 1362 = 0$

$$t^2 = \frac{1362}{16} \Rightarrow t = \frac{\sqrt{1362}}{4} \approx 9.226 \text{ sec}$$

(e) $v\left(\dfrac{\sqrt{1362}}{4}\right) = -32\left(\dfrac{\sqrt{1362}}{4}\right)$

$$= -8\sqrt{1362} \approx -295.242 \text{ ft/sec}$$

**101.**    $s(t) = -4.9t^2 + v_0 t + s_0$

$= -4.9t^2 + 120t$

$v(t) = -9.8t + 120$

$v(5) = -9.8(5) + 120 = 71$ m/sec

$v(10) = -9.8(10) + 120 = 22$ m/sec

**103.** From $(0, 0)$ to $(4, 2)$, $s(t) = \frac{1}{2}t \Rightarrow v(t) = \frac{1}{2}$ mi/min.

$v(t) = \frac{1}{2}(60) = 30$ mi/h for $0 < t < 4$

Similarly, $v(t) = 0$ for $4 < t < 6$. Finally, from $(6, 2)$ to $(10, 6)$,

$s(t) = t - 4 \Rightarrow v(t) = 1$ mi/min. $= 60$ mi/h.

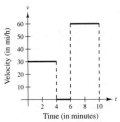

(The velocity has been converted to miles per hour.)

**105.** $v = 40$ mi/h $= \frac{2}{3}$ mi/min

$\left(\frac{2}{3} \text{ mi/min}\right)(6 \text{ min}) = 4$ mi

$v = 0$ mi/h $= 0$ mi/min

$\left(0 \text{ mi/min}\right)(2 \text{ min}) = 0$ mi

$v = 60$ mi/h $= 1$ mi/min

$\left(1 \text{ mi/min}\right)(2 \text{ min}) = 2$ mi

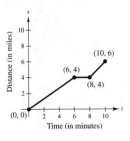

**107.** $V = s^3$, $\dfrac{dV}{ds} = 3s^2$

When $s = 6$ cm, $\dfrac{dV}{ds} = 108$ cm$^3$ per cm change in $s$.

**109. (a)** Using a graphing utility,

$R(v) = 0.417v - 0.02.$

**(b)** Using a graphing utility,

$B(v) = 0.0056v^2 + 0.001v + 0.04.$

**(c)** $T(v) = R(v) + B(v) = 0.0056v^2 + 0.418v + 0.02$

**(d)**

**(e)** $\dfrac{dT}{dv} = 0.0112v + 0.418$

For $v = 40,\ T'(40) \approx 0.866$

For $v = 80,\ T'(80) \approx 1.314$

For $v = 100,\ T'(100) \approx 1.538$

**(f)** For increasing speeds, the total stopping distance increases.

**111.** $s(t) = -\dfrac{1}{2}at^2 + c$ and $s'(t) = -at$

Average velocity: $\dfrac{s(t_0 + \Delta t) - s(t_0 - \Delta t)}{(t_0 + \Delta t) - (t_0 - \Delta t)} = \dfrac{\left[-(1/2)a(t_0 + \Delta t)^2 + c\right] - \left[-(1/2)a(t_0 - \Delta t)^2 + c\right]}{2\Delta t}$

$$= \dfrac{-(1/2)a\left(t_0^2 + 2t_0\Delta t + (\Delta t)^2\right) + (1/2)a\left(t_0^2 - 2t_0\Delta t + (\Delta t)^2\right)}{2\,\Delta t}$$

$$= \dfrac{-2at_0\,\Delta t}{2\,\Delta t} = -at_0 = s'(t_0) \qquad \text{instantaneous velocity at } t = t_0$$

**113.** $y = ax^2 + bx + c$

Because the parabola passes through $(0, 1)$ and $(1, 0)$,

you have:

$(0, 1)\!: 1 = a(0)^2 + b(0) + c \Rightarrow c = 1$

$(1, 0)\!: 0 = a(1)^2 + b(1) + 1 \Rightarrow b = -a - 1$

So, $y = ax^2 + (-a - 1)x + 1.$

From the tangent line $y = x - 1$, you know that the derivative is 1 at the point $(1, 0)$.

$y' = 2ax + (-a - 1)$

$1 = 2a(1) + (-a - 1)$

$1 = a - 1$

$a = 2$

$b = -a - 1 = -3$

Therefore, $y = 2x^2 - 3x + 1.$

**115.** $y = x^3 - 9x$

$y' = 3x^2 - 9$

Tangent lines through $(1, -9)$:     $y + 9 = (3x^2 - 9)(x - 1)$

$$(x^3 - 9x) + 9 = 3x^3 - 3x^2 - 9x + 9$$

$$0 = 2x^3 - 3x^2 = x^2(2x - 3)$$

$$x = 0 \text{ or } x = \tfrac{3}{2}$$

The points of tangency are $(0, 0)$ and $\left(\tfrac{3}{2}, -\tfrac{81}{8}\right)$. At $(0, 0)$, the slope is $y'(0) = -9$. At $\left(\tfrac{3}{2}, -\tfrac{81}{8}\right)$, the slope is $y'\left(\tfrac{3}{2}\right) = -\tfrac{9}{4}$.

Tangent Lines:  $y - 0 = -9(x - 0)$  and          $y + \tfrac{81}{8} = -\tfrac{9}{4}\left(x - \tfrac{3}{2}\right)$

$$y = -9x \qquad\qquad y = -\tfrac{9}{4}x - \tfrac{27}{4}$$

$$9x + y = 0 \qquad\qquad 9x + 4y + 27 = 0$$

**117.** $f(x) = \begin{cases} ax^3, & x \le 2 \\ x^2 + b, & x > 2 \end{cases}$

$f$ must be continuous at $x = 2$ to be differentiable at $x = 2$.

$$\left.\begin{array}{l} \lim\limits_{x \to 2^-} f(x) = \lim\limits_{x \to 2^-} ax^3 = 8a \\[2mm] \lim\limits_{x \to 2^+} f(x) = \lim\limits_{x \to 2^+} (x^2 + b) = 4 + b \end{array}\right\} \quad \begin{array}{l} 8a = 4 + b \\[2mm] 8a - 4 = b \end{array}$$

$$f'(x) = \begin{cases} 3ax^2, & x < 2 \\ 2x, & x > 2 \end{cases}$$

For $f$ to be differentiable at $x = 2$, the left derivative must equal the right derivative.

$$3a(2)^2 = 2(2)$$

$$12a = 4$$

$$a = \tfrac{1}{3}$$

$$b = 8a - 4 = -\tfrac{4}{3}$$

**119.** $f_1(x) = |\sin x|$ is differentiable for all $x \ne n\pi$, $n$ an integer.

$f_2(x) = \sin|x|$ is differentiable for all $x \ne 0$.

You can verify this by graphing $f_1$ and $f_2$ and observing the locations of the sharp turns.

**121.** You are given $f : R \to R$ satisfying

$$(*)f'(x) = \frac{f(x+n) - f(x)}{n} \text{ for all real numbers } x \text{ and all positive integers } n. \text{ You claim that } f(x) = mx + b, m, b \in R.$$

For this case, $f'(x) = m = \dfrac{\left[m(x+n) + b\right] - \left[mx + b\right]}{n} = m.$

Furthermore, these are the only solutions:

Note first that $f'(x+1) = \dfrac{f(x+2) - f(x+1)}{1}$, and $f'(x) = f(x+1) - f(x)$. From $(*)$ you have

$$2f'(x) = f(x+2) - f(x)$$
$$= \left[f(x+2) - f(x+1)\right] + \left[f(x+1) - f(x)\right]$$
$$= f'(x+1) + f'(x).$$

Thus, $f'(x) = f'(x+1).$

Let $g(x) = f(x+1) - f(x).$

Let $m = g(0) = f(1) - f(0).$

Let $b = f(0).$ Then

$$g'(x) = f'(x+1) - f'(x) = 0$$
$$g(x) = \text{constant} = g(0) = m$$
$$f'(x) = f(x+1) - f(x) = g(x) = m \Rightarrow f(x) = mx + b.$$

## Section 3.3  Product and Quotient Rules and Higher-Order Derivatives

**1.** $g(x) = (x^2 + 3)(x^2 - 4x)$

$g'(x) = (x^2 + 3)(2x - 4) + (x^2 - 4x)(2x)$
$= 2x^3 - 4x^2 + 6x - 12 + 2x^3 - 8x^2$
$= 4x^3 - 12x^2 + 6x - 12$
$= 2(2x^3 - 6x^2 + 3x - 6)$

**3.** $h(t) = \sqrt{t}(1 - t^2) = t^{1/2}(1 - t^2)$

$h'(t) = t^{1/2}(-2t) + (1 - t^2)\dfrac{1}{2}t^{-1/2}$
$= -2t^{3/2} + \dfrac{1}{2t^{1/2}} - \dfrac{1}{2}t^{3/2}$
$= -\dfrac{5}{2}t^{3/2} + \dfrac{1}{2t^{1/2}}$
$= \dfrac{1 - 5t^2}{2t^{1/2}} = \dfrac{1 - 5t^2}{2\sqrt{t}}$

**5.** $f(x) = e^x \cos x$

$f'(x) = e^x(-\sin x) + e^x \cos x$
$= e^x(\cos x - \sin x)$

**7.** $f(x) = \dfrac{x}{x^2 + 1}$

$f'(x) = \dfrac{(x^2 + 1)(1) - x(2x)}{(x^2 + 1)^2} = \dfrac{1 - x^2}{(x^2 + 1)^2}$

**9.** $h(x) = \dfrac{\sqrt{x}}{x^3 + 1} = \dfrac{x^{1/2}}{x^3 + 1}$

$h'(x) = \dfrac{(x^3 + 1)\dfrac{1}{2}x^{-1/2} - x^{1/2}(3x^2)}{(x^3 + 1)^2}$
$= \dfrac{x^3 + 1 - 6x^3}{2x^{1/2}(x^3 + 1)^2}$
$= \dfrac{1 - 5x^3}{2\sqrt{x}(x^3 + 1)^2}$

**11.** $g(x) = \dfrac{\sin x}{e^x}$

$g'(x) = \dfrac{e^x \cos x - \sin x(e^x)}{(e^x)^2}$
$= \dfrac{\cos x - \sin x}{e^x}$

**13.** $f(x) = (x^3 + 4x)(3x^2 + 2x - 5)$

$f'(x) = (x^3 + 4x)(6x + 2) + (3x^2 + 2x - 5)(3x^2 + 4)$

$\quad = 6x^4 + 24x^2 + 2x^3 + 8x + 9x^4 + 6x^3 - 15x^2 + 12x^2 + 8x - 20$

$\quad = 15x^4 + 8x^3 + 21x^2 + 16x - 20$

$f'(0) = -20$

**15.** $f(x) = \dfrac{x^2 - 4}{x - 3}$

$f'(x) = \dfrac{(x - 3)(2x) - (x^2 - 4)(1)}{(x - 3)^2}$

$\quad = \dfrac{2x^2 - 6x - x^2 + 4}{(x - 3)^2}$

$\quad = \dfrac{x^2 - 6x + 4}{(x - 3)^2}$

$f'(1) = \dfrac{1 - 6 + 4}{(1 - 3)^2} = -\dfrac{1}{4}$

**17.** $f(x) = x \cos x$

$f'(x) = (x)(-\sin x) + (\cos x)(1) = \cos x - x \sin x$

$f'\left(\dfrac{\pi}{4}\right) = \dfrac{\sqrt{2}}{2} - \dfrac{\pi}{4}\left(\dfrac{\sqrt{2}}{2}\right) = \dfrac{\sqrt{2}}{8}(4 - \pi)$

**19.** $f(x) = e^x \sin x$

$f'(x) = e^x \cos x + e^x \sin x$

$\quad = e^x (\cos x + \sin x)$

$f'(0) = 1$

| *Function* | *Rewrite* | *Differentiate* | *Simplify* |
|---|---|---|---|
| **21.** $y = \dfrac{x^2 + 3x}{7}$ | $y = \dfrac{1}{7}x^2 + \dfrac{3}{7}x$ | $y' = \dfrac{2}{7}x + \dfrac{3}{7}$ | $y' = \dfrac{2x + 3}{7}$ |
| **23.** $y = \dfrac{6}{7x^2}$ | $y = \dfrac{6}{7}x^{-2}$ | $y' = -\dfrac{12}{7}x^{-3}$ | $y' = -\dfrac{12}{7x^3}$ |
| **25.** $y = \dfrac{4x^{3/2}}{x}$ | $y = 4x^{1/2},\ x > 0$ | $y' = 2x^{-1/2}$ | $y' = \dfrac{2}{\sqrt{x}},\ x > 0$ |

**27.** $f(x) = \dfrac{4 - 3x - x^2}{x^2 - 1}$

$f'(x) = \dfrac{(x^2 - 1)(-3 - 2x) - (4 - 3x - x^2)(2x)}{(x^2 - 1)^2}$

$\quad = \dfrac{-3x^2 + 3 - 2x^3 + 2x - 8x + 6x^2 + 2x^3}{(x^2 - 1)^2}$

$\quad = \dfrac{3x^2 - 6x + 3}{(x^2 - 1)^2}$

$\quad = \dfrac{3(x^2 - 2x + 1)}{(x^2 - 1)^2}$

$\quad = \dfrac{3(x - 1)^2}{(x - 1)^2(x + 1)^2} = \dfrac{3}{(x + 1)^2},\ x \neq 1$

**29.** $f(x) = x\left(1 - \dfrac{4}{x + 3}\right) = x - \dfrac{4x}{x + 3}$

$f'(x) = 1 - \dfrac{(x + 3)4 - 4x(1)}{(x + 3)^2}$

$\quad = \dfrac{(x^2 + 6x + 9) - 12}{(x + 3)^2}$

$\quad = \dfrac{x^2 + 6x - 3}{(x + 3)^2}$

**31.** $f(x) = \dfrac{3x-1}{\sqrt{x}} = 3x^{1/2} - x^{-1/2}$

$f'(x) = \dfrac{3}{2}x^{-1/2} + \dfrac{1}{2}x^{-3/2} = \dfrac{3x+1}{2x^{3/2}}$

**Alternate solution:**

$f(x) = \dfrac{3x-1}{\sqrt{x}} = \dfrac{3x-1}{x^{1/2}}$

$f'(x) = \dfrac{x^{1/2}(3) - (3x-1)\left(\dfrac{1}{2}\right)(x^{-1/2})}{x}$

$= \dfrac{\dfrac{1}{2}x^{-1/2}(3x+1)}{x}$

$= \dfrac{3x+1}{2x^{3/2}}$

**33.** $h(s) = (s^3 - 2)^2 = s^6 - 4s^3 + 4$

$h'(s) = 6s^5 - 12s^2 = 6s^2(s^3 - 2)$

**35.** $f(x) = \dfrac{2 - (1/x)}{x-3} = \dfrac{2x-1}{x(x-3)} = \dfrac{2x-1}{x^2 - 3x}$

$f'(x) = \dfrac{(x^2 - 3x)2 - (2x-1)(2x-3)}{(x^2 - 3x)^2}$

$= \dfrac{2x^2 - 6x - 4x^2 + 8x - 3}{(x^2 - 3x)^2}$

$= \dfrac{-2x^2 + 2x - 3}{(x^2 - 3x)^2} = \dfrac{2x^2 - 2x + 3}{x^2(x-3)^2}$

**37.** $f(x) = (2x^3 + 5x)(x - 3)(x + 2)$

$f'(x) = (6x^2 + 5)(x - 3)(x + 2) + (2x^3 + 5x)(1)(x + 2) + (2x^3 + 5x)(x - 3)(1)$

$= (6x^2 + 5)(x^2 - x - 6) + (2x^3 + 5x)(x + 2) + (2x^3 + 5x)(x - 3)$

$= (6x^4 + 5x^2 - 6x^3 - 5x - 36x^2 - 30) + (2x^4 + 4x^3 + 5x^2 + 10x) + (2x^4 + 5x^2 - 6x^3 - 15x)$

$= 10x^4 - 8x^3 - 21x^2 - 10x - 30$

***Note:*** You could simplify first: $f(x) = (2x^3 + 5x)(x^2 - x - 6)$

**39.** $f(x) = \dfrac{x^2 + c^2}{x^2 - c^2}$

$f'(x) = \dfrac{(x^2 - c^2)(2x) - (x^2 + c^2)(2x)}{(x^2 - c^2)^2} = -\dfrac{4xc^2}{(x^2 - c^2)^2}$

**41.** $f(t) = t^2 \sin t$

$f'(t) = t^2 \cos t + 2t \sin t = t(t \cos t + 2 \sin t)$

**43.** $f(t) = \dfrac{\cos t}{t}$

$f'(t) = \dfrac{-t \sin t - \cos t}{t^2} = -\dfrac{t \sin t + \cos t}{t^2}$

**45.** $f(x) = -e^x + \tan x$

$f'(x) = -e^x + \sec^2 x$

**47.** $g(t) = \sqrt[4]{t} + 6 \csc t = t^{1/4} + 6 \csc t$

$g'(t) = \dfrac{1}{4}t^{-3/4} - 6 \csc t \cot t = \dfrac{1}{4t^{3/4}} - 6 \csc t \cot t$

**49.** $y = \dfrac{3(1 - \sin x)}{2 \cos x} = \dfrac{3 - 3 \sin x}{2 \cos x}$

$y' = \dfrac{(-3 \cos x)(2 \cos x) - (3 - 3 \sin x)(-2 \sin x)}{(2 \cos x)^2}$

$= \dfrac{-6 \cos^2 x + 6 \sin x - 6 \sin^2 x}{4 \cos^2 x}$

$= \dfrac{3}{2}(-1 + \tan x \sec x - \tan^2 x)$

$= \dfrac{3}{2} \sec x(\tan x - \sec x)$

**51.** $y = -\csc x - \sin x$

$y' = \csc x \cot x - \cos x$

$= \dfrac{\cos x}{\sin^2 x} - \cos x$

$= \cos x(\csc^2 x - 1)$

$= \cos x \cot^2 x$

**53.** $f(x) = x^2 \tan x$

$f'(x) = x^2 \sec^2 x + 2x \tan x = x(x \sec^2 x + 2 \tan x)$

**55.** $y = 2x \sin x + x^2 e^x$

$y' = 2x(\cos x) + 2 \sin x + x^2 e^x + 2xe^x$

$= 2x \cos x + 2 \sin x + xe^x(x + 2)$

**57.** $y = \dfrac{e^x}{4\sqrt{x}}$

$$y' = \frac{4\sqrt{x}\,e^x - e^x\left(4/2\sqrt{x}\right)}{\left(4\sqrt{x}\right)^2} = \frac{e^x\left[4\sqrt{x} - \left(2/\sqrt{x}\right)\right]}{16x} = \frac{e^x(4x - 2)}{16x^{3/2}} = \frac{e^x(2x - 1)}{8x^{3/2}}$$

**59.** $g(x) = \left(\dfrac{x + 1}{x + 2}\right)(2x - 5)$

$$g'(x) = \left(\frac{x + 1}{x + 2}\right)(2) + (2x - 5)\left[\frac{(x + 2)(1) - (x + 1)(1)}{(x + 2)^2}\right] = \frac{2x^2 + 8x - 1}{(x + 2)^2}$$

(Form of answer may vary.)

**61.** $g(\theta) = \dfrac{\theta}{1 - \sin\theta}$

$$g'(\theta) = \frac{1 - \sin\theta + \theta\cos\theta}{(1 - \sin\theta)^2}$$

(Form of answer may vary.)

**63.**  $y = \dfrac{1 + \csc x}{1 - \csc x}$

$$y' = \frac{(1 - \csc x)(-\csc x \cot x) - (1 + \csc x)(\csc x \cot x)}{(1 - \csc x)^2} = \frac{-2 \csc x \cot x}{(1 - \csc x)^2}$$

$$y'\left(\frac{\pi}{6}\right) = \frac{-2(2)\left(\sqrt{3}\right)}{(1 - 2)^2} = -4\sqrt{3}$$

**65.**  $h(t) = \dfrac{\sec t}{t}$

$$h'(t) = \frac{t(\sec t \tan t) - (\sec t)(1)}{t^2} = \frac{\sec t(t \tan t - 1)}{t^2}$$

$$h'(\pi) = \frac{\sec \pi(\pi \tan \pi - 1)}{\pi^2} = \frac{1}{\pi^2}$$

**67. (a)** $f(x) = \left(x^3 + 4x - 1\right)(x - 2)$,    $(1, -4)$

$$f'(x) = \left(x^3 + 4x - 1\right)(1) + (x - 2)\left(3x^2 + 4\right)$$

$$= x^3 + 4x - 1 + 3x^3 - 6x^2 + 4x - 8$$

$$= 4x^3 - 6x^2 + 8x - 9$$

$f'(1) = -3$; Slope at $(1, -4)$

Tangent line: $y + 4 = -3(x - 1) \Rightarrow y = -3x - 1$

**(b)**

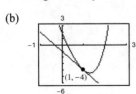

**(c)** Graphing utility confirms $\dfrac{dy}{dx} = -3$ at $(1, -4)$.

**69. (a)**    $f(x) = \dfrac{x}{x + 4}$,    $(-5, 5)$

$$f'(x) = \frac{(x + 4)(1) - x(1)}{(x + 4)^2} = \frac{4}{(x + 4)^2}$$

$$f'(-5) = \frac{4}{(-5 + 4)^2} = 4; \quad \text{Slope at } (-5, 5)$$

Tangent line: $y - 5 = 4(x + 5) \Rightarrow y = 4x + 25$

**(b)**

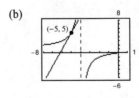

**(c)** Graphing utility confirms $\dfrac{dy}{dx} = 4$ at $(-5, 5)$.

**71. (a)** $f(x) = \tan x$, $\left(\dfrac{\pi}{4}, 1\right)$

$f'(x) = \sec^2 x$

$f'\left(\dfrac{\pi}{4}\right) = 2$; Slope at $\left(\dfrac{\pi}{4}, 1\right)$

Tangent line: $\qquad y - 1 = 2\left(x - \dfrac{\pi}{4}\right)$

$\qquad\qquad\qquad\qquad y - 1 = 2x - \dfrac{\pi}{2}$

$\qquad\qquad 4x - 2y - \pi + 2 = 0$

**(b)**

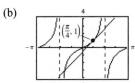

**(c)** Graphing utility confirms $\dfrac{dy}{dx} = 2$ at $\left(\dfrac{\pi}{4}, 1\right)$.

**73. (a)** $f(x) = (x - 1)e^x$, $(1, 0)$

$f'(x) = (x - 1)e^x + e^x = e^x$

$f'(1) = e$

Tangent line: $y - 0 = e(x - 1)$

$\qquad\qquad\qquad y = e(x - 1)$

**(b)**

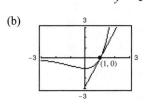

**(c)** Graphing utility confirms $\dfrac{dy}{dx} = e$ at $(1, 0)$.

**75.** $f(x) = \dfrac{8}{x^2 + 4}$; $(2, 1)$

$f'(x) = \dfrac{(x^2 + 4)(0) - 8(2x)}{(x^2 + 4)^2} = \dfrac{-16x}{(x^2 + 4)^2}$

$f'(2) = \dfrac{-16(2)}{(4 + 4)^2} = -\dfrac{1}{2}$

$y - 1 = -\dfrac{1}{2}(x - 2)$

$\qquad y = -\dfrac{1}{2}x + 2$

$2y + x - 4 = 0$

**77.** $f(x) = \dfrac{16x}{x^2 + 16}$; $\left(-2, -\dfrac{8}{5}\right)$

$f'(x) = \dfrac{(x^2 + 16)(16) - 16x(2x)}{(x^2 + 16)^2} = \dfrac{256 - 16x^2}{(x^2 + 16)^2}$

$f'(-2) = \dfrac{256 - 16(4)}{20^2} = \dfrac{12}{25}$

$y + \dfrac{8}{5} = \dfrac{12}{25}(x + 2)$

$\qquad y = \dfrac{12}{25}x - \dfrac{16}{25}$

$25y - 12x + 16 = 0$

**79.** $f(x) = \dfrac{2x - 1}{x^2} = 2x^{-1} - x^{-2}$

$f'(x) = -2x^{-2} + 2x^{-3} = \dfrac{2(-x + 1)}{x^3}$

$f'(x) = 0$ when $x = 1$, and $f(1) = 1$.

Horizontal tangent at $(1, 1)$.

**81.** $g(x) = \dfrac{8(x - 2)}{e^x}$

$g'(x) = \dfrac{e^x(8) - 8(x - 2)e^x}{e^{2x}} = \dfrac{24 - 8x}{e^x}$

$g'(x) = 0$ when $x = 3$.

Horizontal tangent is at $\left(3, 8e^{-3}\right)$.

**83.** $f(x) = \dfrac{x + 1}{x - 1}$

$f'(x) = \dfrac{(x - 1) - (x + 1)}{(x - 1)^2} = \dfrac{-2}{(x - 1)^2}$

$2y + x = 6 \Rightarrow y = -\dfrac{1}{2}x + 3$; Slope: $-\dfrac{1}{2}$

$\dfrac{-2}{(x - 1)^2} = -\dfrac{1}{2}$

$(x - 1)^2 = 4$

$x - 1 = \pm 2$

$\qquad x = -1, 3$; $f(-1) = 0$, $f(3) = 2$

$y - 0 = -\dfrac{1}{2}(x + 1) \Rightarrow y = -\dfrac{1}{2}x - \dfrac{1}{2}$

$y - 2 = -\dfrac{1}{2}(x - 3) \Rightarrow y = -\dfrac{1}{2}x + \dfrac{7}{2}$

**85.** $f'(x) = \dfrac{(x+2)3 - 3x(1)}{(x+2)^2} = \dfrac{6}{(x+2)^2}$

$g'(x) = \dfrac{(x+2)5 - (5x+4)(1)}{(x+2)^2} = \dfrac{6}{(x+2)^2}$

$g(x) = \dfrac{5x+4}{(x+2)} = \dfrac{3x}{(x+2)} + \dfrac{2x+4}{(x+2)} = f(x) + 2$

$f$ and $g$ differ by a constant.

**87.** (a)   $p'(x) = f'(x)g(x) + f(x)g'(x)$

$p'(1) = f'(1)g(1) + f(1)g'(1) = 1(4) + 6\left(-\dfrac{1}{2}\right) = 1$

(b)   $q'(x) = \dfrac{g(x)f'(x) - f(x)g'(x)}{g(x)^2}$

$q'(4) = \dfrac{3(-1) - 7(0)}{3^2} = -\dfrac{1}{3}$

**89.**   Area $= A(t) = (6t+5)\sqrt{t} = 6t^{3/2} + 5t^{1/2}$

$A'(t) = 9t^{1/2} + \dfrac{5}{2}t^{-1/2} = \dfrac{18t+5}{2\sqrt{t}}$ cm$^2$/sec

**91.**   $C = 100\left(\dfrac{200}{x^2} + \dfrac{x}{x+30}\right), \ 1 \le x$

$\dfrac{dC}{dx} = 100\left(-\dfrac{400}{x^3} + \dfrac{30}{(x+30)^2}\right)$

(a)   When $x = 10$:   $\dfrac{dC}{dx} = -\$38.13$ thousand/100 components

(b)   When $x = 15$:   $\dfrac{dC}{dx} = -\$10.37$ thousand/100 components

(c)   When $x = 20$:   $\dfrac{dC}{dx} = -\$3.80$ thousand/100 components

As the order size increases, the cost per item decreases.

**93.** (a)   $\cot x = \dfrac{\cos x}{\sin x}$

$\dfrac{d}{dx}[\cot x] = \dfrac{d}{dx}\left[\dfrac{\cos x}{\sin x}\right] = \dfrac{\sin x(-\sin x) - (\cos x)(\cos x)}{(\sin x)^2} = -\dfrac{\sin^2 x + \cos^2 x}{\sin^2 x} = -\dfrac{1}{\sin^2 x} = -\csc^2 x$

(b)   $\sec x = \dfrac{1}{\cos x}$

$\dfrac{d}{dx}[\sec x] = \dfrac{d}{dx}\left[\dfrac{1}{\cos x}\right] = \dfrac{(\cos x)(0) - (1)(-\sin x)}{(\cos x)^2} = \dfrac{\sin x}{\cos x \cos x} = \dfrac{1}{\cos x} \cdot \dfrac{\sin x}{\cos x} = \sec x \tan x$

(c)   $\csc x = \dfrac{1}{\sin x}$

$\dfrac{d}{dx}[\csc x] = \dfrac{d}{dx}\left[\dfrac{1}{\sin x}\right] = \dfrac{(\sin x)(0) - (1)(\cos x)}{(\sin x)^2} = -\dfrac{\cos x}{\sin x \sin x} = -\dfrac{1}{\sin x} \cdot \dfrac{\cos x}{\sin x} = -\csc x \cot x$

**95.** (a) $h(t) = 112.4t + 1332$

$p(t) = 2.9t + 282$

(b)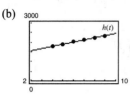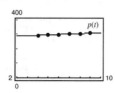

(c) $A = \dfrac{112.4t + 1332}{2.9t + 282}$

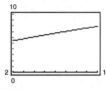

$A$ represents the average health care expenses per person (in thousands of dollars).

(d) $A'(t) \approx \dfrac{3407.5}{(t + 98.53)^2} \approx \dfrac{27{,}834}{8.41t^2 + 1635.6t + 79{,}524}$

$A'(t)$ represents the rate of change of the average health care expenses per person per year $t$.

**97.** $f(x) = x^4 + 2x^3 - 3x^2 - x$

$f'(x) = 4x^3 + 6x^2 - 6x - 1$

$f''(x) = 12x^2 + 12x - 6$

**99.** $f(x) = 4x^{3/2}$

$f'(x) = 6x^{1/2}$

$f''(x) = 3x^{-1/2} = \dfrac{3}{\sqrt{x}}$

**101.** $f(x) = \dfrac{x}{x - 1}$

$f'(x) = \dfrac{(x - 1)(1) - x(1)}{(x - 1)^2} = \dfrac{-1}{(x - 1)^2}$

$f''(x) = \dfrac{2}{(x - 1)^3}$

**103.** $f(x) = x \sin x$

$f'(x) = x \cos x + \sin x$

$f''(x) = x(-\sin x) + \cos x + \cos x$

$= -x \sin x + 2 \cos x$

**105.** $g(x) = \dfrac{e^x}{x}$

$g'(x) = \dfrac{xe^x - e^x}{x^2}$

$g''(x) = \dfrac{x^2(xe^x + e^x - e^x) - 2x(xe^x - e^x)}{x^4}$

$= \dfrac{e^x}{x^3}(x^2 - 2x + 2)$

**107.** $f'(x) = x^2$

$f''(x) = 2x$

**109.** $f'''(x) = 2\sqrt{x}$

$f^{(4)}(x) = \dfrac{1}{2}(2)x^{-1/2} = \dfrac{1}{\sqrt{x}}$

**111.** $f(x) = 2g(x) + h(x)$

$f'(x) = 2g'(x) + h'(x)$

$f'(2) = 2g'(2) + h'(2)$

$= 2(-2) + 4$

$= 0$

**113.** $f(x) = \dfrac{g(x)}{h(x)}$

$f'(x) = \dfrac{h(x)g'(x) - g(x)h'(x)}{[h(x)]^2}$

$f'(2) = \dfrac{h(2)g'(2) - g(2)h'(2)}{[h(2)]^2}$

$= \dfrac{(-1)(-2) - (3)(4)}{(-1)^2}$

$= -10$

**115.** The graph of a differentiable function $f$ such that $f(2) = 0$, $f' < 0$ for $-\infty < x < 2$, and $f' > 0$ for $2 < x < \infty$ would, in general, look like the graph below.

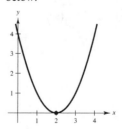

One such function is $f(x) = (x - 2)^2$.

**117.**

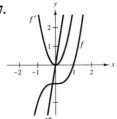

It appears that $f$ is cubic, so $f'$ would be quadratic and $f''$ would be linear.

**121.**

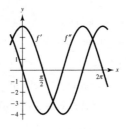

**119.**

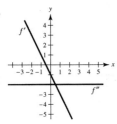

**123.** $v(t) = 36 - t^2, 0 \le t \le 6$

$a(t) = v'(t) = -2t$

$v(3) = 27$ m/sec

$a(3) = -6$ m/sec$^2$

The speed of the object is decreasing.

**125.** $s(t) = -8.25t^2 + 66t$

$v(t) = s'(t) = 16.50t + 66$

$a(t) = v'(t) = -16.50$

| $t$(sec) | 0 | 1 | 2 | 3 | 4 |
|---|---|---|---|---|---|
| $s(t)$ (ft) | 0 | 57.75 | 99 | 123.75 | 132 |
| $v(t) = s'(t)$ (ft/sec) | 66 | 49.5 | 33 | 16.5 | 0 |
| $a(t) = v'(t)$ (ft/sec$^2$) | -16.5 | -16.5 | -16.5 | -16.5 | -16.5 |

Average velocity on:

$[0, 1]$ is $\dfrac{57.75 - 0}{1 - 0} = 57.75$

$[1, 2]$ is $\dfrac{99 - 57.75}{2 - 1} = 41.25$

$[2, 3]$ is $\dfrac{123.75 - 99}{3 - 2} = 24.75$

$[3, 4]$ is $\dfrac{132 - 123.75}{4 - 3} = 8.25$

**127.** $f(x) = x^n$

$f^{(n)}(x) = n(n - 1)(n - 2) \cdots (2)(1) = n!$

**Note:** $n! = n(n - 1) \cdots 3 \cdot 2 \cdot 1$ (read "$n$ factorial")

**129.** $f(x) = g(x)h(x)$

(a) $f'(x) = g(x)h'(x) + h(x)g'(x)$

$f''(x) = g(x)h''(x) + g'(x)h'(x) + h(x)g''(x) + h'(x)g'(x)$

$\qquad = g(x)h''(x) + 2g'(x)h'(x) + h(x)g''(x)$

$f'''(x) = g(x)h'''(x) + g'(x)h''(x) + 2g'(x)h''(x) + 2g''(x)h'(x) + h(x)g'''(x) + h'(x)g''(x)$

$\qquad = g(x)h'''(x) + 3g'(x)h''(x) + 3g''(x)h'(x) + g'''(x)h(x)$

$f^{(4)}(x) = g(x)h^{(4)}(x) + g'(x)h'''(x) + 3g'(x)h'''(x) + 3g''(x)h''(x) + 3g''(x)h''(x) + 3g'''(x)h'(x)$

$\qquad\qquad + g'''(x)h'(x) + g^{(4)}(x)h(x)$

$\qquad = g(x)h^{(4)}(x) + 4g'(x)h'''(x) + 6g''(x)h''(x) + 4g'''(x)h'(x) + g^{(4)}(x)h(x)$

(b) $f^{(n)}(x) = g(x)h^{(n)}(x) + \dfrac{n(n-1)(n-2)\cdots(2)(1)}{1\big[(n-1)(n-2)\cdots(2)(1)\big]}g'(x)h^{(n-1)}(x) + \dfrac{n(n-1)(n-2)\cdots(2)(1)}{(2)(1)\big[(n-2)(n-3)\cdots(2)(1)\big]}g''(x)h^{(n-2)}(x)$

$\qquad\quad + \dfrac{n(n-1)(n-2)\cdots(2)(1)}{(3)(2)(1)\big[(n-3)(n-4)\cdots(2)(1)\big]}g'''(x)h^{(n-3)}(x) + \cdots$

$\qquad\quad + \dfrac{n(n-1)(n-2)\cdots(2)(1)}{\big[(n-1)(n-2)\cdots(2)(1)\big](1)}g^{(n-1)}(x)h'(x) + g^{(n)}(x)h(x)$

$\qquad = g(x)h^{(n)}(x) + \dfrac{n!}{1!(n-1)!}g'(x)h^{(n-1)}(x) + \dfrac{n!}{2!(n-2)!}g''(x)h^{(n-2)}(x) + \cdots$

$\qquad\quad + \dfrac{n!}{(n-1)!1!}g^{(n-1)}(x)h'(x) + g^{(n)}(x)h(x)$

**Note:** $n! = n(n-1)\cdots 3 \cdot 2 \cdot 1$ (read "$n$ factorial")

**131.** $f(x) = x^n \sin x$

$f'(x) = x^n \cos x + nx^{n-1}\sin x$

When $n = 1$: $f'(x) = x \cos x + \sin x$

When $n = 2$: $f'(x) = x^2 \cos x + 2\sin x$

When $n = 3$: $f'(x) = x^3 \cos x + 3x^2 \sin x$

When $n = 4$: $f'(x) = x^4 \cos x + 4x^3 \sin x$

For general $n$, $f'(x) = x^n \cos x + nx^{n-1}\sin x$.

**133.** $y = \dfrac{1}{x},\ y' = -\dfrac{1}{x^2},\ y'' = \dfrac{2}{x^3}$

$x^3 y'' + 2x^2 y' = x^3\left[\dfrac{2}{x^3}\right] + 2x^2\left[-\dfrac{1}{x^2}\right] = 2 - 2 = 0$

**135.** $\qquad y = 2\sin x + 3$

$\qquad\quad y' = 2\cos x$

$\qquad\quad y'' = -2\sin x$

$y'' + y = -2\sin x + (2\sin x + 3) = 3$

**137.** False. If $y = f(x)g(x)$, then

$\dfrac{dy}{dx} = f(x)g'(x) + g(x)f'(x).$

**139.** True

$h'(c) = f(c)g'(c) + g(c)f'(c)$

$\qquad = f(c)(0) + g(c)(0)$

$\qquad = 0$

**141.** True

**143.** $f(x) = x|x| = \begin{cases} x^2, & x \geq 0 \\ -x^2, & x < 0 \end{cases}$

$f'(x) = \begin{cases} 2x, & x > 0 \\ -2x, & x < 0 \end{cases} = 2|x|$

$f''(x) = \begin{cases} 2, & x > 0 \\ -2, & x < 0 \end{cases}$

$f''(0)$ does not exist because the left and right derivatives do not agree at $x = 0$.

**145.** $\dfrac{d}{dx}\Big[f(x)g(x)h(x)\Big] = \dfrac{d}{dx}\Big[\big(f(x)g(x)\big)h(x)\Big]$

$\qquad\qquad\qquad = \dfrac{d}{dx}\Big[f(x)g(x)\Big]h(x) + f(x)g(x)h'(x)$

$\qquad\qquad\qquad = \Big[f(x)g'(x) + g(x)f'(x)\Big]h(x) + f(x)g(x)h'(x)$

$\qquad\qquad\qquad = f'(x)g(x)h(x) + f(x)g'(x)h(x) + f(x)g(x)h'(x)$

## Section 3.4   The Chain Rule

| $y = f\big(g(x)\big)$ | $u = g(x)$ | $y = f(u)$ |
|---|---|---|
| **1.** $y = (5x - 8)^4$ | $u = 5x - 8$ | $y = u^4$ |
| **3.** $y = \csc^3 x$ | $u = \csc x$ | $y = u^3$ |
| **5.** $y = e^{-2x}$ | $u = -2x$ | $y = e^u$ |

**7.** $y = (4x - 1)^3$

$\quad y' = 3(4x - 1)^2(4) = 12(4x - 1)^2$

**9.** $g(x) = 3(4 - 9x)^4$

$\quad g'(x) = 12(4 - 9x)^3(-9) = -108(4 - 9x)^3$

**11.** $f(t) = \sqrt{5 - t} = (5 - t)^{1/2}$

$\quad f'(t) = \dfrac{1}{2}(5 - t)^{-1/2}(-1) = \dfrac{-1}{2\sqrt{5 - t}}$

**13.** $y = \sqrt[3]{6x^2 + 1} = (6x^2 + 1)^{1/3}$

$\quad y' = \dfrac{1}{3}(6x^2 + 1)^{-2/3}(12x) = \dfrac{4x}{(6x^2 + 1)^{2/3}} = \dfrac{4x}{\sqrt[3]{(6x^2 + 1)^2}}$

**15.** $y = 2\sqrt[4]{9 - x^2} = 2(9 - x^2)^{1/4}$

$\quad y' = 2\left(\dfrac{1}{4}\right)(9 - x^2)^{-3/4}(-2x)$

$\qquad = \dfrac{-x}{(9 - x^2)^{3/4}} = \dfrac{-x}{\sqrt[4]{(9 - x^2)^3}}$

**17.** $y = (x - 2)^{-1}$

$\quad y' = -1(x - 2)^{-2}(1) = \dfrac{-1}{(x - 2)^2}$

**19.** $f(t) = (t - 3)^{-2}$

$\quad f'(t) = -2(t - 3)^{-3}(1) = \dfrac{-2}{(t - 3)^3}$

**21.** $y = \dfrac{1}{\sqrt{3x + 5}} = (3x + 5)^{-1/2}$

$\quad y' = -\dfrac{1}{2}(3x + 5)^{-3/2}(3)$

$\qquad = \dfrac{-3}{2(3x + 5)^{3/2}}$

$\qquad = -\dfrac{3}{2\sqrt{(3x + 5)^3}}$

**23.** $f(x) = x^2(x - 2)^4$

$\quad f'(x) = x^2\Big[4(x - 2)^3(1)\Big] + (x - 2)^4(2x)$

$\qquad = 2x(x - 2)^3\Big[2x + (x - 2)\Big]$

$\qquad = 2x(x - 2)^3(3x - 2)$

**25.** $y = x\sqrt{1 - x^2} = x(1 - x^2)^{1/2}$

$\quad y' = x\left[\dfrac{1}{2}(1 - x^2)^{-1/2}(-2x)\right] + (1 - x^2)^{1/2}(1)$

$\qquad = -x^2(1 - x^2)^{-1/2} + (1 - x^2)^{1/2}$

$\qquad = (1 - x^2)^{-1/2}\Big[-x^2 + (1 - x^2)\Big]$

$\qquad = \dfrac{1 - 2x^2}{\sqrt{1 - x^2}}$

**27.** $y = \dfrac{x}{\sqrt{x^2 + 1}} = \dfrac{x}{\left(x^2 + 1\right)^{1/2}}$

$y' = \dfrac{\left(x^2 + 1\right)^{1/2}(1) - x\left(\dfrac{1}{2}\right)\left(x^2 + 1\right)^{-1/2}(2x)}{\left[\left(x^2 + 1\right)^{1/2}\right]^2}$

$= \dfrac{\left(x^2 + 1\right)^{1/2} - x^2\left(x^2 + 1\right)^{-1/2}}{x^2 + 1}$

$= \dfrac{\left(x^2 + 1\right)^{-1/2}\left[x^2 + 1 - x^2\right]}{x^2 + 1}$

$= \dfrac{1}{\left(x^2 + 1\right)^{3/2}} = \dfrac{1}{\sqrt{\left(x^2 + 1\right)^3}}$

**29.** $g(x) = \left(\dfrac{x + 5}{x^2 + 2}\right)^2$

$g'(x) = 2\left(\dfrac{x + 5}{x^2 + 2}\right)\left(\dfrac{\left(x^2 + 2\right) - (x + 5)(2x)}{\left(x^2 + 2\right)^2}\right)$

$= \dfrac{2(x + 5)\left(2 - 10x - x^2\right)}{\left(x^2 + 2\right)^3}$

$= \dfrac{-2(x + 5)\left(x^2 + 10x - 2\right)}{\left(x^2 + 2\right)^3}$

**31.** $f(v) = \left(\dfrac{1 - 2v}{1 + v}\right)^3$

$f'(v) = 3\left(\dfrac{1 - 2v}{1 + v}\right)^2\left(\dfrac{(1 + v)(-2) - (1 - 2v)}{(1 + v)^2}\right)$

$= \dfrac{-9(1 - 2v)^2}{(1 + v)^4}$

**33.** $f(x) = \left(\left(x^2 + 3\right)^5 + x\right)^2$

$f'(x) = 2\left(\left(x^2 + 3\right)^5 + x\right)\left(5\left(x^2 + 3\right)^4(2x) + 1\right)$

$= 2\left[10x\left(x^2 + 3\right)^9 + \left(x^2 + 3\right)^5 + 10x^2\left(x^2 + 3\right)^4 + x\right] = 20x\left(x^2 + 3\right)^9 + 2\left(x^2 + 3\right)^5 + 20x^2\left(x^2 + 3\right)^4 + 2x$

**35.** $y = \dfrac{\sqrt{x} + 1}{x^2 + 1}$

$y' = \dfrac{1 - 3x^2 - 4x^{3/2}}{2\sqrt{x}\left(x^2 + 1\right)^2}$

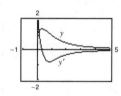

The zero of $y'$ corresponds to the point on the graph of $y$ where the tangent line is horizontal.

**37.** $y = \sqrt{\dfrac{x + 1}{x}}$

$y' = -\dfrac{\sqrt{(x + 1)/x}}{2x(x + 1)}$

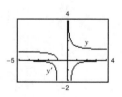

$y'$ has no zeros.

**39.** $y = \dfrac{\cos \pi x + 1}{x}$

$\dfrac{dy}{dx} = \dfrac{-\pi x \sin \pi x - \cos \pi x - 1}{x^2}$

$= -\dfrac{\pi x \sin \pi x + \cos \pi x + 1}{x^2}$

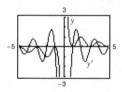

The zeros of $y'$ correspond to the points on the graph of $y$ where the tangent lines are horizontal.

**41. (a)** $\quad y = \sin x$

$\qquad y' = \cos x$

$\quad y'(0) = 1$

$\quad$ 1 cycle in $[0, 2\pi]$

**(b)** $\quad y = \sin 2x$

$\qquad y' = 2 \cos 2x$

$\quad y'(0) = 2$

$\quad$ 2 cycles in $[0, 2\pi]$

The slope of $\sin ax$ at the origin is $a$.

**43.** $y = e^{3x}$

$y' = 3e^{3x}$

At $(0, 1)$, $y' = 3$.

**45.** $y = \ln x^3 = 3 \ln x$

$y' = \dfrac{3}{x}$

At $(1, 0)$, $y' = 3$.

**47.** $y = \cos 4x$

$\dfrac{dy}{dx} = -4 \sin 4x$

**49.** $g(x) = 5 \tan 3x$

$g'(x) = 15 \sec^2 3x$

**51.** $y = \sin(\pi x)^2 = \sin(\pi^2 x^2)$

$y' = \cos(\pi^2 x^2)\left[2\pi^2 x\right] = 2\pi^2 x \cos(\pi^2 x^2)$

$\quad = 2\pi^2 x \cos(\pi x)^2$

**53.** $h(x) = \sin 2x \cos 2x$

$h'(x) = \sin 2x(-2 \sin 2x) + \cos 2x(2 \cos 2x)$

$\quad = 2 \cos^2 2x - 2 \sin^2 2x$

$\quad = 2 \cos 4x$

**Alternate solution:** $h(x) = \frac{1}{2} \sin 4x$

$\qquad\qquad h'(x) = \frac{1}{2} \cos 4x(4) = 2 \cos 4x$

**55.** $f(x) = \dfrac{\cot x}{\sin x} = \dfrac{\cos x}{\sin^2 x}$

$f'(x) = \dfrac{\sin^2 x(-\sin x) - \cos x(2 \sin x \cos x)}{\sin^4 x}$

$\quad = \dfrac{-\sin^2 x - 2 \cos^2 x}{\sin^3 x} = \dfrac{-1 - \cos^2 x}{\sin^3 x}$

**57.** $y = 4 \sec^2 x$

$y' = 8 \sec x \cdot \sec x \tan x = 8 \sec^2 x \tan x$

**59.** $f(\theta) = \tan^2 5\theta = (\tan 5\theta)^2$

$f'(\theta) = 2(\tan 5\theta)(\sec^2 5\theta)5 = 10 \tan 5\theta \sec^2 5\theta$

**61.** $f(\theta) = \frac{1}{4} \sin^2 2\theta = \frac{1}{4}(\sin 2\theta)^2$

$f'(\theta) = 2\left(\frac{1}{4}\right)(\sin 2\theta)(\cos 2\theta)(2)$

$\quad = \sin 2\theta \cos 2\theta = \frac{1}{2} \sin 4\theta$

**63.** $f(t) = 3\sec^2(\pi t - 1)$

$f'(t) = 6 \sec(\pi t - 1) \sec(\pi t - 1) \tan(\pi t - 1)(\pi)$

$\quad = 6\pi \sec^2(\pi t - 1) \tan(\pi t - 1) = \dfrac{6\pi \sin(\pi t - 1)}{\cos^3(\pi t - 1)}$

**65.** $y = \sqrt{x} + \dfrac{1}{4} \sin(2x)^2 = \sqrt{x} + \dfrac{1}{4} \sin(4x^2)$

$\dfrac{dy}{dx} = \dfrac{1}{2}x^{-1/2} + \dfrac{1}{4} \cos(4x^2)(8x) = \dfrac{1}{2\sqrt{x}} + 2x \cos(2x)^2$

**67.** $y = \sin(\tan 2x)$

$y' = \cos(\tan 2x)(\sec^2 2x)(2) = 2 \cos(\tan 2x) \sec^2 2x$

**69.** $f(x) = e^{2x}$

$f'(x) = 2e^{2x}$

**71.** $y = e^{\sqrt{x}}$

$\dfrac{dy}{dx} = \dfrac{e^{\sqrt{x}}}{2\sqrt{x}}$

**73.** $g(t) = \left(e^{-t} + e^{t}\right)^3$

$g'(t) = 3\left(e^{-t} + e^{t}\right)^2\left(e^{t} - e^{-t}\right)$

**75.** $y = \ln e^{x^2} = x^2$

$\dfrac{dy}{dx} = 2x$

**77.** $y = \dfrac{2}{e^x + e^{-x}} = 2\left(e^x + e^{-x}\right)^{-1}$

$\dfrac{dy}{dx} = -2\left(e^x + e^{-x}\right)^{-2}\left(e^x - e^{-x}\right)$

$\quad = \dfrac{-2\left(e^x - e^{-x}\right)}{\left(e^x + e^{-x}\right)^2}$

**79.** $y = x^2 e^x - 2xe^x + 2e^x = e^x\left(x^2 - 2x + 2\right)$

$\dfrac{dy}{dx} = e^x(2x - 2) + e^x\left(x^2 - 2x + 2\right) = x^2 e^x$

**81.** $f(x) = e^{-x} \ln x$

$f'(x) = e^{-x}\left(\dfrac{1}{x}\right) - e^{-x} \ln x = e^{-x}\left(\dfrac{1}{x} - \ln x\right)$

**83.** $y = e^x(\sin x + \cos x)$

$\dfrac{dy}{dx} = e^x(\cos x - \sin x) + (\sin x + \cos x)\left(e^x\right)$

$\quad = e^x(2 \cos x) = 2e^x \cos x$

**85.** $g(x) = \ln x^2 = 2 \ln x$

$g'(x) = \dfrac{2}{x}$

**87.** $y = (\ln x)^4$

$\dfrac{dy}{dx} = 4(\ln x)^3 \left(\dfrac{1}{x}\right) = \dfrac{4(\ln x)^3}{x}$

**89.** $y = \ln x\sqrt{x^2 - 1} = \ln x + \dfrac{1}{2}\ln(x^2 - 1)$

$\dfrac{dy}{dx} = \dfrac{1}{x} + \dfrac{1}{2}\left(\dfrac{2x}{x^2 - 1}\right) = \dfrac{2x^2 - 1}{x(x^2 - 1)}$

**91.** $f(x) = \ln\dfrac{x}{x^2 + 1} = \ln x - \ln(x^2 + 1)$

$f'(x) = \dfrac{1}{x} - \dfrac{2x}{x^2 + 1} = \dfrac{1 - x^2}{x(x^2 + 1)}$

**93.** $g(t) = \dfrac{\ln t}{t^2}$

$g'(t) = \dfrac{t^2(1/t) - 2t \ln t}{t^4} = \dfrac{1 - 2 \ln t}{t^3}$

**95.** $y = \ln\sqrt{\dfrac{x + 1}{x - 1}} = \dfrac{1}{2}\big[\ln(x + 1) - \ln(x - 1)\big]$

$\dfrac{dy}{dx} = \dfrac{1}{2}\left[\dfrac{1}{x + 1} - \dfrac{1}{x - 1}\right] = \dfrac{1}{1 - x^2}$

**97.** $y = \dfrac{-\sqrt{x^2 + 1}}{x} + \ln\left(x + \sqrt{x^2 + 1}\right)$

$\dfrac{dy}{dx} = \dfrac{-x\left(x/\sqrt{x^2 + 1}\right) + \sqrt{x^2 + 1}}{x^2} + \left(\dfrac{1}{x + \sqrt{x^2 + 1}}\right)\left(1 + \dfrac{x}{\sqrt{x^2 + 1}}\right)$

$= \dfrac{1}{x^2\sqrt{x^2 + 1}} + \left(\dfrac{1}{x + \sqrt{x^2 + 1}}\right)\left(\dfrac{\sqrt{x^2 + 1} + x}{\sqrt{x^2 + 1}}\right)$

$= \dfrac{1}{x^2\sqrt{x^2 + 1}} + \dfrac{1}{\sqrt{x^2 + 1}} = \dfrac{1 + x^2}{x^2\sqrt{x^2 + 1}} = \dfrac{\sqrt{x^2 + 1}}{x^2}$

**99.** $y = \ln|\sin x|$

$\dfrac{dy}{dx} = \dfrac{\cos x}{\sin x} = \cot x$

**101.** $y = \ln\left|\dfrac{\cos x}{\cos x - 1}\right|$

$= \ln|\cos x| - \ln|\cos x - 1|$

$\dfrac{dy}{dx} = \dfrac{-\sin x}{\cos x} - \dfrac{-\sin x}{\cos x - 1} = -\tan x + \dfrac{\sin x}{\cos x - 1}$

**103.** $y = \ln\left|\dfrac{-1 + \sin x}{2 + \sin x}\right| = \ln|-1 + \sin x| - \ln|2 + \sin x|$

$\dfrac{dy}{dx} = \dfrac{\cos x}{-1 + \sin x} - \dfrac{\cos x}{2 + \sin x} = \dfrac{3 \cos x}{(\sin x - 1)(\sin x + 2)}$

**105.** $y = \sqrt{x^2 + 8x} = (x^2 + 8x)^{1/2}, \ (1, 3)$

$y' = \dfrac{1}{2}(x^2 + 8x)^{-1/2}(2x + 8) = \dfrac{2(x + 4)}{2(x^2 + 8x)^{1/2}} = \dfrac{x + 4}{\sqrt{x^2 + 8x}}$

$y'(1) = \dfrac{1 + 4}{\sqrt{1^2 + 8(1)}} = \dfrac{5}{\sqrt{9}} = \dfrac{5}{3}$

**107.** $f(x) = \dfrac{5}{x^3 - 2} = 5\left(x^3 - 2\right)^{-1}, \quad \left(-2, -\dfrac{1}{2}\right)$

$f'(x) = -5\left(x^3 - 2\right)^{-2}\left(3x^2\right) = \dfrac{-15x^2}{\left(x^3 - 2\right)^2}$

$f'(-2) = -\dfrac{60}{100} = -\dfrac{3}{5}$

**109.** $f(t) = \dfrac{3t + 2}{t - 1}, \quad (0, -2)$

$f'(t) = \dfrac{(t - 1)(3) - (3t + 2)(1)}{(t - 1)^2}$

$= \dfrac{3t - 3 - 3t - 2}{(t - 1)^2}$

$= \dfrac{-5}{(t - 1)^2}$

$f'(0) = -5$

**111.** $y = 26 - \sec^3 4x, \quad (0, 25)$

$y' = -3\sec^2 4x \sec 4x \tan 4x \cdot 4$

$= -12 \sec^3 4x \tan 4x$

$y'(0) = 0$

**113.** (a) $f(x) = \left(2x^2 - 7\right)^{1/2}, \quad (4, 5)$

$f'(x) = \dfrac{1}{2}\left(2x^2 - 7\right)^{-1/2}(4x) = \dfrac{2x}{\sqrt{2x^2 - 7}}$

$f'(4) = \dfrac{8}{5}$

Tangent line:

$y - 5 = \dfrac{8}{5}(x - 4) \Rightarrow 8x - 5y - 7 = 0$

(b)
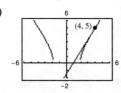

**115.** (a) $f(x) = \sin 2x, \quad (\pi, 0)$

$f'(x) = 2\cos 2x$

$f'(\pi) = 2$

Tangent line:

$y = 2(x - \pi) \Rightarrow 2x - y - 2\pi = 0$

(b)
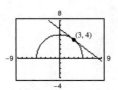

**117.** (a) $f(x) = \tan^2 x, \quad \left(\dfrac{\pi}{4}, 1\right)$

$f'(x) = 2\tan x \sec^2 x$

$f'\left(\dfrac{\pi}{4}\right) = 2(1)(2) = 4$

Tangent line:

$y - 1 = 4\left(x - \dfrac{\pi}{4}\right) \Rightarrow 4x - y + (1 - \pi) = 0$

(b)
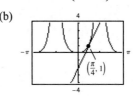

**119.** (a) $y = 4 - x^2 - \ln\left(\dfrac{1}{2}x + 1\right), \quad (0, 4)$

$\dfrac{dy}{dx} = -2x - \dfrac{1}{(1/2)x + 1}\left(\dfrac{1}{2}\right)$

$= -2x - \dfrac{1}{x + 2}$

When $x = 0$, $\dfrac{dy}{dx} = -\dfrac{1}{2}$.

Tangent line: $y - 4 = -\dfrac{1}{2}(x - 0)$

$y = -\dfrac{1}{2}x + 4$

(b)
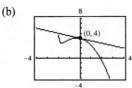

**121.** $f(x) = \sqrt{25 - x^2} = \left(25 - x^2\right)^{1/2}, \quad (3, 4)$

$f'(x) = \dfrac{1}{2}\left(25 - x^2\right)(-2x) = \dfrac{-x}{\sqrt{25 - x^2}}$

$f'(3) = -\dfrac{3}{4}$

Tangent line:

$y - 4 = -\dfrac{3}{4}(x - 3) \Rightarrow 3x + 4y - 25 = 0$

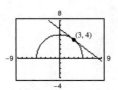

**123.**
$$f(x) = 2\cos x + \sin 2x, \quad 0 < x < 2\pi$$
$$f'(x) = -2\sin x + 2\cos 2x$$
$$= -2\sin x + 2 - 4\sin^2 x = 0$$
$$2\sin^2 x + \sin x - 1 = 0$$
$$(\sin x + 1)(2\sin x - 1) = 0$$
$$\sin x = -1 \Rightarrow x = \frac{3\pi}{2}$$
$$\sin x = \frac{1}{2} \Rightarrow x = \frac{\pi}{6}, \frac{5\pi}{6}$$

Horizontal tangents at $x = \dfrac{\pi}{6}, \dfrac{3\pi}{2}, \dfrac{5\pi}{6}$

Horizontal tangent at the points $\left(\dfrac{\pi}{6}, \dfrac{3\sqrt{3}}{2}\right), \left(\dfrac{3\pi}{2}, 0\right)$, and $\left(\dfrac{5\pi}{6}, -\dfrac{3\sqrt{3}}{2}\right)$

**125.** $f(x) = 5(2 - 7x)^4$
$$f'(x) = 20(2 - 7x)^3(-7) = -140(2 - 7x)^3$$
$$f''(x) = -420(2 - 7x)^2(-7) = 2940(2 - 7x)^2$$

**127.** $f(x) = \dfrac{1}{x - 6} = (x - 6)^{-1}$
$$f'(x) = -(x - 6)^{-2}$$
$$f''(x) = 2(x - 6)^{-3} = \dfrac{2}{(x - 6)^3}$$

**129.** $f(x) = \sin x^2$
$$f'(x) = 2x\cos x^2$$
$$f''(x) = 2x\left[2x(-\sin x^2)\right] + 2\cos x^2$$
$$= 2(\cos x^2 - 2x^2\sin x^2)$$

**131.** $f(x) = (3 + 2x)e^{-3x}$
$$f'(x) = (3 + 2x)(-3e^{-3x}) + 2e^{-3x}$$
$$= (-7 - 6x)e^{-3x}$$
$$f''(x) = (-7 - 6x)(-3e^{-3x}) - 6e^{-3x}$$
$$= 3(6x + 5)e^{-3x}$$

**133.** $h(x) = \frac{1}{9}(3x + 1)^3, \quad \left(1, \frac{64}{9}\right)$
$$h'(x) = \frac{1}{9}3(3x + 1)^2(3) = (3x + 1)^2$$
$$h''(x) = 2(3x + 1)(3) = 18x + 6$$
$$h''(1) = 24$$

**135.** $f(x) = \cos x^2, \quad (0, 1)$
$$f'(x) = -\sin(x^2)(2x) = -2x\sin(x^2)$$
$$f''(x) = -2x\cos(x^2)(2x) - 2\sin(x^2)$$
$$= -4x^2\cos(x^2) - 2\sin(x^2)$$
$$f''(0) = 0$$

**137.** $f(x) = 4^x$
$$f'(x) = (\ln 4)4^x$$

**139.** $y = 5^{x-2}$
$$\frac{dy}{dx} = (\ln 5)5^{x-2}$$

**141.** $g(t) = t^2 2^t$
$$g'(t) = t^2(\ln 2)2^t + (2t)2^t$$
$$= t2^t(t\ln 2 + 2)$$
$$= 2^t t(2 + t\ln 2)$$

**143.** $h(\theta) = 2^{-\theta}\cos \pi\theta$
$$h'(\theta) = 2^{-\theta}(-\pi\sin \pi\theta) - (\ln 2)2^{-\theta}\cos \pi\theta$$
$$= -2^{-\theta}\left[(\ln 2)\cos \pi\theta + \pi\sin \pi\theta\right]$$

**145.** $y = \log_3 x$
$$\frac{dy}{dx} = \frac{1}{x\ln 3}$$

**147.** $f(x) = \log_2 \dfrac{x^2}{x-1}$

$\qquad = 2\log_2 x - \log_2(x-1)$

$\quad f'(x) = \dfrac{2}{x\ln 2} - \dfrac{1}{(x-1)\ln 2}$

$\qquad = \dfrac{x-2}{(\ln 2)x(x-1)}$

**149.** $y = \log_5\sqrt{x^2-1} = \dfrac{1}{2}\log_5(x^2-1)$

$\quad \dfrac{dy}{dx} = \dfrac{1}{2}\cdot\dfrac{2x}{(x^2-1)\ln 5} = \dfrac{x}{(x^2-1)\ln 5}$

**151.** $g(t) = \dfrac{10\log_4 t}{t} = \dfrac{10}{\ln 4}\left(\dfrac{\ln t}{t}\right)$

$\quad g'(t) = \dfrac{10}{\ln 4}\left[\dfrac{t(1/t) - \ln t}{t^2}\right]$

$\qquad = \dfrac{10}{t^2\ln 4}[1 - \ln t] = \dfrac{5}{t^2\ln 2}(1 - \ln t)$

**153.**

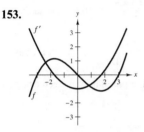

The zeros of $f'$ correspond to the points where the graph of $f$ has horizontal tangents.

**155.**

The zeros of $f'$ correspond to the points where the graph of $f$ has horizontal tangents.

**157.** $g(x) = f(3x)$

$\quad g'(x) = f'(3x)(3) \Rightarrow g'(x) = 3f'(3x)$

**159.** $f(x) = g(x)h(x)$

$\quad f'(x) = g(x)h'(x) + g'(x)h(x)$

$\quad f'(5) = (-3)(-2) + (6)(3) = 24$

**161.** $f(x) = \dfrac{g(x)}{h(x)}$

$\quad f'(x) = \dfrac{h(x)g'(x) - g(x)h'(x)}{\left[h(x)\right]^2}$

$\quad f'(5) = \dfrac{(3)(6) - (-3)(-2)}{(3)^2} = \dfrac{12}{9} = \dfrac{4}{3}$

**163.** (a) $h(x) = f(g(x)),\ g(1) = 4,\ g'(1) = -\frac{1}{2},\ f'(4) = -1$

$\qquad h'(x) = f'(g(x))g'(x)$

$\qquad h'(1) = f'(g(1))g'(1) = f'(4)g'(1) = (-1)\left(-\frac{1}{2}\right) = \frac{1}{2}$

   (b) $s(x) = g(f(x)),\ f(5) = 6,\ f'(5) = -1,\ g'(6)$ does not exist.

$\qquad s'(x) = g'(f(x))f'(x)$

$\qquad s'(5) = g'(f(5))f'(5) = g'(6)(-1)$

$\qquad s'(5)$ does not exist because $g$ is not differentiable at 6.

**165.** (a) $F = 132,400(331 - v)^{-1}$

$$F' = (-1)(132,400)(331 - v)^{-2}(-1) = \frac{132,400}{(331 - v)^2}$$

When $v = 30$, $F' \approx 1.461$.

(b) $F = 132,400(331 + v)^{-1}$

$$F' = (-1)(132,400)(331 + v)^{-2}(-1) = \frac{-132,400}{(331 + v)^2}$$

When $v = 30$, $F' \approx -1.016$.

**167.** $\theta = 0.2 \cos 8t$

The maximum angular displacement is $\theta = 0.2$ (because $-1 \le \cos 8t \le 1$).

$$\frac{d\theta}{dt} = 0.2[-8 \sin 8t] = -1.6 \sin 8t$$

When $t = 3$, $d\theta/dt = -1.6 \sin 24 \approx 1.4489$ rad/sec.

**169.** (a) Using a graphing utility, you obtain a model similar to
$T(t) = 56.1 + 27.6 \sin (0.48t - 1.86)$.

(a)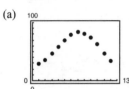

(b)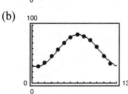

The model is a good fit.

(c)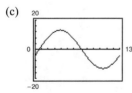

$T'(t) \approx 13.25 \cos (0.48t - 1.86)$

(d) The temperature changes most rapidly around spring (March–May), and fall (Oct–Nov).

**171.** (a)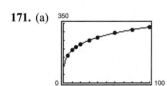

(b) $T'(p) = \dfrac{34.96}{p} + \dfrac{3.955}{\sqrt{p}}$

$T'(10) \approx 4.75$ deg/lb/in.²

$T'(70) \approx 0.97$ deg/lb/in.²

**173.** $S = C(R^2 - r^2)$

$$\frac{dS}{dt} = C\left(2R\frac{dR}{dt} - 2r\frac{dr}{dt}\right)$$

Because $r$ is constant, you have $dr/dt = 0$ and

$$\frac{dS}{dt} = (1.76 \times 10^5)(2)(1.2 \times 10^{-2})(10^{-5})$$

$$= 4.224 \times 10^{-2} = 0.04224 \text{ cm/sec}^2.$$

**175.** $N = 400\left[1 - \dfrac{3}{(t^2 + 2)^2}\right] = 400 - 1200(t^2 + 2)^{-2}$

$$N'(t) = 2400(t^2 + 2)^{-3}(2t) = \frac{4800t}{(t^2 + 2)^3}$$

(a) $N'(0) = 0$ bacteria/day

(b) $N'(1) = \dfrac{4800(1)}{(1 + 2)^3} = \dfrac{4800}{27} \approx 177.8$ bacteria/day

(c) $N'(2) = \dfrac{4800(2)}{(4 + 2)^3} = \dfrac{9600}{216} \approx 44.4$ bacteria/day

(d) $N'(3) = \dfrac{4800(3)}{(9 + 2)^3} = \dfrac{14,400}{1331} \approx 10.8$ bacteria/day

(e) $N'(4) = \dfrac{4800(4)}{(16 + 2)^3} = \dfrac{19,200}{5832} \approx 3.3$ bacteria/day

(f) The rate of change of the population is decreasing as $t \to \infty$.

**177.** $f(x) = \sin \beta x$

(a) $f'(x) = \beta \cos \beta x$

$f''(x) = -\beta^2 \sin \beta x$

$f'''(x) = -\beta^3 \cos \beta x$

$f^{(4)} = \beta^4 \sin \beta x$

(b) $f''(x) + \beta^2 f(x) = -\beta^2 \sin \beta x + \beta^2(\sin \beta x) = 0$

(c) $f^{(2k)}(x) = (-1)^k \beta^{2k} \sin \beta x$

$f^{(2k-1)}(x) = (-1)^{k+1} \beta^{2k-1} \cos \beta x$

**179.** (a) $r'(x) = f'(g(x))g'(x)$

$r'(1) = f'(g(1))g'(1)$

Note that $g(1) = 4$ and $f'(4) = \dfrac{5 - 0}{6 - 2} = \dfrac{5}{4}$.

Also, $g'(1) = 0$. So, $r'(1) = 0$.

(b) $s'(x) = g'(f(x))f'(x)$

$s'(4) = g'(f(4))f'(4)$

Note that $f(4) = \dfrac{5}{2}$, $g'\left(\dfrac{5}{2}\right) = \dfrac{6 - 4}{6 - 2} = \dfrac{1}{2}$ and

$f'(4) = \dfrac{5}{4}$. So, $s'(4) = \dfrac{1}{2}\left(\dfrac{5}{4}\right) = \dfrac{5}{8}$.

**181.** (a) If $f(-x) = -f(x)$, then

$$\frac{d}{dx}[f(-x)] = \frac{d}{dx}[-f(x)]$$
$$f'(-x)(-1) = -f'(x)$$
$$f'(-x) = f'(x).$$

So, $f'(x)$ is even.

(b) If $f(-x) = f(x)$, then

$$\frac{d}{dx}[f(-x)] = \frac{d}{dx}[f(x)]$$
$$f'(-x)(-1) = f'(x)$$
$$f'(-x) = -f'(x).$$

So, $f'$ is odd.

**183.** $g(x) = |3x - 5|$

$$g'(x) = 3\left(\frac{3x - 5}{|3x - 5|}\right), \quad x \neq \frac{5}{3}$$

**185.** $h(x) = |x|\cos x$

$$h'(x) = -|x|\sin x + \frac{x}{|x|}\cos x, \quad x \neq 0$$

**187.** (a)  $f(x) = \tan x$ $\qquad f(\pi/4) = 1$

$\qquad f'(x) = \sec^2 x$ $\qquad f'(\pi/4) = 2$

$\qquad f''(x) = 2\sec^2 x \tan x$ $\qquad f''(\pi/4) = 4$

$\qquad P_1(x) = 2(x - \pi/4) + 1$

$\qquad P_2(x) = \frac{1}{2}(4)(x - \pi/4)^2 + 2(x - \pi/4) + 1$

$\qquad\qquad = 2(x - \pi/4)^2 + 2(x - \pi/4) + 1$

(b)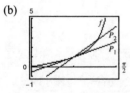

(c)  $P_2$ is a better approximation than $P_1$.

(d)  The accuracy worsens as you move away from $x = \pi/4$.

**189.** (a)  $f(x) = e^x$ $\qquad f(0) = 1$

$\qquad f'(x) = e^x$ $\qquad f'(0) = 1$

$\qquad f''(x) = e^x$ $\qquad f''(0) = 1$

$\qquad P_1(x) = 1(x - 0) + 1 = x + 1$

$\qquad P_2(x) = \frac{1}{2}(1)(x - 0)^2 + 1(x - 0) + 1$

$\qquad\qquad = \frac{1}{2}x^2 + x + 1$

(b)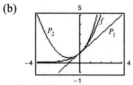

(c)  $P_2$ is a better approximation than $P_1$.

(d)  The accuracy worsens as you move away from $x = 0$.

**191.** False. If $y = (1 - x)^{1/2}$, then $y' = \frac{1}{2}(1 - x)^{-1/2}(-1)$.

**193.** True

**195.**
$$f(x) = a_1 \sin x + a_2 \sin 2x + \cdots + a_n \sin nx$$
$$f'(x) = a_1 \cos x + 2a_2 \cos 2x + \cdots + na_n \cos nx$$
$$f'(0) = a_1 + 2a_2 + \cdots + na_n$$

$$\left| a_1 + 2a_2 + \cdots + na_n \right| = \left| f'(0) \right| = \lim_{x \to 0} \left| \frac{f(x) - f(0)}{x - 0} \right| = \lim_{x \to 0} \left| \frac{f(x)}{\sin x} \right| \cdot \left| \frac{\sin x}{x} \right| = \lim_{x \to 0} \left| \frac{f(x)}{\sin x} \right| \le 1$$

# Section 3.5   Implicit Differentiation

**1.**   $x^2 + y^2 = 9$

$2x + 2yy' = 0$

$$y' = -\frac{x}{y}$$

**3.**   $x^{1/2} + y^{1/2} = 16$

$$\frac{1}{2}x^{-1/2} + \frac{1}{2}y^{-1/2}y' = 0$$

$$y' = -\frac{x^{-1/2}}{y^{-1/2}}$$

$$= -\sqrt{\frac{y}{x}}$$

**5.**   $x^3 - xy + y^2 = 7$

$3x^2 - xy' - y + 2yy' = 0$

$(2y - x)y' = y - 3x^2$

$$y' = \frac{y - 3x^2}{2y - x}$$

**7.**   $x^3y^3 - y - x = 0$

$3x^3y^2y' + 3x^2y^3 - y' - 1 = 0$

$(3x^3y^2 - 1)y' = 1 - 3x^2y^3$

$$y' = \frac{1 - 3x^2y^3}{3x^3y^2 - 1}$$

**9.**   $xe^y - 10x + 3y = 0$

$xe^y \dfrac{dy}{dx} + e^y - 10 + 3\dfrac{dy}{dx} = 0$

$\dfrac{dy}{dx}(xe^y + 3) = 10 - e^y$

$$\frac{dy}{dx} = \frac{10 - e^y}{xe^y + 3}$$

**11.**   $\sin x + 2 \cos 2y = 1$

$\cos x - 4(\sin 2y)y' = 0$

$$y' = \frac{\cos x}{4 \sin 2y}$$

**13.**   $\sin x = x(1 + \tan y)$

$\cos x = x(\sec^2 y)y' + (1 + \tan y)(1)$

$$y' = \frac{\cos x - \tan y - 1}{x \sec^2 y}$$

**15.**   $y = \sin xy$

$y' = [xy' + y] \cos(xy)$

$y' - x \cos(xy)y' = y \cos(xy)$

$$y' = \frac{y \cos(xy)}{1 - x \cos(xy)}$$

**17.**   $x^2 - 3 \ln y + y^2 = 10$

$2x - \dfrac{3}{y}\dfrac{dy}{dx} + 2y\dfrac{dy}{dx} = 0$

$2x = \dfrac{dy}{dx}\left(\dfrac{3}{y} - 2y\right)$

$$\frac{dy}{dx} = \frac{2x}{(3/y) - 2y} = \frac{2xy}{3 - 2y^2}$$

**19.**   $4x^3 + \ln y^2 + 2y = 2x$

$12x^2 + \dfrac{2}{y}y' + 2y' = 2$

$\left(\dfrac{2}{y} + 2\right)y' = 2 - 12x^2$

$$y' = \frac{2 - 12x^2}{2/y + 2}$$

$$y' = \frac{y - 6yx^2}{1 + y} = \frac{y(1 - 6x^2)}{1 + y}$$

**21.** (a) $x^2 + y^2 = 64$

$$y^2 = 64 - x^2$$

$$y = \pm\sqrt{64 - x^2}$$

(b)

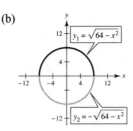

(c) Explicitly: $\dfrac{dy}{dx} = \pm\dfrac{1}{2}\left(64 - x^2\right)^{-1/2}(-2x) = \dfrac{\mp x}{\sqrt{64 - x^2}} = \dfrac{-x}{\pm\sqrt{64 - x^2}} = -\dfrac{x}{y}$

(d) Implicitly: $2x + 2yy' = 0$

$$y' = -\dfrac{x}{y}$$

**23.** (a) $16y^2 - x^2 = 16$

$$16y^2 = x^2 + 16$$

$$y^2 = \dfrac{x^2}{16} + 1 = \dfrac{x^2 + 16}{16}$$

$$y = \dfrac{\pm\sqrt{x^2 + 16}}{4}$$

(b)

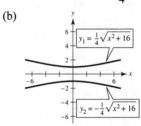

(c) Explicitly: $\dfrac{dy}{dx} = \dfrac{\pm\dfrac{1}{2}\left(x^2 + 16\right)^{-1/2}(-2x)}{4}$

$$= \dfrac{\pm x}{4\sqrt{x^2 + 16}} = \dfrac{\pm x}{4(\pm 4y)} = -\dfrac{x}{16y}$$

(d) Implicitly: $16y^2 - x^2 = 16$

$$32yy' - 2x = 0$$

$$32yy' = 2x$$

$$y' = \dfrac{2x}{32y} = \dfrac{x}{16y}$$

**25.** $xy = 6$

$$xy' + y(1) = 0$$

$$xy' = -y$$

$$y' = -\dfrac{y}{x}$$

At $(-6, -1)$: $y' = -\dfrac{1}{6}$

**27.** $y^2 = \dfrac{x^2 - 49}{x^2 + 49}$

$$2yy' = \dfrac{\left(x^2 + 49\right)(2x) - \left(x^2 - 49\right)(2x)}{\left(x^2 + 49\right)^2}$$

$$2yy' = \dfrac{196x}{\left(x^2 + 49\right)^2}$$

$$y' = \dfrac{98x}{y\left(x^2 + 49\right)^2}$$

At $(7, 0)$: $y'$ is undefined.

**29.** $\tan(x + y) = x$

$$\left(1 + y'\right)\sec^2(x + y) = 1$$

$$y' = \dfrac{1 - \sec^2(x + y)}{\sec^2(x + y)}$$

$$= \dfrac{-\tan^2(x + y)}{\tan^2(x + y) + 1}$$

$$= -\sin^2(x + y)$$

$$= -\dfrac{x^2}{x^2 + 1}$$

At $(0, 0)$: $y' = 0$

**31.** $3e^{xy} - x = 0$

$$3e^{xy}\left[xy' + y\right] - 1 = 0$$

$$3e^{xy}xy' = 1 - 3ye^{xy}$$

$$y' = \dfrac{1 - 3ye^{xy}}{3xe^{xy}}$$

At $(3, 0)$: $y' = \dfrac{1}{9}$

**33.**
$$\left(x^2 + 4\right)y = 8$$

$$\left(x^2 + 4\right)y' + y(2x) = 0$$

$$y' = \frac{-2xy}{x^2 + 4}$$

$$= \frac{-2x\left[8/\left(x^2 + 4\right)\right]}{x^2 + 4}$$

$$= \frac{-16x}{\left(x^2 + 4\right)^2}$$

At $(2, 1)$: $y' = \dfrac{-32}{64} = -\dfrac{1}{2}$

$\left(\text{Or, you could just solve for } y: \ y = \dfrac{8}{x^2 + 4}\right)$

**35.**
$$\left(x^2 + y^2\right)^2 = 4x^2 y$$

$$2\left(x^2 + y^2\right)(2x + 2yy') = 4x^2 y' + y(8x)$$

$$4x^3 + 4x^2 yy' + 4xy^2 + 4y^3 y' = 4x^2 y' + 8xy$$

$$4x^2 yy' + 4y^3 y' - 4x^2 y' = 8xy - 4x^3 - 4xy^2$$

$$4y'\left(x^2 y + y^3 - x^2\right) = 4\left(2xy - x^3 - xy^2\right)$$

$$y' = \frac{2xy - x^3 - xy^2}{x^2 y + y^3 - x^2}$$

At $(1, 1)$: $y' = 0$

**37.** $(y - 3)^2 = 4(x - 5)$,  $(6, 1)$

$$2(y - 3)y' = 4$$

$$y' = \frac{2}{y - 3}$$

At $(6, 1)$: $y' = \dfrac{2}{1 - 3} = -1$

Tangent line: $y - 1 = -1(x - 6)$

$$y = -x + 7$$

**39.** $xy = 1$,  $(1, 1)$

$$xy' + y = 0$$

$$y' = \frac{-y}{x}$$

At $(1, 1)$: $y' = -1$

Tangent line: $y - 1 = -1(x - 1)$

$$y = -x + 2$$

**41.** $x^2 y^2 - 9x^2 - 4y^2 = 0$,  $\left(-4, 2\sqrt{3}\right)$

$$x^2 2yy' + 2xy^2 - 18x - 8yy' = 0$$

$$y' = \frac{18x - 2xy^2}{2x^2 y - 8y}$$

At $\left(-4, 2\sqrt{3}\right)$: $y' = \dfrac{18(-4) - 2(-4)(12)}{2(16)\left(2\sqrt{3}\right) - 16\sqrt{3}}$

$$= \frac{24}{48\sqrt{3}} = \frac{1}{2\sqrt{3}} = \frac{\sqrt{3}}{6}$$

Tangent line: $y - 2\sqrt{3} = \dfrac{\sqrt{3}}{6}(x + 4)$

$$y = \frac{\sqrt{3}}{6}x + \frac{8}{3}\sqrt{3}$$

**43.** $3\left(x^2 + y^2\right)^2 = 100\left(x^2 - y^2\right)$,   $(4, 2)$

$$6\left(x^2 + y^2\right)(2x + 2yy') = 100(2x - 2yy')$$

At $(4, 2)$: $6(16 + 4)(8 + 4y') = 100(8 - 4y')$

$$960 + 480y' = 800 - 400y'$$

$$880y' = -160$$

$$y' = -\frac{2}{11}$$

Tangent line: $\quad y - 2 = -\frac{2}{11}(x - 4)$

$$11y + 2x - 30 = 0$$

$$y = -\frac{2}{11}x + \frac{30}{11}$$

**45.** $4xy = 9$,   $\left(1, \dfrac{9}{4}\right)$

$$4xy' + 4y = 0$$

$$xy' = -y$$

$$y' = \frac{-y}{x}$$

At $\left(1, \dfrac{9}{4}\right)$, $y' = \dfrac{-9/4}{1} = \dfrac{-9}{4}$

Tangent line: $\quad y - \dfrac{9}{4} = \dfrac{-9}{4}(x - 1)$

$$4y - 9 = -9x + 9$$

$$4y + 9x = 18$$

$$y = \frac{-9}{4}x + \frac{9}{2}$$

**47.**
$$x + y - 1 = \ln(x^2 + y^2), \quad (1, 0)$$
$$1 + y' = \frac{2x + 2yy'}{x^2 + y^2}$$
$$x^2 + y^2 + (x^2 + y^2)y' = 2x + 2yy'$$
At $(1, 0)$: $1 + y' = 2$
$$y' = 1$$
Tangent line: $y = x - 1$

**49. (a)** $\dfrac{x^2}{2} + \dfrac{y^2}{8} = 1, \quad (1, 2)$
$$x + \frac{yy'}{4} = 0$$
$$y' = -\frac{4x}{y}$$
At $(1, 2)$: $y' = -2$
Tangent line: $y - 2 = -2(x - 1)$
$$y = -2x + 4$$

**(b)** $\dfrac{x^2}{a^2} + \dfrac{y^2}{b^2} = 1 \Rightarrow \dfrac{2x}{a^2} + \dfrac{2yy'}{b^2} = 0 \Rightarrow y' = \dfrac{-b^2x}{a^2y}$
$$y - y_0 = \frac{-b^2x_0}{a^2y_0}(x - x_0), \text{ Tangent line at } (x_0, y_0)$$
$$\frac{y_0y}{b^2} - \frac{y_0^2}{b^2} = \frac{-x_0x}{a^2} + \frac{x_0^2}{a^2}$$
Because $\dfrac{x_0^2}{a^2} + \dfrac{y_0^2}{b^2} = 1$, you have $\dfrac{y_0y}{b^2} + \dfrac{x_0x}{a^2} = 1$.

**Note:** From part (a),
$$\frac{1(x)}{2} + \frac{2(y)}{8} = 1 \Rightarrow \frac{1}{4}y = -\frac{1}{2}x + 1 \Rightarrow y = -2x + 4,$$
Tangent line.

**51.**
$$\tan y = x$$
$$y' \sec^2 y = 1$$
$$y' = \frac{1}{\sec^2 y} = \cos^2 y, -\frac{\pi}{2} < y < \frac{\pi}{2}$$
$$\sec^2 y = 1 + \tan^2 y = 1 + x^2$$
$$y' = \frac{1}{1 + x^2}$$

**53.**
$$x^2 + y^2 = 4$$
$$2x + 2yy' = 0$$
$$y' = \frac{-x}{y}$$
$$y'' = \frac{y(-1) + xy'}{y^2}$$
$$= \frac{-y + x(-x/y)}{y^2}$$
$$= \frac{-y^2 - x^2}{y^3}$$
$$= -\frac{4}{y^3}$$

**55.**
$$x^2 - y^2 = 36$$
$$2x - 2yy' = 0$$
$$y' = \frac{x}{y}$$
$$x - yy' = 0$$
$$1 - yy'' - (y')^2 = 0$$
$$1 - yy'' - \left(\frac{x}{y}\right)^2 = 0$$
$$y^2 - y^3y'' = x^2$$
$$y'' = \frac{y^2 - x^2}{y^3} = -\frac{36}{y^3}$$

**57.**
$$y^2 = x^3$$
$$2yy' = 3x^2$$
$$y' = \frac{3x^2}{2y} = \frac{3x^2}{2y} \cdot \frac{xy}{xy} = \frac{3y}{2x} \cdot \frac{x^3}{y^2} = \frac{3y}{2x}$$
$$y'' = \frac{2x(3y') - 3y(2)}{4x^2}$$
$$= \frac{2x[3 \cdot (3y/2x)] - 6y}{4x^2} = \frac{3y}{4x^2} = \frac{3x}{4y}$$

**59.** $x^2 + y^2 = 25$

$2x + 2yy' = 0$

$$y' = \frac{-x}{y}$$

At $(4, 3)$:

Tangent line:

$$y - 3 = \frac{-4}{3}(x - 4) \Rightarrow 4x + 3y - 25 = 0$$

Normal line: $y - 3 = \frac{3}{4}(x - 4) \Rightarrow 3x - 4y = 0$

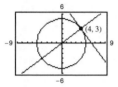

At $(-3, 4)$:

Tangent line:

$$y - 4 = \frac{3}{4}(x + 3) \Rightarrow 3x - 4y + 25 = 0$$

Normal line: $y - 4 = \frac{-4}{3}(x + 3) \Rightarrow 4x + 3y = 0$

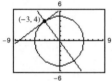

**61.** $x^2 + y^2 = r^2$

$2x + 2yy' = 0$

$$y' = \frac{-x}{y} = \text{slope of tangent line}$$

$$\frac{y}{x} = \text{slope of normal line}$$

Let $(x_0, y_0)$ be a point on the circle. If $x_0 = 0$, then the tangent line is horizontal, the normal line is vertical and, hence, passes through the origin. If $x_0 \neq 0$, then the equation of the normal line is

$$y - y_0 = \frac{y_0}{x_0}(x - x_0)$$

$$y = \frac{y_0}{x_0}x$$

which passes through the origin.

**63.** $25x^2 + 16y^2 + 200x - 160y + 400 = 0$

$50x + 32yy' + 200 - 160y' = 0$

$$y' = \frac{200 + 50x}{160 - 32y}$$

Horizontal tangents occur when $x = -4$:

$$25(16) + 16y^2 + 200(-4) - 160y + 400 = 0$$

$$y(y - 10) = 0 \Rightarrow y = 0, 10$$

Horizontal tangents: $(-4, 0), (-4, 10)$

Vertical tangents occur when $y = 5$:

$$25x^2 + 400 + 200x - 800 + 400 = 0$$

$$25x(x + 8) = 0 \Rightarrow x = 0, -8$$

Vertical tangents: $(0, 5), (-8, 5)$

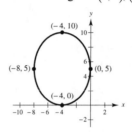

**65.** $y = x\sqrt{x^2 + 1}$

$$\ln y = \ln x + \frac{1}{2}\ln(x^2 + 1)$$

$$\frac{1}{y}\left(\frac{dy}{dx}\right) = \frac{1}{x} + \frac{x}{x^2 + 1}$$

$$\frac{dy}{dx} = y\left[\frac{2x^2 + 1}{x(x^2 + 1)}\right] = \frac{2x^2 + 1}{\sqrt{x^2 + 1}}$$

**67.** $y = \frac{x^2\sqrt{3x - 2}}{(x + 1)^2}$

$$\ln y = 2\ln x + \frac{1}{2}\ln(3x - 2) - 2\ln(x + 1)$$

$$\frac{1}{y}\left(\frac{dy}{dx}\right) = \frac{2}{x} + \frac{3}{2(3x - 2)} - \frac{2}{x + 1}$$

$$\frac{dy}{dx} = y\left[\frac{3x^2 + 15x - 8}{2x(3x - 2)(x + 1)}\right]$$

$$= \frac{3x^3 + 15x^2 - 8x}{2(x + 1)^3\sqrt{3x - 2}}$$

**69.** $y = \dfrac{x(x-1)^{3/2}}{\sqrt{x+1}}$

$\ln y = \ln x + \dfrac{3}{2}\ln(x-1) - \dfrac{1}{2}\ln(x+1)$

$\dfrac{1}{y}\left(\dfrac{dy}{dx}\right) = \dfrac{1}{x} + \dfrac{3}{2}\left(\dfrac{1}{x-1}\right) - \dfrac{1}{2}\left(\dfrac{1}{x+1}\right)$

$\dfrac{dy}{dx} = \dfrac{y}{2}\left[\dfrac{2}{x} + \dfrac{3}{x-1} - \dfrac{1}{x+1}\right]$

$= \dfrac{y}{2}\left[\dfrac{4x^2 + 4x - 2}{x(x^2 - 1)}\right] = \dfrac{(2x^2 + 2x - 1)\sqrt{x-1}}{(x+1)^{3/2}}$

**71.** $y = x^{2/x}$

$\ln y = \dfrac{2}{x}\ln x$

$\dfrac{1}{y}\left(\dfrac{dy}{dx}\right) = \dfrac{2}{x}\left(\dfrac{1}{x}\right) + \ln x\left(-\dfrac{2}{x^2}\right) = \dfrac{2}{x^2}(1 - \ln x)$

$\dfrac{dy}{dx} = \dfrac{2y}{x^2}(1 - \ln x) = 2x^{(2/x)-2}(1 - \ln x)$

**73.** $y = (x-2)^{x+1}$

$\ln y = (x+1)\ln(x-2)$

$\dfrac{1}{y}\left(\dfrac{dy}{dx}\right) = (x+1)\left(\dfrac{1}{x-2}\right) + \ln(x-2)$

$\dfrac{dy}{dx} = y\left[\dfrac{x+1}{x-2} + \ln(x-2)\right]$

$= (x-2)^{x+1}\left[\dfrac{x+1}{x-2} + \ln(x-2)\right]$

**75.** $y = x^{\ln x}, \quad x > 0$

$\ln y = \ln x^{\ln x} = (\ln x)(\ln x) = (\ln x)^2$

$\dfrac{y'}{y} = 2\ln x(1/x)$

$y' = \dfrac{2y\ln x}{x} = \dfrac{2x^{\ln x}\cdot \ln x}{x}$

**77.** Find the points of intersection by letting $y^2 = 4x$ in the equation $2x^2 + y^2 = 6$.

$2x^2 + 4x = 6$ and $(x + 3)(x - 1) = 0$

The curves intersect at $(1, \pm 2)$.

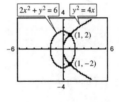

*Ellipse*:        *Parabola*:

$4x + 2yy' = 0$     $2yy' = 4$

$\quad y' = -\dfrac{2x}{y}$     $y' = \dfrac{2}{y}$

At $(1, 2)$, the slopes are:

$\quad y' = -1$       $y' = 1$

At $(1, -2)$, the slopes are:

$\quad y' = 1$       $y' = -1$

Tangents are perpendicular

**79.** $y = -x$ and $x = \sin y$

Point of intersection: $(0, 0)$

$\underline{y = -x}$:        $\underline{x = \sin y}$:

$y' = -1$         $1 = y'\cos y$

                 $y' = \sec y$

At $(0, 0)$, the slopes are:

$\quad y' = -1$       $y' = 1$

Tangents are perpendicular.

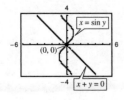

**81.**     $xy = C$        $x^2 - y^2 = K$

      $xy' + y = 0$     $2x - 2yy' = 0$

        $y' = -\dfrac{y}{x}$      $y' = \dfrac{x}{y}$

At any point of intersection $(x, y)$ the product of the slopes is $(-y/x)(x/y) = -1$. The curves are orthogonal.

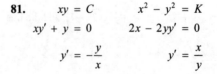

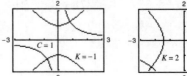

**83.** Answers will vary. *Sample answer:* In the explicit form of a function, the variable is explicitly written as a function of $x$. In an implicit equation, the function is only implied by an equation. An example of an implicit function is $x^2 + xy = 5$. In explicit form it would be $y = \left(5 - x^2\right)/x$.

**85.** (a) True

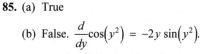

(b) False. $\dfrac{d}{dy}\cos\left(y^2\right) = -2y\sin\left(y^2\right)$.

(c) False. $\dfrac{d}{dx}\cos\left(y^2\right) = -2yy'\sin\left(y^2\right)$.

**87.** (a)  $x^4 = 4\left(4x^2 - y^2\right)$

$$4y^2 = 16x^2 - x^4$$

$$y^2 = 4x^2 - \frac{1}{4}x^4$$

$$y = \pm\sqrt{4x^2 - \frac{1}{4}x^4}$$

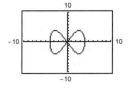

(b)  $y = 3 \Rightarrow 9 = 4x^2 - \dfrac{1}{4}x^4$

$$36 = 16x^2 - x^4$$

$$x^4 - 16x^2 + 36 = 0$$

$$x^2 = \frac{16 \pm \sqrt{256 - 144}}{2} = 8 \pm \sqrt{28}$$

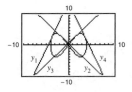

Note that $x^2 = 8 \pm \sqrt{28} = 8 \pm 2\sqrt{7} = \left(1 \pm \sqrt{7}\right)^2$. So, there are four values of $x$:

$$-1 - \sqrt{7}, 1 - \sqrt{7}, -1 + \sqrt{7}, 1 + \sqrt{7}$$

To find the slope, $2yy' = 8x - x^3 \Rightarrow y' = \dfrac{x\left(8 - x^2\right)}{2(3)}$.

For $x = -1 - \sqrt{7}$, $y' = \dfrac{1}{3}\left(\sqrt{7} + 7\right)$, and the line is

$$y_1 = \frac{1}{3}\left(\sqrt{7} + 7\right)\left(x + 1 + \sqrt{7}\right) + 3 = \frac{1}{3}\left[\left(\sqrt{7} + 7\right)x + 8\sqrt{7} + 23\right].$$

For $x = 1 - \sqrt{7}$, $y' = \dfrac{1}{3}\left(\sqrt{7} - 7\right)$, and the line is

$$y_2 = \frac{1}{3}\left(\sqrt{7} - 7\right)\left(x - 1 + \sqrt{7}\right) + 3 = \frac{1}{3}\left[\left(\sqrt{7} - 7\right)x + 23 - 8\sqrt{7}\right].$$

For $x = -1 + \sqrt{7}$, $y' = -\dfrac{1}{3}\left(\sqrt{7} - 7\right)$, and the line is

$$y_3 = -\frac{1}{3}\left(\sqrt{7} - 7\right)\left(x + 1 - \sqrt{7}\right) + 3 = -\frac{1}{3}\left[\left(\sqrt{7} - 7\right)x - \left(23 - 8\sqrt{7}\right)\right].$$

For $x = 1 + \sqrt{7}$, $y' = -\dfrac{1}{3}\left(\sqrt{7} + 7\right)$, and the line is

$$y_4 = -\frac{1}{3}\left(\sqrt{7} + 7\right)\left(x - 1 - \sqrt{7}\right) + 3 = -\frac{1}{3}\left[\left(\sqrt{7} + 7\right)x - \left(8\sqrt{7} + 23\right)\right].$$

(c) Equating $y_3$ and $y_4$:

$$-\frac{1}{3}\left(\sqrt{7} - 7\right)\left(x + 1 - \sqrt{7}\right) + 3 = -\frac{1}{3}\left(\sqrt{7} + 7\right)\left(x - 1 - \sqrt{7}\right) + 3$$

$$\left(\sqrt{7} - 7\right)\left(x + 1 - \sqrt{7}\right) = \left(\sqrt{7} + 7\right)\left(x - 1 - \sqrt{7}\right)$$

$$\sqrt{7}x + \sqrt{7} - 7 - 7x - 7 + 7\sqrt{7} = \sqrt{7}x - \sqrt{7} - 7 + 7x - 7 - 7\sqrt{7}$$

$$16\sqrt{7} = 14x$$

$$x = \frac{8\sqrt{7}}{7}$$

If $x = \dfrac{8\sqrt{7}}{7}$, then $y = 5$ and the lines intersect at $\left(\dfrac{8\sqrt{7}}{7}, 5\right)$.

**89.**   $x^2 + y^2 = 100$, slope $= \dfrac{3}{4}$

$$2x + 2yy' = 0$$

$$y' = -\frac{x}{y} = \frac{3}{4} \Rightarrow y = -\frac{4}{3}x$$

$$x^2 + \left(\frac{16}{9}x^2\right) = 100$$

$$\frac{25}{9}x^2 = 100$$

$$x = \pm 6$$

Points: $(6, -8)$ and $(-6, 8)$

**91.**   $\dfrac{x^2}{4} + \dfrac{y^2}{9} = 1$,   $(4, 0)$

$$\frac{2x}{4} + \frac{2yy'}{9} = 0$$

$$y' = \frac{-9x}{4y}$$

$$\frac{-9x}{4y} = \frac{y - 0}{x - 4}$$

$$-9x(x - 4) = 4y^2$$

But, $9x^2 + 4y^2 = 36 \Rightarrow 4y^2 = 36 - 9x^2$.

So, $-9x^2 + 36x = 4y^2 = 36 - 9x^2 \Rightarrow x = 1$.

Points on ellipse: $\left(1, \pm \dfrac{3}{2}\sqrt{3}\right)$

At $\left(1, \dfrac{3}{2}\sqrt{3}\right)$: $y' = \dfrac{-9x}{4y} = \dfrac{-9}{4\left[(3/2)\sqrt{3}\right]} = -\dfrac{\sqrt{3}}{2}$

At $\left(1, -\dfrac{3}{2}\sqrt{3}\right)$: $y' = \dfrac{\sqrt{3}}{2}$

Tangent lines: $y = -\dfrac{\sqrt{3}}{2}(x - 4) = -\dfrac{\sqrt{3}}{2}x + 2\sqrt{3}$

$$y = \frac{\sqrt{3}}{2}(x - 4) = \frac{\sqrt{3}}{2}x - 2\sqrt{3}$$

**93. (a)**   $\dfrac{x^2}{32} + \dfrac{y^2}{8} = 1$

$$\frac{2x}{32} + \frac{2yy'}{8} = 0 \Rightarrow y' = \frac{-x}{4y}$$

At $(4, 2)$: $y' = \dfrac{-4}{4(2)} = -\dfrac{1}{2}$

Slope of normal line is 2.

$$y - 2 = 2(x - 4)$$

$$y = 2x - 6$$

**(b)**

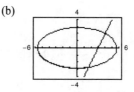

**(c)**   $\dfrac{x^2}{32} + \dfrac{(2x - 6)^2}{8} = 1$

$$x^2 + 4\left(4x^2 - 24x + 36\right) = 32$$

$$17x^2 - 96x + 112 = 0$$

$$(17x - 28)(x - 4) = 0 \Rightarrow x = 4, \frac{28}{17}$$

Second point: $\left(\dfrac{28}{17}, -\dfrac{46}{17}\right)$

## Section 3.6 Derivatives of Inverse Functions

**1.** $f(x) = x^3 - 1$, $\quad a = 26$

$f'(x) = 3x^2$

$f$ is monotonic (increasing) on $(-\infty, \infty)$ therefore $f$ has an inverse.

$$f(3) = 26 \Rightarrow f^{-1}(26) = 3$$

$$\left(f^{-1}\right)'(26) = \frac{1}{f'\left(f^{-1}(26)\right)} = \frac{1}{f'(3)} = \frac{1}{3(3^2)} = \frac{1}{27}$$

**3.** $f(x) = x^3 + 2x - 1$, $\quad a = 2$

$f'(x) = 3x^2 + 2 > 0$

$f$ is monotonic (increasing) on $(-\infty, \infty)$ therefore $f$ has an inverse.

$$f(1) = 2 \Rightarrow f^{-1}(2) = 1$$

$$\left(f^{-1}\right)'(2) = \frac{1}{f'\left(f^{-1}(2)\right)} = \frac{1}{f'(1)} = \frac{1}{3(1^2) + 2} = \frac{1}{5}$$

**5.** $f(x) = \sin x$, $\quad a = 1/2$, $-\dfrac{\pi}{2} \le x \le \dfrac{\pi}{2}$

$f'(x) = \cos x > 0$ on $\left(-\dfrac{\pi}{2}, \dfrac{\pi}{2}\right)$

$f$ is monotonic (increasing) on $\left[-\dfrac{\pi}{2}, \dfrac{\pi}{2}\right]$ therefore $f$ has an inverse.

$$f\left(\frac{\pi}{6}\right) = \sin\frac{\pi}{6} = \frac{1}{2} \Rightarrow f^{-1}\left(\frac{1}{2}\right) = \frac{\pi}{6}$$

$$\left(f^{-1}\right)'\left(\frac{1}{2}\right) = \frac{1}{f'\left(f^{-1}\left(\frac{1}{2}\right)\right)}$$

$$= \frac{1}{f'\left(\frac{\pi}{6}\right)} = \frac{1}{\cos\left(\frac{\pi}{6}\right)} = \frac{2}{\sqrt{3}} = \frac{2\sqrt{3}}{3}$$

**7.** $f(x) = \dfrac{x + 6}{x - 2}$, $\quad x > 0, a = 3$

$$f'(x) = \frac{(x - 2)(1) - (x + 6)(1)}{(x - 2)^2}$$

$$= \frac{-8}{(x - 2)^2} < 0 \text{ on } (2, \infty)$$

$f$ is monotonic (decreasing) on $(2, \infty)$ therefore $f$ has an inverse.

$$f(6) = 3 \Rightarrow f^{-1}(3) = 6$$

$$\left(f^{-1}\right)'(3) = \frac{1}{f'\left(f^{-1}(3)\right)} = \frac{1}{f'(6)} = \frac{1}{-8/(6 - 2)^2} = -2$$

**9.** $f(x) = x^3 - \dfrac{4}{x}$, $\quad a = 6, x > 0$

$$f'(x) = 3x^2 + \frac{4}{x^2} > 0$$

$f$ is monotonic (increasing) on $(0, \infty)$ therefore $f$ has an inverse.

$$f(2) = 6 \Rightarrow f^{-1}(6) = 2$$

$$\left(f^{-1}\right)'(6) = \frac{1}{f'\left(f^{-1}(6)\right)} = \frac{1}{f'(2)} = \frac{1}{3(2^2) + 4/2^2} = \frac{1}{13}$$

**11.** $f(x) = x^3$, $\quad \left(\dfrac{1}{2}, \dfrac{1}{8}\right)$

$f'(x) = 3x^2$

$f'\left(\dfrac{1}{2}\right) = \dfrac{3}{4}$

$f^{-1}(x) = \sqrt[3]{x}$, $\quad \left(\dfrac{1}{8}, \dfrac{1}{2}\right)$

$\left(f^{-1}\right)'(x) = \dfrac{1}{3\sqrt[3]{x}}$

$\left(f^{-1}\right)'\left(\dfrac{1}{8}\right) = \dfrac{4}{3}$

**13.** $f(x) = \sqrt{x - 4}$, $\quad (5, 1)$

$$f'(x) = \frac{1}{2\sqrt{x - 4}}$$

$$f'(5) = \frac{1}{2}$$

$$f^{-1}(x) = x^2 + 4, \quad (1, 5)$$

$$\left(f^{-1}\right)'(x) = 2x$$

$$\left(f^{-1}\right)'(1) = 2$$

**15.** (a) $f(x) = \arccos(x^2)$

$$f'(x) = \frac{-1}{\sqrt{1 - x^4}}(2x) = \frac{-2x}{\sqrt{1 - x^4}}$$

$$f'(0) = 0$$

$$y - \frac{\pi}{2} = 0(x - 0)$$

$$y = \frac{\pi}{2}, \quad \text{tangent line}$$

(b)

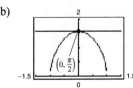

**17.** (a)  $f(x) = \arcsin 3x$

$$f'(x) = \frac{1}{\sqrt{1 - (3x)^2}}(3) = \frac{3}{\sqrt{1 - 9x^2}}$$

$$f'\left(\sqrt{2}/6\right) = \frac{3}{\sqrt{1 - 9(1/18)}} = \frac{3}{\sqrt{1/2}} = 3\sqrt{2}$$

$$y - \frac{\pi}{4} = 3\sqrt{2}\left(x - \sqrt{2}/6\right)$$

$$y = 3\sqrt{2}x + \frac{\pi}{4} - 1, \quad \text{Tangent line}$$

(b)

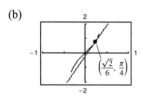

**19.**  $x = y^3 - 7y^2 + 2$

$$1 = 3y^2\frac{dy}{dx} - 14y\frac{dy}{dx}$$

$$\frac{dy}{dx} = \frac{1}{3y^2 - 14y}$$

At $(-4, 1)$: $\dfrac{dy}{dx} = \dfrac{1}{3 - 14} = \dfrac{-1}{11}$.

**Alternate Solution:**

Let $f(x) = x^3 - 7x^2 + 2$. Then

$f'(x) = 3x^2 - 14x$ and $f'(1) = -11$. So,

$$\frac{dy}{dx} = \frac{1}{-11} = \frac{-1}{11}.$$

**21.**  $x \arctan x = e^y$

$$x\frac{1}{1 + x^2} + \arctan x = e^y \cdot \frac{dy}{dx}$$

At $\left(1, \ln\dfrac{\pi}{4}\right)$: $\dfrac{1}{2} + \dfrac{\pi}{4} = \dfrac{\pi}{4}\dfrac{dy}{dx}$

$$\frac{dy}{dx} = \frac{\pi + 2}{\pi}$$

**23.**  $f(x) = \arcsin(x + 1)$

$$f'(x) = \frac{1}{\sqrt{1 - (x + 1)^2}} = \frac{1}{\sqrt{-x^2 - 2x}}$$

**25.**  $g(x) = 3\arccos\dfrac{x}{2}$

$$g'(x) = \frac{-3(1/2)}{\sqrt{1 - (x^2/4)}} = \frac{-3}{\sqrt{4 - x^2}}$$

**27.**  $f(x) = \arctan\left(e^x\right)$

$$f'(x) = \frac{1}{1 + \left(e^x\right)^2}e^x = \frac{e^x}{1 + e^{2x}}$$

**29.**  $g(x) = \dfrac{\arcsin 3x}{x}$

$$g'(x) = \frac{x\left(3/\sqrt{1 - 9x^2}\right) - \arcsin 3x}{x^2}$$

$$= \frac{3x - \sqrt{1 - 9x^2}\arcsin 3x}{x^2\sqrt{1 - 9x^2}}$$

**31.**  $g(x) = e^{2x}\arcsin x$

$$g'(x) = e^{2x}\frac{1}{\sqrt{1 - x^2}} + 2e^{2x}\arcsin x$$

$$= e^{2x}\left[2\arcsin x + \frac{1}{\sqrt{1 - x^2}}\right]$$

**33.**  $h(x) = \text{arccot } 6x$

$$h'(x) = \frac{-6}{1 + 36x^2}$$

**35.**  $h(t) = \sin(\arccos t) = \sqrt{1 - t^2}$

$$h'(t) = \frac{1}{2}\left(1 - t^2\right)^{-1/2}(-2t)$$

$$= \frac{-t}{\sqrt{1 - t^2}}$$

**37.**  $y = 2x\arccos x - 2\sqrt{1 - x^2}$

$$y' = 2\arccos x - 2x\frac{1}{\sqrt{1 - x^2}} - 2\left(\frac{1}{2}\right)\left(1 - x^2\right)^{-1/2}(-2x)$$

$$= 2\arccos x - \frac{2x}{\sqrt{1 - x^2}} + \frac{2x}{\sqrt{1 - x^2}} = 2\arccos x$$

**39.**  $y = \dfrac{1}{2}\left(\dfrac{1}{2}\ln\dfrac{x + 1}{x - 1} + \arctan x\right) = \dfrac{1}{4}\left[\ln(x + 1) - \ln(x - 1)\right] + \dfrac{1}{2}\arctan x$

$$\frac{dy}{dx} = \frac{1}{4}\left(\frac{1}{x + 1} - \frac{1}{x - 1}\right) + \frac{1/2}{1 + x^2} = \frac{1}{1 - x^4}$$

**41.** $g(t) = \tan(\arcsin t) = \dfrac{t}{\sqrt{1 - t^2}}$

$g'(t) = \dfrac{\sqrt{1 - t^2} - t\left(-t/\sqrt{1 - t^2}\right)}{1 - t^2} = \dfrac{1}{\left(1 - t^2\right)^{3/2}}$

**43.** $y = x \arcsin x + \sqrt{1 - x^2}$

$\dfrac{dy}{dx} = x\left(\dfrac{1}{\sqrt{1 - x^2}}\right) + \arcsin x - \dfrac{x}{\sqrt{1 - x^2}} = \arcsin x$

**45.** $y = 8 \arcsin \dfrac{x}{4} - \dfrac{x\sqrt{16 - x^2}}{2}$

$y' = 2\dfrac{1}{\sqrt{1 - (x/4)^2}} - \dfrac{\sqrt{16 - x^2}}{2} - \dfrac{x}{4}\left(16 - x^2\right)^{-1/2}(-2x)$

$= \dfrac{8}{\sqrt{16 - x^2}} - \dfrac{\sqrt{16 - x^2}}{2} + \dfrac{x^2}{2\sqrt{16 - x^2}} = \dfrac{16 - \left(16 - x^2\right) + x^2}{2\sqrt{16 - x^2}} = \dfrac{x^2}{\sqrt{16 - x^2}}$

**47.** $y = \arctan x + \dfrac{x}{1 + x^2}$

$y' = \dfrac{1}{1 + x^2} + \dfrac{\left(1 + x^2\right) - x(2x)}{\left(1 + x^2\right)^2}$

$= \dfrac{\left(1 + x^2\right) + \left(1 - x^2\right)}{\left(1 + x^2\right)^2}$

$= \dfrac{2}{\left(1 + x^2\right)^2}$

**49.** $y = 2 \arcsin x, \quad \left(\dfrac{1}{2}, \dfrac{\pi}{3}\right)$

$y' = \dfrac{2}{\sqrt{1 - x^2}}$

At $\left(\dfrac{1}{2}, \dfrac{\pi}{3}\right), y' = \dfrac{2}{\sqrt{1 - (1/4)}} = \dfrac{4}{\sqrt{3}}.$

Tangent line: $y - \dfrac{\pi}{3} = \dfrac{4}{\sqrt{3}}\left(x - \dfrac{1}{2}\right)$

$y = \dfrac{4}{\sqrt{3}}x + \dfrac{\pi}{3} - \dfrac{2}{\sqrt{3}}$

$y = \dfrac{4\sqrt{3}}{3}x + \dfrac{\pi}{3} - \dfrac{2\sqrt{3}}{3}$

**51.** $y = \arcsin\left(\dfrac{x}{2}\right), \quad \left(2, \dfrac{\pi}{4}\right)$

$y' = \dfrac{1}{1 + (x^2/4)}\left(\dfrac{1}{2}\right) = \dfrac{2}{4 + x^2}$

At $\left(2, \dfrac{\pi}{4}\right), y' = \dfrac{2}{4 + 4} = \dfrac{1}{4}.$

Tangent line: $y - \dfrac{\pi}{4} = \dfrac{1}{4}(x - 2)$

$y = \dfrac{1}{4}x + \dfrac{\pi}{4} - \dfrac{1}{2}$

**53.** $y = 4x \arccos(x - 1), \quad (1, 2\pi)$

$y' = 4x\dfrac{-1}{\sqrt{1 - (x - 1)^2}} + 4 \arccos(x - 1)$

At $(1, 2\pi), y' = -4 + 2\pi.$

Tangent line: $y - 2\pi = (2\pi - 4)(x - 1)$

$y = (2\pi - 4)x + 4$

**55.** $f(x) = \arccos x$

$f'(x) = \dfrac{-1}{\sqrt{1 - x^2}} = -2$ when $x = \pm\dfrac{\sqrt{3}}{2}.$

When $x = \sqrt{3}/2, f\left(\sqrt{3}/2\right) = \pi/6.$ When $x = -\sqrt{3}/2, f\left(-\sqrt{3}/2\right) = 5\pi/6.$

Tangent lines: $y - \dfrac{\pi}{6} = -2\left(x - \dfrac{\sqrt{3}}{2}\right) \Rightarrow y = -2x + \left(\dfrac{\pi}{6} + \sqrt{3}\right)$

$y - \dfrac{5\pi}{6} = -2\left(x + \dfrac{\sqrt{3}}{2}\right) \Rightarrow y = -2x + \left(\dfrac{5\pi}{6} - \sqrt{3}\right)$

**57.** $f(x) = \arctan x, \; a = 0$

$f(0) = 0$

$f'(x) = \dfrac{1}{1 + x^2}, \qquad f'(0) = 1$

$f''(x) = \dfrac{-2x}{\left(1 + x^2\right)^2}, \quad f''(0) = 0$

$P_1(x) = f(0) + f'(0)x = x$

$P_2(x) = f(0) + f'(0)x + \dfrac{1}{2}f''(0)x^2 = x$

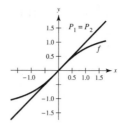

**59.** $f(x) = \arcsin x, \; a = \dfrac{1}{2}$

$f'(x) = \dfrac{1}{\sqrt{1 - x^2}}$

$f''(x) = \dfrac{x}{\left(1 - x^2\right)^{3/2}}$

$P_1(x) = f\!\left(\dfrac{1}{2}\right) + f'\!\left(\dfrac{1}{2}\right)\!\left(x - \dfrac{1}{2}\right) = \dfrac{\pi}{6} + \dfrac{2\sqrt{3}}{3}\!\left(x - \dfrac{1}{2}\right)$

$P_2(x) = f\!\left(\dfrac{1}{2}\right) + f'\!\left(\dfrac{1}{2}\right)\!\left(x - \dfrac{1}{2}\right) + \dfrac{1}{2}f''\!\left(\dfrac{1}{2}\right)\!\left(x - \dfrac{1}{2}\right)^2 = \dfrac{\pi}{6} + \dfrac{2\sqrt{3}}{3}\!\left(x - \dfrac{1}{2}\right) + \dfrac{2\sqrt{3}}{9}\!\left(x - \dfrac{1}{2}\right)^2$

**61.** $\qquad x^2 + x \arctan y = y - 1, \qquad \left(-\dfrac{\pi}{4}, 1\right)$

$2x + \arctan y + \dfrac{x}{1 + y^2}y' = y'$

$\left(1 - \dfrac{x}{1 + y^2}\right)y' = 2x + \arctan y$

$y' = \dfrac{2x + \arctan y}{1 - \dfrac{x}{1 + y^2}}$

At $\left(-\dfrac{\pi}{4}, 1\right)$: $y' = \dfrac{-\dfrac{\pi}{2} + \dfrac{\pi}{4}}{1 - \dfrac{-\pi/4}{2}} = \dfrac{-\dfrac{\pi}{2}}{2 + \dfrac{\pi}{4}} = \dfrac{-2\pi}{8 + \pi}$

Tangent line: $y - 1 = \dfrac{-2\pi}{8 + \pi}\!\left(x + \dfrac{\pi}{4}\right)$

$y = \dfrac{-2\pi}{8 + \pi}x + 1 - \dfrac{\pi^2}{16 + 2\pi}$

**63.** $\qquad \arcsin x + \arcsin y = \dfrac{\pi}{2}, \qquad \left(\dfrac{\sqrt{2}}{2}, \dfrac{\sqrt{2}}{2}\right)$

$\dfrac{1}{\sqrt{1 - x^2}} + \dfrac{1}{\sqrt{1 - y^2}}y' = 0$

$\dfrac{1}{\sqrt{1 - y^2}}y' = \dfrac{-1}{\sqrt{1 - x^2}}$

At $\left(\dfrac{\sqrt{2}}{2}, \dfrac{\sqrt{2}}{2}\right)$: $y' = -1$

Tangent line: $y - \dfrac{\sqrt{2}}{2} = -1\!\left(x - \dfrac{\sqrt{2}}{2}\right)$

$y = -x + \sqrt{2}$

**65.** $f$ is not one-to-one because many different $x$-values yield the same $y$-value.

Example: $f(0) = f(\pi) = 0$

Not continuous at $\dfrac{(2n - 1)\pi}{2}$, where $n$ is an integer.

**67.** Because you know that $f^{-1}$ exists and that

$y_1 = f(x_1)$ by Theorem 3.17, then $\left(f^{-1}\right)'(y_1) = \dfrac{1}{f'(x_1)}$,

provided that $f'(x_1) \neq 0$.

**69.** The derivatives are algebraic. See Theorem 3.18.

**71.** Because the slope of $f$ at $(1, 3)$ is $m = 2$, the slope of $f^{-1}$ at $(3, 1)$ is $1/2$.

**73.** (a) $\cot \theta = \dfrac{x}{5}$

$\theta = \text{arccot}\left(\dfrac{x}{5}\right)$

(b) $\dfrac{d\theta}{dt} = \dfrac{-1/5}{1 + (x/5)^2}\dfrac{dx}{dt} = \dfrac{-5}{x^2 + 25}\dfrac{dx}{dt}$

If $\dfrac{dx}{dt} = -400$ and $x = 10$, $\dfrac{d\theta}{dt} = 16$ rad/h.

If $\dfrac{dx}{dt} = -400$ and $x = 3$, $\dfrac{d\theta}{dt} \approx 58.824$ rad/h.

**75.** (a) $\qquad h(t) = -16t^2 + 256$

$-16t^2 + 256 = 0$ when $t = 4$ sec

(b) $\tan \theta = \dfrac{h}{500} = \dfrac{-16t^2 + 256}{500}$

$\theta = \arctan\left[\dfrac{16}{500}(-t^2 + 16)\right]$

$\dfrac{d\theta}{dt} = \dfrac{-8t/125}{1 + \left[(4/125)(-t^2 + 16)\right]^2}$

$= \dfrac{-1000t}{15{,}625 + 16(16 - t^2)^2}$

When $t = 1$, $d\theta/dt \approx -0.0520$ rad/sec.

When $t = 2$, $d\theta/dt \approx -0.1116$ rad/sec.

**77.** $\tan \theta = \dfrac{h}{300}$

$\dfrac{dh}{dt} = 5$ ft/sec

$\theta = \arctan\left(\dfrac{h}{300}\right)$

$\dfrac{d\theta}{dt} = \dfrac{1/300}{1 + \left(h^2/300^2\right)}\left(\dfrac{dh}{dt}\right)$

$= \dfrac{300}{300^2 + h^2}(5)$

$= \dfrac{1500}{300^2 + h^2} = \dfrac{3}{200}$ rad/sec when $h = 100$

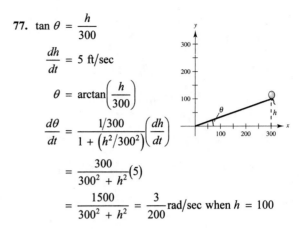

**79.** (a) $\qquad$ Let $y = \arccos u$. Then

$\cos y = u$

$-\sin y \dfrac{dy}{dx} = u'$

$\dfrac{dy}{dx} = -\dfrac{u'}{\sin y} = -\dfrac{u'}{\sqrt{1 - u^2}}.$

(b) $\qquad$ Let $y = \arctan u$. Then

$\tan y = u$

$\sec^2 y \dfrac{dy}{dx} = u'$

$\dfrac{dy}{dx} = \dfrac{u'}{\sec^2 y} = \dfrac{u'}{1 - u^2}.$

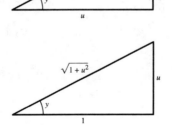

(c)      Let $y = \text{arcsec } u$. Then

$$\sec y = u$$

$$\sec y \tan y \frac{dy}{dx} = u'$$

$$\frac{dy}{dx} = \frac{u'}{\sec y \tan y} = \frac{u'}{|u|\sqrt{u^2 - 1}}.$$

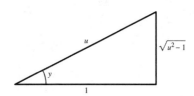

**Note:** The absolute value sign in the formula for the derivative of arcsec $u$ is necessary because the inverse secant function has a positive slope at every value in its domain.

(d)      Let $y = \text{arccot } u$. Then

$$\cot y = u$$

$$-\csc^2 y \frac{dy}{dx} = u'$$

$$\frac{dy}{dx} = \frac{u'}{-\csc^2 y} = -\frac{u'}{1 + u^2}.$$

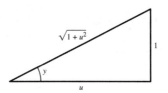

(e)      Let $y = \text{arccsc } u$. Then

$$\csc y = u$$

$$-\csc y \cot y \frac{dy}{dx} = u'$$

$$\frac{dy}{dx} = \frac{u'}{-\csc y \cot y} = -\frac{u'}{|u|\sqrt{u^2 - 1}}.$$

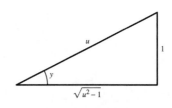

**Note:** The absolute value sign in the formula for the derivative of arccsc $u$ is necessary because the inverse cosecant function has a negative slope at every value in its domain.

**81.** True

$$\frac{d}{dx}[\text{arcsec } u] = \frac{u'}{|u|\sqrt{u^2 - 1}}$$

$$\frac{d}{dx}[\text{arccsc } u] = \frac{-u'}{|u|\sqrt{u^2 - 1}}$$

**83.** True

$$\frac{d}{dx}\left[\text{arctan}(\tan x)\right] = \frac{\sec^2 x}{1 + \tan^2 x} = \frac{\sec^2 x}{\sec^2 x} = 1$$

**85.** Let      $\theta = \arctan\left(\dfrac{x}{\sqrt{1 - x^2}}\right),\quad -1 < x < 1$

$$\tan \theta = \frac{x}{\sqrt{1 - x^2}}$$

$$\sin \theta = \frac{x}{1} = x$$

$$\arcsin x = \theta.$$

So, $\arcsin x = \arctan\left(\dfrac{x}{\sqrt{1 - x^2}}\right)$ for $-1 < x < 1$.

**87.** $f(x) = \sec x, \quad 0 \le x < \dfrac{\pi}{2}, \pi \le x < \dfrac{3\pi}{2}$

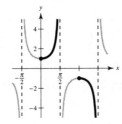

(a) $y = \text{arcsec } x, \quad x \le -1 \quad \text{or} \quad x \ge 1$

$0 \le y < \dfrac{\pi}{2} \quad \text{or} \quad \pi \le y < \dfrac{3\pi}{2}$

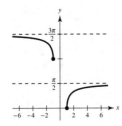

(b)
$$y = \text{arcsec } x$$
$$x = \sec y$$
$$1 = \sec y \tan y \cdot y'$$
$$y' = \frac{1}{\sec y \tan y}$$
$$= \frac{1}{x\sqrt{x^2 - 1}}$$
$$\tan^2 y + 1 = \sec^2 y$$
$$\tan y = \pm\sqrt{\sec^2 y - 1}$$

On $0 \le y < \pi/2$ and $\pi \le y < 3\pi/2$, $\tan y \ge 0$.

# Section 3.7   Related Rates

**1.** $y = \sqrt{x}$

$$\frac{dy}{dt} = \left(\frac{1}{2\sqrt{x}}\right)\frac{dx}{dt}$$

$$\frac{dx}{dt} = 2\sqrt{x}\,\frac{dy}{dt}$$

(a) When $x = 4$ and $dx/dt = 3$:

$$\frac{dy}{dt} = \frac{1}{2\sqrt{4}}(3) = \frac{3}{4}$$

(b) When $x = 25$ and $dy/dt = 2$:

$$\frac{dx}{dt} = 2\sqrt{25}(2) = 20$$

**3.** $xy = 4$

$$x\frac{dy}{dt} + y\frac{dx}{dt} = 0$$

$$\frac{dy}{dt} = \left(-\frac{y}{x}\right)\frac{dx}{dt}$$

$$\frac{dx}{dt} = \left(-\frac{x}{y}\right)\frac{dy}{dt}$$

(a) When $x = 8$, $y = 1/2$, and $dx/dt = 10$:

$$\frac{dy}{dt} = -\frac{1/2}{8}(10) = -\frac{5}{8}$$

(b) When $x = 1$, $y = 4$, and $dy/dt = -6$:

$$\frac{dx}{dt} = -\frac{1}{4}(-6) = \frac{3}{2}$$

**5.** $y = 2x^2 + 1$

$$\frac{dx}{dt} = 2$$

$$\frac{dy}{dt} = 4x\frac{dx}{dt}$$

(a) When $x = -1$:

$$\frac{dy}{dt} = 4(-1)(2) = -8 \text{ cm/sec}$$

(b) When $x = 0$:

$$\frac{dy}{dt} = 4(0)(2) = 0 \text{ cm/sec}$$

(c) When $x = 1$:

$$\frac{dy}{dt} = 4(1)(2) = 8 \text{ cm/sec}$$

**7.**  $y = \tan x, \dfrac{dx}{dt} = 3$

$\dfrac{dy}{dt} = \sec^2 x \cdot \dfrac{dx}{dt} = \sec^2 x(3) = 3 \sec^2 x$

(a)  When $x = -\dfrac{\pi}{3}$:

$\dfrac{dy}{dt} = 3 \sec^2\left(-\dfrac{\pi}{3}\right) = 3(2)^2 = 12 \text{ ft/sec}$

(b)  When $x = -\dfrac{\pi}{4}$:

$\dfrac{dy}{dt} = 3 \sec^2\left(-\dfrac{\pi}{4}\right) = 3\left(\sqrt{2}\right)^2 = 6 \text{ ft/sec}$

(c)  When $x = 0$:

$\dfrac{dy}{dt} = 3 \sec^2(0) = 3 \text{ ft/sec}$

**9.**  Yes, $y$ changes at a constant rate.

$\dfrac{dy}{dt} = a \cdot \dfrac{dx}{dt}$

No, the rate $dy/dt$ is a multiple of $dx/dt$.

**11.**  $A = \pi r^2$

$\dfrac{dr}{dt} = 4$

$\dfrac{dA}{dt} = 2\pi r \dfrac{dr}{dt}$

(a)  When $r = 8, \dfrac{dA}{dt} = 2\pi(8)(4) = 64\pi \text{ cm}^2/\text{min}.$

(b)  When $r = 32, \dfrac{dA}{dt} = 2\pi(32)(4) = 256\pi \text{ cm}^2/\text{min}.$

**13.**  $V = \dfrac{4}{3}\pi r^3$

$\dfrac{dr}{dt} = 3$

$\dfrac{dV}{dt} = 4\pi r^2 \dfrac{dr}{dt}$

(a)  When $r = 9$,

$\dfrac{dV}{dt} = 4\pi(9)^2(3) = 972\pi \text{ in.}^3/\text{min}.$

When $r = 36$,

$\dfrac{dV}{dt} = 4\pi(36)^2(3) = 15{,}552\pi \text{ in.}^3/\text{min}.$

(b)  If $dr/dt$ is constant, $dV/dt$ is proportional to $r^2$.

**15.**  $V = x^3$

$\dfrac{dx}{dt} = 6$

$\dfrac{dV}{dt} = 3x^2 \dfrac{dx}{dt}$

(a)  When $x = 2$,

$\dfrac{dV}{dt} = 3(2)^2(6) = 72 \text{ cm}^3/\text{sec}.$

(b)  When $x = 10$,

$\dfrac{dV}{dt} = 3(10)^2(6) = 1800 \text{ cm}^3/\text{sec}.$

**17.**  $V = \dfrac{1}{3}\pi r^2 h = \dfrac{1}{3}\pi\left(\dfrac{9}{4}h^2\right)h \qquad$ [because $2r = 3h$]

$\qquad = \dfrac{3\pi}{4}h^3$

$\dfrac{dV}{dt} = 10$

$\dfrac{dV}{dt} = \dfrac{9\pi}{4}h^2 \dfrac{dh}{dt} \Rightarrow \dfrac{dh}{dt} = \dfrac{4(dV/dt)}{9\pi h^2}$

When $h = 15$,

$\dfrac{dh}{dt} = \dfrac{4(10)}{9\pi(15)^2} = \dfrac{8}{405\pi} \text{ ft/min}.$

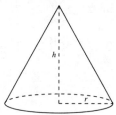

**19.**

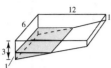

(a) Total volume of pool $= \dfrac{1}{2}(2)(12)(6) + (1)(6)(12) = 144 \text{ m}^3$

Volume of 1 m of water $= \dfrac{1}{2}(1)(6)(6) = 18 \text{ m}^3$    (see similar triangle diagram)

% pool filled $= \dfrac{18}{144}(100\%) = 12.5\%$

(b) Because for $0 \le h \le 2$, $b = 6h$, you have

$$V = \dfrac{1}{2}bh(6) = 3bh = 3(6h)h = 18h^2$$

$$\dfrac{dV}{dt} = 36h\dfrac{dh}{dt} = \dfrac{1}{4} \Rightarrow \dfrac{dh}{dt} = \dfrac{1}{144h} = \dfrac{1}{144(1)} = \dfrac{1}{144} \text{ m/min.}$$

**21.**

$$x^2 + y^2 = 25^2$$

$$2x\dfrac{dx}{dt} + 2y\dfrac{dy}{dt} = 0$$

$$\dfrac{dy}{dt} = \dfrac{-x}{y} \cdot \dfrac{dx}{dt} = \dfrac{-2x}{y} \qquad \text{because } \dfrac{dx}{dt} = 2.$$

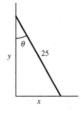

(a) When $x = 7$, $y = \sqrt{576} = 24$, $\dfrac{dy}{dt} = \dfrac{-2(7)}{24} = -\dfrac{7}{12}$ ft/sec.

When $x = 15$, $y = \sqrt{400} = 20$, $\dfrac{dy}{dt} = \dfrac{-2(15)}{20} = -\dfrac{3}{2}$ ft/sec.

When $x = 24$, $y = 7$, $\dfrac{dy}{dt} = \dfrac{-2(24)}{7} = -\dfrac{48}{7}$ ft/sec.

(b) $A = \dfrac{1}{2}xy$

$$\dfrac{dA}{dt} = \dfrac{1}{2}\left(x\dfrac{dy}{dt} + y\dfrac{dx}{dt}\right)$$

From part (a) you have $x = 7$, $y = 24$, $\dfrac{dx}{dt} = 2$, and $\dfrac{dy}{dt} = -\dfrac{7}{12}$. So,

$$\dfrac{dA}{dt} = \dfrac{1}{2}\left[7\left(-\dfrac{7}{12}\right) + 24(2)\right] = \dfrac{527}{24} \text{ ft}^2/\text{sec.}$$

(c) $\tan \theta = \dfrac{x}{y}$

$$\sec^2\theta \dfrac{d\theta}{dt} = \dfrac{1}{y} \cdot \dfrac{dx}{dt} - \dfrac{x}{y^2} \cdot \dfrac{dy}{dt}$$

$$\dfrac{d\theta}{dt} = \cos^2\theta\left[\dfrac{1}{y} \cdot \dfrac{dx}{dt} - \dfrac{x}{y^2} \cdot \dfrac{dy}{dt}\right]$$

Using $x = 7$, $y = 24$, $\dfrac{dx}{dt} = 2$, $\dfrac{dy}{dt} = -\dfrac{7}{12}$ and $\cos \theta = \dfrac{24}{25}$, you have

$$\dfrac{d\theta}{dt} = \left(\dfrac{24}{25}\right)^2\left[\dfrac{1}{24}(2) - \dfrac{7}{(24)^2}\left(-\dfrac{7}{12}\right)\right] = \dfrac{1}{12} \text{ rad/sec.}$$

**23.** When $y = 6$, $x = \sqrt{12^2 - 6^2} = 6\sqrt{3}$, and $s = \sqrt{x^2 + (12 - y)^2} = \sqrt{108 + 36} = 12$.

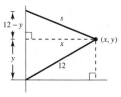

$$x^2 + (12 - y)^2 = s^2$$

$$2x\frac{dx}{dt} + 2(12 - y)(-1)\frac{dy}{dt} = 2s\frac{ds}{dt}$$

$$x\frac{dx}{dt} + (y - 12)\frac{dy}{dt} = s\frac{ds}{dt}$$

Also, $x^2 + y^2 = 12^2$.

$$2x\frac{dx}{dt} + 2y\frac{dy}{dt} = 0 \Rightarrow \frac{dy}{dt} = \frac{-x}{y}\frac{dx}{dt}$$

So, $x\dfrac{dx}{dt} + (y - 12)\left(\dfrac{-x}{y}\dfrac{dx}{dt}\right) = s\dfrac{ds}{dt}$.

$$\frac{dx}{dt}\left[x - x + \frac{12x}{y}\right] = s\frac{ds}{dt} \Rightarrow \frac{dx}{dt} = \frac{sy}{12x} \cdot \frac{ds}{dt} = \frac{(12)(6)}{(12)(6\sqrt{3})}(-0.2) = \frac{-1}{5\sqrt{3}} = \frac{-\sqrt{3}}{15} \text{ m/sec (horizontal)}$$

$$\frac{dy}{dt} = \frac{-x}{y}\frac{dx}{dt} = \frac{-6\sqrt{3}}{6} \cdot \frac{(-\sqrt{3})}{15} = \frac{1}{5} \text{ m/sec (vertical)}$$

**25.** (a)    $s^2 = x^2 + y^2$

$$\frac{dx}{dt} = -450$$

$$\frac{dy}{dt} = -600$$

$$2s\frac{ds}{dt} = 2x\frac{dx}{dt} + 2y\frac{dy}{dt}$$

$$\frac{ds}{dt} = \frac{x(dx/dt) + y(dy/dt)}{s}$$

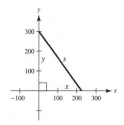

When $x = 225$ and $y = 300$, $s = 375$ and

$$\frac{ds}{dt} = \frac{225(-450) + 300(-600)}{375} = -750 \text{ mi/h}.$$

(b)  $t = \dfrac{375}{750} = \dfrac{1}{2}\text{ h} = 30 \text{ min}$

**27.**
$$s^2 = 90^2 + x^2$$
$$x = 20$$
$$\frac{dx}{dt} = -25$$
$$2s\frac{ds}{dt} = 2x\frac{dx}{dt} \Rightarrow \frac{ds}{dt} = \frac{x}{s} \cdot \frac{dx}{dt}$$

When $x = 20$, $s = \sqrt{90^2 + 20^2} = 10\sqrt{85}$,

$$\frac{ds}{dt} = \frac{20}{10\sqrt{85}}(-25) = \frac{-50}{\sqrt{85}} \approx -5.42 \text{ ft/sec.}$$

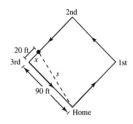

**29. (a)**
$$\frac{15}{6} = \frac{y}{y - x} \Rightarrow 15y - 15x = 6y$$
$$y = \frac{5}{3}x$$
$$\frac{dx}{dt} = 5$$
$$\frac{dy}{dt} = \frac{5}{3} \cdot \frac{dx}{dt} = \frac{5}{3}(5) = \frac{25}{3} \text{ ft/sec}$$

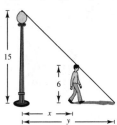

**(b)** $\dfrac{d(y - x)}{dt} = \dfrac{dy}{dt} - \dfrac{dx}{dt} = \dfrac{25}{3} - 5 = \dfrac{10}{3} \text{ ft/sec}$

**31.** $x(t) = \dfrac{1}{2}\sin\dfrac{\pi t}{6}$, $x^2 + y^2 = 1$

**(a)** Period: $\dfrac{2\pi}{\pi/6} = 12$ seconds

**(b)** When $x = \dfrac{1}{2}$, $y = \sqrt{1^2 - \left(\dfrac{1}{2}\right)^2} = \dfrac{\sqrt{3}}{2}$ m.

Lowest point: $\left(0, \dfrac{\sqrt{3}}{2}\right)$

**(c)** When $x = \dfrac{1}{4}$, $y = \sqrt{1 - \left(\dfrac{1}{4}\right)^2} = \dfrac{\sqrt{15}}{4}$ and $t = 1$:

$$\frac{dx}{dt} = \frac{1}{2}\left(\frac{\pi}{6}\right)\cos\frac{\pi t}{6} = \frac{\pi}{12}\cos\frac{\pi t}{6}$$

$$x^2 + y^2 = 1$$

$$2x\frac{dx}{dt} + 2y\frac{dy}{dt} = 0 \Rightarrow \frac{dy}{dt} = \frac{-x}{y}\frac{dx}{dt}$$

So, $\dfrac{dy}{dt} = -\dfrac{1/4}{\sqrt{15}/4} \cdot \dfrac{\pi}{12}\cos\left(\dfrac{\pi}{6}\right)$

$$= \frac{-\pi}{\sqrt{15}}\left(\frac{1}{12}\right)\frac{\sqrt{3}}{2} = \frac{-\pi}{24}\frac{1}{\sqrt{5}} = \frac{-\sqrt{5}\pi}{120}.$$

Speed $= \left|\dfrac{-\sqrt{5}\pi}{120}\right| = \dfrac{\sqrt{5}\pi}{120}$ m/sec

**33.** Because the evaporation rate is proportional to the surface area, $dV/dt = k\left(4\pi r^2\right)$. However, because

$V = (4/3)\pi r^3$, you have

$$\frac{dV}{dt} = 4\pi r^2\frac{dr}{dt}.$$

Therefore, $k\left(4\pi r^2\right) = 4\pi r^2\dfrac{dr}{dt} \Rightarrow k = \dfrac{dr}{dt}.$

**35.**
$$pV^{1.3} = k$$
$$1.3pV^{0.3}\frac{dV}{dt} + V^{1.3}\frac{dp}{dt} = 0$$
$$V^{0.3}\left(1.3p\frac{dV}{dt} + V\frac{dp}{dt}\right) = 0$$
$$1.3p\frac{dV}{dt} = -V\frac{dp}{dt}$$

**37.** $\tan \theta = \dfrac{y}{50}$

$\dfrac{dy}{dt} = 4 \text{ m/sec}$

$\sec^2 \theta \cdot \dfrac{d\theta}{dt} = \dfrac{1}{50} \dfrac{dy}{dt}$

$\dfrac{d\theta}{dt} = \dfrac{1}{50} \cos^2 \theta \cdot \dfrac{dy}{dt}$

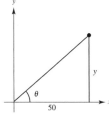

When $y = 50$, $\theta = \dfrac{\pi}{4}$, and $\cos \theta = \dfrac{\sqrt{2}}{2}$.

So, $\dfrac{d\theta}{dt} = \dfrac{1}{50}\left(\dfrac{\sqrt{2}}{2}\right)^2 (4) = \dfrac{1}{25}$ rad/sec.

**39.** $H = \dfrac{4347}{400,000,000} e^{369,444/(50t+19,793)}$

(a) $t = 65° \Rightarrow H \approx 99.79\%$

$t = 80° \Rightarrow H \approx 60.20\%$

(b) $H' = H \cdot \left(\dfrac{-369,444(50)}{(50t + 19,793)^2}\right) t'$

At $t = 75$ and $t' = 2$, $H' \approx -4.7\%/h$.

**41.** $\dfrac{d\theta}{dt} = (10 \text{ rev/sec})(2\pi \text{ rad/rev}) = 20\pi \text{ rad/sec}$

(a) $\cos \theta = \dfrac{x}{30}$

$-\sin \theta \dfrac{d\theta}{dt} = \dfrac{1}{30} \dfrac{dx}{dt}$

$\dfrac{dx}{dt} = -30 \sin \theta \dfrac{d\theta}{dt}$

$= -30 \sin \theta (20\pi)$

$= -600\pi \sin \theta$

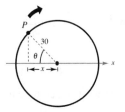

(b)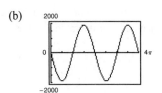

(c) $\left| dx/dt \right| = \left| -600\pi \sin \theta \right|$ is greatest when

$\left| \sin \theta \right| = 1 \Rightarrow \theta = \dfrac{\pi}{2} + n\pi \quad (\text{or } 90° + n \cdot 180°)$.

$\left| dx/dt \right|$ is least when $\theta = n\pi \quad (\text{or } n \cdot 180°)$.

(d) For $\theta = 30°$,

$\dfrac{dx}{dt} = -600\pi \sin(30°) = -600\pi\dfrac{1}{2} = -300\pi$ cm/sec.

For $\theta = 60°$,

$\dfrac{dx}{dt} = -600\pi \sin(60°)$

$= -600\pi\dfrac{\sqrt{3}}{2} = -300\sqrt{3}\pi$ cm/sec.

**43.** $\tan \theta = \dfrac{x}{50} \Rightarrow x = 50 \tan \theta$

$\dfrac{dx}{dt} = 50 \sec^2 \theta \dfrac{d\theta}{dt}$

$2 = 50 \sec^2 \theta \dfrac{d\theta}{dt}$

$\dfrac{d\theta}{dt} = \dfrac{1}{25} \cos^2 \theta, \quad -\dfrac{\pi}{4} \le \theta \le \dfrac{\pi}{4}$

**45.** $x^2 + y^2 = 25$; acceleration of the top of the ladder $= \dfrac{d^2y}{dt^2}$

First derivative:   $2x\dfrac{dx}{dt} + 2y\dfrac{dy}{dt} = 0$

$$x\dfrac{dx}{dt} + y\dfrac{dy}{dt} = 0$$

Second derivative:   $x\dfrac{d^2x}{dt^2} + \dfrac{dx}{dt}\cdot\dfrac{dx}{dt} + y\dfrac{d^2y}{dt^2} + \dfrac{dy}{dt}\cdot\dfrac{dy}{dt} = 0$

$$\dfrac{d^2y}{dt^2} = \left(\dfrac{1}{y}\right)\left[-x\dfrac{d^2x}{dt^2} - \left(\dfrac{dx}{dt}\right)^2 - \left(\dfrac{dy}{dt}\right)^2\right]$$

When $x = 7$, $y = 24$, $\dfrac{dy}{dt} = -\dfrac{7}{12}$, and $\dfrac{dx}{dt} = 2$ (see Exercise 25). Because $\dfrac{dx}{dt}$ is constant, $\dfrac{d^2x}{dt^2} = 0$.

$$\dfrac{d^2y}{dt^2} = \dfrac{1}{24}\left[-7(0) - (2)^2 - \left(-\dfrac{7}{12}\right)^2\right] = \dfrac{1}{24}\left[-4 - \dfrac{49}{144}\right] = \dfrac{1}{24}\left[-\dfrac{625}{144}\right] \approx -0.1808 \text{ ft/sec}^2$$

**47.** (a) $dy/dt = 3(dx/dt)$ means that y changes three times as fast as x changes.

(b) y changes slowly when $x \approx 0$ or $x \approx L$. y changes more rapidly when $x$ is near the middle of the interval.

**49.** (a) $A = (\text{base})(\text{height}) = 2xe^{-x^2/2}$

(b) $\dfrac{dA}{dt} = \left[2x\left(-xe^{-x^2/2}\right) + 2e^{-x^2/2}\right]\dfrac{dx}{dt}$

$$= \left(-2x^2 + 2\right)e^{-x^2/2}\dfrac{dx}{dt}$$

For $x = 2$ and $\dfrac{dx}{dt} = 4$,

$$\dfrac{dA}{dt} = -6e^{-2}(4) = \dfrac{-24}{e^2} \approx -3.25 \text{ cm}^2/\text{min}.$$

## Section 3.8   Newton's Method

**The following solutions may vary depending on the software or calculator used, and on rounding.**

**1.** $f(x) = x^2 - 5$

$f'(x) = 2x$

$x_1 = 2.2$

| $n$ | $x_n$ | $f(x_n)$ | $f'(x_n)$ | $\dfrac{f(x_n)}{f'(x_n)}$ | $x_n - \dfrac{f(x_n)}{f'(x_n)}$ |
|---|---|---|---|---|---|
| 1 | 2.2000 | −0.1600 | 4.4000 | −0.0364 | 2.2364 |
| 2 | 2.2364 | 0.0013 | 4.4727 | 0.0003 | 2.2361 |

**3.** $f(x) = \cos x$

$f'(x) = -\sin x$

$x_1 = 1.6$

| $n$ | $x_n$ | $f(x_n)$ | $f'(x_n)$ | $\dfrac{f(x_n)}{f'(x_n)}$ | $x_n - \dfrac{f(x_n)}{f'(x_n)}$ |
|---|---|---|---|---|---|
| 1 | 1.6000 | −0.0292 | −0.9996 | 0.0292 | 1.5708 |
| 2 | 1.5708 | 0.0000 | −1.0000 | 0.0000 | 1.5708 |

**5.**  $f(x) = x^3 + 4$

  $f'(x) = 3x^2$

  $x_1 = -2$

| $n$ | $x_n$ | $f(x_n)$ | $f'(x_n)$ | $\dfrac{f(x_n)}{f'(x_n)}$ | $x_n - \dfrac{f(x_n)}{f'(x_n)}$ |
|---|---|---|---|---|---|
| 1 | −2.0000 | −4.0000 | 12.0000 | −0.3333 | −1.6667 |
| 2 | −1.6667 | −0.6296 | 8.3333 | −0.0756 | −1.5911 |
| 3 | −1.5911 | −0.0281 | 7.5949 | −0.0037 | −1.5874 |
| 4 | −1.5874 | −0.0000 | 7.5596 | 0.0000 | −1.5874 |

Approximation of the zero of $f$ is −1.587.

**7.**  $f(x) = x^3 + x - 1$

  $f'(x) = 3x^2 + 1$

| $n$ | $x_n$ | $f(x_n)$ | $f'(x_n)$ | $\dfrac{f(x_n)}{f'(x_n)}$ | $x_n - \dfrac{f(x_n)}{f'(x_n)}$ |
|---|---|---|---|---|---|
| 1 | 0.5000 | −0.3750 | 1.7500 | −0.2143 | 0.7143 |
| 2 | 0.7143 | 0.0788 | 2.5307 | 0.0311 | 0.6832 |
| 3 | 0.6832 | 0.0021 | 2.4003 | 0.0009 | 0.6823 |

Approximation of the zero of $f$ is 0.682.

**9.**  $f(x) = 5\sqrt{x - 1} - 2x$

  $f'(x) = \dfrac{5}{2\sqrt{x - 1}} - 2$

From the graph you see that these are two zeros. Begin with $x = 1.2$.

| $n$ | $x_n$ | $f(x_n)$ | $f'(x_n)$ | $\dfrac{f(x_n)}{f'(x_n)}$ | $x_n - \dfrac{f(x_n)}{f'(x_n)}$ |
|---|---|---|---|---|---|
| 1 | 1.2000 | −0.1639 | 3.5902 | −0.0457 | 1.2457 |
| 2 | 1.2457 | −0.0131 | 3.0440 | −0.0043 | 1.2500 |
| 3 | 1.2500 | −0.0001 | 3.0003 | −0.0003 | 1.2500 |

Approximation of the zero of $f$ is 1.250.

Similarly, the other zero is approximately 5.000.

(**Note:** These answers are exact)

**11.** $f(x) = x - e^{-x}$

$f'(x) = 1 + e^{-x}$

$x_1 = 0.5$

| $n$ | $x_n$ | $f(x_n)$ | $f'(x_n)$ | $\dfrac{f(x_n)}{f'(x_n)}$ | $x_n - \dfrac{f(x_n)}{f'(x_n)}$ |
|---|---|---|---|---|---|
| 1 | 0.5 | −0.1065 | 1.6065 | −0.0663 | 0.5663 |
| 2 | 0.5663 | 0.0013 | 1.5676 | 0.0008 | 0.5671 |
| 3 | 0.5671 | 0.0001 | 1.5672 | −0.0000 | 0.5671 |

Approximation of the zero of $f$ is 0.567.

**13.** $f(x) = x^3 - 3.9x^2 + 4.79x - 1.881$

$f'(x) = 3x^2 - 7.8x + 4.79$

| $n$ | $x_n$ | $f(x_n)$ | $f'(x_n)$ | $\dfrac{f(x_n)}{f'(x_n)}$ | $x_n - \dfrac{f(x_n)}{f'(x_n)}$ |
|---|---|---|---|---|---|
| 1 | 0.5000 | −0.3360 | 1.6400 | −0.2049 | 0.7049 |
| 2 | 0.7049 | −0.0921 | 0.7824 | −0.1177 | 0.8226 |
| 3 | 0.8226 | −0.0231 | 0.4037 | −0.0573 | 0.8799 |
| 4 | 0.8799 | −0.0045 | 0.2495 | −0.0181 | 0.8980 |
| 5 | 0.8980 | −0.0004 | 0.2048 | −0.0020 | 0.9000 |
| 6 | 0.9000 | 0.0000 | 0.2000 | 0.0000 | 0.9000 |

Approximation of the zero of $f$ is 0.900.

| $n$ | $x_n$ | $f(x_n)$ | $f'(x_n)$ | $\dfrac{f(x_n)}{f'(x_n)}$ | $x_n - \dfrac{f(x_n)}{f'(x_n)}$ |
|---|---|---|---|---|---|
| 1 | 1.1 | 0.0000 | −0.1600 | −0.0000 | 1.1000 |

Approximation of the zero of $f$ is 1.100.

| $n$ | $x_n$ | $f(x_n)$ | $f'(x_n)$ | $\dfrac{f(x_n)}{f'(x_n)}$ | $x_n - \dfrac{f(x_n)}{f'(x_n)}$ |
|---|---|---|---|---|---|
| 1 | 1.9 | 0.0000 | 0.8000 | 0.0000 | 1.9000 |

Approximation of the zero of $f$ is 1.900.

**15.** $f(x) = 1 - x + \sin x$

$f'(x) = -1 + \cos x$

$x_1 = 2$

| $n$ | $x_n$ | $f(x_n)$ | $f'(x_n)$ | $\dfrac{f(x_n)}{f'(x_n)}$ | $x_n - \dfrac{f(x_n)}{f'(x_n)}$ |
|---|---|---|---|---|---|
| 1 | 2.0000 | −0.0907 | −1.4161 | 0.0640 | 1.9360 |
| 2 | 1.9360 | −0.0019 | −1.3571 | 0.0014 | 1.9346 |
| 3 | 1.9346 | 0.0000 | −1.3558 | 0.0000 | 1.9346 |

Approximate zero: $x \approx 1.935$

**17.** $h(x) = f(x) - g(x) = 2x + 1 - \sqrt{x+4}$

$h'(x) = 2 - \dfrac{1}{2\sqrt{x+4}}$

| $n$ | $x_n$ | $h(x_n)$ | $h'(x_n)$ | $\dfrac{h(x_n)}{h'(x_n)}$ | $x_n - \dfrac{h(x_n)}{h'(x_n)}$ |
|---|---|---|---|---|---|
| 1 | 0.6000 | 0.0552 | 1.7669 | 0.0313 | 0.5687 |
| 2 | 0.5687 | 0.0000 | 1.7661 | 0.0000 | 0.5687 |

Point of intersection of the graphs of $f$ and $g$ occurs when $x \approx 0.569$.

**19.** $h(x) = f(x) - g(x) = x - \tan x$

$h'(x) = 1 - \sec^2 x$

| $n$ | $x_n$ | $h(x_n)$ | $h'(x_n)$ | $\dfrac{h(x_n)}{h'(x_n)}$ | $x_n - \dfrac{h(x_n)}{h'(x_n)}$ |
|---|---|---|---|---|---|
| 1 | 4.5000 | −0.1373 | −21.5048 | 0.0064 | 4.4936 |
| 2 | 4.4936 | −0.0039 | −20.2271 | 0.0002 | 4.4934 |

Point of intersection of the graphs of $f$ and $g$ occurs when $x \approx 4.493$.

**Note:** $f(x) = x$ and $g(x) = \tan x$ intersect infinitely often.

**21.** (a) $f(x) = x^2 - a, a > 0$

$f'(x) = 2x$

$x_{n+1} = x_n - \dfrac{f(x_n)}{f'(x_n)} = x_n - \dfrac{x_n^2 - a}{2x_n} = \dfrac{1}{2}\left(x_n + \dfrac{a}{x_n}\right)$

(b) $\sqrt{5}$: $x_{n+1} = \dfrac{1}{2}\left(x_n + \dfrac{5}{x_n}\right)$, $x_1 = 2$

| $n$ | 1 | 2 | 3 | 4 |
|---|---|---|---|---|
| $x_n$ | 2 | 2.25 | 2.2361 | 2.2361 |

For example, given $x_1 = 2$,

$x_2 = \dfrac{1}{2}\left(2 + \dfrac{5}{2}\right) = \dfrac{9}{4} = 2.25.$

$\sqrt{5} \approx 2.236$

$\sqrt{7}$: $x_{n+1} = \dfrac{1}{2}\left(x_n + \dfrac{7}{x_n}\right)$, $x_1 = 2$

| $n$ | 1 | 2 | 3 | 4 | 5 |
|---|---|---|---|---|---|
| $x_n$ | 2 | 2.75 | 2.6477 | 2.6458 | 2.6458 |

$\sqrt{7} \approx 2.646$

**23.** $y = 2x^3 - 6x^2 + 6x - 1 = f(x)$

$y' = 6x^2 - 12x + 6 = f'(x)$

$x_1 = 1$

$f'(x) = 0$; therefore, the method fails.

| $n$ | $x_n$ | $f(x_n)$ | $f'(x_n)$ |
|---|---|---|---|
| 1 | 1 | 1 | 0 |

**25.** Let $g(x) = f(x) - x = \cos x - x$

$g'(x) = -\sin x - 1$.

| $n$ | $x_n$ | $g(x_n)$ | $g'(x_n)$ | $\dfrac{g(x_n)}{g'(x_n)}$ | $x_n - \dfrac{g(x_n)}{g'(x_n)}$ |
|---|---|---|---|---|---|
| 1 | 1.0000 | −0.4597 | −1.8415 | 0.2496 | 0.7504 |
| 2 | 0.7504 | −0.0190 | −1.6819 | 0.0113 | 0.7391 |
| 3 | 0.7391 | 0.0000 | −1.6736 | 0.0000 | 0.7391 |

The fixed point is approximately 0.74.

**27.** Let $g(x) = e^{x/10} - x$

$g'(x) = \dfrac{1}{10}e^{x/10} - 1$

| $n$ | $x_n$ | $g(x_n)$ | $g'(x_n)$ | $\dfrac{g(x_n)}{g'(x_n)}$ | $x_n - \dfrac{g(x_n)}{g'(x_n)}$ |
|---|---|---|---|---|---|
| 1 | 1.0 | 0.1052 | −0.8895 | −0.1182 | 1.1182 |
| 2 | 1.1182 | 0.0001 | −0.8882 | −0.0001 | 1.1183 |

The fixed point is approximately 1.12.

**29.** $f(x) = \dfrac{1}{x} - a = 0$

$f'(x) = -\dfrac{1}{x^2}$

$x_{n+1} = x_n - \dfrac{(1/x_n) - a}{-1/x_n^2} = x_n + x_n^2\left(\dfrac{1}{x_n} - a\right) = x_n + x_n - x_n^2 a = 2x_n - x_n^2 a = x_n(2 - ax_n)$

**31.** $f(x) = x^3 - 3x^2 + 3$, $f'(x) = 3x^2 - 6x$

(a)

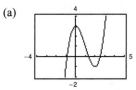

(b) $x_1 = 1$

$x_2 = x_1 - \dfrac{f(x_1)}{f'(x_1)} \approx 1.333$

Continuing, the zero is 1.347.

(c) $x_1 = \dfrac{1}{4}$

$x_2 = x_1 - \dfrac{f(x_1)}{f'(x_1)} \approx 2.405$

Continuing, the zero is 2.532.

(d)

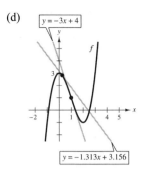

The $x$-intercept of $y = -3x + 4$ is $\frac{4}{3}$. The $x$-intercept of $y = 1.313x + 3.156$ is approximately 2.405.

The $x$-intercepts correspond to the values resulting from the first iteration of Newton's Method.

(e) If the initial guess $x_1$ is not "close to" the desired zero of the function, the $x$-intercept of the tangent line may approximate another zero of the function.

**33.** Answers will vary. See page 229.

If $f$ is a function continuous on $[a, b]$ and differentiable on $(a, b)$ where $c \in [a, b]$ and $f(c) = 0$, Newton's Method uses tangent lines to approximate $c$ such that $f(c) = 0$.

First, estimate an initial $x_1$ close to $c$ (see graph).

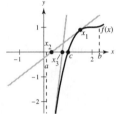

Then determine $x_2$ by $x_2 = x_1 - \dfrac{f(x_1)}{f'(x_1)}$.

Calculate a third estimate by $x_3 = x_2 - \dfrac{f(x_2)}{f'(x_2)}$.

Continue this process until $\left| x_n - x_{n+1} \right|$ is within the desired accuracy.

Let $x_{n+1}$ be the final approximation of $c$.

**35.** $y = f(x) = 4 - x^2, (1, 0)$

$$d = \sqrt{(x - 1)^2 + (y - 0)^2} = \sqrt{(x - 1)^2 + (4 - x^2)^2} = \sqrt{x^4 - 7x^2 - 2x + 17}$$

$d$ is minimized when $D = x^4 - 7x^2 - 2x + 17$ is a minimum.

$g(x) = D' = 4x^3 - 14x - 2$

$g'(x) = 12x^2 - 14$

| $n$ | $x_n$ | $g(x_n)$ | $g'(x_n)$ | $\dfrac{g(x_n)}{g'(x_n)}$ | $x_n - \dfrac{g(x_n)}{g'(x_n)}$ |
|---|---|---|---|---|---|
| 1 | 2.0000 | 2.0000 | 34.0000 | 0.0588 | 1.9412 |
| 2 | 1.9412 | 0.0830 | 31.2191 | 0.0027 | 1.9385 |
| 3 | 1.9385 | −0.0012 | 31.0934 | 0.0000 | 1.9385 |

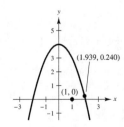

$x \approx 1.939$

Point closest to $(1, 0)$ is $\approx (1.939, 0.240)$.

**37.**

$$\text{Minimize: } T = \frac{\text{Distance rowed}}{\text{Rate rowed}} + \frac{\text{Distance walked}}{\text{Rate walked}}$$

$$T = \frac{\sqrt{x^2 + 4}}{3} + \frac{\sqrt{x^2 - 6x + 10}}{4}$$

$$T' = \frac{x}{3\sqrt{x^2 + 4}} + \frac{x - 3}{4\sqrt{x^2 - 6x + 10}} = 0$$

$$4x\sqrt{x^2 - 6x + 10} = -3(x - 3)\sqrt{x^2 + 4}$$

$$16x^2(x^2 - 6x + 10) = 9(x - 3)^2(x^2 + 4)$$

$$7x^4 - 42x^3 + 43x^2 + 216x - 324 = 0$$

Let $f(x) = 7x^4 - 42x^3 + 43x^2 + 216x - 324$ and $f'(x) = 28x^3 - 126x^2 + 86x + 216$. Becasuse $f(1) = -100$ and $f(2) = 56$, the solution is in the interval $(1, 2)$.

| $n$ | $x_n$ | $f(x_n)$ | $f'(x_n)$ | $\dfrac{f(x_n)}{f'(x_n)}$ | $x_n - \dfrac{f(x_n)}{f'(x_n)}$ |
|---|---|---|---|---|---|
| 1 | 1.7000 | 19.5887 | 135.6240 | 0.1444 | 1.5556 |
| 2 | 1.5556 | -1.0480 | 150.2780 | -0.0070 | 1.5626 |
| 3 | 1.5626 | 0.0014 | 49.5591 | 0.0000 | 1.5626 |

Approximation: $x \approx 1.563$ mi

**39.** False. Let $f(x) = (x^2 - 1)/(x - 1)$. $x = 1$ is a discontinuity. It is not a zero of $f(x)$. This statement would be true if $f(x) = p(x)/q(x)$ was given in **reduced** form.

**41.** True

**43.** $f(x) = -\sin x$

$f'(x) = -\cos x$

Let $(x_0, y_1) = (x_0, -\sin(x_0))$ be a point on the graph of $f$. If $(x_0, y_0)$ is a point of tangency, then

$$-\cos(x_0) = \frac{y_0 - 0}{x_0 - 0} = \frac{y_0}{x_0} = \frac{-\sin(x_0)}{x_0}.$$

So, $x_0 = \tan(x_0)$.

$x_0 \approx 4.4934$

Slope $= -\cos(x_0) \approx 0.217$

You can verify this answer by graphing $y_1 = -\sin x$ and the tangent line $y_2 = 0.217x$.

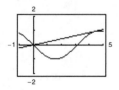

# Review Exercises for Chapter 3

**1.** $f(x) = 12$

$$f'(x) = \lim_{\Delta x \to 0} \frac{f(x + \Delta x) - f(x)}{\Delta x}$$

$$= \lim_{\Delta x \to 0} \frac{12 - 12}{\Delta x}$$

$$= \lim_{\Delta x \to 0} \frac{0}{\Delta x} = 0$$

**3.** $f(x) = x^2 - 4x + 5$

$$f'(x) = \lim_{\Delta x \to 0} \frac{f(x + \Delta x) - f(x)}{\Delta x}$$

$$= \lim_{\Delta x \to 0} \frac{\left[(x + \Delta x)^2 - 4(x + \Delta x) + 5\right] - \left[x^2 - 4x + 5\right]}{\Delta x}$$

$$= \lim_{\Delta x \to 0} \frac{\left(x^2 + 2x(\Delta x) + (\Delta x)^2 - 4x - 4(\Delta x) + 5\right) - \left(x^2 - 4x + 5\right)}{\Delta x}$$

$$= \lim_{\Delta x \to 0} \frac{2x(\Delta x) + (\Delta x)^2 - 4(\Delta x)}{\Delta x} = \lim_{\Delta x \to 0} (2x + \Delta x - 4) = 2x - 4$$

**5.** $g(x) = 2x^2 - 3x,\ c = 2$

$$g'(2) = \lim_{x \to 2} \frac{g(x) - g(2)}{x - 2}$$

$$= \lim_{x \to 2} \frac{(2x^2 - 3x) - 2}{x - 2}$$

$$= \lim_{x \to 2} \frac{(x - 2)(2x + 1)}{x - 2}$$

$$= \lim_{x \to 2} (2x + 1) = 2(2) + 1 = 5$$

**7.** $f$ is differentiable for all $x \neq 3$.

**9.** $y = 25$

$y' = 0$

**11.** $f(x) = x^3 - 11x^2$

$f'(x) = 3x^2 - 22x$

**13.** $h(x) = 6\sqrt{x} + 3\sqrt[3]{x} = 6x^{1/2} + 3x^{1/3}$

$h'(x) = 3x^{-1/2} + x^{-2/3} = \dfrac{3}{\sqrt{x}} + \dfrac{1}{\sqrt[3]{x^2}}$

**15.** $g(t) = \dfrac{2}{3}t^{-2}$

$g'(t) = \dfrac{-4}{3}t^{-3} = -\dfrac{4}{3t^3}$

**17.** $f(\theta) = 4\theta - 5 \sin \theta$

$f'(\theta) = 4 - 5 \cos \theta$

**19.** $f(t) = 3 \cos t - 4e^t$

$f'(t) = -3 \sin t - 4e^t$

**21.** $f(x) = \dfrac{27}{x^3} = 27x^{-3},\ (3, 1)$

$f'(x) = 27(-3)x^{-4} = -\dfrac{81}{x^4}$

$f'(3) = -\dfrac{81}{3^4} = -1$

**23.** $f(x) = 2x^4 - 8,\ (0, -8)$

$f'(x) = 8x^3$

$f'(0) = 0$

**25.** $\quad F = 200\sqrt{T}$

$F'(t) = \dfrac{100}{\sqrt{T}}$

(a) When $T = 4$, $F'(4) = 50$ vibrations/sec/lb.

(b) When $T = 9$, $F'(9) = 33\frac{1}{3}$ vibrations/sec/lb.

**27.** $s(t) = -16t^2 + v_0 t + s_0$; $s_0 = 600$, $v_0 = -30$

   (a)  $s(t) = -16t^2 - 30t + 600$

       $s'(t) = v(t) = -32t - 30$

   (b)  Average velocity $= \dfrac{s(3) - s(1)}{3 - 1}$

                           $= \dfrac{366 - 554}{2}$

                           $= -94$ ft/sec

   (c)  $v(1) = -32(1) - 30 = -62$ ft/sec

       $v(3) = -32(3) - 30 = -126$ ft/sec

   (d)  $s(t) = 0 = -16t^2 - 30t + 600$

       Using a graphing utility or the Quadratic Formula, $t \approx 5.258$ seconds.

   (e)  When

       $t \approx 5.258$, $v(t) \approx -32(5.258) - 30 \approx -198.3$ ft/sec.

**29.** $f(x) = (5x^2 + 8)(x^2 - 4x - 6)$

   $f'(x) = (5x^2 + 8)(2x - 4) + (x^2 - 4x - 6)(10x)$

          $= 10x^3 + 16x - 20x^2 - 32 + 10x^3 - 40x^2 - 60x$

          $= 20x^3 - 60x^2 - 44x - 32$

          $= 4(5x^3 - 15x^2 - 11x - 8)$

**31.** $h(x) = \sqrt{x}\,\sin x = x^{1/2}\sin x$

   $h'(x) = \dfrac{1}{2\sqrt{x}}\sin x + \sqrt{x}\cos x$

**33.** $f(x) = \dfrac{x^2 + x - 1}{x^2 - 1}$

   $f'(x) = \dfrac{(x^2 - 1)(2x + 1) - (x^2 + x - 1)(2x)}{(x^2 - 1)^2}$

       $= \dfrac{-(x^2 + 1)}{(x^2 - 1)^2}$

**35.** $y = \dfrac{x^4}{\cos x}$

   $y' = \dfrac{(\cos x)\,4x^3 - x^4(-\sin x)}{\cos^2 x}$

      $= \dfrac{4x^3\cos x + x^4\sin x}{\cos^2 x}$

**37.** $y = 3x^2 \sec x$

   $y' = 3x^2 \sec x \tan x + 6x \sec x$

**39.** $y = 4xe^x - \cot x$

   $y' = 4xe^x + 4e^x + \csc^2 x$

**41.** $f(x) = (x + 2)(x^2 + 5)$, $(-1, 6)$

   $f'(x) = (x + 2)(2x) + (x^2 + 5)(1)$

       $= 2x^2 + 4x + x^2 + 5 = 3x^2 + 4x + 5$

   $f'(-1) = 3 - 4 + 5 = 4$

   Tangent line: $y - 6 = 4(x + 1)$

                 $y = 4x + 10$

**43.** $f(x) = \dfrac{x + 1}{x - 1}$, $\left(\dfrac{1}{2}, -3\right)$

   $f'(x) = \dfrac{(x - 1) - (x + 1)}{(x - 1)^2} = \dfrac{-2}{(x - 1)^2}$

   $f'\left(\dfrac{1}{2}\right) = \dfrac{-2}{(1/4)} = -8$

   Tangent line: $y + 3 = -8\left(x - \dfrac{1}{2}\right)$

                 $y = -8x + 1$

**45.** $g(t) = -8t^3 - 5t + 12$

   $g'(t) = -24t^2 - 5$

   $g''(t) = -48t$

**47.** $f(x) = 15x^{5/2}$

   $f'(x) = \frac{75}{2}x^{3/2}$

   $f''(x) = \frac{225}{4}x^{1/2} = \frac{225}{4}\sqrt{x}$

**49.** $f(\theta) = 3\tan\theta$

   $f'(\theta) = 3\sec^2\theta$

   $f''(\theta) = 6\sec\theta(\sec\theta\tan\theta) = 6\sec^2\theta\tan\theta$

**51.** $v(t) = 20 - t^2$, $0 \le t \le 6$

   $a(t) = v'(t) = -2t$

   $v(3) = 20 - 3^2 = 11$ m/sec

   $a(3) = -2(3) = -6$ m/sec$^2$

**53.** $y = (7x + 3)^4$

   $y' = 4(7x + 3)^3(7) = 28(7x + 3)^3$

**55.** $y = \dfrac{1}{x^2 + 4} = (x^2 + 4)^{-1}$

   $y' = -1(x^2 + 4)^{-2}(2x) = -\dfrac{2x}{(x^2 + 4)^2}$

**57.** $y = 5\cos(9x + 1)$

$y' = -5\sin(9x + 1)(9) = -45\sin(9x + 1)$

**59.** $y = \dfrac{x}{2} - \dfrac{\sin 2x}{4}$

$y' = \dfrac{1}{2} - \dfrac{1}{4}\cos 2x(2) = \dfrac{1}{2}(1 - \cos 2x) = \sin^2 x$

**61.** $y = x(6x + 1)^5$

$\begin{aligned} y' &= x\,5(6x + 1)^4(6) + (6x + 1)^5(1) \\ &= 30x(6x + 1)^4 + (6x + 1)^5 \\ &= (6x + 1)^4(30x + 6x + 1) \\ &= (6x + 1)^4(36x + 1) \end{aligned}$

**63.** $f(x) = \dfrac{3x}{\sqrt{x^2 + 1}}$

$\begin{aligned} f'(x) &= \dfrac{3(x^2 + 1)^{1/2} - 3x\dfrac{1}{2}(x^2 + 1)^{-1/2}(2x)}{x^2 + 1} \\[2mm] &= \dfrac{3(x^2 + 1) - 3x^2}{(x^2 + 1)^{3/2}} = \dfrac{3}{(x^2 + 1)^{3/2}} \end{aligned}$

**65.** $g(t) = t^2 e^{t/4}$

$\begin{aligned} g'(t) &= \tfrac{1}{4}t^2 e^{t/4} + 2t e^{t/4} \\ &= \tfrac{1}{4}t e^{t/4}[t + 8] \end{aligned}$

**67.** $y = \sqrt{e^{2x} + e^{-2x}} = (e^{2x} + e^{-2x})^{1/2}$

$y' = \dfrac{1}{2}(e^{2x} + e^{-2x})^{-1/2}(2e^{2x} - 2e^{-2x}) = \dfrac{e^{2x} - e^{-2x}}{\sqrt{e^{2x} + e^{-2x}}}$

**69.** $g(x) = \dfrac{x^2}{e^x}$

$g'(x) = \dfrac{e^x(2x) - x^2 e^x}{e^{2x}} = \dfrac{x(2 - x)}{e^x}$

**71.** $g(x) = \ln\sqrt{x} = \dfrac{1}{2}\ln x$

$g'(x) = \dfrac{1}{2x}$

**73.** $f(x) = x\sqrt{\ln x}$

$\begin{aligned} f'(x) &= \left(\dfrac{x}{2}\right)(\ln x)^{-1/2}\left(\dfrac{1}{x}\right) + \sqrt{\ln x} \\ &= \dfrac{1}{2\sqrt{\ln x}} + \sqrt{\ln x} = \dfrac{1 + 2\ln x}{2\sqrt{\ln x}} \end{aligned}$

**75.** $y = \dfrac{1}{b^2}\left[\ln(a + bx) + \dfrac{a}{a + bx}\right]$

$\dfrac{dy}{dx} = \dfrac{1}{b^2}\left[\dfrac{b}{a + bx} - \dfrac{ab}{(a + bx)^2}\right] = \dfrac{x}{(a + bx)^2}$

**77.** $y = -\dfrac{1}{a}\ln\left(\dfrac{a + bx}{x}\right) = -\dfrac{1}{a}\left[\ln(a + bx) - \ln x\right]$

$\dfrac{dy}{dx} = -\dfrac{1}{a}\left(\dfrac{b}{a + bx} - \dfrac{1}{x}\right) = \dfrac{1}{x(a + bx)}$

**79.** $f(x) = \sqrt{1 - x^3},\ (-2, 3)$

$f'(x) = \dfrac{1}{2}(1 - x^3)^{-1/2}(-3x^2) = \dfrac{-3x^2}{2\sqrt{1 - x^3}}$

$f'(-2) = \dfrac{-12}{2(3)} = -2$

**81.** $f(x) = \dfrac{4}{x^2 + 1} = 4(x^2 + 1)^{-1},\ (-1, 2)$

$f'(x) = -4(x^2 + 1)^{-2}(2x) = -\dfrac{8x}{(x^2 + 1)^2}$

$f'(-1) = -\dfrac{8(-1)}{\left[(-1)^2 + 1\right]^2} = \dfrac{8}{4} = 2$

**83.** $y = \dfrac{1}{2}\csc 2x,\ \left(\dfrac{\pi}{4}, \dfrac{1}{2}\right)$

$y' = -\csc 2x \cot 2x$

$y'\left(\dfrac{\pi}{4}\right) = 0$

**85.** $y = (8x + 5)^3$

$y' = 3(8x + 5)^2(8) = 24(8x + 5)^2$

$y'' = 24(2)(8x + 5)(8) = 384(8x + 5)$

**87.** $f(x) = \cot x$

$f'(x) = -\csc^2 x$

$\begin{aligned} f''(x) &= -2\csc x(-\csc x \cdot \cot x) \\ &= 2\csc^2 x \cot x \end{aligned}$

**89.** $T = \dfrac{700}{t^2 + 4t + 10}$

$T = 700\left(t^2 + 4t + 10\right)^{-1}$

$T' = \dfrac{-1400(t + 2)}{\left(t^2 + 4t + 10\right)^2}$

(a) When $t = 1$,

$T' = \dfrac{-1400(1 + 2)}{\left(1 + 4 + 10\right)^2} \approx -18.667$ deg/h.

(b) When $t = 3$,

$T' = \dfrac{-1400(3 + 2)}{\left(9 + 12 + 10\right)^2} \approx -7.284$ deg/h.

(c) When $t = 5$,

$T' = \dfrac{-1400(5 + 2)}{\left(25 + 20 + 10\right)^2} \approx -3.240$ deg/h.

(d) When $t = 10$,

$T' = \dfrac{-1400(10 + 2)}{\left(100 + 40 + 10\right)^2} \approx -0.747$ deg/h.

**91.** (a) You get an error message because $\ln h$ does not exist for $h = 0$.

(b) Reversing the data, you obtain
$h = 0.8627 - 6.4474 \ln p$.

(c)

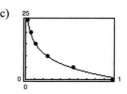

(d) If $p = 0.75$, $h \approx 2.72$ km.

(e) If $h = 13$ km, $p \approx 0.15$ atmosphere.

(f) $h = 0.8627 - 6.4474 \ln p$

$1 = -6.4474 \dfrac{1}{p} \dfrac{dp}{dh}$ (implicit differentiation)

$\dfrac{dp}{dh} = \dfrac{p}{-6.4474}$

For $h = 5$,

$p = 0.5264$ and $\dfrac{dp}{dh} = -0.0816$ atm/km

For $h = 20$,

$p = 0.0514$ and $\dfrac{dp}{dh} = -0.0080$ atm/km

As the altitude increases, the rate of change of pressure decreases.

**93.** $x^2 + y^2 = 64$

$2x + 2yy' = 0$

$2yy' = -2x$

$y' = -\dfrac{x}{y}$

**95.** $x^3 y - xy^3 \overset{!}{=} 4$

$x^3 y' + 3x^2 y - x3y^2 y' - y^3 = 0$

$x^3 y' - 3xy^2 y' = y^3 - 3x^2 y$

$y'\left(x^3 - 3xy^2\right) = y^3 - 3x^2 y$

$y' = \dfrac{y^3 - 3x^2 y}{x^3 - 3xy^2}$

$y' = \dfrac{y\left(y^2 - 3x^2\right)}{x\left(x^2 - 3y^2\right)}$

**97.** $x \sin y = y \cos x$

$\left(x \cos y\right)y' + \sin y = -y \sin x + y' \cos x$

$y'\left(x \cos y - \cos x\right) = -y \sin x - \sin y$

$y' = \dfrac{y \sin x + \sin y}{\cos x - x \cos y}$

**99.** $x^2 + y^2 = 10$

$2x + 2yy' = 0$

$y' = \dfrac{-x}{y}$

At $(3, 1)$, $y' = -3$

Tangent line: $\quad y - 1 = -3(x - 3) \Rightarrow 3x + y - 10 = 0$

Normal line: $\quad y - 1 = \dfrac{1}{3}(x - 3) \Rightarrow x - 3y = 0$

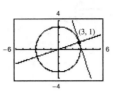

**101.** $\quad y \ln x + y^2 = 0, \quad (e, -1)$

$y' \ln x + \dfrac{y}{x} + 2yy' = 0$

$y'(\ln x + 2y) = \dfrac{-y}{x}$

$y' = \dfrac{-y}{x(\ln x + 2y)}$

At $(e, -1)$: $y' = \dfrac{-1}{e}$

Tangent line: $y + 1 = \dfrac{-1}{e}(x - e)$

$y = \dfrac{-1}{e}x$

Normal line: $y + 1 = e(x - e)$

$y = ex - e^2 - 1$

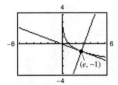

**103.** $\quad y = \dfrac{x\sqrt{x^2 + 1}}{x + 4}$

$\ln y = \ln x + \dfrac{1}{2}\ln(x^2 + 1) - \ln(x + 4)$

$\dfrac{y'}{y} = \dfrac{1}{x} + \dfrac{x}{x^2 + 1} - \dfrac{1}{x + 4}$

$y' = \dfrac{x\sqrt{x^2 + 1}}{x + 4}\left(\dfrac{1}{x} + \dfrac{x}{x^2 + 1} - \dfrac{1}{x + 4}\right)$

$= \dfrac{x^3 + 8x^2 + 4}{(x + 4)^2\sqrt{x^2 + 1}}$

**105.** $f(x) = x^3 + 2, \quad a = -1$

$f'(x) = 3x^2 > 0$

$f$ is monotonic (increasing) on $(-\infty, \infty)$ therefore $f$ has an inverse.

$f(-3^{1/3}) = -1 \Rightarrow f^{-1}(-1) = -3^{1/3}$

$f'(-3^{1/3}) = 3^{2/3}$

$(f^{-1})'(-1) = \dfrac{1}{f'(f^{-1}(-1))} = \dfrac{1}{f'(-3^{1/3})} = \dfrac{1}{3(3^{2/3})} = \dfrac{1}{3^{5/3}}$

**107.** $f(x) = \tan x, \quad a = \dfrac{\sqrt{3}}{3}, -\dfrac{\pi}{4} \le x \le \dfrac{\pi}{4}$

$f'(x) = \sec^2 x > 0$ on $\left(-\dfrac{\pi}{4}, \dfrac{\pi}{4}\right)$

$f$ is monotonic (increasing) on $\left[-\dfrac{\pi}{4}, \dfrac{\pi}{4}\right]$ therefore $f$ has an inverse.

$f\left(\dfrac{\pi}{6}\right) = \dfrac{\sqrt{3}}{3} \Rightarrow f^{-1}\left(\dfrac{\sqrt{3}}{3}\right) = \dfrac{\pi}{6}$

$f'\left(\dfrac{\pi}{6}\right) = \dfrac{4}{3}$

$(f^{-1})'\left(\dfrac{\sqrt{3}}{3}\right) = \dfrac{1}{f'\left(f^{-1}\left(\dfrac{\sqrt{3}}{3}\right)\right)} = \dfrac{1}{f'\left(\dfrac{\pi}{6}\right)} = \dfrac{1}{\left(\dfrac{4}{3}\right)} = \dfrac{3}{4}$

**109.** $y = \tan(\arcsin x) = \dfrac{x}{\sqrt{1 - x^2}}$

$y' = \dfrac{(1 - x^2)^{1/2} + x^2(1 - x^2)^{-1/2}}{1 - x^2} = (1 - x^2)^{-3/2}$

**111.** $y = x \operatorname{arcsec} x$

$y' = \dfrac{x}{|x|\sqrt{x^2 - 1}} + \operatorname{arcsec} x$

**113.**  $y = x(\arcsin x)^2 - 2x + 2\sqrt{1 - x^2}\, \arcsin x$

$y' = \dfrac{2x \arcsin x}{\sqrt{1 - x^2}} + (\arcsin x)^2 - 2 + \dfrac{2\sqrt{1 - x^2}}{\sqrt{1 - x^2}} - \dfrac{2x}{\sqrt{1 - x^2}}\, \arcsin x = (\arcsin x)^2$

**115.**  $y = \sqrt{x}$

$\dfrac{dy}{dt} = 2$ units/sec

$\dfrac{dy}{dt} = \dfrac{1}{2\sqrt{x}}\dfrac{dx}{dt} \Rightarrow \dfrac{dx}{dt} = 2\sqrt{x}\dfrac{dy}{dt} = 4\sqrt{x}$

(a)  When $x = \dfrac{1}{2}, \dfrac{dx}{dt} = 2\sqrt{2}$ units/sec.

(b)  When $x = 1, \dfrac{dx}{dt} = 4$ units/sec.

(c)  When $x = 4, \dfrac{dx}{dt} = 8$ units/sec.

**117.**  $\tan\theta = x$

$\dfrac{d\theta}{dt} = 3(2\pi)$ rad/min

$\sec^2\theta\left(\dfrac{d\theta}{dt}\right) = \dfrac{dx}{dt}$

$\dfrac{dx}{dt} = (\tan^2\theta + 1)(6\pi) = 6\pi(x^2 + 1)$

When $x = \dfrac{1}{2}$,

$\dfrac{dx}{dt} = 6\pi\left(\dfrac{1}{4} + 1\right) = \dfrac{15\pi}{2}$ km/min $= 450\pi$ km/h.

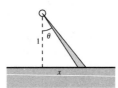

**119.**  $f(x) = x^3 - 3x - 1$

From the graph you can see that $f(x)$ has three real zeros.

$f'(x) = 3x^2 - 3$

| $n$ | $x_n$ | $f(x_n)$ | $f'(x_n)$ | $\dfrac{f(x_n)}{f'(x_n)}$ | $x_n - \dfrac{f(x_n)}{f'(x_n)}$ |
|---|---|---|---|---|---|
| 1 | −1.5000 | 0.1250 | 3.7500 | 0.0333 | −1.5333 |
| 2 | −1.5333 | −0.0049 | 4.0530 | −0.0012 | −1.5321 |

| $n$ | $x_n$ | $f(x_n)$ | $f'(x_n)$ | $\dfrac{f(x_n)}{f'(x_n)}$ | $x_n - \dfrac{f(x_n)}{f'(x_n)}$ |
|---|---|---|---|---|---|
| 1 | −0.5000 | 0.3750 | −2.2500 | −0.1667 | −0.3333 |
| 2 | −0.3333 | −0.0371 | −2.6667 | 0.0139 | −0.3472 |
| 3 | −0.3472 | −0.0003 | −2.6384 | 0.0001 | −0.3473 |

| $n$ | $x_n$ | $f(x_n)$ | $f'(x_n)$ | $\dfrac{f(x_n)}{f'(x_n)}$ | $x_n - \dfrac{f(x_n)}{f'(x_n)}$ |
|---|---|---|---|---|---|
| 1 | 1.9000 | 0.1590 | 7.8300 | 0.0203 | 1.8797 |
| 2 | 1.8797 | 0.0024 | 7.5998 | 0.0003 | 1.8794 |

The three real zeros of $f(x)$ are $x \approx -1.532$, $x \approx -0.347$, and $x \approx 1.879$.

**121.** $g(x) = xe^x - 4$

$g'(x) = (x + 1)e^x$

From the graph, there is one zero near 1.

| $n$ | $x_n$ | $g(x_n)$ | $g'(x_n)$ | $\dfrac{g(x_n)}{g'(x_n)}$ | $x_n - \dfrac{g(x_n)}{g'(x_n)}$ |
|---|---|---|---|---|---|
| 1 | 1.0 | −1.2817 | 5.4366 | −0.2358 | 1.2358 |
| 2 | 1.2358 | 0.2525 | 7.6937 | 0.0328 | 1.2030 |
| 3 | 1.2030 | 0.0059 | 7.3359 | 0.0008 | 1.2022 |

To three decimal places, $x = 1.202$.

**123.** $f(x) = x^4 + x^3 - 3x^2 + 2$

From the graph you can see that $f(x)$ has two real zeros.

$f'(x) = 4x^3 + 3x^2 - 6x$

| $n$ | $x_n$ | $f(x_n)$ | $f'(x_n)$ | $\dfrac{f(x_n)}{f'(x_n)}$ | $x_n - \dfrac{f(x_n)}{f'(x_n)}$ |
|---|---|---|---|---|---|
| 1 | −2.0 | −2.0 | −8.0 | 0.25 | −2.25 |
| 2 | −2.25 | 1.0508 | −16.875 | −0.0623 | −2.1877 |
| 3 | −2.1877 | 0.0776 | −14.3973 | −0.0054 | −2.1823 |
| 4 | −2.1823 | 0.0004 | −14.3911 | −0.00003 | −2.1873 |

| $n$ | $x_n$ | $f(x_n)$ | $f'(x_n)$ | $\dfrac{f(x_n)}{f'(x_n)}$ | $x_n - \dfrac{f(x_n)}{f'(x_n)}$ |
|---|---|---|---|---|---|
| 1 | −1.0 | −1.0 | 5.0 | −0.2 | −0.8 |
| 2 | −0.8 | −0.0224 | 4.6720 | −0.0048 | −0.7952 |
| 3 | −0.7952 | −0.00001 | 4.6569 | −0.0000 | −0.7952 |

The two zeros of $f(x)$ are $x \approx -2.1823$ and $x \approx -0.7952$.

**125.** Find the zeros of $f(x) = x^4 - x - 3$.

$f'(x) = 4x^3 - 1$

From the graph you can see that $f(x)$ has two real zeros.

$f$ changes sign in $[-2, -1]$.

| $n$ | $x_n$ | $f(x_n)$ | $f'(x_n)$ | $\dfrac{f(x_n)}{f'(x_n)}$ | $x_n - \dfrac{f(x_n)}{f'(x_n)}$ |
|---|---|---|---|---|---|
| 1 | $-1.2000$ | $0.2736$ | $-7.9120$ | $-0.0346$ | $-1.1654$ |
| 2 | $-1.1654$ | $0.0100$ | $-7.3312$ | $-0.0014$ | $-1.1640$ |

On the interval $[-2, -1]$: $x \approx -1.164$.

$f$ changes sign in $[1, 2]$.

| $n$ | $x_n$ | $f(x_n)$ | $f'(x_n)$ | $\dfrac{f(x_n)}{f'(x_n)}$ | $x_n - \dfrac{f(x_n)}{f'(x_n)}$ |
|---|---|---|---|---|---|
| 1 | $1.5000$ | $0.5625$ | $12.5000$ | $0.0450$ | $1.4550$ |
| 2 | $1.4550$ | $0.0268$ | $11.3211$ | $0.0024$ | $1.4526$ |
| 3 | $1.4526$ | $-0.0003$ | $11.2602$ | $0.0000$ | $1.4526$ |

On the interval $[1, 2]$: $x \approx 1.453$.

**127.** Find the zeros of $f(x) = \ln x + x$.

$f'(x) = \dfrac{1}{x} + 1$

From the graph you can see that $f(x)$ has one real zero.

| $n$ | $x_n$ | $f(x_n)$ | $f'(x_n)$ | $\dfrac{f(x_n)}{f'(x_n)}$ | $x_n - \dfrac{f(x_n)}{f'(x_n)}$ |
|---|---|---|---|---|---|
| 1 | $0.5$ | $-0.1931$ | $3.0000$ | $-0.0644$ | $0.5644$ |
| 2 | $0.5644$ | $-0.0076$ | $2.7718$ | $-0.0027$ | $0.5671$ |
| 3 | $0.5671$ | $0.0001$ | $2.7634$ | $-0.0000$ | $0.5671$ |

The real zero of $f(x)$ is $x \approx 0.567$.

# Problem Solving for Chapter 3

**1. (a)** $x^2 + (y - r)^2 = r^2$, Circle

$\qquad x^2 = y$, Parabola

Substituting:

$$(y - r)^2 = r^2 - y$$

$$y^2 - 2ry + r^2 = r^2 - y$$

$$y^2 - 2ry + y = 0$$

$$y(y - 2r + 1) = 0$$

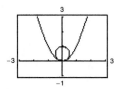

Because you want only one solution, let $1 - 2r = 0 \Rightarrow r = \dfrac{1}{2}$. Graph $y = x^2$ and $x^2 + \left(y - \dfrac{1}{2}\right)^2 = \dfrac{1}{4}$.

**(b)** Let $(x, y)$ be a point of tangency:

$$x^2 + (y - b)^2 = 1 \Rightarrow 2x + 2(y - b)y' = 0 \Rightarrow y' = \frac{x}{b - y}, \text{ Circle}$$

$$y = x^2 \Rightarrow y' = 2x, \text{ Parabola}$$

Equating:

$$2x = \frac{x}{b - y}$$

$$2(b - y) = 1$$

$$b - y = \frac{1}{2} \Rightarrow b = y + \frac{1}{2}$$

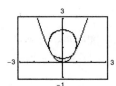

Also, $x^2 + (y - b)^2 = 1$ and $y = x^2$ imply:

$$y + (y - b)^2 = 1 \Rightarrow y + \left[y - \left(y + \frac{1}{2}\right)\right]^2 = 1 \Rightarrow y + \frac{1}{4} = 1 \Rightarrow y = \frac{3}{4} \text{ and } b = \frac{5}{4}$$

Center: $\left(0, \dfrac{5}{4}\right)$

Graph $y = x^2$ and $x^2 + \left(y - \dfrac{5}{4}\right)^2 = 1$.

**3. (a)**

| $f(x) = \cos x$ | $P_1(x) = a_0 + a_1 x$ |
|---|---|
| $f(0) = 1$ | $P_1(0) = a_0 \Rightarrow a_0 = 1$ |
| $f'(0) = 0$ | $P_1'(0) = a_1 \Rightarrow a_1 = 0$ |
| $P_1(x) = 1$ | |

**(b)**

| $f(x) = \cos x$ | $P_2(x) = a_0 + a_1 x + a_2 x^2$ |
|---|---|
| $f(0) = 1$ | $P_2(0) = a_0 \Rightarrow a_0 = 1$ |
| $f'(0) = 0$ | $P_2'(0) = a_1 \Rightarrow a_1 = 0$ |
| $f''(0) = -1$ | $P_2''(0) = 2a_2 \Rightarrow a_2 = -\frac{1}{2}$ |
| $P_2(x) = 1 - \frac{1}{2}x^2$ | |

**(c)**

| $x$ | $-1.0$ | $-0.1$ | $-0.001$ | $0$ | $0.001$ | $0.1$ | $1.0$ |
|---|---|---|---|---|---|---|---|
| $\cos x$ | $0.5403$ | $0.9950$ | $\approx 1$ | $1$ | $\approx 1$ | $0.9950$ | $0.5403$ |
| $P_2(x)$ | $0.5$ | $0.9950$ | $\approx 1$ | $1$ | $\approx 1$ | $0.9950$ | $0.5$ |

$P_2(x)$ is a good approximation of $f(x) = \cos x$ when $x$ is near $0$.

(d) $f(x) = \sin x$       $P_3(x) = a_0 + a_1 x + a_2 x^2 + a_3 x^3$

    $f(0) = 0$        $P_3(0) = a_0 \Rightarrow a_0 = 0$

    $f'(0) = 1$       $P_3'(0) = a_1 \Rightarrow a_1 = 1$

    $f''(0) = 0$      $P_3''(0) = 2a_2 \Rightarrow a_2 = 0$

    $f'''(0) = -1$     $P_3'''(0) = 6a_3 \Rightarrow a_3 = -\frac{1}{6}$

    $P_3(x) = x - \frac{1}{6}x^3$

**5.** Let $p(x) = Ax^3 + Bx^2 + Cx + D$

    $p'(x) = 3Ax^2 + 2Bx + C.$

At $(1, 1)$:                             At $(-1, -3)$:

$A + B + C + D = 1$    Equation 1         $A + B - C + D = -3$    Equation 3

$3A + 2B + C = 14$      Equation 2         $3A + 2B + C = -2$    Equation 4

Adding Equations 1 and 3: $2B + 2D = -2$

Subtracting Equations 1 and 3: $2A + 2C = 4$

Adding Equations 2 and 4: $6A + 2C = 12$

Subtracting Equations 2 and 4: $4B = 16$

So, $B = 4$ and $D = \frac{1}{2}(-2 - 2B) = -5$. Subtracting $2A + 2C = 4$ and $6A + 2C = 12$,

you obtain $4A = 8 \Rightarrow A = 2$. Finally, $C = \frac{1}{2}(4 - 2A) = 0$. So, $p(x) = 2x^3 + 4x^2 - 5$.

**7.** (a)     $x^4 = a^2 x^2 - a^2 y^2$

         $a^2 y^2 = a^2 x^2 - x^4$

             $y = \dfrac{\pm\sqrt{a^2 x^2 - x^4}}{a}$

    Graph: $y_1 = \dfrac{\sqrt{a^2 x^2 - x^4}}{a}$ and $y_2 = -\dfrac{\sqrt{a^2 x^2 - x^4}}{a}$.

(b)

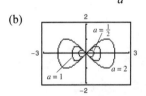

$(\pm a, 0)$ are the $x$-intercepts, along with $(0, 0)$.

(c) Differentiating implicitly:

       $4x^3 = 2a^2 x - 2a^2 y y'$

        $y' = \dfrac{2a^2 x - 4x^3}{2a^2 y}$

          $= \dfrac{x(a^2 - 2x^2)}{a^2 y} = 0 \Rightarrow 2x^2 = a^2 \Rightarrow x = \dfrac{\pm a}{\sqrt{2}}$

        $\left(\dfrac{a^2}{2}\right)^2 = a^2\left(\dfrac{a^2}{2}\right) - a^2 y^2$

           $\dfrac{a^4}{4} = \dfrac{a^4}{2} - a^2 y^2$

         $a^2 y^2 = \dfrac{a^4}{4}$

            $y^2 = \dfrac{a^2}{4}$

             $y = \pm\dfrac{a}{2}$

Four points: $\left(\dfrac{a}{\sqrt{2}}, \dfrac{a}{2}\right), \left(\dfrac{a}{\sqrt{2}}, -\dfrac{a}{2}\right), \left(-\dfrac{a}{\sqrt{2}}, \dfrac{a}{2}\right), \left(\dfrac{-a}{\sqrt{2}}, -\dfrac{a}{2}\right)$

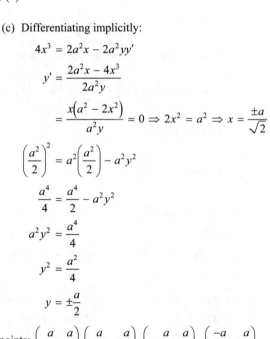

**9. (a)**

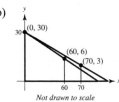

Not drawn to scale

Line determined by $(0, 30)$ and $(90, 6)$:

$$y - 30 = \frac{30 - 6}{0 - 90}(x - 0) = -\frac{24}{90}x = -\frac{4}{15}x \Rightarrow y = -\frac{4}{15}x + 30$$

When $x = 100$: $y = -\frac{4}{15}(100) + 30 = \frac{10}{3} > 3$

As you can see from the figure, the shadow determined by the man extends beyond the shadow determined by the child.

**(b)**

Line determined by $(0, 30)$ and $(60, 6)$:

$$y - 30 = \frac{30 - 6}{0 - 60}(x - 0) = -\frac{2}{5}x \Rightarrow y = -\frac{2}{5}x + 30$$

When $x = 70$: $y = -\frac{2}{5}(70) + 30 = 2 < 3$

Not drawn to scale

As you can see from the figure, the shadow determined by the child extends beyond the shadow determined by the man.

**(c)** Need $(0, 30), (d, 6), (d + 10, 3)$ collinear.

$$\frac{30 - 6}{0 - d} = \frac{6 - 3}{d - (d + 10)} \Rightarrow \frac{24}{d} = \frac{3}{10} \Rightarrow d = 80 \text{ feet}$$

**(d)** Let $y$ be the distance from the base of the street light to the tip of the shadow. You know that $dx/dt = -5$.

For $x > 80$, the shadow is determined by the man.

$$\frac{y}{30} = \frac{y - x}{6} \Rightarrow y = \frac{5}{4}x \text{ and } \frac{dy}{dt} = \frac{5}{4}\frac{dx}{dt} = \frac{-25}{4}$$

For $x < 80$, the shadow is determined by the child.

$$\frac{y}{30} = \frac{y - x - 10}{3} \Rightarrow y = \frac{10}{9}x + \frac{100}{9} \text{ and } \frac{dy}{dt} = \frac{10}{9}\frac{dx}{dt} = -\frac{50}{9}$$

Therefore:

$$\frac{dy}{dt} = \begin{cases} -\dfrac{25}{4}, & x > 80 \\ -\dfrac{50}{9}, & 0 < x < 80 \end{cases}$$

$dy/dt$ is not continuous at $x = 80$.

**ALTERNATE SOLUTION for parts (a) and (b):**

**(a)** As before, the line determined by the man's shadow is

$$y_m = -\frac{4}{15}x + 30$$

The line determined by the child's shadow is obtained by finding the line through $(0, 30)$ and $(100, 3)$:

$$y - 30 = \frac{30 - 3}{0 - 100}(x - 0) \Rightarrow y_c = -\frac{27}{100}x + 30$$

By setting $y_m = y_c = 0$, you can determine how far the shadows extend:

Man: $y_m = 0 \Rightarrow \dfrac{4}{15}x = 30 \Rightarrow x = 112.5 = 112\dfrac{1}{2}$

Child: $y_c = 0 \Rightarrow \dfrac{27}{100}x = 30 \Rightarrow x = 111.\overline{11} = 111\dfrac{1}{9}$

The man's shadow is $112\dfrac{1}{2} - 111\dfrac{1}{9} = 1\dfrac{7}{18}$ ft beyond the child's shadow.

(b) As before, the line determined by the man's shadow is

$$y_m = -\frac{2}{5}x + 30$$

For the child's shadow,

$$y - 30 = \frac{30 - 3}{0 - 70}(x - 0) \Rightarrow y_c = -\frac{27}{70}x + 30$$

Man: $y_m = 0 \Rightarrow \frac{2}{5}x = 30 \Rightarrow x = 75$

Child: $y_c = 0 \Rightarrow \frac{27}{70}x = 30 \Rightarrow x = \frac{700}{9} = 77\frac{7}{9}$

So the child's shadow is $77\frac{7}{9} - 75 = 2\frac{7}{9}$ ft beyond the man's shadow.

**11.** (a) $v(t) = -\frac{27}{5}t + 27$ ft/sec

$a(t) = -\frac{27}{5}$ ft/sec$^2$

(b) $v(t) = -\frac{27}{5}t + 27 = 0 \Rightarrow \frac{27}{5}t = 27 \Rightarrow t = 5$ seconds

$S(5) = -\frac{27}{10}(5)^2 + 27(5) + 6 = 73.5$ feet

(c) The acceleration due to gravity on Earth is greater in magnitude than that on the moon.

**13.** $f(x) = \dfrac{a + bx}{1 + cx}$

$f(0) = a = e^0 = 1 \Rightarrow a = 1$

$f'(x) = \dfrac{(1 + cx)(b) - (a + bx)c}{(1 + cx)^2} = \dfrac{b - ac}{(1 + cx)^2}$

$f'(0) = b - ac = 1 \Rightarrow b = 1 + c$

$f''(x) = \dfrac{(1 + cx)^2(0) - (b - ac)2c(1 + cx)}{(1 + cx)^4} = \dfrac{2c(ac - b)}{(1 + cx)^3}$

$f''(0) = 2c(ac - b) = 2c(c - (1 + c)) = 2c(-1) = 1 \Rightarrow c = -\dfrac{1}{2}$

So, $b = 1 + c = 1 - \dfrac{1}{2} = \dfrac{1}{2}$.

$$f(x) = \frac{1 + \dfrac{1}{2}x}{1 - \dfrac{1}{2}x}$$

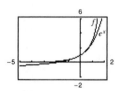

**15.** $j(t) = a'(t)$

(a) $j(t)$ is the rate of change of acceleration.

(b) $s(t) = -8.25t^2 + 66t$

$v(t) = -16.5t + 66$

$a(t) = -16.5$

$a'(t) = j(t) = 0$

The acceleration is constant, so $j(t) = 0$.

(c) $a$ is position.

$b$ is acceleration.

$c$ is jerk.

$d$ is velocity.

# C H A P T E R  4
# Applications of Differentiation

# C H A P T E R   4
## Applications of Differentiation

### Section 4.1  Extrema on an Interval

**1.** $f(x) = \dfrac{x^2}{x^2 + 4}$

$f'(x) = \dfrac{(x^2 + 4)(2x) - (x^2)(2x)}{(x^2 + 4)^2} = \dfrac{8x}{(x^2 + 4)^2}$

$f'(0) = 0$

**3.** $f(x) = x + \dfrac{4}{x^2} = x + 4x^{-2}$

$f'(x) = 1 - 8x^{-3} = 1 - \dfrac{8}{x^3}$

$f'(2) = 0$

**5.** $f(x) = (x + 2)^{2/3}$

$f'(x) = \tfrac{2}{3}(x + 2)^{-1/3}$

$f'(-2)$ is undefined.

**7.** Critical number: $x = 2$

$x = 2$: absolute maximum (and relative maximum)

**9.** Critical numbers: $x = 1, 2, 3$

$x = 1, 3$: absolute maxima (and relative maxima)

$x = 2$: absolute minimum (and relative minimum)

**11.** $f(x) = x^3 - 3x^2$

$f'(x) = 3x^2 - 6x = 3x(x - 2)$

Critical numbers: $x = 0, 2$

**13.** $g(t) = t\sqrt{4 - t}, \ t < 3$

$g'(t) = t\left[\dfrac{1}{2}(4 - t)^{-1/2}(-1)\right] + (4 - t)^{1/2}$

$\quad = \dfrac{1}{2}(4 - t)^{-1/2}\left[-t + 2(4 - t)\right]$

$\quad = \dfrac{8 - 3t}{2\sqrt{4 - t}}$

Critical number: $t = \dfrac{8}{3}$

**15.** $h(x) = \sin^2 x + \cos x, \ 0 < x < 2\pi$

$h'(x) = 2\sin x \cos x - \sin x = \sin x(2 \cos x - 1)$

Critical numbers in $(0, 2\pi)$: $x = \dfrac{\pi}{3}, \pi, \dfrac{5\pi}{3}$

**17.** $f(t) = te^{-2t}$

$f'(t) = e^{-2t} - 2te^{-2t} = e^{-2t}(1 - 2t)$

Critical number: $t = \tfrac{1}{2}$

**19.** $f(x) = x^2\log_2(x^2 + 1) = x^2\dfrac{\ln(x^2 + 1)}{\ln 2}$

$f'(x) = 2x\dfrac{\ln(x^2 + 1)}{\ln 2} + x^2\dfrac{2x}{(\ln 2)(x^2 + 1)}$

$\quad = \dfrac{2x}{\ln 2}\left[\ln(x^2 + 1) + \dfrac{x^2}{x^2 + 1}\right]$

Critical number: $x = 0$

**21.** $f(x) = 3 - x, \quad [-1, 2]$

$f'(x) = -1 \Rightarrow$ no critical numbers

Left endpoint: $(-1, 4)$ Maximum

Right endpoint: $(2, 1)$ Minimum

**23.** $g(x) = 2x^2 - 8x, \ [0, 6]$

$g'(x) = 4x - 8 = 4(x - 2)$

Critical number: $x = 2$

Left endpoint: $(0, 0)$

Critical number: $(2, -8)$ Minimum

Right endpoint: $(6, 24)$ Maximum

**25.** $f(x) = x^3 - \tfrac{3}{2}x^2, [-1, 2]$

$f'(x) = 3x^2 - 3x = 3x(x - 1)$

Left endpoint: $\left(-1, -\tfrac{5}{2}\right)$ Minimum

Right endpoint: $(2, 2)$ Maximum

Critical number: $(0, 0)$

Critical number: $\left(1, -\tfrac{1}{2}\right)$

**27.** $f(x) = 3x^{2/3} - 2x$, $[-1, 1]$

$$f'(x) = 2x^{-1/3} - 2 = \frac{2\left(1 - \sqrt[3]{x}\right)}{\sqrt[3]{x}}$$

Left endpoint: $(-1, 5)$ Maximum

Critical number: $(0, 0)$ Minimum

Right endpoint: $(1, 1)$

**29.** $h(s) = \dfrac{1}{s - 2} = (s - 2)^{-1}$, $[0, 1]$

$$h'(s) = \frac{-1}{(s - 2)^2}$$

Left endpoint: $\left(0, -\dfrac{1}{2}\right)$ Maximum

Right endpoint: $(1, -1)$ Minimum

**31.** $y = 3 - |t - 3|$, $[-1, 5]$

For $x < 3$, $y = 3 + (t - 3) = t$

and $y' = 1 \neq 0$ on $[-1, 3)$

For $x > 3$, $y = 3 - (t - 3) = 6 - t$

and $y' = -1 \neq 0$ on $(3, 5]$

So, $x = 3$ is the only critical number.

Left endpoint: $(-1, -1)$ Minimum

Right endpoint: $(5, 1)$

Critical number: $(3, 3)$ Maximum

**33.** $f(x) = [\![x]\!]$, $[-2, 2]$

From the graph of $f$, you see that the maximum value of $f$ is 2 for $x = 2$, and the minimum value is $-2$ for $-2 \leq x < -1$.

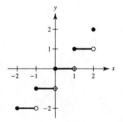

**35.** $f(x) = \sin x$, $\left[\dfrac{5\pi}{6}, \dfrac{11\pi}{6}\right]$

$$f'(x) = \cos x$$

Critical number: $x = \dfrac{3\pi}{2}$

Left endpoint: $\left(\dfrac{5\pi}{6}, \dfrac{1}{2}\right)$ Maximum

Critical number: $\left(\dfrac{3\pi}{2}, -1\right)$ Minimum

Right endpoint: $\left(\dfrac{11\pi}{6}, -\dfrac{1}{2}\right)$

**37.** $y = 3 \cos x$, $[0, 2\pi]$

$$y' = -3 \sin x$$

Critical number in $(0, 2\pi)$: $x = \pi$

Left endpoint: $(0, 3)$ Maximum

Critical number: $(\pi, -3)$ Minimum

Right endpoint: $(2\pi, 3)$ Maximum

**39.** $f(x) = \arctan x^2$, $[-2, 1]$

$$f'(x) = \frac{2x}{1 + x^4}$$

Critical number: $x = 0$

Left endpoint: $(-2, \arctan 4) \approx (-2, 1.326)$ Maximum

Right endpoint: $(1, \arctan 1) = \left(1, \dfrac{\pi}{4}\right) \approx (1, 0.785)$

Critical number: $(0, 0)$ Minimum

**41.** $h(x) = 5e^x - e^{2x}$, $[-1, 2]$

$$h'(x) = 5e^x - 2e^{2x} = e^x\left(5 - 2e^x\right)$$

$$5 - 2e^x = 0 \Rightarrow e^x = \frac{5}{2} \Rightarrow x = \ln\left(\frac{5}{2}\right) \approx 0.916$$

Critical number: $x = \ln\left(\dfrac{5}{2}\right)$

Left endpoint: $\left(-1, \dfrac{5}{e} - \dfrac{1}{e^2}\right) \approx (-1, 1.704)$

Right endpoint: $\left(2, 5e^2 - e^4\right) \approx (2, -17.653)$ Minimum

Critical number: $\left(\ln\left(\dfrac{5}{2}\right), \dfrac{25}{4}\right)$ Maximum

**Note:** $h\left(\ln\left(\dfrac{5}{2}\right)\right) = 5e^{\ln(5/2)} - e^{2\ln(5/2)}$

$$= 5\left(\frac{5}{2}\right) - \left(\frac{5}{2}\right)^2 = \frac{25}{4}$$

**43.** $y = e^x \sin x$, $[0, \pi]$

$$y' = e^x \sin x + e^x \cos x = e^x(\sin x + \cos x)$$

Left endpoint: $(0, 0)$ Minimum

Critical number:

$$\left(\frac{3\pi}{4}, \frac{\sqrt{2}}{2}e^{3\pi/4}\right) \approx \left(\frac{3\pi}{4}, 7.46\right)$$ Maximum

Right endpoint: $(\pi, 0)$ Minimum

**45.** $f(x) = 2x - 3$

  (a)  Minimum: $(0, -3)$

       Maximum: $(2, 1)$

  (b)  Minimum: $(0, -3)$

  (c)  Maximum: $(2, 1)$

  (d)  No extrema

**47.** $f(x) = \dfrac{3}{x - 1}, \quad (1, 4]$

Right endpoint: $(4, 1)$ Minimum

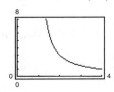

**49.** $f(x) = \sqrt{x + 4}\,e^{x^2/10}, \quad [-2, 2]$

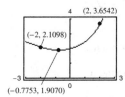

$$f'(x) = \frac{\left(2x^2 + 8x + 5\right)e^{x^2/10}}{10\sqrt{x + 4}}$$

Right endpoint: $(2, 3.6542)$ Maximum

Critical point: $(-0.7753, 1.9070)$ Minimum

**51.** (a)

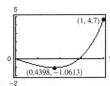

    Minimum: $(0.4398, -1.0613)$

  (b) $\qquad f(x) = 3.2x^5 + 5x^3 - 3.5x, \quad [0, 1]$

$\qquad\qquad f'(x) = 16x^4 + 15x^2 - 3.5$

$16x^4 + 15x^2 - 3.5 = 0$

$$x^2 = \frac{-15 \pm \sqrt{(15)^2 - 4(16)(-3.5)}}{2(16)} = \frac{-15 \pm \sqrt{449}}{32}$$

$$x = \sqrt{\frac{-15 + \sqrt{449}}{32}} \approx 0.4398$$

Left endpoint: $(0, 0)$

Critical point: $(0.4398, -1.0613)$ Minimum

Right endpoint: $(1, 4.7)$ Maximum

**53.** (a)

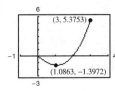

    Minimum: $(1.0863, -1.3972)$

  (b) $f(x) = \left(x^2 - 2x\right)\ln(x + 3), \quad [0, 3]$

$$f'(x) = \left(x^2 - 2x\right) \cdot \frac{1}{x + 3} + (2x - 2)\ln(x + 3) = \frac{x^2 - 2x + \left(2x^2 + 4x - 6\right)\ln(x + 3)}{x + 3}$$

Left endpoint: $(0, 0)$

Critical point: $(1.0863, -1.3972)$ Minimum

Right endpoint: $(3, 5.3753)$ Maximum

**55.** $f(x) = \left(1 + x^3\right)^{1/2}, \quad [0, 2]$

$f'(x) = \frac{3}{2}x^2\left(1 + x^3\right)^{-1/2}$

$f''(x) = \frac{3}{4}\left(x^4 + 4x\right)\left(1 + x^3\right)^{-3/2}$

$f'''(x) = -\frac{3}{8}\left(x^6 + 20x^3 - 8\right)\left(1 + x^3\right)^{-5/2}$

Setting $f''' = 0$, you have $x^6 + 20x^3 - 8 = 0$.

$x^3 = \dfrac{-20 \pm \sqrt{400 - 4(1)(-8)}}{2}$

$x = \sqrt[3]{-10 \pm \sqrt{108}} = \sqrt{3} - 1$

In the interval $[0, 2]$, choose

$x = \sqrt[3]{-10 \pm \sqrt{108}} = \sqrt{3} - 1 \approx 0.732.$

$\left| f''\left(\sqrt[3]{-10 + \sqrt{108}}\right) \right| \approx 1.47$ is the maximum value.

**57.** $f(x) = e^{-x^2/2}, [0, 1]$

$f'(x) = -xe^{-x^2/2}$

$f''(x) = -x\left(-xe^{-x^2/2}\right) - e^{-x^2/2}$

$\quad = e^{-x^2/2}\left(x^2 - 1\right)$

$f'''(x) = e^{-x^2/2}(2x) + \left(x^2 - 1\right)\left(-xe^{-x^2/2}\right)$

$\quad = xe^{-x^2/2}\left(3 - x^2\right)$

$\left| f''(0) \right| = 1$ is the maximum value.

**59.** $f(x) = (x + 1)^{2/3}, \quad [0, 2]$

$f'(x) = \frac{2}{3}(x + 1)^{-1/3}$

$f''(x) = -\frac{2}{9}(x + 1)^{-4/3}$

$f'''(x) = \frac{8}{27}(x + 1)^{-7/3}$

$f^{(4)}(x) = -\frac{56}{81}(x + 1)^{-10/3}$

$f^{(5)}(x) = \frac{560}{243}(x + 1)^{-13/3}$

$\left| f^{(4)}(0) \right| = \frac{56}{81}$ is the maximum value.

**61.** $f(x) = \tan x$

$f$ is continuous on $[0, \pi/4]$ but not on $[0, \pi]$.

$\displaystyle\lim_{x \to (\pi/2)^-} \tan x = \infty.$

**63.**

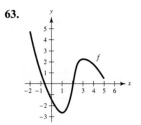

**65.** (a) Yes

(b) No

**67.** (a) No

(b) Yes

**69.** $P = VI - RI^2 = 12I - 0.5I^2, 0 \le I \le 15$

$P = 0$ when $I = 0$.

$P = 67.5$ when $I = 15$.

$P' = 12 - I = 0$

Critical number: $I = 12$ amps

When $I = 12$ amps, $P = 72$, the maximum output.

No, a 20-amp fuse would not increase the power output. $P$ is decreasing for $I > 12$.

**71.**
$$S = 6hs + \frac{3s^2}{2}\left(\frac{\sqrt{3} - \cos \theta}{\sin \theta}\right), \frac{\pi}{6} \le \theta \le \frac{\pi}{2}$$

$$\frac{dS}{d\theta} = \frac{3s^2}{2}\left(-\sqrt{3}\csc \theta \cot \theta + \csc^2 \theta\right)$$

$$= \frac{3s^2}{2}\csc \theta\left(-\sqrt{3}\cot \theta + \csc \theta\right) = 0$$

$$\csc \theta = \sqrt{3}\cot \theta$$

$$\sec \theta = \sqrt{3}$$

$$\theta = \operatorname{arcsec}\sqrt{3} \approx 0.9553 \text{ radians}$$

$$S\left(\frac{\pi}{6}\right) = 6hs + \frac{3s^2}{2}\left(\sqrt{3}\right)$$

$$S\left(\frac{\pi}{6}\right) = 6hs + \frac{3s^2}{2}\left(\sqrt{3}\right)$$

$$S\left(\operatorname{arcsec}\sqrt{3}\right) = 6hs + \frac{3s^2}{2}\left(\sqrt{2}\right)$$

$S$ is minimum when $\theta = \operatorname{arcsec}\sqrt{3} \approx 0.9553$ radian.

**73.** True. See Exercise 37.

**75.** True

**77.** If $f$ has a maximum value at $x = c$, then $f(c) \ge f(x)$ for all $x$ in $I$. So, $-f(c) \le -f(x)$ for all $x$ in $I$. So, $-f$ has a minimum value at $x = c$.

**79.** First do an example: Let $a = 4$ and $f(x) = 4$.

Then $R$ is the square $0 \le x \le 4, 0 \le y \le 4$.

Its area and perimeter are both $k = 16$.

Claim that all real numbers $a > 2$ work. On the one hand, if $a > 2$ is given, then let $f(x) = 2a/(a - 2)$.

Then the rectangle

$$R = \left\{(x, y): 0 \le x \le a, 0 \le y \le \frac{2a}{a - 2}\right\} \text{ has } k = \frac{2a^2}{a - 2}:$$

$$\text{Area} = a\left(\frac{2a}{a - 2}\right) = \frac{2a^2}{a - 2}$$

$$\text{Perimeter} = 2a + 2\left(\frac{2a}{a - 2}\right) = \frac{2a(a - 2) + 2(2a)}{a - 2} = \frac{2a^2}{a - 2}.$$

To see that $a$ must be greater than 2, consider

$$R = \{(x, y): 0 \le x \le a, 0 \le y \le f(x)\}.$$

$f$ attains its maximum value on $[0, a]$ at some point $P(x_0, y_0)$, as indicated in the figure.

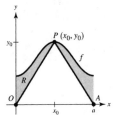

Draw segments $\overline{OP}$ and $\overline{PA}$. The region $R$ is bounded by the rectangle $0 \le x \le a, 0 \le y \le y_0$, so $\text{area}(R) = k \le ay_0$.

Furthermore, from the figure, $y_0 < \overline{OP}$ and $y_0 < \overline{PA}$. So, $k = \text{Perimeter}(R) > \overline{OP} + \overline{PA} > 2y_0$. Combining,

$$2y_0 < k \le ay_0 \Rightarrow a > 2.$$

# Section 4.2   Rolle's Theorem and the Mean Value Theorem

**1.** $f(x) = \left| \dfrac{1}{x} \right|$

$f(-1) = f(1) = 1$. But, $f$ is not continuous on $[-1, 1]$.

**3.** Rolle's Theorem does not apply to $f(x) = 1 - |x - 1|$ over $[0, 2]$ because $f$ is not differentiable at $x = 1$.

**5.** $f(x) = x^2 - x - 2 = (x - 2)(x + 1)$

$x$-intercepts: $(-1, 0), (2, 0)$

$f'(x) = 2x - 1 = 0$ at $x = \dfrac{1}{2}$.

**7.** $f(x) = x\sqrt{x + 4}$

$x$-intercepts: $(-4, 0), (0, 0)$

$f'(x) = x\dfrac{1}{2}(x + 4)^{-1/2} + (x + 4)^{1/2}$

$\quad = (x + 4)^{-1/2}\left( \dfrac{x}{2} + (x + 4) \right)$

$f'(x) = \left( \dfrac{3}{2}x + 4 \right)(x + 4)^{-1/2} = 0$ at $x = -\dfrac{8}{3}$

**9.** $f(x) = -x^2 + 3x, \quad [0, 3]$

$f(0) = -(0)^2 + 3(0)$

$f(3) = -(3)^2 + 3(3) = 0$

$f$ is continuous on $[0, 3]$ and differentiable on $(0, 3)$. Rolle's Theorem applies.

$f'(x) = -2x + 3 = 0$

$\qquad -2x = -3 \Rightarrow x = \dfrac{3}{2}$

$c$-value: $\dfrac{3}{2}$

**11.** $f(x) = (x - 1)(x - 2)(x - 3), [1, 3]$

$f(1) = (1 - 1)(1 - 2)(1 - 3) = 0$

$f(3) = (3 - 1)(3 - 2)(3 - 3) = 0$

$f$ is continuous on $[1, 3]$. $f$ is differentiable on $(1, 3)$. Rolle's Theorem applies.

$f(x) = x^3 - 6x^2 + 11x - 6$

$f'(x) = 3x^2 - 12x + 11 = 0$

$\qquad\qquad x = \dfrac{6 \pm \sqrt{3}}{3}$

$c$-values: $\dfrac{6 - \sqrt{3}}{3}, \dfrac{6 + \sqrt{3}}{3}$

**13.**  $f(x) = x^{2/3} - 1, [-8, 8]$

$f(-8) = (-8)^{2/3} - 1 = 3$

$f(8) = (8)^{2/3} - 1 = 3$

$f$ is continuous on $[-8, 8]$. $f$ is not differentiable

on $(-8, 8)$ because $f'(0)$ does not exist. Rolle's Theorem does not apply.

**15.**  $f(x) = \dfrac{x^2 - 2x}{x + 2}, \quad [-1, 6]$

$f(-1) = \dfrac{1 + 2}{1} = 3$

$f(6) = \dfrac{36 - 12}{8} = 3$

$f$ is continuous on $[-1, 6]$. $f$ is differentiable on $(-1, 6)$.  Rolle's Theorem applies.

$f'(x) = \dfrac{(x + 2)(2x - 2) - (x^2 - 2x)(1)}{(x + 2)^2} = \dfrac{2x^2 + 4x - 2x - 4 - x^2 + 2x}{(x + 2)^2} = \dfrac{x^2 + 4x - 4}{(x + 2)^2}$

$f'(x) = x^2 + 4x - 4 = 0 \Rightarrow x = \dfrac{-4 \pm\sqrt{16 + 16}}{2} = -2 + 2\sqrt{2}$

$\left( \text{Note: } -2 - 2\sqrt{2} \text{ is not in the interval.} \right)$

$c$-value: $-2 + 2\sqrt{2}$

**17.** $f(x) = \sin x, [0, 2\pi]$

$f(0) = \sin 0 = 0$

$f(2\pi) = \sin(2\pi) = 0$

$f$ is continuous on $[0, 2\pi]$. $f$ is differentiable on $(0, 2\pi)$. Rolle's Theorem applies.

$f'(x) = \cos x = 0 \Rightarrow x = \dfrac{\pi}{2}, \dfrac{3\pi}{2}$

$c$-values: $\dfrac{\pi}{2}, \dfrac{3\pi}{2}$

**19.** $f(x) = \tan x, [0, \pi]$

$f(0) = \tan 0 = 0$

$f(\pi) = \tan \pi = 0$

$f$ is not continuous on $[0, \pi]$ because $f(\pi/2)$ does not exist. Rolle's Theorem does not apply.

**21.** $f(x) = (x^2 - 2x)e^x, [0, 2]$

$f(0) = f(2) = 0$

$f$ is continuous on $[0, 2]$ and differentiable on $(0, 2)$, so Rolle's Theorem applies.

$f'(x) = (x^2 - 2x)e^x + (2x - 2)e^x = e^x(x^2 - 2)$

$= 0 \Rightarrow x = \sqrt{2}$

$c$-value: $\sqrt{2} \approx 1.414$

**23.** $f(x) = |x| - 1, [-1, 1]$

$f(-1) = f(1) = 0$

$f$ is continuous on $[-1, 1]$. $f$ is not differentiable on $(-1, 1)$ because $f'(0)$ does not exist. Rolle's Theorem does not apply.

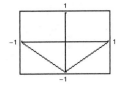

**25.** $f(x) = x - \tan \pi x, \left[-\frac{1}{4}, \frac{1}{4}\right]$

$f\left(-\frac{1}{4}\right) = -\frac{1}{4} + 1 = \frac{3}{4}$

$f\left(\frac{1}{4}\right) = \frac{1}{4} - 1 = -\frac{3}{4}$

Rolle's Theorem does not apply.

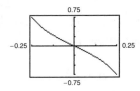

**27.** $f(x) = 2 + \arcsin(x^2 - 1), [-1, 1]$

$f(-1) = f(1) = 2$

$f'(x) = \dfrac{2x}{\sqrt{1 - (x^2 - 1)^2}} = \dfrac{2x}{\sqrt{2x^2 - x^4}}$

$f'(0)$ does not exist. Rolle's Theorem does not apply.

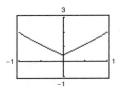

**29.** $f(t) = -16t^2 + 48t + 6$

(a) $f(1) = f(2) = 38$

(b) $v = f'(t)$ must be 0 at some time in $(1, 2)$.

$f'(t) = -32t + 48 = 0$

$t = \frac{3}{2}$ sec

**31.**

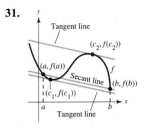

**33.** $f$ is not continuous on the interval $[0, 6]$. ($f$ is not continuous at $x = 2$.)

**35.** $f(x) = \dfrac{1}{x - 3}, [0, 6]$

$f$ has a discontinuity at $x = 3$.

**37.** $f(x) = -x^2 + 5$

   **(a)** Slope $= \dfrac{1-4}{2+1} = -1$

      Secant line:     $y - 4 = -(x+1)$

                              $y = -x + 3$

                    $x + y - 3 = 0$

   **(b)** $f'(x) = -2x = -1 \;\Rightarrow\; x = c = \dfrac{1}{2}$

   **(c)** $f(c) = f\!\left(\dfrac{1}{2}\right) = -\dfrac{1}{4} + 5 = \dfrac{19}{4}$

      Tangent line:     $y - \dfrac{19}{4} = -\left(x - \dfrac{1}{2}\right)$

                           $4y - 19 = -4x + 2$

                    $4x + 4y - 21 = 0$

   **(d)**

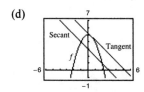

**39.** $f(x) = x^2$ is continuous on $[-2, 1]$ and differentiable on $(-2, 1)$.

$$\frac{f(1) - f(-2)}{1 - (-2)} = \frac{1 - 4}{3} = -1$$

$$f'(x) = 2x = -1$$

$$x = -\frac{1}{2}$$

$$c = -\frac{1}{2}$$

**41.** $f(x) = x^3 + 2x$ is continuous on $[-1, 1]$ and differentiable on $(-1, 1)$.

$$\frac{f(1) - f(-1)}{1 - (-1)} = \frac{3 - (-3)}{2} = 3$$

$$f'(x) = 3x^2 + 2 = 3$$

$$3x^2 = 1$$

$$x = \pm\frac{1}{\sqrt{3}}$$

$$c = \pm\frac{\sqrt{3}}{3}$$

**43.** $f(x) = x^{2/3}$ is continuous on $[0, 1]$ and differentiable on $(0, 1)$.

$$\frac{f(1) - f(0)}{1 - 0} = 1$$

$$f'(x) = \frac{2}{3}x^{-1/3} = 1$$

$$x = \left(\frac{2}{3}\right)^3 = \frac{8}{27}$$

$$c = \frac{8}{27}$$

**45.** $f(x) = |2x + 1|$ is not differentiable at $x = -1/2$. The Mean Value Theorem does not apply.

**47.** $f(x) = \sin x$ is continuous on $[0, \pi]$ and differentiable on $(0, \pi)$.

$$\frac{f(\pi) - f(0)}{\pi - 0} = \frac{0 - 0}{\pi} = 0$$

$$f'(x) = \cos x = 0$$

$$x = \pi/2$$

$$c = \frac{\pi}{2}$$

**49.** $f(x) = \cos x + \tan x$ is not continuous at $x = \pi/2$. The Mean Value Theorem does not apply.

**51.** $f(x) = x \log_2 x = x\dfrac{\ln x}{\ln 2}$

  $f$ is continuous on $[1, 2]$ and differentiable on $(1, 2)$.

$$\frac{f(2) - f(1)}{2 - 1} = \frac{2 - 0}{2 - 1} = 2$$

$$f'(x) = x\frac{1}{x \ln 2} + \frac{\ln x}{\ln 2} = \frac{1 + \ln x}{\ln 2} = 2$$

$$1 + \ln x = 2\ln 2 = \ln 4$$

$$xe = 4$$

$$x = \frac{4}{e}$$

$$c = \frac{4}{e}$$

**53.** $f(x) = \dfrac{x}{x+1}, \left[-\dfrac{1}{2}, 2\right]$

(a)–(c)

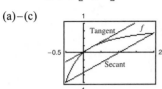

(b) Secant line:

$$\text{slope} = \frac{f(2) - f(-1/2)}{2 - (-1/2)} = \frac{2/3 - (-1)}{5/2} = \frac{2}{3}$$

$$y - \frac{2}{3} = \frac{2}{3}(x - 2)$$

$$y = \frac{2}{3}(x - 1)$$

(c) $f'(x) = \dfrac{1}{(x+1)^2} = \dfrac{2}{3}$

$$(x+1)^2 = \frac{3}{2}$$

$$x = -1 \pm \sqrt{\frac{3}{2}} = -1 \pm \frac{\sqrt{6}}{2}$$

In the interval $[-1/2, 2]$: $c = -1 + \left(\sqrt{6}/2\right)$

$$f(c) = \frac{-1 + \left(\sqrt{6}/2\right)}{\left[-1 + \left(\sqrt{6}/2\right)\right] + 1}$$

$$= \frac{-2 + \sqrt{6}}{\sqrt{6}}$$

$$= \frac{-2}{\sqrt{6}} + 1$$

Tangent line: $y - 1 + \dfrac{2}{\sqrt{6}} = \dfrac{2}{3}\left(x - \dfrac{\sqrt{6}}{2} + 1\right)$

$$y - 1 + \frac{\sqrt{6}}{3} = \frac{2}{3}x - \frac{\sqrt{6}}{3} + \frac{2}{3}$$

$$y = \frac{1}{3}\left(2x + 5 - 2\sqrt{6}\right)$$

**55.** $f(x) = \sqrt{x}, [1, 9]$

(a)–(c)

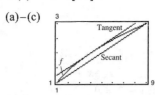

(b) Secant line:

$$\text{slope} = \frac{f(9) - f(1)}{9 - 1} = \frac{3 - 1}{8} = \frac{1}{4}$$

$$y - 1 = \frac{1}{4}(x - 1)$$

$$y = \frac{1}{4}x + \frac{3}{4}$$

(c) $f'(x) = \dfrac{1}{2\sqrt{x}} = \dfrac{1}{4}$

$$x = c = 4$$

$$f(4) = 2$$

Tangent line: $y - 2 = \dfrac{1}{4}(x - 4)$

$$y = \frac{1}{4}x + 1$$

**57.** $f(x) = 2e^{x/4} \cos \dfrac{\pi x}{4}, \; 0 \le x \le 2$

(a)–(c)

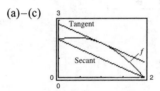

(b) Secant line:

$$\text{slope} = \frac{f(2) - f(0)}{2 - 0} = \frac{0 - 2}{2 - 0} = -1$$

$$y - 2 = -1(x - 0)$$

$$y = -x + 2$$

(c) $f'(x) = 2\left(\dfrac{1}{4}e^{x/4} \cos \dfrac{\pi x}{4}\right) + 2e^{x/4}\left(-\sin \dfrac{\pi x}{4}\right)\dfrac{\pi}{4}$

$$= e^{x/4}\left[\frac{1}{2} \cos \frac{\pi x}{4} - \frac{\pi}{2} \sin \frac{\pi x}{4}\right]$$

$$f'(c) = -1 \Rightarrow c \approx 1.0161, f(c) \approx 1.8$$

Tangent line: $y - 1.8 = -1(x - 1.0161)$

$$y = -x + 2.8161$$

**59.** $s(t) = -4.9t^2 + 300$

(a) $v_{avg} = \dfrac{s(3) - s(0)}{3 - 0} = \dfrac{255.9 - 300}{3} = -14.7$ m/sec

(b) $s(t)$ is continuous on $[0, 3]$ and differentiable on $(0, 3)$. Therefore, the Mean Value Theorem applies.

$$v(t) = s'(t) = -9.8t = -14.7 \text{ m/sec}$$

$$t = \dfrac{-14.7}{-9.8} = 1.5 \text{ sec}$$

**61.** No. Let $f(x) = x^2$ on $[-1, 2]$.

$$f'(x) = 2x$$

$f'(0) = 0$ and zero is in the interval $(-1, 2)$ but $f(-1) \neq f(2)$.

**63.** $f(x) = \begin{cases} 0, & x = 0 \\ 1 - x, & 0 < x \leq 1 \end{cases}$

No, this does not contradict Rolle's Theorem. $f$ is not continuous on $[0, 1]$.

**65.** Let $S(t)$ be the position function of the plane. If $t = 0$ corresponds to 2 P.M., $S(0) = 0$, $S(5.5) = 2500$ and the Mean Value Theorem says that there exists a time $t_0$, $0 < t_0 < 5.5$, such that

$$S'(t_0) = v(t_0) = \dfrac{2500 - 0}{5.5 - 0} \approx 454.54.$$

Applying the Intermediate Value Theorem to the velocity function on the intervals $[0, t_0]$ and $[t_0, 5.5]$, you see that there are at least two times during the flight when the speed was 400 miles per hour. $(0 < 400 < 454.54)$

**67.** Let $S(t)$ be the difference in the positions of the 2 bicyclists, $S(t) = S_1(t) - S_2(t)$. Because $S(0) = S(2.25) = 0$, there must exist a time $t_0 \in (0, 2.25)$ such that $S'(t_0) = v(t_0) = 0$. At this time, $v_1(t_0) = v_2(t_0)$.

**69.** $f(x) = 3 \cos^2\left(\dfrac{\pi x}{2}\right)$, $f'(x) = 6 \cos\left(\dfrac{\pi x}{2}\right)\left(-\sin\left(\dfrac{\pi x}{2}\right)\right)\left(\dfrac{\pi}{2}\right)$

$$= -3\pi \cos\left(\dfrac{\pi x}{2}\right) \sin\left(\dfrac{\pi x}{2}\right)$$

(a)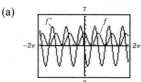

(b) $f$ and $f'$ are both continuous on the entire real line.

(c) Because $f(-1) = f(1) = 0$, Rolle's Theorem applies on $[-1, 1]$. Because $f(1) = 0$ and $f(2) = 3$, Rolle's Theorem does not apply on $[1, 2]$.

(d) $\lim\limits_{x \to 3^-} f'(x) = 0$

$\lim\limits_{x \to 3^+} f'(x) = 0$

**71.** $f$ is continuous on $[-5, 5]$ and does not satisfy the conditions of the Mean Value Theorem. $\Rightarrow$ $f$ is not differentiable on $(-5, 5)$. Example: $f(x) = |x|$

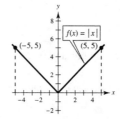

**73.** $f(x) = x^5 + x^3 + x + 1$

$f$ is differentiable for all $x$.

$f(-1) = -2$ and $f(0) = 1$, so the Intermediate Value Theorem implies that $f$ has at least one zero $c$ in $[-1, 0]$, $f(c) = 0$.

Suppose $f$ had 2 zeros, $f(c_1) = f(c_2) = 0$. Then Rolle's Theorem would guarantee the existence of a number $a$ such that

$$f'(a) = f(c_2) - f(c_1) = 0.$$

But, $f'(x) = 5x^4 + 3x^2 + 1 > 0$ for all $x$. So, $f$ has exactly one real solution.

**75.** $f(x) = 3x + 1 - \sin x$

$f$ is differentiable for all $x$.

$f(-\pi) = -3\pi + 1 < 0$ and $f(0) = 1 > 0$, so the Intermediate Value Theorem implies that $f$ has at least one zero $c$ in $[-\pi, 0]$, $f(c) = 0$.

Suppose $f$ had 2 zeros, $f(c_1) = f(c_2) = 0$. Then Rolle's Theorem would guarantee the existence of a number $a$ such that

$f'(a) = f(c_2) - f(c_1) = 0$.

But $f'(x) = 3 - \cos x > 0$ for all $x$. So, $f(x) = 0$ has exactly one real solution.

**77.** $f'(x) = 0$

$f(x) = c$

$f(2) = 5$

So, $f(x) = 5$.

**79.** $f'(x) = 2x$

$f(x) = x^2 + c$

$f(1) = 0 \Rightarrow 0 = 1 + c \Rightarrow c = -1$

So, $f(x) = x^2 - 1$.

**81.** False. $f(x) = 1/x$ has a discontinuity at $x = 0$.

**83.** True. A polynomial is continuous and differentiable everywhere.

**85.** Suppose that $p(x) = x^{2n+1} + ax + b$ has two real roots $x_1$ and $x_2$. Then by Rolle's Theorem, because $p(x_1) = p(x_2) = 0$, there exists $c$ in $(x_1, x_2)$ such that $p'(c) = 0$. But $p'(x) = (2n + 1)x^{2n} + a \ne 0$, because $n > 0$, $a > 0$. Therefore, $p(x)$ cannot have two real roots.

**87.** If $p(x) = Ax^2 + Bx + C$, then

$$p'(x) = 2Ax + B = \frac{f(b) - f(a)}{b - a}$$
$$= \frac{(Ab^2 + Bb + C) - (Aa^2 + Ba + C)}{b - a}$$
$$= \frac{A(b^2 - a^2) + B(b - a)}{b - a}$$
$$= \frac{(b - a)[A(b + a) + B]}{b - a}$$
$$= A(b + a) + B.$$

So, $2Ax = A(b + a)$ and $x = (b + a)/2$ which is the midpoint of $[a, b]$.

**89.** Suppose $f(x)$ has two fixed points $c_1$ and $c_2$. Then, by the Mean Value Theorem, there exists $c$ such that

$$f'(c) = \frac{f(c_2) - f(c_1)}{c_2 - c_1} = \frac{c_2 - c_1}{c_2 - c_1} = 1.$$

This contradicts the fact that $f'(x) < 1$ for all $x$.

**91.** Let $f(x) = \cos x$. $f$ is continuous and differentiable for all real numbers. By the Mean Value Theorem, for any interval $[a, b]$, there exists $c$ in $(a, b)$ such that

$$\frac{f(b) - f(a)}{b - a} = f'(c)$$
$$\frac{\cos b - \cos a}{b - a} = -\sin c$$
$$\cos b - \cos a = (-\sin c)(b - a)$$
$$|\cos b - \cos a| = |-\sin c||b - a|$$
$$|\cos b - \cos a| \le |b - a| \text{ since } |-\sin c| \le 1.$$

**93.** Let $0 < a < b$. $f(x) = \sqrt{x}$ satisfies the hypotheses of the Mean Value Theorem on $[a, b]$. Hence, there exists $c$ in $(a, b)$ such that

$$f'(c) = \frac{1}{2\sqrt{c}} = \frac{f(b) - f(a)}{b - a} = \frac{\sqrt{b} - \sqrt{a}}{b - a}.$$

So, $\sqrt{b} - \sqrt{a} = (b - a)\frac{1}{2\sqrt{c}} < \frac{b - a}{2\sqrt{a}}$.

## Section 4.3   Increasing and Decreasing Functions and the First Derivative Test

**1.** (a) Increasing: $(0, 6)$ and $(8, 9)$. Largest: $(0, 6)$

(b) Decreasing: $(6, 8)$ and $(9, 10)$. Largest: $(6, 8)$

**3.** $f(x) = x^2 - 6x + 8$

From the graph, $f$ is decreasing on $(-\infty, 3)$ and increasing on $(3, \infty)$.

Analytically, $f'(x) = 2x - 6$.

Critical number: $x = 3$

| Test intervals: | $-\infty < x < 3$ | $3 < x < \infty$ |
|---|---|---|
| Sign of $f'(x)$: | $f' < 0$ | $f' > 0$ |
| Conclusion: | Decreasing | Increasing |

**5.** $y = \dfrac{x^3}{4} - 3x$

From the graph, $y$ is increasing on $(-\infty, -2)$ and $(2, \infty)$, and decreasing on $(-2, 2)$.

Analytically, $y' = \dfrac{3x^2}{4} - 3 = \dfrac{3}{4}(x^2 - 4) = \dfrac{3}{4}(x - 2)(x + 2)$

Critical numbers: $x = \pm 2$

| Test intervals: | $-\infty < x < -2$ | $-2 < x < 2$ | $2 < x < \infty$ |
|---|---|---|---|
| Sign of $y'$: | $y' > 0$ | $y' < 0$ | $y' > 0$ |
| Conclusion: | Increasing | Decreasing | Increasing |

**7.** $f(x) = \dfrac{1}{(x + 1)^2}$

From the graph, $f$ is increasing on $(-\infty, -1)$ and decreasing on $(-1, \infty)$.

Analytically, $f'(x) = \dfrac{-2}{(x + 1)^3}$.

No critical numbers. Discontinuity: $x = -1$

| Test intervals: | $-\infty < x < -1$ | $-1 < x < \infty$ |
|---|---|---|
| Sign of $f'(x)$: | $f' > 0$ | $f' < 0$ |
| Conclusion: | Increasing | Decreasing |

**9.** $g(x) = x^2 - 2x - 8$

$g'(x) = 2x - 2$

Critical number: $x = 1$

| Test intervals: | $-\infty < x < 1$ | $1 < x < \infty$ |
|---|---|---|
| Sign of $g'(x)$: | $g' < 0$ | $g' > 0$ |
| Conclusion: | Decreasing | Increasing |

Increasing on: $(1, \infty)$

Decreasing on: $(-\infty, 1)$

**11.** $y = x\sqrt{16 - x^2}$   Domain: $[-4, 4]$

$y' = \dfrac{-2(x^2 - 8)}{\sqrt{16 - x^2}} = \dfrac{-2}{\sqrt{16 - x^2}}(x - 2\sqrt{2})(x + 2\sqrt{2})$

Critical numbers: $x = \pm 2\sqrt{2}$

| Test intervals: | $-4 < x < -2\sqrt{2}$ | $-2\sqrt{2} < x < 2\sqrt{2}$ | $2\sqrt{2} < x < 4$ |
|---|---|---|---|
| Sign of $y'$: | $y' < 0$ | $y' > 0$ | $y' < 0$ |
| Conclusion: | Decreasing | Increasing | Decreasing |

Increasing on: $\left(-2\sqrt{2}, 2\sqrt{2}\right)$

Decreasing on: $\left(-4, -2\sqrt{2}\right), \left(2\sqrt{2}, 4\right)$

**13.** $f(x) = \sin x - 1, \quad 0 < x < 2\pi$

$f'(x) = \cos x$

Critical numbers: $x = \dfrac{\pi}{2}, \dfrac{3\pi}{2}$

| Test intervals: | $0 < x < \dfrac{\pi}{2}$ | $\dfrac{\pi}{2} < x < \dfrac{3\pi}{2}$ | $\dfrac{3\pi}{2} < x < 2\pi$ |
|---|---|---|---|
| Sign of $f'(x)$: | $f' > 0$ | $f' < 0$ | $f' > 0$ |
| Conclusion: | Increasing | Decreasing | Increasing |

Increasing on: $\left(0, \dfrac{\pi}{2}\right), \left(\dfrac{3\pi}{2}, 2\pi\right)$

Decreasing on: $\left(\dfrac{\pi}{2}, \dfrac{3\pi}{2}\right)$

**15.** $y = x - 2\cos x, \quad 0 < x < 2\pi$

$y' = 1 + 2\sin x$

$y' = 0: \sin x = -\dfrac{1}{2}$

Critical numbers: $x = \dfrac{7\pi}{6}, \dfrac{11\pi}{6}$

| Test intervals: | $0 < x < \dfrac{7\pi}{6}$ | $\dfrac{7\pi}{6} < x < \dfrac{11\pi}{6}$ | $\dfrac{11\pi}{6} < x < 2\pi$ |
|---|---|---|---|
| Sign of $y'$: | $y' > 0$ | $y' < 0$ | $y' > 0$ |
| Conclusion: | Increasing | Decreasing | Increasing |

Increasing on: $\left(0, \dfrac{7\pi}{6}\right), \left(\dfrac{11\pi}{6}, 2\pi\right)$

Decreasing on: $\left(\dfrac{7\pi}{6}, \dfrac{11\pi}{6}\right)$

**17.** $g(x) = e^{-x} + e^{3x}$

$g'(x) = -e^{-x} + 3e^{3x}$

Critical number: $x = -\frac{1}{4} \ln 3$

| Test intervals: | $-\infty < x < -\frac{1}{4} \ln 3$ | $-\frac{1}{4} \ln 3 < x < \infty$ |
|---|---|---|
| Sign of $g'(x)$: | $g' < 0$ | $g' > 0$ |
| Conclusion: | Decreasing | Increasing |

Increasing on: $\left(-\frac{1}{4} \ln 3, \infty\right)$

Decreasing on: $\left(-\infty, -\frac{1}{4} \ln 3\right)$

**19.** $f(x) = x^2 \ln\left(\frac{x}{2}\right), \quad x > 0$

$f'(x) = 2x \ln\left(\frac{x}{2}\right) + \frac{x^2}{x} = 2x \ln\left(\frac{x}{2}\right) + x$

Critical number: $x = \dfrac{2}{\sqrt{e}}$

| Test intervals: | $0 < x < \dfrac{2}{\sqrt{e}}$ | $\dfrac{2}{\sqrt{e}} < x < \infty$ |
|---|---|---|
| Sign of $f'(x)$: | $f' < 0$ | $f' > 0$ |
| Conclusion: | Decreasing | Increasing |

Increasing on: $\left(\dfrac{2}{\sqrt{e}}, \infty\right)$

Decreasing on: $\left(0, \dfrac{2}{\sqrt{e}}\right)$

**21.** (a) $f(x) = x^2 - 4x$

$f'(x) = 2x - 4$

Critical number: $x = 2$

(b)

| Test intervals: | $-\infty < x < 2$ | $2 < x < \infty$ |
|---|---|---|
| Sign of $f'$: | $f' < 0$ | $f' > 0$ |
| Conclusion: | Decreasing | Increasing |

Decreasing on: $(-\infty, 2)$

Increasing on: $(2, \infty)$

(c) Relative minimum: $(2, -4)$

**23.** (a) $f(x) = -2x^2 + 4x + 3$

$f'(x) = -4x + 4 = 0$

Critical number: $x = 1$

(b)

| Test intervals: | $-\infty < x < 1$ | $1 < x < \infty$ |
|---|---|---|
| Sign of $f'(x)$: | $f' > 0$ | $f' < 0$ |
| Conclusion: | Increasing | Decreasing |

Increasing on: $(-\infty, 1)$

Decreasing on: $(1, \infty)$

(c) Relative maximum: $(1, 5)$

**25.** (a) $f(x) = 2x^3 + 3x^2 - 12x$

$f'(x) = 6x^2 + 6x - 12 = 6(x + 2)(x - 1) = 0$

Critical numbers: $x = -2, 1$

(b)

| Test intervals: | $-\infty < x < -2$ | $-2 < x < 1$ | $1 < x < \infty$ |
|---|---|---|---|
| Sign of $f'(x)$: | $f' > 0$ | $f' < 0$ | $f' > 0$ |
| Conclusion: | Increasing | Decreasing | Increasing |

Increasing on: $(-\infty, -2), (1, \infty)$

Decreasing on: $(-2, 1)$

(c) Relative maximum: $(-2, 20)$

Relative minimum: $(1, -7)$

**27.** (a) $f(x) = (x - 1)^2(x + 3) = x^3 + x^2 - 5x + 3$

$f'(x) = 3x^2 + 2x - 5 = (x - 1)(3x + 5)$

Critical numbers: $x = 1, -\frac{5}{3}$

(b)

| Test intervals: | $-\infty < x < -\frac{5}{3}$ | $-5/3 < x < 1$ | $1 < x < \infty$ |
|---|---|---|---|
| Sign of $f'$: | $f' > 0$ | $f' < 0$ | $f' > 0$ |
| Conclusion: | Increasing | Decreasing | Increasing |

Increasing on: $\left(-\infty, -\frac{5}{3}\right)$ and $(1, \infty)$

Decreasing on: $\left(-\frac{5}{3}, 1\right)$

(c) Relative maximum: $\left(-\frac{5}{3}, \frac{256}{27}\right)$

Relative minimum: $(1, 0)$

**29.** (a) $f(x) = \dfrac{x^5 - 5x}{5}$

$f'(x) = x^4 - 1$

Critical numbers: $x = -1, 1$

(b)

| Test intervals: | $-\infty < x < -1$ | $-1 < x < 1$ | $1 < x < \infty$ |
|---|---|---|---|
| Sign of $f'(x)$: | $f' > 0$ | $f' < 0$ | $f' > 0$ |
| Conclusion: | Increasing | Decreasing | Increasing |

Increasing on: $(-\infty, -1), (1, \infty)$

Decreasing on: $(-1, 1)$

(c) Relative maximum: $\left(-1, \dfrac{4}{5}\right)$

Relative minimum: $\left(1, -\dfrac{4}{5}\right)$

**31.** (a) $f(x) = x^{1/3} + 1$

$f'(x) = \dfrac{1}{3}x^{-2/3} = \dfrac{1}{3x^{2/3}}$

Critical number: $x = 0$

(b)

| Test intervals: | $-\infty < x < 0$ | $0 < x < \infty$ |
|---|---|---|
| Sign of $f'(x)$: | $f' > 0$ | $f' > 0$ |
| Conclusion: | Increasing | Increasing |

Increasing on: $(-\infty, \infty)$

(c) No relative extrema

**33.** (a) $f(x) = (x + 2)^{2/3}$

$f'(x) = \dfrac{2}{3}(x + 2)^{-1/3} = \dfrac{2}{3(x + 2)^{1/3}}$

Critical number: $x = -2$

(b)

| Test intervals: | $-\infty < x < -2$ | $-2 < x < \infty$ |
|---|---|---|
| Sign of $f'$: | $f' < 0$ | $f' > 0$ |
| Conclusion: | Decreasing | Increasing |

Decreasing on: $(-\infty, -2)$

Increasing on: $(-2, \infty)$

(c) Relative minimum: $(-2, 0)$

**35.** (a) $f(x) = 5 - |x - 5|$

$f'(x) = -\dfrac{x - 5}{|x - 5|} = \begin{cases} 1, & x < 5 \\ -1, & x > 5 \end{cases}$

Critical number: $x = 5$

(b)

| Test intervals: | $-\infty < x < 5$ | $5 < x < \infty$ |
|---|---|---|
| Sign of $f'(x)$: | $f' > 0$ | $f' < 0$ |
| Conclusion: | Increasing | Decreasing |

Increasing on: $(-\infty, 5)$

Decreasing on: $(5, \infty)$

(c) Relative maximum: $(5, 5)$

**37.** (a) $f(x) = 2x + \dfrac{1}{x}$

$f'(x) = 2 - \dfrac{1}{x^2} = \dfrac{2x^2 - 1}{x^2}$

Critical numbers: $x = \pm\dfrac{\sqrt{2}}{2}$

Discontinuity: $x = 0$

(b)

| Test intervals: | $-\infty < x < -\dfrac{\sqrt{2}}{2}$ | $-\dfrac{\sqrt{2}}{2} < x < 0$ | $0 < x < \dfrac{\sqrt{2}}{2}$ | $\dfrac{\sqrt{2}}{2} < x < \infty$ |
|---|---|---|---|---|
| Sign of $f'$: | $f' > 0$ | $f' < 0$ | $f' < 0$ | $f' > 0$ |
| Conclusion: | Increasing | Decreasing | Decreasing | Increasing |

Increasing on: $\left(-\infty, -\dfrac{\sqrt{2}}{2}\right)$ and $\left(\dfrac{\sqrt{2}}{2}, \infty\right)$

Decreasing on: $\left(-\dfrac{\sqrt{2}}{2}, 0\right)$ and $\left(0, \dfrac{\sqrt{2}}{2}\right)$

(c) Relative maximum: $\left(-\dfrac{\sqrt{2}}{2}, -2\sqrt{2}\right)$

Relative minimum: $\left(\dfrac{\sqrt{2}}{2}, 2\sqrt{2}\right)$

**39.** (a) $f(x) = \dfrac{x^2}{x^2 - 9}$

$$f'(x) = \dfrac{(x^2 - 9)(2x) - (x^2)(2x)}{(x^2 - 9)^2} = \dfrac{-18x}{(x^2 - 9)^2}$$

Critical number: $x = 0$

Discontinuities: $x = -3, 3$

(b)

| Test intervals: | $-\infty < x < -3$ | $-3 < x < 0$ | $0 < x < 3$ | $3 < x < \infty$ |
|---|---|---|---|---|
| Sign of $f'(x)$: | $f' > 0$ | $f' > 0$ | $f' < 0$ | $f' < 0$ |
| Conclusion: | Increasing | Increasing | Decreasing | Decreasing |

Increasing on: $(-\infty, -3), (-3, 0)$

Decreasing on: $(0, 3), (3, \infty)$

(c) Relative maximum: $(0, 0)$

**41.** (a) $f(x) = \begin{cases} 4 - x^2, & x \le 0 \\ -2x, & x > 0 \end{cases}$

$f'(x) = \begin{cases} -2x, & x < 0 \\ -2, & x > 0 \end{cases}$

Critical number: $x = 0$

(b)

| Test intervals: | $-\infty < x < 0$ | $0 < x < \infty$ |
|---|---|---|
| Sign of $f'$: | $f' > 0$ | $f' < 0$ |
| Conclusion: | Increasing | Decreasing |

Increasing on: $(-\infty, 0)$

Decreasing on: $(0, \infty)$

(c) Relative maximum: $(0, 4)$

**43.** (a) $f(x) = \begin{cases} 3x + 1, & x \le 1 \\ 5 - x^2, & x > 1 \end{cases}$

$f'(x) = \begin{cases} 3, & x < 1 \\ -2x, & x > 1 \end{cases}$

Critical number: $x = 1$

(b)

| Test intervals: | $-\infty < x < 1$ | $1 < x < \infty$ |
|---|---|---|
| Sign of $f'$: | $f' > 0$ | $f' < 0$ |
| Conclusion: | Increasing | Decreasing |

Increasing on: $(-\infty, 1)$

Decreasing on: $(1, \infty)$

(c) Relative maximum: $(1, 4)$

**45.** $f(x) = (3 - x)e^{x-3}$

$f'(x) = (3 - x)e^{x-3} - e^{x-3}$

$\quad = e^{x-3}(2 - x)$

Critical number: $x = 2$

| Test intervals: | $-\infty < x < 2$ | $2 < x < \infty$ |
|---|---|---|
| Sign of $f'(x)$: | $f' > 0$ | $f' < 0$ |
| Conclusion: | Increasing | Decreasing |

Increasing on: $(-\infty, 2)$

Decreasing on: $(2, \infty)$

Relative minimum: $(2, e^{-1})$

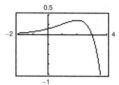

**47.** $f(x) = 4(x - \arcsin x), -1 \le x \le 1$

$f'(x) = 4 - \dfrac{4}{\sqrt{1 - x^2}}$

Critical number: $x = 0$

| Test intervals: | $-1 \le x < 0$ | $0 < x \le 1$ |
|---|---|---|
| Sign of $f'(x)$: | $f' < 0$ | $f' < 0$ |
| Conclusion: | Decreasing | Decreasing |

Decreasing on: $[-1, 1]$

No relative extrema

(Absolute maximum at $x = -1$, absolute minimum at $x = 1$)

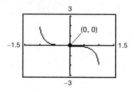

**49.** $g(x) = (x)3^{-x}$

$g'(x) = (1 - x \ln 3)3^{-x}$

Critical number: $x = \dfrac{1}{\ln 3} \approx 0.9102$

| Test intervals: | $-\infty < x < \dfrac{1}{\ln 3}$ | $\dfrac{1}{\ln 3} < x < \infty$ |
|---|---|---|
| Sign of $f'(x)$: | $f' > 0$ | $f' < 0$ |
| Conclusion: | Increasing | Decreasing |

Increasing on: $\left(-\infty, \dfrac{1}{\ln 3}\right)$

Decreasing on: $\left(\dfrac{1}{\ln 3}, \infty\right)$

Relative maximum: $\left(\dfrac{1}{\ln 3}, \dfrac{1}{e \ln 3}\right) \approx (0.9102, 0.3349)$

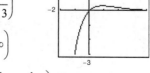

**51.** $f(x) = x - \log_4 x = x - \dfrac{\ln x}{\ln 4}$

$f'(x) = 1 - \dfrac{1}{x \ln 4} = 0 \Rightarrow x \ln 4 = 1 \Rightarrow x = \dfrac{1}{\ln 4}$

Critical number: $x = \dfrac{1}{\ln 4}$

| Test intervals: | $0 < x < \dfrac{1}{\ln 4}$ | $\dfrac{1}{\ln 4} < x < \infty$ |
|---|---|---|
| Sign of $f'(x)$: | $f' < 0$ | $f' > 0$ |
| Conclusion: | Decreasing | Increasing |

Increasing on: $\left(\dfrac{1}{\ln 4}, \infty\right)$

Decreasing on: $\left(0, \dfrac{1}{\ln 4}\right)$

Relative maximum:

$\left(\dfrac{1}{\ln 4}, \dfrac{1}{\ln 4} - \log_4\left(\dfrac{1}{\ln 4}\right)\right) = \left(\dfrac{1}{\ln 4}, \dfrac{\ln(\ln 4) + 1}{\ln 4}\right)$

$\approx (0.7213, 0.9570)$

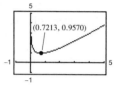

**53.** $g(x) = \dfrac{e^{2x}}{e^{2x} + 1}$

$g'(x) = \dfrac{(e^{2x} + 1)2e^{2x} - e^{2x}(2e^{2x})}{(e^{2x} + 1)^2} = \dfrac{2e^{2x}}{(e^{2x} + 1)^2}$

No critical numbers.

Increasing on: $(-\infty, \infty)$

No relative extrema.

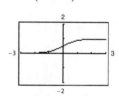

**55.** $f(x) = e^{-1/(x-2)} = e^{1/(2-x)}, x \neq 2$

$f'(x) = e^{1/(2-x)}\left(\dfrac{1}{(2 - x)^2}\right)$

No critical numbers.

$x = 2$ is a vertical asymptote.

| Test intervals: | $-\infty < x < 2$ | $2 < x < \infty$ |
|---|---|---|
| Sign of $f'(x)$: | $f' > 0$ | $f' > 0$ |
| Conclusion: | Increasing | Increasing |

Increasing on: $(-\infty, 2), (2, \infty)$

No relative extrema.

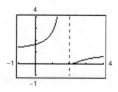

**57.** (a) $f(x) = \dfrac{x}{2} + \cos x, \; 0 < x < 2\pi$

$f'(x) = \dfrac{1}{2} - \sin x = 0$

Critical numbers: $x = \dfrac{\pi}{6}, \dfrac{5\pi}{6}$

| Test intervals: | $0 < x < \dfrac{\pi}{4}$ | $\dfrac{\pi}{4} < x < \dfrac{5\pi}{4}$ | $\dfrac{5\pi}{4} < x < 2\pi$ |
|---|---|---|---|
| Sign of $f'(x)$: | $f' > 0$ | $f' < 0$ | $f' > 0$ |
| Conclusion: | Increasing | Decreasing | Increasing |

Increasing on: $\left(0, \dfrac{\pi}{6}\right), \left(\dfrac{5\pi}{6}, 2\pi\right)$

Decreasing on: $\left(\dfrac{\pi}{6}, \dfrac{5\pi}{6}\right)$

(b) Relative maximum: $\left(\dfrac{\pi}{6}, \dfrac{\pi + 6\sqrt{3}}{12}\right)$

Relative minimum: $\left(\dfrac{5\pi}{6}, \dfrac{5\pi - 6\sqrt{3}}{12}\right)$

(c)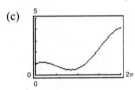

**59.** (a) $f(x) = \sin x + \cos x, \quad 0 < x < 2\pi$

$f'(x) = \cos x - \sin x = 0 \Rightarrow \sin x = \cos x$

Critical numbers: $x = \dfrac{\pi}{4}, \dfrac{5\pi}{4}$

| Test intervals: | $0 < x < \dfrac{\pi}{4}$ | $\dfrac{\pi}{4} < x < \dfrac{5\pi}{4}$ | $\dfrac{5\pi}{4} < x < 2\pi$ |
|---|---|---|---|
| Sign of $f'(x)$: | $f' > 0$ | $f' < 0$ | $f' > 0$ |
| Conclusion: | Increasing | Decreasing | Increasing |

Increasing on: $\left(0, \dfrac{\pi}{4}\right), \left(\dfrac{5\pi}{4}, 2\pi\right)$

Decreasing on: $\left(\dfrac{\pi}{4}, \dfrac{5\pi}{4}\right)$

(b) Relative maximum: $\left(\dfrac{\pi}{4}, \sqrt{2}\right)$

Relative minimum: $\left(\dfrac{5\pi}{4}, -\sqrt{2}\right)$

(c)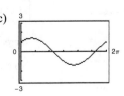

**61.** (a)  $f(x) = \cos^2(2x),$    $0 < x < 2\pi$

$f'(x) = -4 \cos 2x \sin 2x = 0 \Rightarrow \cos 2x = 0$ or $\sin 2x = 0$

Critical numbers:  $x = \dfrac{\pi}{4}, \dfrac{3\pi}{4}, \dfrac{5\pi}{4}, \dfrac{7\pi}{4}, \dfrac{\pi}{2}, \pi, \dfrac{3\pi}{2}$

| Test intervals: | $0 < x < \dfrac{\pi}{4}$ | $\dfrac{\pi}{4} < x < \dfrac{\pi}{2}$ | $\dfrac{\pi}{2} < x < \dfrac{3\pi}{4}$ | $\dfrac{3\pi}{4} < x < \pi$ |
|---|---|---|---|---|
| Sign of $f'(x)$: | $f' < 0$ | $f' > 0$ | $f' < 0$ | $f' > 0$ |
| Conclusion: | Decreasing | Increasing | Decreasing | Increasing |

| Test intervals: | $\pi < x < \dfrac{5\pi}{4}$ | $\dfrac{5\pi}{4} < x < \dfrac{3\pi}{2}$ | $\dfrac{3\pi}{2} < x < \dfrac{7\pi}{4}$ | $\dfrac{7\pi}{4} < x < 2\pi$ |
|---|---|---|---|---|
| Sign of $f'(x)$: | $f' < 0$ | $f' > 0$ | $f' < 0$ | $f' > 0$ |
| Conclusion: | Decreasing | Increasing | Decreasing | Increasing |

Increasing on:  $\left(\dfrac{\pi}{4}, \dfrac{\pi}{2}\right), \left(\dfrac{3\pi}{4}, \pi\right), \left(\dfrac{5\pi}{4}, \dfrac{3\pi}{2}\right), \left(\dfrac{7\pi}{4}, 2\pi\right)$

Decreasing on:  $\left(0, \dfrac{\pi}{4}\right), \left(\dfrac{\pi}{2}, \dfrac{3\pi}{4}\right), \left(\pi, \dfrac{5\pi}{4}\right), \left(\dfrac{3\pi}{2}, \dfrac{7\pi}{4}\right)$

(b) Relative maxima:  $\left(\dfrac{\pi}{2}, 1\right), (\pi, 1), \left(\dfrac{3\pi}{2}, 1\right)$

Relative minima:  $\left(\dfrac{\pi}{4}, 0\right), \left(\dfrac{3\pi}{4}, 0\right), \left(\dfrac{5\pi}{4}, 0\right), \left(\dfrac{7\pi}{4}, 0\right)$

(c)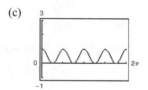

**63.** (a)  $f(x) = \sin^2 x + \sin x,$    $0 < x < 2\pi$

$f'(x) = 2 \sin x \cos x + \cos x = \cos x (2 \sin x + 1) = 0$

Critical numbers:  $x = \dfrac{\pi}{2}, \dfrac{7\pi}{6}, \dfrac{3\pi}{2}, \dfrac{11\pi}{6}$

| Test intervals: | $0 < x < \dfrac{\pi}{2}$ | $\dfrac{\pi}{2} < x < \dfrac{7\pi}{6}$ | $\dfrac{7\pi}{6} < x < \dfrac{3\pi}{2}$ | $\dfrac{3\pi}{2} < x < \dfrac{11\pi}{6}$ | $\dfrac{11\pi}{6} < x < 2\pi$ |
|---|---|---|---|---|---|
| Sign of $f'(x)$: | $f' > 0$ | $f' < 0$ | $f' > 0$ | $f' < 0$ | $f' > 0$ |
| Conclusion: | Increasing | Decreasing | Increasing | Decreasing | Increasing |

Increasing on:  $\left(0, \dfrac{\pi}{2}\right), \left(\dfrac{7\pi}{6}, \dfrac{3\pi}{2}\right), \left(\dfrac{11\pi}{6}, 2\pi\right)$

Decreasing on:  $\left(\dfrac{\pi}{2}, \dfrac{7\pi}{6}\right), \left(\dfrac{3\pi}{2}, \dfrac{11\pi}{6}\right)$

(b) Relative minima:  $\left(\dfrac{7\pi}{6}, -\dfrac{1}{4}\right), \left(\dfrac{11\pi}{6}, -\dfrac{1}{4}\right)$

Relative maxima:  $\left(\dfrac{\pi}{2}, 2\right), \left(\dfrac{3\pi}{2}, 0\right)$

(c)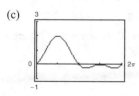

**65.** $f(x) = 2x\sqrt{9 - x^2}, [-3, 3]$

(a) $f'(x) = \dfrac{2(9 - 2x^2)}{\sqrt{9 - x^2}}$

(b)

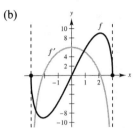

(c) $\dfrac{2(9 - 2x^2)}{\sqrt{9 - x^2}} = 0$

Critical numbers: $x = \pm\dfrac{3}{\sqrt{2}} = \pm\dfrac{3\sqrt{2}}{2}$

(d) Intervals:

$$\left(-3, -\dfrac{3\sqrt{2}}{2}\right) \quad \left(-\dfrac{3\sqrt{2}}{2}, \dfrac{3\sqrt{2}}{2}\right) \quad \left(\dfrac{3\sqrt{2}}{2}, 3\right)$$

$\quad\ f'(x) < 0 \qquad\ \ f'(x) > 0 \qquad\ \ f'(x) < 0$

Decreasing    Increasing    Decreasing

$f$ is increasing when $f'$ is positive and decreasing when $f'$ is negative.

**67.** $f(t) = t^2 \sin t, [0, 2\pi]$

(a) $f'(t) = t^2 \cos t + 2t \sin t = t(t \cos t + 2 \sin t)$

(b)

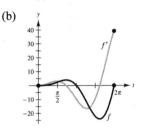

(c) $t(t \cos t + 2 \sin t) = 0$

$t = 0 \text{ or } t = -2 \tan t$

$t \cot t = -2$

$t \approx 2.2889, 5.0870 \text{ (graphing utility)}$

Critical numbers: $t = 2.2889, 5.0870$

(d) Intervals:

$(0, 2.2889) \quad (2.2889, 5.0870) \quad (5.0870, 2\pi)$

$\ f'(t) > 0 \qquad\ \ f'(t) < 0 \qquad\ \ f'(t) > 0$

Increasing    Decreasing    Increasing

$f$ is increasing when $f'$ is positive and decreasing when $f'$ is negative.

**69.** (a) $f(x) = -3 \sin \dfrac{x}{3}, [0, 6\pi]$

$f'(x) = -\cos \dfrac{x}{3}$

(b)

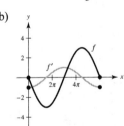

(c) Critical numbers: $x = \dfrac{3\pi}{2}, \dfrac{9\pi}{2}$

(d) Intervals:

$$\left(0, \dfrac{3\pi}{2}\right) \quad \left(\dfrac{3\pi}{2}, \dfrac{9\pi}{2}\right) \quad \left(\dfrac{9\pi}{2}, 6\pi\right)$$

$\ \ f' < 0 \qquad\ \ f' > 0 \qquad\ \ f' < 0$

Decreasing    Increasing    Decreasing

$f$ is increasing when $f'$ is positive and decreasing when $f'$ is negative.

**71.** $f(x) = \dfrac{1}{2}(x^2 - \ln x), (0, 3]$

(a) $f'(x) = \dfrac{2x^2 - 1}{2x}$

(b)

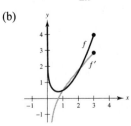

(c) $\dfrac{2x^2 - 1}{2x} = 0$

Critical number: $x = \dfrac{1}{\sqrt{2}} = \dfrac{\sqrt{2}}{2}$

(d) Intervals: $\left(0, \dfrac{\sqrt{2}}{2}\right) \qquad \left(\dfrac{\sqrt{2}}{2}, 3\right)$

$\quad\ \ f'(x) < 0 \qquad\qquad f'(x) > 0$

Decreasing    Increasing

(e) $f$ is increasing when $f'$ is positive, and decreasing when $f'$ is negative.

**73.** $f(x) = \dfrac{x^5 - 4x^3 + 3x}{x^2 - 1} = \dfrac{(x^2 - 1)(x^3 - 3x)}{x^2 - 1} = x^3 - 3x, \; x \neq \pm 1$

$f(x) = g(x) = x^3 - 3x$ for all $x \neq \pm 1$.

$f'(x) = 3x^2 - 3 = 3(x^2 - 1), \; x \neq \pm 1 \Rightarrow f'(x) \neq 0$

$f$ symmetric about origin

zeros of $f: (0, 0), (\pm\sqrt{3}, 0)$

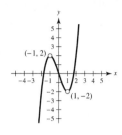

$g(x)$ is continuous on $(-\infty, \infty)$ and $f(x)$ has holes at $(-1, 2)$ and $(1, -2)$.

**75.** $f(x) = c$ is constant $\Rightarrow f'(x) = 0$.

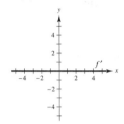

**77.** $f$ is quadratic $\Rightarrow f'$ is a line.

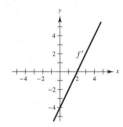

**79.** $f$ has positive, but decreasing slope.

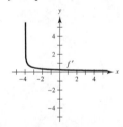

**In Exercises 81–86,** $f'(x) > 0$ **on** $(-\infty, -4)$, $f'(x) < 0$ **on** $(-4, 6)$ **and** $f'(x) > 0$ **on** $(6, \infty)$.

**81.** $g(x) = f(x) + 5$

$g'(x) = f'(x)$

$g'(0) = f'(0) < 0$

**83.**   $g(x) = -f(x)$

$g'(x) = -f'(x)$

$g'(-6) = -f'(-6) < 0$

**85.** $g(x) = f(x - 10)$

$g'(x) = f'(x - 10)$

$g'(0) = f'(-10) > 0$

**87.** No. $f$ does have a horizontal tangent line at $x = c$, but $f$ could be increasing (or decreasing) on both sides of the point. For example, $f(x) = x^3$ at $x = 0$.

**89.** $f'(x) \begin{cases} > 0, & x < 4 \Rightarrow f \text{ is increasing on } (-\infty, 4). \\ \text{undefined}, & x = 4 \\ < 0, & x > 4 \Rightarrow f \text{ is decreasing on } (4 \, \infty). \end{cases}$

Two possibilities for $f(x)$ are given below.

(a)

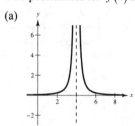

(b)

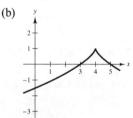

**91.** Critical number: $x = 5$

$f'(4) = -2.5 \Rightarrow f$ is decreasing at $x = 4$.

$f'(6) = 3 \Rightarrow f$ is increasing at $x = 6$.

$(5, f(5))$ is a relative minimum.

**In Exercise 93, answers will vary.**

*Sample answers:*

**93.** (a)

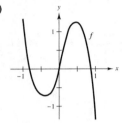

(b) The critical numbers are in intervals $(-0.50, -0.25)$

and $(0.25, 0.50)$ because the sign of $f'$ changes in these intervals. $f$ is decreasing on approximately

$(-1, -0.40), (0.48, 1)$, and increasing on $(-0.40, 0.48)$.

(c) Relative minimum when $x \approx -0.40$: $(-0.40, 0.75)$

Relative maximum when $x \approx 0.48$: $(0.48, 1.25)$

**95.** $s(t) = 4.9(\sin \theta)t^2$

(a) $s'(t) = 4.9(\sin \theta)(2t) = 9.8(\sin \theta)t$

speed $= |s'(t)| = |9.8(\sin \theta)t|$

(b)

| $\theta$ | $0$ | $\dfrac{\pi}{4}$ | $\dfrac{\pi}{3}$ | $\dfrac{\pi}{2}$ | $\dfrac{2\pi}{3}$ | $\dfrac{3\pi}{4}$ | $\pi$ |
|---|---|---|---|---|---|---|---|
| $\lvert s'(t) \rvert$ | $0$ | $4.9\sqrt{2}t$ | $4.9\sqrt{3}t$ | $9.8t$ | $4.9\sqrt{3}t$ | $4.9\sqrt{2}t$ | $0$ |

The speed is maximum for $\theta = \dfrac{\pi}{2}$.

**97.** $C = \dfrac{3t}{27 + t^3}, \ t \geq 0$

(a)

| $t$ | $0$ | $0.5$ | $1$ | $1.5$ | $2$ | $2.5$ | $3$ |
|---|---|---|---|---|---|---|---|
| $C(t)$ | $0$ | $0.055$ | $0.107$ | $0.148$ | $0.171$ | $0.176$ | $0.167$ |

The concentration seems greatest near $t = 2.5$ hours.

(b)

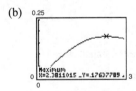

The concentration is greatest when $t \approx 2.38$ hours.

(c) $C' = \dfrac{(27 + t^3)(3) - (3t)(3t^2)}{(27 + t^3)^2} = \dfrac{3(27 - 2t^3)}{(27 + t^3)^2}$

$C' = 0$ when $t = 3/\sqrt[3]{2} \approx 2.38$ hours.

By the First Derivative Test, this is a maximum.

**99.** $v = k(R - r)r^2 = k(Rr^2 - r^3)$

$v' = k(2Rr - 3r^2)$

$\quad = kr(2R - 3r) = 0$

$r = 0$ or $\frac{2}{3}R$

Maximum when $r = \frac{2}{3}R$.

**101.** (a)  $s(t) = 6t - t^2, t \geq 0$

$v(t) = 6 - 2t$

(b)  $v(t) = 0$ when $t = 3$.

Moving in positive direction for $0 \leq t < 3$ because $v(t) > 0$ on $0 \leq t < 3$.

(c)  Moving in negative direction when $t > 3$.

(d)  The particle changes direction at $t = 3$.

**103.** (a)  $s(t) = t^3 - 5t^2 + 4t, t \geq 0$

$v(t) = 3t^2 - 10t + 4$

(b)  $v(t) = 0$ for $t = \dfrac{10 \pm \sqrt{100 - 48}}{6} = \dfrac{5 \pm \sqrt{13}}{3}$

Particle is moving in a positive direction on

$\left[0, \dfrac{5 - \sqrt{13}}{3}\right) \approx [0, 0.4648)$ and $\left(\dfrac{5 + \sqrt{13}}{3}, \infty\right) \approx (2.8685, \infty)$ because $v > 0$ on these intervals.

(c)  Particle is moving in a negative direction on

$\left(\dfrac{5 - \sqrt{13}}{3}, \dfrac{5 + \sqrt{13}}{3}\right) \approx (0.4648, 2.8685)$

(d)  The particle changes direction at $t = \dfrac{5 \pm \sqrt{13}}{3}$.

**105.** Answers will vary.

**107.** (a)  Use a cubic polynomial

$f(x) = a_3x^3 + a_2x^2 + a_1x + a_0$

(b)  $f'(x) = 3a_3x^2 + 2a_2x + a_1$.

$f(0) = 0$:  $a_3(0)^3 + a_2(0)^2 + a_1(0) + a_0 = 0 \Rightarrow \qquad a_0 = 0$

$f'(0) = 0$:  $\qquad 3a_3(0)^2 + 2a_2(0) + a_1 = 0 \Rightarrow \qquad a_1 = 0$

$f(2) = 2$:  $a_3(2)^3 + a_2(2)^2 + a_1(2) + a_0 = 2 \Rightarrow \quad 8a_3 + 4a_2 = 2$

$f'(2) = 0$:  $\qquad 3a_3(2)^2 + 2a_2(2) + a_1 = 0 \Rightarrow \quad 12a_3 + 4a_2 = 0$

(c)  The solution is $a_0 = a_1 = 0, a_2 = \frac{3}{2}, a_3 = -\frac{1}{2}$:

$f(x) = -\frac{1}{2}x^3 + \frac{3}{2}x^2$.

(d)

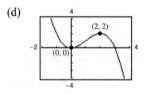

**109.** (a)  Use a fourth degree polynomial

$$f(x) = a_4 x^4 + a_3 x^3 + a_2 x^2 + a_1 x + a_0.$$

(b)  $f'(x) = 4a_4 x^3 + 3a_3 x^2 + 2a_2 x + a_1$

$f(0) = 0$:  $a_4(0)^4 + a_3(0)^3 + a_2(0)^2 + a_1(0) + a_0 = 0 \Rightarrow$ $\qquad a_0 = 0$

$f'(0) = 0$:  $\qquad 4a_4(0)^3 + 3a_3(0)^2 + 2a_2(0) + a_1 = 0 \Rightarrow$ $\qquad a_1 = 0$

$f(4) = 0$:  $a_4(4)^4 + a_3(4)^3 + a_2(4)^2 + a_1(4) + a_0 = 0 \Rightarrow$ $256a_4 + 64a_3 + 16a_2 = 0$

$f'(4) = 0$:  $\qquad 4a_4(4)^3 + 3a_3(4)^2 + 2a_2(4) + a_1 = 0 \Rightarrow$ $256a_4 + 48a_3 + 8a_2 = 0$

$f(2) = 4$:  $a_4(2)^4 + a_3(2)^3 + a_2(2)^2 + a_1(2) + a_0 = 4 \Rightarrow$ $16a_4 + 8a_3 + 4a_2 = 4$

$f'(2) = 0$:  $\qquad 4a_4(2)^3 + 3a_3(2)^2 + 2a_2(2) + a_1 = 0 \Rightarrow$ $32a_4 + 12a_3 + 4a_2 = 0$

(c)  The solution is $a_0 = a_1 = 0, a_2 = 4, a_3 = -2, \ a_4 = \frac{1}{4}$.

$$f(x) = \tfrac{1}{4}x^4 - 2x^3 + 4x^2$$

(d)

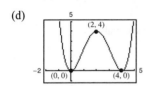

**111.** True.

Let $h(x) = f(x) + g(x)$ where $f$ and $g$ are increasing.

Then $h'(x) = f'(x) + g'(x) > 0$ because

$f'(x) > 0$ and $g'(x) > 0$.

**113.** False.

Let $f(x) = x^3$, then $f'(x) = 3x^2$ and $f$ only has one

critical number. Or, let $f(x) = x^3 + 3x + 1$, then

$f'(x) = 3(x^2 + 1)$ has no critical numbers.

**115.** False. For example, $f(x) = x^3$ does not have a relative

extrema at the critical number $x = 0$.

**117.** Assume that $f'(x) < 0$ for all $x$ in the interval $(a, b)$ and

let $x_1 < x_2$ be any two points in the interval. By the

Mean Value Theorem, you know there exists a number $c$

such that $x_1 < c < x_2$, and

$$f'(c) = \frac{f(x_2) - f(x_1)}{x_2 - x_1}$$

Because $f'(c) < 0$ and $x_2 - x_1 > 0$, then

$f(x_2) - f(x_1) < 0$, which implies that

$f(x_2) < f(x_1)$. So, $f$ is decreasing on the interval.

**119.** Let $f(x) = (1 + x)^n - nx - 1$. Then

$$f'(x) = n(1 + x)^{n-1} - n = n\left[(1 + x)^{n-1} - 1\right] > 0$$

because $x > 0$ and $n > 1$.

So, $f(x)$ is increasing on $(0, \infty)$. Because

$f(0) = 0 \Rightarrow f(x) > 0$ on $(0, \infty)$

$(1 + x)^n - nx - 1 > 0 \Rightarrow (1 + x)^n > 1 + nx$.

**121.** Let $x_1$ and $x_2$ be two positive real numbers,

$0 < x_1 < x_2$. Then

$$\frac{1}{x_1} > \frac{1}{x_2}$$

$$f(x_1) > f(x_2)$$

So, $f$ is decreasing on $(0, \infty)$.

**123.** First observe that

$$\tan x + \cot x + \sec x + \csc x = \frac{\sin x}{\cos x} + \frac{\cos x}{\sin x} + \frac{1}{\cos x} + \frac{1}{\sin x}$$

$$= \frac{\sin^2 x + \cos^2 x + \sin x + \cos x}{\sin x \cos x}$$

$$= \frac{1 + \sin x + \cos x}{\sin x \cos x}\left(\frac{\sin x + \cos x - 1}{\sin x + \cos x - 1}\right)$$

$$= \frac{(\sin x + \cos x)^2 - 1}{\sin x \cos x(\sin x + \cos x - 1)}$$

$$= \frac{2 \sin x \cos x}{\sin x \cos x(\sin x + \cos x - 1)}$$

$$= \frac{2}{\sin x + \cos x - 1}$$

Let $t = \sin x + \cos x - 1$. The expression inside the absolute value sign is

$$f(t) = \sin x + \cos x + \frac{2}{\sin x + \cos x - 1} = (\sin x + \cos x - 1) + 1 + \frac{2}{\sin x + \cos x - 1} = t + 1 + \frac{2}{t}$$

Because $\sin\left(x + \dfrac{\pi}{4}\right) = \sin x \cos\dfrac{\pi}{4} + \cos x \sin\dfrac{\pi}{4} = \dfrac{\sqrt{2}}{2}(\sin x + \cos x)$,

$\sin x + \cos x \in \left[-\sqrt{2}, \sqrt{2}\right]$ and $t = \sin x + \cos x - 1 \in \left[-1 - \sqrt{2}, -1 + \sqrt{2}\right]$.

$$f'(t) = 1 - \frac{2}{t^2} = \frac{t^2 - 2}{t^2} = \frac{\left(t + \sqrt{2}\right)\left(t - \sqrt{2}\right)}{t^2}$$

$$f\left(-1 + \sqrt{2}\right) = -1 + \sqrt{2} + 1 + \frac{2}{-1 + \sqrt{2}} = \sqrt{2} + \frac{2}{\sqrt{2} - 1}$$

$$= \frac{4 - \sqrt{2}}{\sqrt{2} - 1}\left(\frac{\sqrt{2} + 1}{\sqrt{2} + 1}\right) = \frac{4\sqrt{2} - 2 + 4 - \sqrt{2}}{1} = 2 + 3\sqrt{2}$$

For $t > 0$, $f$ is decreasing and $f(t) > f\left(-1 + \sqrt{2}\right) = 2 + 3\sqrt{2}$

For $t < 0$, $f$ is increasing on $\left(-\sqrt{2} - 1, -\sqrt{2}\right)$, then decreasing on $\left(-\sqrt{2}, 0\right)$. So $f(t) < f\left(-\sqrt{2}\right) = 1 - 2\sqrt{2}$.

Finally, $\left|f(t)\right| \geq 2\sqrt{2} - 1$.

(You can verify this easily with a graphing utility.)

# Section 4.4   Concavity and the Second Derivative Test

**1.** $y = x^2 - x - 2$

$y' = 2x - 1$

$y'' = 2$

$y'' > 0$ for all $x$.

Concave upward: $(-\infty, \infty)$

**3.** $f(x) = -x^3 + 6x^2 - 9x - 1$

$f'(x) = -3x^2 + 12x - 9$

$f''(x) = -6x + 12 = -6(x - 2)$

$f''(x) = 0$ when $x = 2$.

Concave upward: $(-\infty, 2)$

Concave downward: $(2, \infty)$

| Intervals: | $-\infty < x < 2$ | $2 < x < \infty$ |
|---|---|---|
| Sign of $f''$: | $f'' > 0$ | $f'' < 0$ |
| Conclusion: | Concave upward | Concave downward |

**5.** $f(x) = \dfrac{24}{x^2 + 12}$

$f'(x) = \dfrac{-48x}{(x^2 + 12)^2}$

$f''(x) = \dfrac{-144(4 - x^2)}{(x^2 + 12)^3}$

$f''(x) = 0$ when $x = \pm 2$.

| Intervals: | $-\infty < x < -2$ | $-2 < x < 2$ | $2 < x < \infty$ |
|---|---|---|---|
| Sign of $f''$: | $f'' > 0$ | $f'' < 0$ | $f'' > 0$ |
| Conclusion: | Concave upward | Concave downward | Concave upward |

Concave upward: $(-\infty, -2), (2, \infty)$

Concave downward: $(-2, 2)$

**7.** $f(x) = \dfrac{x^2 + 1}{x^2 - 1}$

$f' = \dfrac{-4x}{(x^2 - 1)^2}$

$f'' = \dfrac{4(3x^2 + 1)}{(x^2 - 1)^3}$

$f$ is not continuous at $x = \pm 1$.

| Intervals: | $-\infty < x < -1$ | $-1 < x < 1$ | $1 < x < \infty$ |
|---|---|---|---|
| Sign of $f''$: | $f'' > 0$ | $f'' < 0$ | $f'' > 0$ |
| Conclusion: | Concave upward | Concave downward | Concave upward |

Concave upward: $(-\infty, -1), (1, \infty)$

Concave downward: $(-1, 1)$

**9.** $g(x) = \dfrac{x^2 + 4}{4 - x^2}$

$g'(x) = \dfrac{16x}{(4 - x^2)^2}$

$g''(x) = \dfrac{16(3x^2 + 4)}{(4 - x^2)^3} = \dfrac{16(3x^2 + 4)}{(2 - x)^3(2 + x)^3}$

$f$ is not continuous at $x = \pm 2$.

| Intervals: | $-\infty < x < -2$ | $-2 < x < 2$ | $2 < x < \infty$ |
|---|---|---|---|
| Sign of $g''$: | $g'' < 0$ | $g'' > 0$ | $g'' < 0$ |
| Conclusion: | Concave downward | Concave upward | Concave downward |

Concave upward: $(-2, 2)$

Concave downward: $(-\infty, -2), (2, \infty)$

**11.** $y = 2x - \tan x, \left(-\dfrac{\pi}{2}, \dfrac{\pi}{2}\right)$

$y' = 2 - \sec^2 x$

$y'' = -2 \sec^2 x \tan x$

$y'' = 0$ when $x = 0$.

| Intervals: | $-\dfrac{\pi}{2} < x < 0$ | $0 < x < \dfrac{\pi}{2}$ |
|---|---|---|
| Sign of $y''$: | $y'' > 0$ | $y'' < 0$ |
| Conclusion: | Concave upward | Concave downward |

Concave upward: $\left(-\dfrac{\pi}{2}, 0\right)$

Concave downward: $\left(0, \dfrac{\pi}{2}\right)$

**13.** $f(x) = x^3 - 6x^2 + 12x$

$f'(x) = 3x^2 - 12x + 12$

$f''(x) = 6(x - 2) = 0$ when $x = 2$.

| Intervals: | $-\infty < x < 2$ | $2 < x < \infty$ |
|---|---|---|
| Sign of $f''$: | $f'' < 0$ | $f'' > 0$ |
| Conclusion: | Concave downward | Concave upward |

Concave upward: $(2, \infty)$

Concave downward: $(-\infty, 2)$

Point of inflection: $(2, 8)$

**15.** $f(x) = \frac{1}{2}x^4 + 2x^3$

$f'(x) = 2x^3 + 6x^2$

$f''(x) = 6x^2 + 12x = 6x(x + 2)$

$f''(x) = 0$ when $x = 0, -2$

| Intervals: | $-\infty < x < -2$ | $-2 < x < 2$ | $0 < x < \infty$ |
|---|---|---|---|
| Sign of $f''$: | $f'' > 0$ | $f'' < 0$ | $f'' > 0$ |
| Conclusion: | Concave upward | Concave downward | Concave upward |

Concave upward: $(-\infty, -2), (0, \infty)$

Concave downward: $(-2, 0)$

Points of inflection: $(-2, -8)$ and $(0, 0)$

**17.** $f(x) = x(x - 4)^3$

$f'(x) = x\left[3(x - 4)^2\right] + (x - 4)^3 = (x - 4)^2(4x - 4)$

$f''(x) = 4(x - 1)\left[2(x - 4)\right] + 4(x - 4)^2 = 4(x - 4)\left[2(x - 1) + (x - 4)\right] = 4(x - 4)(3x - 6) = 12(x - 4)(x - 2)$

$f''(x) = 12(x - 4)(x - 2) = 0$ when $x = 2, 4$.

| Intervals: | $-\infty < x < 2$ | $2 < x < 4$ | $4 < x < \infty$ |
|---|---|---|---|
| Sign of $f''(x)$: | $f''(x) > 0$ | $f''(x) < 0$ | $f''(x) > 0$ |
| Conclusion: | Concave upward | Concave downward | Concave upward |

Concave upward: $(-\infty, 2), (4, \infty)$

Concave downward: $(2, 4)$

Points of inflection: $(2, -16), (4, 0)$

**19.** $f(x) = x\sqrt{x + 3}$, Domain: $[-3, \infty)$

$$f'(x) = x\left(\frac{1}{2}\right)(x + 3)^{-1/2} + \sqrt{x + 3} = \frac{3(x + 2)}{2\sqrt{x + 3}}$$

$$f''(x) = \frac{6\sqrt{x + 3} - 3(x + 2)(x + 3)^{-1/2}}{4(x + 3)}$$

$$= \frac{3(x + 4)}{4(x + 3)^{3/2}} = 0 \text{ when } x = -4.$$

$x = -4$ is not in the domain. $f''$ is not continuous at $x = -3$.

| Interval: | $-3 < x < \infty$ |
|---|---|
| Sign of $f''$: | $f'' > 0$ |
| Conclusion: | Concave upward |

Concave upward: $(-3, \infty)$

There are no points of inflection.

**21.** $f(x) = \frac{4}{x^2 + 1}$

$$f'(x) = \frac{-8x}{(x^2 + 1)^2}$$

$$f''(x) = \frac{8(3x^2 - 1)}{(x^2 + 1)^3}$$

$$f''(x) = 0 \text{ for } x = \pm\frac{\sqrt{3}}{3}$$

| Intervals: | $-\infty < x < -\frac{\sqrt{3}}{3}$ | $-\frac{\sqrt{3}}{3} < x < \frac{\sqrt{3}}{3}$ | $\frac{\sqrt{3}}{3} < x < \infty$ |
|---|---|---|---|
| Sign of $f''$: | $f'' > 0$ | $f'' < 0$ | $f'' > 0$ |
| Conclusion: | Concave upward | Concave downward | Concave upward |

Concave upward: $\left(-\infty, -\frac{\sqrt{3}}{3}\right), \left(\frac{\sqrt{3}}{3}, \infty\right)$

Concave downward: $\left(-\frac{\sqrt{3}}{3}, \frac{\sqrt{3}}{3}\right)$

Points of inflection: $\left(-\frac{\sqrt{3}}{3}, 3\right)$ and $\left(\frac{\sqrt{3}}{3}, 3\right)$

**23.** $f(x) = \sin \dfrac{x}{2}, 0 \le x \le 4\pi$

$f'(x) = \dfrac{1}{2} \cos\left(\dfrac{x}{2}\right)$

$f''(x) = -\dfrac{1}{4} \sin\left(\dfrac{x}{2}\right)$

$f''(x) = 0$ when $x = 0, 2\pi, 4\pi$.

| Intervals: | $0 < x < 2\pi$ | $2\pi < x < 4\pi$ |
|---|---|---|
| Sign of $f''$: | $f'' < 0$ | $f'' > 0$ |
| Conclusion: | Concave downward | Concave upward |

Concave upward: $(2\pi, 4\pi)$

Concave downward: $(0, 2\pi)$

Point of inflection: $(2\pi, 0)$

**25.** $f(x) = \sec\left(x - \dfrac{\pi}{2}\right), 0 < x < 4\pi$

$f'(x) = \sec\left(x - \dfrac{\pi}{2}\right) \tan\left(x - \dfrac{\pi}{2}\right)$

$f''(x) = \sec^3\left(x - \dfrac{\pi}{2}\right) + \sec\left(x - \dfrac{\pi}{2}\right) \tan^2\left(x - \dfrac{\pi}{2}\right) \ne 0$ for any $x$ in the domain of $f$.

$f''$ is not continuous at $x = \pi, x = 2\pi$, and $x = 3\pi$.

| Intervals: | $0 < x < \pi$ | $\pi < x < 2\pi$ | $2\pi < x < 3\pi$ | $3\pi < x < 4\pi$ |
|---|---|---|---|---|
| Sign of $f''$: | $f'' > 0$ | $f'' < 0$ | $f'' > 0$ | $f'' < 0$ |
| Conclusion: | Concave upward | Concave downward | Concave upward | Concave upward |

Concave upward: $(0, \pi), (2\pi, 3\pi)$

Concave downward: $(\pi, 2\pi), (3\pi, 4\pi)$

No point of inflection

**27.** $f(x) = 2 \sin x + \sin 2x, 0 \le x \le 2\pi$

$f'(x) = 2 \cos x + 2 \cos 2x$

$f''(x) = -2 \sin x - 4 \sin 2x = -2 \sin x(1 + 4 \cos x)$

$f''(x) = 0$ when $x = 0, 1.823, \pi, 4.460$.

| Intervals: | $0 < x < 1.823$ | $1.823 < x < \pi$ | $\pi < x < 4.460$ | $4.460 < x < 2\pi$ |
|---|---|---|---|---|
| Sign of $f''$: | $f'' < 0$ | $f'' > 0$ | $f'' < 0$ | $f'' > 0$ |
| Conclusion: | Concave downward | Concave upward | Concave downward | Concave upward |

Concave upward: $(1.823, \pi), (4.460, 2\pi)$

Concave downward: $(0, 1.823), (\pi, 4.460)$

Points of inflection: $(1.823, 1.452), (\pi, 0), (4.46, -1.452)$

**29.** $y = e^{-3/x}$

$$y' = \frac{3}{x^2} e^{-3/x}$$

$$y'' = \frac{e^{-3/x}(9 - 6x)}{x^4}$$

$y'' = 0$ when $x = \dfrac{3}{2}$. $y$ is not defined at $x = 0$.

| Test intervals: | $-\infty < x < 0$ | $0 < x < \dfrac{3}{2}$ | $\dfrac{3}{2} < x < \infty$ |
|---|---|---|---|
| Sign of $y''$: | $y'' > 0$ | $y'' > 0$ | $y'' < 0$ |
| Conclusion: | Concave upward | Concave upward | Concave downward |

Point of inflection: $\left(\dfrac{3}{2}, e^{-2}\right)$

Concave upward: $(-\infty, 0), \left(0, \dfrac{3}{2}\right)$

Concave downward: $\left(\dfrac{3}{2}, \infty\right)$

**31.** $f(x) = x - \ln x$, Domain: $x > 0$

$$f'(x) = 1 - \frac{1}{x}$$

$$f''(x) = \frac{1}{x^2}$$

$f''(x) > 0$ on the entire domain of $f$. There are no points of inflection.

Concave upward: $(0, \infty)$

**33.** $f(x) = \arcsin x^{4/5}, \quad -1 \le x \le 1$

$$f'(x) = \frac{4}{5x^{1/5}\sqrt{1 - x^{8/5}}}$$

$$f''(x) = \frac{20x^{8/5} - 4}{25x^{6/5}\left(1 - x^{8/5}\right)^{3/2}}$$

$f''(x) = 0$ when $20x^{8/5} = 4 \Rightarrow x^{8/5} = \dfrac{1}{5} \Rightarrow x = \pm\left(\dfrac{1}{5}\right)^{5/8} \approx \pm 0.3657$.

$f''$ is undefined at $x = 0$.

| Test intervals: | $-1 < x < -\left(\dfrac{1}{5}\right)^{5/8}$ | $-\left(\dfrac{1}{5}\right)^{5/8} < x < 0$ | $0 < x < \left(\dfrac{1}{5}\right)^{5/8}$ | $\left(\dfrac{1}{5}\right)^{5/8} < x < 1$ |
|---|---|---|---|---|
| Sign of $f''$: | $f'' > 0$ | $f'' < 0$ | $f'' < 0$ | $f'' > 0$ |
| Conclusion: | Concave upward | Concave downward | Concave downward | Concave upward |

Points of inflection: $\left(\pm\left(\dfrac{1}{5}\right)^{5/8}, \arcsin\sqrt{\dfrac{1}{5}}\right) \approx (\pm 0.3657, 0.4636)$

Concave upward: $\left(-1, -\left(\dfrac{1}{5}\right)^{5/8}\right), \left(\left(\dfrac{1}{5}\right)^{5/8}, 1\right)$

Concave downward: $\left(-\left(\dfrac{1}{5}\right)^{5/8}, 0\right), \left(0, \left(\dfrac{1}{5}\right)^{5/8}\right)$

**35.** $f(x) = 6x - x^2$

$f'(x) = 6 - 2x$

$f''(x) = -2$

Critical number: $x = 3$

$f''(3) = -2 < 0$

Therefore, $(3, 9)$ is a relative maximum.

**37.** $f(x) = x^3 - 3x^2 + 3$

$f'(x) = 3x^2 - 6x = 3x(x - 2)$

$f''(x) = 6x - 6 = 6(x - 1)$

Critical numbers: $x = 0, x = 2$

$f''(0) = -6 < 0$

Therefore, $(0, 3)$ is a relative maximum.

$f''(2) = 6 > 0$

Therefore, $(2, -1)$ is a relative minimum.

**39.** $f(x) = x^4 - 4x^3 + 2$

$f'(x) = 4x^3 - 12x^2 = 4x^2(x - 3)$

$f''(x) = 12x^2 - 24x = 12x(x - 2)$

Critical numbers: $x = 0, x = 3$

However, $f''(0) = 0$, so you must use the First Derivative Test. $f'(x) < 0$ on the intervals $(-\infty, 0)$ and $(0, 3)$; so, $(0, 2)$ is not an extremum. $f''(3) > 0$ so $(3, -25)$ is a relative minimum.

**41.** $f(x) = x^{2/3} - 3$

$f'(x) = \dfrac{2}{3x^{1/3}}$

$f''(x) = -\dfrac{2}{9x^{4/3}}$

Critical number: $x = 0$

However, $f''(0)$ is undefined, so you must use the First Derivative Test. Because $f'(x) < 0$ on $(-\infty, 0)$ and $f'(x) > 0$ on $(0, \infty)$, $(0, -3)$ is a relative minimum.

**43.** $f(x) = x + \dfrac{4}{x}$

$f'(x) = 1 - \dfrac{4}{x^2} = \dfrac{x^2 - 4}{x^2}$

$f''(x) = \dfrac{8}{x^3}$

Critical numbers: $x = \pm 2$

$f''(-2) = -1 < 0$

Therefore, $(-2, -4)$ is a relative maximum.

$f''(2) = 1 > 0$

Therefore, $(2, 4)$ is a relative minimum.

**45.** $f(x) = \cos x - x, 0 \le x \le 4\pi$

$f'(x) = -\sin x - 1 \le 0$

Therefore, $f$ is non-increasing and there are no relative extrema.

**47.** $y = f(x) = 8x^2 - \ln x$

$f'(x) = 16x - \dfrac{1}{x}$

$f''(x) = 16 + \dfrac{1}{x^2}$

$f'(x) = 0 \Rightarrow 16x = \dfrac{1}{x} \Rightarrow 16x^2 = 1 \Rightarrow x = \pm\dfrac{1}{4}$

Critical number:

$x = \dfrac{1}{4}$   $\left( x = -\dfrac{1}{4} \text{ is not in the domain.} \right)$

$f''\left(\dfrac{1}{4}\right) > 0$

Therefore, $\left(\dfrac{1}{4}, \dfrac{1}{2} - \ln\dfrac{1}{4}\right) = \left(\dfrac{1}{4}, \dfrac{1}{2} + \ln 4\right)$ is a relative minimum.

**49.** $y = f(x) = \dfrac{x}{\ln x}$

Domain: $0 < x < 1, x > 1$

$f'(x) = \dfrac{(\ln x)(1) - (x)(1/x)}{(\ln x)^2} = \dfrac{\ln x - 1}{(\ln x)^2}$

$f''(x) = \dfrac{2 - \ln x}{x(\ln x)}$

Critical number: $x = e$

$f''(e) > 0$

Therefore, $(e, e)$ is a relative minimum.

**51.** $f(x) = \dfrac{e^x + e^{-x}}{2}$

$f'(x) = \dfrac{e^x - e^{-x}}{2}$

$f''(x) = \dfrac{e^x + e^{-x}}{2}$

Critical number: $x = 0$

$f''(0) > 0$

Therefore, $(0, 1)$ is a relative minimum.

**53.** $f(x) = x^2 e^{-x}$

$f'(x) = -x^2 e^{-x} + 2xe^{-x} = xe^{-x}(2 - x)$

$f''(x) = -e^{-x}(2x - x^2) + e^{-x}(2 - 2x)$

$\quad = e^{-x}(x^2 - 4x + 2)$

Critical numbers: $x = 0, 2$

$f''(0) > 0$

Therefore, $(0, 0)$ is a relative minimum.

$f''(2) < 0$

Therefore, $(2, 4e^{-2})$ is a relative maximum.

**55.** $f(x) = 8x(4^{-x})$

$f'(x) = -8(4^{-x})(x \ln 4 - 1)$

$f''(x) = 8(4^{-x}) \ln 4(x \ln 4 - 2)$

Critical number: $x = \dfrac{1}{\ln 4} = \dfrac{1}{2 \ln 2}$

$f''\left(\dfrac{1}{2 \ln 2}\right) < 0$

Therefore, $\left(\dfrac{1}{2 \ln 2}, \dfrac{4e^{-1}}{\ln 2}\right)$ is a relative maximum.

**57.** $f(x) = \operatorname{arcsec} x - x$

$f'(x) = \dfrac{1}{|x| \sqrt{x^2 - 1}} - 1 = 0$ when $|x| \sqrt{x^2 - 1} = 1$.

$x^2(x^2 - 1) = 1$

$x^4 - x^2 - 1 = 0$ when $x^2 = \dfrac{1 + \sqrt{5}}{2}$

or $x = \pm \sqrt{\dfrac{1 + \sqrt{5}}{2}} = \pm 1.272$.

$f''(x) = -\dfrac{1}{x\sqrt{x^2 - 1}\,|x|} - \dfrac{x}{(x^2 - 1)^{3/2}\,|x|}$

$f''(1.272) < 0$

Therefore, $(1.272, -0.606)$ is a relative maximum.

$f''(-1.272) > 0$

Therefore, $(-1.272, 3.747)$ is a relative minimum.

**59.** $f(x) = 0.2x^2(x - 3)^3, [-1, 4]$

(a)  $f'(x) = 0.2x(5x - 6)(x - 3)^2$

$f''(x) = (x - 3)(4x^2 - 9.6x + 3.6)$

$\quad = 0.4(x - 3)(10x^2 - 24x + 9)$

(b)  $f''(0) < 0 \Rightarrow (0, 0)$ is a relative maximum.

$f''\left(\frac{6}{5}\right) > 0 \Rightarrow (1.2, -1.6796)$ is a relative minimum.

Points of inflection:
$(3, 0), (0.4652, -0.7048), (1.9348, -0.9049)$

(c)

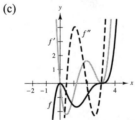

$f$ is increasing when $f' > 0$ and decreasing when $f' < 0$. $f$ is concave upward when $f'' > 0$ and concave downward when $f'' < 0$.

**61.** $f(x) = \sin x - \dfrac{1}{3}\sin 3x + \dfrac{1}{5}\sin 5x, \quad [0, \pi]$

(a) $f'(x) = \cos x - \cos 3x + \cos 5x$

$f'(x) = 0$ when $x = \dfrac{\pi}{6}, x = \dfrac{\pi}{2}, x = \dfrac{5\pi}{6}.$

$f''(x) = -\sin x + 3\sin 3x - 5\sin 5x$

$f''(x) = 0$ when $x = \dfrac{\pi}{6}, x = \dfrac{5\pi}{6},$

$x \approx 1.1731, x \approx 1.9685$

(b) $f''\left(\dfrac{\pi}{2}\right) < 0 \Rightarrow \left(\dfrac{\pi}{2}, 1.53333\right)$ is a relative

maximum.

Points of inflection: $\left(\dfrac{\pi}{6}, 0.2667\right), (1.1731, 0.9638),$

$(1.9685, 0.9637), \left(\dfrac{5\pi}{6}, 0.2667\right)$

**Note:** $(0, 0)$ and $(\pi, 0)$ are not points of inflection
because they are endpoints.

(c)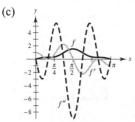

The graph of $f$ is increasing when $f' > 0$ and
decreasing when $f' < 0$. $f$ is concave upward when
$f'' > 0$ and concave downward when $f'' < 0$.

**63.** (a)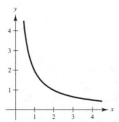

$f' < 0$ means $f$ decreasing

$f'$ increasing means concave upward

(b)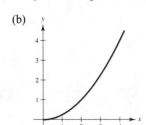

$f' > 0$ means $f$ increasing

$f'$ increasing means concave upward

**65.** Answers will vary. *Sample answer*:

Let $f(x) = x^4.$

$f''(x) = 12x^2$

$f''(0) = 0$, but $(0, 0)$ is not a point of inflection.

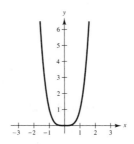

**67.** (a)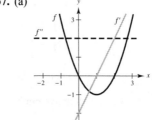

(b)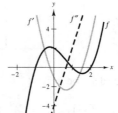

**69.**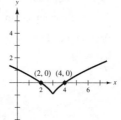

**71.**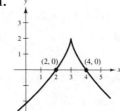

**73.**

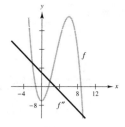

$f''$ is linear.

$f'$ is quadratic.

$f$ is cubic.

$f$ concave upward on $(-\infty, 3)$, downward on $(3, \infty)$.

**75.** (a)

| $n = 1$: | $n = 2$: | $n = 3$: | $n = 4$: |
|---|---|---|---|
| $f(x) = x - 2$ | $f(x) = (x - 2)^2$ | $f(x) = (x - 2)^3$ | $f(x) = (x - 2)^4$ |
| $f'(x) = 1$ | $f'(x) = 2(x - 2)$ | $f'(x) = 3(x - 2)^2$ | $f'(x) = 4(x - 2)^3$ |
| $f''(x) = 0$ | $f''(x) = 2$ | $f''(x) = 6(x - 2)$ | $f''(x) = 12(x - 2)^2$ |
| No point of inflection | No point of inflection | Point of inflection: $(2, 0)$ | No point of inflection |
| | Relative minimum: $(2, 0)$ | | Relative minimum: $(2, 0)$ |

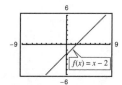 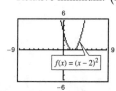 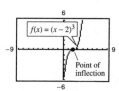 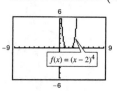

**Conclusion:** If $n \geq 3$ and $n$ is odd, then $(2, 0)$ is point of inflection. If $n \geq 2$ and $n$ is even, then $(2, 0)$ is a relative minimum.

(b) Let $f(x) = (x - 2)^n$, $f'(x) = n(x - 2)^{n-1}$, $f''(x) = n(n - 1)(x - 2)^{n-2}$.

For $n \geq 3$ and odd, $n - 2$ is also odd and the concavity changes at $x = 2$.

For $n \geq 4$ and even, $n - 2$ is also even and the concavity does not change at $x = 2$.

So, $x = 2$ is point of inflection if and only if $n \geq 3$ is odd.

**77.** $f(x) = ax^3 + bx^2 + cx + d$

Relative maximum: $(3, 3)$

Relative minimum: $(5, 1)$

Point of inflection: $(4, 2)$

$f'(x) = 3ax^2 + 2bx + c$, $f''(x) = 6ax + 2b$

$\left.\begin{array}{l} f(3) = 27a + 9b + 3c + d = 3 \\ f(5) = 125a + 25b + 5c + d = 1 \end{array}\right\}$ $98a + 16b + 2c = -2 \Rightarrow 49a + 8b + c = -1$

$f'(3) = 27a + 6b + c = 0$, $f''(4) = 24a + 2b = 0$

$\begin{array}{lll} 49a + 8b + c = -1 & \quad & 24a + 2b = \phantom{-}0 \\ \underline{27a + 6b + c = \phantom{-}0} & \quad & \underline{22a + 2b = -1} \\ 22a + 2b \phantom{+ c} = -1 & \quad & 2a \phantom{+ 2b} = \phantom{-}1 \end{array}$

$a = \frac{1}{2}, b = -6, c = \frac{45}{2}, d = -24$

$f(x) = \frac{1}{2}x^3 - 6x^2 + \frac{45}{2}x - 24$

**79.** $f(x) = ax^3 + bx^2 + cx + d$

Maximum: $(-4, 1)$

Minimum: $(0, 0)$

(a)  $f'(x) = 3ax^2 + 2bx + c, \qquad f''(x) = 6ax + 2b$

$$
\begin{aligned}
f(0) &= 0 \Rightarrow d = 0 \\
f(-4) &= 1 \Rightarrow -64a + 16b - 4c = 1 \\
f'(-4) &= 0 \Rightarrow \quad 48a - 8b + c = 0 \\
f'(0) &= 0 \Rightarrow \qquad\qquad c = 0
\end{aligned}
$$

Solving this system yields $a = \frac{1}{32}$ and $b = 6a = \frac{3}{16}$.

$$f(x) = \frac{1}{32}x^3 + \frac{3}{16}x^2$$

(b)  The plane would be descending at the greatest rate at the point of inflection.

$$f''(x) = 6ax + 2b = \frac{3}{16}x + \frac{3}{8} = 0 \Rightarrow x = -2.$$

Two miles from touchdown.

**81.**  $C = 0.5x^2 + 15x + 5000$

$$\overline{C} = \frac{C}{x} = 0.5x + 15 + \frac{5000}{x}$$

$\overline{C}$ = average cost per unit

$$\frac{d\overline{C}}{dx} = 0.5 - \frac{5000}{x^2} = 0 \text{ when } x = 100$$

By the First Derivative Test, $\overline{C}$ is minimized when $x = 100$ units.

**83.**  $S = \dfrac{5000t^2}{8 + t^2}, \; 0 \le t \le 3$

(a)

| $t$ | 0.5 | 1 | 1.5 | 2 | 2.5 | 3 |
|---|---|---|---|---|---|---|
| $S$ | 151.5 | 555.6 | 1097.6 | 1666.7 | 2193.0 | 2647.1 |

Increasing at greatest rate when $1.5 < t < 2$

(b)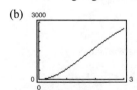

Increasing at greatest rate when $t \approx 1.5$.

(c)  $S = \dfrac{5000t^2}{8 + t^2}$

$$S'(t) = \frac{80,000t}{\left(8 + t^2\right)^2}$$

$$S''(t) = \frac{80,000\left(8 - 3t^2\right)}{\left(8 + t^2\right)^3}$$

$$S''(t) = 0 \text{ for } t = \pm\sqrt{\frac{8}{3}}. \text{ So, } t = \frac{2\sqrt{6}}{3} \approx 1.633 \text{ yrs.}$$

**85.** $f(x) = 2(\sin x + \cos x), \qquad f\left(\dfrac{\pi}{4}\right) = 2\sqrt{2}$

$f'(x) = 2(\cos x - \sin x), \qquad f'\left(\dfrac{\pi}{4}\right) = 0$

$f''(x) = 2(-\sin x - \cos x), \qquad f''\left(\dfrac{\pi}{4}\right) = -2\sqrt{2}$

$P_1(x) = 2\sqrt{2} + 0\left(x - \dfrac{\pi}{4}\right) = 2\sqrt{2}$

$P_1'(x) = 0$

$P_2(x) = 2\sqrt{2} + 0\left(x - \dfrac{\pi}{4}\right) + \dfrac{1}{2}(-2\sqrt{2})\left(x - \dfrac{\pi}{4}\right)^2 = 2\sqrt{2} - \sqrt{2}\left(x - \dfrac{\pi}{4}\right)^2$

$P_2'(x) = -2\sqrt{2}\left(x - \dfrac{\pi}{4}\right)$

$P_2''(x) = -2\sqrt{2}$

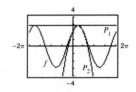

The values of $f$, $P_1$, $P_2$, and their first derivatives are equal at $x = \pi/4$. The values of the second derivatives of $f$ and $P_2$ are equal at $x = \pi/4$. The approximations worsen as you move away from $x = \pi/4$.

**87.** $f(x) = \arctan x, \; a = -1, \qquad f(-1) = -\dfrac{\pi}{4}$

$f'(x) = \dfrac{1}{1 + x^2}, \qquad\qquad f'(-1) = \dfrac{1}{2}$

$f''(x) = -\dfrac{2x}{(1 + x^2)^2}, \qquad f''(-1) = \dfrac{1}{2}$

$P_1(x) = f(-1) + f'(-1)(x + 1) = -\dfrac{\pi}{4} + \dfrac{1}{2}(x + 1)$

$P_1'(x) = \dfrac{1}{2}$

$P_2(x) = f(-1) + f'(-1)(x + 1) + \dfrac{1}{2}f''(-1)(x + 1)^2 = -\dfrac{\pi}{4} + \dfrac{1}{2}(x + 1) + \dfrac{1}{4}(x + 1)^2$

$P_2'(x) = \dfrac{1}{2} + \dfrac{1}{2}(x + 1)$

$P_2''(x) = \dfrac{1}{2}$

The values of $f$, $P_1$, $P_2$, and their first derivatives are equal when $x = -1$. The approximations worsen as you move away from $x = -1$.

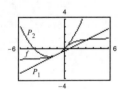

**89.**  $f(x) = x \sin\left(\dfrac{1}{x}\right)$

$f'(x) = x\left[-\dfrac{1}{x^2}\cos\left(\dfrac{1}{x}\right)\right] + \sin\left(\dfrac{1}{x}\right) = -\dfrac{1}{x}\cos\left(\dfrac{1}{x}\right) + \sin\left(\dfrac{1}{x}\right)$

$f''(x) = -\dfrac{1}{x}\left[\dfrac{1}{x^2}\sin\left(\dfrac{1}{x}\right)\right] + \dfrac{1}{x^2}\cos\left(\dfrac{1}{x}\right) - \dfrac{1}{x^2}\cos\left(\dfrac{1}{x}\right) = -\dfrac{1}{x^3}\sin\left(\dfrac{1}{x}\right) = 0$

$x = \dfrac{1}{\pi}$

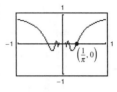

Point of inflection: $\left(\dfrac{1}{\pi}, 0\right)$

When $x > 1/\pi$, $f'' < 0$, so the graph is concave downward.

**91.** True. Let $y = ax^3 + bx^2 + cx + d,\ a \ne 0$. Then
$y'' = 6ax + 2b = 0$ when $x = -(b/3a)$, and the concavity changes at this point.

**93.** False. Concavity is determined by $f''$. For example, let
$f(x) = x$ and $c = 2$. $f'(c) = f'(2) > 0$, but $f$ is not concave upward at $c = 2$.

**95.** $f$ and $g$ are concave upward on $(a, b)$ implies that $f'$ and $g'$ are increasing on $(a, b)$, and $f'' > 0$ and $g'' > 0$.

So, $(f + g)'' > 0 \Rightarrow f + g$ is concave upward on $(a, b)$ by Theorem 4.7.

# Section 4.5    Limits at Infinity

**1.**  $f(x) = \dfrac{2x^2}{x^2 + 2}$

No vertical asymptotes

Horizontal asymptote: $y = 2$

Matches (f).

**3.**  $f(x) = \dfrac{x}{x^2 + 2}$

No vertical asymptotes

Horizontal asymptote: $y = 0$

$f(1) < 1$

Matches (d).

**5.**  $f(x) = \dfrac{4 \sin x}{x^2 + 1}$

No vertical asymptotes

Horizontal asymptote: $y = 0$

$f(1) > 1$

Matches (b).

**7.**  $f(x) = \dfrac{4x + 3}{2x - 1}$

| $x$ | $10^0$ | $10^1$ | $10^2$ | $10^3$ | $10^4$ | $10^5$ | $10^6$ |
|-----|--------|--------|--------|--------|--------|--------|--------|
| $f(x)$ | 7 | 2.26 | 2.025 | 2.0025 | 2.0003 | 2 | 2 |

$\lim\limits_{x \to \infty} f(x) = 2$

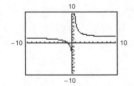

**9.** $f(x) = \dfrac{-6x}{\sqrt{4x^2 + 5}}$

| $x$ | $10^0$ | $10^1$ | $10^2$ | $10^3$ | $10^4$ | $10^5$ | $10^6$ |
|---|---|---|---|---|---|---|---|
| $f(x)$ | $-2$ | $-2.98$ | $-2.9998$ | $-3$ | $-3$ | $-3$ | $-3$ |

$\displaystyle\lim_{x\to\infty} f(x) = -3$

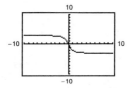

**11.** $f(x) = 5 - \dfrac{1}{x^2 + 1}$

| $x$ | $10^0$ | $10^1$ | $10^2$ | $10^3$ | $10^4$ | $10^5$ | $10^6$ |
|---|---|---|---|---|---|---|---|
| $f(x)$ | $4.5$ | $4.99$ | $4.9999$ | $4.999999$ | $5$ | $5$ | $5$ |

$\displaystyle\lim_{x\to\infty} f(x) = 5$

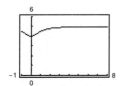

**13.** (a) $h(x) = \dfrac{f(x)}{x^2} = \dfrac{5x^3 - 3x^2 + 10x}{x^2} = 5x - 3 + \dfrac{10}{x}$

$\displaystyle\lim_{x\to\infty} h(x) = \infty$   (Limit does not exist)

(b) $h(x) = \dfrac{f(x)}{x^3} = \dfrac{5x^3 - 3x^2 + 10x}{x^3} = 5 - \dfrac{3}{x} + \dfrac{10}{x^2}$

$\displaystyle\lim_{x\to\infty} h(x) = 5$

(c) $h(x) = \dfrac{f(x)}{x^4} = \dfrac{5x^3 - 3x^2 + 10x}{x^4} = \dfrac{5}{x} - \dfrac{3}{x^2} + \dfrac{10}{x^3}$

$\displaystyle\lim_{x\to\infty} h(x) = 0$

**15.** (a) $\displaystyle\lim_{x\to\infty} \dfrac{x^2 + 2}{x^3 - 1} = 0$

(b) $\displaystyle\lim_{x\to\infty} \dfrac{x^2 + 2}{x^2 - 1} = 1$

(c) $\displaystyle\lim_{x\to\infty} \dfrac{x^2 + 2}{x - 1} = \infty$   (Limit does not exist)

**17.** (a) $\displaystyle\lim_{x\to\infty} \dfrac{5 - 2x^{3/2}}{3x^2 - 4} = 0$

(b) $\displaystyle\lim_{x\to\infty} \dfrac{5 - 2x^{3/2}}{3x^{3/2} - 4} = -\dfrac{2}{3}$

(c) $\displaystyle\lim_{x\to\infty} \dfrac{5 - 2x^{3/2}}{3x - 4} = -\infty$   (Limit does not exist)

**19.** $\displaystyle\lim_{x\to\infty}\left(4 + \dfrac{3}{x}\right) = 4 + 0 = 4$

**21.** $\displaystyle\lim_{x\to\infty} \dfrac{2x - 1}{3x + 2} = \lim_{x\to\infty} \dfrac{2 - (1/x)}{3 + (2/x)} = \dfrac{2 - 0}{3 + 0} = \dfrac{2}{3}$

**23.** $\displaystyle\lim_{x\to\infty} \dfrac{x}{x^2 - 1} = \lim_{x\to\infty} \dfrac{1/x}{1 - (1/x^2)} = \dfrac{0}{1} = 0$

**25.** $\displaystyle \lim_{x \to -\infty} \frac{x}{\sqrt{x^2 - x}}$

$= \displaystyle \lim_{x \to -\infty} \frac{1}{\left( \dfrac{\sqrt{x^2 - x}}{-\sqrt{x^2}} \right)}$

$= \displaystyle \lim_{x \to -\infty} \frac{-1}{\sqrt{1 - (1/x)}}$

$= -1, \left(\text{for } x < 0 \text{ we have } x = -\sqrt{x^2}\right)$

**27.** $\displaystyle \lim_{x \to -\infty} \frac{2x + 1}{\sqrt{x^2 - x}}$

$= \displaystyle \lim_{x \to -\infty} \frac{2 + \dfrac{1}{x}}{\left( \dfrac{\sqrt{x^2 - x}}{-\sqrt{x^2}} \right)}$

$= \displaystyle \lim_{x \to -\infty} \frac{-2 - \left(\dfrac{1}{x}\right)}{\sqrt{1 - \dfrac{1}{x}}}$

$= -2, \left(\text{for } x < 0, \ x = -\sqrt{x^2}\right)$

**29.** $\displaystyle \lim_{x \to \infty} \frac{\sqrt{x^2 - 1}}{2x - 1} = \lim_{x \to \infty} \frac{\sqrt{x^2 - 1}/\sqrt{x^2}}{2 - 1/x}$

$= \displaystyle \lim_{x \to \infty} \frac{\sqrt{1 - 1/x^2}}{2 - 1/x} = \frac{1}{2}$

**31.** $\displaystyle \lim_{x \to \infty} \frac{x + 1}{\left(x^2 + 1\right)^{1/3}} = \lim_{x \to \infty} \frac{x + 1}{\left(x^2 + 1\right)^{1/3}} \left( \frac{1/x^{2/3}}{1/\left(x^2\right)^{1/3}} \right)$

$= \displaystyle \lim_{x \to \infty} \frac{x^{1/3} + 1/x^{2/3}}{\left(1 + 1/x^2\right)^{1/3}} = \infty$

Limit does not exist.

**33.** $\displaystyle \lim_{x \to \infty} \frac{1}{2x + \sin x} = 0$

**35.** Because $(-1/x) \le (\sin 2x)/x \le (1/x)$ for all $x \ne 0$, you have by the Squeeze Theorem,

$\displaystyle \lim_{x \to \infty} -\frac{1}{x} \le \lim_{x \to \infty} \frac{\sin 2x}{x} \le \lim_{x \to \infty} \frac{1}{x}$

$0 \le \displaystyle \lim_{x \to \infty} \frac{\sin 2x}{x} \le 0.$

Therefore, $\displaystyle \lim_{x \to \infty} \frac{\sin 2x}{x} = 0.$

**37.** $\displaystyle \lim_{x \to \infty} \left(2 - 5e^{-x}\right) = 2$

**39.** $\displaystyle \lim_{x \to \infty} \log_{10}\left(1 + 10^{-x}\right) = 0$

**41.** $\displaystyle \lim_{t \to \infty} \left(8t^{-1} - \arctan t\right) = \lim_{t \to \infty} \left(\frac{8}{t}\right) - \lim_{t \to \infty} \arctan t$

$= 0 - \dfrac{\pi}{2} = -\dfrac{\pi}{2}$

**43.** $f(x) = \dfrac{|x|}{x + 1}$

$\displaystyle \lim_{x \to \infty} \frac{|x|}{x + 1} = 1$

$\displaystyle \lim_{x \to -\infty} \frac{|x|}{x + 1} = -1$

Therefore, $y = 1$ and $y = -1$ are both horizontal asymptotes.

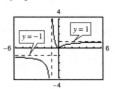

**45.** $f(x) = \dfrac{3x}{\sqrt{x^2 + 2}}$

$\displaystyle \lim_{x \to \infty} f(x) = 3$

$\displaystyle \lim_{x \to -\infty} f(x) = -3$

Therefore, $y = 3$ and $y = -3$ are both horizontal asymptotes.

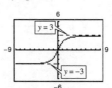

**47.** $\displaystyle \lim_{x \to \infty} x \sin \frac{1}{x} = \lim_{t \to 0^+} \frac{\sin t}{t} = 1$

(Let $x = 1/t$.)

**49.** $\lim\limits_{x \to -\infty} \left( x + \sqrt{x^2 + 3} \right) = \lim\limits_{x \to -\infty} \left[ \left( x + \sqrt{x^2 + 3} \right) \cdot \dfrac{x - \sqrt{x^2 + 3}}{x - \sqrt{x^2 + 3}} \right] = \lim\limits_{x \to -\infty} \dfrac{-3}{x - \sqrt{x^2 + 3}} = 0$

**51.** $\lim\limits_{x \to -\infty} \left( 3x + \sqrt{9x^2 - x} \right) = \lim\limits_{x \to -\infty} \left[ \left( 3x + \sqrt{9x^2 - x} \right) \cdot \dfrac{3x - \sqrt{9x^2 - x}}{3x - \sqrt{9x^2 - x}} \right]$

$$= \lim\limits_{x \to -\infty} \dfrac{x}{3x - \sqrt{9x^2 - x}}$$

$$= \lim\limits_{x \to -\infty} \dfrac{1}{3 - \dfrac{\sqrt{9x^2 - x}}{-\sqrt{x^2}}} \quad \left( \text{for } x < 0 \text{ you have } x = -\sqrt{x^2} \right)$$

$$= \lim\limits_{x \to -\infty} \dfrac{1}{3 + \sqrt{9 - (1/x)}} = \dfrac{1}{6}$$

**53.**

| $x$ | $10^0$ | $10^1$ | $10^2$ | $10^3$ | $10^4$ | $10^5$ | $10^6$ |
|---|---|---|---|---|---|---|---|
| $f(x)$ | 1 | 0.513 | 0.501 | 0.500 | 0.500 | 0.500 | 0.500 |

$$\lim\limits_{x \to \infty} \left( x - \sqrt{x(x-1)} \right) = \lim\limits_{x \to \infty} \dfrac{x - \sqrt{x^2 - x}}{1} \cdot \dfrac{x + \sqrt{x^2 - x}}{x + \sqrt{x^2 - x}} = \lim\limits_{x \to \infty} \dfrac{x}{x + \sqrt{x^2 - x}} = \lim\limits_{x \to \infty} \dfrac{1}{1 + \sqrt{1 - (1/x)}} = \dfrac{1}{2}$$

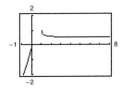

**55.**

| $x$ | $10^0$ | $10^1$ | $10^2$ | $10^3$ | $10^4$ | $10^5$ | $10^6$ |
|---|---|---|---|---|---|---|---|
| $f(x)$ | 0.479 | 0.500 | 0.500 | 0.500 | 0.500 | 0.500 | 0.500 |

Let $x = 1/t$.

$$\lim\limits_{x \to \infty} x \sin\left( \dfrac{1}{2x} \right) = \lim\limits_{t \to 0^+} \dfrac{\sin(t/2)}{t} = \lim\limits_{t \to 0^+} \dfrac{1}{2} \dfrac{\sin(t/2)}{t/2} = \dfrac{1}{2}$$

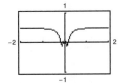

**57.** (a) $\lim\limits_{x \to \infty} f(x) = 4$ means that $f(x)$ approaches 4 as $x$ becomes large.

(b) $\lim\limits_{x \to -\infty} f(x) = 2$ means that $f(x)$ approaches 2 as $x$ becomes very large (in absolute value) and negative.

**59.** $x = 2$ is a critical number.

$f'(x) < 0$ for $x < 2$.

$f'(x) > 0$ for $x > 2$.

$\lim\limits_{x \to -\infty} f(x) = \lim\limits_{x \to \infty} f(x) = 6$

For example, let

$$f(x) = \dfrac{-6}{0.1(x-2)^2 + 1} + 6.$$

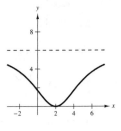

**61. (a)**

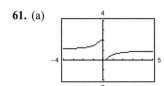

(b) When $x$ increases without bound, $1/x$ approaches zero and $e^{1/x}$ approaches 1. Therefore, $f(x)$ approaches $2/(1 + 1) = 1$. So, $f(x)$ has a horizontal asymptote at $y = 1$. As $x$ approaches zero from the right, $1/x$ approaches $\infty$, $e^{1/x}$ approaches $\infty$, and $f(x)$ approaches zero. As $x$ approaches zero from the left, $1/x$ approaches $-\infty$, $e^{1/x}$ approaches zero, and $f(x)$ approaches 2. The limit does not exist because the left limit does not equal the right limit. Therefore, $x = 0$ is a nonremovable discontinuity.

**63.** $y = \dfrac{x}{1 - x}$

Intercept: $(0, 0)$

Symmetry: none

Horizontal asymptote: $y = -1$

Vertical asymptote: $x = 1$

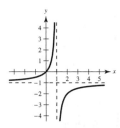

**65.** $y = \dfrac{x + 1}{x^2 - 4}$

Intercepts: $(0, -1/4), (-1, 0)$

Symmetry: none

Horizontal asymptote: $y = 0$

Vertical asymptotes: $x = \pm 2$

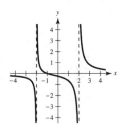

**67.** $y = \dfrac{x^2}{x^2 + 16}$

Intercept: $(0, 0)$

Symmetry: $y$-axis

Horizontal asymptote: $y = 1$

$y' = \dfrac{32x}{\left(x^2 + 16\right)^2}$

Relative minimum: $(0, 0)$

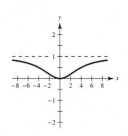

**69.** $xy^2 = 9$

Domain: $x > 0$

Intercepts: none

Symmetry: $x$-axis

$y = \pm \dfrac{3}{\sqrt{x}}$

Horizontal asymptote: $y = 0$

Vertical asymptote: $x = 0$

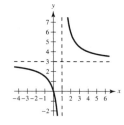

**71.** $y = \dfrac{3x}{x - 1}$

Intercept: $(0, 0)$

Symmetry: none

Horizontal asymptote: $y = 3$

Vertical asymptote: $x = 1$

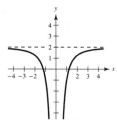

**73.** $y = 2 - \dfrac{3}{x^2} = \dfrac{2x^2 - 3}{x^2}$

Intercepts: $\left(\pm\sqrt{\dfrac{3}{2}}, 0\right)$

Symmetry: $y$-axis

Horizontal asymptote: $y = 2$

Vertical asymptote: $x = 0$

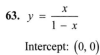

**75.** $y = 3 + \dfrac{2}{x}$

Intercept:

$y = 0 = 3 + \dfrac{2}{x} \Rightarrow \dfrac{2}{x} = -3 \Rightarrow x = -\dfrac{2}{3}; \left(-\dfrac{2}{3}, 0\right)$

Symmetry: none

Horizontal asymptote: $y = 3$

Vertical asymptote: $x = 0$

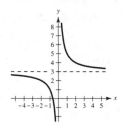

**77.** $y = \dfrac{x^3}{\sqrt{x^2 - 4}}$

Domain: $(-\infty, -2), (2, \infty)$

Intercepts: none

Symmetry: origin

Horizontal asymptote: none

Vertical asymptotes: $x = \pm 2$ (discontinuities)

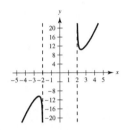

**79.** $f(x) = 9 - \dfrac{5}{x^2}$

Domain: all $x \neq 0$

$f'(x) = \dfrac{10}{x^3} \Rightarrow$ No relative extrema

$f''(x) = -\dfrac{30}{x^4} \Rightarrow$ No points of inflection

Vertical asymptote: $x = 0$

Horizontal asymptote: $y = 9$

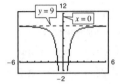

**81.** $f(x) = \dfrac{x - 2}{x^2 - 4x + 3} = \dfrac{x - 2}{(x - 1)(x - 3)}$

$f'(x) = \dfrac{(x^2 - 4x + 3) - (x - 2)(2x - 4)}{(x^2 - 4x + 3)^2} = \dfrac{-x^2 + 4x - 5}{(x^2 - 4x + 3)^2} \neq 0$

$f''(x) = \dfrac{(x^2 - 4x + 3)^2(-2x + 4) - (-x^2 + 4x - 5)(2)(x^2 - 4x + 3)(2x - 4)}{(x^2 - 4x + 3)^4}$

$= \dfrac{2(x^3 - 6x^2 + 15x - 14)}{(x^2 - 4x + 3)^3} = \dfrac{2(x - 2)(x^2 - 4x + 7)}{(x^2 - 4x + 3)^3} = 0$ when $x = 2$.

Because $f''(x) > 0$ on $(1, 2)$ and $f''(x) < 0$ on $(2, 3)$, then $(2, 0)$ is a point of inflection.

Vertical asymptotes: $x = 1, x = 3$

Horizontal asymptote: $y = 0$

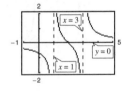

**83.** $f(x) = \dfrac{3x}{\sqrt{4x^2 + 1}}$

$f'(x) = \dfrac{3}{(4x^2 + 1)^{3/2}} \Rightarrow$ No relative extrema

$f''(x) = \dfrac{-36x}{(4x^2 + 1)^{5/2}} = 0$ when $x = 0$.

Point of inflection: $(0, 0)$

Horizontal asymptotes: $y = \pm\dfrac{3}{2}$

No vertical asymptotes

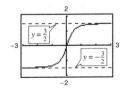

**85.** $g(x) = \sin\left(\dfrac{x}{x - 2}\right), 3 < x < \infty$

$g'(x) = \dfrac{-2\cos\left(\dfrac{x}{x - 2}\right)}{(x - 2)^2}$

Horizontal asymptote: $y = \sin(1)$

Relative maximum:

$\dfrac{x}{x - 2} = \dfrac{\pi}{2} \Rightarrow x = \dfrac{2\pi}{\pi - 2} \approx 5.5039$

No vertical asymptotes

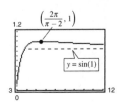

**87.** $f(x) = 2 + (x^2 - 3)e^{-x}$

$f'(x) = -e^{-x}(x + 1)(x - 3)$

Critical numbers: $x = -1, x = 3$

Relative minimum: $(-1, 2 - 2e) \approx (-1, -3.4366)$

Relative maximum: $(3, 2 + 6e^{-3}) \approx (3, 2.2987)$

Horizontal asymptote: $y = 2$

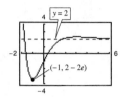

**89.** $f(x) = \dfrac{x^3 - 3x^2 + 2}{x(x - 3)}$, $g(x) = x + \dfrac{2}{x(x - 3)}$

(a)

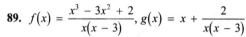

(b) $f(x) = \dfrac{x^3 - 3x^2 + 2}{x(x - 3)}$

$= \dfrac{x^2(x - 3)}{x(x - 3)} + \dfrac{2}{x(x - 3)}$

$= x + \dfrac{2}{x(x - 3)} = g(x)$

(c)

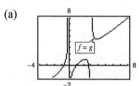

The graph appears as the slant asymptote $y = x$.

**91.** $\displaystyle\lim_{v_1/v_2 \to \infty} 100\left[1 - \dfrac{1}{(v_1/v_2)^c}\right] = 100[1 - 0] = 100\%$

**93.** $\displaystyle\lim_{t \to \infty} N(t) = \infty$

$\displaystyle\lim_{t \to \infty} E(t) = c$

**95.** (a) $\displaystyle\lim_{n \to \infty} \dfrac{0.83}{1 + e^{-0.2n}} = 0.83 = 83\%$

(b) $P' = \dfrac{0.166 e^{-0.2n}}{(1 + e^{-0.2n})^2}$

$P'(3) \approx 0.038$

$P'(10) \approx 0.017$

**97.** $f(x) = \dfrac{2x^2}{x^2 + 2}$

(a) $\displaystyle\lim_{x \to \infty} f(x) = 2 = L$

(b) $f(x_1) + \varepsilon = \dfrac{2x_1^2}{x_1^2 + 2} + \varepsilon = 2$

$2x_1^2 + \varepsilon x_1^2 + 2\varepsilon = 2x_1^2 + 4$

$x_1^2 \varepsilon = 4 - 2\varepsilon$

$x_1 = \sqrt{\dfrac{4 - 2\varepsilon}{\varepsilon}}$

$x_2 = -x_1$ by symmetry

(c) Let $M = \sqrt{\dfrac{4 - 2\varepsilon}{\varepsilon}} > 0$. For $x > M$:

$x > \sqrt{\dfrac{4 - 2\varepsilon}{\varepsilon}}$

$x^2 \varepsilon > 4 - 2\varepsilon$

$2x^2 + x^2 \varepsilon + 2\varepsilon > 2x^2 + 4$

$\dfrac{2x^2}{x^2 + 2} + \varepsilon > 2$

$\left|\dfrac{2x^2}{x^2 + 2} - 2\right| > |-\varepsilon| = \varepsilon$

$|f(x) - L| > \varepsilon$

(d) Similarly, $N = -\sqrt{\dfrac{4 - 2\varepsilon}{\varepsilon}}$.

**99.** $\displaystyle\lim_{x \to \infty} \dfrac{3x}{\sqrt{x^2 + 3}} = 3$

$f(x_1) + \varepsilon = \dfrac{3x_1}{\sqrt{x_1^2 + 3}} + \varepsilon = 3$

$3x_1 = (3 - \varepsilon)\sqrt{x_1^2 + 3}$

$9x_1^2 = (3 - \varepsilon)^2(x_1^2 + 3)$

$9x_1^2 - (3 - \varepsilon)^2 x_1^2 = 3(3 - \varepsilon)^2$

$x_1^2(9 - 9 + 6\varepsilon - \varepsilon^2) = 3(3 - \varepsilon)^2$

$x_1^2 = \dfrac{3(3 - \varepsilon)^2}{6\varepsilon - \varepsilon^2}$

$x_1 = (3 - \varepsilon)\sqrt{\dfrac{3}{6\varepsilon - \varepsilon^2}}$

Let $M = x_1 = (3 - \varepsilon)\sqrt{\dfrac{3}{6\varepsilon - \varepsilon^2}}$

(a) When $\varepsilon = 0.5$:

$M = (3 - 0.5)\sqrt{\dfrac{3}{6(0.5) - (0.5)^2}} = \dfrac{5\sqrt{33}}{11}$

(b) When $\varepsilon = 0.1$:

$M = (3 - 0.1)\sqrt{\dfrac{3}{6(0.1) - (0.1)^2}} = \dfrac{29\sqrt{177}}{59}$

**101.** $\lim\limits_{x\to\infty} \dfrac{1}{x^2} = 0$. Let $\varepsilon > 0$ be given. You need

$M > 0$ such that

$$\left| f(x) - L \right| = \left| \dfrac{1}{x^2} - 0 \right| = \dfrac{1}{x^2} < \varepsilon \text{ whenever } x > M.$$

$$x^2 > \dfrac{1}{\varepsilon} \Rightarrow x > \dfrac{1}{\sqrt{\varepsilon}}$$

Let $M = \dfrac{1}{\sqrt{\varepsilon}}$.

For $x > M$, you have

$$x > \dfrac{1}{\sqrt{\varepsilon}} \Rightarrow x^2 > \dfrac{1}{\varepsilon} \Rightarrow \dfrac{1}{x^2} < \varepsilon \Rightarrow \left| f(x) - L \right| < \varepsilon.$$

**103.** $\lim\limits_{x\to-\infty} \dfrac{1}{x^3} = 0$. Let $\varepsilon > 0$. You need $N < 0$ such that

$$\left| f(x) - L \right| = \left| \dfrac{1}{x^3} - 0 \right| = \dfrac{-1}{x^3} < \varepsilon \text{ whenever } x < N.$$

$$\dfrac{-1}{x^3} < \varepsilon \Rightarrow -x^3 > \dfrac{1}{\varepsilon} \Rightarrow x < \dfrac{-1}{\varepsilon^{1/3}}$$

Let $N = \dfrac{-1}{\sqrt[3]{\varepsilon}}$.

For $x < N = \dfrac{-1}{\sqrt[3]{\varepsilon}}$,

$$\dfrac{1}{x} > -\sqrt[3]{\varepsilon}$$

$$-\dfrac{1}{x} < \sqrt[3]{\varepsilon}$$

$$-\dfrac{1}{x^3} < \varepsilon$$

$$\Rightarrow \left| f(x) - L \right| < \varepsilon.$$

**107.** $\lim\limits_{x\to\infty} \dfrac{p(x)}{q(x)} = \lim\limits_{x\to\infty} \dfrac{a_n x^n + \cdots + a_1 x + a_0}{b_m x^m + \cdots + b_1 x + b_0}$

Divide $p(x)$ and $q(x)$ by $x^m$.

**Case 1:** If $n < m$: $\lim\limits_{x\to\infty} \dfrac{p(x)}{q(x)} = \lim\limits_{x\to\infty} \dfrac{\dfrac{a_n}{x^{m-n}} + \cdots + \dfrac{a_1}{x^{m-1}} + \dfrac{a_0}{x^m}}{b_m + \cdots + \dfrac{b_1}{x^{m-1}} + \dfrac{b_0}{x^m}} = \dfrac{0 + \cdots + 0 + 0}{b_m + \cdots + 0 + 0} = \dfrac{0}{b_m} = 0.$

**Case 2:** If $m = n$: $\lim\limits_{x\to\infty} \dfrac{p(x)}{q(x)} = \lim\limits_{x\to\infty} \dfrac{a_n + \cdots + \dfrac{a_1}{x^{m-1}} + \dfrac{a_0}{x^m}}{b_m + \cdots + \dfrac{b_1}{x^{m-1}} + \dfrac{b_0}{x^m}} = \dfrac{a_n + \cdots + 0 + 0}{b_m + \cdots + 0 + 0} = \dfrac{a_n}{b_m}.$

**Case 3:** If $n > m$: $\lim\limits_{x\to\infty} \dfrac{p(x)}{q(x)} = \lim\limits_{x\to\infty} \dfrac{a_n x^{n-m} + \cdots + \dfrac{a_1}{x^{m-1}} + \dfrac{a_0}{x^m}}{b_m + \cdots + \dfrac{b_1}{x^{m-1}} + \dfrac{b_0}{x^m}} = \dfrac{\pm\infty + \cdots + 0}{b_m + \cdots + 0} = \pm\infty.$

**109.** False. Let $f(x) = \dfrac{2x}{\sqrt{x^2 + 2}}$. (See Exercise 57(b).)

---

**105.** line: $mx - y + 4 = 0$

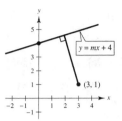

(a) $d = \dfrac{\left| A x_1 + B y_1 + C \right|}{\sqrt{A^2 + B^2}} = \dfrac{\left| m(3) - 1(1) + 4 \right|}{\sqrt{m^2 + 1}}$

$\qquad = \dfrac{\left| 3m + 3 \right|}{\sqrt{m^2 + 1}}$

(b)

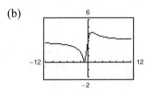

(c) $\lim\limits_{m\to\infty} d(m) = 3 = \lim\limits_{m\to-\infty} d(m)$

The line approaches the vertical line $x = 0$. So, the distance from $(3, 1)$ approaches 3.

# Section 4.6   A Summary of Curve Sketching

**1.**  $y = \dfrac{1}{x-2} - 3$

$y' = \dfrac{1}{(x-2)^2} \Rightarrow$ undefined when $x = 2$

$y'' = \dfrac{2}{(x-2)^3} \Rightarrow$ undefined when $x = 2$

Intercepts: $\left( \dfrac{7}{3}, 0 \right), \left( 0, -\dfrac{7}{2} \right)$

Vertical asymptote: $x = 2$

Horizontal asymptote: $y = -3$

|  | $y$ | $y'$ | $y''$ | Conclusion |
|---|---|---|---|---|
| $-\infty < x < 2$ |  | − | − | Decreasing, concave down |
| $2 < x < \infty$ |  | − | + | Decreasing, concave up |

No relative extrema, no points of inflection

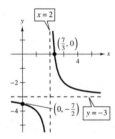

**3.**  $y = \dfrac{x^2}{x^2 + 3}$

$y' = \dfrac{6x}{\left(x^2 + 3\right)^2} = 0$ when $x = 0$.

$y'' = \dfrac{18\left(1 - x^2\right)}{\left(x^2 + 3\right)^3} = 0$ when $x = \pm 1$.

Horizontal asymptote: $y = 1$

|  | $y$ | $y'$ | $y''$ | Conclusion |
|---|---|---|---|---|
| $-\infty < x < -1$ |  | − | − | Decreasing, concave down |
| $x = -1$ | $\dfrac{1}{4}$ | − | 0 | Point of inflection |
| $-1 < x < 0$ |  | − | + | Decreasing, concave up |
| $x = 0$ | 0 | 0 | + | Relative minimum |
| $0 < x < 1$ |  | + | + | Increasing, concave up |
| $x = 1$ | $\dfrac{1}{4}$ | + | 0 | Point of inflection |
| $1 < x < \infty$ |  | + | − | Increasing, concave down |

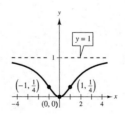

**5.** $y = \dfrac{3x}{x^2 - 1}$

$y' = \dfrac{-3(x^2 + 1)}{(x^2 - 1)^2}$ undefined when $x = \pm 1$

$y'' = \dfrac{6x(x^2 + 3)}{(x^2 - 1)^3}$

Intercept: $(0, 0)$

Symmetry with respect to origin

Vertical asymptotes: $x = \pm 1$

Horizontal asymptote: $y = 0$

|  | $y$ | $y'$ | $y''$ | Conclusion |
|---|---|---|---|---|
| $-\infty < x < -1$ |  | $-$ | $-$ | Decreasing, concave down |
| $-1 < x < 0$ |  | $-$ | $+$ | Decreasing, concave up |
| $x = 0$ | 0 | $-3$ | 0 | Point of inflection |
| $0 < x < 1$ |  | $-$ | $-$ | Decreasing, concave down |
| $1 < x < \infty$ |  | $-$ | $+$ | Decreasing, concave up |

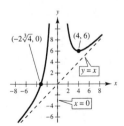

**7.** $f(x) = x + \dfrac{32}{x^2}$

$f'(x) = 1 - \dfrac{64}{x^3} = \dfrac{(x - 4)(x^2 + 4x + 16)}{x^3} = 0$ when $x = 4$ and undefined when $x = 0$.

$f''(x) = \dfrac{192}{x^4}$

Intercept: $\left(-2\sqrt[3]{4}, 0\right)$

Vertical asymptote: $x = 0$

Slant asymptote: $y = x$

|  | $y$ | $y'$ | $y''$ | Conclusion |
|---|---|---|---|---|
| $-\infty < x < 0$ |  | $+$ | $+$ | Increasing, concave up |
| $0 < x < 4$ |  | $-$ | $+$ | Decreasing, concave up |
| $x = 4$ | 6 | 0 | $+$ | Relative minimum |
| $4 < x < \infty$ |  | $+$ | $+$ | Increasing, concave up |

**9.** $y = \dfrac{x^2 - 6x + 12}{x - 4} = x - 2 + \dfrac{4}{x - 4}$

$y' = 1 - \dfrac{4}{(x - 4)^2} = \dfrac{(x - 2)(x - 6)}{(x - 4)^2} = 0$ when $x = 2, 6$ and is undefined when $x = 4$.

$y'' = \dfrac{8}{(x - 4)^3}$

Vertical asymptote: $x = 4$

Slant asymptote: $y = x - 2$

| | $y$ | $y'$ | $y''$ | Conclusion |
|---|---|---|---|---|
| $-\infty < x < 2$ | | + | − | Increasing, concave down |
| $x = 2$ | −2 | 0 | − | Relative maximum |
| $2 < x < 4$ | | − | − | Decreasing, concave down |
| $4 < x < 6$ | | − | + | Decreasing, concave up |
| $x = 6$ | 6 | 0 | + | Relative minimum |
| $6 < x < \infty$ | | + | + | Increasing, concave up |

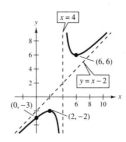

**11.** $y = x\sqrt{4 - x}$,  Domain: $(-\infty, 4]$

$y' = \dfrac{8 - 3x}{2\sqrt{4 - x}} = 0$ when $x = \dfrac{8}{3}$ and undefined when $x = 4$.

$y'' = \dfrac{3x - 16}{4(4 - x)^{3/2}} = 0$ when $x = \dfrac{16}{3}$ and undefined when $x = 4$.

**Note:** $x = \dfrac{16}{3}$ is not in the domain.

| | $y$ | $y'$ | $y''$ | Conclusion |
|---|---|---|---|---|
| $-\infty < x < \dfrac{8}{3}$ | | + | − | Increasing, concave down |
| $x = \dfrac{8}{3}$ | $\dfrac{16}{3\sqrt{3}}$ | 0 | − | Relative maximum |
| $\dfrac{8}{3} < x < 4$ | | − | − | Decreasing, concave down |
| $x = 4$ | 0 | Undefined | Undefined | Endpoint |

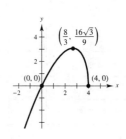

**13.** $y = 3x^{2/3} - 2x$

$y' = 2x^{-1/3} - 2 = \dfrac{2\left(1 - x^{1/3}\right)}{x^{1/3}} = 0$ when $x = 1$ and undefined when $x = 0$.

$y'' = \dfrac{-2}{3x^{4/3}} < 0$ when $x \neq 0$.

|  | $y$ | $y'$ | $y''$ | Conclusion |
|---|---|---|---|---|
| $-\infty < x < 0$ |  | $-$ | $-$ | Decreasing, concave down |
| $x = 0$ | 0 | Undefined | Undefined | Relative minimum |
| $0 < x < 1$ |  | $+$ | $-$ | Increasing, concave down |
| $x = 1$ | 1 | 0 | $-$ | Relative maximum |
| $1 < x < \infty$ |  | $-$ | $-$ | Decreasing, concave down |

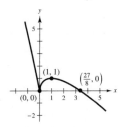

**15.** $y = 2 - x - x^3$

$y' = -1 - 3x^2$

No critical numbers

$y'' = -6x = 0$ when $x = 0$.

|  | $y$ | $y'$ | $y''$ | Conclusion |
|---|---|---|---|---|
| $-\infty < x < 0$ |  | $-$ | $+$ | Decreasing, concave up |
| $x = 0$ | 2 | $-$ | 0 | Point of inflection |
| $0 < x < \infty$ |  | $-$ | $-$ | Decreasing, concave down |

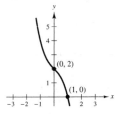

**17.** $y = 3x^4 + 4x^3$

$y' = 12x^3 + 12x^2 = 12x^2(x + 1) = 0$ when $x = 0$, $x = -1$.

$y'' = 36x^2 + 24x = 12x(3x + 2) = 0$ when $x = 0$, $x = -\frac{2}{3}$.

|  | $y$ | $y'$ | $y''$ | Conclusion |
|---|---|---|---|---|
| $-\infty < x < -1$ |  | $-$ | $+$ | Decreasing, concave up |
| $x = -1$ | $-1$ | 0 | $+$ | Relative minimum |
| $-1 < x < -\frac{2}{3}$ |  | $+$ | $+$ | Increasing, concave up |
| $x = -\frac{2}{3}$ | $-\frac{16}{27}$ | $+$ | 0 | Point of inflection |
| $-\frac{2}{3} < x < 0$ |  | $+$ | $-$ | Increasing, concave down |
| $x = 0$ | 0 | 0 | 0 | Point of inflection |
| $0 < x < \infty$ |  | $+$ | $+$ | Increasing, concave up |

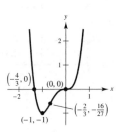

**19.**   $y = x^5 - 5x$

$y' = 5x^4 - 5 = 5(x^4 - 1) = 0$ when $x = \pm 1$.

$y'' = 20x^3 = 0$ when $x = 0$.

|  | $y$ | $y'$ | $y''$ | Conclusion |
|---|---|---|---|---|
| $-\infty < x < -1$ |  | + | − | Increasing, concave down |
| $x = -1$ | 4 | 0 | − | Relative maximum |
| $-1 < x < 0$ |  | − | − | Decreasing, concave down |
| $x = 0$ | 0 | − | 0 | Point of inflection |
| $0 < x < 1$ |  | − | + | Decreasing, concave up |
| $x = 1$ | −4 | 0 | + | Relative minimum |
| $1 < x < \infty$ |  | + | + | Increasing, concave up |

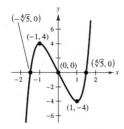

**21.**   $y = |2x - 3|$

$y' = \dfrac{2(2x - 3)}{|2x - 3|}$ undefined at $x = \dfrac{3}{2}$.

$y'' = 0$

|  | $y$ | $y'$ | Conclusion |
|---|---|---|---|
| $-\infty < x < \frac{3}{2}$ |  | − | Decreasing |
| $x = \frac{3}{2}$ | 0 | Undefined | Relative minimum |
| $\frac{3}{2} < x < \infty$ |  | + | Increasing |

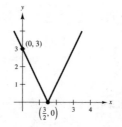

**23.** $f(x) = e^{3x}(2 - x)$

$f'(x) = -e^{3x} + 2(2 - x)e^{3x} = e^{3x}(5 - 3x) = 0$ when $x = \dfrac{5}{3}$.

$f''(x) = -3e^{3x}(-4 + 3x) = 0$ when $x = \dfrac{4}{3}$.

|  | $f(x)$ | $f'(x)$ | $f''(x)$ | Conclusion |
|---|---|---|---|---|
| $-\infty < x < \dfrac{4}{3}$ |  | $+$ | $+$ | Increasing, concave up |
| $x = \dfrac{4}{3}$ | $\dfrac{2e^4}{3}$ | 54.6 | 0 | Point of inflection |
| $\dfrac{4}{3} < x < \dfrac{5}{3}$ |  | $+$ | $-$ | Increasing, concave down |
| $x = \dfrac{5}{3}$ | $\dfrac{e^5}{3}$ | 0 | $-445.2$ | Relative maximum |
| $\dfrac{5}{3} < x < \infty$ |  | $-$ | $-$ | Decreasing, concave down |

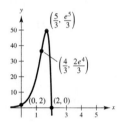

**25.** $g(t) = \dfrac{10}{1 + 4e^{-t}}$

$g'(t) = \dfrac{40e^{-t}}{\left(1 + 4e^{-t}\right)^2} > 0$ for all $t$.

$g''(t) = \dfrac{40e^{-t}\left(4e^{-t} - 1\right)}{\left(1 + 4e^{-t}\right)^3} = 0$ at $t \approx 1.386$.

$\lim\limits_{t \to \infty} g(t) = 10 \Rightarrow t = 10$ is a horizontal asymptote.

$\lim\limits_{t \to -\infty} g(t) = 0 \Rightarrow t = 0$ is a horizontal asymptote.

|  | $g(t)$ | $g'(t)$ | $g''(t)$ | Conclusion |
|---|---|---|---|---|
| $-\infty < t < 1.386$ |  | $+$ | $+$ | Increasing, concave up |
| $t = 1.386$ | 5 | 2.5 | 0 | Point of inflection |
| $1.386 < t < \infty$ |  | $+$ | $-$ | Increasing, concave down |

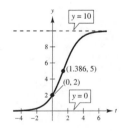

**27.** $y = (x - 1) \ln(x - 1)$, Domain: $x > 1$

$y' = 1 + \ln(x - 1) = 0$ when $\ln(x - 1) = -1 \Rightarrow (x - 1) = e^{-1} \Rightarrow x = 1 + e^{-1}$

$y'' = \dfrac{1}{x - 1}$

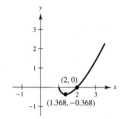

|  | $y$ | $y'$ | $y''$ | Conclusion |
|---|---|---|---|---|
| $1 < x < 1 + e^{-1}$ |  | $-$ | $+$ | Decreasing, concave up |
| $x = 1 + e^{-1}$ | $-e^{-1}$ | $0$ | $e$ | Relative minimum |
| $1 + e^{-1} < x < \infty$ |  | $+$ | $+$ | Increasing, concave up |

**29.** $g(x) = 6 \arcsin\left(\dfrac{x - 2}{2}\right)^2$, Domain: $[0, 4]$

$g'(x) = \dfrac{12(x - 2)}{\sqrt{(4x - x^2)(x^2 - 4x + 8)}} = 0$ when $x = 2$.

$g''(x) = \dfrac{12(x^4 - 8x^3 + 24x^2 - 32x + 32)}{\left[(4x - x^2)(x^2 - 4x + 8)\right]^{3/2}}$

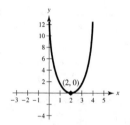

|  | $g(x)$ | $g'(x)$ | $g''(x)$ | Conclusion |
|---|---|---|---|---|
| $0 < x < 2$ |  | $-$ | $+$ | Decreasing, concave up |
| $x = 2$ | $0$ | $0$ | $+$ | Relative minimum |
| $2 < x < 4$ |  | $+$ | $+$ | Increasing, concave down |

**31.** $f(x) = \dfrac{x}{3^{x-3}} = \dfrac{27x}{3^x}$

$f'(x) = \dfrac{27(1 - x \ln 3)}{3^x} = 0 \Rightarrow x = \dfrac{1}{\ln 3} \approx 0.910$

$f''(x) = \dfrac{27 \ln 3(x \ln 3 - 2)}{3^x} = 0 \Rightarrow x = \dfrac{2}{\ln 3} \approx 1.820$

$\lim\limits_{x \to \infty} f(x) = 0$, $\lim\limits_{x \to -\infty} f(x) = -\infty$

Horizontal symptote: $y = 0$

Intercept: $(0, 0)$

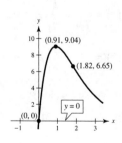

|  | $f(x)$ | $f'(x)$ | $f''(x)$ | Conclusion |
|---|---|---|---|---|
| $-\infty < x < 0.910$ |  | $+$ | $-$ | Increasing, concave down |
| $x = 0.910$ | $9.041$ | $0$ | $-$ | Relative maximum |
| $0.910 < x < 1.820$ |  | $-$ | $-$ | Decreasing, concave down |
| $x = 1.820$ |  | $-$ | $0$ | Point of inflection |
| $1.820 < x < \infty$ | $6.652$ | $-$ | $+$ | Decreasing, concave up |

**33.**  $g(x) = \log_4(x - x^2) = \dfrac{\ln(x - x^2)}{\ln 4}$,    Domain: $0 < x < 1$

$g'(x) = \dfrac{2x - 1}{\ln 4 \cdot x(x - 1)} = 0$ when $x = \dfrac{1}{2}$.

$g''(x) = \dfrac{-2x^2 + 2x - 1}{\ln 4 \cdot x^2(x - 1)^2}$

|  | $g(x)$ | $g'(x)$ | $g''(x)$ | Conclusion |
|---|---|---|---|---|
| $0 < x < \dfrac{1}{2}$ |  | + | – | Increasing, concave down |
| $x = \dfrac{1}{2}$ | –1 | 0 | – | Relative maximum |
| $\dfrac{1}{2} < x < 1$ |  | – | – | Decreasing, concave down |

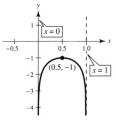

**35.**  $f(x) = \dfrac{20x}{x^2 + 1} - \dfrac{1}{x} = \dfrac{19x^2 - 1}{x(x^2 + 1)}$

$f'(x) = \dfrac{-(19x^4 - 22x^2 - 1)}{x^2(x^2 + 1)^2} = 0$ for $x \approx \pm 1.10$

$f''(x) = \dfrac{2(19x^6 - 63x^9 - 3x^2 - 1)}{x^3(x^2 + 1)^3} = 0$ for $x \approx \pm 1.84$

Vertical asymptote: $x = 0$

Horizontal asymptote: $y = 0$

Minimum: $(-1.10, -9.05)$

Maximum: $(1.10, 9.05)$

Points of inflection: $(-1.84, -7.86)$, $(1.84, 7.86)$

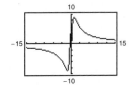

**39.**  $y = \dfrac{x}{2} + \ln\!\left(\dfrac{x}{x + 3}\right)$

$y' = \dfrac{1}{2} + \dfrac{3}{x(x + 3)}$

$y'' = \dfrac{-3(2x + 3)}{x^2(x + 3)^2}$

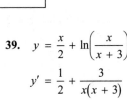

Vertical asymptotes: $x = -3, x = 0$

Slant asymptote: $y = \dfrac{x}{2}$

**37.**  $f(x) = \dfrac{-2x}{\sqrt{x^2 + 7}}$

$f'(x) = \dfrac{-14}{(x^2 + 7)^{3/2}} < 0$

$f''(x) = \dfrac{42x}{(x^2 + 7)^{5/2}} = 0$ at $x = 0$

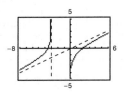

Horizontal asymptotes: $y = \pm 2$

Point of inflection: $(0, 0)$

**41.**  $f(x) = 2x - 4 \sin x, \, 0 \le x \le 2\pi$

$f'(x) = 2 - 4 \cos x$

$f''(x) = 4 \sin x$

$f'(x) = 0 \Rightarrow \cos x = \dfrac{1}{2} \Rightarrow x = \dfrac{\pi}{3}, \dfrac{5\pi}{3}$

$f''(x) = 0 \Rightarrow x = 0, \pi, 2\pi$

Relative minimum: $\left(\dfrac{\pi}{3}, \dfrac{2\pi}{3} - 2\sqrt{3}\right)$

Relative maximum: $\left(\dfrac{5\pi}{3}, \dfrac{10\pi}{3} + 2\sqrt{3}\right)$

Points of inflection: $(0, 0), (\pi, 2\pi), (2\pi, 4\pi)$

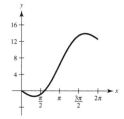

**43.**  $y = \sin x - \dfrac{1}{18} \sin 3x, \, 0 \le x \le 2\pi$

$y' = \cos x - \dfrac{1}{6} \cos 3x$

$\phantom{y'} = \cos x - \dfrac{1}{6}\left[\cos 2x \cos x - \sin 2x \sin x\right]$

$\phantom{y'} = \cos x - \dfrac{1}{6}\left[\left(1 - 2\sin^2 x\right)\cos x - 2\sin^2 x \cos x\right]$

$\phantom{y'} = \cos x\left[1 - \dfrac{1}{6}\left(1 - 2\sin^2 x - 2\sin^2 x\right)\right] = \cos x\left[\dfrac{5}{6} + \dfrac{2}{3}\sin^2 x\right]$

$y' = 0: \qquad \cos x = 0 \Rightarrow x = \pi/2, 3\pi/2$

$\qquad \dfrac{5}{6} + \dfrac{2}{3}\sin^2 x = 0 \Rightarrow \sin^2 x = -5/4, \text{impossible}$

$y'' = -\sin x + \dfrac{1}{2}\sin 3x = 0 \Rightarrow 2\sin x = \sin 3x$

$\qquad\qquad\qquad\qquad\qquad\quad = \sin 2x \cos x + \cos 2x \sin x$

$\qquad\qquad\qquad\qquad\qquad\quad = 2\sin x \cos^2 x + \left(2\cos^2 x - 1\right)\sin x$

$\qquad\qquad\qquad\qquad\qquad\quad = \sin x\left(2\cos^2 x + 2\cos^2 x - 1\right)$

$\qquad\qquad\qquad\qquad\qquad\quad = \sin x\left(4\cos^2 x - 1\right)$

$\sin x = 0 \Rightarrow x = 0, \pi, 2\pi$

$2 = 4\cos^2 x - 1 \Rightarrow \cos x = \pm\sqrt{3}/2 \Rightarrow x = \dfrac{\pi}{6}, \dfrac{5\pi}{6}, \dfrac{7\pi}{6}, \dfrac{11\pi}{6}$

Relative maximum: $\left(\dfrac{\pi}{2}, \dfrac{19}{18}\right)$

Relative minimum: $\left(\dfrac{3\pi}{2}, -\dfrac{19}{18}\right)$

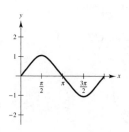

Points of inflection: $\left(\dfrac{\pi}{6}, \dfrac{4}{9}\right), \left(\dfrac{5\pi}{6}, \dfrac{4}{9}\right), (\pi, 0), \left(\dfrac{7\pi}{6}, -\dfrac{4}{9}\right), \left(\dfrac{11\pi}{6}, -\dfrac{4}{9}\right)$

**45.** $y = 2x - \tan x, -\dfrac{\pi}{2} < x < \dfrac{\pi}{2}$

$y' = 2 - \sec^2 x = 0$ when $x = \pm\dfrac{\pi}{4}$.

$y'' = -2 \sec^2 x \tan x = 0$ when $x = 0$.

Relative maximum: $\left(\dfrac{\pi}{4}, \dfrac{\pi}{2} - 1\right)$

Relative minimum: $\left(-\dfrac{\pi}{4}, 1 - \dfrac{\pi}{2}\right)$

Point of inflection: $(0, 0)$

Vertical asymptotes: $x = \pm\dfrac{\pi}{2}$

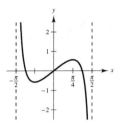

**47.** $y = 2(\csc x + \sec x), 0 < x < \dfrac{\pi}{2}$

$y' = 2(\sec x \tan x - \csc x \cot x) = 0 \Rightarrow x = \dfrac{\pi}{4}$

Relative minimum: $\left(\dfrac{\pi}{4}, 4\sqrt{2}\right)$

Vertical asymptotes: $x = 0, \dfrac{\pi}{2}$

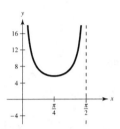

**49.** $g(x) = x \tan x, -\dfrac{3\pi}{2} < x < \dfrac{3\pi}{2}$

$g'(x) = \dfrac{x + \sin x \cos x}{\cos^2 x} = 0$ when $x = 0$.

$g''(x) = \dfrac{2(\cos x + x \sin x)}{\cos^3 x}$

Vertical asymptotes: $x = -\dfrac{3\pi}{2}, -\dfrac{\pi}{2}, \dfrac{\pi}{2}, \dfrac{3\pi}{2}$

Intercepts: $(-\pi, 0), (0, 0), (\pi, 0)$

Symmetric with respect to $y$-axis.

Increasing on $\left(0, \dfrac{\pi}{2}\right)$ and $\left(\dfrac{\pi}{2}, \dfrac{3\pi}{2}\right)$

Points of inflection: $(\pm 2.80, -1)$

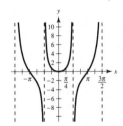

**51.** Because the slope is negative, the function is decreasing on $(2, 8)$, and so $f(3) > f(5)$.

**53.** $f$ is cubic.

$f'$ is quadratic.

$f''$ is linear.

The zeros of $f'$ correspond to the points where the graph of $f$ has horizontal tangents. The zero of $f''$ corresponds to the point where the graph of $f'$ has a horizontal tangent.

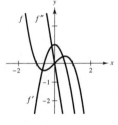

**55.** $f(x) = \dfrac{4(x - 1)^2}{x^2 - 4x + 5}$

Vertical asymptote: none

Horizontal asymptote: $y = 4$

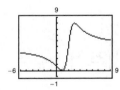

The graph crosses the horizontal asymptote $y = 4$. If a function has a vertical asymptote at $x = c$, the graph would not cross it because $f(c)$ is undefined.

**57.** $h(x) = \dfrac{\sin 2x}{x}$

Vertical asymptote: none

Horizontal asymptote: $y = 0$

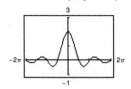

Yes, it is possible for a graph to cross its horizontal asymptote.

It is not possible to cross a vertical asymptote because the function is not continuous there.

**59.** $h(x) = \dfrac{6 - 2x}{3 - x}$

$= \dfrac{2(3 - x)}{3 - x} = \begin{cases} 2, & \text{if } x \neq 3 \\ \text{Undefined}, & \text{if } x = 3 \end{cases}$

The rational function is not reduced to lowest terms.

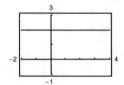

There is a hole at $(3, 2)$.

**61.** $f(x) = -\dfrac{x^2 - 3x - 1}{x - 2} = -x + 1 + \dfrac{3}{x - 2}$

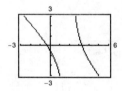

The graph appears to approach the slant asymptote $y = -x + 1$.

**69.** $f(x) = \dfrac{\cos^2 \pi x}{\sqrt{x^2 + 1}}, (0, 4)$

(a)

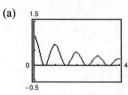

On $(0, 4)$ there seem to be 7 critical numbers: 0.5, 1.0, 1.5, 2.0, 2.5, 3.0, 3.5

**63.** $f(x) = \dfrac{2x^3}{x^2 + 1} = 2x - \dfrac{2x}{x^2 + 1}$

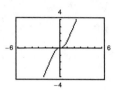

The graph appears to approach the slant asymptote $y = 2x$.

**65.**

 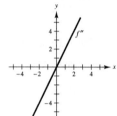

(or any vertical translation of $f$)

**67.**

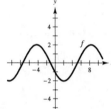 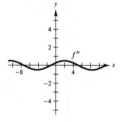

(or any vertical translation of $f$)

(b) $f'(x) = \dfrac{-\cos \pi x \left( x \cos \pi x + 2\pi \left( x^2 + 1 \right) \sin \pi x \right)}{\left( x^2 + 1 \right)^{3/2}} = 0$

Critical numbers $\approx \dfrac{1}{2}, 0.97, \dfrac{3}{2}, 1.98, \dfrac{5}{2}, 2.98, \dfrac{7}{2}$.

The critical numbers where maxima occur appear to be integers in part (a), but approximating them using $f'$ shows that they are not integers.

**71.** Vertical asymptote: $x = 3$

Horizontal asymptote: $y = 0$

$y = \dfrac{1}{x - 3}$

**73.** Vertical asymptote: $x = 3$

Slant asymptote: $y = 3x + 2$

$y = 3x + 2 + \dfrac{1}{x - 3} = \dfrac{3x^2 - 7x - 5}{x - 3}$

**75.** (a)  $f(x) = \ln x, \, g(x) = \sqrt{x}$

$f'(x) = \dfrac{1}{x}, \, g'(x) = \dfrac{1}{2\sqrt{x}}$

For $x > 4, g'(x) > f'(x)$. $g$ is increasing at a higher rate than $f$ for "large" values of $x$.

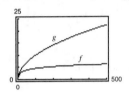

(b)  $f(x) = \ln x, \, g(x) = \sqrt[4]{x}$

$f'(x) = \dfrac{1}{x}, \, g'(x) = \dfrac{1}{4\sqrt[4]{x^3}}$

For $x > 256, g'(x) > f'(x)$. $g$ is increasing at a higher rate than $f$ for "large" values of $x$. $f(x) = \ln x$ increases very slowly for "large" values of $x$.

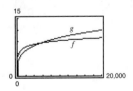

**77.** (a)  $f'(x) = 0$ at $x_0, x_2$ and $x_4$ (horizontal tangent).

(b)  $f''(x) = 0$ at $x_2$ and $x_3$ (point of inflection).

(c)  $f'(x)$ does not exist at $x_1$ (sharp corner).

(d)  $f$ has a relative maximum at $x_1$.

(e)  $f$ has a point of inflection at $x_2$ and $x_3$ (change in concavity).

**79.** Tangent line at $P$: $y - y_0 = f'(x_0)(x - x_0)$

(a)  Let $y = 0$: $-y_0 = f'(x_0)(x - x_0)$

$f'(x_0)x = x_0 f'(x_0) - y_0$

$x = x_0 - \dfrac{y_0}{f'(x_0)} = x_0 - \dfrac{f(x_0)}{f'(x_0)}$

$x$-intercept: $\left( x_0 - \dfrac{f(x_0)}{f'(x_0)}, 0 \right)$

(b)  Let $x = 0$: $y - y_0 = f'(x_0)(-x_0)$

$y = y_0 - x_0 f'(x_0)$

$y = f(x_0) - x_0 f'(x_0)$

$y$-intercept: $\left( 0, f(x_0) - x_0 f'(x_0) \right)$

(c)  Normal line: $y - y_0 = -\dfrac{1}{f'(x_0)}(x - x_0)$

Let $y = 0$: $-y_0 = -\dfrac{1}{f'(x_0)}(x - x_0)$

$-y_0 f'(x_0) = -x + x_0$

$x = x_0 + y_0 f'(x_0) = x_0 + f(x_0) f'(x_0)$

$x$-intercept: $\left( x_0 + f(x_0) f'(x_0), 0 \right)$

(d)  Let $x = 0$: $y - y_0 = \dfrac{-1}{f'(x_0)}(-x_0)$

$y = y_0 + \dfrac{x_0}{f'(x_0)}$

$y$-intercept: $\left( 0, y_0 + \dfrac{x_0}{f'(x_0)} \right)$

(e)  $\left| BC \right| = \left| x_0 - \dfrac{f(x_0)}{f'(x_0)} - x_0 \right| = \left| \dfrac{f(x_0)}{f'(x_0)} \right|$

(f)  $\left| PC \right|^2 = y_0^2 + \left( \dfrac{f(x_0)}{f'(x_0)} \right) = \dfrac{f(x_0)^2 f'(x_0)^2 + f(x_0)^2}{f'(x_0)^2}$

$\left| PC \right| = \left| \dfrac{f(x_0)\sqrt{1 + \left[ f'(x_0) \right]^2}}{f'(x_0)} \right|$

(g)  $\left| AB \right| = \left| x_0 - \left( x_0 + f(x_0) f'(x_0) \right) \right| = \left| f(x_0) f'(x_0) \right|$

(h)  $\left| AP \right|^2 = f(x_0)^2 f'(x_0)^2 + y_0^2$

$\left| AP \right| = \left| f(x_0) \right| \sqrt{1 + \left[ f'(x_0) \right]^2}$

**81.** $f(x) = \dfrac{ax}{(x-b)^2}$

Answers will vary. *Sample answer*: The graph has a vertical asymptote at $x = b$. If $a$ and $b$ are both positive, or both negative, then the graph of $f$ approaches $\infty$ as $x$ approaches $b$, and the graph has a minimum at $x = -b$. If $a$ and $b$ have opposite signs, then the graph of $f$ approaches $-\infty$ as $x$ approaches $b$, and the graph has a maximum at $x = -b$.

**83.** $y = \sqrt{4 + 16x^2}$

As $x \to \infty$, $y \to 4x$. As $x \to -\infty$, $y \to -4x$.

Slant asymptotes: $y = \pm 4x$

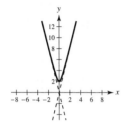

**85.** Let $\lambda = \dfrac{\dfrac{f(x) - f(a)}{x - a} - \dfrac{f(b) - f(a)}{b - a}}{x - b}$, $a < x < b$.

$$\lambda(x - b) = \frac{f(x) - f(a)}{x - a} - \frac{f(b) - f(a)}{b - a}$$

$$\lambda(x - b)(x - a) = f(x) - f(a) - \frac{f(b) - f(a)}{b - a}(x - a)$$

$$f(x) = f(a) + \frac{f(b) - f(a)}{b - a}(x - a) + \lambda(x - b)(x - a)$$

Let $h(t) = f(t) - \left\{ f(a) + \dfrac{f(b) - f(a)}{b - a}(t - a) + \lambda(t - a)(t - b) \right\}$.

$h(a) = 0$, $h(b) = 0$, $h(x) = 0$

By Rolle's Theorem, there exist numbers $\alpha_1$ and $\alpha_2$ such that $a < \alpha_1 < x < \alpha_2 < b$ and $h'(\alpha_1) = h'(\alpha_2) = 0$.

By Rolle's Theorem, there exists $\beta$ in $(a, b)$ such that $h''(\beta) = 0$.

Finally,

$$0 = h''(\beta) = f''(\beta) - \{2\lambda\} \Rightarrow \lambda = \tfrac{1}{2} f''(\beta).$$

# Section 4.7   Optimization Problems

**1.** (a)

| First Number, $x$ | Second Number | Product, $P$ |
|---|---|---|
| 10 | $110 - 10$ | $10(110 - 10) = 1000$ |
| 20 | $110 - 20$ | $20(110 - 20) = 1800$ |
| 30 | $110 - 30$ | $30(110 - 30) = 2400$ |
| 40 | $110 - 40$ | $40(110 - 40) = 2800$ |
| 50 | $110 - 50$ | $50(110 - 50) = 3000$ |
| 60 | $110 - 60$ | $60(110 - 60) = 3000$ |

(b)

| First Number, $x$ | Second Number | Product, $P$ |
|---|---|---|
| 10 | 110 – 10 | $10(110 - 10) = 1000$ |
| 20 | 110 – 20 | $20(110 - 20) = 1800$ |
| 30 | 110 – 30 | $30(110 - 30) = 2400$ |
| 40 | 110 – 40 | $40(110 - 40) = 2800$ |
| 50 | 110 – 50 | $50(110 - 50) = 3000$ |
| 60 | 110 – 60 | $60(110 - 60) = 3000$ |
| 70 | 110 – 70 | $70(110 - 70) = 2800$ |
| 80 | 110 – 80 | $80(110 - 80) = 2400$ |
| 90 | 110 – 90 | $90(110 - 90) = 1800$ |
| 100 | 110 – 100 | $100(110 - 100) = 1000$ |

The maximum is attained near $x = 50$ and $60$.

(c)  $P = x(110 - x) = 110x - x^2$

(d)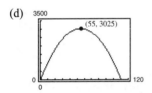

The solution appears to be $x = 55$.

(e)  $\dfrac{dP}{dx} = 110 - 2x = 0$ when $x = 55$.

$\dfrac{d^2P}{dx^2} = -2 < 0$

$P$ is a maximum when $x = 110 - x = 55$. The two numbers are 55 and 55.

**3.** Let $x$ and $y$ be two positive numbers such that $x + y = S$.

$$P = xy = x(S - x) = Sx - x^2$$

$$\frac{dP}{dx} = S - 2x = 0 \text{ when } x = \frac{S}{2}.$$

$$\frac{d^2P}{dx^2} = -2 < 0 \text{ when } x = \frac{S}{2}.$$

$P$ is a maximum when $x = y = S/2$.

**5.** Let $x$ and $y$ be two positive numbers such that $xy = 147$.

$$S = x + 3y = \frac{147}{y} + 3y$$

$$\frac{dS}{dy} = 3 - \frac{147}{y^2} = 0 \text{ when } y = 7.$$

$$\frac{d^2S}{dy^2} = \frac{294}{y^3} > 0 \text{ when } y = 7.$$

$S$ is minimum when $y = 7$ and $x = 21$.

**7.** Let $x$ and $y$ be two positive numbers such that $x + 2y = 108$.

$$P = xy = y(108 - 2y) = 108y - 2y^2$$

$$\frac{dP}{dy} = 108 - 4y = 0 \text{ when } y = 27.$$

$$\frac{d^2P}{dy^2} = -4 < 0 \text{ when } y = 27.$$

$P$ is a maximum when $x = 54$ and $y = 27$.

**9.** Let $x$ be the length and $y$ the width of the rectangle.

$$2x + 2y = 80$$

$$y = 40 - x$$

$$A = xy = x(40 - x) = 40x - x^2$$

$$\frac{dA}{dx} = 40 - 2x = 0 \text{ when } x = 20.$$

$$\frac{d^2A}{dx^2} = -2 < 0 \text{ when } x = 20.$$

$A$ is maximum when $x = y = 20$ m.

**11.** Let $x$ be the length and $y$ the width of the rectangle.

$$xy = 32$$

$$y = \frac{32}{x}$$

$$P = 2x + 2y = 2x + 2\left(\frac{32}{x}\right) = 2x + \frac{64}{x}$$

$$\frac{dP}{dx} = 2 - \frac{64}{x^2} = 0 \text{ when } x = 4\sqrt{2}.$$

$$\frac{d^2P}{dx^2} = \frac{128}{x^3} > 0 \text{ when } x = 4\sqrt{2}.$$

$P$ is minimum when $x = y = 4\sqrt{2}$ ft.

**13.** $d = \sqrt{(x - 2)^2 + \left[x^2 - (1/2)\right]^2}$

$\quad = \sqrt{x^4 - 4x + (17/4)}$

Because $d$ is smallest when the expression inside the radical is smallest, you need only find the critical numbers of

$$f(x) = x^4 - 4x + \frac{17}{4}.$$

$$f'(x) = 4x^3 - 4 = 0$$

$$x = 1$$

By the First Derivative Test, the point nearest to $\left(2, \frac{1}{2}\right)$ is $(1, 1)$.

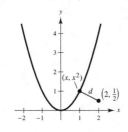

**15.** $d = \sqrt{(x - 4)^2 + \left(\sqrt{x} - 0\right)^2}$

$\quad = \sqrt{x^2 - 7x + 16}$

Because $d$ is smallest when the expression inside the radical is smallest, you need only find the critical numbers of

$$f(x) = x^2 - 7x + 16.$$

$$f'(x) = 2x - 7 = 0$$

$$x = \frac{7}{2}$$

By the First Derivative Test, the point nearest to $(4, 0)$ is $\left(7/2, \sqrt{7/2}\right)$.

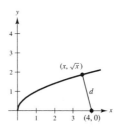

**17.** $xy = 30 \Rightarrow y = \dfrac{30}{x}$

$$A = (x + 2)\left(\frac{30}{x} + 2\right) \text{ (see figure)}$$

$$\frac{dA}{dx} = (x + 2)\left(\frac{-30}{x^2}\right) + \left(\frac{30}{x} + 2\right)$$

$$= \frac{2(x^2 - 30)}{x^2} = 0 \text{ when } x = \sqrt{30}.$$

$$y = \frac{30}{\sqrt{30}} = \sqrt{30}$$

By the First Derivative Test, the dimensions $(x + 2)$ by $(y + 2)$ are $\left(2 + \sqrt{30}\right)$ by $\left(2 + \sqrt{30}\right)$ (approximately 7.477 by 7.477). These dimensions yield a minimum area.

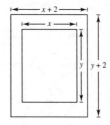

**19.** $xy = 245,000$ (see figure)

$S = x + 2y$

$= \left(x + \dfrac{490,000}{x}\right)$ where $S$ is the length

of fence needed.

$\dfrac{dS}{dx} = 1 - \dfrac{490,000}{x^2} = 0$ when $x = 700$.

$\dfrac{d^2S}{dx^2} = \dfrac{980,000}{x^3} > 0$ when $x = 700$.

$S$ is a minimum when $x = 700$ m and $y = 350$ m.

**21.** $16 = 2y + x + \pi\left(\dfrac{x}{2}\right)$

$32 = 4y + 2x + \pi x$

$y = \dfrac{32 - 2x - \pi x}{4}$

$A = xy + \dfrac{\pi}{2}\left(\dfrac{x}{2}\right)^2 = \left(\dfrac{32 - 2x - \pi x}{4}\right)x + \dfrac{\pi x^2}{8} = 8x - \dfrac{1}{2}x^2 - \dfrac{\pi}{4}x^2 + \dfrac{\pi}{8}x^2$

$\dfrac{dA}{dx} = 8 - x - \dfrac{\pi}{2}x + \dfrac{\pi}{4}x = 8 - x\left(1 + \dfrac{\pi}{4}\right) = 0$ when $x = \dfrac{8}{1 + (\pi/4)} = \dfrac{32}{4 + \pi}$.

$\dfrac{d^2A}{dx^2} = -\left(1 + \dfrac{\pi}{4}\right) < 0$ when $x = \dfrac{32}{4 + \pi}$.

$y = \dfrac{32 - 2[32/(4 + \pi)] - \pi[32/(4 + \pi)]}{4} = \dfrac{16}{4 + \pi}$

The area is maximum when $y = \dfrac{16}{4 + \pi}$ ft and $x = \dfrac{32}{4 + \pi}$ ft.

**23. (a)** $\dfrac{y - 2}{0 - 1} = \dfrac{0 - 2}{x - 1}$

$y = 2 + \dfrac{2}{x - 1}$

$L = \sqrt{x^2 + y^2} = \sqrt{x^2 + \left(2 + \dfrac{2}{x - 1}\right)^2} = \sqrt{x^2 + 4 + \dfrac{8}{x - 1} + \dfrac{4}{(x - 1)^2}}, \quad x > 1$

**(b)**

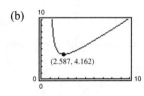

$L$ is minimum when $x \approx 2.587$ and $L \approx 4.162$.

**(c)** Area $= A(x) = \dfrac{1}{2}xy = \dfrac{1}{2}x\left(2 + \dfrac{2}{x - 1}\right) = x + \dfrac{x}{x - 1}$

$A'(x) = 1 + \dfrac{(x - 1) - x}{(x - 1)^2} = 1 - \dfrac{1}{(x - 1)^2} = 0$

$(x - 1)^2 = 1$

$x - 1 = \pm 1$

$x = 0, 2$ (select $x = 2$)

They $y = 4$ and $A = 4$.

Vertices: $(0, 0), (2, 0), (0, 4)$

**25.**  $A = 2xy = 2x\sqrt{25 - x^2}$ (see figure)

$$\frac{dA}{dx} = 2x\left(\frac{1}{2}\right)\left(\frac{-2x}{\sqrt{25 - x^2}}\right) + 2\sqrt{25 - x^2} = 2\left(\frac{25 - 2x^2}{\sqrt{25 - x^2}}\right) = 0 \text{ when } x = y = \frac{5\sqrt{2}}{2} \approx 3.54.$$

By the First Derivative Test, the inscribed rectangle of maximum area has vertices

$$\left(\pm\frac{5\sqrt{2}}{2}, 0\right), \left(\pm\frac{5\sqrt{2}}{2}, \frac{5\sqrt{2}}{2}\right).$$

Width: $\dfrac{5\sqrt{2}}{2}$; Length: $5\sqrt{2}$

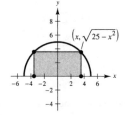

**27. (a)**  $P = 2x + 2\pi r = 2x + 2\pi\left(\dfrac{y}{2}\right) = 2x + \pi y = 200 \Rightarrow y = \dfrac{200 - 2x}{\pi} = \dfrac{2}{\pi}(100 - x)$

**(b)**

| Length, $x$ | Width, $y$ | Area, $xy$ |
|---|---|---|
| 10 | $\dfrac{2}{\pi}(100 - 10)$ | $(10)\dfrac{2}{\pi}(100 - 10) \approx 573$ |
| 20 | $\dfrac{2}{\pi}(100 - 20)$ | $(20)\dfrac{2}{\pi}(100 - 20) \approx 1019$ |
| 30 | $\dfrac{2}{\pi}(100 - 30)$ | $(30)\dfrac{2}{\pi}(100 - 30) \approx 1337$ |
| 40 | $\dfrac{2}{\pi}(100 - 40)$ | $(40)\dfrac{2}{\pi}(100 - 40) \approx 1528$ |
| 50 | $\dfrac{2}{\pi}(100 - 50)$ | $(50)\dfrac{2}{\pi}(100 - 50) \approx 1592$ |
| 60 | $\dfrac{2}{\pi}(100 - 60)$ | $(60)\dfrac{2}{\pi}(100 - 60) \approx 1528$ |

The maximum area of the rectangle is approximately 1592 m².

**(c)**  $A = xy = x\dfrac{2}{\pi}(100 - x) = \dfrac{2}{\pi}(100x - x^2)$

**(d)**  $A' = \dfrac{2}{\pi}(100 - 2x)$. $A' = 0$ when $x = 50$.

Maximum value is approximately 1592 when length $= 50$ m and width $= \dfrac{100}{\pi}$.

**(e)**

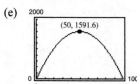

Maximum area is approximately

1591.55 m² $(x = 50$ m$)$.

**29.** Let $x$ be the sides of the square ends and $y$ the length of the package.

$$P = 4x + y = 108 \Rightarrow y = 108 - 4x$$

$$V = x^2 y = x^2(108 - 4x) = 108x^2 - 4x^3$$

$$\frac{dV}{dx} = 216x - 12x^2$$

$$= 12x(18 - x) = 0 \text{ when } x = 18.$$

$$\frac{d^2V}{dx^2} = 216 - 24x = -216 < 0 \text{ when } x = 18.$$

The volume is maximum when $x = 18$ in. and $y = 108 - 4(18) = 36$ in.

**31.** No. The volume will change because the shape of the container changes when squeezed.

**33.**    $V = 14 = \dfrac{4}{3}\pi r^3 + \pi r^2 h$

$$h = \frac{14 - (4/3)\pi r^3}{\pi r^2} = \frac{14}{\pi r^2} - \frac{4}{3}r$$

$$S = 4\pi r^2 + 2\pi rh = 4\pi r^2 + 2\pi r\left(\frac{14}{\pi r^2} - \frac{4}{3}r\right) = 4\pi r^2 + \frac{28}{r} - \frac{8}{3}\pi r^2 = \frac{4}{3}\pi r^2 + \frac{28}{r}$$

$$\frac{dS}{dr} = \frac{8}{3}\pi r - \frac{28}{r^2} = 0 \text{ when } r = \sqrt[3]{\frac{21}{2\pi}} \approx 1.495 \text{ cm.}$$

$$\frac{d^2S}{dr^2} = \frac{8}{3}\pi + \frac{56}{r^3} > 0 \text{ when } r = \sqrt[3]{\frac{21}{2\pi}}.$$

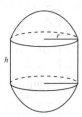

The surface area is minimum when $r = \sqrt[3]{\dfrac{21}{2\pi}}$ cm and $h = 0$.

The resulting solid is a sphere of radius $r \approx 1.495$ cm.

**35.** Let $x$ be the length of a side of the square and $y$ the length of a side of the triangle.

$$4x + 3y = 10$$

$$A = x^2 + \frac{1}{2}y\left(\frac{\sqrt{3}}{2}y\right)$$

$$= \frac{(10 - 3y)^2}{16} + \frac{\sqrt{3}}{4}y^2$$

$$\frac{dA}{dy} = \frac{1}{8}(10 - 3y)(-3) + \frac{\sqrt{3}}{2}y = 0$$

$$-30 + 9y + 4\sqrt{3}y = 0$$

$$y = \frac{30}{9 + 4\sqrt{3}}$$

$$\frac{d^2A}{dy^2} = \frac{9 + 4\sqrt{3}}{8} > 0$$

$A$ is minimum when $y = \dfrac{30}{9 + 4\sqrt{3}}$ and $x = \dfrac{10\sqrt{3}}{9 + 4\sqrt{3}}$.

**37.** Let $S$ be the strength and $k$ the constant of proportionality. Given

$$h^2 + w^2 = 20^2, h^2 = 20^2 - w^2,$$

$$S = kwh^2$$

$$S = kw(400 - w^2) = k(400w - w^3)$$

$$\frac{dS}{dw} = k(400 - 3w^2) = 0 \text{ when } w = \frac{20\sqrt{3}}{3} \text{ in.}$$

and $h = \dfrac{20\sqrt{6}}{3}$ in.

$$\frac{d^2S}{dw^2} = -6kw < 0 \text{ when } w = \frac{20\sqrt{3}}{3}.$$

These values yield a maximum.

**39.**

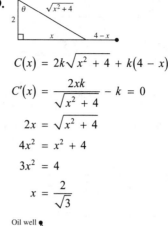

$$C(x) = 2k\sqrt{x^2 + 4} + k(4 - x)$$

$$C'(x) = \frac{2xk}{\sqrt{x^2 + 4}} - k = 0$$

$$2x = \sqrt{x^2 + 4}$$

$$4x^2 = x^2 + 4$$

$$3x^2 = 4$$

$$x = \frac{2}{\sqrt{3}}$$

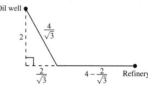

The path of the pipe should go underwater from the oil well to the coast following the hypotenuse of a right triangle with leg lengths of 2 kilometers and $2/\sqrt{3}$ kilometers for a distance of $4/\sqrt{3}$ kilometers. Then the pipe should go down the coast to the refinery for a distance of $\left(4 - 2/\sqrt{3}\right)$ kilometers.

**41.** (a)

$$S = \sqrt{x^2 + 4}, L = \sqrt{1 + (3 - x)^2}$$

$$\text{Time} = T = \frac{\sqrt{x^2 + 4}}{2} + \frac{\sqrt{x^2 - 6x + 10}}{4}$$

$$\frac{dT}{dx} = \frac{x}{2\sqrt{x^2 + 4}} + \frac{x - 3}{4\sqrt{x^2 - 6x + 10}} = 0$$

$$\frac{x^2}{x^2 + 4} = \frac{9 - 6x + x^2}{4(x^2 - 6x + 10)}$$

$$x^4 - 6x^3 + 9x^2 + 8x - 12 = 0$$

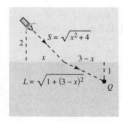

You need to find the roots of this equation in the interval $[0, 3]$. By using a computer or graphing utility you can determine that this equation has only one root in this interval $(x = 1)$. Testing at this value and at the endpoints, you see that $x = 1$ yields the minimum time. So, the man should row to a point 1 mile from the nearest point on the coast.

(b)   $T = \dfrac{\sqrt{x^2 + 4}}{v_1} + \dfrac{\sqrt{x^2 - 6x + 10}}{v_2}$

$\dfrac{dT}{dx} = \dfrac{x}{v_1 \sqrt{x^2 + 4}} + \dfrac{x - 3}{v_2 \sqrt{x^2 - 6x + 10}} = 0$

Because $\dfrac{x}{\sqrt{x^2 + 4}} = \sin \theta_1$ and $\dfrac{x - 3}{\sqrt{x^2 - 6x + 10}} = -\sin \theta_2$

you have $\dfrac{\sin \theta_1}{v_1} - \dfrac{\sin \theta_2}{v_2} = 0 \Rightarrow \dfrac{\sin \theta_1}{v_1} = \dfrac{\sin \theta_2}{v_2}.$

Because $\dfrac{d^2T}{dx^2} = \dfrac{4}{v_1(x^2 + 4)^{3/2}} + \dfrac{1}{v_2(x^2 - 6x + 10)^{3/2}} > 0$

this condition yields a minimum time.

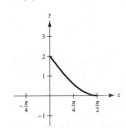

**43.** $f(x) = 2 - 2 \sin x$

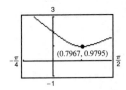

(a)  Distance from origin to $y$-intercept is 2.

Distance from origin to $x$-intercept is $\pi/2 \approx 1.57$.

(b)  $d = \sqrt{x^2 + y^2} = \sqrt{x^2 + (2 - 2 \sin x)^2}$

(0.7967, 0.9795)

Minimum distance = 0.9795 at $x = 0.7967$.

(c)  Let $f(x) = d^2(x) = x^2 + (2 - 2 \sin x)^2.$

$f'(x) = 2x + 2(2 - 2 \sin x)(-2 \cos x)$

Setting $f'(x) = 0$, you obtain $x \approx 0.7967$, which corresponds to $d = 0.9795$.

**45.**   $V = \dfrac{1}{3}\pi r^2 h = \dfrac{1}{3}\pi r^2 \sqrt{144 - r^2}$

$\dfrac{dV}{dr} = \dfrac{1}{3}\pi \left[ r^2 \left( \dfrac{1}{2} \right)(144 - r^2)^{-1/2}(-2r) + 2r\sqrt{144 - r^2} \right]$

$= \dfrac{1}{3}\pi \left[ \dfrac{288r - 3r^3}{\sqrt{144 - r^2}} \right]$

$= \pi \left[ \dfrac{r(96 - r^2)}{\sqrt{144 - r^2}} \right] = 0$ when $r = 0, 4\sqrt{6}.$

By the First Derivative Test, $V$ is maximum when $r = 4\sqrt{6}$ and $h = 4\sqrt{3}.$

Area of circle: $A = \pi(12)^2 = 144\pi$

Lateral surface area of cone:

$S = \pi(4\sqrt{6})\sqrt{(4\sqrt{6})^2 + (4\sqrt{3})^2} = 48\sqrt{6}\pi$

Area of sector:

$144\pi - 48\sqrt{6}\pi = \dfrac{1}{2}\theta r^2 = 72\theta$

$\theta = \dfrac{144\pi - 48\sqrt{6}\pi}{72}$

$= \dfrac{2\pi}{3}(3 - \sqrt{6}) \approx 1.153$ radians or $66°$

**47.**  Let $d$ be the amount deposited in the bank, $i$ be the interest rate paid by the bank, and $P$ be the profit.

$P = (0.12)d - id$

$d = ki^2 \left( \text{because } d \text{ is proportional to } i^2 \right)$

$P = (0.12)(ki^2) - i(ki^2) = k(0.12i^2 - i^3)$

$\dfrac{dP}{di} = k(0.24i - 3i^2) = 0$ when $i = \dfrac{0.24}{3} = 0.08.$

$\dfrac{d^2P}{di^2} = k(0.24 - 6i) < 0$ when $i = 0.08 \, (\textbf{Note:} \, k > 0).$

The profit is a maximum when $i = 8\%$.

**49.** $y = \dfrac{L}{1 + ae^{-x/b}},\ a > 0,\ b > 0,\ L > 0$

$y' = \dfrac{-L\left(-\dfrac{a}{b}e^{-x/b}\right)}{\left(1 + ae^{-x/b}\right)^2} = \dfrac{\dfrac{aL}{b}e^{-x/b}}{\left(1 + ae^{-x/b}\right)^2}$

$y'' = \dfrac{\left(1 + ae^{-x/b}\right)^2\left(\dfrac{-aL}{b^2}e^{-x/b}\right) - \left(\dfrac{aL}{b}e^{-x/b}\right)2\left(1 + ae^{-x/b}\right)\left(\dfrac{-a}{b}e^{-x/b}\right)}{\left(1 + ae^{-x/b}\right)^4}$

$= \dfrac{\left(1 + ae^{-x/b}\right)\left(\dfrac{-aL}{b^2}e^{-x/b}\right) + 2\left(\dfrac{aL}{b}e^{-x/b}\right)\left(\dfrac{a}{b}e^{-x/b}\right)}{\left(1 + ae^{-x/b}\right)^3} = \dfrac{Lae^{-x/b}\left(ae^{-x/b} - 1\right)}{\left(1 + ae^{-x/b}\right)^3 b^2}$

$y'' = 0$ if $ae^{-x/b} = 1 \Rightarrow \dfrac{-x}{b} = \ln\left(\dfrac{1}{a}\right) \Rightarrow x = b\ln a$

$y(b\ln a) = \dfrac{L}{1 + ae^{-(b\ln a)/b}} = \dfrac{L}{1 + a(1/a)} = \dfrac{L}{2}$

Therefore, the $y$-coordinate of the inflection point is $L/2$.

**51.** $S_1 = (4m - 1)^2 + (5m - 6)^2 + (10m - 3)^2$

$\dfrac{dS_1}{dm} = 2(4m - 1)(4) + 2(5m - 6)(5) + 2(10m - 3)(10)$

$= 282m - 128 = 0$ when $m = \dfrac{64}{141}$.

Line: $y = \dfrac{64}{141}x$

$S = \left|4\left(\dfrac{64}{141}\right) - 1\right| + \left|5\left(\dfrac{64}{141}\right) - 6\right| + \left|10\left(\dfrac{64}{141}\right) - 3\right|$

$= \left|\dfrac{256}{141} - 1\right| + \left|\dfrac{320}{141} - 6\right| + \left|\dfrac{640}{141} - 3\right| = \dfrac{858}{141} \approx 6.1$ mi

**53.** $S_3 = \dfrac{|4m - 1|}{\sqrt{m^2 + 1}} + \dfrac{|5m - 6|}{\sqrt{m^2 + 1}} + \dfrac{|10m - 3|}{\sqrt{m^2 + 1}}$

Using a graphing utility, you can see that the minimum occurs when $x \approx 0.3$.

Line: $y \approx 0.3x$

$S_3 = \dfrac{|4(0.3) - 1| + |5(0.3) - 6| + |10(0.3) - 3|}{\sqrt{(0.3)^2 + 1}} \approx 4.5$ mi.

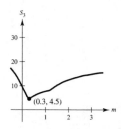

**55.** $f(x) = x^3 - 3x;\ x^4 + 36 \le 13x^2$

$x^4 - 13x^2 + 36 = (x^2 - 9)(x^2 - 4)$

$\qquad\qquad = (x - 3)(x - 2)(x + 2)(x + 3) \le 0$

So, $-3 \le x \le -2$ or $2 \le x \le 3$.

$f'(x) = 3x^2 - 3 = 3(x + 1)(x - 1)$

$f$ is increasing on $(-\infty, -1)$ and $(1, \infty)$.

So, $f$ is increasing on $[-3, -2]$ and $[2, 3]$.

$f(-2) = -2,\ f(3) = 18$. The maximum value of $f$ is 18.

## Section 4.8 Differentials

**1.** $f(x) = x^2$

$f'(x) = 2x$

Tangent line at $(2, 4)$: $y - f(2) = f'(2)(x - 2)$

$$y - 4 = 4(x - 2)$$
$$y = 4x - 4$$

| $x$ | 1.9 | 1.99 | 2 | 2.01 | 2.1 |
|---|---|---|---|---|---|
| $f(x) = x^2$ | 3.6100 | 3.9601 | 4 | 4.0401 | 4.4100 |
| $T(x) = 4x - 4$ | 3.6000 | 3.9600 | 4 | 4.0400 | 4.4000 |

**3.** $f(x) = x^5$

$f'(x) = 5x^4$

Tangent line at $(2, 32)$:

$$y - f(2) = f'(2)(x - 2)$$
$$y - 32 = 80(x - 2)$$
$$y = 80x - 128$$

| $x$ | 1.9 | 1.99 | 2 | 2.01 | 2.1 |
|---|---|---|---|---|---|
| $f(x) = x^5$ | 24.7610 | 31.2080 | 32 | 32.8080 | 40.8410 |
| $T(x) = 80x - 128$ | 24.0000 | 31.2000 | 32 | 32.8000 | 40.0000 |

**5.** $f(x) = \sin x$

$f'(x) = \cos x$

Tangent line at $(2, \sin 2)$:

$$y - f(2) = f'(2)(x - 2)$$
$$y - \sin 2 = (\cos 2)(x - 2)$$
$$y = (\cos 2)(x - 2) + \sin 2$$

| $x$ | 1.9 | 1.99 | 2 | 2.01 | 2.1 |
|---|---|---|---|---|---|
| $f(x) = \sin x$ | 0.9463 | 0.9134 | 0.9093 | 0.9051 | 0.8632 |
| $T(x) = (\cos 2)(x - 2) + \sin 2$ | 0.9509 | 0.9135 | 0.9093 | 0.9051 | 0.8677 |

**7.** $y = f(x) = x^3$, $f'(x) = 3x^2$, $x = 1$, $\Delta x = dx = 0.1$

$\Delta y = f(x + \Delta x) - f(x)$      $dy = f'(x)\, dx$

$\quad = f(1.1) - f(1)$          $= f'(1)(0.1)$

$\quad = 0.331$                $= 3(0.1)$

                            $= 0.3$

**9.** $y = f(x) = x^4 + 1$, $f'(x) = 4x^3$, $x = -1$, $\Delta x = dx = 0.01$

$$\begin{aligned}
\Delta y &= f(x + \Delta x) - f(x) & dy &= f'(x)\, dx \\
&= f(-0.99) - f(-1) & &= f'(-1)(0.01) \\
&= \left[(-0.99)^4 + 1\right] - \left[(-1)^4 + 1\right] \approx -0.0394 & &= (-4)(0.01) = -0.04
\end{aligned}$$

**11.** $y = 3x^2 - 4$

$dy = 6x\, dx$

**13.** $y = x \tan x$

$dy = \left(x \sec^2 x + \tan x\right) dx$

**15.** $y = \dfrac{x + 1}{2x - 1}$

$dy = -\dfrac{3}{(2x - 1)^2}\, dx$

**17.** $y = \sqrt{9 - x^2}$

$dy = \dfrac{1}{2}\left(9 - x^2\right)^{-1/2}(-2x)\, dx = \dfrac{-x}{\sqrt{9 - x^2}}\, dx$

**19.** $y = 3x - \sin^2 x$

$dy = (3 - 2\sin x \cos x)\, dx = (3 - \sin 2x)\, dx$

**21.** $y = \ln\sqrt{4 - x^2} = \dfrac{1}{2}\ln\left(4 - x^2\right)$

$dy = \dfrac{1}{2}\left(\dfrac{-2x}{4 - x^2}\right) dx = \dfrac{-x}{4 - x^2}\, dx$

**23.** $y = x \arcsin x$

$dy = \left(\dfrac{x}{\sqrt{1 - x^2}} + \arcsin x\right) dx$

**25.** (a) $f(1.9) = f(2 - 0.1) \approx f(2) + f'(2)(-0.1)$

$\qquad\qquad\qquad \approx 1 + (1)(-0.1) = 0.9$

(b) $f(2.04) = f(2 + 0.04) \approx f(2) + f'(2)(0.04)$

$\qquad\qquad\qquad \approx 1 + (1)(0.04) = 1.04$

**27.** (a) $g(2.93) = g(3 - 0.07) \approx g(3) + g'(3)(-0.07)$

$\qquad\qquad\qquad \approx 8 + \left(-\tfrac{1}{2}\right)(-0.07) = 8.035$

(b) $g(3.1) = g(3 + 0.1) \approx g(3) + g'(3)(0.1)$

$\qquad\qquad\qquad \approx 8 + \left(-\tfrac{1}{2}\right)(0.1) = 7.95$

**29.** $x = 10$ in., $\Delta x = dx = \pm\dfrac{1}{32}$ in.

(a) $A = x^2$

$dA = 2x\, dx$

$\Delta A \approx dA = 2(10)\left(\pm\dfrac{1}{32}\right) = \pm\dfrac{5}{8}$ in.$^2$

(b) Percent error:

$\dfrac{dA}{A} = \dfrac{5/8}{100} = \dfrac{5}{800} = \dfrac{1}{100} = 0.00625 = 0.625\%$

**31.** $b = 36$ cm, $h = 50$ cm,

$\Delta b = \Delta h = db = dh = \pm 0.25$ cm

(a) $A = \dfrac{1}{2}bh$

$dA = \dfrac{1}{2}b\, dh + \dfrac{1}{2}h\, db$

$\Delta A \approx dA = \dfrac{1}{2}(36)(\pm 0.25) + \dfrac{1}{2}(50)(\pm 0.25)$

$\qquad\quad = \pm 10.75$ cm$^2$

(b) Percent error:

$\dfrac{dA}{A} = \dfrac{10.75}{\frac{1}{2}(36)(50)} \approx 0.011944 = 1.19\%$

**33.** $x = 15$ in., $\Delta x = dx = \pm 0.03$ in.

(a) $V = x^3$

$dV = 3x^2 dx$

$\Delta V \approx dV = 3(15)^2(\pm 0.03) = \pm 20.25$ in.$^3$

(b) $S = 6x^2$

$dS = 12x\, dx$

$\Delta S \approx dS = 12(15)(\pm 0.03) = \pm 5.4$ in.$^2$

(c) Percent error of volume:

$\dfrac{dV}{V} = \dfrac{20.25}{15^3} = 0.006$ or $0.6\%$

Percent error of surface area:

$\dfrac{dS}{S} = \dfrac{5.4}{6(15)^2} = 0.004$ or $0.4\%$

**35.** $T = 2.5x + 0.5x^2$, $\Delta x = dx = 26 - 25 = 1$, $x = 25$

$dT = (2.5 + x)dx = (2.5 + 25)(1) = 27.5$ mi

Percentage change $= \dfrac{dT}{T} = \dfrac{27.5}{375} \approx 7.3\%$

**37.** (a)  $T = 2\pi\sqrt{L/g}$

$$dT = \dfrac{\pi}{g\sqrt{L/g}}dL$$

Relative error:

$$\dfrac{dT}{T} = \dfrac{(\pi\, dL)\big/\big(g\sqrt{L/g}\big)}{2\pi\sqrt{L/g}}$$

$$= \dfrac{dL}{2L}$$

$$= \dfrac{1}{2}\left(\text{relative error in } L\right)$$

$$= \dfrac{1}{2}(0.005) = 0.0025$$

Percentage error: $\dfrac{dT}{T}(100) = 0.25\% = \dfrac{1}{4}\%$

(b)  $(0.0025)(3600)(24) = 216 \text{ sec} = 3.6 \text{ min}$

**39.** $dH = -\dfrac{401{,}493{,}267}{2{,}000{,}000}\dfrac{e^{369{,}444/(50t+19{,}793)}}{(50t+19{,}793)^2}dt$

At $t = 72$ and $dt = 1$, $dH \approx -2.65$.

**41.** Let $f(x) = \sqrt{x}$, $x = 100$, $dx = -0.6$.

$$f(x + \Delta x) \approx f(x) + f'(x)\, dx$$

$$= \sqrt{x} + \dfrac{1}{2\sqrt{x}}dx$$

$$f(x + \Delta x) = \sqrt{99.4}$$

$$\approx \sqrt{100} + \dfrac{1}{2\sqrt{100}}(-0.6) = 9.97$$

Using a calculator: $\sqrt{99.4} \approx 9.96995$

**43.** Let $f(x) = \sqrt[4]{x}$, $x = 625$, $dx = -1$.

$$f(x + \Delta x) \approx f(x) + f'(x)\, dx = \sqrt[4]{x} + \dfrac{1}{4\sqrt[4]{x^3}}dx$$

$$f(x + \Delta x) = \sqrt[4]{624} \approx \sqrt[4]{625} + \dfrac{1}{4\left(\sqrt[4]{625}\right)^3}(-1)$$

$$= 5 - \dfrac{1}{500} = 4.998$$

Using a calculator, $\sqrt[4]{624} \approx 4.9980$.

**45.** $f(x) = \sqrt{x + 4}$

$$f'(x) = \dfrac{1}{2\sqrt{x + 4}}$$

At $(0, 2)$, $f(0) = 2$, $f'(0) = \dfrac{1}{4}$

Tangent line: $y - 2 = \dfrac{1}{4}(x - 0)$

$$y = \dfrac{1}{4}x + 2$$

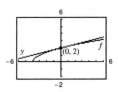

**47.** In general, when $\Delta x \to 0$, $dy$ approaches $\Delta y$.

**49.** (a)  Let $f(x) = \sqrt{x}$, $x = 4$, $dx = 0.02$,

$f'(x) = 1/\left(2\sqrt{x}\right)$.

Then

$$f(4.02) \approx f(4) + f'(4)\, dx$$

$$\sqrt{4.02} \approx \sqrt{4} + \dfrac{1}{2\sqrt{4}}(0.02) = 2 + \dfrac{1}{4}(0.02).$$

(b)  Let

$f(x) = \tan x$, $x = 0$, $dx = 0.05$, $f'(x) = \sec^2 x$.

Then

$$f(0.05) \approx f(0) + f'(0)\, dx$$

$$\tan 0.05 \approx \tan 0 + \sec^2 0(0.05) = 0 + 1(0.05).$$

**51.** True

**53.** True

# Review Exercises for Chapter 4

**1.** $f(x) = x^2 + 5x, \quad [-4, 0]$

$f'(x) = 2x + 5 = 0$ when $x = -5/2$

Critical number: $x = -5/2$

Left endpoint: $(-4, -4)$

Critical number: $(-5/2, -25/4)$    Minimum

Right endpoint: $(0, 0)$         Maximum

**3.** $f(x) = \sqrt{x} - 2, [0, 4]$

$f'(x) = \dfrac{1}{2\sqrt{x}}$

No critical numbers on $(0, 4)$

Left endpoint: $(0, -2)$      Minimum

Right endpoint: $(4, 0)$      Maximum

**5.** $f(x) = \dfrac{4x}{x^2 + 9}, [-4, 4]$

$f'(x) = \dfrac{(x^2 + 9)4 - 4x(2x)}{(x^2 + 9)^2} = \dfrac{36 - 4x^2}{(x^2 + 9)^2}$

$= 0 \Rightarrow 36 - 4x^2 = 0 \Rightarrow x = \pm 3$

Critical numbers: $x = \pm 3$

Left endpoint: $\left(-4, -\dfrac{16}{25}\right)$

Critical number: $\left(-3, -\dfrac{2}{3}\right)$      Minimum

Critical number: $\left(3, \dfrac{2}{3}\right)$      Maximum

Right endpoint: $\left(4, \dfrac{16}{25}\right)$

**7.** $g(x) = 2x + 5 \cos x, [0, 2\pi]$

$g'(x) = 2 - 5 \sin x = 0$ when $\sin x = \frac{2}{5}$.

Critical numbers: $x \approx 0.41, x \approx 2.73$

Left endpoint: $(0, 5)$

Critical number: $(0.41, 5.41)$

Critical number: $(2.73, 0.88)$      Minimum

Right endpoint: $(2\pi, 17.57)$      Maximum

**9.** No, Rolle's Theorem cannot be applied.

$f(0) = -7 \neq 25 = f(4)$

**11.** No. $f(x) = \dfrac{x^2}{1 - x^2}$ is not continuous on $[-2, 2]$. $f(-1)$ is not defined.

**13.** $f(x) = x^{2/3}, 1 \le x \le 8$

$f'(x) = \dfrac{2}{3} x^{-1/3}$

$\dfrac{f(b) - f(a)}{b - a} = \dfrac{4 - 1}{8 - 1} = \dfrac{3}{7}$

$f'(c) = \dfrac{2}{3} c^{-1/3} = \dfrac{3}{7}$

$c = \left(\dfrac{14}{9}\right)^3 = \dfrac{2744}{729} \approx 3.764$

**15.** The Mean Value Theorem cannot be applied. $f$ is not differentiable at $x = 5$ in $[2, 6]$.

**17.** $f(x) = x - \cos x, -\dfrac{\pi}{2} \le x \le \dfrac{\pi}{2}$

$f'(x) = 1 + \sin x$

$\dfrac{f(b) - f(a)}{b - a} = \dfrac{(\pi/2) - (-\pi/2)}{(\pi/2) - (-\pi/2)} = 1$

$f'(c) = 1 + \sin c = 1$

$c = 0$

**19.** No; the function is discontinuous at $x = 0$ which is in the interval $[-2, 1]$.

**21.** $f(x) = x^2 + 3x - 12$

$f'(x) = 2x + 3$

Critical number: $x = -\dfrac{3}{2}$

| Intervals: | $-\infty < x < -\frac{3}{2}$ | $-\frac{3}{2} < x < \infty$ |
| --- | --- | --- |
| Sign of $f'(x)$: | $f'(x) < 0$ | $f'(x) > 0$ |
| Conclusion: | Decreasing | Increasing |

**23.** $f(x) = (x - 1)^2(x - 3)$

$f'(x) = (x - 1)^2(1) + (x - 3)(2)(x - 1)$

$\quad = (x - 1)(3x - 7)$

Critical numbers: $x = 1$ and $x = \frac{7}{3}$

| Intervals: | $-\infty < x < 1$ | $1 < x < \frac{7}{3}$ | $\frac{7}{3} < x < \infty$ |
|---|---|---|---|
| Sign of $f'(x)$: | $f'(x) > 0$ | $f'(x) < 0$ | $f'(x) > 0$ |
| Conclusion: | Increasing | Decreasing | Increasing |

**25.** $h(x) = \sqrt{x}(x - 3) = x^{3/2} - 3x^{1/2}$

Domain: $(0, \infty)$

$h'(x) = \frac{3}{2}x^{3/2} - \frac{3}{2}x^{-1/2} = \frac{3}{2}x^{-1/2}(x - 1) = \frac{3(x - 1)}{2\sqrt{x}}$

Critical number: $x = 1$

| Intervals: | $0 < x < 1$ | $1 < x < \infty$ |
|---|---|---|
| Sign of $h'(x)$: | $h'(x) < 0$ | $h'(x) > 0$ |
| Conclusion: | Decreasing | Increasing |

**27.** $f(t) = (2 - t)2^t$

$f'(t) = (2 - t)2^t \ln 2 - 2^t = 2^t[(2 - t) \ln 2 - 1]$

$f'(t) = 0: (2 - t)\ln 2 = 1$

$2 - t = \frac{1}{\ln 2}$

$t = 2 - \frac{1}{\ln 2} \approx 0.5573$, Critical number

Increasing on: $\left(-\infty, 2 - \frac{1}{\ln 2}\right)$

Decreasing on: $\left(2 - \frac{1}{\ln 2}, \infty\right)$

| Intervals: | $-\infty < t < 2 - \frac{1}{\ln 2}$ | $2 - \frac{1}{\ln 2} < t < \infty$ |
|---|---|---|
| Sign of $f'(t)$: | $f'(t) > 0$ | $f'(t) < 0$ |
| Conclusion: | Increasing | Decreasing |

**29.** (a) $f(x) = x^2 - 6x + 5$

$f'(x) = 2x - 6 = 0$ when $x = 3$.

(b)

| Intervals: | $-\infty < x < 3$ | $3 < x < \infty$ |
|---|---|---|
| Sign of $f'(x)$: | $f'(x) < 0$ | $f'(x) > 0$ |
| Conclusion: | Decreasing | Increasing |

(c) Relative minimum: $(3, -4)$

(d)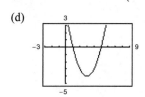

**31.** (a) $h(t) = \frac{1}{4}t^4 - 8t$

$h'(t) = t^3 - 8 = 0$ when $t = 2$.

(c) Relative minimum: $(2, -12)$

(b)

| Intervals: | $-\infty < t < 2$ | $2 < t < \infty$ |
|---|---|---|
| Sign of $h'(t)$: | $h'(t) < 0$ | $h'(t) > 0$ |
| Conclusion: | Decreasing | Increasing |

(d)

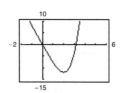

**33.** (a) $f(x) = \dfrac{x + 4}{x^2}$

$f'(x) = \dfrac{x^2(1) - (x + 4)(2x)}{x^4} = -\dfrac{x^2 + 8x}{x^4} = -\dfrac{x + 8}{x^3}$

$f'(x) = 0$ when $x = -8$.

Discontinuity at: $x = 0$

(b)

| Intervals: | $-\infty < x < -8$ | $-8 < x < 0$ | $0 < x < \infty$ |
|---|---|---|---|
| Sign of $f'(x)$: | $f'(x) < 0$ | $f'(x) > 0$ | $f'(x) < 0$ |
| Conclusion: | Decreasing | Increasing | Decreasing |

(c) Relative minimum: $\left(-8, -\frac{1}{16}\right)$

(d)

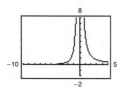

**35.** (a) $f(x) = \cos x - \sin x, \ (0, 2\pi)$

$f'(x) = -\sin x - \cos x = 0 \Rightarrow -\cos x = \sin x \Rightarrow \tan x = -1$

Critical numbers: $x = \dfrac{3\pi}{4}, \dfrac{7\pi}{4}$

(b)

| Intervals: | $0 < x < \dfrac{3\pi}{4}$ | $\dfrac{3\pi}{4} < x < \dfrac{7\pi}{4}$ | $\dfrac{7\pi}{4} < x < 2\pi$ |
|---|---|---|---|
| Sign of $f'(x)$: | $f'(x) < 0$ | $f'(x) > 0$ | $f'(x) < 0$ |
| Conclusion: | Decreasing | Increasing | Decreasing |

(c) Relative minimum: $\left(\dfrac{3\pi}{4}, -\sqrt{2}\right)$

Relative maximum: $\left(\dfrac{7\pi}{4}, \sqrt{2}\right)$

(d)

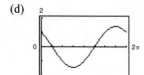

**37.** $f(x) = x^3 - 9x^2$

$f'(x) = 3x^2 - 18x$

$f''(x) = 6x - 18 = 0$ when $x = 3$.

| Intervals: | $-\infty < x < 3$ | $3 < x < \infty$ |
|---|---|---|
| Sign of $f''(x)$: | $f''(x) < 0$ | $f''(x) > 0$ |
| Conclusion: | Concave downward | Concave upward |

Point of inflection: $(3, -54)$

**39.** $g(x) = x\sqrt{x + 5}$, Domain: $x \geq -5$

$g'(x) = x\left(\dfrac{1}{2}\right)(x + 5)^{-1/2} + (x + 5)^{1/2} = \dfrac{1}{2}(x + 5)^{-1/2}(x + 2(x + 5)) = \dfrac{3x + 10}{2\sqrt{x + 5}}$

$g''(x) = \dfrac{2\sqrt{x + 5}(3) - (3x + 10)(x + 5)^{-1/2}}{4(x + 5)} = \dfrac{6(x + 5) - (3x + 10)}{4(x + 5)^{3/2}} = \dfrac{3x + 20}{4(x + 5)^{3/2}} > 0$ on $(-5, \infty)$.

Concave upward on $(-5, \infty)$

No point of inflection

**41.** $f(x) = x + \cos x, 0 \leq x \leq 2\pi$

$f'(x) = 1 - \sin x$

$f''(x) = -\cos x = 0$ when $x = \dfrac{\pi}{2}, \dfrac{3\pi}{2}$.

| Intervals: | $0 < x < \dfrac{\pi}{2}$ | $\dfrac{\pi}{2} < x < \dfrac{3\pi}{2}$ | $\dfrac{3\pi}{2} < x < 2\pi$ |
|---|---|---|---|
| Sign of $f''(x)$: | $f''(x) < 0$ | $f''(x) > 0$ | $f''(x) < 0$ |
| Conclusion: | Concave downward | Concave upward | Concave downward |

Points of inflection: $\left(\dfrac{\pi}{2}, \dfrac{\pi}{2}\right), \left(\dfrac{3\pi}{2}, \dfrac{3\pi}{2}\right)$

**43.** $f(x) = (x + 9)^2$

$f'(x) = 2(x + 9) = 0 \Rightarrow x = -9$

$f''(x) = 2 > 0 \Rightarrow (-9, 0)$ is a relative minimum.

**45.** $g(x) = 2x^2(1 - x^2)$

$g'(x) = -4x(2x^2 - 1) = 0 \Rightarrow x = 0, \pm\dfrac{1}{\sqrt{2}}$

$g''(x) = 4 - 24x^2$

$g''(0) = 4 > 0 \quad (0, 0)$ is a relative minimum.

$g''\left(\pm\dfrac{1}{\sqrt{2}}\right) = -8 < 0 \left(\pm\dfrac{1}{\sqrt{2}}, \dfrac{1}{2}\right)$ are relative maxima.

**47.** $f(x) = 2x + \dfrac{18}{x}$

$f'(x) = 2 - \dfrac{18}{x^2} = 0 \Rightarrow 2x^2 = 18 \Rightarrow x = \pm3$

Critical numbers: $x = \pm3$

$f''(x) = \dfrac{36}{x^3}$

$f''(-3) < 0 \Rightarrow (-3, -12)$ is a relative maximum.

$f''(3) > 0 \Rightarrow (3, 12)$ is a relative minimum.

**49.**

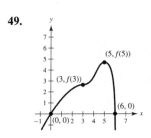

**51.** The first derivative is positive and the second derivative is negative. The graph is increasing and is concave down.

**53. (a)**

$$D = 0.00188t^4 - 0.1273t^3 + 2.672t^2 - 7.81t + 77.1,$$
$$0 \le t \le 40$$

**(b)**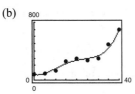

**(c)** Maximum occurs at $t = 40$ (2010).

Minimum occurs at $t \approx 1.6$ (1970).

**(d)** $D'(t)$ is greatest at $t = 40$ (2010).

**55.** $\lim\limits_{x \to \infty} \left( 8 + \dfrac{1}{x} \right) = 8 + 0 = 8$

**57.** $\lim\limits_{x \to \infty} \dfrac{2x^2}{3x^2 + 5} = \lim\limits_{x \to \infty} \dfrac{2}{3 + 5/x^2} = \dfrac{2}{3}$

**59.** $\lim\limits_{x \to -\infty} \dfrac{3x^2}{x + 5} = -\infty$

**61.** $\lim\limits_{x \to \infty} \dfrac{5 \cos x}{x} = 0$, because $\left| 5 \cos x \right| \le 5.$

**63.** $\lim\limits_{x \to -\infty} \dfrac{6x}{x + \cos x} = 6$

**65.** $f(x) = \dfrac{3}{x} - 2$

Discontinuity: $x = 0$

$\lim\limits_{x \to \infty} \left( \dfrac{3}{x} - 2 \right) = -2$

Vertical asymptote: $x = 0$

Horizontal asymptote: $y = -2$

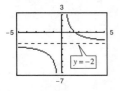

**67.** $h(x) = \dfrac{2x + 3}{x - 4}$

Discontinuity: $x = 4$

$\lim\limits_{x \to \infty} \dfrac{2x + 3}{x - 4} = \lim\limits_{x \to \infty} \dfrac{2 + (3/x)}{1 - (4/x)} = 2$

Vertical asymptote: $x = 4$

Horizontal asymptote: $y = 2$

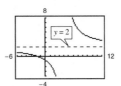

**69.** $f(x) = \dfrac{5}{3 + 2e^{-x}}$

$\lim\limits_{x \to \infty} \dfrac{5}{3 + 2e^{-x}} = \dfrac{5}{3}$

$\lim\limits_{x \to -\infty} \dfrac{5}{3 + 2e^{-x}} = 0$

Horizontal asymptotes: $y = 0, \ y = \dfrac{5}{3}$

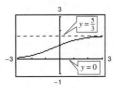

**71.** $g(x) = 3 \ln\left( 1 + e^{-x/4} \right)$

$\lim\limits_{x \to \infty} 3 \ln\left( 1 + e^{-x/4} \right) = 0$

Horizontal asymptote: $y = 0$

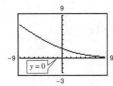

**73.** $f(x) = 4x - x^2 = x(4 - x)$

Domain: $(-\infty, \infty)$; Range: $(-\infty, 4]$

$f'(x) = 4 - 2x = 0$ when $x = 2$.

$f''(x) = -2$

Therefore, $(2, 4)$ is a relative maximum.

Intercepts: $(0, 0), (4, 0)$

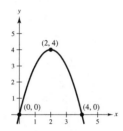

**75.** $f(x) = x\sqrt{16 - x^2}$

Domain: $[-4, 4]$; Range: $[-8, 8]$

$f'(x) = \dfrac{16 - 2x^2}{\sqrt{16 - x^2}} = 0$ when $x = \pm 2\sqrt{2}$ and

undefined when $x = \pm 4$.

$f''(x) = \dfrac{2x(x^2 - 24)}{(16 - x^2)^{3/2}}$

$f''(-2\sqrt{2}) > 0$

Therefore, $(-2\sqrt{2}, -8)$ is a relative minimum.

$f''(2\sqrt{2}) < 0$

Therefore, $(2\sqrt{2}, 8)$ is a relative maximum.

Point of inflection: $(0, 0)$

Intercepts: $(-4, 0), (0, 0), (4, 0)$

Symmetry with respect to origin

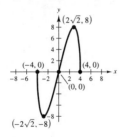

**77.** $f(x) = x^{1/3}(x + 3)^{2/3}$

Domain: $(-\infty, \infty)$; Range: $(-\infty, \infty)$

$f'(x) = \dfrac{x + 1}{(x + 3)^{1/3} x^{2/3}} = 0$ when $x = -1$ and

undefined when $x = -3, 0$.

$f''(x) = \dfrac{-2}{x^{5/3}(x + 3)^{4/3}}$ is undefined when $x = 0, -3$.

By the First Derivative Test $(-3, 0)$ is a relative

maximum and $\left(-1, -\sqrt[3]{4}\right)$ is a relative minimum. $(0, 0)$

is a point of inflection.

Intercepts: $(-3, 0), (0, 0)$

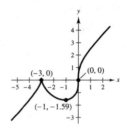

**79.** $f(x) = \dfrac{5 - 3x}{x - 2}$

$f'(x) = \dfrac{1}{(x - 2)^2} > 0$ for all $x \neq 2$

$f''(x) = \dfrac{-2}{(x - 2)^3}$

Concave upward on $(-\infty, 2)$

Concave downward on $(2, \infty)$

Vertical asymptote: $x = 2$

Horizontal asymptote: $y = -3$

Intercepts: $\left(\dfrac{5}{3}, 0\right), \left(0, -\dfrac{5}{2}\right)$

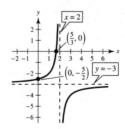

**81.** $f(x) = x^3 + x + \dfrac{4}{x}$

Domain: $(-\infty, 0), (0, \infty)$; Range: $(-\infty, -6], [6, \infty)$

$f'(x) = 3x^2 + 1 - \dfrac{4}{x^2}$

$\qquad = \dfrac{3x^4 + x^2 - 4}{x^2} = \dfrac{(3x^2 + 4)(x^2 - 1)}{x^2} = 0$

when $x = \pm 1$.

$f''(x) = 6x + \dfrac{8}{x^3} = \dfrac{6x^4 + 8}{x^3} \neq 0$

$f''(-1) < 0$

Therefore, $(-1, -6)$ is a relative maximum.

$f''(1) > 0$

Therefore, $(1, 6)$ is a relative minimum.

Vertical asymptote: $x = 0$

Symmetric with respect to origin

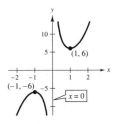

**83.** $4x + 3y = 400$ is the perimeter.

$$A = 2xy = 2x\left(\dfrac{400 - 4x}{3}\right) = \dfrac{8}{3}(100x - x^2)$$

$\dfrac{dA}{dx} = \dfrac{8}{3}(100 - 2x) = 0$ when $x = 50$.

$\dfrac{d^2A}{dx^2} = -\dfrac{16}{3} < 0$ when $x = 50$.

$A$ is a maximum when $x = 50$ ft and $y = \dfrac{200}{3}$ ft.

**85.** You have points $(0, y), (x, 0)$, and $(1, 8)$. So,

$$m = \dfrac{y - 8}{0 - 1} = \dfrac{0 - 8}{x - 1} \text{ or } y = \dfrac{8x}{x - 1}.$$

Let $f(x) = L^2 = x^2 + \left(\dfrac{8x}{x - 1}\right)^2.$

$$f'(x) = 2x + 128\left(\dfrac{x}{x - 1}\right)\left[\dfrac{(x - 1) - x}{(x - 1)^2}\right] = 0$$

$$x - \dfrac{64x}{(x - 1)^3} = 0$$

$x\left[(x - 1)^3 - 64\right] = 0$ when $x = 0, 5$ (minimum).

Vertices of triangle: $(0, 0), (5, 0), (0, 10)$

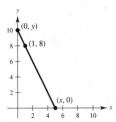

**87.**

| $h$ | 0 | 5 | 10 | 15 | 20 |
|---|---|---|---|---|---|
| $P$ | 10,332 | 5,583 | 2,376 | 1,240 | 517 |
| $\ln P$ | 9.243 | 8.627 | 7.773 | 7.123 | 6.248 |

(a)

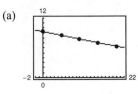

$y = -0.1499h + 9.3018$ is the regression line for data $(h, \ln P)$.

(b)  $\ln P = ah + b$

$\quad\quad P = e^{ah+b} = e^b e^{ah}$

$\quad\quad P = Ce^{ah}, \; C = e^b$

For our data, $a = -0.1499$ and $C = e^{9.3018} = 10{,}957.7$.

$\quad\quad P = 10{,}957.7e^{-0.1499h}$

(c)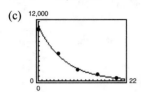

(d)  $\dfrac{dP}{dh} = (10{,}957.71)(-0.1499)e^{-0.1499h}$

$\quad\quad\quad = -1{,}642.56e^{-0.1499h}$

For $h = 5$, $\dfrac{dP}{dh} \approx -776.3$. For $h = 18$, $\dfrac{dP}{dh} \approx -110.6$.

**89.** You can form a right triangle with vertices $(0, 0)$, $(x, 0)$ and $(0, y)$. Assume that the hypotenuse of length $L$ passes through $(4, 6)$.

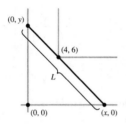

$$m = \frac{y - 6}{0 - 4} = \frac{6 - 0}{4 - x} \text{ or } y = \frac{6x}{x - 4}$$

Let $f(x) = L^2 = x^2 + y^2 = x^2 + \left(\dfrac{6x}{x - 4}\right)^2$.

$$f'(x) = 2x + 72\left(\frac{x}{x - 4}\right)\left[\frac{-4}{(x - 4)^2}\right] = 0$$

$x\left[(x - 4)^3 - 144\right] = 0$ when $x = 0$ or $x = 4 + \sqrt[3]{144}$.

$$L \approx 14.05 \text{ ft}$$

**91.** $V = \frac{1}{3}\pi x^2 h = \frac{1}{3}\pi x^2\left(r + \sqrt{r^2 - x^2}\right)$ (see figure)

$$\frac{dV}{dx} = \frac{1}{3}\pi\left[\frac{-x^3}{\sqrt{r^2 - x^2}} + 2x\left(r + \sqrt{r^2 - x^2}\right)\right] = \frac{\pi x}{3\sqrt{r^2 - x^2}}\left(2r^2 + 2r\sqrt{r^2 - x^2} - 3x^2\right) = 0$$

$$2r^2 + 2r\sqrt{r^2 - x^2} - 3x^2 = 0$$

$$2r\sqrt{r^2 - x^2} = 3x^2 - 2r^2$$

$$4r^2\left(r^2 - x^2\right) = 9x^4 - 12x^2r^2 + 4r^4$$

$$0 = 9x^4 - 8x^2r^2 = x^2\left(9x^2 - 8r^2\right)$$

$$x = 0, \frac{2\sqrt{2}r}{3}$$

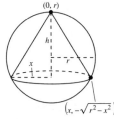

By the First Derivative Test, the volume is a maximum when

$x = \frac{2\sqrt{2}r}{3}$ and $h = r + \sqrt{r^2 - x^2} = \frac{4r}{3}$.

Thus, the maximum volume is $V = \frac{1}{3}\pi\left(\frac{8r^2}{9}\right)\left(\frac{4r}{3}\right) = \frac{32\pi r^3}{81}$ cubic units.

**93.** $y = f(x) = 0.5x^2$, $f'(x) = x$, $x = 3$, $\Delta x = dx = 0.01$

$$\begin{aligned}\Delta y &= f(x + \Delta x) - f(x) \\ &= f(3.01) - f(3) \\ &= 4.53005 - 4.5 \\ &= 0.03005\end{aligned} \qquad \begin{aligned}dy &= f'(x)dx \\ &= f'(3)dx \\ &= 3(0.01) \\ &= 0.03\end{aligned}$$

**95.** $y = x(1 - \cos x) = x - x\cos x$

$$\frac{dy}{dx} = 1 + x\sin x - \cos x$$

$$dy = (1 + x\sin x - \cos x)dx$$

**97.** $r = 9$ cm, $dr = \Delta r = \pm 0.025$

(a) $V = \frac{4}{3}\pi r^3$

$dV = 4\pi r^2\, dr$

$\Delta V \approx dV = 4\pi(9)^2(\pm 0.025) = \pm 8.1\pi$ cm$^3$

(b) $S = 4\pi r^2$

$dS = 8\pi r\, dr$

$\Delta S \approx dS = 8\pi(9)(\pm 0.025) = \pm 1.8\pi$ cm$^2$

(c) Percent error of volume:

$$\frac{dV}{V} = \frac{8.1\pi}{\frac{4}{3}\pi(9)^3} = 0.0083, \text{ or } 0.83\%$$

Percent error of surface area:

$$\frac{dS}{S} = \frac{1.8\pi}{4\pi(9)^2} = 0.0056, \text{ or } 0.56\%$$

**99.** $P = 100xe^{-x/400}$, $x$ changes from 115 to 120.

$$dP = 100\left(e^{-x/400} - \frac{x}{400}e^{-x/400}\right)dx$$

$$= e^{-115/400}\left(100 - \frac{115}{4}\right)(120 - 115)$$

$$\approx 267.24$$

Approximate percentage change: $\frac{dP}{P}(100) = \frac{267.24}{8626.57}(100) \approx 3.1\%$

# Problem Solving for Chapter 4

**1.** $p(x) = x^4 + ax^2 + 1$

(a)  $p'(x) = 4x^3 + 2ax = 2x(2x^2 + a)$

$p''(x) = 12x^2 + 2a$

For $a \geq 0$, there is one relative minimum at $(0, 1)$.

(b) For $a < 0$, there is a relative maximum at $(0, 1)$.

(c) For $a < 0$, there are two relative minima at $x = \pm\sqrt{-\dfrac{a}{2}}$.

(d) If $a < 0$, there are three critical points; if $a > 0$, there is only one critical point.

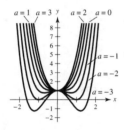

**3.** $f(x) = \dfrac{c}{x} + x^2$

$$f'(x) = -\frac{c}{x^2} + 2x = 0 \Rightarrow \frac{c}{x^2} = 2x \Rightarrow x^3 = \frac{c}{2} \Rightarrow x = \sqrt[3]{\frac{c}{2}}$$

$$f''(x) = \frac{2c}{x^3} + 2$$

If $c = 0$, $f(x) = x^2$ has a relative minimum, but no relative maximum.

If $c > 0$, $x = \sqrt[3]{\dfrac{c}{2}}$ is a relative minimum, because $f''\left(\sqrt[3]{\dfrac{c}{2}}\right) > 0$.

If $c < 0$, $x = \sqrt[3]{\dfrac{c}{2}}$ is a relative minimum, too.

Answer: All $c$.

**5.** Set $\dfrac{f(b) - f(a) - f'(a)(b - a)}{(b - a)^2} = k.$

Define $F(x) = f(x) - f(a) - f'(a)(x - a) - k(x - a)^2.$

$F(a) = 0, F(b) = f(b) - f(a) - f'(a)(b - a) - k(b - a)^2 = 0$

$F$ is continuous on $[a, b]$ and differentiable on $(a, b)$.

There exists $c_1, a < c_1 < b$, satisfying $F'(c_1) = 0$.

$F'(x) = f'(x) - f'(a) - 2k(x - a)$ satisfies the hypothesis of Rolle's Theorem on $[a, c_1]$:

$F'(a) = 0, F'(c_1) = 0.$

There exists $c_2, a < c_2 < c_1$ satisfying $F''(c_2) = 0$.

Finally, $F''(x) = f''(x) - 2k$ and $F''(c_2) = 0$ implies that

$k = \dfrac{f''(c_2)}{2}.$

So, $k = \dfrac{f(b) - f(a) - f'(a)(b - a)}{(b - a)^2} = \dfrac{f''(c_2)}{2} \Rightarrow f(b) = f(a) + f'(a)(b - a) + \dfrac{1}{2}f''(c_2)(b - a)^2.$

**7.** Distance $= \sqrt{4^2 + x^2} + \sqrt{(4 - x)^2 + 4^2} = f(x)$

$f'(x) = \dfrac{x}{\sqrt{4^2 + x^2}} - \dfrac{4 - x}{\sqrt{(4 - x)^2 + 4^2}} = 0$

$x\sqrt{(4 - x)^2 + 4^2} = -(x - 4)\sqrt{4^2 + x^2}$

$x^2\left[16 - 8x + x^2 + 16\right] = (x^2 - 8x + 16)(16 + x^2)$

$32x^2 - 8x^3 + x^4 = x^4 - 8x^3 + 32x^2 - 128x + 256$

$128x = 256$

$x = 2$

The bug should head towards the midpoint of the opposite side.

Without Calculus: Imagine opening up the cube:

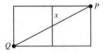

The shortest distance is the line $PQ$, passing through the midpoint.

**9.** $f$ continuous at $x = 0$: $1 = b$

$f$ continuous at $x = 1$: $a + 1 = 5 + c$

$f$ differentiable at $x = 1$: $a = 2 + 4 = 6$. So, $c = 2$.

$$f(x) = \begin{cases} 1, & x = 0 \\ 6x + 1, & 0 < x \le 1 \\ x^2 + 4x + 2, & 1 < x \le 3 \end{cases}$$

$$= \begin{cases} 6x + 1, & 0 \le x \le 1 \\ x^2 + 4x + 2, & 1 < x \le 3 \end{cases}$$

**11.** Let $h(x) = g(x) - f(x)$, which is continuous on $[a, b]$ and differentiable on $(a, b)$. $h(a) = 0$ and $h(b) = g(b) - f(b)$.

By the Mean Value Theorem, there exists $c$ in $(a, b)$ such that

$$h'(c) = \frac{h(b) - h(a)}{b - a} = \frac{g(b) - f(b)}{b - a}.$$

Because $h'(c) = g'(c) - f'(c) > 0$ and $b - a > 0$,

$g(b) - f(b) > 0 \Rightarrow g(b) > f(b)$.

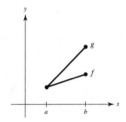

**13.** $y = \left(1 + x^2\right)^{-1}$

$$y' = \frac{-2x}{\left(1 + x^2\right)^2}$$

$$y'' = \frac{2\left(3x^2 - 1\right)}{\left(x^2 + 1\right)^3} = 0 \Rightarrow x = \pm\frac{1}{\sqrt{3}} = \pm\frac{\sqrt{3}}{3}$$

$$y: \quad \begin{array}{ccccc} +++ & ---- & ---- & +++ \\ \hline & -\frac{\sqrt{3}}{3} & 0 & \frac{\sqrt{3}}{3} & \end{array}$$

The tangent line has greatest slope at $\left(-\frac{\sqrt{3}}{3}, \frac{3}{4}\right)$ and

least slope at $\left(\frac{\sqrt{3}}{3}, \frac{3}{4}\right)$.

**15.** Assume $y_1 < d < y_2$. Let $g(x) = f(x) - d(x - a)$. $g$ is continuous on $[a, b]$ and therefore has a minimum $(c, g(c))$ on $[a, b]$. The point $c$ cannot be an endpoint of $[a, b]$ because

$g'(a) = f'(a) - d = y_1 - d < 0$

$g'(b) = f'(b) - d = y_2 - d > 0$.

So, $a < c < b$ and $g'(c) = 0 \Rightarrow f'(c) = d$.

**17.**    $p(x) = ax^3 + bx^2 + cx + d$

   $p'(x) = 3ax^2 + 2bx + c$

   $p''(x) = 6ax + 2b$

$6ax + 2b = 0$

   $x = -\dfrac{b}{3a}$

The sign of $p''(x)$ changes at $x = -b/3a$. Therefore, $\left(-b/3a,\, p(-b/3a)\right)$ is a point of inflection.

$$p\left(-\frac{b}{3a}\right) = a\left(-\frac{b^3}{27a^3}\right) + b\left(\frac{b^2}{9a^2}\right) + c\left(-\frac{b}{3a}\right) + d = \frac{2b^3}{27a^2} - \frac{bc}{3a} + d$$

When $p(x) = x^3 - 3x^2 + 2$, $a = 1$, $b = -3$, $c = 0$, and $d = 2$.

$$x_0 = \frac{-(-3)}{3(1)} = 1$$

$$y_0 = \frac{2(-3)^3}{27(1)^2} - \frac{(-3)(0)}{3(1)} + 2 = -2 - 0 + 2 = 0$$

The point of inflection of $p(x) = x^3 - 3x^2 + 2$ is $(x_0, y_0) = (1, 0)$.

**19.** $f(x) = \sin(\ln x)$

(a) Domain: $x > 0$  or  $(0, \infty)$

(b) $f(x) = 1 = \sin(\ln x) \Rightarrow \ln x = \dfrac{\pi}{2} + 2k\pi$.

Two values are $x = e^{\pi/2},\ e^{(\pi/2)+2\pi}$.

(c) $f(x) = -1 = \sin(\ln x) \Rightarrow \ln x = \dfrac{3\pi}{2} + 2k\pi$.

Two values are $x = e^{-\pi/2},\ e^{3\pi/2}$.

(d) Because the range of the sine function is $[-1, 1]$,  parts (b) and (c) show that the range of $f$ is $[-1, 1]$.

(e) $f'(x) = \dfrac{1}{x}\cos(\ln x)$

$f'(x) = 0 \Rightarrow \cos(\ln x) = 0 \Rightarrow \ln x = \dfrac{\pi}{2} + k\pi \Rightarrow x = e^{\pi/2}$ on $[1, 10]$.

$\left.\begin{array}{l} f\!\left(e^{\pi/2}\right) = 1 \\[4pt] f(1) = 0 \\[4pt] f(10) \approx 0.7440 \end{array}\right\}$ Maximum is 1 at $x = e^{\pi/2} \approx 4.8105$.

(f)

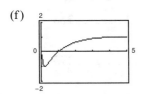

$\displaystyle\lim_{x \to 0^{+}} f(x)$ seems to be $-\dfrac{1}{2}$. (This is incorrect.)

(g) For the points $x = e^{\pi/2},\ x = e^{-3\pi/2},\ x = e^{-7\pi/2}, \ldots$ you have $f(x) = 1$.

For the points $x = e^{-\pi/2},\ x = e^{-5\pi/2},\ x = e^{-9\pi/2}, \ldots$ you have $f(x) = -1$.

That is, as $x \to 0^{+}$, there is an infinite number of points where $f(x) = 1$,  and an infinite number where $f(x) = -1$.

So, $\displaystyle\lim_{x \to 0^{+}} \sin(\ln x)$ does not exist.

You can verify this by graphing $f(x)$ on small intervals close to the origin.

# C H A P T E R  5
# Integration

# C H A P T E R 5
# Integration

## Section 5.1   Antiderivatives and Indefinite Integration

**1.** $\dfrac{d}{dx}\left(\dfrac{2}{x^3} + C\right) = \dfrac{d}{dx}\left(2x^{-3} + C\right) = -6x^{-4} = \dfrac{-6}{x^4}$

**3.** $\dfrac{dy}{dt} = 9t^2$

$y = 3t^3 + C$

**Check:** $\dfrac{d}{dt}\left[3t^3 + C\right] = 9t^2$

**5.** $\dfrac{dy}{dx} = x^{3/2}$

$y = \dfrac{2}{5}x^{5/2} + C$

**Check:** $\dfrac{d}{dx}\left[\dfrac{2}{5}x^{5/2} + C\right] = x^{3/2}$

| *Given* | *Rewrite* | *Integrate* | *Simplify* |
|---|---|---|---|
| **7.** $\displaystyle\int \sqrt[3]{x}\,dx$ | $\displaystyle\int x^{1/3}\,dx$ | $\dfrac{x^{4/3}}{4/3} + C$ | $\dfrac{3}{4}x^{4/3} + C$ |
| **9.** $\displaystyle\int \dfrac{1}{x\sqrt{x}}\,dx$ | $\displaystyle\int x^{-3/2}\,dx$ | $\dfrac{x^{-1/2}}{-1/2} + C$ | $-\dfrac{2}{\sqrt{x}} + C$ |

**11.** $\displaystyle\int (x + 7)\,dx = \dfrac{x^2}{2} + 7x + C$

**Check:** $\dfrac{d}{dx}\left[\dfrac{x^2}{2} + 7x + C\right] = x + 7$

**13.** $\displaystyle\int \left(x^{3/2} + 2x + 1\right)dx = \dfrac{2}{5}x^{5/2} + x^2 + x + C$

**Check:** $\dfrac{d}{dx}\left(\dfrac{2}{5}x^{5/2} + x^2 + x + C\right) = x^{3/2} + 2x + 1$

**15.** $\displaystyle\int \sqrt[3]{x^2}\,dx = \int x^{2/3}\,dx = \dfrac{x^{5/3}}{5/3} + C = \dfrac{3}{5}x^{5/3} + C$

**Check:** $\dfrac{d}{dx}\left(\dfrac{3}{5}x^{5/3} + C\right) = x^{2/3} = \sqrt[3]{x^2}$

**17.** $\displaystyle\int \dfrac{1}{x^5}\,dx = \int x^{-5}\,dx = \dfrac{x^{-4}}{-4} + C = \dfrac{-1}{4x^4} + C$

**Check:** $\dfrac{d}{dx}\left(\dfrac{-1}{4x^4} + C\right) = \dfrac{d}{dx}\left(-\dfrac{1}{4}x^{-4} + C\right)$

$= -\dfrac{1}{4}\left(-4x^{-5}\right) = \dfrac{1}{x^5}$

**19.** $\displaystyle\int \dfrac{x + 6}{\sqrt{x}}\,dx = \int \left(x^{1/2} + 6x^{-1/2}\right)dx$

$= \dfrac{x^{3/2}}{3/2} + 6\dfrac{x^{1/2}}{1/2} + C$

$= \dfrac{2}{3}x^{3/2} + 12x^{1/2} + C$

$= \dfrac{2}{3}x^{1/2}(x + 18) + C$

**Check:** $\dfrac{d}{dx}\left(\dfrac{2}{3}x^{3/2} + 12x^{1/2} + C\right)$

$= \dfrac{2}{3}\left(\dfrac{3}{2}x^{1/2}\right) + 12\left(\dfrac{1}{2}x^{-1/2}\right)$

$= x^{1/2} + 6x^{-1/2} = \dfrac{x + 6}{\sqrt{x}}$

**21.** $\displaystyle\int (x + 1)(3x - 2)\,dx = \int \left(3x^2 + x - 2\right)dx$

$= x^3 + \dfrac{1}{2}x^2 - 2x + C$

**Check:** $\dfrac{d}{dx}\left(x^3 + \dfrac{1}{2}x^2 - 2x + C\right) = 3x^2 + x - 2$

$= (x + 1)(3x - 2)$

**23.** $\displaystyle\int (5\cos x + 4\sin x)\,dx = 5\sin x - 4\cos x + C$

**Check:**

$\dfrac{d}{dx}(5\sin x - 4\cos x + C) = 5\cos x + 4\sin x$

**25.** $\int \left(2 \sin x - 5e^x\right) dx = -2 \cos x - 5e^x + C$

**Check:** $\dfrac{d}{dx}\left(-2 \cos x - 5e^x + C\right) = 2 \sin x - 5e^x$

**27.** $\int \left(\tan^2 y + 1\right) dy = \int \sec^2 y \, dy = \tan y + C$

**Check:** $\dfrac{d}{dy}(\tan y + C) = \sec^2 y = \tan^2 y + 1$

**29.** $\int \left(2x - 4^x\right) dx = x^2 - \dfrac{4^x}{\ln 4} + C$

**Check:** $\dfrac{d}{dx}\left(x^2 - \dfrac{4^x}{\ln 4} + C\right) = 2x - 4^x$

**31.** $\int \left(x - \dfrac{5}{x}\right) dx = \dfrac{x^2}{2} - 5 \ln|x| + C$

**Check:** $\dfrac{d}{dx}\left(\dfrac{x^2}{2} - 5 \ln|x| + C\right) = x - \dfrac{5}{x}$

**33.** $f'(x) = 4$

$f(x) = 4x + C$

Answers will vary.

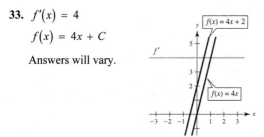

**35.** $f'(x) = 6x, \ f(0) = 8$

$f(x) = \int 6x \, dx = 3x^2 + C$

$f(0) = 8 = 3(0)^2 + C \Rightarrow C = 8$

$f(x) = 3x^2 + 8$

**37.** $f''(x) = 2$

$f'(2) = 5$

$f(2) = 10$

$f'(x) = \int 2 \, dx = 2x + C_1$

$f'(2) = 4 + C_1 = 5 \Rightarrow C_1 = 1$

$f'(x) = 2x + 1$

$f(x) = \int (2x + 1) \, dx = x^2 + x + C_2$

$f(2) = 6 + C_2 = 10 \Rightarrow C_2 = 4$

$f(x) = x^2 + x + 4$

**39.** $f''(x) = x^{-3/2}$

$f'(4) = 2$

$f(0) = 0$

$f'(x) = \int x^{-3/2} \, dx = -2x^{-1/2} + C_1 = -\dfrac{2}{\sqrt{x}} + C_1$

$f'(4) = -\dfrac{2}{2} + C_1 = 2 \Rightarrow C_1 = 3$

$f'(x) = -\dfrac{2}{\sqrt{x}} + 3$

$f(x) = \int \left(-2x^{-1/2} + 3\right) dx = -4x^{1/2} + 3x + C_2$

$f(0) = 0 + 0 + C_2 = 0 \Rightarrow C_2 = 0$

$f(x) = -4x^{1/2} + 3x = -4\sqrt{x} + 3x$

**41.** $f''(x) = e^x$

$f'(0) = 2$

$f(0) = 5$

$f'(x) = \int e^x \, dx = e^x + C_1$

$f'(0) = 2 = e^0 + C_1 \Rightarrow C_1 = 1$

$f'(x) = e^x + 1$

$f(x) = \int \left(e^x + 1\right) dx = e^x + x + C_2$

$f(0) = 5 = e^0 + 0 + C_2 \Rightarrow C_2 = 4$

$f(x) = e^x + x + 4$

**43.** (a) Answers will vary. *Sample answer.*

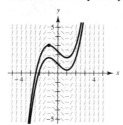

(b) $\dfrac{dy}{dx} = x^2 - 1, \ (-1, 3)$

$y = \dfrac{x^3}{3} - x + C$

$3 = \dfrac{(-1)^3}{3} - (-1) + C$

$C = \dfrac{7}{3}$

$y = \dfrac{x^3}{3} - x + \dfrac{7}{3}$

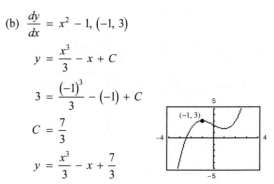

**45.** (a)

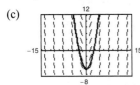

(b) $\dfrac{dy}{dx} = 2x$, $(-2, -2)$

$$y = \int 2x \, dx = x^2 + C$$

$$-2 = (-2)^2 + C = 4 + C \Rightarrow C = -6$$

$$y = x^2 - 6$$

(c)

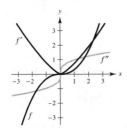

**47.** They are the same. In both cases you are finding a function $F(x)$ such that $F'(x) = f(x)$.

**49.** Because $f''$ is negative on $(-\infty, 0)$, $f'$ is decreasing on $(-\infty, 0)$. Because $f''$ is positive on $(0, \infty)$, $f'$ is increasing on $(0, \infty)$. $f'$ has a relative minimum at $(0, 0)$. Because $f'$ is positive on $(-\infty, \infty)$, $f$ is increasing on $(-\infty, \infty)$.

**51.** (a) $h(t) = \int (1.5t + 5) \, dt = 0.75t^2 + 5t + C$

$$h(0) = 0 + 0 + C = 12 \Rightarrow C = 12$$

$$h(t) = 0.75t^2 + 5t + 12$$

(b) $h(6) = 0.75(6)^2 + 5(6) + 12 = 69$ cm

**53.** $a(t) = -32$ ft/sec$^2$

$$v(t) = \int -32 \, dt = -32t + C_1$$

$$v(0) = 60 = C_1$$

$$s(t) = \int (-32t + 60) \, dt = -16t^2 + 60t + C_2$$

$$s(0) = 6 = C_2$$

$$s(t) = -16t^2 + 60t + 6, \text{ Position function}$$

The ball reaches its maximum height when

$$v(t) = -32t + 60 = 0$$

$$32t = 60$$

$$t = \tfrac{15}{8} \text{ seconds.}$$

$$s\left(\tfrac{15}{8}\right) = -16\left(\tfrac{15}{8}\right)^2 + 60\left(\tfrac{15}{8}\right) + 6 = 62.25 \text{ feet}$$

**55.** $v_0 = 16$ ft/sec

$s_0 = 64$ ft

(a) $\qquad s(t) = -16t^2 + 16t + 64 = 0$

$$-16(t^2 - t - 4) = 0$$

$$t = \frac{1 \pm \sqrt{17}}{2}$$

Choosing the positive value,

$$t = \frac{1 + \sqrt{17}}{2} \approx 2.562 \text{ seconds.}$$

(b) $\qquad v(t) = s'(t) = -32t + 16$

$$v\left(\frac{1 + \sqrt{17}}{2}\right) = -32\left(\frac{1 + \sqrt{17}}{2}\right) + 16$$

$$= -16\sqrt{17} \approx -65.970 \text{ ft/sec}$$

**57.** From Exercise 56, $f(t) = -4.9t^2 + v_0 t + 2$. If

$$f(t) = 200 = -4.9t^2 + v_0 t + 2,$$

Then $v(t) = -9.8t + v_0 = 0$

for this $t$ value. So, $t = v_0/9.8$ and you solve

$$-4.9\left(\frac{v_0}{9.8}\right)^2 + v_0\left(\frac{v_0}{9.8}\right) + 2 = 200$$

$$\frac{-4.9v_0^2}{(9.8)^2} + \left(\frac{v_0^2}{9.8}\right) = 198$$

$$-4.9v_0^2 + 9.8v_0^2 = (9.8)^2 198$$

$$4.9v_0^2 = (9.8)^2 198$$

$$v_0^2 = 3880.8$$

$$v_0 \approx 62.3 \text{ m/sec.}$$

**59.** $a = -1.6$

$v(t) = \int -1.6\, dt = -1.6t + v_0 = -1.6t$, because the stone was dropped, $v_0 = 0$.

$s(t) = \int (-1.6t)\, dt = -0.8t^2 + s_0$

$s(20) = 0 \Rightarrow -0.8(20)^2 + s_0 = 0$

$$s_0 = 320$$

So, the height of the cliff is 320 meters.

$v(t) = -1.6t$

$v(20) = -32$ m/sec

**61.** $x(t) = t^3 - 6t^2 + 9t - 2, \ 0 \le t \le 5$

(a) $v(t) = x'(t) = 3t^2 - 12t + 9$

$\qquad = 3(t^2 - 4t + 3) = 3(t-1)(t-3)$

$\qquad a(t) = v'(t) = 6t - 12 = 6(t-2)$

(b) $v(t) > 0$ when $0 < t < 1$ or $3 < t < 5$.

(c) $a(t) = 6(t-2) = 0$ when $t = 2$.

$\qquad v(2) = 3(1)(-1) = -3$

**63.** $v(t) = \dfrac{1}{\sqrt{t}} = t^{-1/2} \quad t > 0$

$x(t) = \int v(t)\, dt = 2t^{1/2} + C$

$x(1) = 4 = 2(1) + C \Rightarrow C = 2$

Position function: $x(t) = 2t^{1/2} + 2$

Acceleration function: $a(t) = v'(t) = -\dfrac{1}{2}t^{-3/2} = \dfrac{-1}{2t^{3/2}}$

**65.** (a) $v(0) = 25$ km/h $= 25 \cdot \dfrac{1000}{3600} = \dfrac{250}{36}$ m/sec

$\qquad v(13) = 80$ km/h $= 80 \cdot \dfrac{1000}{3600} = \dfrac{800}{36}$ m/sec

$\qquad a(t) = a$ (constant acceleration)

$\qquad v(t) = at + C$

$\qquad v(0) = \dfrac{250}{36} \Rightarrow v(t) = at + \dfrac{250}{36}$

$\qquad v(13) = \dfrac{800}{36} = 13a + \dfrac{250}{36}$

$\qquad \dfrac{550}{36} = 13a$

$\qquad a = \dfrac{550}{468} = \dfrac{275}{234} \approx 1.175$ m$/$sec$^2$

(b) $s(t) = a\dfrac{t^2}{2} + \dfrac{250}{36}t \quad (s(0) = 0)$

$s(13) = \dfrac{275}{234}\dfrac{(13)^2}{2} + \dfrac{250}{36}(13) \approx 189.58$ m

**67.** Truck: $v(t) = 30$

$\qquad s(t) = 30t \ \big(\text{Let } s(0) = 0.\big)$

Automobile: $a(t) = 6$

$\qquad v(t) = 6t \ \big(\text{Let } v(0) = 0.\big)$

$\qquad s(t) = 3t^2 \ \big(\text{Let } s(0) = 0.\big)$

At the point where the automobile overtakes the truck:

$30t = 3t^2$

$\quad 0 = 3t^2 - 30t$

$\quad 0 = 3t(t - 10)$ when $t = 10$ sec.

(a) $s(10) = 3(10)^2 = 300$ ft

(b) $v(10) = 6(10) = 60$ ft/sec $\approx 41$ mi/h

**69.** False. $f$ has an infinite number of antiderivatives, each differing by a constant.

**71.** $f''(x) = 2x$

$f'(x) = x^2 + C$

$f'(2) = 0 \Rightarrow 4 + C = 0 \Rightarrow C = -4$

$f(x) = \dfrac{x^3}{3} - 4x + C_1$

$f(2) = 0 \Rightarrow \dfrac{8}{3} - 8 + C_1 = 0 \Rightarrow C_1 = \dfrac{16}{3}$

$f(x) = \dfrac{x^3}{3} - 4x + \dfrac{16}{3}$

**73.** $\dfrac{d}{dx}\Big[ [s(x)]^2 + [c(x)]^2 \Big] = 2s(x)s'(x) + 2c(x)c'(x)$

$\qquad\qquad\qquad\qquad = 2s(x)c(x) - 2c(x)s(x) = 0$

So, $[s(x)]^2 + [c(x)]^2 = k$ for some constant $k$.

Because, $s(0) = 0$ and $c(0) = 1, k = 1$.

Therefore, $[s(x)]^2 + [c(x)]^2 = 1$.

[Note that $s(x) = \sin x$ and $c(x) = \cos x$ satisfy these properties.]

**75.** $\dfrac{d}{dx}\big(\ln|x| + C\big) = \dfrac{1}{x} + 0 = \dfrac{1}{x}$

## Section 5.2  Area

**1.** $\displaystyle\sum_{i=1}^{6}(3i + 2) = 3\sum_{i=1}^{6}i + \sum_{i=1}^{6}2 = 3(1 + 2 + 3 + 4 + 5 + 6) + 12 = 75$

**3.** $\displaystyle\sum_{k=0}^{4}\dfrac{1}{k^2 + 1} = 1 + \dfrac{1}{2} + \dfrac{1}{5} + \dfrac{1}{10} + \dfrac{1}{17} = \dfrac{158}{85}$

**5.** $\displaystyle\sum_{k=1}^{4} c = c + c + c + c = 4c$

**7.** $\displaystyle\sum_{i=1}^{11} \frac{1}{5i}$

**15.** $\displaystyle\sum_{i=1}^{24} 4i = 4\sum_{i=1}^{24} i = 4\left[\frac{24(25)}{2}\right] = 1200$

**9.** $\displaystyle\sum_{j=1}^{6} \left[7\left(\frac{j}{6}\right) + 5\right]$

**17.** $\displaystyle\sum_{i=1}^{20} (i-1)^2 = \sum_{i=1}^{19} i^2 = \left[\frac{19(20)(39)}{6}\right] = 2470$

**11.** $\displaystyle\frac{2}{n}\sum_{i=1}^{n} \left[\left(\frac{2i}{n}\right)^3 - \left(\frac{2i}{n}\right)\right]$

**13.** $\displaystyle\sum_{i=1}^{12} 7 = 7(12) = 84$

**19.** $\displaystyle\sum_{i=1}^{15} i(i-1)^2 = \sum_{i=1}^{15} i^3 - 2\sum_{i=1}^{15} i^2 + \sum_{i=1}^{15} i$

$$= \frac{15^2(16)^2}{4} - 2\frac{15(16)(31)}{6} + \frac{15(16)}{2}$$

$$= 14{,}400 - 2480 + 120 = 12{,}040$$

**21.** $\displaystyle\sum_{i=1}^{n} \frac{2i+1}{n^2} = \frac{1}{n^2}\sum_{i=1}^{n} (2i+1) = \frac{1}{n^2}\left[2\frac{n(n+1)}{2} + n\right] = \frac{n+2}{n} = 1 + \frac{2}{n} = S(n)$

$$S(10) = \frac{12}{10} = 1.2$$

$$S(100) = 1.02$$

$$S(1000) = 1.002$$

$$S(10{,}000) = 1.0002$$

**23.** $\displaystyle\sum_{k=1}^{n} \frac{6k(k-1)}{n^3} = \frac{6}{n^3}\sum_{k=1}^{n} (k^2 - k) = \frac{6}{n^3}\left[\frac{n(n+1)(2n+1)}{6} - \frac{n(n+1)}{2}\right]$

$$= \frac{6}{n^2}\left[\frac{2n^2 + 3n + 1 - 3n - 3}{6}\right] = \frac{1}{n^2}\left[2n^2 - 2\right] = 2 - \frac{2}{n^2} = S(n)$$

$$S(10) = 1.98$$

$$S(100) = 1.9998$$

$$S(1000) = 1.999998$$

$$S(10{,}000) = 1.99999998$$

**25.**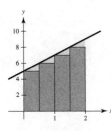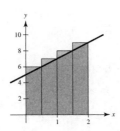

$$\Delta x = \frac{2-0}{4} = \frac{1}{2}$$

Left endpoints: Area $\approx \frac{1}{2}[5 + 6 + 7 + 8] = \frac{26}{2} = 13$

Right endpoints: Area $\approx \frac{1}{2}[6 + 7 + 8 + 9] = \frac{30}{2} = 15$

$13 < \text{Area} < 15$

**27.**

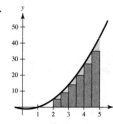

$$\Delta x = \frac{5-2}{6} = \frac{1}{2}$$

Left endpoints: Area $\approx \frac{1}{2}[5 + 9 + 14 + 20 + 27 + 35] = 55$

Right endpoints: Area $\approx \frac{1}{2}[9 + 14 + 20 + 27 + 35 + 44] = \frac{149}{2} = 74.5$

$55 < \text{Area} < 74.5$

**29.**

$$\Delta x = \frac{\frac{\pi}{2} - 0}{4} = \frac{\pi}{8}$$

Left endpoints: Area $\approx \frac{\pi}{8}\left[\cos(0) + \cos\left(\frac{\pi}{8}\right) + \cos\left(\frac{\pi}{4}\right) + \cos\left(\frac{3\pi}{8}\right)\right] \approx 1.1835$

Right endpoints: Area $\approx \frac{\pi}{8}\left[\cos\left(\frac{\pi}{8}\right) + \cos\left(\frac{\pi}{4}\right) + \cos\left(\frac{3\pi}{8}\right) + \cos\left(\frac{\pi}{2}\right)\right] \approx 0.7908$

$0.7908 < \text{Area} < 1.1835$

**31.** $S = \left[3 + 4 + \frac{9}{2} + 5\right](1) = \frac{33}{2} = 16.5$

$s = \left[1 + 3 + 4 + \frac{9}{2}\right](1) = \frac{25}{2} = 12.5$

**33.** $S(4) = \sqrt{\frac{1}{4}}\left(\frac{1}{4}\right) + \sqrt{\frac{1}{2}}\left(\frac{1}{4}\right) + \sqrt{\frac{3}{4}}\left(\frac{1}{4}\right) + \sqrt{1}\left(\frac{1}{4}\right) = \frac{1 + \sqrt{2} + \sqrt{3} + 2}{8} \approx 0.768$

$s(4) = 0\left(\frac{1}{4}\right) + \sqrt{\frac{1}{4}}\left(\frac{1}{4}\right) + \sqrt{\frac{1}{2}}\left(\frac{1}{4}\right) + \sqrt{\frac{3}{4}}\left(\frac{1}{4}\right) = \frac{1 + \sqrt{2} + \sqrt{3}}{8} \approx 0.518$

**35.** $S(5) = 1\left(\frac{1}{5}\right) + \frac{1}{6/5}\left(\frac{1}{5}\right) + \frac{1}{7/5}\left(\frac{1}{5}\right) + \frac{1}{8/5}\left(\frac{1}{5}\right) + \frac{1}{9/5}\left(\frac{1}{5}\right) = \frac{1}{5} + \frac{1}{6} + \frac{1}{7} + \frac{1}{8} + \frac{1}{9} \approx 0.746$

$s(5) = \frac{1}{6/5}\left(\frac{1}{5}\right) + \frac{1}{7/5}\left(\frac{1}{5}\right) + \frac{1}{8/5}\left(\frac{1}{5}\right) + \frac{1}{9/5}\left(\frac{1}{5}\right) + \frac{1}{2}\left(\frac{1}{5}\right) = \frac{1}{6} + \frac{1}{7} + \frac{1}{8} + \frac{1}{9} + \frac{1}{10} \approx 0.646$

**37.** $\displaystyle\lim_{n\to\infty}\sum_{i=1}^{n}\left(\frac{24i}{n^2}\right) = \lim_{n\to\infty}\frac{24}{n^2}\sum_{i=1}^{n}i = \lim_{n\to\infty}\frac{24}{n^2}\left(\frac{n(n+1)}{2}\right) = \lim_{n\to\infty}\left[12\left(\frac{n^2+n}{n^2}\right)\right] = 12\lim_{n\to\infty}\left(1 + \frac{1}{n}\right) = 12$

**39.** $\lim\limits_{n\to\infty} \sum\limits_{i=1}^{n} \dfrac{1}{n^3}(i-1)^2 = \lim\limits_{n\to\infty} \dfrac{1}{n^3} \sum\limits_{i=1}^{n-1} i^2 = \lim\limits_{n\to\infty} \dfrac{1}{n^3}\left[\dfrac{(n-1)(n)(2n-1)}{6}\right]$

$\qquad\qquad = \lim\limits_{n\to\infty} \dfrac{1}{6}\left[\dfrac{2n^3 - 3n^2 + n}{n^3}\right] = \lim\limits_{n\to\infty} \left[\dfrac{1}{6}\left(\dfrac{2 - (3/n) + (1/n^2)}{1}\right)\right] = \dfrac{1}{3}$

**41.** $\lim\limits_{n\to\infty} \sum\limits_{i=1}^{n} \left(1 + \dfrac{i}{n}\right)\!\left(\dfrac{2}{n}\right) = 2\lim\limits_{n\to\infty} \dfrac{1}{n}\left[\sum\limits_{i=1}^{n} 1 + \dfrac{1}{n}\sum\limits_{i=1}^{n} i\right] = 2\lim\limits_{n\to\infty} \dfrac{1}{n}\left[n + \dfrac{1}{n}\left(\dfrac{n(n+1)}{2}\right)\right] = 2\lim\limits_{n\to\infty}\left[1 + \dfrac{n^2 + n}{2n^2}\right] = 2\left(1 + \dfrac{1}{2}\right) = 3$

**43. (a)**

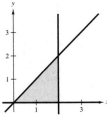

**(b)** $\Delta x = \dfrac{2 - 0}{n} = \dfrac{2}{n}$

$\qquad$ Endpoints: $0 < 1\!\left(\dfrac{2}{n}\right) < 2\!\left(\dfrac{2}{n}\right) < \ldots < (n-1)\!\left(\dfrac{2}{n}\right) < n\!\left(\dfrac{2}{n}\right) = 2$

**(c)** Because $y = x$ is increasing, $f(m_i) = f(x_{i-1})$ on $[x_{i-1}, x_i]$.

$\qquad s(n) = \sum\limits_{i=1}^{n} f(x_{i-1})\Delta x = \sum\limits_{i=1}^{n} f\!\left(\dfrac{2i-2}{n}\right)\!\left(\dfrac{2}{n}\right) = \sum\limits_{i=1}^{n}\left[(i-1)\!\left(\dfrac{2}{n}\right)\right]\!\left(\dfrac{2}{n}\right)$

**(d)** $f(M_i) = f(x_i)$ on $[x_{i-1}, x_i]$

$\qquad S(n) = \sum\limits_{i=1}^{n} f(x_i)\,\Delta x = \sum\limits_{i=1}^{n} f\!\left(\dfrac{2i}{n}\right)\dfrac{2}{n} = \sum\limits_{i=1}^{n}\left[i\!\left(\dfrac{2}{n}\right)\right]\!\left(\dfrac{2}{n}\right)$

**(e)**

| $x$ | 5 | 10 | 50 | 100 |
|---|---|---|---|---|
| $s(n)$ | 1.6 | 1.8 | 1.96 | 1.98 |
| $S(n)$ | 2.4 | 2.2 | 2.04 | 2.02 |

**(f)** $\lim\limits_{n\to\infty} \sum\limits_{i=1}^{n}\left[(i-1)\!\left(\dfrac{2}{n}\right)\right]\!\left(\dfrac{2}{n}\right) = \lim\limits_{n\to\infty} \dfrac{4}{n^2}\sum\limits_{i=1}^{n}(i-1) = \lim\limits_{n\to\infty} \dfrac{4}{n^2}\left[\dfrac{n(n+1)}{2} - n\right] = \lim\limits_{n\to\infty}\left[\dfrac{2(n+1)}{n} - \dfrac{4}{n}\right] = 2$

$\qquad \lim\limits_{n\to\infty} \sum\limits_{i=1}^{n}\left[i\!\left(\dfrac{2}{n}\right)\right]\!\left(\dfrac{2}{n}\right) = \lim\limits_{n\to\infty} \dfrac{4}{n^2}\sum\limits_{i=1}^{n} i = \lim\limits_{n\to\infty}\left(\dfrac{4}{n^2}\right)\dfrac{n(n+1)}{2} = \lim\limits_{n\to\infty} \dfrac{2(n+1)}{n} = 2$

**45.** $y = -4x + 5$ on $[0, 1]$. $\left(\textbf{Note: } \Delta x = \dfrac{1}{n}\right)$

$$s(n) = \sum_{i=1}^{n} f\left(\frac{i}{n}\right)\left(\frac{1}{n}\right) = \sum_{i=1}^{n}\left[-4\left(\frac{i}{n}\right) + 5\right]\left(\frac{1}{n}\right)$$

$$= -\frac{4}{n^2}\sum_{i=1}^{n} i + 5$$

$$= -\frac{4}{n^2}\frac{n(n+1)}{2} + 5$$

$$= -2\left(1 + \frac{1}{n}\right) + 5$$

$$\text{Area} = \lim_{n\to\infty} s(n) = 3$$

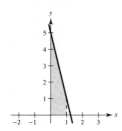

**47.** $y = x^2 + 2$ on $[0, 1]$. $\left(\textbf{Note: } \Delta x = \dfrac{1}{n}\right)$

$$S(n) = \sum_{i=1}^{n} f\left(\frac{i}{n}\right)\left(\frac{1}{n}\right)$$

$$= \sum_{i=1}^{n}\left[\left(\frac{i}{n}\right)^2 + 2\right]\left(\frac{1}{n}\right)$$

$$= \left[\frac{1}{n^3}\sum_{i=1}^{n} i^2\right] + 2$$

$$= \frac{n(n+1)(2n+1)}{6n^3} + 2 = \frac{1}{6}\left(2 + \frac{3}{n} + \frac{1}{n^2}\right) + 2$$

$$\text{Area} = \lim_{n\to\infty} S(n) = \frac{7}{3}$$

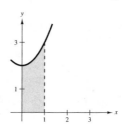

**49.** $y = 25 - x^2$ on $[1, 4]$. $\left(\textbf{Note: } \Delta x = \dfrac{3}{n}\right)$

$$s(n) = \sum_{i=1}^{n} f\left(1 + \frac{3i}{n}\right)\left(\frac{3}{n}\right) = \sum_{i=1}^{n}\left[25 - \left(1 + \frac{3i}{n}\right)^2\right]\left(\frac{3}{n}\right)$$

$$= \frac{3}{n}\sum_{i=1}^{n}\left[24 - \frac{9i^2}{n^2} - \frac{6i}{n}\right]$$

$$= \frac{3}{n}\left[24n - \frac{9}{n^2}\frac{n(n+1)(2n+1)}{6} - \frac{6}{n}\frac{n(n+1)}{2}\right]$$

$$= 72 - \frac{9}{2n^2}(n+1)(2n+1) - \frac{9}{n}(n+1)$$

$$\text{Area} = \lim_{n\to\infty} s(n) = 72 - 9 - 9 = 54$$

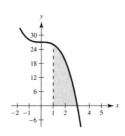

**51.** $y = 27 - x^3$ on $[1, 3]$. $\left(\textbf{Note: } \Delta x = \dfrac{3-1}{n} = \dfrac{2}{n}\right)$

$$s(n) = \sum_{i=1}^{n} f\left(1 + \frac{2i}{n}\right)\left(\frac{2}{n}\right) = \sum_{i=1}^{n}\left[27 - \left(1 + \frac{2i}{n}\right)^3\right]\left(\frac{2}{n}\right)$$

$$= \frac{2}{n}\sum_{i=1}^{n}\left[26 - \frac{8i^3}{n^3} - \frac{12i^2}{n^2} - \frac{6i}{n}\right]$$

$$= \frac{2}{n}\left[26n - \frac{8}{n^3}\frac{n^2(n+1)^2}{4} - \frac{12}{n^2}\frac{n(n+1)(2n+1)}{6} - \frac{6}{n}\frac{n(n+1)}{2}\right]$$

$$= 52 - \frac{4}{n^2}(n+1)^2 - \frac{4}{n^2}(n+1)(2n+1) - \frac{6n+1}{n}$$

$$\text{Area} = \lim_{n\to\infty} s(n) = 52 - 4 - 8 - 6 = 34$$

**53.** $y = x^2 - x^3$ on $[-1, 1]$. $\left(\textbf{Note: } \Delta x = \dfrac{1 - (-1)}{n} = \dfrac{2}{n}\right)$

Because $y$ both increases and decreases on $[-1, 1]$, $T(n)$ is neither an upper nor a lower sum.

$$T(n) = \sum_{i=1}^{n} f\left(-1 + \frac{2i}{n}\right)\left(\frac{2}{n}\right) = \sum_{i=1}^{n}\left[\left(-1 + \frac{2i}{n}\right)^2 - \left(-1 + \frac{2i}{n}\right)^3\right]\left(\frac{2}{n}\right)$$

$$= \sum_{i=1}^{n}\left[\left(1 - \frac{4i}{n} + \frac{4i^2}{n^2}\right) - \left(-1 + \frac{6i}{n} - \frac{12i^2}{n^2} + \frac{8i^3}{n^3}\right)\right]\left(\frac{2}{n}\right)$$

$$= \sum_{i=1}^{n}\left[2 - \frac{10i}{n} + \frac{16i^2}{n^2} - \frac{8i^3}{n^3}\right]\left(\frac{2}{n}\right) = \frac{4}{n}\sum_{i=1}^{n}1 - \frac{20}{n^2}\sum_{i=1}^{n}i + \frac{32}{n^3}\sum_{i=1}^{n}i^2 - \frac{16}{n^4}\sum_{i=1}^{n}i^3$$

$$= \frac{4}{n}(n) - \frac{20}{n^2}\cdot\frac{n(n+1)}{2} + \frac{32}{n^3}\cdot\frac{n(n+1)(2n+1)}{6} - \frac{16}{n^4}\cdot\frac{n^2(n+1)^2}{4}$$

$$= 4 - 10\left(1 + \frac{1}{n}\right) + \frac{16}{3}\left(2 + \frac{3}{n} + \frac{1}{n^2}\right) - 4\left(1 + \frac{2}{n} + \frac{1}{n^2}\right)$$

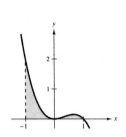

$$\text{Area} = \lim_{n\to\infty} T(n) = 4 - 10 + \frac{32}{3} - 4 = \frac{2}{3}$$

**55.** $f(y) = 4y$, $0 \le y \le 2$ $\left(\textbf{Note: } \Delta y = \dfrac{2 - 0}{n} = \dfrac{2}{n}\right)$

$$S(n) = \sum_{i=1}^{n} f(m_i)\Delta y$$

$$= \sum_{i=1}^{n} f\left(\frac{2i}{n}\right)\left(\frac{2}{n}\right)$$

$$= \sum_{i=1}^{n} 4\left(\frac{2i}{n}\right)\left(\frac{2}{n}\right)$$

$$= \frac{16}{n^2}\sum_{i=1}^{n} i$$

$$= \left(\frac{16}{n^2}\right)\cdot\frac{n(n+1)}{2} = \frac{8(n+1)}{n} = 8 + \frac{8}{n}$$

$$\text{Area} = \lim_{n\to\infty} S(n) = \lim_{n\to\infty}\left(8 + \frac{8}{n}\right) = 8$$

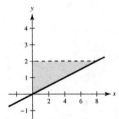

**57.** $f(y) = y^2$, $0 \le y \le 5$ $\left(\textbf{Note: } \Delta y = \dfrac{5 - 0}{n} = \dfrac{5}{n}\right)$

$$S(n) = \sum_{i=1}^{n} f\left(\frac{5i}{n}\right)\left(\frac{5}{n}\right)$$

$$= \sum_{i=1}^{n}\left(\frac{5i}{n}\right)^2\left(\frac{5}{n}\right)$$

$$= \frac{125}{n^3}\sum_{i=1}^{n} i^2$$

$$= \frac{125}{n^3}\cdot\frac{n(n+1)(2n+1)}{6}$$

$$= \frac{125}{n^2}\left(\frac{2n^2 + 3n + 1}{6}\right) = \frac{125}{3} + \frac{125}{2n} + \frac{125}{6n^2}$$

$$\text{Area } \lim_{n\to\infty} S(n) = \lim_{n\to\infty}\left(\frac{125}{3} + \frac{125}{2n} + \frac{125}{6n^2}\right) = \frac{125}{3}$$

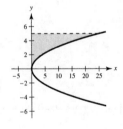

**59.** $g(y) = 4y^2 - y^3$, $1 \le y \le 3$. $\left( \textbf{Note:} \; \Delta y = \dfrac{3-1}{n} = \dfrac{2}{n} \right)$

$$S(n) = \sum_{i=1}^{n} g\left( 1 + \frac{2i}{n} \right)\left( \frac{2}{n} \right)$$

$$= \sum_{i=1}^{n} \left[ 4\left( 1 + \frac{2i}{n} \right)^2 - \left( 1 + \frac{2i}{n} \right)^3 \right]\frac{2}{n}$$

$$= \frac{2}{n}\sum_{i=1}^{n} 4\left[ 1 + \frac{4i}{n} + \frac{4i^2}{n^2} \right] - \left[ 1 + \frac{6i}{n} + \frac{12i^2}{n^2} + \frac{8i^3}{n^3} \right]$$

$$= \frac{2}{n}\sum_{i=1}^{n} \left[ 3 + \frac{10i}{n} + \frac{4i^2}{n^2} - \frac{8i^3}{n^3} \right]$$

$$= \frac{2}{n}\left[ 3n + \frac{10}{n}\frac{n(n+1)}{2} + \frac{4}{n^2}\frac{n(n+1)(2n+1)}{6} - \frac{8}{n^2}\frac{n^2(n+1)^2}{4} \right]$$

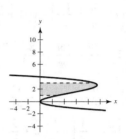

Area $= \displaystyle\lim_{n\to\infty} S(n) = 6 + 10 + \frac{8}{3} - 4 = \frac{44}{3}$

**61.** $f(x) = x^2 + 3$, $0 \le x \le 2$, $n = 4$

Let $c_i = \dfrac{x_i + x_{i-1}}{2}$.

$\Delta x = \dfrac{1}{2}$, $c_1 = \dfrac{1}{4}$, $c_2 = \dfrac{3}{4}$, $c_3 = \dfrac{5}{4}$, $c_4 = \dfrac{7}{4}$

Area $\approx \displaystyle\sum_{i=1}^{n} f(c_i)\Delta x = \sum_{i=1}^{4} \left[ c_i^2 + 3 \right]\left( \frac{1}{2} \right) = \frac{1}{2}\left[ \left( \frac{1}{16} + 3 \right) + \left( \frac{9}{16} + 3 \right) + \left( \frac{25}{16} + 3 \right) + \left( \frac{49}{16} + 3 \right) \right] = \frac{69}{8}$

**63.** $f(x) = \tan x$, $0 \le x \le \dfrac{\pi}{4}$, $n = 4$

Let $c_i = \dfrac{x_i + x_{i-1}}{2}$.

$\Delta x = \dfrac{\pi}{16}$, $c_1 = \dfrac{\pi}{32}$, $c_2 = \dfrac{3\pi}{32}$, $c_3 = \dfrac{5\pi}{32}$, $c_4 = \dfrac{7\pi}{32}$

Area $\approx \displaystyle\sum_{i=1}^{n} f(c_i)\Delta x = \sum_{i=1}^{4} (\tan c_i)\left( \frac{\pi}{16} \right) = \frac{\pi}{16}\left( \tan\frac{\pi}{32} + \tan\frac{3\pi}{32} + \tan\frac{5\pi}{32} + \tan\frac{7\pi}{32} \right) \approx 0.345$

**65.** $f(x) = \ln x$, $1 \le x \le 5$, $n = 4$

Let $c_i = \dfrac{x_i + x_{i-1}}{2}$, $\Delta x = 1$

$c_1 = \dfrac{3}{2}$, $c_2 = \dfrac{5}{2}$, $c_3 = \dfrac{7}{2}$, $c_4 = \dfrac{9}{2}$

Area $\approx \displaystyle\sum_{i=1}^{n} f(c_i)\Delta x = \sum_{i=1}^{4} \left[ \ln(c_i) \right](1) \approx 0.40547 + 0.91629 + 1.25276 + 1.50408 \approx 4.0786$

**67.**

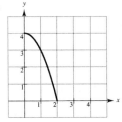

(b)   $A \approx 6$ square units

**69.** You can use the line $y = x$ bounded by $x = a$ and $x = b$. The sum of the areas of these inscribed rectangles is the lower sum.

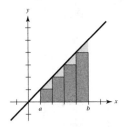

The sum of the areas of these circumscribed rectangles is the upper sum.

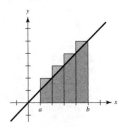

You can see that the rectangles do not contain all of the area in the first graph and the rectangles in the second graph cover more than the area of the region. The exact value of the area lies between these two sums.

**71. (a)**

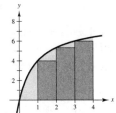

Lower sum:

$s(4) = 0 + 4 + 5\frac{1}{3} + 6 = 15\frac{1}{3} = \frac{46}{3} \approx 15.333$

**(b)**

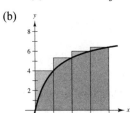

Upper sum:

$S(4) = 4 + 5\frac{1}{3} + 6 + 6\frac{2}{5} = 21\frac{11}{15} = \frac{326}{15} \approx 21.733$

**(c)**

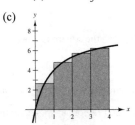

Midpoint Rule:

$M(4) = 2\frac{2}{3} + 4\frac{4}{5} + 5\frac{5}{7} + 6\frac{2}{9} = \frac{6112}{315} \approx 19.403$

**(d)** In each case, $\Delta x = 4/n$. The lower sum uses left end-points, $(i - 1)(4/n)$. The upper sum uses right endpoints, $(i)(4/n)$. The Midpoint Rule uses midpoints, $(i - \frac{1}{2})(4/n)$.

**(e)**

| $N$ | 4 | 8 | 20 | 100 | 200 |
|-----|-----|-----|-----|-----|-----|
| $s(n)$ | 15.333 | 17.368 | 18.459 | 18.995 | 19.06 |
| $S(n)$ | 21.733 | 20.568 | 19.739 | 19.251 | 19.188 |
| $M(n)$ | 19.403 | 19.201 | 19.137 | 19.125 | 19.125 |

**(f)** $s(n)$ increases because the lower sum approaches the exact value as $n$ increases. $S(n)$ decreases because the upper sum approaches the exact value as $n$ increases. Because of the shape of the graph, the lower sum is always smaller than the exact value, whereas the upper sum is always larger.

**73.** True. (Theorem 5.2 (2))

**75.** Suppose there are $n$ rows and $n + 1$ columns in the figure. The stars on the left total $1 + 2 + \cdots + n$, as do the stars on the right. There are $n(n + 1)$ stars in total, so

$$2[1 + 2 + \cdots + n] = n(n + 1)$$
$$1 + 2 + \cdots + n = \tfrac{1}{2}(n)(n + 1).$$

**77.** For $n$ odd,

| | | |
|---|---|---|
| $n = 1,$ | 1 row, | 1 block |
| $n = 3,$ | 2 rows, | 4 blocks |
| $n = 5,$ | 3 rows, | 9 blocks |
| $n,$ | $\dfrac{n + 1}{2}$ rows, | $\left(\dfrac{n + 1}{2}\right)^2$ blocks, |

For $n$ even,

| | | |
|---|---|---|
| $n = 2,$ | 1 row, | 2 block |
| $n = 4,$ | 2 rows, | 6 blocks |
| $n = 6,$ | 3 rows, | 12 blocks |
| $n,$ | $\dfrac{n}{2}$ rows, | $\dfrac{n^2 + 2n}{4}$ blocks, |

**79.** Assume that the dartboard has corners at $(\pm 1, \pm 1)$.

A point $(x, y)$ in the square is closer to the center than the top edge if

$$\sqrt{x^2 + y^2} \le 1 - y$$
$$x^2 + y^2 \le 1 - 2y + y^2$$
$$y \le \tfrac{1}{2}(1 - x^2).$$

By symmetry, a point $(x, y)$ in the square is closer to the center than the right edge if

$$x \le \tfrac{1}{2}(1 - y^2).$$

In the first quadrant, the parabolas $y = \tfrac{1}{2}(1 - x^2)$ and $x = \tfrac{1}{2}(1 - y^2)$ intersect at $\left(\sqrt{2} - 1, \sqrt{2} - 1\right)$. There are 8 equal regions that make up the total region, as indicated in the figure.

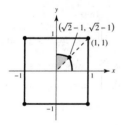

Area of shaded region $S = \displaystyle\int_0^{\sqrt{2}-1} \left[\frac{1}{2}(1 - x^2) - x\right] dx = \frac{2\sqrt{2}}{3} - \frac{5}{6}$

Probability $= \dfrac{8S}{\text{Area square}} = 2\left[\dfrac{2\sqrt{2}}{3} - \dfrac{5}{6}\right] = \dfrac{4\sqrt{2}}{3} - \dfrac{5}{3}$

## Section 5.3    Riemann Sums and Definite Integrals

**1.** $f(x) = \sqrt{x}$, $y = 0$, $x = 0$, $x = 3$, $c_i = \dfrac{3i^2}{n^2}$

$$\Delta x_i = \frac{3i^2}{n^2} - \frac{3(i-1)^2}{n^2} = \frac{3}{n^2}(2i - 1)$$

$$\lim_{n \to \infty} \sum_{i=1}^{n} f(c_i)\Delta x_i = \lim_{n \to \infty} \sum_{i=1}^{n} \sqrt{\frac{3i^2}{n^2}} \frac{3}{n^2}(2i - 1)$$

$$= \lim_{n \to \infty} \frac{3\sqrt{3}}{n^3} \sum_{i=1}^{n} \left(2i^2 - i\right)$$

$$= \lim_{n \to \infty} \frac{3\sqrt{3}}{n^3}\left[2\frac{n(n+1)(2n+1)}{6} - \frac{n(n+1)}{2}\right]$$

$$= \lim_{n \to \infty} 3\sqrt{3}\left[\frac{(n+1)(2n+1)}{3n^2} - \frac{n+1}{2n^2}\right]$$

$$= 3\sqrt{3}\left[\frac{2}{3} - 0\right] = 2\sqrt{3} \approx 3.464$$

**3.** $y = 8$ on $[2, 6]$. $\left(\textbf{Note: } \Delta x = \dfrac{6-2}{n} = \dfrac{4}{n}, \ \|\Delta\| \to 0 \text{ as } n \to \infty\right)$

$$\sum_{i=1}^{n} f(c_i)\,\Delta x_i = \sum_{i=1}^{n} f\left(2 + \frac{4i}{n}\right)\left(\frac{4}{n}\right) = \sum_{i=1}^{n} 8\left(\frac{4}{n}\right) = \sum_{i=1}^{n} \frac{32}{n} = \frac{1}{n}\sum_{i=1}^{n} 32 = \frac{1}{n}(32n) = 32$$

$$\int_{2}^{6} 8\,dx = \lim_{n \to \infty} 32 = 32$$

**5.** $y = x^3$ on $[-1, 1]$. $\left(\textbf{Note:} \Delta x = \dfrac{1-(-1)}{n} = \dfrac{2}{n}, \ \|\Delta\| \to 0 \text{ as } n \to \infty\right)$

$$\sum_{i=1}^{n} f(c_i)\Delta x_i = \sum_{i=1}^{n} f\left(-1 + \frac{2i}{n}\right)\left(\frac{2}{n}\right)$$

$$= \sum_{i=1}^{n}\left(-1 + \frac{2i}{n}\right)^3\left(\frac{2}{n}\right)$$

$$= \sum_{i=1}^{n}\left[-1 + \frac{6i}{n} - \frac{12i^2}{n^2} + \frac{8i^3}{n^3}\right]\left(\frac{2}{n}\right)$$

$$= -2 + \frac{12}{n^2}\sum_{i=1}^{n} i - \frac{24}{n^3}\sum_{i=1}^{n} i^2 + \frac{16}{n^4}\sum_{i=1}^{n} i^3$$

$$= -2 + 6\left(1 + \frac{1}{n}\right) - 4\left(2 + \frac{3}{n} + \frac{1}{n^2}\right) + 4\left(1 + \frac{2}{n} + \frac{1}{n^2}\right) = \frac{2}{n}$$

$$\int_{-1}^{1} x^3\,dx = \lim_{n \to \infty} \frac{2}{n} = 0$$

**7.** $y = x^2 + 1$ on [1, 2]. $\left(\text{Note: } \Delta x = \dfrac{2-1}{n} = \dfrac{1}{n}, \|\Delta\| \to 0 \text{ as } n \to \infty\right)$

$$\sum_{i=1}^{n} f(c_i)\,\Delta x_i = \sum_{i=1}^{n} f\left(1 + \frac{i}{n}\right)\left(\frac{1}{n}\right)$$

$$= \sum_{i=1}^{n}\left[\left(1 + \frac{i}{n}\right)^2 + 1\right]\left(\frac{1}{n}\right)$$

$$= \sum_{i=1}^{n}\left[1 + \frac{2i}{n} + \frac{i^2}{n^2} + 1\right]\left(\frac{1}{n}\right)$$

$$= 2 + \frac{2}{n^2}\sum_{i=1}^{n} i + \frac{1}{n^3}\sum_{i=1}^{n} i^2 = 2 + \left(1 + \frac{1}{n}\right) + \frac{1}{6}\left(2 + \frac{3}{n} + \frac{1}{n^2}\right) = \frac{10}{3} + \frac{3}{2n} + \frac{1}{6n^2}$$

$$\int_1^2 (x^2 + 1)\,dx = \lim_{n \to \infty}\left(\frac{10}{3} + \frac{3}{2n} + \frac{1}{6n^2}\right) = \frac{10}{3}$$

**9.** $\displaystyle\lim_{\|\Delta\| \to 0} \sum_{i=1}^{n} (3c_i + 10)\,\Delta x_i = \int_{-1}^{5} (3x + 10)\,dx$

    on the interval $[-1, 5]$.

**11.** $\displaystyle\lim_{\|\Delta\| \to 0} \sum_{i=1}^{n} \sqrt{c_i^2 + 4}\,\Delta x_i = \int_0^3 \sqrt{x^2 + 4}\,dx$

    on the interval $[0, 3]$.

**13.** $\displaystyle\lim_{\|\Delta\| \to 0} \sum_{i=1}^{n}\left(1 + \frac{3}{c_i}\right)\Delta x_i = \int_1^5\left(1 + \frac{3}{x}\right)dx$

    on the interval $[1, 5]$.

**15.** $\displaystyle\int_0^4 5\,dx$

**17.** $\displaystyle\int_{-4}^4 (4 - |x|)\,dx$

**19.** $\displaystyle\int_{-5}^5 (25 - x^2)\,dx$

**21.** $\displaystyle\int_0^{\pi/2} \cos x\,dx$

**23.** $\displaystyle\int_0^2 y^3\,dy$

**25.** $\displaystyle\int_1^4 \frac{2}{x}\,dx$

**27.** Rectangle

$$A = bh = 3(4)$$

$$A = \int_0^3 4\,dx = 12$$

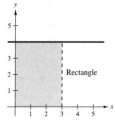

**29.** Triangle

$$A = \tfrac{1}{2}bh = \tfrac{1}{2}(4)(4) = 8$$

$$A = \int_0^4 x\,dx = 8$$

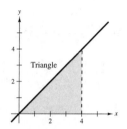

**31.** Trapezoid

$$A = \frac{b_1 + b_2}{2}h = \left(\frac{4 + 10}{2}\right)2 = 14$$

$$A = \int_0^2 (3x + 4)\,dx = 14$$

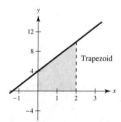

**33.** Triangle

$$A = \tfrac{1}{2}bh = \tfrac{1}{2}(2)(1) = 1$$

$$A = \int_{-1}^1 (1 - |x|)\,dx = 1$$

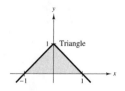

**35.** Semicircle

$$A = \frac{1}{2}\pi r^2 = \frac{1}{2}\pi(7)^2 = \frac{49\pi}{2}$$

$$A = \int_{-7}^{7} \sqrt{49 - x^2}\, dx = \frac{49\pi}{2}$$

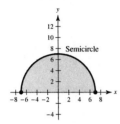

In Exercises 37 – 43, $\int_{2}^{4} x^3\, dx = 60$, $\int_{2}^{4} x\, dx = 6$,

$\int_{2}^{4} dx = 2$

**37.** $\int_{4}^{2} x\, dx = -\int_{2}^{4} x\, dx = -6$

**39.** $\int_{2}^{4} 8x\, dx = 8\int_{2}^{4} x\, dx = 8(6) = 48$

**41.** $\int_{2}^{4} (x - 9)\, dx = \int_{2}^{4} x\, dx - 9\int_{2}^{4} dx = 6 - 9(2) = -12$

**43.** $\int_{2}^{4} \left(\frac{1}{2}x^3 - 3x + 2\right) dx = \frac{1}{2}\int_{2}^{4} x^3\, dx - 3\int_{2}^{4} x\, dx + 2\int_{2}^{4} dx$

$$= \frac{1}{2}(60) - 3(6) + 2(2) = 16$$

**45.** (a) $\int_{0}^{7} f(x)\, dx = \int_{0}^{5} f(x)\, dx + \int_{5}^{7} f(x)\, dx = 10 + 3 = 13$

(b) $\int_{5}^{0} f(x)\, dx = -\int_{0}^{5} f(x)\, dx = -10$

(c) $\int_{5}^{5} f(x)\, dx = 0$

(d) $\int_{0}^{5} 3f(x)\, dx = 3\int_{0}^{5} f(x)\, dx = 3(10) = 30$

**47.** (a) $\int_{2}^{6} \left[f(x) + g(x)\right] dx = \int_{2}^{6} f(x)\, dx + \int_{2}^{6} g(x)\, dx$

$$= 10 + (-2) = 8$$

(b) $\int_{2}^{6} \left[g(x) - f(x)\right] dx = \int_{2}^{6} g(x)\, dx - \int_{2}^{6} f(x)\, dx$

$$= -2 - 10 = -12$$

(c) $\int_{2}^{6} 2g(x)\, dx = 2\int_{2}^{6} g(x)\, dx = 2(-2) = -4$

(d) $\int_{2}^{6} 3f(x)\, dx = 3\int_{2}^{6} f(x)\, dx = 3(10) = 30$

**49.** Lower estimate: $\left[24 + 12 - 4 - 20 - 36\right](2) = -48$

Upper estimate: $\left[32 + 24 + 12 - 4 - 20\right](2) = 88$

**51.** (a) Quarter circle below *x*-axis:

$$-\tfrac{1}{4}\pi r^2 = -\tfrac{1}{4}\pi(2)^2 = -\pi$$

(b) Triangle: $\frac{1}{2}bh = \frac{1}{2}(4)(2) = 4$

(c) Triangle + Semicircle below *x*-axis:

$$-\tfrac{1}{2}(2)(1) - \tfrac{1}{2}\pi(2)^2 = -(1 + 2\pi)$$

(d) Sum of parts (b) and (c): $4 - (1 + 2\pi) = 3 - 2\pi$

(e) Sum of absolute values of (b) and (c):

$$4 + (1 + 2\pi) = 5 + 2\pi$$

(f) Answers to (d) plus

$$2(10) = 20: (3 - 2\pi) + 20 = 23 - 2\pi$$

**53.** (a) $\int_{0}^{5} \left[f(x) + 2\right] dx = \int_{0}^{5} f(x)\, dx + \int_{0}^{5} 2\, dx$

$$= 4 + 10 = 14$$

(b) $\int_{-2}^{3} f(x + 2)\, dx = \int_{0}^{5} f(x)\, dx = 4$ (Let $u = x + 2$.)

(c) $\int_{-5}^{5} f(x)\, dx = 2\int_{0}^{5} f(x)\, dx = 2(4) = 8$ ($f$ even)

(d) $\int_{-5}^{5} f(x)\, dx = 0$ ($f$ odd)

**55.** $f(x) = \begin{cases} 4, & x < 4 \\ x, & x \geq 4 \end{cases}$

$$\int_{0}^{8} f(x)\, dx = 4(4) + 4(4) + \tfrac{1}{2}(4)(4) = 40$$

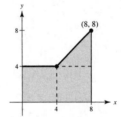

**57.**

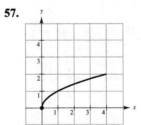

(a) $A \approx 5$ square units

**59.**

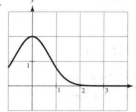

(c) $A \approx 2$ square units

**61.** $f(x) = \dfrac{1}{x - 4}$

is not integrable on the interval $[3, 5]$ because $f$ has a discontinuity at $x = 4$.

**63.** $\displaystyle\int_{-2}^{1} f(x)\,dx + \int_{1}^{5} f(x)\,dx = \int_{-2}^{5} f(x)\,dx$

$a = -2,\ b = 5$

**73.** $f(x) = x^2 + 3x,\ [0, 8]$

$x_0 = 0,\ x_1 = 1,\ x_2 = 3,\ x_3 = 7,\ x_4 = 8$

$\Delta x_1 = 1,\ \Delta x_2 = 2,\ \Delta x_3 = 4,\ \Delta x_4 = 1$

$c_1 = 1,\ c_2 = 2,\ c_3 = 5,\ c_4 = 8$

$\displaystyle\sum_{i=1}^{4} f(c_i)\Delta x = f(1)\Delta x_1 + f(2)\Delta x_2 + f(5)\Delta x_3 + f(8)\Delta x_4$

$\qquad\qquad = (4)(1) + (10)(2) + (40)(4) + (88)(1) = 272$

**75.** $\Delta x = \dfrac{b - a}{n},\ c_i = a + i(\Delta x) = a + i\left(\dfrac{b - a}{n}\right)$

$\displaystyle\int_{0}^{b} x\,dx = \lim_{\|\Delta\| \to 0} \sum_{i=1}^{n} f(c_i)\Delta x$

$\qquad\qquad = \lim_{n \to \infty} \sum_{i=1}^{n} \left[a + i\left(\dfrac{b - a}{n}\right)\right]\left(\dfrac{b - a}{n}\right)$

$\qquad\qquad = \lim_{n \to \infty} \left[\left(\dfrac{b - a}{n}\right)\sum_{i=1}^{n} a + \left(\dfrac{b - a}{n}\right)^2 \sum_{i=1}^{n} i\right]$

$\qquad\qquad = \lim_{n \to \infty} \left[\dfrac{b - a}{n}(an) + \left(\dfrac{b - a}{n}\right)^2 \dfrac{n(n + 1)}{2}\right]$

$\qquad\qquad = \lim_{n \to \infty} \left[a(b - a) + \dfrac{(b - a)^2}{n} \dfrac{n + 1}{2}\right]$

$\qquad\qquad = a(b - a) + \dfrac{(b - a)^2}{2}$

$\qquad\qquad = (b - a)\left[a + \dfrac{b - a}{2}\right]$

$\qquad\qquad = \dfrac{(b - a)(a + b)}{2} = \dfrac{b^2 - a^2}{2}$

**65.** Answers will vary. *Sample answer:* $a = \pi,\ b = 2\pi$

$\displaystyle\int_{\pi}^{2\pi} \sin x\,dx < 0$

**67.** True

**69.** True

**71.** False

$\displaystyle\int_{0}^{2} (-x)\,dx = -2$

**77.** $f(x) = \begin{cases} 1, & x \text{ is rational} \\ 0, & x \text{ is irrational} \end{cases}$

is not integrable on the interval $[0, 1]$. As

$\|\Delta\| \to 0$, $f(c_i) = 1$ or $f(c_i) = 0$ in each subinterval

because there are an infinite number of both rational and irrational numbers in any interval, no matter how small.

**79.** The function $f$ is nonnegative between $x = -1$ and $x = 1$.

So, $\int_a^b \left(1 - x^2\right) dx$

is a maximum for

$a = -1$ and $b = 1$.

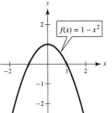

**81.** Let $f(x) = x^2$, $0 \le x \le 1$, and $\Delta x_i = 1/n$. The appropriate Riemann Sum is

$$\sum_{i=1}^{n} f(c_i) \Delta x_i = \sum_{i=1}^{n} \left(\frac{i}{n}\right)^2 \frac{1}{n} = \frac{1}{n^3} \sum_{i=1}^{n} i^2.$$

$$\lim_{n \to \infty} \frac{1}{n^3}\left[1^2 + 2^2 + 3^2 + \cdots + n^2\right] = \lim_{n \to \infty} \frac{1}{n^3} \cdot \frac{n(2n+1)(n+1)}{6} = \lim_{n \to \infty} \frac{2n^2 + 3n + 1}{6n^2} = \lim_{n \to \infty}\left(\frac{1}{3} + \frac{1}{2n} + \frac{1}{6n^2}\right) = \frac{1}{3}$$

## Section 5.4   The Fundamental Theorem of Calculus

**1.** $f(x) = \dfrac{4}{x^2 + 1}$

$\int_0^\pi \dfrac{4}{x^2 + 1}\, dx$ is positive.

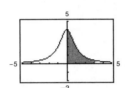

**3.** $f(x) = x\sqrt{x^2 + 1}$

$\int_{-2}^{2} x\sqrt{x^2 + 1}\, dx = 0$

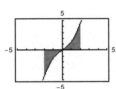

**5.** $\int_0^2 6x\, dx = \left[3x^2\right]_0^2 = 3(2)^2 - 0$

**7.** $\int_{-1}^{0} (2x - 1)\, dx = \left[x^2 - x\right]_{-1}^{0}$

$\qquad = 0 - \left((-1)^2 - (-1)\right) = -(1 + 1) = -2$

**9.** $\int_{-1}^{1} \left(t^2 - 2\right) dt = \left[\dfrac{t^3}{3} - 2t\right]_{-1}^{1}$

$\qquad = \left(\dfrac{1}{3} - 2\right) - \left(-\dfrac{1}{3} + 2\right) = -\dfrac{10}{3}$

**11.** $\int_0^1 (2t - 1)^2\, dt = \int_0^1 \left(4t^2 - 4t + 1\right) dt = \left[\frac{4}{3}t^3 - 2t^2 + t\right]_0^1 = \frac{4}{3} - 2 + 1 = \frac{1}{3}$

**13.** $\int_1^2 \left(\dfrac{3}{x^2} - 1\right) dx = \left[-\dfrac{3}{x} - x\right]_1^2 = \left(-\dfrac{3}{2} - 2\right) - (-3 - 1) = \dfrac{1}{2}$

**15.** $\int_1^4 \dfrac{u - 2}{\sqrt{u}}\, du = \int_1^4 \left(u^{1/2} - 2u^{-1/2}\right) du = \left[\frac{2}{3}u^{3/2} - 4u^{1/2}\right]_1^4 = \left[\frac{2}{3}\left(\sqrt{4}\right)^3 - 4\sqrt{4}\right] - \left[\frac{2}{3} - 4\right] = \dfrac{2}{3}$

**17.** $\int_{-1}^{1}\left(\sqrt[3]{t} - 2\right) dt = \left[\frac{3}{4}t^{4/3} - 2t\right]_{-1}^{1} = \left(\frac{3}{4} - 2\right) - \left(\frac{3}{4} + 2\right) = -4$

**19.** $\int_0^1 \dfrac{x - \sqrt{x}}{3}\, dx = \dfrac{1}{3}\int_0^1 \left(x - x^{1/2}\right) dx = \dfrac{1}{3}\left[\dfrac{x^2}{2} - \dfrac{2}{3}x^{3/2}\right]_0^1 = \dfrac{1}{3}\left(\dfrac{1}{2} - \dfrac{2}{3}\right) = -\dfrac{1}{18}$

**21.** $\int_{-1}^{0}\left(t^{1/3} - t^{2/3}\right) dt = \left[\frac{3}{4}t^{4/3} - \frac{3}{5}t^{5/3}\right]_{-1}^{0} = 0 - \left(\frac{3}{4} + \frac{3}{5}\right) = -\dfrac{27}{20}$

**23.** $\int_0^5 |2x - 5|\, dx = \int_0^{5/2} (5 - 2x)\, dx + \int_{5/2}^{5} (2x - 5)\, dx$ $\left(\text{split up the integral at the zero } x = \frac{5}{2}\right)$

$\qquad = \left[5x - x^2\right]_0^{5/2} + \left[x^2 - 5x\right]_{5/2}^{5} = \left(\frac{25}{2} - \frac{25}{4}\right) - 0 + (25 - 25) - \left(\frac{25}{4} - \frac{25}{2}\right) = 2\left(\frac{25}{2} - \frac{25}{4}\right) = \dfrac{25}{2}$

**Note:** By Symmetry, $\int_0^5 |2x - 5|\, dx = 2\int_{5/2}^{5} (2x - 5)\, dx.$

**25.** $\int_0^4 |x^2 - 9|\, dx = \int_0^3 (9 - x^2)\, dx + \int_3^4 (x^2 - 9)\, dx$ (split up integral at the zero $x = 3$)

$$= \left[9x - \frac{x^3}{3}\right]_0^3 + \left[\frac{x^3}{3} - 9x\right]_3^4 = (27 - 9) + \left(\frac{64}{3} - 36\right) - (9 - 27) = \frac{64}{3}$$

**27.** $\int_0^\pi (1 + \sin x)\, dx = \left[x - \cos x\right]_0^\pi = (\pi + 1) - (0 - 1) = 2 + \pi$

**29.** $\int_0^{\pi/4} \frac{1 - \sin^2 \theta}{\cos^2 \theta}\, d\theta = \int_0^{\pi/4} d\theta = \left[\theta\right]_0^{\pi/4} = \frac{\pi}{4}$

**31.** $\int_{-\pi/6}^{\pi/6} \sec^2 x\, dx = \left[\tan x\right]_{-\pi/6}^{\pi/6} = \frac{\sqrt{3}}{3} - \left(-\frac{\sqrt{3}}{3}\right) = \frac{2\sqrt{3}}{3}$

**33.** $\int_{-\pi/3}^{\pi/3} 4 \sec \theta \tan \theta\, d\theta = \left[4 \sec \theta\right]_{-\pi/3}^{\pi/3} = 4(2) - 4(2) = 0$

**35.** $\int_0^2 (2^x + 6)\, dx = \left[\frac{2^x}{\ln 2} + 6x\right]_0^2 = \left(\frac{4}{\ln 2} + 12\right) - \left(\frac{1}{\ln 2} + 0\right) = \frac{3}{\ln 2} + 12$

**37.** $\int_{-1}^1 (e^\theta + \sin \theta)\, d\theta = \left[e^\theta - \cos \theta\right]_{-1}^1 = (e - \cos 1) - \left[e^{-1} - \cos(-1)\right] = e - \frac{1}{e}$

**39.** $A = \int_0^1 (x - x^2)\, dx = \left[\frac{x^2}{2} - \frac{x^3}{3}\right]_0^1 = \frac{1}{6}$

**41.** $A = \int_0^{\pi/2} \cos x\, dx = \left[\sin x\right]_0^{\pi/2} = 1$

**43.** Because $y > 0$ on $[0, 2]$,

$$\text{Area} = \int_0^2 (5x^2 + 2)\, dx = \left[\tfrac{5}{3}x^3 + 2x\right]_0^2 = \tfrac{40}{3} + 4 = \tfrac{52}{3}.$$

**45.** Because $y > 0$ on $[0, 8]$,

$$\text{Area} = \int_0^8 \left(1 + x^{1/3}\right) dx = \left[x + \frac{3}{4}x^{4/3}\right]_0^8 = 8 + \frac{3}{4}(16) = 20.$$

**47.** Because $y > 0$ on $[1, e]$,

$$\text{Area} = \int_1^e \frac{4}{x}\, dx = \left[4 \ln x\right]_1^e = 4 \ln e - 4 \ln 1 = 4.$$

**49.** $\int_0^3 x^3\, dx = \left[\frac{x^4}{4}\right]_0^3 = \frac{81}{4}$

$$f(c)(3 - 0) = \frac{81}{4}$$

$$f(c) = \frac{27}{4}$$

$$c^3 = \frac{27}{4}$$

$$c = \frac{3}{\sqrt[3]{4}} = \frac{3}{2}\sqrt[3]{2} \approx 1.8899$$

**51.** $\int_1^4 \left(5 - \frac{1}{x}\right) dx = \left[5x - \ln x\right]_1^4$

$$= (20 - \ln 4) - (5 - 0) = 15 - \ln 4$$

$$f(c)(4 - 1) = 15 - \ln 4$$

$$\left(5 - \frac{1}{c}\right)(3) = 15 - \ln 4$$

$$15 - \frac{3}{c} = 15 - \ln 4$$

$$\frac{3}{c} = \ln 4$$

$$c = \frac{3}{\ln 4} \approx 2.1640$$

**53.** $\int_{-\pi/4}^{\pi/4} 2\sec^2 x\, dx = \left[2\tan x\right]_{-\pi/4}^{\pi/4} = 2(1) - 2(-1) = 4$

$$f(c)\left[\frac{\pi}{4} - \left(-\frac{\pi}{4}\right)\right] = 4$$

$$2\sec^2 c = \frac{8}{\pi}$$

$$\sec^2 c = \frac{4}{\pi}$$

$$\sec c = \pm\frac{2}{\sqrt{\pi}}$$

$$c = \pm\operatorname{arcsec}\left(\frac{2}{\sqrt{\pi}}\right)$$

$$= \pm\arccos\frac{\sqrt{\pi}}{2} \approx \pm 0.4817$$

**55.** $\dfrac{1}{3 - (-3)} \int_{-3}^{3} \left(9 - x^2\right) dx = \dfrac{1}{6}\left[9x - \dfrac{1}{3}x^3\right]_{-3}^{3}$

$$= \frac{1}{6}\left[(27 - 9) - (-27 + 9)\right]$$

$$= 6$$

Average value $= 6$

$9 - x^2 = 6$ when $x^2 = 9 - 6$ or $x = \pm\sqrt{3} \approx \pm 1.7321$.

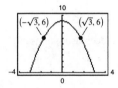

**57.** $\dfrac{1}{1 - (-1)} \int_{-1}^{1} 2e^x\, dx = \int_{-1}^{1} e^x\, dx$

$$= \left[e^x\right]_{-1}^{1} = e - e^{-1} \approx 2.3504$$

Average value $= e - e^{-1} \approx 2.3504$

$$2e^x = e - e^{-1}$$

$$e^x = \frac{1}{2}\left(e - e^{-1}\right)$$

$$x = \ln\left(\frac{e - e^{-1}}{2}\right) \approx 0.1614$$

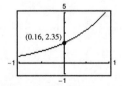

**59.** $\dfrac{1}{\pi - 0} \int_{0}^{\pi} \sin x\, dx = \left[-\dfrac{1}{\pi}\cos x\right]_{0}^{\pi} = \dfrac{2}{\pi}$

Average value $= \dfrac{2}{\pi}$

$$\sin x = \frac{2}{\pi}$$

$$x \approx 0.690,\ 2.451$$

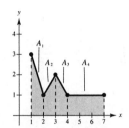

**61.** The distance traveled is $\int_{0}^{8} v(t)\, dt$. The area under the curve from $0 \le t \le 8$ is approximately (18 squares) $(30) \approx 540$ ft.

**63.** (a) $\int_{1}^{7} f(x)\, dx = $ Sum of the areas

$$= A_1 + A_2 + A_3 + A_4$$

$$= \frac{1}{2}(3 + 1) + \frac{1}{2}(1 + 2) + \frac{1}{2}(2 + 1) + (3)(1)$$

$$= 8$$

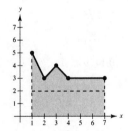

(b) Average value $= \dfrac{\int_{1}^{7} f(x)\, dx}{7 - 1} = \dfrac{8}{6} = \dfrac{4}{3}$

(c) $A = 8 + (6)(2) = 20$

Average value $= \dfrac{20}{6} = \dfrac{10}{3}$

**65.** (a) $F(x) = k\sec^2 x$

$$F(0) = k = 500$$

$$F(x) = 500\sec^2 x$$

(b) $\dfrac{1}{\pi/3 - 0} \int_{0}^{\pi/3} 500\sec^2 x\, dx = \dfrac{1500}{\pi}\left[\tan x\right]_{0}^{\pi/3}$

$$= \frac{1500}{\pi}\left(\sqrt{3} - 0\right)$$

$$\approx 826.99 \text{ newtons}$$

$$\approx 827 \text{ newtons}$$

**67.** $\frac{1}{5-0}\int_0^5 \left(0.1729t + 0.1522t^2 - 0.0374t^3\right) dt \approx \frac{1}{5}\left[0.08645t^2 + 0.05073t^3 - 0.00935t^4\right]_0^5 \approx 0.5318$ liter

**69.** (a) $v = -0.00086t^3 + 0.0782t^2 - 0.208t + 0.10$

(b)

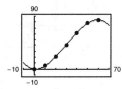

(c) $\int_0^{60} v(t)\, dt = \left[\dfrac{-0.00086t^4}{4} + \dfrac{0.0782t^3}{3} - \dfrac{0.208t^2}{2} + 0.10t\right]_0^{60} \approx 2476$ meters

**71.** $F(x) = \int_0^x (4t - 7)\, dt = \left[2t^2 - 7t\right]_0^x = 2x^2 - 7x$

$F(2) = 2(2^2) - 7(2) = -6$

$F(5) = 2(5^2) - 7(5) = 15$

$F(8) = 2(8^2) - 7(8) = 72$

**73.** $F(x) = \int_1^x \dfrac{20}{v^2}\, dv = \int_1^x 20v^{-2}\, dv = -\dfrac{20}{v}\Big]_1^x$

$\quad = -\dfrac{20}{x} + 20 = 20\left(1 - \dfrac{1}{x}\right)$

$F(2) = 20\left(\tfrac{1}{2}\right) = 10$

$F(5) = 20\left(\tfrac{4}{5}\right) = 16$

$F(8) = 20\left(\tfrac{7}{8}\right) = \tfrac{35}{2}$

**75.** $F(x) = \int_1^x \cos\theta\, d\theta = \sin\theta\Big]_1^x = \sin x - \sin 1$

$F(2) = \sin 2 - \sin 1 \approx 0.0678$

$F(5) = \sin 5 - \sin 1 \approx -1.8004$

$F(8) = \sin 8 - \sin 1 \approx 0.1479$

**77.** $g(x) = \int_0^x f(t)\, dt$

(a) $g(0) = \int_0^0 f(t)\, dt = 0$

$g(2) = \int_0^2 f(t)\, dt \approx 4 + 2 + 1 = 7$

$g(4) = \int_0^4 f(t)\, dt \approx 7 + 2 = 9$

$g(6) = \int_0^6 f(t)\, dt \approx 9 + (-1) = 8$

$g(8) = \int_0^8 f(t)\, dt \approx 8 - 3 = 5$

(b) $g$ increasing on $(0, 4)$ and decreasing on $(4, 8)$

(c) $g$ is a maximum of 9 at $x = 4$.

(d)

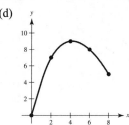

**79.** (a) $\int_0^x (t + 2)\, dt = \left[\dfrac{t^2}{2} + 2t\right]_0^x = \dfrac{1}{2}x^2 + 2x$

(b) $\dfrac{d}{dx}\left[\dfrac{1}{2}x^2 + 2x\right] = x + 2$

**81.** (a) $\int_8^x \sqrt[3]{t}\, dt = \left[\dfrac{3}{4}t^{4/3}\right]_8^x = \dfrac{3}{4}\left(x^{4/3} - 16\right) = \dfrac{3}{4}x^{4/3} - 12$

(b) $\dfrac{d}{dx}\left[\dfrac{3}{4}x^{4/3} - 12\right] = x^{1/3} = \sqrt[3]{x}$

**83.** (a) $\int_{\pi/4}^x \sec^2 t\, dt = \left[\tan t\right]_{\pi/4}^x = \tan x - 1$

(b) $\dfrac{d}{dx}\left[\tan x - 1\right] = \sec^2 x$

**85.** (a) $F(x) = \int_{-1}^{x} e^{t}\, dt = e^{t}\Big]_{-1}^{x} = e^{x} - e^{-1}$

(b) $\dfrac{d}{dx}\left(e^{x} - e^{-1}\right) = e^{x}$

**87.** $F(x) = \int_{-2}^{x} \left(t^{2} - 2t\right) dt$

$F'(x) = x^{2} - 2x$

**89.** $F(x) = \int_{-1}^{x} \sqrt{t^{4} + 1}\, dt$

$F'(x) = \sqrt{x^{4} + 1}$

**91.** $F(x) = \int_{0}^{x} t \cos t\, dt$

$F'(x) = x \cos x$

**93.** $F(x) = \int_{x}^{x+2} \left(4t + 1\right) dt$

$= \left[2t^{2} + t\right]_{x}^{x+2}$

$= \left[2(x + 2)^{2} + (x + 2)\right] - \left[2x^{2} + x\right]$

$= 8x + 10$

$F'(x) = 8$

**Alternate solution:**

$F(x) = \int_{x}^{x+2} \left(4t + 1\right) dt$

$= \int_{x}^{0} \left(4t + 1\right) dt + \int_{0}^{x+2} \left(4t + 1\right) dt$

$= -\int_{0}^{x} \left(4t + 1\right) dt + \int_{0}^{x+2} \left(4t + 1\right) dt$

$F'(x) = -(4x + 1) + 4(x + 2) + 1 = 8$

**95.** $F(x) = \int_{0}^{\sin x} \sqrt{t}\, dt = \left[\tfrac{2}{3}t^{3/2}\right]_{0}^{\sin x} = \tfrac{2}{3}(\sin x)^{3/2}$

$F'(x) = (\sin x)^{1/2} \cos x = \cos x \sqrt{\sin x}$

**Alternate solution:**

$F(x) = \int_{0}^{\sin x} \sqrt{t}\, dt$

$F'(x) = \sqrt{\sin x}\,\dfrac{d}{dx}(\sin x) = \sqrt{\sin x}\,(\cos x)$

**97.** $F(x) = \int_{0}^{x^{3}} \sin t^{2}\, dt$

$F'(x) = \sin\left(x^{3}\right)^{2} \cdot 3x^{2} = 3x^{2} \sin x^{6}$

**99.** $g(x) = \int_{0}^{x} f(t)\, dt$

$g(0) = 0,\ g(1) \approx \tfrac{1}{2},\ g(2) \approx 1,\ g(3) \approx \tfrac{1}{2},\ g(4) = 0$

$g$ has a relative maximum at $x = 2$.

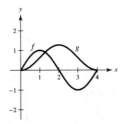

**101.** (a) $v(t) = 5t - 7,\ 0 \le t \le 3$

Displacement $= \int_{0}^{3} \left(5t - 7\right) dt = \left[\dfrac{5t^{2}}{2} - 7t\right]_{0}^{3} = \dfrac{45}{2} - 21 = \dfrac{3}{2}$ ft to the right

(b) Total distance traveled $= \int_{0}^{3} \left|5t - 7\right| dt$

$= \int_{0}^{7/5} \left(7 - 5t\right) dt + \int_{7/5}^{3} \left(5t - 7\right) dt$

$= \left[7t - \dfrac{5t^{2}}{2}\right]_{0}^{7/5} + \left[\dfrac{5t^{2}}{2} - 7t\right]_{7/5}^{3}$

$= 7\left(\dfrac{7}{5}\right) - \dfrac{5}{2}\left(\dfrac{7}{2}\right)^{2} + \left(\dfrac{5}{2}(9) - 21\right) - \left(\dfrac{5}{2}\left(\dfrac{7}{5}\right)^{2} - 7\left(\dfrac{7}{5}\right)\right)$

$= \dfrac{49}{5} - \dfrac{49}{10} + \dfrac{45}{2} - 21 - \dfrac{49}{10} + \dfrac{49}{5} = \dfrac{113}{10}$ ft

**103.** (a) $v(t) = t^3 - 10t^2 + 27t - 18 = (t-1)(t-3)(t-6)$, $1 \le t \le 7$

Displacement $= \displaystyle\int_1^7 \left( t^3 - 10t^2 + 27t - 18 \right) dt$

$$= \left[ \frac{t^4}{4} - \frac{10t^3}{3} + \frac{27t^2}{2} - 18t \right]_1^7$$

$$= \left[ \frac{7^4}{4} - \frac{10(7^3)}{3} + \frac{27(7^2)}{2} - 18(7) \right] - \left[ \frac{1}{4} - \frac{10}{3} + \frac{27}{2} - 18 \right]$$

$$= -\frac{91}{12} - \left( -\frac{91}{12} \right) = 0$$

(b) Total distance traveled $= \displaystyle\int_1^7 \left| v(t) \right| dt$

$$= \int_1^3 \left( t^3 - 10t^2 + 27t - 18 \right) dt - \int_3^6 \left( t^3 - 10t^2 + 27t - 18 \right) dt + \int_6^7 \left( t^3 - 10t^2 + 27t - 18 \right) dt$$

Evaluating each of these integrals, you obtain

Total distance $= \frac{16}{3} - \left( -\frac{63}{4} \right) + \frac{125}{12} = \frac{63}{2}$ ft

**105.** (a) $v(t) = \dfrac{1}{\sqrt{t}}$, $1 \le t \le 4$

Because $v(t) > 0$,

Displacement = Total Distance

Displacement $= \displaystyle\int_1^4 t^{-1/2}\, dt = \left[ 2t^{1/2} \right]_1^4 = 4 - 2 = 2$ ft to the right

(b) Total distance $= 2$ ft

**107.** $x(t) = t^3 - 6t^2 + 9t - 2$

$x'(t) = 3t^2 - 12t + 9 = 3(t^2 - 4t + 3) = 3(t-3)(t-1)$

Total distance $= \displaystyle\int_0^5 \left| x'(t) \right| dt$

$$= \int_0^5 3 \left| (t-3)(t-1) \right| dt$$

$$= 3\int_0^1 \left( t^2 - 4t + 3 \right) dt - 3\int_1^3 \left( t^2 - 4t + 3 \right) dt + 3\int_3^5 \left( t^2 - 4t + 3 \right) dt = 4 + 4 + 20 = 28 \text{ units}$$

**109.** Let $c(t)$ be the amount of water that is flowing out of the tank. Then $c'(t) = 500 - 5t$ L/min is the rate of flow.

$$\int_0^{18} c'(t)\,dt = \int_0^{18} (500 - 5t)\,dt = \left[ 500t - \frac{5t^2}{2} \right]_0^{18} = 9000 - 810 = 8190 \text{ L}$$

**111.** The function $f(x) = x^{-2}$ is not continuous on $[-1, 1]$.

$$\int_{-1}^1 x^{-2}\,dx = \int_{-1}^0 x^{-2}\,dx + \int_0^1 x^{-2}\,dx$$

Each of these integrals is infinite. $f(x) = x^{-2}$ has a nonremovable discontinuity at $x = 0$.

**113.** The function $f(x) = \sec^2 x$ is not continuous on $\left[\dfrac{\pi}{4}, \dfrac{3\pi}{4}\right]$.

$$\int_{\pi/4}^{3\pi/4} \sec^2 x \, dx = \int_{\pi/4}^{\pi/2} \sec^2 x \, dx + \int_{\pi/2}^{3\pi/4} \sec^2 x \, dx$$

Each of these integrals is infinite. $f(x) = \sec^2 x$ has a nonremovable discontinuity at $x = \dfrac{\pi}{2}$

**115.** $P = \dfrac{2}{\pi} \int_0^{\pi/2} \sin \theta \, d\theta = \left[ -\dfrac{2}{\pi} \cos \theta \right]_0^{\pi/2} = -\dfrac{2}{\pi}(0 - 1) = \dfrac{2}{\pi} \approx 63.7\%$

**117.** True

**119.** $f(x) = \int_0^{1/x} \dfrac{1}{t^2 + 1} \, dt + \int_0^x \dfrac{1}{t^2 + 1} \, dt$

By the Second Fundamental Theorem of Calculus, you have $f'(x) = \dfrac{1}{(1/x)^2 + 1}\left(-\dfrac{1}{x^2}\right) + \dfrac{1}{x^2 + 1} = -\dfrac{1}{1 + x^2} + \dfrac{1}{x^2 + 1} = 0$.

Because $f'(x) = 0$, $f(x)$ must be constant.

**121.** $G(x) = \int_0^x \left[ s \int_0^s f(t) \, dt \right] ds$

(a)  $G(0) = \int_0^0 \left[ s \int_0^s f(t) \, dt \right] ds = 0$

(b)  Let $F(s) = s \int_0^s f(t) \, dt$.

$G(x) = \int_0^x F(s) \, ds$

$G'(x) = F(x) = x \int_0^x f(t) \, dt$

$G'(0) = 0 \int_0^0 f(t) \, dt = 0$

(c)  $G''(x) = x \cdot f(x) + \int_0^x f(t) \, dt$

(d)  $G''(0) = 0 \cdot f(0) + \int_0^0 f(t) \, dt = 0$

## Section 5.5   Integration by Substitution

| $\int f(g(x))g'(x)\,dx$ | $u = g(x)$ | $du = g'(x)\,dx$ |
|---|---|---|
| **1.** $\int \left(8x^2 + 1\right)^2 (16x)\,dx$ | $8x^2 + 1$ | $16x\,dx$ |
| **3.** $\int \tan^2 x \sec^2 x \, dx$ | $\tan x$ | $\sec^2 x \, dx$ |

**5.** $\displaystyle\int (1+6x)^4 (6)\, dx = \frac{(1+6x)^5}{5} + C$

**Check:** $\dfrac{d}{dx}\left[\dfrac{(1+6x)^5}{5} + C\right] = 6(1+6x)^4$

**7.** $\displaystyle\int \sqrt{25-x^2}\,(-2x)\, dx = \frac{\left(25-x^2\right)^{3/2}}{3/2} + C = \frac{2}{3}\left(25-x^2\right)^{3/2} + C$

**Check:** $\dfrac{d}{dx}\left[\dfrac{2}{3}\left(25-x^2\right)^{3/2} + C\right] = \dfrac{2}{3}\left(\dfrac{3}{2}\right)\left(25-x^2\right)^{1/2}(-2x) = \sqrt{25-x^2}\,(-2x)$

**9.** $\displaystyle\int x^3 \left(x^4+3\right)^2 dx = \frac{1}{4}\int \left(x^4+3\right)^2 \left(4x^3\right) dx = \frac{1}{4}\frac{\left(x^4+3\right)^3}{3} + C = \frac{\left(x^4+3\right)^3}{12} + C$

**Check:** $\dfrac{d}{dx}\left[\dfrac{\left(x^4+3\right)^3}{12} + C\right] = \dfrac{3\left(x^4+3\right)^2}{12}\left(4x^3\right) = \left(x^4+3\right)^2 \left(x^3\right)$

**11.** $\displaystyle\int x^2 \left(x^3-1\right)^4 dx = \frac{1}{3}\int \left(x^3-1\right)^4 \left(3x^2\right) dx = \frac{1}{3}\left[\frac{\left(x^3-1\right)^5}{5}\right] + C = \frac{\left(x^3-1\right)^5}{15} + C$

**Check:** $\dfrac{d}{dx}\left[\dfrac{\left(x^3-1\right)^5}{15} + C\right] = \dfrac{5\left(x^3-1\right)^4 \left(3x^2\right)}{15} = x^2 \left(x^3-1\right)^4$

**13.** $\displaystyle\int t\sqrt{t^2+2}\, dt = \frac{1}{2}\int \left(t^2+2\right)^{1/2} (2t)\, dt = \frac{1}{2}\frac{\left(t^2+2\right)^{3/2}}{3/2} + C = \frac{\left(t^2+2\right)^{3/2}}{3} + C$

**Check:** $\dfrac{d}{dt}\left[\dfrac{\left(t^2+2\right)^{3/2}}{3} + C\right] = \dfrac{3/2\left(t^2+2\right)^{1/2}(2t)}{3} = \left(t^2+2\right)^{1/2} t$

**15.** $\displaystyle\int 5x\left(1-x^2\right)^{1/3} dx = -\frac{5}{2}\int \left(1-x^2\right)^{1/3}(-2x)\, dx = -\frac{5}{2}\cdot\frac{\left(1-x^2\right)^{4/3}}{4/3} + C = -\frac{15}{8}\left(1-x^2\right)^{4/3} + C$

**Check:** $\dfrac{d}{dx}\left[-\dfrac{15}{8}\left(1-x^2\right)^{4/3} + C\right] = -\dfrac{15}{8}\cdot\dfrac{4}{3}\left(1-x^2\right)^{1/3}(-2x) = 5x\left(1-x^2\right)^{1/3} = 5x\sqrt[3]{1-x^2}$

**17.** $\displaystyle\int \frac{x}{\left(1-x^2\right)^3}\, dx = -\frac{1}{2}\int \left(1-x^2\right)^{-3}(-2x)\, dx = -\frac{1}{2}\frac{\left(1-x^2\right)^{-2}}{-2} + C = \frac{1}{4\left(1-x^2\right)^2} + C$

**Check:** $\dfrac{d}{dx}\left[\dfrac{1}{4\left(1-x^2\right)^2} + C\right] = \dfrac{1}{4}(-2)\left(1-x^2\right)^{-3}(-2x) = \dfrac{x}{\left(1-x^2\right)^3}$

**19.** $\displaystyle\int \frac{x^2}{\left(1+x^3\right)^2}\, dx = \frac{1}{3}\int \left(1+x^3\right)^{-2}\left(3x^2\right) dx = \frac{1}{3}\left[\frac{\left(1+x^3\right)^{-1}}{-1}\right] + C = -\frac{1}{3\left(1+x^3\right)} + C$

**Check:** $\dfrac{d}{dx}\left[-\dfrac{1}{3\left(1+x^3\right)} + C\right] = -\dfrac{1}{3}(-1)\left(1+x^3\right)^{-2}\left(3x^2\right) = \dfrac{x^2}{\left(1+x^3\right)^2}$

**21.** $\int \dfrac{x}{\sqrt{1-x^2}}\,dx = -\dfrac{1}{2}\int\left(1-x^2\right)^{-1/2}(-2x)\,dx = -\dfrac{1}{2}\dfrac{\left(1-x^2\right)^{1/2}}{1/2} + C = -\sqrt{1-x^2} + C$

**Check:** $\dfrac{d}{dx}\left[-\left(1-x^2\right)^{1/2} + C\right] = -\dfrac{1}{2}\left(1-x^2\right)^{-1/2}(-2x) = \dfrac{x}{\sqrt{1-x^2}}$

**23.** $\int\left(1+\dfrac{1}{t}\right)^3\left(\dfrac{1}{t^2}\right)dt = -\int\left(1+\dfrac{1}{t}\right)^3\left(-\dfrac{1}{t^2}\right)dt = -\dfrac{\left[1+\left(\dfrac{1}{t}\right)\right]^4}{4} + C$

**Check:** $\dfrac{d}{dt}\left[-\dfrac{\left[1+(1/t)\right]^4}{4} + C\right] = -\dfrac{1}{4}(4)\left(1+\dfrac{1}{t}\right)^3\left(-\dfrac{1}{t^2}\right) = \dfrac{1}{t^2}\left(1+\dfrac{1}{t}\right)^3$

**25.** $\int\dfrac{1}{\sqrt{2x}}\,dx = \dfrac{1}{2}\int(2x)^{-1/2}\,2\,dx = \dfrac{1}{2}\left[\dfrac{(2x)^{1/2}}{1/2}\right] + C = \sqrt{2x} + C$

**Alternate Solution:** $\int\dfrac{1}{\sqrt{2x}}\,dx = \dfrac{1}{\sqrt{2}}\int x^{-1/2}\,dx = \dfrac{1}{\sqrt{2}}\dfrac{x^{1/2}}{(1/2)} + C = \sqrt{2x} + C$

**Check:** $\dfrac{d}{dx}\left[\sqrt{2x} + C\right] = \dfrac{1}{2}(2x)^{-1/2}(2) = \dfrac{1}{\sqrt{2x}}$

**27.** $y = \int\left[4x + \dfrac{4x}{\sqrt{16-x^2}}\right]dx = 4\int x\,dx - 2\int\left(16-x^2\right)^{-1/2}(-2x)\,dx = 4\left(\dfrac{x^2}{2}\right) - 2\left[\dfrac{\left(16-x^2\right)^{1/2}}{1/2}\right] + C = 2x^2 - 4\sqrt{16-x^2} + C$

**29.** $y = \int\dfrac{x+1}{\left(x^2+2x-3\right)^2}\,dx$

$= \dfrac{1}{2}\int\left(x^2+2x-3\right)^{-2}(2x+2)\,dx$

$= \dfrac{1}{2}\left[\dfrac{\left(x^2+2x-3\right)^{-1}}{-1}\right] + C$

$= -\dfrac{1}{2\left(x^2+2x-3\right)} + C$

**31.** (a)  Answers will vary. *Sample answer:*

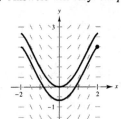

(b)  $\dfrac{dy}{dx} = x\sqrt{4-x^2},\ (2,2)$

$y = \int x\sqrt{4-x^2}\,dx = -\dfrac{1}{2}\int\left(4-x^2\right)^{1/2}(-2x\,dx)$

$= -\dfrac{1}{2}\cdot\dfrac{2}{3}\left(4-x^2\right)^{3/2} + C = -\dfrac{1}{3}\left(4-x^2\right)^{3/2} + C$

$(2,2):\ 2 = -\dfrac{1}{3}\left(4-2^2\right)^{3/2} + C \Rightarrow C = 2$

$y = -\dfrac{1}{3}\left(4-x^2\right)^{3/2} + 2$

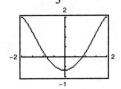

**33.** $\int \pi \sin \pi x \, dx = -\cos \pi x + C$

**35.** $\int \cos 8x \, dx = \frac{1}{8}\int (\cos 8x)(8) \, dx = \frac{1}{8}\sin 8x + C$

**37.** $\int \frac{1}{\theta^2} \cos \frac{1}{\theta} \, d\theta = -\int \cos \frac{1}{\theta}\left(-\frac{1}{\theta^2}\right) d\theta = -\sin \frac{1}{\theta} + C$

**39.** $\int \sin 2x \cos 2x \, dx = \frac{1}{2}\int (\sin 2x)(2\cos 2x) \, dx = \frac{1}{2}\frac{(\sin 2x)^2}{2} + C = \frac{1}{4}\sin^2 2x + C$ OR

$\int \sin 2x \cos 2x \, dx = -\frac{1}{2}\int (\cos 2x)(-2\sin 2x) \, dx = -\frac{1}{2}\frac{(\cos 2x)^2}{2} + C_1 = -\frac{1}{4}\cos^2 2x + C_1$ OR

$\int \sin 2x \cos 2x \, dx = \frac{1}{2}\int 2\sin 2x \cos 2x \, dx = \frac{1}{2}\int \sin 4x \, dx = -\frac{1}{8}\cos 4x + C_2$

**41.** $\int \frac{\csc^2 x}{\cot^3 x} \, dx = -\int (\cot x)^{-3}\left(-\csc^2 x\right) dx$

$= -\frac{(\cot x)^{-2}}{-2} + C = \frac{1}{2\cot^2 x} + C = \frac{1}{2}\tan^2 x + C = \frac{1}{2}\left(\sec^2 x - 1\right) + C = \frac{1}{2}\sec^2 x + C_1$

**43.** $\int e^{-x^3}\left(-3x^2\right) dx = e^{-x^3} + C$

**45.** $\int e^x\left(e^x + 1\right)^2 dx = \frac{\left(e^x + 1\right)^3}{3} + C$

**47.** $\int \frac{5 - e^x}{e^{2x}} \, dx = \int 5e^{-2x} \, dx - \int e^{-x} \, dx$

$= -\frac{5}{2}e^{-2x} + e^{-x} + C$

**49.** $\int e^{\sin \pi x} \cos \pi x \, dx = \frac{1}{\pi}\int e^{\sin \pi x}(\pi \cos \pi x) \, dx$

$= \frac{1}{\pi}e^{\sin \pi x} + C$

**51.** $\int e^{-x} \sec^2\left(e^{-x}\right) dx = -\int \sec^2\left(e^{-x}\right)\left(-e^{-x}\right) dx$

$= -\tan\left(e^{-x}\right) + C$

**53.** $\int 3^{x/2} \, dx = 2\int 3^{x/2}\left(\frac{1}{2}\right) dx = 2\frac{3^{x/2}}{\ln 3} + C = \frac{2}{\ln 3}3^{x/2} + C$

**55.** $f(x) = \int -\sin \frac{x}{2} \, dx = 2\cos \frac{x}{2} + C$

Because $f(0) = 6 = 2\cos\left(\frac{0}{2}\right) + C$, $C = 4$. So,

$f(x) = 2\cos \frac{x}{2} + 4.$

**57.** $f(x) = \int 2e^{-x/4} \, dx = -8\int e^{-x/4}\left(-\frac{1}{4}\right) dx$

$= -8e^{-x/4} + C$

$f(0) = 1 = -8 + C \Rightarrow C = 9$

$f(x) = -8e^{-x/4} + 9$

**59.** $f'(x) = 2x\left(4x^2 - 10\right)^2, \ (2, 10)$

$f(x) = \frac{\left(4x^2 - 10\right)^3}{12} + C = \frac{2\left(2x^2 - 5\right)^3}{3} + C$

$f(2) = \frac{2(8 - 5)^3}{3} + C = 18 + C = 10 \Rightarrow C = -8$

$f(x) = \frac{2}{3}\left(2x^2 - 5\right)^3 - 8$

**61.** $u = x + 6, \ x = u - 6, \ dx = du$

$\int x\sqrt{x + 6} \, dx = \int (u - 6)\sqrt{u} \, du$

$= \int \left(u^{3/2} - 6u^{1/2}\right) du$

$= \frac{2}{5}u^{5/2} - 4u^{3/2} + C$

$= \frac{2u^{3/2}}{5}(u - 10) + C$

$= \frac{2}{5}(x + 6)^{3/2}\left[(x + 6) - 10\right] + C$

$= \frac{2}{5}(x + 6)^{3/2}(x - 4) + C$

**63.** $u = 1 - x$, $x = 1 - u$, $dx = -du$

$$\int x^2 \sqrt{1 - x}\, dx = -\int (1 - u)^2 \sqrt{u}\, du$$

$$= -\int \left( u^{1/2} - 2u^{3/2} + u^{5/2} \right) du$$

$$= -\left( \frac{2}{3} u^{3/2} - \frac{4}{5} u^{5/2} + \frac{2}{7} u^{7/2} \right) + C$$

$$= -\frac{2u^{3/2}}{105} \left( 35 - 42u + 15u^2 \right) + C$$

$$= -\frac{2}{105} (1 - x)^{3/2} \left[ 35 - 42(1 - x) + 15(1 - x)^2 \right] + C$$

$$= -\frac{2}{105} (1 - x)^{3/2} \left( 15x^2 + 12x + 8 \right) + C$$

**65.** $u = 2x - 1$, $x = \frac{1}{2}(u + 1)$, $dx = \frac{1}{2}\, du$

$$\int \frac{x^2 - 1}{\sqrt{2x - 1}}\, dx = \int \frac{\left[ (1/2)(u + 1) \right]^2 - 1}{\sqrt{u}} \frac{1}{2}\, du$$

$$= \frac{1}{8} \int u^{-1/2} \left[ \left( u^2 + 2u + 1 \right) - 4 \right] du$$

$$= \frac{1}{8} \int \left( u^{3/2} + 2u^{1/2} - 3u^{-1/2} \right) du$$

$$= \frac{1}{8} \left( \frac{2}{5} u^{5/2} + \frac{4}{3} u^{3/2} - 6u^{1/2} \right) + C$$

$$= \frac{u^{1/2}}{60} \left( 3u^2 + 10u - 45 \right) + C$$

$$= \frac{\sqrt{2x - 1}}{60} \left[ 3(2x - 1)^2 + 10(2x - 1) - 45 \right] + C$$

$$= \frac{1}{60} \sqrt{2x - 1} \left( 12x^2 + 8x - 52 \right) + C$$

$$= \frac{1}{15} \sqrt{2x - 1} \left( 3x^2 + 2x - 13 \right) + C$$

**67.** $u = x + 1$, $x = u - 1$, $dx = du$

$$\int \frac{-x}{(x + 1) - \sqrt{x + 1}}\, dx = \int \frac{-(u - 1)}{u - \sqrt{u}}\, du$$

$$= -\int \frac{\left( \sqrt{u} + 1 \right)\left( \sqrt{u} - 1 \right)}{\sqrt{u}\left( \sqrt{u} - 1 \right)}\, du$$

$$= -\int \left( 1 + u^{-1/2} \right) du$$

$$= -\left( u + 2u^{1/2} \right) + C$$

$$= -u - 2\sqrt{u} + C$$

$$= -(x + 1) - 2\sqrt{x + 1} + C$$

$$= -x - 2\sqrt{x + 1} - 1 + C$$

$$= -\left( x + 2\sqrt{x + 1} \right) + C_1$$

where $C_1 = -1 + C$.

**69.** Let $u = x^2 + 1$, $du = 2x\,dx$.

$$\int_{-1}^{1} x(x^2 + 1)^3\,dx = \frac{1}{2}\int_{-1}^{1}(x^2 + 1)^3(2x)\,dx = \left[\frac{1}{8}(x^2 + 1)^4\right]_{-1}^{1} = 0$$

**71.** Let $u = x^3 + 1$, $du = 3x^2\,dx$.

$$\int_{1}^{2} 2x^2\sqrt{x^3 + 1}\,dx = 2 \cdot \frac{1}{3}\int_{1}^{2}(x^3 + 1)^{1/2}(3x^2)\,dx = \left[\frac{(x^3 + 1)^{3/2}}{3/2}\right]_{1}^{2} = \frac{4}{9}\left[(x^3 + 1)^{3/2}\right]_{1}^{2} = \frac{4}{9}\left[27 - 2\sqrt{2}\right] = 12 - \frac{8}{9}\sqrt{2}$$

**73.** Let $u = 2x + 1$, $du = 2\,dx$.

$$\int_{0}^{4} \frac{1}{\sqrt{2x + 1}}\,dx = \frac{1}{2}\int_{0}^{4}(2x + 1)^{-1/2}(2)\,dx = \left[\sqrt{2x + 1}\right]_{0}^{4} = \sqrt{9} - \sqrt{1} = 2$$

**75.** Let $u = 1 + \sqrt{x}$, $du = \frac{1}{2\sqrt{x}}\,dx$.

$$\int_{1}^{9} \frac{1}{\sqrt{x}\left(1 + \sqrt{x}\right)^2}\,dx = 2\int_{1}^{9}\left(1 + \sqrt{x}\right)^{-2}\left(\frac{1}{2\sqrt{x}}\right)dx = \left[-\frac{2}{1 + \sqrt{x}}\right]_{1}^{9} = -\frac{1}{2} + 1 = \frac{1}{2}$$

**77.** $\displaystyle \int_{0}^{1} e^{-2x}\,dx = -\frac{1}{2}\int_{0}^{1}e^{-2x}(-2)\,dx = \left[-\frac{1}{2}e^{-2x}\right]_{0}^{1} = -\frac{1}{2}e^{-2} + \frac{1}{2}$

**79.** $\displaystyle \int_{1}^{3} \frac{e^{3/x}}{x^2}\,dx = -\frac{1}{3}\int_{1}^{3}e^{3/x}\left(-\frac{3}{x^2}\right)dx = \left[-\frac{1}{3}e^{3/x}\right]_{1}^{3} = -\frac{1}{3}\left(e - e^3\right) = \frac{e}{3}\left(e^2 - 1\right)$

**81.** $u = x + 1$, $x = u - 1$, $dx = du$

When $x = 0$, $u = 1$. When $x = 7$, $u = 8$.

$$\text{Area} = \int_{0}^{7} x\sqrt[3]{x + 1}\,dx = \int_{1}^{8}(u - 1)\sqrt[3]{u}\,du$$

$$= \int_{1}^{8}\left(u^{4/3} - u^{1/3}\right)du$$

$$= \left[\frac{3}{7}u^{7/3} - \frac{3}{4}u^{4/3}\right]_{1}^{8}$$

$$= \left(\frac{384}{7} - 12\right) - \left(\frac{3}{7} - \frac{3}{4}\right)$$

$$= \frac{1209}{28}$$

**83.** $\displaystyle \text{Area} = \int_{\pi/2}^{2\pi/3} \sec^2\left(\frac{x}{2}\right)dx$

$$= 2\int_{\pi/2}^{2\pi/3} \sec^2\left(\frac{x}{2}\right)\left(\frac{1}{2}\right)dx$$

$$= \left[2\tan\left(\frac{x}{2}\right)\right]_{\pi/2}^{2\pi/3} = 2\left(\sqrt{3} - 1\right)$$

**85.** $\displaystyle \int_{0}^{5} e^x\,dx = \left[e^x\right]_{0}^{5} = e^5 - 1 \approx 147.413$

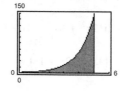

**87.** $\displaystyle \int_{0}^{\sqrt{6}} xe^{-x^2/4}\,dx = \left[-2e^{-x^2/4}\right]_{0}^{\sqrt{6}}$

$$= -2e^{-3/2} + 2 \approx 1.554$$

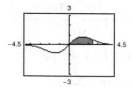

**89.** $f(x) = x^2(x^2 + 1)$ is even.

$$\int_{-2}^{2} x^2(x^2 + 1)\,dx = 2\int_{0}^{2}(x^4 + x^2)\,dx = 2\left[\frac{x^5}{5} + \frac{x^3}{3}\right]_{0}^{2}$$

$$= 2\left[\frac{32}{5} + \frac{8}{3}\right] = \frac{272}{15}$$

**91.** $f(x) = \sin^2 x \cos x$ is even.

$$\int_{-\pi/2}^{\pi/2} \sin^2 x \cos x \, dx = 2\int_0^{\pi/2} \sin^2 x (\cos x) \, dx$$

$$= 2\left[\frac{\sin^3 x}{3}\right]_0^{\pi/2}$$

$$= \frac{2}{3}$$

**93.** $\int_0^4 x^2 \, dx = \left[\frac{x^3}{3}\right]_0^4 = \frac{64}{3}$; the function $x^2$ is an even function.

(a) $\int_{-4}^0 x^2 \, dx = \int_0^4 x^2 \, dx = \frac{64}{3}$

(b) $\int_{-4}^4 x^2 \, dx = 2\int_0^4 x^2 \, dx = \frac{128}{3}$

(c) $\int_0^4 (-x^2) \, dx = -\int_0^4 x^2 \, dx = -\frac{64}{3}$

(d) $\int_{-4}^0 3x^2 \, dx = 3\int_0^4 x^2 \, dx = 64$

**95.** $\int_{-3}^3 \left(x^3 + 4x^2 - 3x - 6\right) dx = \int_{-3}^3 \left(x^3 - 3x\right) dx + \int_{-3}^3 \left(4x^2 - 6\right) dx = 0 + 2\int_0^3 \left(4x^2 - 6\right) dx = 2\left[\frac{4}{3}x^3 - 6x\right]_0^3 = 36$

**97.** If $u = 5 - x^2$, then $du = -2x \, dx$ and $\int x\left(5 - x^2\right)^3 dx = -\frac{1}{2}\int \left(5 - x^2\right)^3 (-2x) \, dx = -\frac{1}{2}\int u^3 \, du$.

**99.** (a) The second integral is easier. Use substitution with $u = x^3 + 1$ and $du = 3x^2 dx$. The answer is

$$\int x^2 \sqrt{x^3 + 1} \, dx = \frac{1}{3}\int \left(x^3 + 1\right)^{1/2} 3x^2 dx$$

$$= \frac{2}{9}\left(x^3 + 1\right)^{3/2} + C.$$

(b) The first integral is easier. Use substitution with $u = \tan 3x$ and $du = 3\sec^2(3x) dx$. The answer is

$$\int \tan(3x)\sec^2(3x) \, dx = \frac{1}{3}\int \tan(3x) 3\sec^2(3x) \, dx = \frac{1}{6}\tan^2 3x + C.$$

**101.** $\dfrac{dV}{dt} = \dfrac{k}{(t+1)^2}$

$V(t) = \int \dfrac{k}{(t+1)^2} \, dt = -\dfrac{k}{t+1} + C$

$V(0) = -k + C = 500{,}000$

$V(1) = -\dfrac{1}{2}k + C = 400{,}000$

Solving this system yields $k = -200{,}000$ and $C = 300{,}000$. So, $V(t) = \dfrac{200{,}000}{t+1} + 300{,}000$.

When $t = 4, V(4) = \$340{,}000$.

**103.** $\dfrac{1}{b-a}\int_a^b \left[74.50 + 43.75\sin\dfrac{\pi t}{6}\right] dt = \dfrac{1}{b-a}\left[74.50t - \dfrac{262.5}{\pi}\cos\dfrac{\pi t}{6}\right]_a^b$

(a) $\dfrac{1}{3}\left[74.50t - \dfrac{262.5}{\pi}\cos\dfrac{\pi t}{6}\right]_0^3 = \dfrac{1}{3}\left(223.5 + \dfrac{262.5}{\pi}\right) \approx 102.352$ thousand units

(b) $\dfrac{1}{3}\left[74.50t - \dfrac{262.5}{\pi}\cos\dfrac{\pi t}{6}\right]_3^6 = \dfrac{1}{3}\left(447 + \dfrac{262.5}{\pi} - 223.5\right) \approx 102.352$ thousand units

(c) $\dfrac{1}{12}\left[74.50t - \dfrac{262.5}{\pi}\cos\dfrac{\pi t}{6}\right]_0^{12} = \dfrac{1}{12}\left(894 - \dfrac{262.5}{\pi} + \dfrac{262.5}{\pi}\right) = 74.5$ thousand units

**105.** $u = 1 - x, \; x = 1 - u, \; dx = -du$

When $x = a, \; u = 1 - a$. When $x = b, \; u = 1 - b$.

$$P_{a,b} = \int_a^b \frac{15}{4} x\sqrt{1-x} \, dx = \frac{15}{4} \int_{1-a}^{1-b} -(1-u)\sqrt{u} \, du$$

$$= \frac{15}{4} \int_{1-a}^{1-b} \left(u^{3/2} - u^{1/2}\right) du = \frac{15}{4}\left[\frac{2}{5}u^{5/2} - \frac{2}{3}u^{3/2}\right]_{1-a}^{1-b} = \frac{15}{4}\left[\frac{2u^{3/2}}{15}(3u-5)\right]_{1-a}^{1-b} = \left[-\frac{(1-x)^{3/2}}{2}(3x+2)\right]_a^b$$

(a) $P_{0.50,\,0.75} = \left[-\dfrac{(1-x)^{3/2}}{2}(3x+2)\right]_{0.50}^{0.75} = 0.353 = 35.3\%$

(b) $P_{0,b} = \left[-\dfrac{(1-x)^{3/2}}{2}(3x+2)\right]_0^b = -\dfrac{(1-b)^{3/2}}{2}(3b+2) + 1 = 0.5$

$$(1-b)^{3/2}(3b+2) = 1$$

$$b \approx 0.586 = 58.6\%$$

**107.** (a)

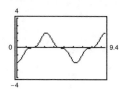

(b) $g$ is nonnegative because the graph of $f$ is positive at the beginning, and generally has more positive sections than negative ones.

(c) The points on $g$ that correspond to the extrema of $f$ are points of inflection of $g$.

(d) No, some zeros of $f$, like $x = \pi/2$, do not correspond to an extrema of $g$. The graph of $g$ continues to increase after $x = \pi/2$ because $f$ remains above the $x$-axis.

(e) The graph of $h$ is that of $g$ shifted 2 units downward.

$$g(t) = \int_0^t f(x)\,dx = \int_0^{\pi/2} f(x)\,dx + \int_{\pi/2}^t f(x)\,dx = 2 + h(t).$$

**109.** (a) Let $u = 1 - x, \; du = -dx, \; x = 1 - u$

$x = 0 \Rightarrow u = 1, \; x = 1 \Rightarrow u = 0$

$$\int_0^1 x^2(1-x)^5 \, dx = \int_1^0 (1-u)^2 u^5 (-du)$$

$$= \int_0^1 u^5(1-u)^2 \, du$$

$$= \int_0^1 x^5(1-x)^2 \, dx$$

(b) Let $u = 1 - x, \; du = -dx, \; x = 1 - u$

$x = 0 \Rightarrow u = 1, \; x = 1 \Rightarrow u = 0$

$$\int_0^1 x^a(1-x)^b \, dx = \int_1^0 (1-u)^a u^b (-du)$$

$$= \int_0^1 u^b(1-u)^a \, du$$

$$= \int_0^1 x^b(1-x)^a \, dx$$

**111.** False

$$\int (2x+1)^2 \, dx = \frac{1}{2}\int (2x+1)^2 \, 2\, dx = \frac{1}{6}(2x+1)^3 + C$$

**113.** True

$$\int_{-10}^{10} \left(ax^3 + bx^2 + cx + d\right) dx = \underbrace{\int_{-10}^{10}\left(ax^3 + cx\right) dx}_{\text{Odd}} + \underbrace{\int_{-10}^{10}\left(bx^2 + d\right) dx}_{\text{Even}} = 0 + 2\int_0^{10}\left(bx^2 + d\right) dx$$

**115.** True

$$4 \int \sin x \cos x \, dx = 2 \int \sin 2x \, dx = -\cos 2x + C$$

**117.** Let $u = cx, du = c \, dx$:

$$c \int_a^b f(cx) \, dx = c \int_{ca}^{cb} f(u) \frac{du}{c}$$

$$= \int_{ca}^{cb} f(u) \, du$$

$$= \int_{ca}^{cb} f(x) \, dx$$

**119.** Because $f$ is odd, $f(-x) = -f(x)$. Then

$$\int_{-a}^a f(x) \, dx = \int_{-a}^0 f(x) \, dx + \int_0^a f(x) \, dx$$

$$= -\int_0^{-a} f(x) \, dx + \int_0^a f(x) \, dx.$$

Let $x = -u$, $dx = -du$ in the first integral.

When $x = 0$, $u = 0$. When $x = -a$, $u = a$.

$$\int_{-a}^1 f(x) \, dx = -\int_0^a f(-u)(-du) + \int_0^a f(x) \, dx$$

$$= -\int_0^a f(u) \, du + \int_0^a f(x) \, dx = 0$$

**121.** Let $f(x) = a_0 + a_1 x + a_2 x^2 + \cdots + a_n x^n$.

$$\int_0^1 f(x) \, dx = \left[ a_0 x + a_1 \frac{x^2}{2} + a_2 \frac{x^3}{3} + \cdots + a_n \frac{x^{n+1}}{n+1} \right]_0^1$$

$$= a_0 + \frac{a_1}{2} + \frac{a_2}{3} + \cdots + \frac{a_n}{n+1} = 0 \text{ (Given)}$$

By the Mean Value Theorem for Integrals, there exists $c$ in $[0, 1]$ such that

$$\int_0^1 f(x) \, dx = f(c)(1 - 0)$$

$$0 = f(c).$$

So the equation has at least one real zero.

# Section 5.6   Numerical Integration

**1.** Exact: $\int_0^2 x^2 \, dx = \left[ \frac{1}{3} x^3 \right]_0^2 = \frac{8}{3} \approx 2.6667$

Trapezoidal: $\int_0^2 x^2 \, dx \approx \frac{1}{4} \left[ 0 + 2\left(\frac{1}{2}\right)^2 + 2(1)^2 + 2\left(\frac{3}{2}\right)^2 + (2)^2 \right] = \frac{11}{4} = 2.7500$

Simpson's: $\int_0^2 x^2 \, dx \approx \frac{1}{6} \left[ 0 + 4\left(\frac{1}{2}\right)^2 + 2(1)^2 + 4\left(\frac{3}{2}\right)^2 + (2)^2 \right] = \frac{8}{3} \approx 2.6667$

**3.** Exact: $\int_0^2 x^3 \, dx = \left[ \frac{x^4}{4} \right]_0^2 = 4.0000$

Trapezoidal: $\int_0^2 x^3 \, dx \approx \frac{1}{4} \left[ 0 + 2\left(\frac{1}{2}\right)^3 + 2(1)^3 + 2\left(\frac{3}{2}\right)^3 + (2)^3 \right] = \frac{17}{4} = 4.2500$

Simpson's: $\int_0^2 x^3 \, dx \approx \frac{1}{6} \left[ 0 + 4\left(\frac{1}{2}\right)^3 + 2(1)^3 + 4\left(\frac{3}{2}\right)^3 + (2)^3 \right] = \frac{24}{6} = 4.0000$

**5.** Exact: $\int_1^3 x^3 \, dx = \left[ \frac{x^4}{4} \right]_1^3 = \frac{81}{4} - \frac{1}{4} = 20$

Trapezoidal: $\int_1^3 x^3 \, dx \approx \frac{1}{6} \left[ 1 + 2\left(\frac{4}{3}\right)^3 + 2\left(\frac{5}{3}\right)^3 + 2(2)^3 + 2\left(\frac{7}{3}\right)^3 + 2\left(\frac{8}{3}\right)^3 + 27 \right] \approx 20.2222$

Simpson's: $\int_1^3 x^3 \, dx \approx \frac{1}{9} \left[ 1 + 4\left(\frac{4}{3}\right)^3 + 2\left(\frac{5}{3}\right)^3 + 4(2)^3 + 2\left(\frac{7}{3}\right)^3 + 4\left(\frac{8}{3}\right)^3 + 27 \right] = 20.0000$

**7.** Exact: $\int_4^9 \sqrt{x}\, dx = \left[\frac{2}{3}x^{3/2}\right]_4^9 = 18 - \frac{16}{3} = \frac{38}{3} \approx 12.6667$

Trapezoidal: $\int_4^9 \sqrt{x}\, dx \approx \frac{5}{16}\left[2 + 2\sqrt{\frac{37}{8}} + 2\sqrt{\frac{21}{4}} + 2\sqrt{\frac{47}{8}} + 2\sqrt{\frac{26}{4}} + 2\sqrt{\frac{57}{8}} + 2\sqrt{\frac{31}{4}} + 2\sqrt{\frac{67}{8}} + 3\right] \approx 12.6640$

Simpson's: $\int_4^9 \sqrt{x}\, dx \approx \frac{5}{24}\left[2 + 4\sqrt{\frac{37}{8}} + \sqrt{21} + 4\sqrt{\frac{47}{8}} + \sqrt{26} + 4\sqrt{\frac{57}{8}} + \sqrt{31} + 4\sqrt{\frac{67}{8}} + 3\right] \approx 12.6667$

**9.** Exact: $\int_0^1 \frac{2}{(x+2)^2}\, dx = \left[\frac{-2}{(x+2)}\right]_0^1 = \frac{-2}{3} + \frac{2}{2} = \frac{1}{3}$

Trapezoidal: $\int_0^1 \frac{2}{(x+2)^2}\, dx \approx \frac{1}{8}\left[\frac{1}{2} + 2\left(\frac{2}{((1/4)+2)^2}\right) + 2\left(\frac{2}{((1/2)+2)^2}\right) + 2\left(\frac{2}{((3/4)+2)^2}\right) + \frac{2}{9}\right]$

$= \frac{1}{8}\left[\frac{1}{2} + 2\left(\frac{32}{81}\right) + 2\left(\frac{8}{25}\right) + 2\left(\frac{32}{121}\right) + \frac{2}{9}\right] \approx 0.3352$

Simpson's: $\int_0^1 \frac{2}{(x+2)^2}\, dx \approx \frac{1}{12}\left[\frac{1}{2} + 4\left(\frac{2}{((1/4)+2)^2}\right) + 2\left(\frac{2}{((1/2)+2)^2}\right) + 4\left(\frac{2}{((3/4)+2)^2}\right) + \frac{2}{9}\right]$

$= \frac{1}{12}\left[\frac{1}{2} + 4\left(\frac{32}{81}\right) + 2\left(\frac{8}{25}\right) + 4\left(\frac{32}{121}\right) + \frac{2}{9}\right] \approx 0.3334$

**11.** Trapezoidal: $\int_0^2 \sqrt{1+x^3}\, dx \approx \frac{1}{4}\left[1 + 2\sqrt{1+\left(\frac{1}{8}\right)} + 2\sqrt{2} + 2\sqrt{1+\left(\frac{27}{8}\right)} + 3\right] \approx 3.283$

Simpson's: $\int_0^2 \sqrt{1+x^3}\, dx \approx \frac{1}{6}\left[1 + 4\sqrt{1+\left(\frac{1}{8}\right)} + 2\sqrt{2} + 4\sqrt{1+\left(\frac{27}{8}\right)} + 3\right] \approx 3.240$

Graphing utility: 3.241

**13.** $\int_0^1 \sqrt{x}\sqrt{1-x}\, dx = \int_0^1 \sqrt{x(1-x)}\, dx$

Trapezoidal: $\int_0^1 \sqrt{x(1-x)}\, dx \approx \frac{1}{8}\left[0 + 2\sqrt{\frac{1}{4}\left(1-\frac{1}{4}\right)} + 2\sqrt{\frac{1}{2}\left(1-\frac{1}{2}\right)} + 2\sqrt{\frac{3}{4}\left(1-\frac{3}{4}\right)}\right] \approx 0.342$

Simpson's: $\int_0^1 \sqrt{x(1-x)}\, dx \approx \frac{1}{12}\left[0 + 4\sqrt{\frac{1}{4}\left(1-\frac{1}{4}\right)} + 2\sqrt{\frac{1}{2}\left(1-\frac{1}{2}\right)} + 4\sqrt{\frac{3}{4}\left(1-\frac{3}{4}\right)}\right] \approx 0.372$

Graphing utility: 0.393

**15.** Trapezoidal: $\int_0^{\sqrt{\pi/2}} \sin(x^2)\, dx \approx \frac{\sqrt{\pi/2}}{8}\left[\sin 0 + 2\sin\left(\frac{\sqrt{\pi/2}}{4}\right)^2 + 2\sin\left(\frac{\sqrt{\pi/2}}{2}\right)^2 + 2\sin\left(\frac{3\sqrt{\pi/2}}{4}\right)^2 + \sin\left(\sqrt{\frac{\pi}{2}}\right)^2\right] \approx 0.550$

Simpson's: $\int_0^{\sqrt{\pi/2}} \sin(x^2)\, dx \approx \frac{\sqrt{\pi/2}}{12}\left[\sin 0 + 4\sin\left(\frac{\sqrt{\pi/2}}{4}\right)^2 + 2\sin\left(\frac{\sqrt{\pi/2}}{2}\right)^2 + 4\sin\left(\frac{3\sqrt{\pi/2}}{4}\right)^2 + \sin\left(\sqrt{\frac{\pi}{2}}\right)^2\right] \approx 0.548$

Graphing utility: 0.549

**17.** Trapezoidal: $\int_3^{3.1} \cos x^2\, dx \approx \frac{0.1}{8}\left[\cos(3)^2 + 2\cos(3.025)^2 + 2\cos(3.05)^2 + 2\cos(3.075)^2 + \cos(3.1)^2\right] \approx -0.098$

Simpson's: $\int_3^{3.1} \cos x^2\, dx \approx \frac{0.1}{12}\left[\cos(3)^2 + 4\cos(3.025)^2 + 2\cos(3.05)^2 + 4\cos(3.075)^2 + \cos(3.1)^2\right] \approx -0.098$

Graphing utility: $-0.098$

**19.** Trapezoidal: $\int_0^2 x \ln(x+1)\, dx \approx \frac{1}{4}\left[0 + 2(0.5)\ln(1.5) + 2\ln(2) + 2(1.5)\ln(2.5) + 2\ln(3)\right] \approx 1.684$

Simpson's: $\int_0^2 x \ln(x+1)\, dx \approx \frac{1}{6}\left[0 + 4(0.5)\ln(1.5) + 2\ln(2) + 4(1.5)\ln(2.5) + 2\ln(3)\right] \approx 1.649$

Graphing utility: 1.648

**21.** Trapezoidal: $\int_0^2 xe^{-x}\, dx \approx \frac{1}{4}\left[0 + e^{-1/2} + 2e^{-1} + 3e^{-3/2} + 2e^{-2}\right] \approx \frac{2.2824}{4} \approx 0.5706$

Simpson's: $\int_0^2 2xe^{-x}\, dx \approx \frac{1}{6}\left[0 + 2e^{-1/2} + 2e^{-1} + 6e^{-3/2} + 2e^{-2}\right] \approx \frac{3.5583}{6} \approx 0.5930$

Graphing utility: 0.594

**23.** Trapezoidal: Linear polynomials

Simpson's: Quadratic polynomials

**25.**  $f(x) = 2x^3$

$f'(x) = 6x^2$

$f''(x) = 12x$

$f'''(x) = 12$

$f^{(4)}(x) = 0$

(a)  Trapezoidal: Error $\leq \dfrac{(3-1)^3}{12(4^2)}(36) = 1.5$ because

$\left| f''(x) \right|$ is maximum in $[1, 3]$ when $x = 3$.

(b)  Simpson's: Error $\leq \dfrac{(3-1)^5}{180(4^4)}(0) = 0$ because

$f^{(4)}(x) = 0$.

**27.**  $f(x) = (x-1)^{-2}$

$f'(x) = -2(x-1)^{-3}$

$f''(x) = 6(x-1)^{-4}$

$f'''(x) = -24(x-1)^{-5}$

$f^{(4)}(x) = 120(x-1)^{-6}$

(a)  Trapezoidal: Error $\leq \dfrac{(4-2)^3}{12(4^2)}(6) = \dfrac{1}{4}$ because

$\left| f''(x) \right|$ is a maximum of 6 at $x = 2$.

(b)  Simpson's: Error $\leq \dfrac{(4-2)^5}{180(4^4)}(120) = \dfrac{1}{12}$ because

$\left| f^{(4)}(x) \right|$ is a maximum of 120 at $x = 2$.

**29.**   $f(x) = x^{-1}, \quad 1 \leq x \leq 3$

$f'(x) = -x^{-2}$

$f''(x) = 2x^{-3}$

$f'''(x) = -6x^{-4}$

$f^{(4)}(x) = 24x^{-5}$

(a)  Maximum of $\left| f''(x) \right| = \left| 2x^{-3} \right|$ is 2.

Trapezoidal: Error

$\leq \dfrac{2^3}{12n^2}(2) \leq 0.00001, \ n^2 \geq 133,333.33,$

$n \geq 365.15$ Let $n = 366$.

(b)  Maximum of $\left| f^{(4)}(x) \right| = \left| 24x^{-5} \right|$ is 24.

Simpson's: Error $\leq \dfrac{2^5}{180n^4}(24) \leq 0.00001,$

$n^4 \geq 426,666.67, \ n \geq 25.56$ Let $n = 26$.

**31.** $f(x) = (x + 2)^{1/2}, \quad 0 \le x \le 2$

$f'(x) = \dfrac{1}{2}(x + 2)^{-1/2}$

$f''(x) = -\dfrac{1}{4}(x + 2)^{-3/2}$

$f'''(x) = \dfrac{3}{8}(x + 2)^{-5/2}$

$f^{(4)}(x) = \dfrac{-15}{16}(x + 2)^{-7/2}$

(a) Maximum of $\left| f''(x) \right| = \left| \dfrac{-1}{4(x + 2)^{3/2}} \right|$ is

$\dfrac{\sqrt{2}}{16} \approx 0.0884.$

Trapezoidal:

Error $\le \dfrac{(2 - 0)^3}{12n^2}\left(\dfrac{\sqrt{2}}{16}\right) \le 0.00001$

$n^2 \ge \dfrac{8\sqrt{2}}{12(16)}10^5 = \dfrac{\sqrt{2}}{24}10^5$

$n \ge 76.8.$ Let $n = 77.$

(b) Maximum of $\left| f^{(4)}(x) \right| = \left| \dfrac{-15}{16(x + 2)^{7/2}} \right|$ is

$\dfrac{15\sqrt{2}}{256} \approx 0.0829.$

Simpson's:

Error

$\le \dfrac{2^5}{180n^4}\left(\dfrac{15\sqrt{2}}{256}\right) \le 0.00001$

$n^4 \ge \dfrac{32(15)\sqrt{2}}{180(256)}10^5 = \dfrac{\sqrt{2}}{96}10^5$

$n \ge 6.2.$ Let $n = 8$ (even).

**33.** $f(x) = \sqrt{1 + x}$

(a) $f''(x) = -\dfrac{1}{4(1 + x)^{3/2}}$ in $[0, 2]$.

$\left| f''(x) \right|$ is maximum when $x = 0$ and $\left| f''(0) \right| = \dfrac{1}{4}.$

Trapezoidal: Error $\le \dfrac{8}{12n^2}\left(\dfrac{1}{4}\right) \le 0.00001,$

$n^2 \ge 16{,}666.67,\ n \ge 129.10;$ let $n = 130.$

(b) $f^{(4)}(x) = -\dfrac{15}{16(1 + x)^{7/2}}$ in $[0, 2]$

$\left| f^{(4)}(x) \right|$ is maximum when $x = 0$ and

$\left| f^{(4)}(0) \right| = \dfrac{15}{16}.$

Simpson's: Error $\le \dfrac{32}{180n^4}\left(\dfrac{15}{16}\right) \le 0.00001,$

$n^4 \ge 16{,}666.67,\ n \ge 11.36;$ let $n = 12.$

**35.** $f(x) = \tan(x^2)$

(a) $f''(x) = 2\sec^2(x^2)\left[1 + 4x^2\tan(x^2)\right]$ in $[0, 1]$.

$\left| f''(x) \right|$ is maximum when $x = 1$ and $\left| f''(1) \right| \approx 49.5305.$

Trapezoidal: Error $\le \dfrac{(1 - 0)^3}{12n^2}(49.5305) \le 0.00001,\ n^2 \ge 412{,}754.17,\ n \ge 642.46;$ let $n = 643.$

(b) $f^{(4)}(x) = 8\sec^2(x^2)\left[12x^2 + (3 + 32x^4)\tan(x^2) + 36x^2\tan^2(x^2) + 48x^4\tan^3(x^2)\right]$ in $[0, 1]$

$\left| f^{(4)}(x) \right|$ is maximum when $x = 1$ and $\left| f^{(4)}(1) \right| \approx 9184.4734.$

Simpson's: Error $\le \dfrac{(1 - 0)^5}{180n^4}(9184.4734) \le 0.00001,\ n^4 \ge 5{,}102{,}485.22,\ n \ge 47.53;$ let $n = 48.$

**37.** $n = 4$, $b - a = 4 - 0 = 4$

(a) $\displaystyle\int_0^4 f(x)\,dx \approx \frac{4}{8}\big[3 + 2(7) + 2(9) + 2(7) + 0\big] = \frac{1}{2}(49) = \frac{49}{2} = 24.5$

(b) $\displaystyle\int_0^4 f(x)\,dx \approx \frac{4}{12}\big[3 + 4(7) + 2(9) + 4(7) + 0\big] = \frac{77}{3} \approx 25.67$

**39.** $A = \displaystyle\int_0^{\pi/2} \sqrt{x}\,\cos x\,dx$

Simpson's Rule: $n = 14$

$\displaystyle\int_0^{\pi/2} \sqrt{x}\,\cos x\,dx \approx \frac{\pi}{84}\left[\sqrt{0}\,\cos 0 + 4\sqrt{\frac{\pi}{28}}\,\cos\frac{\pi}{28} + 2\sqrt{\frac{\pi}{14}}\,\cos\frac{\pi}{14} + 4\sqrt{\frac{3\pi}{28}}\,\cos\frac{3\pi}{28} + \cdots + \sqrt{\frac{\pi}{2}}\,\cos\frac{\pi}{2}\right] \approx 0.701$

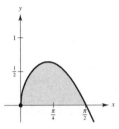

**41.** Area $\approx \dfrac{1000}{2(10)}\big[125 + 2(125) + 2(120) + 2(112) + 2(90) + 2(90) + 2(95) + 2(88) + 2(75) + 2(35)\big] = 89{,}250\ \text{m}^2$

**43.** $W = \displaystyle\int_0^5 100x\sqrt{125 - x^3}\,dx$

Simpson's Rule: $n = 12$

$\displaystyle\int_0^5 100x\sqrt{125 - x^3}\,dx \approx \frac{5}{3(12)}\left[0 + 400\left(\frac{5}{12}\right)\sqrt{125 - \left(\frac{5}{12}\right)^3} + 200\left(\frac{10}{12}\right)\sqrt{125 - \left(\frac{10}{12}\right)^3}\right.$

$\left. + 400\left(\frac{15}{12}\right)\sqrt{125 - \left(\frac{15}{12}\right)^3} + \cdots + 0\right] \approx 10{,}233.58\ \text{ft-lb}$

**45.** $\displaystyle\int_0^{1/2} \frac{6}{\sqrt{1 - x^2}}\,dx$   Simpson's Rule, $n = 6$

$\pi \approx \dfrac{\left(\frac{1}{2} - 0\right)}{3(6)}\big[6 + 4(6.0209) + 2(6.0851) + 4(6.1968) + 2(6.3640) + 4(6.6002) + 6.9282\big] \approx \dfrac{1}{36}[113.098] \approx 3.1416$

**47.** $\displaystyle\int_0^t \sin\sqrt{x}\,dx = 2$, $n = 10$

By trial and error, you obtain $t \approx 2.477$.

**49.** The quadratic polynomial

$p(x) = \dfrac{(x - x_2)(x - x_3)}{(x_1 - x_2)(x_1 - x_3)}y_1 + \dfrac{(x - x_1)(x - x_3)}{(x_2 - x_1)(x_2 - x_3)}y_2 + \dfrac{(x - x_1)(x - x_2)}{(x_3 - x_1)(x_3 - x_2)}y_3$

passes through the three points.

## Section 5.7 The Natural Logarithmic Function: Integration

**1.** $\int \dfrac{5}{x}\,dx = 5\int \dfrac{1}{x}\,dx = 5\ln|x| + C$

**3.** $u = x + 1,\ du = dx$

$\quad \int \dfrac{1}{x+1}\,dx = \ln|x+1| + C$

**5.** $u = 2x + 5,\ du = 2\,dx$

$\quad \int \dfrac{1}{2x+5}\,dx = \dfrac{1}{2}\int \dfrac{1}{2x+5}(2)\,dx$

$\qquad\qquad\qquad = \dfrac{1}{2}\ln|2x+5| + C$

**7.** $u = x^2 - 3,\ du = 2x\,dx$

$\quad \int \dfrac{x}{x^2-3}\,dx = \dfrac{1}{2}\int \dfrac{1}{x^2-3}(2x)\,dx$

$\qquad\qquad\qquad = \dfrac{1}{2}\ln|x^2-3| + C$

**9.** $u = x^4 + 3x,\ du = \left(4x^3 + 3\right)dx$

$\quad \int \dfrac{4x^3+3}{x^4+3x}\,dx = \int \dfrac{1}{x^4+3x}\left(4x^3+3\right)dx$

$\qquad\qquad\qquad = \ln|x^4+3x| + C$

**11.** $\int \dfrac{x^2-4}{x}\,dx = \int\left(x - \dfrac{4}{x}\right)dx$

$\qquad\qquad\quad = \dfrac{x^2}{2} - 4\ln|x| + C$

$\qquad\qquad\quad = \dfrac{x^2}{2} - \ln\left(x^4\right) + C$

**13.** $u = x^3 + 3x^2 + 9x,\ du = 3\left(x^2 + 2x + 3\right)dx$

$\quad \int \dfrac{x^2+2x+3}{x^3+3x^2+9x}\,dx = \dfrac{1}{3}\int \dfrac{3\left(x^2+2x+3\right)}{x^3+3x^2+9x}\,dx$

$\qquad\qquad\qquad\qquad = \dfrac{1}{3}\ln|x^3+3x^2+9x| + C$

**15.** $\int \dfrac{x^2-3x+2}{x+1}\,dx = \int\left(x - 4 + \dfrac{6}{x+1}\right)dx$

$\qquad\qquad\qquad = \dfrac{x^2}{2} - 4x + 6\ln|x+1| + C$

**17.** $\int \dfrac{x^3-3x^2+5}{x-3}\,dx = \int\left(x^2 + \dfrac{5}{x-3}\right)dx$

$\qquad\qquad\qquad = \dfrac{x^3}{3} + 5\ln|x-3| + C$

**19.** $\int \dfrac{x^4+x-4}{x^2+2}\,dx = \int\left(x^2 - 2 + \dfrac{x}{x^2+2}\right)dx$

$\qquad\qquad\qquad = \dfrac{x^3}{3} - 2x + \dfrac{1}{2}\ln\left(x^2+2\right) + C$

$\qquad\qquad\qquad = \dfrac{x^3}{3} - 2x + \ln\sqrt{x^2+2} + C$

**21.** $u = \ln x,\ du = \dfrac{1}{x}dx$

$\quad \int \dfrac{(\ln x)^2}{x}\,dx = \dfrac{1}{3}(\ln x)^3 + C$

**23.** $u = 1 - 3\sqrt{x},\ du = \dfrac{-3}{2\sqrt{x}}$

$\quad \int \dfrac{1}{\sqrt{x}\left(1-3\sqrt{x}\right)}\,dx = -\dfrac{2}{3}\int \dfrac{1}{1-3\sqrt{x}}\left(\dfrac{-3}{2\sqrt{x}}\right)dx$

$\qquad\qquad\qquad\qquad = -\dfrac{2}{3}\ln|1-3\sqrt{x}| + C$

**25.** $\int \dfrac{2x}{(x-1)^2}\,dx = \int \dfrac{2x-2+2}{(x-1)^2}\,dx$

$\qquad\qquad\quad = \int \dfrac{2(x-1)}{(x-1)^2}\,dx + 2\int \dfrac{1}{(x-1)^2}\,dx$

$\qquad\qquad\quad = 2\int \dfrac{1}{x-1}\,dx + 2\int \dfrac{1}{(x-1)^2}\,dx$

$\qquad\qquad\quad = 2\ln|x-1| - \dfrac{2}{(x-1)} + C$

**27.** $u = 1 + \sqrt{2x},\ du = \dfrac{1}{\sqrt{2x}}\,dx \Rightarrow (u-1)\,du = dx$

$\quad \int \dfrac{1}{1+\sqrt{2x}}\,dx = \int \dfrac{(u-1)}{u}\,du = \int\left(1 - \dfrac{1}{u}\right)du$

$\qquad\qquad\qquad = u - \ln|u| + C_1$

$\qquad\qquad\qquad = \left(1+\sqrt{2x}\right) - \ln|1+\sqrt{2x}| + C_1$

$\qquad\qquad\qquad = \sqrt{2x} - \ln\left(1+\sqrt{2x}\right) + C$

where $C = C_1 + 1$.

**29.** $u = \sqrt{x} - 3,\ du = \dfrac{1}{2\sqrt{x}}\,dx \Rightarrow 2(u + 3)\,du = dx$

$$\int \dfrac{\sqrt{x}}{\sqrt{x} - 3}\,dx = 2\int \dfrac{(u + 3)^2}{u}\,du$$

$$= 2\int \dfrac{u^2 + 6u + 9}{u}\,du = 2\int \left(u + 6 + \dfrac{9}{u}\right)du$$

$$= 2\left[\dfrac{u^2}{2} + 6u + 9\ln|u|\right] + C_1$$

$$= u^2 + 12u + 18\ln|u| + C_1$$

$$= \left(\sqrt{x} - 3\right)^2 + 12\left(\sqrt{x} - 3\right) + 18\ln\left|\sqrt{x} - 3\right| + C_1$$

$$= x + 6\sqrt{x} + 18\ln\left|\sqrt{x} - 3\right| + C$$

where $C = C_1 - 27$.

**31.** $\displaystyle\int \cot\left(\dfrac{\theta}{3}\right)d\theta = 3\int \cot\left(\dfrac{\theta}{3}\right)\left(\dfrac{1}{3}\right)d\theta = 3\ln\left|\sin\dfrac{\theta}{3}\right| + C$

**33.** $\displaystyle\int \csc 2x\,dx = \dfrac{1}{2}\int (\csc 2x)(2)\,dx$

$$= -\dfrac{1}{2}\ln|\csc 2x + \cot 2x| + C$$

**35.** $\displaystyle\int (\cos 3\theta - 1)\,d\theta = \dfrac{1}{3}\int \cos 3\theta(3)\,d\theta - \int d\theta$

$$= \dfrac{1}{3}\sin 3\theta - \theta + C$$

**37.** $u = 1 + \sin t,\ du = \cos t\,dt$

$$\int \dfrac{\cos t}{1 + \sin t}\,dt = \ln|1 + \sin t| + C$$

**39.** $u = \sec x - 1,\ du = \sec x \tan x\,dx$

$$\int \dfrac{\sec x \tan x}{\sec x - 1}\,dx = \ln|\sec x - 1| + C$$

**41.** $\displaystyle\int e^{-x}\tan\left(e^{-x}\right)dx = -\int \tan\left(e^{-x}\right)\left(-e^{-x}\right)dx$

$$= -\left(-\ln\left|\cos\left(e^{-x}\right)\right|\right) + C$$

$$= \ln\left|\cos\left(e^{-x}\right)\right| + C$$

**43.** $y = \displaystyle\int \dfrac{3}{2 - x}\,dx = -3\int \dfrac{1}{x - 2}\,dx = -3\ln|x - 2| + C$

$(1, 0):\ 0 = -3\ln|1 - 2| + C \Rightarrow C = 0$

$y = -3\ln|x - 2|$

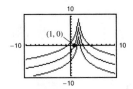

**45.** $y = \displaystyle\int \dfrac{2x}{x^2 - 9}\,dx = \ln|x^2 - 9| + C$

$(0, 4):\ 4 = \ln|0 - 9| + C \Rightarrow C = 4 - \ln 9$

$y = \ln|x^2 - 9| + 4 - \ln 9$

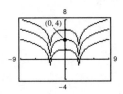

**47.** $f''(x) = \dfrac{2}{x^2} = 2x^{-2},\ x > 0$

$f'(x) = \dfrac{-2}{x} + C$

$f'(1) = 1 = -2 + C \Rightarrow C = 3$

$f'(x) = \dfrac{-2}{x} + 3$

$f(x) = -2\ln x + 3x + C_1$

$f(1) = 1 = -2(0) + 3 + C_1 \Rightarrow C_1 = -2$

$f(x) = -2\ln x + 3x - 2$

**49.** $\dfrac{dy}{dx} = \dfrac{1}{x+2}, (0,1)$

(a)

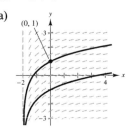

(b) $\quad y = \displaystyle\int \dfrac{1}{x+2}\,dx = \ln|x+2| + C$

$y(0) = 1 \Rightarrow 1 = \ln 2 + C \Rightarrow C = 1 - \ln 2$

So, $y = \ln|x+2| + 1 - \ln 2 = \ln\left(\dfrac{x+2}{2}\right) + 1.$

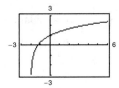

**51.** $\displaystyle\int_0^4 \dfrac{5}{3x+1}\,dx = \left[\dfrac{5}{3}\ln|3x+1|\right]_0^4 = \dfrac{5}{3}\ln 13 \approx 4.275$

**53.** $u = 1 + \ln x, du = \dfrac{1}{x}\,dx$

$\displaystyle\int_1^e \dfrac{(1 + \ln x)^2}{x}\,dx = \left[\dfrac{1}{3}(1 + \ln x)^3\right]_1^e = \dfrac{7}{3}$

**55.** $\displaystyle\int_0^2 \dfrac{x^2 - 2}{x+1}\,dx = \int_0^2\left(x - 1 - \dfrac{1}{x+1}\right)dx$

$\qquad = \left[\dfrac{1}{2}x^2 - x - \ln|x+1|\right]_0^2 = -\ln 3$

$\qquad \approx -1.099$

**57.** $\displaystyle\int_1^2 \dfrac{1 - \cos\theta}{\theta - \sin\theta}\,d\theta = \Big[\ln|\theta - \sin\theta|\Big]_1^2$

$\qquad = \ln\left|\dfrac{2 - \sin 2}{1 - \sin 1}\right| \approx 1.929$

**59.** $\displaystyle\int \dfrac{1}{1 + \sqrt{x}}\,dx = 2\sqrt{x} - 2\ln\left(1 + \sqrt{x}\right) + C$

**61.** $\displaystyle\int \dfrac{\sqrt{x}}{x - 1}\,dx = \ln\left(\dfrac{\sqrt{x} - 1}{\sqrt{x} + 1}\right) + 2\sqrt{x} + C$

**63.** $\displaystyle\int_{\pi/4}^{\pi/2} (\csc x - \sin x)\,dx = \ln\left(\sqrt{2} + 1\right) - \dfrac{\sqrt{2}}{2} \approx 0.174$

**Note: In Exercises 65–67, you can use the Second Fundamental Theorem of Calculus or integrate the function.**

**65.** $F(x) = \displaystyle\int_1^x \dfrac{1}{t}\,dt$

$F'(x) = \dfrac{1}{x}$

**67.** $F(x) = \displaystyle\int_1^{3x} \dfrac{1}{t}\,dt$

$F'(x) = \dfrac{1}{3x}(3) = \dfrac{1}{x}$

(by Second Fundamental Theorem of Calculus)

**Alternate Solution:**

$F(x) = \displaystyle\int_1^{3x} \dfrac{1}{t}\,dt = \Big[\ln|t|\Big]_1^{3x} = \ln|3x|$

$F'(x) = \dfrac{1}{3x}(3) = \dfrac{1}{x}$

**69.** $A = \displaystyle\int_1^3 \dfrac{6}{x}\,dx = \Big[6\ln|x|\Big]_1^3 = 6\ln 3$

**71.** $A = \displaystyle\int_0^{\pi/4} \tan x\,dx = -\ln|\cos x|\Big]_0^{\pi/4} = -\ln\dfrac{\sqrt{2}}{2} + 0 = \ln\sqrt{2} = \dfrac{\ln 2}{2}$

**73.** $A = \displaystyle\int_1^4 \dfrac{x^2 + 4}{x}\,dx = \int_1^4\left(x + \dfrac{4}{x}\right)dx = \left[\dfrac{x^2}{2} + 4\ln x\right]_1^4 = (8 + 4\ln 4) - \dfrac{1}{2} = \dfrac{15}{2} + 8\ln 2 \approx 13.045$

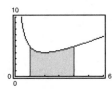

**75.** $\displaystyle\int_0^2 2\sec\dfrac{\pi x}{6}\,dx = \dfrac{12}{\pi}\int_0^2 \sec\left(\dfrac{\pi x}{6}\right)\dfrac{\pi}{6}\,dx = \dfrac{12}{\pi}\left[\ln\left|\sec\dfrac{\pi x}{6} + \tan\dfrac{\pi x}{6}\right|\right]_0^2$

$\qquad = \dfrac{12}{\pi}\left(\ln\left|\sec\dfrac{\pi}{3} + \tan\dfrac{\pi}{3}\right| - \ln|1 + 0|\right) = \dfrac{12}{\pi}\ln\left(2 + \sqrt{3}\right) \approx 5.03041$

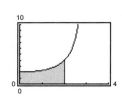

**77.** $f(x) = \dfrac{12}{x}, b - a = 5 - 1 = 4, n = 4$

Trapezoid: $\dfrac{4}{2(4)}\left[f(1) + 2f(2) + 2f(3) + 2f(4) + f(5)\right] = \dfrac{1}{2}[12 + 12 + 8 + 6 + 2.4] = 20.2$

Simpson: $\dfrac{4}{3(4)}\left[f(1) + 4f(2) + 2f(3) + 4f(4) + f(5)\right] = \dfrac{1}{3}[12 + 24 + 8 + 12 + 2.4] \approx 19.4667$

Calculator: $\displaystyle\int_1^5 \dfrac{12}{x}\,dx \approx 19.3133$

Exact: $12 \ln 5$

**79.** $f(x) = \ln x, b - a = 6 - 2 = 4, n = 4$

Trapezoid: $\dfrac{4}{2(4)}\left[f(2) + 2f(3) + 2f(4) + 2f(5) + f(6)\right] = \dfrac{1}{2}[0.6931 + 2.1972 + 2.7726 + 3.2189 + 1.7918] \approx 5.3368$

Simpson: $\dfrac{4}{3(4)}\left[f(2) + 4f(3) + 2f(4) + 4f(5) + f(6)\right] \approx 5.3632$

Calculator: $\displaystyle\int_2^6 \ln x\,dx \approx 5.3643$

**81.** Power Rule

**83.** Substitution: $(u = x^2 + 4)$ and Log Rule

**85.**

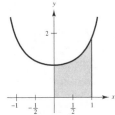

$A \approx 1.25$; Matches (d)

**87.** $\displaystyle\int_1^x \dfrac{3}{t}\,dt = \int_{1/4}^x \dfrac{1}{t}\,dt$

$\left[3 \ln|t|\right]_1^x = \left[\ln|t|\right]_{1/4}^x$

$3 \ln x = \ln x - \ln\left(\dfrac{1}{4}\right)$

$2 \ln x = -\ln\left(\dfrac{1}{4}\right) = \ln 4$

$\ln x = \dfrac{1}{2}\ln 4 = \ln 2$

$x = 2$

**89.** $\displaystyle\int \cot u\,du = \int \dfrac{\cos u}{\sin u}\,du = \ln|\sin u| + C$

**Alternate solution:**

$\dfrac{d}{du}\left[\ln|\sin u| + C\right] = \dfrac{1}{\sin u}\cos u + C = \cot u + C$

**91.** $-\ln|\cos x| + C = \ln\left|\dfrac{1}{\cos x}\right| + C = \ln|\sec x| + C$

**93.** $\ln|\sec x + \tan x| + C = \ln\left|\dfrac{(\sec x + \tan x)(\sec x - \tan x)}{(\sec x - \tan x)}\right| + C$

$= \ln\left|\dfrac{\sec^2 x - \tan^2 x}{\sec x - \tan x}\right| + C$

$= \ln\left|\dfrac{1}{\sec x - \tan x}\right| + C = -\ln|\sec x - \tan x| + C$

**95.** Average value $= \dfrac{1}{4-2} \displaystyle\int_2^4 \dfrac{8}{x^2}\, dx$

$= 4\displaystyle\int_2^4 x^{-2}\, dx$

$= \left[-4\dfrac{1}{x}\right]_2^4$

$= -4\left(\dfrac{1}{4} - \dfrac{1}{2}\right) = 1$

**97.** Average value $= \dfrac{1}{e-1} \displaystyle\int_1^e \dfrac{2\ln x}{x}\, dx$

$= \dfrac{2}{e-1}\left[\dfrac{(\ln x)^2}{2}\right]_1^e$

$= \dfrac{1}{e-1}(1-0)$

$= \dfrac{1}{e-1} \approx 0.582$

**99.** $P(t) = \displaystyle\int \dfrac{3000}{1+0.25t}\, dt = (3000)(4)\displaystyle\int \dfrac{0.25}{1+0.25t}\, dt$

$= 12{,}000\ln|1+0.25t| + C$

$P(0) = 12{,}000\ln|1+0.25(0)| + C = 1000$

$C = 1000$

$P(t) = 12{,}000\ln|1+0.25t| + 1000$

$= 1000\big[12\ln|1+0.25t| + 1\big]$

$P(3) = 1000\big[12(\ln 1.75) + 1\big] \approx 7715$

**101.** $t = \dfrac{10}{\ln 2}\displaystyle\int_{250}^{300} \dfrac{1}{T-100}\, dT$

$= \dfrac{10}{\ln 2}\big[\ln(T-100)\big]_{250}^{300} = \dfrac{10}{\ln 2}\big[\ln 200 - \ln 150\big]$

$= \dfrac{10}{\ln 2}\left[\ln\left(\dfrac{4}{3}\right)\right] \approx 4.1504 \text{ min}$

**103.** $f(x) = \dfrac{x}{1+x^2}$

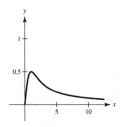

(a) $y = \dfrac{1}{2}x$ intersects $f(x) = \dfrac{x}{1+x^2}$:

$\dfrac{1}{2}x = \dfrac{x}{1+x^2}$

$1+x^2 = 2$

$x = 1$

$A = \displaystyle\int_0^1 \left(\left[\dfrac{x}{1+x^2}\right] - \dfrac{1}{2}x\right) dx = \left[\dfrac{1}{2}\ln(x^2+1) - \dfrac{x^2}{4}\right]_0^1 = \dfrac{1}{2}\ln 2 - \dfrac{1}{4}$

(b) $f'(x) = \dfrac{(1+x^2) - x(2x)}{(1+x^2)^2} = \dfrac{1-x^2}{(1+x^2)^2}$

$f'(0) = 1$

So, for $0 < m < 1$, the graphs of $f$ and $y = mx$ enclose a finite region.

(c)

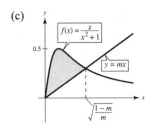

$f(x) = \dfrac{x}{x^2 + 1}$ intersects $y = mx$:

$$\frac{x}{1 + x^2} = mx$$

$$1 = m + mx^2$$

$$x^2 = \frac{1 - m}{m}$$

$$x = \sqrt{\frac{1 - m}{m}}$$

$$A = \int_0^{\sqrt{(1-m)/m}} \left( \frac{x}{1 + x^2} - mx \right) dx, \quad 0 < m < 1$$

$$= \left[ \frac{1}{2} \ln(1 + x^2) - \frac{mx^2}{2} \right]_0^{\sqrt{(1-m)/m}} = \frac{1}{2} \ln\left( 1 + \frac{1 - m}{m} \right) - \frac{1}{2} m\left( \frac{1 - m}{m} \right) = \frac{1}{2} \ln\left( \frac{1}{m} \right) - \frac{1}{2}(1 - m) = \frac{1}{2}\left[ m - \ln(m) - 1 \right]$$

**105.** False

$\frac{1}{2}(\ln x) = \ln\left(x^{1/2}\right) \neq (\ln x)^{1/2}$

**107.** True

$\int \frac{1}{x} dx = \ln|x| + C_1 = \ln|x| + \ln|C| = \ln|Cx|, C \neq 0$

**109.** Let $f(t) = \ln t$ on $[x, y], \quad 0 < x < y.$

By the Mean Value Theorem,

$$\frac{f(y) - f(x)}{y - x} = f'(c), \quad x < c < y,$$

$$\frac{\ln y - \ln x}{y - x} = \frac{1}{c}.$$

Because $0 < x < c < y, \dfrac{1}{x} > \dfrac{1}{c} > \dfrac{1}{y}.$ So,

$$\frac{1}{y} < \frac{\ln y - \ln x}{y - x} < \frac{1}{x}.$$

**111.** $\dfrac{d}{dx} \ln|x| = \dfrac{1}{x}$ implies that

$$\int \frac{1}{x} dx = \ln|x| + C.$$

The second formula follows by the Chain Rule.

# Section 5.8   Inverse Trigonometric Functions: Integration

**1.** $\int \dfrac{dx}{\sqrt{9 - x^2}} = \arcsin\left( \dfrac{x}{3} \right) + C$

**3.** $\int \dfrac{1}{x\sqrt{4x^2 - 1}} dx = \int \dfrac{2}{2x\sqrt{(2x)^2 - 1}} dx = \text{arcsec}|2x| + C$

**5.** $\int \dfrac{1}{\sqrt{1 - (x + 1)^2}} dx = \arcsin(x + 1) + C$

**7.** Let $u = t^2, du = 2t\, dt.$

$$\int \frac{t}{\sqrt{1 - t^4}} dt = \frac{1}{2} \int \frac{1}{\sqrt{1 - (t^2)^2}} (2t)\, dt = \frac{1}{2} \arcsin t^2 + C$$

**9.** $\displaystyle\int \frac{t}{t^4 + 25}\, dt = \frac{1}{2}\int \frac{1}{(t^2)^2 + 5^2}(2)\, dt$

$\displaystyle = \frac{1}{2}\frac{1}{5}\arctan\left(\frac{t^2}{5}\right) + C$

$\displaystyle = \frac{1}{10}\arctan\left(\frac{t^2}{5}\right) + C$

**11.** Let $u = e^{2x}$, $du = 2e^{2x}\, dx$.

$\displaystyle\int \frac{e^{2x}}{4 + e^{4x}}\, dx = \frac{1}{2}\int \frac{2e^{2x}}{4 + (e^{2x})^2}\, dx = \frac{1}{4}\arctan\frac{e^{2x}}{2} + C$

**13.** $\displaystyle\int \frac{\sec^2 x}{\sqrt{25 - \tan^2 x}}\, dx = \int \frac{\sec^2 x}{\sqrt{5^2 - (\tan x)^2}}\, dx$

$\displaystyle = \arcsin\left(\frac{\tan x}{5}\right) + C$

**15.** $\displaystyle\int \frac{1}{\sqrt{x}\sqrt{1 - x}}\, dx,\; u = \sqrt{x},\; x = u^2,\; dx = 2u\, du$

$\displaystyle\int \frac{1}{u\sqrt{1 - u^2}}(2u\, du) = 2\int \frac{du}{\sqrt{1 - u^2}} = 2\arcsin u + C = 2\arcsin\sqrt{x} + C$

**17.** $\displaystyle\int \frac{x - 3}{x^2 + 1}\, dx = \frac{1}{2}\int \frac{2x}{x^2 + 1}\, dx - 3\int \frac{1}{x^2 + 1}\, dx = \frac{1}{2}\ln(x^2 + 1) - 3\arctan x + C$

**19.** $\displaystyle\int \frac{x + 5}{\sqrt{9 - (x - 3)^2}}\, dx = \int \frac{(x - 3)}{\sqrt{9 - (x - 3)^2}}\, dx + \int \frac{8}{\sqrt{9 - (x - 3)^2}}\, dx$

$\displaystyle = -\sqrt{9 - (x - 3)^2} + 8\arcsin\left(\frac{x - 3}{3}\right) + C = -\sqrt{6x - x^2} + 8\arcsin\left(\frac{x}{3} - 1\right) + C$

**21.** Let $u = 3x$, $du = 3\, dx$.

$\displaystyle\int_0^{1/6} \frac{3}{\sqrt{1 - 9x^2}}\, dx = \int_0^{1/6} \frac{1}{\sqrt{1 - (3x)^2}}(3)\, dx$

$\displaystyle = \left[\arcsin(3x)\right]_0^{1/6} = \frac{\pi}{6}$

**23.** Let $u = 2x$, $du = 2\, dx$.

$\displaystyle\int_0^{\sqrt{3}/2} \frac{1}{1 + 4x^2}\, dx = \frac{1}{2}\int_0^{\sqrt{3}/2} \frac{2}{1 + (2x)^2}\, dx$

$\displaystyle = \left[\frac{1}{2}\arctan(2x)\right]_0^{\sqrt{3}/2} = \frac{\pi}{6}$

**25.** $\displaystyle\int_3^6 \frac{1}{25 + (x - 3)^2}\, dx = \left[\frac{1}{5}\arctan\left(\frac{x - 3}{5}\right)\right]_3^6$

$\displaystyle = \frac{1}{5}\arctan(3/5)$

$\displaystyle \approx 0.108$

**27.** Let $u = e^x$, $du = e^x\, dx$

$\displaystyle\int_0^{\ln 5} \frac{e^x}{1 + e^{2x}}\, dx = \left[\arctan(e^x)\right]_0^{\ln 5} = \arctan 5 - \frac{\pi}{4} \approx 0.588$

**29.** Let $u = \cos x$, $du = -\sin x\, dx$.

$\displaystyle\int_{\pi/2}^{\pi} \frac{\sin x}{1 + \cos^2 x}\, dx = -\int_{\pi/2}^{\pi} \frac{-\sin x}{1 + \cos^2 x}\, dx = \left[-\arctan(\cos x)\right]_{\pi/2}^{\pi} = \frac{\pi}{4}$

**31.** Let $u = \arcsin x$, $du = \dfrac{1}{\sqrt{1 - x^2}}\, dx$.

$$\int_0^{1/\sqrt{2}} \frac{\arcsin x}{\sqrt{1 - x^2}}\, dx = \left[\frac{1}{2}\arcsin^2 x\right]_0^{1/\sqrt{2}} = \frac{\pi^2}{32} \approx 0.308$$

**33.** $\displaystyle\int_0^2 \frac{dx}{x^2 - 2x + 2} = \int_0^2 \frac{1}{1 + (x - 1)^2}\, dx = \Big[\arctan(x - 1)\Big]_0^2 = \frac{\pi}{2}$

**35.** $\displaystyle\int \frac{2x}{x^2 + 6x + 13}\, dx = \int \frac{2x + 6}{x^2 + 6x + 13}\, dx - 6\int \frac{1}{x^2 + 6x + 13}\, dx$

$$= \int \frac{2x + 6}{x^2 + 6x + 13}\, dx - 6\int \frac{1}{4 + (x + 3)^2}\, dx = \ln\left|x^2 + 6x + 13\right| - 3\arctan\left(\frac{x + 3}{2}\right) + C$$

**37.** $\displaystyle\int \frac{1}{\sqrt{-x^2 - 4x}}\, dx = \int \frac{1}{\sqrt{4 - (x + 2)^2}}\, dx = \arcsin\left(\frac{x + 2}{2}\right) + C$

**39.** $\displaystyle\int_2^3 \frac{2x - 3}{\sqrt{4x - x^2}}\, dx = \int_2^3 \frac{2x - 4}{\sqrt{4x - x^2}}\, dx + \int_2^3 \frac{1}{\sqrt{4x - x^2}}\, dx$

$$= -\int_2^3 \left(4x - x^2\right)^{-1/2}(4 - 2x)\, dx + \int_2^3 \frac{1}{\sqrt{4 - (x - 2)^2}}\, dx$$

$$= \left[-2\sqrt{4x - x^2} + \arcsin\left(\frac{x - 2}{2}\right)\right]_2^3 = 4 - 2\sqrt{3} + \frac{\pi}{6} \approx 1.059$$

**41.** Let $u = x^2 + 1$, $du = 2x\, dx$.

$$\int \frac{x}{x^4 + 2x^2 + 2}\, dx = \frac{1}{2}\int \frac{2x}{\left(x^2 + 1\right)^2 + 1}\, dx = \frac{1}{2}\arctan\left(x^2 + 1\right) + C$$

**43.** Let $u = \sqrt{e^t - 3}$. Then $u^2 + 3 = e^t$, $2u\, du = e^t\, dt$, and $\dfrac{2u\, du}{u^2 + 3} = dt$.

$$\int \sqrt{e^t - 3}\, dt = \int \frac{2u^2}{u^2 + 3}\, du = \int 2\, du - \int 6\frac{1}{u^2 + 3}\, du$$

$$= 2u - 2\sqrt{3}\arctan\frac{u}{\sqrt{3}} + C = 2\sqrt{e^t - 3} - 2\sqrt{3}\arctan\sqrt{\frac{e^t - 3}{3}} + C$$

**45.** $\displaystyle\int_1^3 \frac{dx}{\sqrt{x}(1 + x)}$

Let $u = \sqrt{x}$, $u^2 = x$, $2u\, du = dx$, $1 + x = 1 + u^2$.

$$\int_1^{\sqrt{3}} \frac{2u\, du}{u(1 + u^2)} = \int_1^{\sqrt{3}} \frac{2}{1 + u^2}\, du$$

$$= \Big[2\arctan(u)\Big]_1^{\sqrt{3}}$$

$$= 2\left(\frac{\pi}{3} - \frac{\pi}{4}\right) = \frac{\pi}{6}$$

**47.** (a) $\displaystyle\int \frac{1}{\sqrt{1 - x^2}}\, dx = \arcsin x + C$, $\quad u = x$

(b) $\displaystyle\int \frac{x}{\sqrt{1 - x^2}}\, dx = -\sqrt{1 - x^2} + C$, $\quad u = 1 - x^2$

(c) $\displaystyle\int \frac{1}{x\sqrt{1 - x^2}}\, dx$ cannot be evaluated using the basic integration rules.

**49.** (a) $\int \sqrt{x-1}\, dx = \dfrac{2}{3}(x-1)^{3/2} + C, \quad u = x - 1$

(b) Let $u = \sqrt{x-1}$. Then $x = u^2 + 1$ and $dx = 2u\, du$.

$$\int x\sqrt{x-1}\, dx = \int (u^2 + 1)(u)(2u)\, du$$
$$= 2\int (u^4 + u^2)\, du$$
$$= 2\left(\frac{u^5}{5} + \frac{u^3}{3}\right) + C$$
$$= \frac{2}{15}u^3(3u^2 + 5) + C$$
$$= \frac{2}{15}(x-1)^{3/2}\left[3(x-1) + 5\right] + C$$
$$= \frac{2}{15}(x-1)^{3/2}(3x + 2) + C$$

(c) Let $u = \sqrt{x-1}$. Then $x = u^2 + 1$ and $dx = 2u\, du$.

$$\int \frac{x}{\sqrt{x-1}}\, dx = \int \frac{u^2 + 1}{u}(2u)\, du$$
$$= 2\int (u^2 + 1)\, du$$
$$= 2\left(\frac{u^3}{3} + u\right) + C$$
$$= \frac{2}{3}u(u^2 + 3) + C$$
$$= \frac{2}{3}\sqrt{x-1}(x + 2) + C$$

**Note:** In (b) and (c), substitution was necessary *before* the basic integration rules could be used.

**51.** No. This integral does not correspond to any of the basic differentiation rules.

**53.** $y' = \dfrac{1}{\sqrt{4 - x^2}}, \quad (0, \pi)$

$y = \int \dfrac{1}{\sqrt{4 - x^2}}\, dx = \arcsin\left(\dfrac{x}{2}\right) + C$

When $x = 0$, $y = \pi \Rightarrow C = \pi$

$y = \arcsin\left(\dfrac{x}{2}\right) + \pi$

**55.** (a)

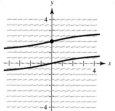

(b) $y' = \dfrac{2}{9 + x^2}, \quad (0, 2)$

$y = \int \dfrac{2}{9 + x^2}\, dx = \dfrac{2}{3}\arctan\left(\dfrac{x}{3}\right) + C$

$2 = C$

$y = \dfrac{2}{3}\arctan\left(\dfrac{x}{3}\right) + 2$

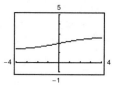

**57.** $\dfrac{dy}{dx} = \dfrac{10}{x\sqrt{x^2 - 1}}$,   $(3, 0)$

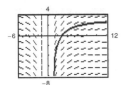

**59.** $\dfrac{dy}{dx} = \dfrac{2y}{\sqrt{16 - x^2}}$,   $(0, 2)$

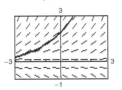

**61.** Area $= \displaystyle\int_0^1 \dfrac{2}{\sqrt{4 - x^2}}\, dx$

$= \left[\, 2 \arcsin\!\left(\dfrac{x}{2}\right) \right]_0^1$

$= 2 \arcsin\!\left(\dfrac{1}{2}\right) - 2 \arcsin(0)$

$= 2\!\left(\dfrac{\pi}{6}\right) = \dfrac{\pi}{3}$

**63.** Area $= \displaystyle\int_1^3 \dfrac{1}{x^2 - 2x + 5}\, dx = \int_1^3 \dfrac{1}{(x-1)^2 + 4}\, dx$

$= \left[\, \dfrac{1}{2} \arctan\!\left(\dfrac{x-1}{2}\right) \right]_1^3$

$= \dfrac{1}{2}\arctan(1) - \dfrac{1}{2}\arctan(0)$

$= \dfrac{\pi}{8}$

**65.** Area $= \displaystyle\int_{-\pi/2}^{\pi/2} \dfrac{3\cos x}{1 + \sin^2 x}\, dx = 3\int_{-\pi/2}^{\pi/2} \dfrac{1}{1 + \sin^2 x}(\cos x\, dx)$

$= \left[\, 3\arctan(\sin x) \right]_{-\pi/2}^{\pi/2}$

$= 3\arctan(1) - 3\arctan(-1)$

$= \dfrac{3\pi}{4} + \dfrac{3\pi}{4} = \dfrac{3\pi}{2}$

**67.** (a)

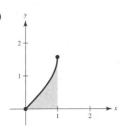

Shaded area is given by $\displaystyle\int_0^1 \arcsin x\, dx$.

(b) $\displaystyle\int_0^1 \arcsin x\, dx \approx 0.5708$

(c) Divide the rectangle into two regions.

Area rectangle $= (\text{base})(\text{height}) = 1\!\left(\dfrac{\pi}{2}\right) = \dfrac{\pi}{2}$

Area rectangle $= \displaystyle\int_0^1 \arcsin x\, dx + \int_0^{\pi/2} \sin y\, dy$

$\dfrac{\pi}{2} = \displaystyle\int_0^1 \arcsin x\, dx + \left(-\cos y\right)\Big]_0^{\pi/2} = \int_0^1 \arcsin x\, dx + 1$

So, $\displaystyle\int_0^1 \arcsin x\, dx = \dfrac{\pi}{2} - 1$, $(\approx 0.5708)$.

**69.** $F(x) = \dfrac{1}{2}\displaystyle\int_x^{x+2} \dfrac{2}{t^2+1}\, dt$

(a) $F(x)$ represents the average value of $f(x)$ over the interval $[x, x+2]$. Maximum at $x = -1$, because the graph is greatest on $[-1, 1]$.

(b) $F(x) = \left[\arctan t\right]_x^{x+2} = \arctan(x+2) - \arctan x$

$$F'(x) = \frac{1}{1+(x+2)^2} - \frac{1}{1+x^2} = \frac{(1+x^2) - (x^2+4x+5)}{(x^2+1)(x^2+4x+5)} = \frac{-4(x+1)}{(x^2+1)(x^2+4x+5)} = 0 \text{ when } x = -1.$$

**71.** False, $\displaystyle\int \dfrac{dx}{3x\sqrt{9x^2-16}} = \dfrac{1}{12}\operatorname{arcsec}\dfrac{|3x|}{4} + C$

**73.** True

$$\frac{d}{dx}\left[-\arccos\frac{x}{2} + C\right] = \frac{1/2}{\sqrt{1-(x/2)^2}} = \frac{1}{\sqrt{4-x^2}}$$

**75.** $\dfrac{d}{dx}\left[\arcsin\left(\dfrac{u}{a}\right) + C\right] = \dfrac{1}{\sqrt{1-(u^2/a^2)}}\left(\dfrac{u'}{a}\right) = \dfrac{u'}{\sqrt{a^2-u^2}}$

So, $\displaystyle\int \dfrac{du}{\sqrt{a^2-u^2}} = \arcsin\left(\dfrac{u}{a}\right) + C.$

**77.** Assume $u > 0$.

$$\frac{d}{dx}\left[\frac{1}{a}\operatorname{arcsec}\frac{u}{a} + C\right] = \frac{1}{a}\left[\frac{u'/a}{(u/a)\sqrt{(u/a)^2-1}}\right] = \frac{1}{a}\left[\frac{u'}{u\sqrt{(u^2-a^2)/a^2}}\right] = \frac{u'}{u\sqrt{u^2-a^2}}.$$

The case $u < 0$ is handled in a similar manner.

So, $\displaystyle\int \dfrac{du}{u\sqrt{u^2-a^2}} = \int \dfrac{u'}{u\sqrt{u^2-a^2}}\, dx = \dfrac{1}{a}\operatorname{arcsec}\dfrac{|u|}{a} + C.$

**79.** (a) Area $= \displaystyle\int_0^1 \dfrac{1}{1+x^2}\, dx$

(b) Trapezoidal Rule: $n = 8, b - a = 1 - 0 = 1$

Area $\approx 0.7847$

(c) Because

$$\int_0^1 \frac{1}{1+x^2}\, dx = \left[\arctan x\right]_0^1 = \frac{\pi}{4},$$

you can use the Trapezoidal Rule to approximate $\pi/4$, and therefore, $\pi$. For example, using $n = 200$, you obtain $\pi \approx 4(0.785397) = 3.141588.$

## Section 5.9  Hyperbolic Functions

**1.** (a) $\sinh 3 = \dfrac{e^3 - e^{-3}}{2} \approx 10.018$

(b) $\tanh(-2) = \dfrac{\sinh(-2)}{\cosh(-2)} = \dfrac{e^{-2} - e^2}{e^{-2} + e^2} \approx -0.964$

**3.** (a) $\operatorname{csch}(\ln 2) = \dfrac{2}{e^{\ln 2} - e^{-\ln 2}} = \dfrac{2}{2 - (1/2)} = \dfrac{4}{3}$

(b) $\coth(\ln 5) = \dfrac{\cosh(\ln 5)}{\sinh(\ln 5)} = \dfrac{e^{\ln 5} + e^{-\ln 5}}{e^{\ln 5} - e^{-\ln 5}}$

$= \dfrac{5 + (1/5)}{5 - (1/5)} = \dfrac{13}{12}$

**5.** (a)  $\cosh^{-1} 2 = \ln\left(2 + \sqrt{3}\right) \approx 1.317$

  (b)  $\operatorname{sech}^{-1}\dfrac{2}{3} = \ln\left(\dfrac{1 + \sqrt{1 - (4/9)}}{2/3}\right) \approx 0.962$

**7.**  $\tanh^2 x + \operatorname{sech}^2 x = \left(\dfrac{e^x - e^{-x}}{e^x + e^{-x}}\right)^2 + \left(\dfrac{2}{e^x + e^{-x}}\right)^2$

$$= \dfrac{e^{2x} - 2 + e^{-2x} + 4}{\left(e^x + e^{-x}\right)^2}$$

$$= \dfrac{e^{2x} + 2 + e^{-2x}}{e^{2x} + 2 + e^{-2x}} = 1$$

**9.**  $\dfrac{1 + \cosh 2x}{2} = \dfrac{1 + \left(e^{2x} + e^{-2x}\right)/2}{2}$

$$= \dfrac{e^{2x} + 2 + e^{-2x}}{4}$$

$$= \left(\dfrac{e^x + e^{-x}}{2}\right)^2 = \cosh^2 x$$

**11.**  $2 \sinh x \cosh x = 2\left(\dfrac{e^x - e^{-x}}{2}\right)\left(\dfrac{e^x + e^{-x}}{2}\right) = \dfrac{e^{2x} - e^{-2x}}{2} = \sinh 2x$

**13.**  $\sinh x \cosh y + \cosh x \sinh y = \left(\dfrac{e^x - e^{-x}}{2}\right)\left(\dfrac{e^y + e^{-y}}{2}\right) + \left(\dfrac{e^x + e^{-x}}{2}\right)\left(\dfrac{e^y - e^{-y}}{2}\right)$

$$= \dfrac{1}{4}\left[e^{x+y} - e^{-x+y} + e^{x-y} - e^{-(x+y)} + e^{x+y} + e^{-x+y} - e^{x-y} - e^{-(x+y)}\right]$$

$$= \dfrac{1}{4}\left[2\left(e^{x+y} - e^{-(x+y)}\right)\right] = \dfrac{e^{(x+y)} - e^{-(x+y)}}{2} = \sinh(x + y)$$

**15.**  $\sinh x = \dfrac{3}{2}$

  $\cosh^2 x - \left(\dfrac{3}{2}\right)^2 = 1 \Rightarrow \cosh^2 x = \dfrac{13}{4} \Rightarrow \cosh x = \dfrac{\sqrt{13}}{2}$

  $\tanh x = \dfrac{3/2}{\sqrt{13}/2} = \dfrac{3\sqrt{13}}{13}$

  $\operatorname{csch} x = \dfrac{1}{3/2} = \dfrac{2}{3}$

  $\operatorname{sech} x = \dfrac{1}{\sqrt{13}/2} = \dfrac{2\sqrt{13}}{13}$

  $\coth x = \dfrac{1}{3/\sqrt{13}} = \dfrac{\sqrt{13}}{3}$

**17.**  $\displaystyle\lim_{x \to \infty} \sinh x = \infty$

**19.**  $\displaystyle\lim_{x \to \infty} \operatorname{sech} x = 0$

**21.**  $\displaystyle\lim_{x \to 0} \dfrac{\sinh x}{x} = \lim_{x \to 0} \dfrac{e^x - e^{-x}}{2x} = 1$

**23.**  $f(x) = \sinh(3x)$

  $f'(x) = 3\cosh(3x)$

**25.**  $y = \operatorname{sech}\left(5x^2\right)$

  $y' = -\operatorname{sech}\left(5x^2\right)\tanh\left(5x^2\right)(10x)$

  $= -10x\,\operatorname{sech}\left(5x^2\right)\tanh\left(5x^2\right)$

**27.**  $f(x) = \ln(\sinh x)$

  $f'(x) = \dfrac{1}{\sinh x}(\cosh x) = \coth x$

**29.**  $h(x) = \dfrac{1}{4}\sinh 2x - \dfrac{x}{2}$

  $h'(x) = \dfrac{1}{2}\cosh(2x) - \dfrac{1}{2} = \dfrac{\cosh(2x) - 1}{2} = \sinh^2 x$

**31.** $f(t) = \arctan(\sinh t)$

$$f'(t) = \frac{1}{1 + \sinh^2 t}(\cosh t) = \frac{\cosh t}{\cosh^2 t} = \operatorname{sech} t$$

**33.** $y = \sinh(1 - x^2), \quad (1, 0)$

$$y' = \cosh(1 - x^2)(-2x)$$

$$y'(1) = -2$$

Tangent line: $y - 0 = -2(x - 1)$

$$y = -2x + 2$$

**35.** $y = (\cosh x - \sinh x)^2, \quad (0, 1)$

$$y' = 2(\cosh x - \sinh x)(\sinh x - \cosh x)$$

At $(0, 1)$, $y' = 2(1)(-1) = -2$.

Tangent line: $y - 1 = -2(x - 0)$

$$y = -2x + 1$$

**37.** $f(x) = \sin x \sinh x - \cos x \cosh x, \quad -4 \le x \le 4$

$$f'(x) = \sin x \cosh x + \cos x \sinh x - \cos x \sinh x + \sin x \cosh x$$

$$= 2 \sin x \cosh x = 0 \text{ when } x = 0, \pm\pi.$$

Relative maxima: $(\pm\pi, \cosh \pi)$

Relative minimum: $(0, -1)$

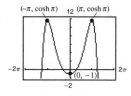

**39.** $g(x) = x \operatorname{sech} x$

$$g'(x) = \operatorname{sech} x - x \operatorname{sech} x \tanh x$$

$$= \operatorname{sech} x(1 - x \tanh x) = 0$$

$x \tanh x = 1$

Using a graphing utility, $x \approx \pm 1.1997$.

By the First Derivative Test, $(1.1997, 0.6627)$ is a relative maximum and $(-1.1997, -0.6627)$ is a relative minimum.

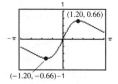

**41.** (a) $y = 10 + 15 \cosh \dfrac{x}{15}, \quad -15 \le x \le 15$

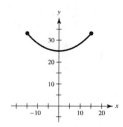

(b) At $x = \pm 15$, $y = 10 + 15 \cosh(1) \approx 33.146$.

At $x = 0$, $y = 10 + 15 \cosh(0) = 25$.

(c) $y' = \sinh \dfrac{x}{15}$. At $x = 15$, $y' = \sinh(1) \approx 1.175$.

**43.** $\displaystyle\int \cosh 2x \, dx = \frac{1}{2}\int \cosh(2x)(2) \, dx$

$$= \frac{1}{2} \sinh 2x + C$$

**45.** Let $u = 1 - 2x$, $du = -2 \, dx$.

$$\int \sinh(1 - 2x) \, dx = -\frac{1}{2}\int \sinh(1 - 2x)(-2) \, dx$$

$$= -\frac{1}{2}\cosh(1 - 2x) + C$$

**47.** Let $u = \cosh(x - 1)$, $du = \sinh(x - 1)\, dx$.

$$\int \cosh^2(x - 1) \sinh(x - 1)\, dx = \tfrac{1}{3}\cosh^3(x - 1) + C$$

**49.** Let $u = \sinh x$, $du = \cosh x\, dx$.

$$\int \frac{\cosh x}{\sinh x}\, dx = \ln|\sinh x| + C$$

**51.** Let $u = \dfrac{x^2}{2}$, $du = x\, dx$.

$$\int x\, \mathrm{csch}^2 \frac{x^2}{2}\, dx = \int \left(\mathrm{csch}^2 \frac{x^2}{2}\right) x\, dx = -\coth \frac{x^2}{2} + C$$

**53.** Let $u = \dfrac{1}{x}$, $du = -\dfrac{1}{x^2}\, dx$.

$$\int \frac{\mathrm{csch}(1/x)\, \coth(1/x)}{x^2}\, dx = -\int \mathrm{csch}\frac{1}{x}\coth\frac{1}{x}\left(-\frac{1}{x^2}\right) dx$$

$$= \mathrm{csch}\frac{1}{x} + C$$

**55.** $\displaystyle\int_0^{\ln 2} \tanh x\, dx = \int_0^{\ln 2} \frac{\sinh x}{\cosh x}\, dx, \quad (u = \cosh x)$

$$= \Big[\ln(\cosh x)\Big]_0^{\ln 2}$$

$$= \ln\big(\cosh(\ln 2) - \ln(\cosh(0)\big)$$

$$= \ln\left(\frac{5}{4}\right) - 0 = \ln\left(\frac{5}{4}\right)$$

**Note:** $\cosh(\ln 2) = \dfrac{e^{\ln 2} + e^{-\ln 2}}{2} = \dfrac{2 + (1/2)}{2} = \dfrac{5}{4}$

**57.** $\displaystyle\int_0^4 \frac{1}{25 - x^2}\, dx = \frac{1}{10}\int \frac{1}{5 - x}\, dx + \frac{1}{10}\int \frac{1}{5 + x}\, dx$

$$= \left[\frac{1}{10}\ln\left|\frac{5 + x}{5 - x}\right|\right]_0^4 = \frac{1}{10}\ln 9 = \frac{1}{5}\ln 3$$

**59.** Let $u = 2x$, $du = 2\, dx$.

$$\int_0^{\sqrt{2}/4} \frac{2}{\sqrt{1 - 4x^2}}\, dx = \int_0^{\sqrt{2}/4} \frac{1}{\sqrt{1 - (2x)^2}}(2)\, dx$$

$$= \Big[\arcsin(2x)\Big]_0^{\sqrt{2}/4} = \frac{\pi}{4}$$

**61.** Answers will vary.

**63.** The derivatives of $f(x) = \cosh x$ and $f(x) = \mathrm{sech}\, x$ differ by a minus sign.

**65.** $y = \cosh^{-1}(3x)$

$$y' = \frac{3}{\sqrt{9x^2 - 1}}$$

**67.** $y = \tanh^{-1}\sqrt{x}$

$$y' = \frac{1}{1 - \left(\sqrt{x}\right)^2}\left(\frac{1}{2}x^{-1/2}\right)$$

$$= \frac{1}{2\sqrt{x}(1 - x)}$$

**69.** $y = \sinh^{-1}(\tan x)$

$$y' = \frac{1}{\sqrt{\tan^2 x + 1}}(\sec^2 x) = |\sec x|$$

**71.** $y = \left(\mathrm{csch}^{-1}\, x\right)^2$

$$y' = 2\,\mathrm{csch}^{-1}x\left(\frac{-1}{|x|\sqrt{1 + x^2}}\right) = \frac{-2\,\mathrm{csch}^{-1}\, x}{|x|\sqrt{1 + x^2}}$$

**73.** $y = 2x \sinh^{-1}(2x) - \sqrt{1 + 4x^2}$

$$y' = 2x\left(\frac{2}{\sqrt{1 + 4x^2}}\right) + 2\sinh^{-1}(2x) - \frac{4x}{\sqrt{1 + 4x^2}}$$

$$= 2\sinh^{-1}(2x)$$

**75.** $\displaystyle\int \frac{1}{3 - 9x^2}\, dx = \frac{1}{3}\int \frac{1}{3 - (3x)^2}(3)\, dx$

$$= \frac{1}{3}\frac{1}{2\sqrt{3}}\ln\left|\frac{\sqrt{3} + 3x}{\sqrt{3} - 3x}\right| + C$$

$$= \frac{\sqrt{3}}{18}\ln\left|\frac{1 + \sqrt{3}x}{1 - \sqrt{3}x}\right| + C$$

**77.** $\displaystyle\int \frac{1}{\sqrt{1 + e^{2x}}}\, dx = \int \frac{e^x}{e^x\sqrt{1 + (e^x)^2}}\, dx$

$$= -\mathrm{csch}^{-1}(e^x) + C$$

$$= -\ln\left(\frac{1 + \sqrt{1 + e^{2x}}}{e^x}\right) + C$$

$$= \ln\left(\frac{e^x}{1 + \sqrt{1 + e^{2x}}}\right) + C$$

$$= \ln\left(\frac{-e^x + e^x\sqrt{1 + e^{2x}}}{e^{2x}}\right) + C$$

$$= \ln\left(\sqrt{1 + e^{2x}} - 1\right) - x + C$$

**79.** Let $u = \sqrt{x}$, $du = \dfrac{1}{2\sqrt{x}}\, dx$.

$$\int \frac{1}{\sqrt{x}\sqrt{1 + x}}\, dx = 2\int \frac{1}{\sqrt{1 + \left(\sqrt{x}\right)^2}}\left(\frac{1}{2\sqrt{x}}\right) dx$$

$$= 2\sinh^{-1}\sqrt{x} + C$$

$$= 2\ln\left(\sqrt{x} + \sqrt{1 + x}\right) + C$$

**81.** $\displaystyle\int \frac{-1}{4x - x^2}\,dx = \int \frac{1}{(x-2)^2 - 4}\,dx$

$$= \frac{1}{4}\ln\left|\frac{(x-2) - 2}{(x-2) + 2}\right|$$

$$= \frac{1}{4}\ln\left|\frac{x-4}{x}\right| + C$$

**83.** $\displaystyle\int_3^7 \frac{1}{\sqrt{x^2 - 4}}\,dx = \left[\ln\left(x + \sqrt{x^2 - 4}\right)\right]_3^7 = \ln\left(7 + \sqrt{45}\right) - \ln\left(3 + \sqrt{5}\right) = \ln\left(\frac{7 + \sqrt{45}}{3 + \sqrt{5}}\right) = \ln\left(\frac{\sqrt{5} + 3}{2}\right)$

**85.** $\displaystyle\int_{-1}^1 \frac{1}{16 - 9x^2}\,dx = \frac{1}{3}\int_{-1}^1 \frac{1}{4^2 - (3x)^2}(3)\,dx$

$$= \left[\frac{1}{3}\frac{1}{4}\frac{1}{2}\ln\left|\frac{4 + 3x}{4 - 3x}\right|\right]_{-1}^1$$

$$= \frac{1}{24}\left[\ln(7) - \ln\left(\frac{1}{7}\right)\right]$$

$$= \frac{1}{24}[\ln 7 - \ln 1 + \ln 7] = \frac{1}{12}\ln 7$$

**87.** Let $u = 4x - 1$, $du = 4\,dx$.

$$y = \int \frac{1}{\sqrt{80 + 8x - 16x^2}}\,dx$$

$$= \frac{1}{4}\int \frac{4}{\sqrt{81 - (4x - 1)^2}}\,dx$$

$$= \frac{1}{4}\arcsin\left(\frac{4x - 1}{9}\right) + C$$

**89.** $\displaystyle y = \int \frac{x^3 - 21x}{5 + 4x - x^2}\,dx = \int\left(-x - 4 + \frac{20}{5 + 4x - x^2}\right)dx$

$$= \int(-x - 4)\,dx + 20\int \frac{1}{3^2 - (x - 2)^2}\,dx$$

$$= -\frac{x^2}{2} - 4x + \frac{20}{6}\ln\left|\frac{3 + (x - 2)}{3 - (x - 2)}\right| + C$$

$$= -\frac{x^2}{2} - 4x + \frac{10}{3}\ln\left|\frac{1 + x}{5 - x}\right| + C$$

$$= \frac{-x^2}{2} - 4x - \frac{10}{3}\ln\left|\frac{5 - x}{x + 1}\right| + C$$

**91.** $\displaystyle A = 2\int_0^4 \operatorname{sech}\frac{x}{2}\,dx$

$$= 2\int_0^4 \frac{2}{e^{x/2} + e^{-x/2}}\,dx$$

$$= 4\int_0^4 \frac{e^{x/2}}{\left(e^{x/2}\right)^2 + 1}\,dx$$

$$= \left[8\arctan\left(e^{x/2}\right)\right]_0^4$$

$$= 8\arctan\left(e^2\right) - 2\pi \approx 5.207$$

**93.** $\displaystyle A = \int_0^2 \frac{5x}{\sqrt{x^4 + 1}}\,dx$

$$= \frac{5}{2}\int_0^2 \frac{2x}{\sqrt{\left(x^2\right)^2 + 1}}\,dx$$

$$= \left[\frac{5}{2}\ln\left(x^2 + \sqrt{x^4 + 1}\right)\right]_0^2$$

$$= \frac{5}{2}\ln\left(4 + \sqrt{17}\right) \approx 5.237$$

**95.** $\int \frac{3k}{16}\, dt = \int \frac{1}{x^2 - 12x + 32}\, dx$

$\frac{3kt}{16} = \int \frac{1}{\left(x - 6\right)^2 - 4}\, dx = \frac{1}{2(2)} \ln \left| \frac{(x - 6) - 2}{(x - 6) + 2} \right| + C = \frac{1}{4} \ln \left| \frac{x - 8}{x - 4} \right| + C$

When $x = 0$:  $t = 0$

$C = -\frac{1}{4} \ln(2)$

When $x = 1$:    $t = 10$

$\frac{30k}{16} = \frac{1}{4} \ln \left| \frac{-7}{-3} \right| - \frac{1}{4} \ln(2) = \frac{1}{4} \ln \left( \frac{7}{6} \right)$

$k = \frac{2}{15} \ln \left( \frac{7}{6} \right)$

When $t = 20$:  $\left( \frac{3}{16} \right)\left( \frac{2}{15} \right) \ln \left( \frac{7}{6} \right)(20) = \frac{1}{4} \ln \frac{x - 8}{2x - 8}$

$\ln \left( \frac{7}{6} \right)^2 = \ln \frac{x - 8}{2x - 8}$

$\frac{49}{36} = \frac{x - 8}{2x - 8}$

$62x = 104$

$x = \frac{104}{62} = \frac{52}{31} \approx 1.677 \text{ kg}$

**97.** (a) $y = a \operatorname{sech}^{-1} \frac{x}{a} - \sqrt{a^2 - x^2}, \quad a > 0$

$\frac{dy}{dx} = \frac{-1}{(x/a)\sqrt{1 - \left(x^2/a^2\right)}} + \frac{x}{\sqrt{a^2 - x^2}} = \frac{-a^2}{x\sqrt{a^2 - x^2}} + \frac{x}{\sqrt{a^2 - x^2}} = \frac{x^2 - a^2}{x\sqrt{a^2 - x^2}} = \frac{-\sqrt{a^2 - x^2}}{x}$

(b) Equation of tangent line through $P = \left( x_0, y_0 \right)$: $y - a \operatorname{sech}^{-1} \frac{x_0}{a} + \sqrt{a^2 - x_0^2} = -\frac{\sqrt{a^2 - x_0^2}}{x_0}(x - x_0)$

When $x = 0$, $y = a \operatorname{sech}^{-1} \frac{x_0}{a} - \sqrt{a^2 - x_0^2} + \sqrt{a^2 - x_0^2} = a \operatorname{sech}^{-1} \frac{x_0}{a}$.

So, $Q$ is the point $\left[ 0, a \operatorname{sech}^{-1}(x_0/a) \right]$.

Distance from P to Q: $d = \sqrt{\left( x_0 - 0 \right)^2 + \left( y_0 - a\operatorname{sech}^{-1}(x_0/a) \right)} = \sqrt{x_0^2 + \left( -\sqrt{a^2 - x_0^2} \right)^2} = \sqrt{a^2} = a$

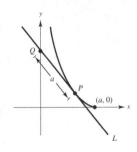

**99.** Let $\quad u = \tanh^{-1}x, \quad -1 < x < 1$

$\tanh u = x.$

$$\frac{\sinh u}{\cosh u} = \frac{e^u - e^{-u}}{e^u + e^{-u}} = x$$

$$e^u - e^{-u} = xe^u + xe^{-u}$$

$$e^{2u} - 1 = xe^{2u} + x$$

$$e^{2u}(1 - x) = 1 + x$$

$$e^{2u} = \frac{1 + x}{1 - x}$$

$$2u = \ln\left(\frac{1 + x}{1 - x}\right)$$

$$u = \frac{1}{2}\ln\left(\frac{1 + x}{1 - x}\right), \quad -1 < x < 1$$

**101.** Let $y = \arcsin(\tanh x)$. Then, $\sin y = \tanh x = \dfrac{e^x - e^{-x}}{e^x + e^{-x}}$ and $\tan y = \dfrac{e^x - e^{-x}}{2} = \sinh x.$

So, $y = \arctan(\sinh x)$. Therefore, $\arctan(\sinh x) = \arcsin(\tanh x).$

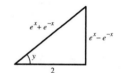

**103.** $y = \cosh x = \dfrac{e^x + e^{-x}}{2}$

$y' = \dfrac{e^x - e^{-x}}{2} = \sinh x$

**105.** $y = \operatorname{sech} x = \dfrac{2}{e^x + e^{-x}}$

$$y' = -2\left(e^x + e^{-x}\right)^{-2}\left(e^x - e^{-x}\right)$$

$$= \left(\frac{-2}{e^x + e^{-x}}\right)\left(\frac{e^x - e^{-x}}{e^x + e^{-x}}\right) = -\operatorname{sech} x \tanh x$$

**107.** $\qquad y = \sinh^{-1} x$

$\qquad \sinh y = x$

$(\cosh y)y' = 1$

$$y' = \frac{1}{\cosh y} = \frac{1}{\sqrt{\sinh^2 y + 1}} = \frac{1}{\sqrt{x^2 + 1}}$$

**109.** $y = c \cosh \dfrac{x}{c}$

Let $P(x_1, y_1)$ be a point on the catenary.

$y' = \sinh \dfrac{x}{c}$

The slope at $P$ is $\sinh(x_1/c)$. The equation of line $L$ is $y - c = \dfrac{-1}{\sinh(x_1/c)}(x - 0).$

When $y = 0, c = \dfrac{x}{\sinh(x_1/c)} \Rightarrow x = c \sinh\left(\dfrac{x_1}{c}\right).$ The length of $L$ is

$$\sqrt{c^2 \sinh^2\left(\frac{x_1}{c}\right) + c^2} = c \cdot \cosh\frac{x_1}{c} = y_1, \text{ the ordinate } y_1 \text{ of the point } P.$$

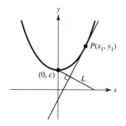

# Review Exercises for Chapter 5

**1.** $\int \left(4x^2 + x + 3\right) dx = \frac{4}{3}x^3 + \frac{1}{2}x^2 + 3x + C$

**3.** $\int \frac{x^4 + 8}{x^3} dx = \int \left(x + 8x^{-3}\right) dx = \frac{1}{2}x^2 - \frac{4}{x^2} + C$

**5.** $\int \left(5 - e^x\right) dx = 5x - e^x + C$

**7.** $f'(x) = -6x, \; f(1) = -2$

$f(x) = -3x^2 + C$

$f(1) = -2 = -3(1)^2 + C \Rightarrow C = 1$

$f(x) = -3x^2 + 1$

**9.** $f''(x) = 24x, \; f'(-1) = 7, \; f(1) = -4$

$f'(x) = 12x^2 + C_1$

$f'(-1) = 7 = 12(-1)^2 + C_1 \Rightarrow C_1 = -5$

$f'(x) = 12x^2 - 5$

$f(x) = 4x^3 - 5x + C_2$

$f(1) = -4 = 4(1)^3 - 5(1) + C_2 \Rightarrow C_2 = -3$

$f(x) = 4x^3 - 5x - 3$

**11.** $a(t) = -32$

$v(t) = -32t + 96$

$s(t) = -16t^2 + 96t$

(a) $v(t) = -32t + 96 = 0$ when $t = 3$ sec.

$s(3) = -144 + 288 = 144$ ft

(b) $v(t) = -32t + 96 = \frac{96}{2}$ when $t = \frac{3}{2}$ sec.

(c) $s\left(\frac{3}{2}\right) = -16\left(\frac{9}{4}\right) + 96\left(\frac{3}{2}\right) = 108$ ft

**13.** $\displaystyle\sum_{i=1}^{5} (5i - 3) = 2 + 7 + 12 + 17 + 22 = 60$

**15.** $\displaystyle\sum_{i=1}^{10} \frac{1}{3i} = \frac{1}{3(1)} + \frac{1}{3(2)} + \cdots + \frac{1}{3(10)}$

**17.** $\displaystyle\sum_{i=1}^{20} 2i = 2\left(\frac{20(21)}{2}\right) = 420$

**19.** $\displaystyle\sum_{i=1}^{20} (i + 1)^2 = \sum_{i=1}^{20} \left(i^2 + 2i + 1\right)$

$= \dfrac{20(21)(41)}{6} = 2\dfrac{20(21)}{2} + 20$

$= 2870 + 420 + 20 = 3310$

**21.** $y = 8 - 2x, \; \Delta x = \dfrac{3}{n},$ right endpoints

Area $= \displaystyle\lim_{n \to \infty} \sum_{i=1}^{n} f(ci)\Delta x$

$= \displaystyle\lim_{n \to \infty} \sum_{i=1}^{n} \left(8 - 2\left(\frac{3i}{n}\right)\right)\frac{3}{n}$

$= \displaystyle\lim_{n \to \infty} \frac{3}{n}\sum_{i=1}^{n} \left(8 - \frac{6i}{n}\right)$

$= \displaystyle\lim_{n \to \infty} \frac{3}{n}\left[8n - \frac{6}{n}\frac{n(n + 1)}{2}\right]$

$= \displaystyle\lim_{n \to \infty} \left[24 - 9\frac{n + 1}{n}\right] = 24 - 9 = 15$

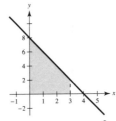

**23.** $y = 5 - x^2, \; \Delta x = \dfrac{3}{n}$

Area $= \displaystyle\lim_{n \to \infty} \sum_{i=1}^{n} f(c_i)\Delta x$

$= \displaystyle\lim_{n \to \infty} \sum_{i=1}^{n} \left[5 - \left(-2 + \frac{3i}{n}\right)^2\right]\left(\frac{3}{n}\right)$

$= \displaystyle\lim_{n \to \infty} \frac{3}{n}\sum_{i=1}^{n} \left[1 + \frac{12i}{n} - \frac{9i^2}{n^2}\right]$

$= \displaystyle\lim_{n \to \infty} \frac{3}{n}\left[n + \frac{12}{n}\frac{n(n + 1)}{2} - \frac{9}{n^2}\frac{n(n + 1)(2n + 1)}{6}\right]$

$= \displaystyle\lim_{n \to \infty} \left[3 + 18\frac{n + 1}{n} - \frac{9}{2}\frac{(n + 1)(2n + 1)}{n^2}\right]$

$= 3 + 18 - 9 = 12$

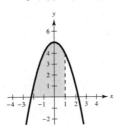

**25.** $x = 5y - y^2, 2 \le y \le 5, \Delta y = \dfrac{3}{n}$

$$\text{Area} = \lim_{n \to \infty} \sum_{i=1}^{n} \left[ 5\left(2 + \dfrac{3i}{n}\right) - \left(2 + \dfrac{3i}{n}\right)^2 \right]\left(\dfrac{3}{n}\right)$$

$$= \lim_{n \to \infty} \dfrac{3}{n} \sum_{i=1}^{n} \left[ 10 + \dfrac{15i}{n} - 4 - 12\dfrac{i}{n} - \dfrac{9i^2}{n^2} \right]$$

$$= \lim_{n \to \infty} \dfrac{3}{n} \sum_{i=1}^{n} \left[ 6 + \dfrac{3i}{n} - \dfrac{9i^2}{n^2} \right]$$

$$= \lim_{n \to \infty} \dfrac{3}{n} \left[ 6n + \dfrac{3}{n}\dfrac{n(n+1)}{2} - \dfrac{9}{n^2}\dfrac{n(n+1)(2n+1)}{6} \right]$$

$$= \left[ 18 + \dfrac{9}{2} - 9 \right] = \dfrac{27}{2}$$

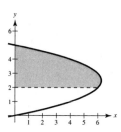

**27.**

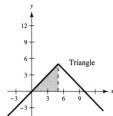

$\int_0^5 \left(5 - |x - 5|\right) dx = \frac{1}{2}(5)(5) = \frac{25}{2}$

(triangle)

**29.** (a) $\displaystyle\int_4^8 \left[ f(x) + g(x) \right] dx = \int_4^8 f(x)\, dx + \int_4^8 g(x)\, dx = 12 + 5 = 17$

(b) $\displaystyle\int_4^8 \left[ f(x) - g(x) \right] dx = \int_4^8 f(x)\, dx - \int_4^8 g(x)\, dx = 12 - 5 = 7$

(c) $\displaystyle\int_4^8 \left[ 2f(x) - 3g(x) \right] dx = 2\int_4^8 f(x)\, dx - 3\int_4^8 g(x)\, dx = 2(12) - 3(5) = 9$

(d) $\displaystyle\int_4^8 7 f(x)\, dx = 7\int_4^8 f(x)\, dx = 7(12) = 84$

**31.** $\displaystyle\int_0^8 (3 + x)\, dx = \left[ 3x + \dfrac{x^2}{2} \right]_0^8 = 24 + \dfrac{64}{2} = 56$

**33.** $\displaystyle\int_4^9 x\sqrt{x}\, dx = \int_4^9 x^{3/2}\, dx = \left[ \dfrac{2}{5}x^{5/2} \right]_4^9 = \dfrac{2}{5}\left[ \left(\sqrt{9}\right)^5 - \left(\sqrt{4}\right)^5 \right] = \dfrac{2}{5}(243 - 32) = \dfrac{422}{5}$

**35.** $\displaystyle\int_0^2 \left(x + e^x\right) dx = \left[ \dfrac{x^2}{2} + e^x \right]_0^2 = 2 + e^2 - 1 = 1 + e^2$

**37.** $A = \displaystyle\int_0^6 (8 - x)\, dx$

$= \left[ 8x - \dfrac{x^2}{2} \right]_0^6$

$= (48 - 18) - 0$

$= 30$

**39.** $A = \displaystyle\int_1^3 \dfrac{2}{x}\, dx$

$= \left[ 2\ln x \right]_1^3$

$= 2\ln 3 - 2\ln 1$

$= \ln 9$

**41.** Average value: $\dfrac{1}{9-4}\displaystyle\int_{4}^{9}\dfrac{1}{\sqrt{x}}\,dx = \left[\dfrac{1}{5}2\sqrt{x}\right]_{4}^{9}$

$$= \dfrac{2}{5}(3-2) = \dfrac{2}{5}$$

$\dfrac{2}{5} = \dfrac{1}{\sqrt{x}}$

$\sqrt{x} = \dfrac{5}{2}$

$x = \dfrac{25}{4}$

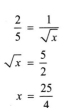

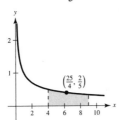

$\left(\dfrac{25}{4}, \dfrac{2}{5}\right)$

**43.** $F'(x) = x^2\sqrt{1+x^3}$

**45.** $F'(x) = x^2 + 3x + 2$

**47.** $u = x^3 + 3,\ du = 3x^2\,dx$

$$\int \dfrac{x^2}{\sqrt{x^3+3}}\,dx = \int \left(x^3+3\right)^{-1/2} x^2\,dx$$

$$= \dfrac{1}{3}\int \left(x^3+3\right)^{-1/2} 3x^2\,dx$$

$$= \dfrac{2}{3}\left(x^3+3\right)^{1/2} + C$$

**49.** $u = 1 - 3x^2,\ du = -6x\,dx$

$$\int x\left(1-3x^2\right)^4\,dx = -\dfrac{1}{6}\int \left(1-3x^2\right)^4(-6x\,dx)$$

$$= -\dfrac{1}{30}\left(1-3x^2\right)^5 + C$$

$$= \dfrac{1}{30}\left(3x^2-1\right)^5 + C$$

**51.** $\displaystyle\int \sin^3 x \cos x\,dx = \dfrac{1}{4}\sin^4 x + C$

**53.** $\displaystyle\int \dfrac{\cos\theta}{\sqrt{1-\sin\theta}}\,d\theta = -\int \left(1-\sin\theta\right)^{-1/2}(-\cos\theta)\,d\theta$

$$= -2\left(1-\sin\theta\right)^{1/2} + C$$

$$= -2\sqrt{1-\sin\theta} + C$$

**55.** $\displaystyle\int xe^{-3x^2}\,dx = -\dfrac{1}{6}\int e^{-3x^2}(-6x)\,dx = -\dfrac{1}{6}e^{-3x^2} + C$

**57.** $\displaystyle\int (x+1)5^{(x+1)^2}\,dx = \dfrac{1}{2}\int 5^{(x+1)^2} 2(x+1)\,dx$

$$= \dfrac{1}{2\ln 5}5^{(x+1)^2} + C$$

**59.** $\displaystyle\int \left(1+\sec\pi x\right)^2 \sec\pi x \tan\pi x\,dx = \dfrac{1}{\pi}\int \left(1+\sec\pi x\right)^2(\pi\sec\pi x \tan\pi x)\,dx = \dfrac{1}{3\pi}\left(1+\sec\pi x\right)^3 + C$

**61. (a)** Answers will vary. *Sample answer:*

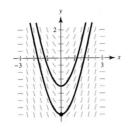

**(b)** $\dfrac{dy}{dx} = x\sqrt{9-x^2},\ (0,-4)$

$$y = \int \left(9-x^2\right)^{1/2} x\,dx = \dfrac{-1}{2}\dfrac{\left(9-x^2\right)^{3/2}}{3/2} + C = -\dfrac{1}{3}\left(9-x^2\right)^{3/2} + C$$

$$-4 = -\dfrac{1}{3}\left(9-0\right)^{3/2} + C = -\dfrac{1}{3}(27) + C \Rightarrow C = 5$$

$$y = -\dfrac{1}{3}\left(9-x^2\right)^{3/2} + 5$$

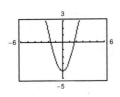

**63.** $\displaystyle\int_{0}^{3}\dfrac{1}{\sqrt{1+x}}\,dx = \int_{0}^{3}\left(1+x\right)^{-1/2}\,dx = \left[2\left(1+x\right)^{1/2}\right]_{0}^{3} = 4 - 2 = 2$

**65.** $u = 1 - y,\ y = 1 - u,\ dy = -du$

When $y = 0, u = 1$. When $y = 1, u = 0$.

$$2\pi \int_0^1 (y + 1)\sqrt{1 - y}\ dy = 2\pi \int_1^0 -\big[(1 - u) + 1\big]\sqrt{u}\ du = 2\pi \int_1^0 \big(u^{3/2} - 2u^{1/2}\big)\ du = 2\pi \left[\frac{2}{5}u^{5/2} - \frac{4}{3}u^{3/2}\right]_1^0 = \frac{28\pi}{15}$$

**67.** $\displaystyle \int_0^\pi \cos\left(\frac{x}{2}\right) dx = 2\int_0^\pi \cos\left(\frac{x}{2}\right)\frac{1}{2}\ dx = \left[2\sin\left(\frac{x}{2}\right)\right]_0^\pi = 2$

**69.** Trapezoidal Rule $(n = 4)$: $\displaystyle \int_2^3 \frac{2}{1 + x^2}\ dx$

$$\approx \frac{1}{8}\left[\frac{2}{1 + 2^2} + 2\left(\frac{2}{1 + (9/4)^2}\right) + 2\left(\frac{2}{1 + (5/2)^2}\right) + 2\left(\frac{2}{1 + (11/4)^2}\right) + \frac{2}{1 + 3^2}\right] \approx 0.285$$

Simpson's Rule $(n = 4)$: $\displaystyle \int_2^3 \frac{2}{1 + x^2}\ dx$

$$\approx \frac{1}{12}\left[\frac{2}{1 + 2^2} + 4\left(\frac{2}{1 + (9/4)^2}\right) + 2\left(\frac{2}{1 + (5/2)^2}\right) + 4\left(\frac{2}{1 + (11/4)^2}\right) + \frac{2}{1 + 3^2}\right] \approx 0.284$$

Graphing utility: 0.284

**71.** Trapezoidal Rule $(n = 4)$: $\displaystyle \int_0^3 \sqrt{x}\ \ln(x + 1)\ dx$

$$\approx \frac{3}{2(4)}\left[\sqrt{0}\ \ln(0 + 1) + 2\sqrt{\frac{3}{4}}\ \ln\left(\frac{3}{4} + 1\right) + 2\sqrt{\frac{3}{2}}\ \ln\left(\frac{3}{2} + 1\right) + 2\sqrt{\frac{9}{4}}\ \ln\left(\frac{9}{4} + 1\right) + \sqrt{3}\ \ln(3 + 1)\right] \approx 3.432$$

Simpson's Rule $(n = 4)$: $\displaystyle \int_0^3 \sqrt{x}\ \ln(x + 1)\ dx$

$$\approx \frac{3}{3(4)}\left[\sqrt{0}\ \ln(0 + 1) + 4\sqrt{\frac{3}{4}}\ \ln\left(\frac{3}{4} + 1\right) + 2\sqrt{\frac{3}{2}}\ \ln\left(\frac{3}{2} + 1\right) + 4\sqrt{\frac{9}{4}}\ \ln\left(\frac{9}{4} + 1\right) + \sqrt{3}\ \ln(3 + 1)\right] \approx 3.414$$

Graphing utility: 3.406

**73.** $u = 7x - 2,\ du = 7\ dx$

$$\int \frac{1}{7x - 2}\ dx = \frac{1}{7}\int \frac{1}{7x - 2}(7)\ dx = \frac{1}{7}\ln|7x - 2| + C$$

**75.** $\displaystyle \int \frac{\sin x}{1 + \cos x}\ dx = -\int \frac{-\sin x}{1 + \cos x}\ dx = -\ln|1 + \cos x| + C$

**77.** Let $u = e^{2x} + e^{-2x},\ du = \big(2e^{2x} - e^{-2x}\big)\ dx$.

$$\int \frac{e^{2x} - e^{-2x}}{e^{2x} + e^{-2x}}\ dx = \frac{1}{2}\int \frac{2e^{2x} - 2e^{-2x}}{e^{2x} + e^{-2x}}\ dx$$
$$= \frac{1}{2}\ln\big(e^{2x} + e^{-2x}\big) + C$$

**79.** $\displaystyle \int_1^4 \frac{2x + 1}{2x}\ dx = \int_1^4\left(1 + \frac{1}{2x}\right)dx$

$$= \left[x + \frac{1}{2}\ln|x|\right]_1^4$$
$$= 4 + \frac{1}{2}\ln 4 - 1 = 3 + \ln 2$$

**81.** $\displaystyle \int_0^{\pi/3} \sec\theta\ d\theta = \Big[\ln|\sec\theta + \tan\theta|\Big]_0^{\pi/3} = \ln\big(2 + \sqrt{3}\big)$

**83.** Let $u = e^{2x},\ du = 2e^{2x}\ dx$.

$$\int \frac{1}{e^{2x} + e^{-2x}}\ dx = \int \frac{e^{2x}}{1 + e^{4x}}\ dx = \frac{1}{2}\int \frac{1}{1 + \big(e^{2x}\big)^2}\big(2e^{2x}\big)\ dx = \frac{1}{2}\arctan\big(e^{2x}\big) + C$$

**85.** Let $u = x^2$, $du = 2x\,dx$.

$$\int \frac{x}{\sqrt{1 - x^4}}\,dx = \frac{1}{2}\int \frac{1}{\sqrt{1 - (x^2)^2}}(2x)\,dx = \frac{1}{2}\arcsin x^2 + C$$

**87.** Let $u = \arctan\left(\dfrac{x}{2}\right)$, $du = \dfrac{2}{4 + x^2}\,dx$.

$$\int \frac{\arctan(x/2)}{4 + x^2}\,dx = \frac{1}{2}\int \left(\arctan\frac{x}{2}\right)\left(\frac{2}{4 + x^2}\right)dx = \frac{1}{4}\left(\arctan\frac{x}{2}\right)^2 + C$$

**89.** $y = \operatorname{sech}(4x - 1)$

$\quad y' = -\operatorname{sech}(4x - 1)\tanh(4x - 1)(4)$

$\quad\quad = -4\operatorname{sech}(4x - 1)\tanh(4x - 1)$

**91.** $y = \sinh^{-1}(4x)$

$$y' = \frac{4}{\sqrt{(4x)^2 + 1}} = \frac{4}{\sqrt{16x^2 + 1}}$$

**93.** Let $u = x^3$, $du = 3x^2\,dx$.

$$\int x^2\left(\operatorname{sech} x^3\right)^2 dx = \frac{1}{3}\int \left(\operatorname{sech} x^3\right)^2\left(3x^2\right)dx$$

$$= \frac{1}{3}\tanh x^3 + C$$

**95.** Let $u = \dfrac{2}{3}x$, $du = \dfrac{2}{3}\,dx$.

$$\int \frac{1}{9 - 4x^2}\,dx = \int \frac{1/9}{1 - \left(\frac{4}{9}x^2\right)}\,dx = \frac{1}{6}\tanh^{-1}\left(\frac{2}{3}x\right) + C$$

**Alternate solution:**

$$\int \frac{1}{3^2 - (2x)^2}\,dx = \frac{1}{12}\ln\left|\frac{3 + 2x}{3 - 2x}\right| + C$$

## Problem Solving for Chapter 5

**1.** (a) $L(1) = \displaystyle\int_1^1 \frac{1}{t}\,dt = 0$

(b) $L'(x) = \dfrac{1}{x}$ by the Second Fundamental Theorem of Calculus.

$\quad L'(1) = 1$

(c) $L(x) = 1 = \displaystyle\int_1^x \frac{1}{t}\,dt$ for $x \approx 2.718$

$$\int_1^{2.718} \frac{1}{t}\,dt = 0.999896$$

(**Note:** The exact value of $x$ is $e$, the base of the natural logarithm function.)

(d) First show that $\displaystyle\int_1^{x_1} \frac{1}{t}\,dt = \int_{1/x_1}^1 \frac{1}{t}\,dt$.

To see this, let $u = \dfrac{t}{x_1}$ and $du = \dfrac{1}{x_1}\,dt$.

Then $\displaystyle\int_1^{x_1} \frac{1}{t}\,dt = \int_{1/x_1}^1 \frac{1}{ux_1}(x_1\,du) = \int_{1/x_1}^1 \frac{1}{u}\,du = \int_{1/x_1}^1 \frac{1}{t}\,dt$.

Now,

$$L(x_1 x_2) = \int_1^{x_1 x_2} \frac{1}{t}\,dt = \int_{1/x_1}^{x_2} \frac{1}{u}\,du \left(\text{using } u = \frac{t}{x_1}\right) = \int_{1/x_1}^1 \frac{1}{u}\,du + \int_1^{x_2} \frac{1}{u}\,du = \int_1^{x_1} \frac{1}{u}\,du + \int_1^{x_2} \frac{1}{u}\,du = L(x_1) + L(x_2).$$

**3. (a)** Let $A = \displaystyle\int_0^b \frac{f(x)}{f(x) + f(b - x)}\,dx$.

Let $u = b - x$, $du = -dx$.

$$A = \int_b^0 \frac{f(b - u)}{f(b - u) + f(u)}(-du) = \int_0^b \frac{f(b - u)}{f(b - u) + f(u)}\,du = \int_0^b \frac{f(b - x)}{f(b - x) + f(x)}\,dx$$

Then, $2A = \displaystyle\int_0^b \frac{f(x)}{f(x) + f(b - x)}\,dx + \int_0^b \frac{f(b - x)}{f(b - x) + f(x)}\,dx = \int_0^b 1\,dx = b$.

So, $A = \dfrac{b}{2}$.

**(b)** $b = 1 \Rightarrow \displaystyle\int_0^1 \frac{\sin x}{\sin(1 - x) + \sin x}\,dx = \frac{1}{2}$

**(c)** $b = 3$, $f(x) = \sqrt{x}$

$$\int_0^3 \frac{\sqrt{x}}{\sqrt{x} + \sqrt{3 - x}}\,dx = \frac{3}{2}$$

**5. (a)** $\displaystyle\int_{-1}^1 \cos x\,dx \approx \cos\left(-\frac{1}{\sqrt{3}}\right) + \cos\left(\frac{1}{\sqrt{3}}\right) = 2\cos\left(\frac{1}{\sqrt{3}}\right) \approx 1.6758$

$\displaystyle\int_{-1}^1 \cos x\,dx = \sin x\Big]_{-1}^1 = 2\sin(1) \approx 1.6829$

Error: $|1.6829 - 1.6758| = 0.0071$

**(b)** $\displaystyle\int_{-1}^1 \frac{1}{1 + x^2}\,dx \approx \frac{1}{1 + (1/3)} + \frac{1}{1 + (1/3)} = \frac{3}{2}$

(**Note:** exact answer is $\pi/2 \approx 1.5708$)

**(c)** Let $p(x) = ax^3 + bx^2 + cx + d$.

$$\int_{-1}^1 p(x)\,dx = \left[\frac{ax^4}{4} + \frac{bx^3}{3} + \frac{cx^2}{2} + dx\right]_{-1}^1 = \frac{2b}{3} + 2d$$

$$p\left(-\frac{1}{\sqrt{3}}\right) + p\left(\frac{1}{\sqrt{3}}\right) = \left(\frac{b}{3} + d\right) + \left(\frac{b}{3} + d\right) = \frac{2b}{3} + 2d$$

**7.** Let $d$ be the distance traversed and $a$ be the uniform acceleration. You can assume that $v(0) = 0$ and $s(0) = 0$. Then

$a(t) = a$

$v(t) = at$

$s(t) = \dfrac{1}{2}at^2$.

$s(t) = d$ when $t = \sqrt{\dfrac{2d}{a}}$.

The highest speed is $v = a\sqrt{\dfrac{2d}{a}} = \sqrt{2ad}$.

The lowest speed is $v = 0$.

The mean speed is $\dfrac{1}{2}\left(\sqrt{2ad} + 0\right) = \sqrt{\dfrac{ad}{2}}$.

The time necessary to traverse the distance $d$ at the mean speed is $t = \dfrac{d}{\sqrt{ad/2}} = \sqrt{\dfrac{2d}{a}}$

which is the same as the time calculated above.

**9.** Consider $F(x) = \left[f(x)\right]^2 \Rightarrow F'(x) = 2f(x)f'(x).$ So,

$$\int_a^b f(x)\, f'(x)\, dx = \int_a^b \tfrac{1}{2} F'(x)\, dx = \left[\tfrac{1}{2}F(x)\right]_a^b = \tfrac{1}{2}\left[F(b) - F(a)\right] = \tfrac{1}{2}\left[f(b)^2 - f(a)^2\right].$$

**11.** Consider $\displaystyle\int_0^1 x^5\, dx = \left.\dfrac{x^6}{6}\right]_0^1 = \dfrac{1}{6}.$

The corresponding Riemann Sum using right endpoints is

$$S(n) = \dfrac{1}{n}\left[\left(\dfrac{1}{n}\right)^5 + \left(\dfrac{2}{n}\right)^5 + \cdots + \left(\dfrac{n}{n}\right)^5\right] = \dfrac{1}{n^6}\left[1^5 + 2^5 + \cdots + n^5\right]. \text{ So, } \lim_{n\to\infty} S(n) = \lim_{n\to\infty}\dfrac{1^5 + 2^5 + \cdots + n^5}{n^6} = \dfrac{1}{6}.$$

**13. (a)**

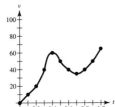

**(b)** $v$ is increasing (positive acceleration) on $(0, 0.4)$ and $(0.7, 1.0)$.

**(c)** Average acceleration $= \dfrac{v(0.4) - v(0)}{0.4 - 0} = \dfrac{60 - 0}{0.4} = 150 \text{ mi/h}^2$

**(d)** This integral is the total distance traveled in miles.

$$\int_0^1 v(t)\, dt \approx \tfrac{1}{10}\left[0 + 2(20) + 2(60) + 2(40) + 2(40) + 65\right] = \tfrac{385}{10} = 38.5 \text{ miles}$$

**(e)** One approximation is

$$a(0.8) \approx \dfrac{v(0.9) - v(0.8)}{0.9 - 0.8} = \dfrac{50 - 40}{0.1} = 100 \text{ mi/h}^2$$

(other answers possible)

**15. (a)** $(1 + i)^3 = 1 + 3i + 3i^2 + i^3 \Rightarrow (1 + i)^3 - i^3 = 3i^2 + 3i + 1$

**(b)** $3i^2 + 3i + 1 = (i + 1)^3 - i^3$

$$\sum_{i=1}^n \left(3i^2 + 3i + 1\right) = \sum_{i=1}^n \left[(i + 1)^3 - i^3\right] = \left(2^3 - 1^3\right) + \left(3^3 - 2^3\right) + \cdots + \left[\left((n + 1)^3 - n^3\right)\right] = (n + 1)^3 - 1$$

So, $(n + 1)^3 = \displaystyle\sum_{i=1}^n \left(3i^2 + 3i + 1\right) + 1.$

**(c)** $(n + 1)^3 - 1 = \displaystyle\sum_{i=1}^n \left(3i^2 + 3i + 1\right) = \sum_{i=1}^n 3i^2 + \dfrac{3(n)(n + 1)}{2} + n$

$$\Rightarrow \sum_{i=1}^n 3i^2 = n^3 + 3n^2 + 3n - \dfrac{3n(n + 1)}{2} - n$$

$$= \dfrac{2n^3 + 6n^2 + 6n - 3n^2 - 3n - 2n}{2}$$

$$= \dfrac{2n^3 + 3n^2 + n}{2}$$

$$= \dfrac{n(n + 1)(2n + 1)}{2}$$

$$\Rightarrow \sum_{i=1}^n i^2 = \dfrac{n(n + 1)(2n + 1)}{6}$$

**17.** Let $u = 1 + \sqrt{x}$, $\sqrt{x} = u - 1$, $x = u^2 - 2u + 1$, $dx = (2u - 2)\,du$.

$$\text{Area} = \int_1^4 \frac{1}{\sqrt{x} + x}\,dx = \int_2^3 \frac{2u - 2}{(u - 1) + (u^2 - 2u + 1)}\,du$$

$$= \int_2^3 \frac{2(u - 1)}{u^2 - u}\,du$$

$$= \int_2^3 \frac{2}{u}\,du = \left[2 \ln |u|\right]_2^3$$

$$= 2 \ln 3 - 2 \ln 2 = 2 \ln\left(\frac{3}{2}\right)$$

$$\approx 0.8109$$

**19.** (a) (i) $y = e^x$

$y_1 = 1 + x$

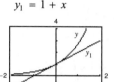

(ii) $y = e^x$

$y_2 = 1 + x + \left(\frac{x^2}{2}\right)$

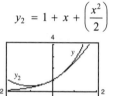

(iii) $y = e^x$

$y_3 = 1 + x + \frac{x^2}{2} + \frac{x^3}{6}$

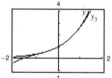

(b) $n^{\text{th}}$ term is $x^n/n!$ in polynomial: $y_4 = 1 + x + \frac{x^2}{2!} + \frac{x^3}{3!} + \frac{x^4}{4!}$

(c) Conjecture: $e^x = 1 + x + \frac{x^2}{2!} + \frac{x^3}{3!} + \cdots$

# C H A P T E R  6
## Differential Equations

# CHAPTER 6
# Differential Equations

## Section 6.1   Slope Fields and Euler's Method

**1.** Differential equation: $y' = 4y$

Solution: $y = Ce^{4x}$

Check: $y' = 4Ce^{4x} = 4y$

**3.** Differential equation: $y' = \dfrac{2xy}{x^2 - y^2}$

Solution: $x^2 + y^2 = Cy$

Check: $2x + 2yy' = Cy'$

$$y' = \frac{-2x}{(2y - C)}$$

$$y' = \frac{-2xy}{2y^2 - Cy}$$

$$= \frac{-2xy}{2y^2 - (x^2 + y^2)}$$

$$= \frac{-2xy}{y^2 - x^2}$$

$$= \frac{2xy}{x^2 - y^2}$$

**5.** Differential equation: $y'' + y = 0$

Solution:   $y = C_1 \sin x - C_2 \cos x$

$y' = C_1 \cos x + C_2 \sin x$

$y'' = -C_1 \sin x + C_2 \cos x$

Check: $y'' + y = \left(-C_1 \sin x + C_2 \cos x\right) + \left(C_1 \sin x - C_2 \cos x\right) = 0$

**7.** Differential equation: $y'' + y = \tan x$

Solution:   $y = -\cos x \ln|\sec x + \tan x|$

$y' = (-\cos x)\dfrac{1}{\sec x + \tan x}\left(\sec x \cdot \tan x + \sec^2 x\right) + \sin x \ln|\sec x + \tan x|$

$\phantom{y'} = \dfrac{(-\cos x)}{\sec x + \tan x}(\sec x)(\tan x + \sec x) + \sin x \ln|\sec x + \tan x|$

$\phantom{y'} = -1 + \sin x \ln|\sec x + \tan x|$

$y'' = (\sin x)\dfrac{1}{\sec x + \tan x}\left(\sec x \cdot \tan x + \sec^2 x\right) + \cos x \ln|\sec x + \tan x|$

$\phantom{y''} = (\sin x)(\sec x) + \cos x \ln|\sec x + \tan x|$

Check: $y'' + y = (\sin x)(\sec x) + \cos x \ln|\sec x + \tan x| - \cos x \ln|\sec x + \tan x| = \tan x$.

**9.**  $y = \sin x \cos x - \cos^2 x$

$\quad y' = -\sin^2 x + \cos^2 x + 2\cos x \sin x$

$\qquad = -1 + 2\cos^2 x + \sin 2x$

Differential equation: $2y + y' = 2\left(\sin x \cos x - \cos^2 x\right) + \left(-1 + 2\cos^2 x + \sin 2x\right)$

$\qquad\qquad\qquad\qquad\qquad = 2\sin x \cos x - 1 + \sin 2x$

$\qquad\qquad\qquad\qquad\qquad = 2\sin 2x - 1$

Initial condition $\left(\dfrac{\pi}{4}, 0\right)$:  $\sin\dfrac{\pi}{4}\cos\dfrac{\pi}{4} - \cos^2\dfrac{\pi}{4} = \dfrac{\sqrt{2}}{2}\cdot\dfrac{\sqrt{2}}{2} - \left(\dfrac{\sqrt{2}}{2}\right)^2 = 0$

**11.**  $y = 4e^{-6x^2}$

$\quad y' = 4e^{-6x^2}(-12\,x) = -48\,xe^{-6x^2}$

Differential equation: $y' = -12xy = -12x\left(4e^{-6x^2}\right) = -48xe^{-6x^2}$

Initial condition $(0, 4)$: $4e^0 = 4$

**In Exercises 13–19, the differential equation is $y^{(4)} - 16y = 0$.**

**13.**  $\qquad\qquad y = 3\cos x$

$\qquad\qquad\quad y^{(4)} = 3\cos x$

$\quad y^{(4)} - 16y = -45\cos x \neq 0,$

No

**15.**  $\qquad\qquad y = 3\cos 2x$

$\qquad\qquad\quad y^{(4)} = 48\cos 2x$

$\quad y^{(4)} - 16y = 48\cos 2x - 48\cos 2x = 0,$

Yes

**17.**  $\qquad\qquad y = e^{-2x}$

$\qquad\qquad\quad y^{(4)} = 16e^{-2x}$

$\quad y^{(4)} - 16y = 16e^{-2x} - 16e^{-2x} = 0,$

Yes

**19.**  $\qquad y = C_1e^{2x} + C_2e^{-2x} + C_3\sin 2x + C_4\cos 2x$

$\qquad\quad y^{(4)} = 16C_1e^{2x} + 16C_2e^{-2x} + 16C_3\sin 2x + 16C_4\cos 2x$

$\quad y^{(4)} - 16y = 0,$

Yes

**In Exercises 21–27, the differential equation is $xy' - 2y = x^3e^x$.**

**21.**  $y = x^2,\ y' = 2x$

$\quad xy' - 2y = x(2x) - 2\left(x^2\right) = 0 \neq x^3e^x,$

No

**23.**  $y = x^2e^x,\ y' = x^2e^x + 2xe^x = e^x\left(x^2 + 2x\right)$

$\quad xy' - 2y = x\left(e^x\left(x^2 + 2x\right)\right) - 2\left(x^2e^x\right) = x^3e^x,$

Yes

**25.**  $y = \sin x,\ y' = \cos x$

$\quad xy' - 2y = x(\cos x) - 2(\sin x) \neq x^3e^x,$

No

**27.**  $y = \ln x,\ y' = \dfrac{1}{x}$

$\quad xy' - 2y = x\left(\dfrac{1}{x}\right) - 2\ln x \neq x^3e^x,$

No

**29.** $y = Ce^{-x/2}$ passes through $(0, 3)$.

$3 = Ce^0 = C \Rightarrow C = 3$

Particular solution: $y = 3e^{-x/2}$

**31.** $y^2 = Cx^3$ passes through $(4, 4)$.

$16 = C(64) \Rightarrow C = \frac{1}{4}$

Particular solution: $y^2 = \frac{1}{4}x^3$ or $4y^2 = x^3$

**33.** Differential equation: $4yy' - x = 0$

General solution: $4y^2 - x^2 = C$

Particular solutions: $C = 0$, Two intersecting lines

$\qquad\qquad\qquad C = \pm 1, C = \pm 4$, Hyperbolas

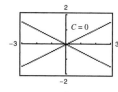 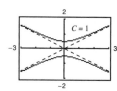 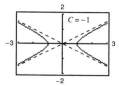

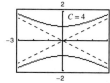 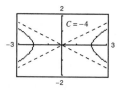

**35.** Differential equation: $y' + 2y = 0$

General solution: $y = Ce^{-2x}$

$y' + 2y = C(-2)e^{-2x} + 2(Ce^{-2x}) = 0$

Initial condition $(0, 3)$: $3 = Ce^0 = C$

Particular solution: $y = 3e^{-2x}$

**37.** Differential equation: $y'' + 9y = 0$

General solution: $y = C_1 \sin 3x + C_2 \cos 3x$

$\qquad y' = 3C_1 \cos 3x - 3C_2 \sin 3x,$

$\qquad y'' = -9C_1 \sin 3x - 9C_2 \cos 3x$

$y'' + 9y = (-9C_1 \sin 3x - 9C_2 \cos 3x) + 9(C_1 \sin 3x + C_2 \cos 3x) = 0$

Initial conditions $\left(\frac{\pi}{6}, 2\right)$ and $y' = 1$ when $x = \frac{\pi}{6}$:

$2 = C_1 \sin\left(\frac{\pi}{2}\right) + C_2 \cos\left(\frac{\pi}{2}\right) \Rightarrow C_1 = 2$

$y' = 3C_1 \cos 3x - 3C_2 \sin 3x$

$1 = 3C_1 \cos\left(\frac{\pi}{2}\right) - 3C_2 \sin\left(\frac{\pi}{2}\right) = -3C_2 \Rightarrow C_2 = -\frac{1}{3}$

Particular solution: $y = 2 \sin 3x - \frac{1}{3} \cos 3x$

**39.** Differential equation: $x^2y'' - 3xy' + 3y = 0$

General solution: $y = C_1x + C_2x^3$

$y' = C_1 + 3C_2x^2$, $y'' = 6C_2x$

$x^2y'' - 3xy' + 3y = x^2(6C_2x) - 3x(C_1 + 3C_2x^2) + 3(C_1x + C_2x^3) = 0$

Initial conditions $(2, 0)$ and $y' = 4$ when $x = 2$:

$0 = 2C_1 + 8C_2$

$y' = C_1 + 3C_2x^2$

$4 = C_1 + 12C_2$

$\left.\begin{array}{l} C_1 + 4C_2 = 0 \\ C_1 + 12C_2 = 4 \end{array}\right\}$ $C_2 = \frac{1}{2}, C_1 = -2$

Particular solution: $y = -2x + \frac{1}{2}x^3$

**41.** $\dfrac{dy}{dx} = 6x^2$

$y = \displaystyle\int 6x^2\,dx = 2x^3 + C$

**43.** $\dfrac{dy}{dx} = \dfrac{x}{1 + x^2}$

$y = \displaystyle\int \dfrac{x}{1 + x^2}\,dx = \dfrac{1}{2}\ln(1 + x^2) + C$

$\left(u = 1 + x^2,\, du = 2x\,dx\right)$

**45.** $\dfrac{dy}{dx} = \dfrac{x - 2}{x} = 1 - \dfrac{2}{x}$

$y = \displaystyle\int \left(1 - \dfrac{2}{x}\right) dx$

$= x - 2\ln|x| + C = x - \ln x^2 + C$

**47.** $\dfrac{dy}{dx} = \sin 2x$

$y = \displaystyle\int \sin 2x\,dx = -\dfrac{1}{2}\cos 2x + C$

$(u = 2x,\, du = 2\,dx)$

**49.** $\dfrac{dy}{dx} = x\sqrt{x - 6}$

Let $u = \sqrt{x - 6}$, then $x = u^2 + 6$ and $dx = 2u\,du$.

$y = \displaystyle\int x\sqrt{x - 6}\,dx = \displaystyle\int (u^2 + 6)(u)(2u)\,du$

$= 2\displaystyle\int (u^4 + 6u^2)\,du$

$= 2\left(\dfrac{u^5}{5} + 2u^3\right) + C$

$= \dfrac{2}{5}(x - 6)^{5/2} + 4(x - 6)^{3/2} + C$

$= \dfrac{2}{5}(x - 6)^{3/2}(x - 6 + 10) + C$

$= \dfrac{2}{5}(x - 6)^{3/2}(x + 4) + C$

**51.** $\dfrac{dy}{dx} = xe^{x^2}$

$y = \displaystyle\int xe^{x^2}\,dx = \dfrac{1}{2}e^{x^2} + C$

$(u = x^2,\, du = 2x\,dx)$

**53.**

| $x$ | $-4$ | $-2$ | $0$ | $2$ | $4$ | $8$ |
|---|---|---|---|---|---|---|
| $y$ | $2$ | $0$ | $4$ | $4$ | $6$ | $8$ |
| $dy/dx$ | $-4$ | Undef. | $0$ | $1$ | $\frac{4}{3}$ | $2$ |

**55.**

| $x$ | $-4$ | $-2$ | $0$ | $2$ | $4$ | $8$ |
|---|---|---|---|---|---|---|
| $y$ | $2$ | $0$ | $4$ | $4$ | $6$ | $8$ |
| $dy/dx$ | $-2\sqrt{2}$ | $-2$ | $0$ | $0$ | $-2\sqrt{2}$ | $-8$ |

**57.** $\dfrac{dy}{dx} = \sin 2x$

For $x = 0$, $\dfrac{dy}{dx} = 0$. Matches (b).

**59.** $\dfrac{dy}{dx} = e^{-2x}$

As $x \to \infty$, $\dfrac{dy}{dx} \to 0$. Matches (d).

**61.** (a), (b)

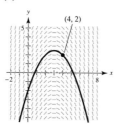

(c) As $x \to \infty$, $y \to -\infty$

As $x \to -\infty$, $y \to -\infty$

**63.** (a), (b)

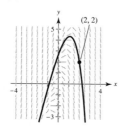

(c) As $x \to \infty$, $y \to -\infty$

As $x \to -\infty$, $y \to -\infty$

**65.** (a) $y' = \dfrac{1}{x}$, $(1, 0)$

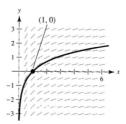

As $x \to \infty$, $y \to \infty$

[Note: The solution is $y = \ln x$.]

(b) $y' = \dfrac{1}{x}$, $(2, -1)$

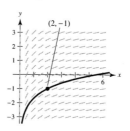

As $x \to \infty$, $y \to \infty$

**67.** $\dfrac{dy}{dx} = 0.25y$, $y(0) = 4$

(a), (b)

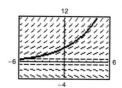

**69.** $\dfrac{dy}{dx} = 0.02y(10 - y)$, $y(0) = 2$

(a), (b)

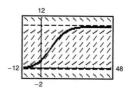

**71.** $\dfrac{dy}{dx} = 0.4y(3 - x)$, $y(0) = 1$

(a), (b)

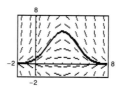

**73.** $y' = x + y$,  $y(0) = 2$,  $n = 10$,  $h = 0.1$

$y_1 = y_0 + hF(x_0, y_0) = 2 + (0.1)(0 + 2) = 2.2$

$y_2 = y_1 + hF(x_1, y_1) = 2.2 + (0.1)(0.1 + 2.2) = 2.43$, etc.

| $n$ | 0 | 1 | 2 | 3 | 4 | 5 | 6 | 7 | 8 | 9 | 10 |
|---|---|---|---|---|---|---|---|---|---|---|---|
| $x_n$ | 0 | 0.1 | 0.2 | 0.3 | 0.4 | 0.5 | 0.6 | 0.7 | 0.8 | 0.9 | 1.0 |
| $y_n$ | 2 | 2.2 | 2.43 | 2.693 | 2.992 | 3.332 | 3.715 | 4.146 | 4.631 | 5.174 | 5.781 |

**75.** $y' = 3x - 2y$,  $y(0) = 3$,  $n = 10$,  $h = 0.05$

$y_1 = y_0 + hF(x_0, y_0) = 3 + (0.05)(3(0) - 2(3)) = 2.7$

$y_2 = y_1 + hF(x_1, y_1) = 2.7 + (0.05)(3(0.05) + 2(2.7)) = 2.4375$, etc.

| $n$ | 0 | 1 | 2 | 3 | 4 | 5 | 6 | 7 | 8 | 9 | 10 |
|---|---|---|---|---|---|---|---|---|---|---|---|
| $x_n$ | 0 | 0.05 | 0.1 | 0.15 | 0.2 | 0.25 | 0.3 | 0.35 | 0.4 | 0.45 | 0.5 |
| $y_n$ | 3 | 2.7 | 2.438 | 2.209 | 2.010 | 1.839 | 1.693 | 1.569 | 1.464 | 1.378 | 1.308 |

**77.** $y' = e^{xy}$,  $y(0) = 1$,  $n = 10$,  $h = 0.1$

$y_1 = y_0 + hF(x_0, y_0) = 1 + (0.1)e^{0(1)} = 1.1$

$y_2 = y_1 + hF(x_1, y_1) = 1.1 + (0.1)e^{(0.1)(1.1)} \approx 1.2116$, etc.

| $n$ | 0 | 1 | 2 | 3 | 4 | 5 | 6 | 7 | 8 | 9 | 10 |
|---|---|---|---|---|---|---|---|---|---|---|---|
| $x_n$ | 0 | 0.1 | 0.2 | 0.3 | 0.4 | 0.5 | 0.6 | 0.7 | 0.8 | 0.9 | 1.0 |
| $y_n$ | 1 | 1.1 | 1.212 | 1.339 | 1.488 | 1.670 | 1.900 | 2.213 | 2.684 | 3.540 | 5.958 |

**79.** $\dfrac{dy}{dx} = y$, $y = 3e^x$, $(0, 3)$

| $x$ | 0 | 0.2 | 0.4 | 0.6 | 0.8 | 1 |
|---|---|---|---|---|---|---|
| $y(x)$ (exact) | 3 | 3.6642 | 4.4755 | 5.4664 | 6.6766 | 8.1548 |
| $y(x)$ $(h = 0.2)$ | 3 | 3.6000 | 4.3200 | 5.1840 | 6.2208 | 7.4650 |
| $y(x)$ $(h = 0.1)$ | 3 | 3.6300 | 4.3923 | 5.3147 | 6.4308 | 7.7812 |

**81.** $\dfrac{dy}{dx} = y + \cos x$, $y = \dfrac{1}{2}(\sin x - \cos x + e^x)$,  $(0, 0)$

| $x$ | 0 | 0.2 | 0.4 | 0.6 | 0.8 | 1 |
|---|---|---|---|---|---|---|
| $y(x)$ (exact) | 0 | 0.2200 | 0.4801 | 0.7807 | 1.1231 | 1.5097 |
| $y(x)$ $(h = 0.2)$ | 0 | 0.2000 | 0.4360 | 0.7074 | 1.0140 | 1.3561 |
| $y(x)$ $(h = 0.1)$ | 0 | 0.2095 | 0.4568 | 0.7418 | 1.0649 | 1.4273 |

**83.** $\dfrac{dy}{dt} = -\dfrac{1}{2}(y - 72)$, $(0, 140)$, $h = 0.1$

(a)

| $t$ | 0 | 1 | 2 | 3 |
|---|---|---|---|---|
| Euler | 140 | 112.7 | 96.4 | 86.6 |

(b) $y = 72 + 68e^{-t/2}$   exact

| $t$ | 0 | 1 | 2 | 3 |
|---|---|---|---|---|
| Exact | 140 | 113.24 | 97.016 | 87.173 |

(c) $\dfrac{dy}{dt} = -\dfrac{1}{2}(y - 72)$, $(0, 140)$, $h = 0.05$

| $t$ | 0 | 1 | 2 | 3 |
|---|---|---|---|---|
| Euler | 140 | 112.98 | 96.7 | 86.9 |

The approximations are better using $h = 0.05$.

**85.** The general solution is a family of curves that satisfies the differential equation. A particular solution is one member of the family that satisfies given conditions.

**87.** Consider $y' = F(x, y)$, $y(x_0) = y_0$. Begin with a point $(x_0, y_0)$ that satisfies the initial condition, $y(x_0) = y_0$. Then, using a step size of $h$, find the point $(x_1, y_1) = (x_0 + h, y_0 + hF(x_0, y_0))$. Continue generating the sequence of points $(x_{n+1}, y_{n+1}) = (x_n + h, y_n + hF(x_n, y_n))$.

**89.** False. Consider Example 2. $y = x^3$ is a solution to $xy' - 3y = 0$, but $y = x^3 + 1$ is not a solution.

**91.** True

**93.** $\dfrac{dy}{dx} = -2y$, $y(0) = 4$, $y = 4e^{-2x}$

(a)

| $x$ | 0 | 0.2 | 0.4 | 0.6 | 0.8 | 1 |
|---|---|---|---|---|---|---|
| $y$ | 4 | 2.6813 | 1.7973 | 1.2048 | 0.8076 | 0.5413 |
| $y_1$ | 4 | 2.5600 | 1.6384 | 1.0486 | 0.6711 | 0.4295 |
| $y_2$ | 4 | 2.4000 | 1.4400 | 0.8640 | 0.5184 | 0.3110 |
| $e_1$ | 0 | 0.1213 | 0.1589 | 0.1562 | 0.1365 | 0.1118 |
| $e_2$ | 0 | 0.2813 | 0.3573 | 0.3408 | 0.2892 | 0.2303 |
| $r$ |  | 0.4312 | 0.4447 | 0.4583 | 0.4720 | 0.4855 |

(b) If $h$ is halved, then the error is approximately halved $(r \approx 0.5)$.

(c) When $h = 0.05$, the errors will again be approximately halved.

**95.** (a) $L\dfrac{dI}{dt} + RI = E(t)$

$4\dfrac{dI}{dt} + 12I = 24$

$\dfrac{dI}{dt} = \dfrac{1}{4}(24 - 12I) = 6 - 3I$

(b) As $t \to \infty$, $I \to 2$. That is, $\displaystyle\lim_{t \to \infty} I(t) = 2$. In fact, $I = 2$ is a solution to the differential equation.

**97.** $y = A \sin \omega t$

$y' = A\omega \cos \omega t$

$y'' = -A\omega^2 \sin \omega t$

$y'' + 16y = 0$

$-A\omega^2 \sin \omega t + 16A \sin \omega t = 0$

$A \sin \omega t \left[16 - \omega^2\right] = 0$

If $A \neq 0$, then $\omega = \pm 4$

**99.** Let the vertical line $x = k$ cut the graph of the solution $y = f(x)$ at $(k, t)$. The tangent line at $(k, t)$ is

$$y - t = f'(k)(x - k)$$

Because $y' + p(x)y = q(x)$, you have

$$y - t = [q(k) - p(k)t](x - k)$$

For any value of $t$, this line passes through the point $\left(k + \dfrac{1}{p(k)}, \dfrac{q(k)}{p(k)}\right)$.

To see this, note that

$$\frac{q(k)}{p(k)} - t \overset{?}{=} [q(k) - p(k)t]\left(k + \frac{1}{p(k)} - k\right)$$

$$\overset{?}{=} q(k)k - p(k)tk + \frac{q(k)}{p(k)} - t - kq(k) + p(k)kt = \frac{q(k)}{p(k)} - t.$$

# Section 6.2   Differential Equations: Growth and Decay

**1.** $\dfrac{dy}{dx} = x + 3$

$$y = \int (x + 3)\, dx = \frac{x^2}{2} + 3x + C$$

**3.** $\dfrac{dy}{dx} = y + 3$

$$\frac{dy}{y + 3} = dx$$

$$\int \frac{1}{y + 3}\, dy = \int dx$$

$$\ln|y + 3| = x + C_1$$

$$y + 3 = e^{x + C_1} = Ce^x$$

$$y = Ce^x - 3$$

**5.** $y' = \dfrac{5x}{y}$

$$yy' = 5x$$

$$\int yy'\, dx = \int 5x\, dx$$

$$\int y\, dy = \int 5x\, dx$$

$$\frac{1}{2}y^2 = \frac{5}{2}x^2 + C_1$$

$$y^2 - 5x^2 = C$$

**7.** $y' = \sqrt{x}\,y$

$$\frac{y'}{y} = \sqrt{x}$$

$$\int \frac{y'}{y}\, dx = \int \sqrt{x}\, dx$$

$$\int \frac{dy}{y} = \int \sqrt{x}\, dx$$

$$\ln|y| = \frac{2}{3}x^{3/2} + C_1$$

$$y = e^{(2/3)x^{3/2} + C_1}$$

$$= e^{C_1}e^{(2/3)x^{3/2}}$$

$$= Ce^{\left(2x^{3/2}\right)/3}$$

**9.** $(1 + x^2)y' - 2xy = 0$

$$y' = \frac{2xy}{1 + x^2}$$

$$\frac{y'}{y} = \frac{2x}{1 + x^2}$$

$$\int \frac{y'}{y}\, dx = \int \frac{2x}{1 + x^2}\, dx$$

$$\int \frac{dy}{y} = \int \frac{2x}{1 + x^2}\, dx$$

$$\ln|y| = \ln(1 + x^2) + C_1$$

$$\ln|y| = \ln(1 + x^2) + \ln C$$

$$\ln|y| = \ln[C(1 + x^2)]$$

$$y = C(1 + x^2)$$

**11.** $\dfrac{dQ}{dt} = \dfrac{k}{t^2}$

$\displaystyle\int \dfrac{dQ}{dt}\,dt = \int \dfrac{k}{t^2}\,dt$

$\displaystyle\int dQ = -\dfrac{k}{t} + C$

$Q = -\dfrac{k}{t} + C$

**13. (a)**

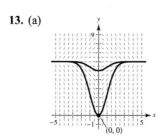

(0, 0)

**(b)** $\dfrac{dy}{dx} = x(6 - y),\quad (0, 0)$

$\dfrac{dy}{y - 6} = -x\,dx$

$\ln|y - 6| = \dfrac{-x^2}{2} + C$

$y - 6 = e^{-x^2/2 + C} = C_1 e^{-x^2/2}$

$y = 6 + C_1 e^{-x^2/2}$

$(0, 0):\ 0 = 6 + C_1 \Rightarrow C_1 = -6$

$y = 6 - 6e^{-x^2/2}$

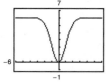

**15.** $\dfrac{dy}{dt} = \dfrac{1}{2}t,\quad (0, 10)$

$\displaystyle\int dy = \int \dfrac{1}{2}t\,dt$

$y = \dfrac{1}{4}t^2 + C$

$10 = \dfrac{1}{4}(0)^2 + C \Rightarrow C = 10$

$y = \dfrac{1}{4}t^2 + 10$

**17.** $\dfrac{dy}{dt} = -\dfrac{1}{2}y,\quad (0, 10)$

$\displaystyle\int \dfrac{dy}{y} = \int -\dfrac{1}{2}\,dt$

$\ln|y| = -\dfrac{1}{2}t + C_1$

$y = e^{-(t/2) + C_1} = e^{C_1}e^{-t/2} = Ce^{-t/2}$

$10 = Ce^0 \Rightarrow C = 10$

$y = 10e^{-t/2}$

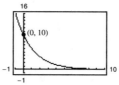

**19.** $\dfrac{dN}{dt} = kN$

$N = Ce^{kt}$   (Theorem 6.1)

$(0, 250):\ C = 250$

$(1, 400):\ 400 = 250e^k \Rightarrow k = \ln \dfrac{400}{250} = \ln \dfrac{8}{5}$

$N = 250e^{\ln(8/5)t} \approx 250e^{0.4700t}$

When $t = 4$, $N = 250e^{4\ln(8/5)} = 250e^{\ln(8/5)^4}$

$= 250\left(\dfrac{8}{5}\right)^4 = \dfrac{8192}{5}.$

**21.** $y = Ce^{kt},\quad \left(0, \dfrac{1}{2}\right), (5, 5)$

$C = \dfrac{1}{2}$

$y = \dfrac{1}{2}e^{kt}$

$5 = \dfrac{1}{2}e^{5k}$

$k = \dfrac{\ln 10}{5}$

$y = \dfrac{1}{2}e^{[(\ln 10)/5]t} = \dfrac{1}{2}\left(10^{t/5}\right)$ or $y \approx \dfrac{1}{2}e^{0.4605t}$

**23.** $y = Ce^{kt}$,   $(1, 5), (5, 2)$

$5 = Ce^{k} \Rightarrow 10 = 2Ce^{k}$

$2 = Ce^{5k} \Rightarrow 10 = 5Ce^{k}$

$2Ce^{k} = 5Ce^{5k}$

$2e^{k} = 5e^{5k}$

$\dfrac{2}{5} = e^{4k}$

$k = \dfrac{1}{4}\ln\left(\dfrac{2}{5}\right) = \ln\left(\dfrac{2}{5}\right)^{1/4}$

$C = 5e^{-k} = 5e^{-1/4\ln(2/5)} = 5\left(\dfrac{2}{5}\right)^{-1/4} = 5\left(\dfrac{5}{2}\right)^{1/4}$

$y = 5\left(\dfrac{5}{2}\right)^{1/4} e^{[1/4\ln(2/5)]t} \approx 6.2872\, e^{-0.2291t}$

**25.** In the model $y = Ce^{kt}$, $C$ represents the initial value of $y$ (when $t = 0$). $k$ is the proportionality constant.

**27.** $\dfrac{dy}{dx} = \dfrac{1}{2}xy$

$\dfrac{dy}{dx} > 0$ when $xy > 0$. Quadrants I and III.

**29.** Because the initial quantity is 20 grams,

$y = 20e^{kt}$.

Because the half-life is 1599 years,

$10 = 20e^{k(1599)}$

$k = \dfrac{1}{1599}\ln\left(\dfrac{1}{2}\right)$.

So, $y = 20e^{[\ln(1/2)/1599]t}$.

When $t = 1000$, $y = 20e^{[\ln(1/2)/1599](1000)} \approx 12.96g$.

When $t = 10,000$, $y \approx 0.26g$.

**31.** Because the half-life is 1599 years,

$\dfrac{1}{2} = 1e^{k(1599)}$

$k = \dfrac{1}{1599}\ln\left(\dfrac{1}{2}\right)$.

Because there are 0.1 gram after 10,000 years,

$0.1 = Ce^{[\ln(1/2)/1599](10,000)}$

$C \approx 7.63$.

So, the initial quantity is approximately 7.63 g.

When $t = 1000$, $y = 7.63e^{[\ln(1/2)/1599](1000)}$

$\approx 4.95$ g.

**33.** Because the initial quantity is 5 grams, $C = 5$.

Because the half-life is 5715 years,

$2.5 = 5e^{k(5715)}$

$k = \dfrac{1}{5715}\ln\left(\dfrac{1}{2}\right)$.

When $t = 1000$ years, $y = 5e^{[\ln(1/2)/5715](1000)} \approx 4.43$ g.

When $t = 10,000$ years, $y = 5e^{[\ln(1/2)/5715](10,000)}$

$\approx 1.49$ g.

**35.** Because the half-life is 24,100 years,

$\dfrac{1}{2} = 1e^{k(24,100)}$

$k = \dfrac{1}{24,100}\ln\left(\dfrac{1}{2}\right)$.

Because there are 2.1 grams after 1000 years,

$2.1 = Ce^{[\ln(1/2)/24,100](1000)}$

$C \approx 2.161$.

So, the initial quantity is approximately 2.161 g.

When $t = 10,000$, $y = 2.161e^{[\ln(1/2)/24,100](10,000)}$

$\approx 1.62$ g.

**37.** $y = Ce^{kt}$

$\dfrac{1}{2}C = Ce^{k(1599)}$

$k = \dfrac{1}{1599}\ln\left(\dfrac{1}{2}\right)$

When $t = 100$, $y = Ce^{[\ln(1/2)/1599](100)}$

$\approx 0.9576\, C$

Therefore, 95.76% remains after 100 years.

**39.** Because $A = 4000e^{0.06t}$, the time to double is given by

$8000 = 4000e^{0.06t}$

$2 = e^{0.06t}$

$\ln 2 = 0.06t$

$t = \dfrac{\ln 2}{0.06} \approx 11.55$ years.

Amount after 10 years: $A = 4000e^{(0.06)(10)} \approx \$7288.48$

**41.** Because $A = 750e^{rt}$ and $A = 1500$ when $t = 7.75$, you have the following.

$1500 = 750e^{7.75r}$

$2 = e^{7.75r}$

$\ln 2 = 7.75r$

$r = \dfrac{\ln 2}{7.75} \approx 0.0894 = 8.94\%$

Amount after 10 years: $A = 750e^{0.0894(10)} \approx \$1833.67$

**43.** Because $A = 500e^{rt}$ and $A = 1292.85$ when $t = 10$, you have the following.

$$1292.85 = 500e^{10r}$$
$$2.5857 = e^{10r}$$
$$\ln(2.5857) = 10r$$
$$r = \frac{\ln(2.5857)}{10} \approx 0.0950 = 9.50\%$$

The time to double is given by

$$1000 = 500e^{0.0950t}$$
$$2 = e^{0.0950t}$$
$$\ln 2 = 0.0950t$$
$$t = \frac{\ln 2}{0.095} \approx 7.30 \text{ years.}$$

**45.** $1{,}000{,}000 = P\left(1 + \dfrac{0.075}{12}\right)^{(12)(20)}$

$$P = 1{,}000{,}000\left(1 + \frac{0.075}{12}\right)^{-240}$$
$$\approx \$224{,}174.18$$

**47.** $1{,}000{,}000 = P\left(1 + \dfrac{0.08}{12}\right)^{(12)(35)}$

$$P = 1{,}000{,}000\left(1 + \frac{0.08}{12}\right)^{-420}$$
$$= \$61{,}377.75$$

**49.** (a) $2000 = 1000(1 + 0.07)^t$

$$2 = 1.07^t$$
$$\ln 2 = t \ln 1.07$$
$$t = \frac{\ln 2}{\ln 1.07} \approx 10.24 \text{ years}$$

(b) $2000 = 1000\left(1 + \dfrac{0.07}{12}\right)^{12t}$

$$2 = \left(1 + \frac{0.007}{12}\right)^{12t}$$
$$\ln 2 = 12t \ln\left(1 + \frac{0.07}{12}\right)$$
$$t = \frac{\ln 2}{12 \ln\left(1 + (0.07/12)\right)} \approx 9.93 \text{ years}$$

(c) $2000 = 1000\left(1 + \dfrac{0.07}{365}\right)^{365t}$

$$2 = \left(1 + \frac{0.07}{365}\right)^{365t}$$
$$\ln 2 = 365t \ln\left(1 + \frac{0.07}{365}\right)$$
$$t = \frac{\ln 2}{365 \ln\left(1 + (0.07/365)\right)} \approx 9.90 \text{ years}$$

(d) $2000 = 1000e^{(0.07)t}$

$$2 = e^{0.07t}$$
$$\ln 2 = 0.07t$$
$$t = \frac{\ln 2}{0.07} \approx 9.90 \text{ years}$$

**51.** (a) $P = Ce^{kt} = Ce^{-0.006t}$

$$P(1) = 2.2 = Ce^{-0.006(1)} \Rightarrow C \approx 2.21$$
$$P = 2.21e^{-0.006t}$$

(b) For 2020, $t = 10$ and
$$P = 2.21e^{-0.006(10)} \approx 2.08 \text{ million.}$$

(c) Because $k < 0$, the population is decreasing.

**53.** (a) $P = Ce^{kt} = Ce^{0.036t}$

$$P(1) = 34.6 = Ce^{0.036(1)} \Rightarrow C \approx 33.38$$
$$P = 33.38e^{0.036t}$$

(b) For 2020, $t = 10$ and
$$P = 33.38e^{0.036(10)} \approx 47.84 \text{ million.}$$

(c) Because $k > 0$, the population is increasing.

**55.** (a) $N = 100.1596(1.2455)^t$

(b) $N = 400$ when $t = 6.3$ hours (graphing utility)

Analytically,

$$400 = 100.1596(1.2455)^t$$
$$1.2455^t = \frac{400}{100.1596} = 3.9936$$
$$t \ln 1.2455 = \ln 3.9936$$
$$t = \frac{\ln 3.9936}{\ln 1.2455} \approx 6.3 \text{ hours}$$

**57.** (a)   $P_1 = Ce^{kt} = 181e^{kt}$

$205 = 181e^{10k} \Rightarrow k = \frac{1}{10} \ln\left(\frac{205}{181}\right) \approx 0.01245$

$P_1 \approx 181e^{0.01245t} \approx 181(1.01253)^t$

(b) Using a graphing utility, $P_2 \approx 182.3248(1.01091)^t$

(c)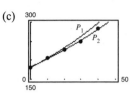

The model $P_2$ fits the data better.

(d) Using the model $P_2$,

$320 = 182.3248(1.01091)^t$

$\frac{320}{182.3248} = (1.01091)^t$

$t = \frac{\ln(320/182.3248)}{\ln(1.01091)}$

$\approx 51.8$ years, or 2011.

**59.** (a) Because the population increases by a constant each month, the rate of change from month to month will always be the same. So, the slope is constant, and the model is linear.

(b) Although the percentage increase is constant each month, the rate of growth is not constant. The rate of change of $y$ is given by

$\frac{dy}{dt} = ry$

which is an exponential model.

**61.** (a)   $P_1 = Ce^{kt} = 106e^{kt} \ (t = 0 \leftrightarrow 1920)$

$123 = 106e^{k(10)} \Rightarrow \frac{123}{106} = e^{10k}$

$\Rightarrow k = \frac{1}{10} \ln\left(\frac{123}{106}\right) \approx 0.01487$

$P_1 = 106e^{0.01487t} = 106e^{\frac{1}{10}\ln\left(\frac{123}{106}\right)t} = 106(1.01499)^t$

(b) Using a graphing utility, $P_2 \approx 107.2727(1.01215)^t$.

(c)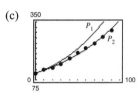

The model $P_2$ fits the data better.

(d)   $P_2 = 400 = 107.2727(1.01215)^t$

$\frac{400}{107.2727} = (1.01215)^t$

$t = \frac{\ln(400/107.2727)}{\ln(1.01215)}$

$\approx 109$, or 2029.

**63.** $\beta(I) = 10 \log_{10}\frac{I}{I_0}, \ I_0 = 10^{-16}$

(a) $\beta(10^{-14}) = 10 \log_{10}\frac{10^{-14}}{10^{-16}} = 20$ decibels

(b) $\beta(10^{-9}) = 10 \log_{10}\frac{10^{-9}}{10^{-16}} = 70$ decibels

(c) $\beta(10^{-6.5}) = 10 \log_{10}\frac{10^{-6.5}}{10^{-16}} = 95$ decibels

(d) $\beta(10^{-4}) = 10 \log_{10}\frac{10^{-4}}{10^{-16}} = 120$ decibels

**65.** Because $\frac{dy}{dt} = k(y - 80)$

$\int \frac{1}{y - 80}\,dy = \int k\,dt$

$\ln(y - 80) = kt + C.$

When $t = 0$, $y = 1500$. So, $C = \ln 1420$.

When $t = 1$, $y = 1120$. So,

$k(1) + \ln 1420 = \ln(1120 - 80)$

$k = \ln 1040 - \ln 1420 = \ln\frac{104}{142}.$

So, $y = 1420e^{\left[\ln(104/142)\right]t} + 80.$

When $t = 5$, $y \approx 379.2°F.$

**67.** False. If $y = Ce^{kt}$, $y' = Cke^{kt} \neq$ constant.

**69.** False. The prices are rising at a rate of 6.2% per year.

## Section 6.3   Differential Equations: Separation of Variables

**1.**   $\dfrac{dy}{dx} = \dfrac{x}{y}$

$\displaystyle\int y\,dy = \int x\,dx$

$\dfrac{y^2}{2} = \dfrac{x^2}{2} + C_1$

$y^2 - x^2 = C$

**3.**   $x^2 + 5y\dfrac{dy}{dx} = 0$

$5y\dfrac{dy}{dx} = -x^2$

$\displaystyle\int 5y\,dy = \int -x^2\,dx$

$\dfrac{5y^2}{2} = \dfrac{-x^3}{3} + C_1$

$15y^2 + 2x^3 = C$

**5.**   $\dfrac{dr}{ds} = 0.75\,r$

$\displaystyle\int \dfrac{dr}{r} = \int 0.75\,ds$

$\ln|r| = 0.75\,s + C_1$

$r = e^{0.75\,s + C_1}$

$r = Ce^{0.75\,s}$

**7.**   $(2 + x)y' = 3y$

$\displaystyle\int \dfrac{dy}{y} = \int \dfrac{3}{2 + x}\,dx$

$\ln|y| = 3\ln|2 + x| + \ln C = \ln\left|C(2 + x)^3\right|$

$y = C(x + 2)^3$

**9.**   $yy' = 4\sin x$

$y\dfrac{dy}{dx} = 4\sin x$

$\displaystyle\int y\,dy = \int 4\sin x\,dx$

$\dfrac{y^2}{2} = -4\cos x + C_1$

$y^2 = C - 8\cos x$

**11.**   $\sqrt{1 - 4x^2}\,y' = x$

$dy = \dfrac{x}{\sqrt{1 - 4x^2}}\,dx$

$\displaystyle\int dy = \int \dfrac{x}{\sqrt{1 - 4x^2}}\,dx$

$\qquad = -\dfrac{1}{8}\int \left(1 - 4x^2\right)^{-1/2}\left(-8x\,dx\right)$

$y = -\dfrac{1}{4}\sqrt{1 - 4x^2} + C$

**13.**   $y\ln x - xy' = 0$

$\displaystyle\int \dfrac{dy}{y} = \int \dfrac{\ln x}{x}\,dx \quad \left(u = \ln x,\ du = \dfrac{dx}{x}\right)$

$\ln|y| = \dfrac{1}{2}(\ln x)^2 + C_1$

$y = e^{(1/2)(\ln x)^2 + C_1} = Ce^{(\ln x)^2/2}$

**15.**   $yy' - 2e^x = 0$

$y\dfrac{dy}{dx} = 2e^x$

$\displaystyle\int y\,dy = \int 2e^x\,dx$

$\dfrac{y^2}{2} = 2e^x + C$

Initial condition $(0, 3)$:   $\dfrac{9}{2} = 2 + C \Rightarrow C = \dfrac{5}{2}$

Particular solution:   $\dfrac{y^2}{2} = 2e^x + \dfrac{5}{2}$

$\qquad\qquad y^2 = 4e^x + 5$

**17.**   $y(x + 1) + y' = 0$

$\displaystyle\int \dfrac{dy}{y} = -\int (x + 1)\,dx$

$\ln|y| = -\dfrac{(x + 1)^2}{2} + C_1$

$y = Ce^{-(x+1)^2/2}$

Initial condition $(-2, 1)$:   $1 = Ce^{-1/2}$, $C = e^{1/2}$

Particular solution:   $y = e^{\left[1 - (x+1)^2\right]/2} = e^{-\left(x^2 + 2x\right)/2}$

**19.** $y(1 + x^2)y' = x(1 + y^2)$

$$\frac{y}{1 + y^2}\, dy = \frac{x}{1 + x^2}\, dx$$

$$\frac{1}{2}\ln(1 + y^2) = \frac{1}{2}\ln(1 + x^2) + C_1$$

$$\ln(1 + y^2) = \ln(1 + x^2) + \ln C = \ln\left[C(1 + x^2)\right]$$

$$1 + y^2 = C(1 + x^2)$$

Initial condition $(0, \sqrt{3})$:   $1 + 3 = C \Rightarrow C = 4$

Particular solution:  $1 + y^2 = 4(1 + x^2)$

$$y^2 = 3 + 4x^2$$

**21.**   $\dfrac{du}{dv} = uv \sin v^2$

$$\int \frac{du}{u} = \int v \sin v^2\, dv$$

$$\ln|u| = -\frac{1}{2}\cos v^2 + C_1$$

$$u = Ce^{-(\cos v^2)/2}$$

Initial condition:  $u(0) = 1$:  $C = \dfrac{1}{e^{-1/2}} = e^{1/2}$

Particular solution:  $u = e^{(1 - \cos v^2)/2}$

**23.** $dP - kP\, dt = 0$

$$\int \frac{dP}{P} = k\int dt$$

$$\ln|P| = kt + C_1$$

$$P = Ce^{kt}$$

Initial condition: $P(0) = P_0$, $P_0 = Ce^0 = C$

Particular solution: $P = P_0 e^{kt}$

**25.**    $y' = \dfrac{dy}{dx} = \dfrac{x}{4y}$

$$\int 4y\, dy = \int x\, dx$$

$$2y^2 = \frac{x^2}{2} + C$$

Initial condition $(0, 2)$:  $2(2^2) = 0 + C \Rightarrow C = 8$

Particular solution:     $2y^2 = \dfrac{x^2}{2} + 8$

$$4y^2 - x^2 = 16$$

**27.**     $y' = \dfrac{dy}{dx} = \dfrac{y}{2x}$

$$\int \frac{2}{y}\, dy = \int \frac{1}{x}\, dx$$

$$2\ln|y| = \ln|x| + C_1 = \ln|x| + \ln C$$

$$y^2 = Cx$$

Initial condition $(9, 1)$:  $1 = 9C \Rightarrow C = \dfrac{1}{9}$

Particular solution:      $y^2 = \dfrac{1}{9}x$

$$9y^2 - x = 0$$

$$y = \frac{1}{3}\sqrt{x}$$

**29.**   $m = \dfrac{dy}{dx} = \dfrac{0 - y}{(x + 2) - x} = -\dfrac{y}{2}$

$$\int \frac{dy}{y} = \int -\frac{1}{2}\, dx$$

$$\ln|y| = -\frac{1}{2}x + C_1$$

$$y = Ce^{-x/2}$$

**31.** $\dfrac{dy}{dx} = x$

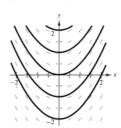

$$y = \int x\, dx = \frac{1}{2}x^2 + C$$

**33.** $\dfrac{dy}{dx} = 4 - y$

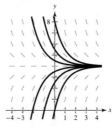

$$\int \frac{dy}{4 - y} = \int dx$$

$$\ln|4 - y| = -x + C_1$$

$$4 - y = e^{-x + C_1}$$

$$y = 4 + Ce^{-x}$$

**35.** (a) Euler's Method gives $y \approx 0.1602$ when $x = 1$.

(b) $\dfrac{dy}{dx} = -6xy$

$\displaystyle\int \dfrac{dy}{y} = \int -6x$

$\ln|y| = -3x^2 + C_1$

$y = Ce^{-3x^2}$

$y(0) = 5 \Rightarrow C = 5$

$y = 5e^{-3x^2}$

(c) At $x = 1$, $y = 5e^{-3(1)} \approx 0.2489$.

Error: $0.2489 - 0.1602 \approx 0.0887$

**37.** (a) Euler's Method gives $y \approx 3.0318$ when $x = 2$.

(b) $\dfrac{dy}{dx} = \dfrac{2x + 12}{3y^2 - 4}$

$\displaystyle\int (3y^2 - 4)\, dy = \int (2x + 12)\, dx$

$y^3 - 4y = x^2 + 12x + C$

$y(1) = 2: 2^3 - 4(2) = 1 + 12 + C \Rightarrow C = -13$

$y^3 - 4y = x^2 + 12x - 13$

(c) At $x = 2$,

$y^3 - 4y = 2^2 + 12(2) - 13 = 15$

$y^3 - 4y - 15 = 0$

$(y - 3)(y^2 + 3y + 5) = 0 \Rightarrow y = 3$.

Error: $3.0318 - 3 = 0.0318$

**39.** $\dfrac{dy}{dt} = ky$, $\quad y = Ce^{kt}$

Initial amount: $y(0) = y_0 = C$

Half-life: $\dfrac{y_0}{2} = y_0 e^{k(1599)}$

$k = \dfrac{1}{1599} \ln\left(\dfrac{1}{2}\right)$

$y = Ce^{[\ln(1/2)/1599]t}$

When $t = 50$, $y = 0.9786C$ or $97.86\%$.

**41.** (a) $\dfrac{dy}{dx} = k(y - 4)$

(b) The direction field satisfies $(dy/dx) = 0$ along $y = 4$; but not along $y = 0$. Matches (a).

**43.** (a) $\dfrac{dy}{dx} = ky(y - 4)$

(b) The direction field satisfies $(dy/dx) = 0$ along $y = 0$ and $y = 4$. Matches (c).

**45.** (a) $\qquad \dfrac{dw}{dt} = k(1200 - w)$

$\displaystyle\int \dfrac{dw}{1200 - w} = \int k\, dt$

$\ln|1200 - w| = -kt + C_1$

$1200 - w = e^{-kt + C_1} = Ce^{-kt}$

$w = 1200 - Ce^{-kt}$

$w(0) = 60 = 1200 - C \Rightarrow C = 1200 - 60 = 1140$

$w = 1200 - 1140e^{-kt}$

(b)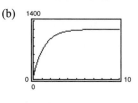

$k = 0.8$

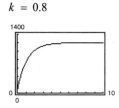

$k = 0.9$

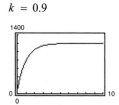

$k = 1$

(c) $k = 0.8$: $\quad t = 1.31$ years

$k = 0.9$: $\quad t = 1.16$ years

$k = 1.0$: $\quad t = 1.05$ years

(d) Maximum weight: 1200 pounds

$\displaystyle\lim_{x \to \infty} w = 1200$

**47.** Given family (circles): $\quad x^2 + y^2 = C$

$2x + 2yy' = 0$

$y' = -\dfrac{x}{y}$

Orthogonal trajectory (lines): $\quad y' = \dfrac{y}{x}$

$\displaystyle\int \dfrac{dy}{y} = \int \dfrac{dx}{x}$

$\ln|y| = \ln|x| + \ln K$

$y = Kx$

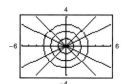

**49.** Given family (parabolas):  $x^2 = Cy$

$$2x = Cy'$$

$$y' = \frac{2x}{C} = \frac{2x}{x^2/y} = \frac{2y}{x}$$

Orthogonal trajectory (ellipses):    $y' = -\dfrac{x}{2y}$

$$2\int y \, dy = -\int x \, dx$$

$$y^2 = -\frac{x^2}{2} + K_1$$

$$x^2 + 2y^2 = K$$

**51.** Given family:   $y^2 = Cx^3$

$$2yy' = 3Cx^2$$

$$y' = \frac{3Cx^2}{2y} = \frac{3x^2}{2y}\left(\frac{y^2}{x^3}\right) = \frac{3y}{2x}$$

Orthogonal trajectory (ellipses):    $y' = -\dfrac{2x}{3y}$

$$3\int y \, dy = -2\int x \, dx$$

$$\frac{3y^2}{2} = -x^2 + K_1$$

$$3y^2 + 2x^2 = K$$

**53.**

$$\frac{dN}{dt} = kN(500 - N)$$

$$\int \frac{dN}{N(500 - N)} = \int k \, dt$$

$$\frac{1}{500}\int \left[\frac{1}{N} + \frac{1}{500 - N}\right] dN = \int k \, dt$$

$$\ln|N| - \ln|500 - N| = 500(kt + C_1)$$

$$\frac{N}{500 - N} = e^{500kt + C_2} = Ce^{500kt}$$

$$N = \frac{500Ce^{500kt}}{1 + Ce^{500kt}}$$

When $t = 0$, $N = 100$. So, $100 = \dfrac{500C}{1 + C} \Rightarrow C = 0.25$. Therefore, $N = \dfrac{125e^{500kt}}{1 + 0.25e^{500kt}}$.

When $t = 4$, $N = 200$. So, $200 = \dfrac{125e^{2000k}}{1 + 0.25e^{2000k}} \Rightarrow k = \dfrac{\ln(8/3)}{2000} \approx 0.00049$.

Therefore, $N = \dfrac{125e^{0.2452t}}{1 + 0.25e^{0.2452t}} = \dfrac{500}{1 + 4e^{-0.2452t}}$.

**55.** The general solution is $y = 1 - Ce^{-kt}$. Because $y = 0$ when $t = 0$, it follows that $C = 1$.

Because $y = 0.75$ when $t = 1$, you have

$$0.75 = 1 - e^{-k(1)}$$

$$-0.25 = -e^{-k}$$

$$0.25 = e^{-k}$$

$$\ln 0.25 = -k$$

$$k = \ln 0.25 = \ln 4 \approx 1.386.$$

So, $y \approx 1 - e^{-1.386t}$.

**Note:** This can be written as $y = 1 - 4^{-x}$.

**57.** The general solution is $y = -\dfrac{1}{kt + C}$.

Because $y = 45$ when $t = 0$, it follows that

$$45 = -\frac{1}{C} \text{ and } C = -\frac{1}{45}.$$

Therefore, $y = -\dfrac{1}{kt - (1/45)} = \dfrac{45}{1 - 45kt}$.

Because $y = 4$ when $t = 2$, you have

$$4 = \frac{45}{1 - 45k(2)} \Rightarrow k = -\frac{41}{360}.$$

So, $y = \dfrac{45}{1 + (41/8)t} = \dfrac{360}{8 + 41t}$.

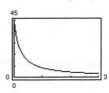

**59.** Because $y = 100$ when $t = 0$, it follows that

$100 = 500e^{-C}$, which implies that $C = \ln 5$. So,

you have $y = 500e^{(-\ln 5)e^{-kt}}$. Because $y = 150$ when

$t = 2$, it follows that

$$150 = 500e^{(-\ln 5)e^{-2k}}$$

$$e^{-2k} = \frac{\ln 0.3}{\ln 0.2}$$

$$k = -\frac{1}{2}\ln\frac{\ln 0.3}{\ln 0.2} \approx 0.1452.$$

So, $y$ is given by

$$y = 500e^{-1.6904e^{-0.1451t}}.$$

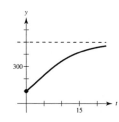

**61.** From Example 8, the general solution is $y = 60e^{-Ce^{-kt}}$.

Because $y = 8$ when $t = 0$,

$$8 = 60e^{-C} \Rightarrow C = \ln\frac{15}{2} \approx 2.0149.$$

Because $y = 15$ when $t = 3$,

$$15 = 60e^{-2.0149e^{-3k}}$$

$$\frac{1}{4} = e^{-2.0149e^{-3k}}$$

$$\ln\frac{1}{4} = -2.0149e^{-3k}$$

$$k = -\frac{1}{3}\ln\left(\frac{\ln(1/4)}{-2.0149}\right) \approx 0.1246.$$

So, $y = 60e^{-2.0149e^{-0.1246t}}$.

When $t = 10$, $y \approx 34$ beavers.

**63.** Following Example 9, the differential equation is

$$\frac{dy}{dt} = ky(1-y)(2-y)$$

and its general solution is $\dfrac{y(2-y)}{(1-y)^2} = Ce^{2kt}$.

$$y = \frac{1}{2} \text{ when } t = 0 \Rightarrow \frac{(1/2)(3/2)}{(1/2)^2} = C \Rightarrow C = 3$$

$$y = 0.75 = \frac{3}{4} \text{ when}$$

$$t = 4 \Rightarrow \frac{(3/4)(5/4)}{(1/4)^2} = 15 = 3e^{2k(4)}$$

$$\Rightarrow 5 = e^{8k}$$

$$\Rightarrow k = \frac{1}{8}\ln 5 \approx 0.2012.$$

So, the particular solution is $\dfrac{y(2-y)}{(1-y)^2} = 3e^{0.4024t}$.

Using a symbolic algebra utility or graphing utility, you find that when $t = 10$,

$$\frac{y(2-y)}{(1-y)^2} = 3e^{0.4024(10)}$$

and $y \approx 0.92$, or 92%.

**65.** (a) $\dfrac{dQ}{dt} = -\dfrac{Q}{20}$

$$\int\frac{dQ}{Q} = \int -\frac{1}{20}dt$$

$$\ln|Q| = -\frac{1}{20}t + C_1$$

$$Q = e^{-(1/20)t + C_1} = Ce^{-(1/20)t}$$

Because $Q = 25$ when $t = 0$, you have $25 = C$.

So, the particular solution is $Q = 25e^{-(1/20)t}$.

(b) When $Q = 15$, you have $15 = 25e^{-(1/20)t}$.

$$\frac{3}{5} = e^{-(1/20)t}$$

$$\ln\left(\frac{3}{5}\right) = -\frac{1}{20}t$$

$$-20\ln\left(\frac{3}{5}\right) = t$$

$$t \approx 10.217 \text{ minutes}$$

**67. (a)**  $\dfrac{dy}{dt} = ky$

$$\int \dfrac{dy}{y} = \int k\,dt$$

$$\ln y = kt + C_1$$

$$y = e^{kt + C_1} = Ce^{kt}$$

**(b)**  $y(0) = 20 \Rightarrow C = 20$

$$y(1) = 16 = 20e^k \Rightarrow k = \ln\dfrac{16}{20} = \ln\left(\dfrac{4}{5}\right)$$

$$y = 20e^{t\,\ln(4/5)}$$

When 75% has changed:

$$5 = 20e^{t\,\ln(4/5)}$$

$$\dfrac{1}{4} = e^{t\,\ln(4/5)}$$

$$t = \dfrac{\ln(1/4)}{\ln(4/5)} \approx 6.2 \text{ hours}$$

$$\dfrac{3}{5} = e^{-(1/20)t}$$

$$\ln\left(\dfrac{3}{5}\right) = -\dfrac{1}{20}t$$

$$-20\ln\left(\dfrac{3}{5}\right) = t$$

$$t \approx 10.217 \text{ minutes}$$

**69.** The general solution is $y = Ce^{kt}$. Because $y = 0.60C$ when $t = 1$, you have

$$0.60C = Ce^k \Rightarrow k = \ln 0.60 \approx -0.5108.$$

So, $y = Ce^{-0.5108t}$. When $y = 0.20C$, you have

$$0.20C = Ce^{-0.5108t}$$

$$\ln 0.20 = -0.5108t$$

$$t \approx 3.15 \text{ hours.}$$

**71.**  $\displaystyle\int \dfrac{1}{kP + N}\,dP = \int dt$

$$\dfrac{1}{k}\ln|kP + N| = t + C_1$$

$$kP + N = C_2 e^{kt}$$

$$P = Ce^{kt} - \dfrac{N}{k}$$

**73.**  $\dfrac{dA}{dt} = rA + P$

$$\dfrac{dA}{rA + P} = dt$$

$$\int \dfrac{dA}{rA + P} = \int dt$$

$$\dfrac{1}{r}\ln(rA + P) = t + C_1$$

$$\ln(rA + P) = rt + C_2$$

$$rA + P = e^{rt + C_2}$$

$$A = \dfrac{C_3 e^{rt} - P}{r}$$

$$A = Ce^{rt} - \dfrac{P}{r}$$

When $t = 0$: $A = 0$

$$0 = C - \dfrac{P}{r} \Rightarrow C = \dfrac{P}{r}$$

$$A = \dfrac{P}{r}\left(e^{rt} - 1\right)$$

**75.** From Exercise 73,

$$A = \dfrac{P}{r}\left(e^{rt} - 1\right).$$

Because $A = 260{,}000{,}000$ when $t = 8$ and $r = 0.0725$, you have

$$P = \dfrac{Ar}{e^{rt} - 1}$$

$$= \dfrac{(260{,}000{,}000)(0.0725)}{e^{(0.0725)(8)} - 1}$$

$$\approx \$23{,}981{,}015.77.$$

**77.**  $\dfrac{dy}{dt} = 0.02y\,\ln\left(\dfrac{5000}{y}\right)$

**(a)**

**(b)** As $t \to \infty$, $y \to L = 5000$.

**(c)** Using a computer algebra system or separation of variables, the general solution is

$$y = 5000e^{-Ce^{-kt}} = 5000e^{-Ce^{-0.02t}}.$$

Using the initial condition $y(0) = 500$, you obtain

$$500 = 5000e^{-C} \Rightarrow C = \ln 10 \approx 2.3026.$$

So, $y = 5000e^{-2.3026e^{-0.02t}}.$

(d)

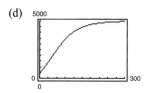

The graph is concave upward on $(0, 41.7)$ and concave downward on $(41.7, \infty)$.

**79.** A differential equation can be solved by separation of variables if it can be written in the form

$$M(x) + N(y)\frac{dy}{dx} = 0.$$

To solve a separable equation, rewrite as,

$$M(x)\,dx = -N(y)\,dy$$

and integrate both sides.

**81.** $y(1 + x)\,dx + x\,dy = 0$

$$x\,dy = -y(1 + x)\,dx$$

$$\frac{1}{y}\,dy = -\frac{(1 + x)}{x}\,dx$$

Separable

**83.** $\dfrac{dy}{dx} + xy = 5$

Not separable

**85.** (a) $\qquad \dfrac{dv}{dt} = k(W - v)$

$$\int \frac{dv}{W - v} = \int k\,dt$$

$$-\ln|W - v| = kt + C_1$$

$$v = W - Ce^{-kt}$$

Initial conditions:

$W = 20, v = 0$ when $t = 0$ and $v = 10$

when $t = 0.5$ so, $C = 20, k = \ln 4$.

Particular solution:

$$v = 20\left(1 - e^{-(\ln 4)t}\right) = 20\left(1 - \left(\frac{1}{4}\right)^{t}\right)$$

or

$$v = 20\left(1 - e^{-1.386t}\right)$$

(b) $s = \displaystyle\int 20\left(1 - e^{-1.386t}\right) dt \approx 20\left(t + 0.7215e^{-1.386t}\right) + C$

Because $s(0) = 0, C \approx -14.43$ and you have

$$s \approx 20t + 14.43\left(e^{-1.386t} - 1\right).$$

**87.** $f(x, y) = x^3 - 4xy^2 + y^3$

$$f(tx, ty) = t^3x^3 - 4txt^2y^2 + t^3y^3$$

$$= t^3\left(x^3 - 4xy^2 + y^3\right)$$

Homogeneous of degree 3

**89.** $f(x, y) = \dfrac{x^2y^2}{\sqrt{x^2 + y^2}}$

$$f(tx, ty) = \frac{t^4x^2y^2}{\sqrt{t^2x^2 + t^2y^2}} = t^3\frac{x^2y^2}{\sqrt{x^2 + y^2}}$$

Homogeneous of degree 3

**91.** $f(x, y) = 2\ln xy$

$$f(tx, ty) = 2\ln[txty]$$

$$= 2\ln\left[t^2xy\right] = 2\left(\ln t^2 + \ln xy\right)$$

Not homogeneous

**93.** $f(x, y) = 2\ln\dfrac{x}{y}$

$$f(tx, ty) = 2\ln\frac{tx}{ty} = 2\ln\frac{x}{y}$$

Homogeneous of degree 0

**95.** $(x + y)dx - 2x\,dy = 0, \ y = ux, \ dy = x\,du + u\,dx$

$$(x + ux)dx - 2x(x\,du + u\,dx) = 0$$

$$(1 + u)dx - 2x\,du - 2u\,dx = 0$$

$$(1 - u)dx = 2x\,du$$

$$\frac{1}{x}\,dx = \frac{2}{1 - u}\,du$$

$$\int \frac{1}{x}\,dx = 2\int \frac{1}{1 - u}\,du$$

$$\ln|x| + \ln C = -2\ln|1 - u|$$

$$\ln|Cx| = \ln|1 - u|^{-2}$$

$$|Cx| = \frac{1}{(1 - u)^2}$$

$$= \frac{1}{\left[1 - (y/x)\right]^2}$$

$$|Cx| = \frac{x^2}{(x - y)^2}$$

$$|x| = C(x - y)^2$$

**97.** $(x - y)dx - (x + y)dy = 0, y = ux, dy = x\,du + u\,dx$

$$(x - ux)dx - (x + ux)(x\,du + u\,dx) = 0$$

$$(1 - u)dx - (1 + u)(x\,du + u\,dx) = 0$$

$$(1 - 2u - u^2)dx = x(1 + u)du$$

$$-\frac{dx}{x} = \frac{1 + u}{u^2 + 2u - 1}du$$

$$-\int\frac{dx}{x} = \int\frac{u + 1}{u^2 + 2u - 1}du$$

$$-\ln|x| + \ln C = \frac{1}{2}\ln|u^2 + 2u - 1|$$

$$\ln\left|\frac{C}{x}\right| = \ln|u^2 + 2u - 1|^{1/2}$$

$$\frac{C^2}{x^2} = |u^2 + 2u - 1|$$

$$\frac{C}{x^2} = \left|\left(\frac{y}{x}\right)^2 + 2\left(\frac{y}{x}\right) - 1\right|$$

$$C = |y^2 + 2yx - x^2|$$

**99.** $xy\,dx + (y^2 - x^2)dy = 0, y = ux, dy = x\,du + u\,dx$

$$x(ux)\,dx + \left[(ux)^2 - x^2\right](x\,du + u\,dx) = 0$$

$$u\,dx + (u^2 - 1)(x\,du + u\,dx) = 0$$

$$u^3\,dx = -(u^2 - 1)x\,du$$

$$\frac{dx}{x} = \frac{1 - u^2}{u^3}du$$

$$\int\frac{dx}{x} = \int\left(u^{-3} - \frac{1}{u}\right)du$$

$$\ln|x| + \ln|C_1| = -\frac{1}{2u^2} - \ln|u|$$

$$\ln|C_1\,xu| = -\frac{1}{2u^2}$$

$$\ln|C_1\,y| = -\frac{1}{2(y/x)^2}$$

$$= -\frac{x^2}{2y^2}$$

$$y = Ce^{-x^2/(2y^2)}$$

**101.** False. $\dfrac{dy}{dx} = \dfrac{x}{y}$ is separable, but $y = 0$ is not a solution.

**103.** True

$$x^2 + y^2 = 2Cy \qquad x^2 + y^2 = 2Kx$$

$$\frac{dy}{dx} = \frac{x}{C - y} \qquad \frac{dy}{dx} = \frac{K - x}{y}$$

$$\frac{x}{C - y} \cdot \frac{K - x}{y} = \frac{Kx - x^2}{Cy - y^2}$$

$$= \frac{2Kx - 2x^2}{2Cy - 2y^2}$$

$$= \frac{x^2 + y^2 - 2x^2}{x^2 + y^2 - 2y^2}$$

$$= \frac{y^2 - x^2}{x^2 - y^2}$$

$$= -1$$

# Section 6.4   The Logistic Equation

**1.** $y = \dfrac{12}{1 + e^{-x}}$

Because $y(0) = 6$, it matches (c) or (d).

Because (d) approaches its horizontal asymptote slower than (c), it matches (d).

**3.** $y = \dfrac{12}{1 + \frac{1}{2}e^{-x}}$

Because $y(0) = \dfrac{12}{\left(\frac{3}{2}\right)} = 8$, it matches (b).

**5.** $y = \dfrac{8}{1 + e^{-2t}} = 8\left(1 + e^{-2t}\right)^{-1}$; $L = 8, k = 2, b = 1$

$$\frac{dy}{dt} = -8\left(1 + e^{-2t}\right)^{-2}\left(-2e^{-2t}\right)$$

$$= \frac{8}{\left(1 + e^{-2t}\right)} \cdot \frac{2e^{-2t}}{\left(1 + e^{-2t}\right)}$$

$$= 2y\left(\frac{e^{-2t}}{1 + e^{-2t}}\right)$$

$$= 2y\left(1 - \frac{8}{8\left(1 + e^{-2t}\right)}\right)$$

$$= 2y\left(1 - \frac{y}{8}\right)$$

$$y(0) = \frac{8}{1 + e^0} = 4$$

**7.** $y = 12\left(1 + 6e^{-t}\right)^{-1}$; $L = 12, k = 1, b = 6$

$$y' = -12\left(1 + 6e^{-t}\right)^{-2}\left(-6e^{-t}\right)$$

$$= \left(\frac{12}{1 + 6e^{-t}}\right)\left(\frac{6e^{-t}}{1 + 6e^{-t}}\right)$$

$$= y\left(1 - \frac{1}{1 + 6e^{-t}}\right)$$

$$= y\left(1 - \frac{12}{12\left(1 + 6e^{-t}\right)}\right)$$

$$= y\left(1 - \frac{y}{12}\right)$$

$$y(0) = \frac{12}{1 + 6} = \frac{12}{7}$$

**9.** $P(t) = \dfrac{2100}{1 + 29e^{-0.75t}}$

(a) $k = 0.75$

(b) $L = 2100$

(c) $P(0) = \dfrac{2100}{1 + 29} = 70$

(d) $\qquad 1050 = \dfrac{2100}{1 + 29e^{-0.75t}}$

$$1 + 29e^{-0.75t} = 2$$

$$e^{-0.75t} = \frac{1}{29}$$

$$-0.75t = \ln\left(\frac{1}{29}\right) = -\ln 29$$

$$t = \frac{\ln 29}{0.75} \approx 4.4897 \text{ years}$$

(e) $\dfrac{dP}{dt} = 0.75P\left(1 - \dfrac{P}{2100}\right)$

**11.** $P(t) = \dfrac{6000}{1 + 4999e^{-0.8t}}$

(a) $k = 0.8$

(b) $L = 6000$

(c) $P(0) = \dfrac{6000}{1 + 4999} = \dfrac{6}{5}$

(d) $\qquad 3000 = \dfrac{6000}{1 + 4999e^{-0.8t}}$

$$1 + 4999e^{-0.8t} = 2$$

$$e^{-0.8t} = \frac{1}{4999}$$

$$-0.8t = \ln\left(\frac{1}{4999}\right) = -\ln 4999$$

$$t = \frac{\ln 4999}{0.8} \approx 10.65 \text{ years}$$

(e) $\dfrac{dP}{dt} = 0.8P\left(1 - \dfrac{P}{6000}\right)$

**13.** $\dfrac{dP}{dt} = 3P\left(1 - \dfrac{P}{100}\right)$

(a) $k = 3$

(b) $L = 100$

(c)

(d) $\dfrac{d^2P}{dt^2} = 3P'\left(1 - \dfrac{P}{100}\right) + 3P\left(\dfrac{-P'}{100}\right)$

$= 3\left[3P\left(1 - \dfrac{P}{100}\right)\right]\left(1 - \dfrac{P}{100}\right) - \dfrac{3P}{100}\left[3P\left(1 - \dfrac{P}{100}\right)\right] = 9P\left(1 - \dfrac{P}{100}\right)\left(1 - \dfrac{P}{100} - \dfrac{P}{100}\right) = 9P\left(1 - \dfrac{P}{100}\right)\left(1 - \dfrac{2P}{100}\right)$

$\dfrac{d^2P}{dt^2} = 0$ for $P = 50$, and by the first Derivative Test, this is a maximum. $\left(\text{Note: } P = 50 = \dfrac{L}{2} = \dfrac{100}{2}\right)$

**15.** $\dfrac{dP}{dt} = 0.1P - 0.0004P^2$

$= 0.1P(1 - 0.004P)$

$= 0.1P\left(1 - \dfrac{P}{250}\right)$

(a) $k = 0.1 = \dfrac{1}{10}$

(b) $L = 250$

(c)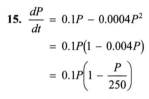

(d) $P = \dfrac{250}{2} = 125$

(Same argument as in Exercise 13)

**17.** $\dfrac{dy}{dt} = y\left(1 - \dfrac{y}{36}\right), \quad y(0) = 4$

$k = 1, L = 36$

$y = \dfrac{L}{1 + be^{-kt}} = \dfrac{36}{1 + be^{-t}}$

$(0, 4): 4 = \dfrac{36}{1 + b} \Rightarrow b = 8$

Solution: $y = \dfrac{36}{1 + 8e^{-t}}$

$y(5) = \dfrac{36}{1 + 8e^{-5}} \approx 34.16$

$y(100) = \dfrac{36}{1 + 8e^{-100}} \approx 36.00$

**19.** $\dfrac{dy}{dt} = \dfrac{4y}{5} - \dfrac{y^2}{150} = \dfrac{4}{5}y\left(1 - \dfrac{y}{120}\right), \quad y(0) = 8$

$k = \dfrac{4}{5} = 0.8, L = 120$

$y = \dfrac{L}{1 + be^{-kt}} = \dfrac{120}{1 + be^{-0.8t}}$

$y(0) = 8: 8 = \dfrac{120}{1 + b} \Rightarrow b = 14$

Solution: $y = \dfrac{120}{1 + 14e^{-0.8t}}$

$y(5) = \dfrac{120}{1 + 14e^{-0.8(5)}} \approx 95.51$

$y(100) = \dfrac{120}{1 + 14e^{-0.8(100)}} \approx 120.0$

**21.** $L = 250$ and $y(0) = 350$

Matches (c).

**23.** $L = 250$ and $y(0) = 50$

Matches (b).

**25.** $\dfrac{dy}{dt} = 0.2y\left(1 - \dfrac{y}{1000}\right)$

(a)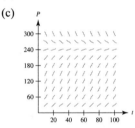

(b) $k = 0.2, L = 1000$

$y = \dfrac{1000}{1 + be^{-0.2t}}$

$y(0) = 105 = \dfrac{1000}{1 + b}$

$1 + b = \dfrac{1000}{105} = \dfrac{200}{21}$

$b = \dfrac{179}{21} \approx 8.524$

$y = \dfrac{1000}{1 + (179/21)e^{-0.2t}}$

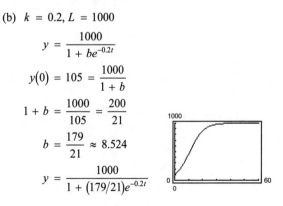

**27.** $L$ represents the value that $y$ approaches as $t$ approaches infinity. $L$ is the carrying capacity.

**29.** Yes, the logistic differential equation is separable. See Example 1.

**31.** (a) $P = \dfrac{L}{1 + be^{-kt}}$, $L = 200$, $P(0) = 25$

$$25 = \frac{200}{1 + b} \Rightarrow b = 7$$

$$39 = \frac{200}{1 + 7e^{-k(2)}}$$

$$1 + 7e^{-2k} = \frac{200}{39}$$

$$e^{-2k} = \frac{23}{39}$$

$$k = -\frac{1}{2}\ln\left(\frac{23}{39}\right) = \frac{1}{2}\ln\left(\frac{39}{23}\right) \approx 0.2640$$

$$P = \frac{200}{1 + 7e^{-0.2640t}}$$

(b) For $t = 5$, $P \approx 70$ panthers.

(c)
$$100 = \frac{200}{1 + 7e^{-0.264t}}$$

$$1 + 7e^{-0.264t} = 2$$

$$-0.264t = \ln\left(\frac{1}{7}\right)$$

$$t \approx 7.37 \text{ years}$$

(d) $\dfrac{dP}{dt} = kP\left(1 - \dfrac{P}{L}\right)$

$$= 0.264P\left(1 - \frac{P}{200}\right), \quad P(0) = 25$$

Using Euler's Method, $P \approx 65.6$ when $t = 5$.

(e) $P$ is increasing most rapidly where $P = 200/2 = 100$, corresponds to $t \approx 7.37$ years.

**33.** False. If $y > L$, then $dy/dt < 0$ and the population decreases.

**35.** $y = \dfrac{1}{1 + be^{-kt}}$

$$y' = \frac{-1}{\left(1 + be^{-kt}\right)^2}\left(-bke^{-kt}\right)$$

$$= \frac{k}{\left(1 + be^{-kt}\right)} \cdot \frac{be^{-kt}}{\left(1 + be^{-kt}\right)}$$

$$= \frac{k}{\left(1 + be^{-kt}\right)} \cdot \frac{1 + be^{-kt} - 1}{\left(1 + be^{-kt}\right)}$$

$$= \frac{k}{\left(1 + be^{-kt}\right)} \cdot \left(1 - \frac{1}{1 + be^{-kt}}\right)$$

$$= ky(1 - y)$$

# Section 6.5  First-Order Linear Differential Equations

**1.** $x^3 y' + xy = e^x + 1$

$$y' + \frac{1}{x^2}y = \frac{1}{x^3}(e^x + 1)$$

Linear

**3.** $y' - y \sin x = xy^2$

Not linear, because of the $xy^2$-term.

**5.** $\dfrac{dy}{dx} + \left(\dfrac{1}{x}\right)y = 6x + 2$

Integrating factor: $e^{\int (1/x)\,dx} = e^{\ln x} = x$

$$xy = \int x(6x + 2)\,dx = 2x^3 + x^2 + C$$

$$y = 2x^2 + x + \frac{C}{x}$$

**7.** $y' - y = 16$

Integrating factor: $e^{\int -1\,dx} = e^{-x}$

$$e^{-x}y' - e^{-x}y = 16e^{-x}$$

$$ye^{-x} = \int 16e^{-x}\,dx = -16e^{-x} + C$$

$$y = -16 + Ce^x$$

**9.** $(y + 1)\cos x\,dx = dy$

$$y' = (y + 1)\cos x = y\cos x + \cos x$$

$$y' - (\cos x)y = \cos x$$

Integrating factor: $e^{\int -\cos x\,dx} = e^{-\sin x}$

$$y'e^{-\sin x} - (\cos x)e^{-\sin x}y = (\cos x)e^{-\sin x}$$

$$ye^{-\sin x} = \int (\cos x)e^{-\sin x}\,dx$$

$$= -e^{-\sin x} + C$$

$$y = -1 + Ce^{\sin x}$$

**11.** $(x - 1)y' + y = x^2 - 1$

$$y' + \left(\frac{1}{x - 1}\right)y = x + 1$$

Integrating factor: $e^{\int [1/(x-1)]\,dx} = e^{\ln|x-1|} = x - 1$

$$y(x - 1) = \int(x^2 - 1)\,dx = \frac{1}{3}x^3 - x + C_1$$

$$y = \frac{x^3 - 3x + C}{3(x - 1)}$$

**13.** $y' - 3x^2y = e^{x^3}$

Integrating factor: $e^{-\int 3x^2\,dx} = e^{-x^3}$

$$ye^{-x^3} = \int e^{x^3}e^{-x^3}\,dx = \int dx = x + C$$

$$y = (x + C)e^{x^3}$$

**15.** (a)  Answers will vary.

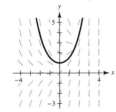

(b) $\dfrac{dy}{dx} = e^x - y$

$\dfrac{dy}{dx} + y = e^x$    Integrating factor: $e^{\int dx} = e^x$

$e^x y' + e^x y = e^{2x}$

$\left(ye^x\right) = \int e^{2x}\,dx$

$ye^x = \dfrac{1}{2}e^{2x} + C$

$y(0) = 1 \Rightarrow 1 = \dfrac{1}{2} + C \Rightarrow C = \dfrac{1}{2}$

$ye^x = \dfrac{1}{2}e^{2x} + \dfrac{1}{2}$

$y = \dfrac{1}{2}e^x + \dfrac{1}{2}e^{-x} = \dfrac{1}{2}\left(e^x + e^{-x}\right)$

(c)

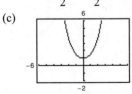

**17.** $y' \cos^2 x + y - 1 = 0$

$$y' + \left(\sec^2 x\right)y = \sec^2 x$$

Integrating factor: $e^{\int \sec^2 x\,dx} = e^{\tan x}$

$$ye^{\tan x} = \int \sec^2 x\, e^{\tan x}\,dx = e^{\tan x} + C$$

$$y = 1 + Ce^{-\tan x}$$

Initial condition: $y(0) = 5, C = 4$

Particular solution: $y = 1 + 4e^{-\tan x}$

**19.** $y' + y\tan x = \sec x + \cos x$

Integrating factor: $e^{\int \tan x\,dx} = e^{\ln|\sec x|} = \sec x$

$$y\sec x = \int \sec x(\sec x + \cos x)\,dx = \tan x + x + C$$

$$y = \sin x + x\cos x + C\cos x$$

Initial condition: $y(0) = 1, 1 = C$

Particular solution: $y = \sin x + (x + 1)\cos x$

**21.** $y' + \left(\dfrac{1}{x}\right)y = 0$

Integrating factor: $e^{\int (1/x)\,dx} = e^{\ln|x|} = x$

Separation of variables:

$$\frac{dy}{dx} = -\frac{y}{x}$$

$$\int \frac{1}{y}\,dy = \int -\frac{1}{x}\,dx$$

$$\ln y = -\ln x + \ln C$$

$$\ln xy = \ln C$$

$$xy = C$$

Initial condition: $y(2) = 2, C = 4$

Particular solution: $xy = 4$

**23.** $x\,dy = (x + y + 2)\,dx$

$$\frac{dy}{dx} = \frac{x + y + 2}{x} = \frac{y}{x} + 1 + \frac{2}{x}$$

$$\frac{dy}{dx} - \frac{1}{x}y = 1 + \frac{2}{x}\qquad \text{Linear}$$

$$u(x) = e^{\int -(1/x)\,dx} = \frac{1}{x}$$

$$y = x\int\left(1 + \frac{2}{x}\right)\frac{1}{x}\,dx = x\int\left(\frac{1}{x} + \frac{2}{x^2}\right)dx$$

$$= x\left[\ln|x| + \frac{-2}{x} + C\right]$$

$$= -2 + x\ln|x| + Cx$$

$$y(1) = 10 = -2 + C \Rightarrow C = 12$$

$$y = -2 + x\ln|x| + 12x$$

**25.**

$$\frac{dP}{dt} = kP + N, \; N \text{ constant}$$

$$\frac{dP}{kP + N} = dt$$

$$\int \frac{1}{kP + N} \, dP = \int dt$$

$$\frac{1}{k} \ln(kP + N) = t + C_1$$

$$\ln(kP + N) = kt + C_2$$

$$kP + N = e^{kt + C_2}$$

$$P = \frac{C_3 e^{kt} - N}{k}$$

$$P = Ce^{kt} - \frac{N}{k}$$

When $t = 0$: $P = P_0$

$$P_0 = C - \frac{N}{k} \Rightarrow C = P_0 + \frac{N}{k}$$

$$P = \left( P_0 + \frac{N}{k} \right) e^{kt} - \frac{N}{k}$$

**27. (a)** $A = \dfrac{P}{r}\left( e^{rt} - 1 \right)$

$$A = \frac{275{,}000}{0.06}\left( e^{0.08(10)} - 1 \right) \approx \$4{,}212{,}796.94$$

**(b)** $A = \dfrac{550{,}000}{0.05}\left( e^{0.059(25)} - 1 \right) \approx \$31{,}424{,}909.75$

**29. (a)** $\dfrac{dN}{dt} = k\left( 75 - N \right)$

**(b)** $N' + kN = 75k$

Integrating factor: $e^{\int k \, dt} = e^{kt}$

$$N'e^{kt} + kNe^{kt} = 75\, ke^{kt}$$

$$\left( Ne^{kt} \right)' = 75\, ke^{kt}$$

$$Ne^{kt} = \int 75\, ke^{kt} = 75\, e^{kt} + C$$

$$N = 75 + Ce^{-kt}$$

**(c)** For $t = 1, N = 20$:

$$20 = 75 + Ce^{-k} \Rightarrow -55 = Ce^{-k}$$

For $t = 20, N = 35$:

$$35 = 75 + Ce^{-20k} \Rightarrow -40 = Ce^{-20k}$$

$$\frac{55}{40} = \frac{Ce^{-k}}{Ce^{-20k}} \Rightarrow e^{19k} = \frac{11}{8} \Rightarrow k = \frac{1}{19}\ln\!\left(\frac{11}{8}\right)$$

$$\approx 0.0168$$

$$Ce^{-k} = -55$$

$$C = -55e^{k} \approx -55.9296$$

$$N = 75 - 55.9296\, e^{-0.0168t}$$

**31.** From Example 3,

$$\frac{dv}{dt} + \frac{kv}{m} = g$$

$$v = \frac{mg}{k}\left( 1 - e^{-kt/m} \right), \quad \text{Solution}$$

$$g = -32, \; mg = -8, \; v(5) = -101, \; m = \frac{-8}{g} = \frac{1}{4}$$

implies that $-101 = \dfrac{-8}{k}\left( 1 - e^{-5k/(1/4)} \right)$.

Using a graphing utility, $k \approx 0.050165$, and

$$v = -159.47\left( 1 - e^{-0.2007t} \right).$$

As $t \to \infty, \, v \to -159.47$ ft/sec. The graph of $v$ is shown below.

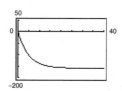

**33.** $L\dfrac{dI}{dt} + RI = E_0, \; I' + \dfrac{R}{L}I = \dfrac{E_0}{L}$

Integrating factor: $e^{\int (R/L)\, dt} = e^{Rt/L}$

$$I\, e^{Rt/L} = \int \frac{E_0}{L} e^{Rt/L}\, dt = \frac{E_0}{R} e^{Rt/L} + C$$

$$I = \frac{E_0}{R} + Ce^{-Rt/L}$$

**35.** Let $Q$ be the number of pounds of concentrate in the solution at any time $t$. Because the number of gallons of solution in the tank at any time $t$ is $v_0 + \left( r_1 - r_2 \right)t$ and because the tank loses $r_2$ gallons of solution per minute, it must lose concentrate at the rate

$$\left[ \frac{Q}{v_0 + \left( r_1 - r_2 \right)t} \right] r_2.$$

The solution gains concentrate at the rate $r_1 q_1$. Therefore, the net rate of change is

$$\frac{dQ}{dt} = q_1 r_1 - \left[ \frac{Q}{v_0 + \left( r_1 - r_2 \right)t} \right] r_2$$

or

$$\frac{dQ}{dt} + \frac{r_2 Q}{v_0 + \left( r_1 - r_2 \right)t} = q_1 r_1.$$

**37.** (a) From Exercise 35,

$$\frac{dQ}{dt} + \frac{r_2 Q}{u_0 + (r_1 - r_2)t} = q_1 r_1$$

You have $Q(0)q_0 = 25$, $q_1 = 0$, $u_0 = 200$, and $r_1 = r_2 = 10$. Hence, the linear differential

equation is $\dfrac{dQ}{dt} + \dfrac{1}{20}Q = 0$.

By separating variables,

$$\int \frac{dQ}{Q} = -\int \frac{1}{20}\, dt$$

$$\ln Q = -\frac{1}{20}t + \ln C_1$$

$$Q = Ce^{-\frac{1}{20}t}.$$

The initial condition $Q(0) = 25$ implies that

$C = 25$. Hence, $Q = 25e^{-\frac{1}{20}t}$.

(b) $15 = 25e^{-\frac{1}{20}t} \Rightarrow \dfrac{3}{5} = e^{-\frac{1}{20}t} \Rightarrow \ln\!\left(\dfrac{3}{5}\right) = -\dfrac{1}{20}t$

$\Rightarrow t = -20 \ln\!\left(\dfrac{3}{5}\right) \approx 10.2$ minutes

(c) $\displaystyle \lim_{t \to \infty} Q = \lim_{t \to \infty} 25e^{-\frac{1}{20}t} = 0$

**39.** $y' + P(x)y = Q(x)$

Integrating factor: $u = e^{\int P(x)\, dx}$

$y'u + P(x)yu = Q(x)u$

$(uy)' = Q(x)u$

so $u'(x) = P(x)u$

Answer (a)

**41.** $\dfrac{dy}{dx} + P(x)y = Q(x)$      Standard form

     $u(x) = e^{\int P(x)\, dx}$      Integrating factor

**43.** $y' - 2x = 0$

$$\int dy = \int 2x\, dx$$

$$y = x^2 + C$$

Matches (c).

**44.** $y' - 2y = 0$

$$\int \frac{dy}{y} = \int 2\, dx$$

$$\ln y = 2x + C_1$$

$$y = Ce^{2x}$$

Matches (d).

**45.** $y' - 2xy = 0$

$$\int \frac{dy}{y} = \int 2x\, dx$$

$$\ln y = x^2 + C_1$$

$$y = Ce^{x^2}$$

Matches (a).

**46.** $y' - 2xy = x$

$$\int \frac{dy}{2y + 1} = \int x\, dx$$

$$\frac{1}{2}\ln(2y + 1) = \frac{1}{2}x^2 + C_1$$

$$2y + 1 = C_2 e^{x^2}$$

$$y = -\frac{1}{2} + Ce^{x^2}$$

Matches (b).

**47.** (a)

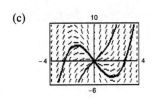

(b) $\dfrac{dy}{dx} - \dfrac{1}{x}y = x^2$

     Integrating factor: $e^{-\int 1/x\, dx} = e^{-\ln x} = \dfrac{1}{x}$

$$\frac{1}{x}y' - \frac{1}{x^2}y = x$$

$$\left(\frac{1}{x}y\right) = \int x\, dx = \frac{x^2}{2} + C$$

$$y = \frac{x^3}{2} + Cx$$

$(-2, 4):\ 4 = \dfrac{-8}{2} - 2C \Rightarrow C = -4 \Rightarrow y = \dfrac{x^3}{2} - 4x = \dfrac{1}{2}x(x^2 - 8)$

$(2, 8):\ 8 = \dfrac{8}{2} + 2C \Rightarrow C = 2 \Rightarrow y = \dfrac{x^3}{2} + 2x = \dfrac{1}{2}x(x^2 + 4)$

(c)

**49.** (a)

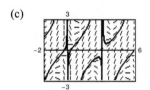

(b) $y' + (\cot x)y = 2$

Integrating factor: $e^{\int \cot x \, dx} = e^{\ln|\sin x|} = \sin x$

$$y'\sin x + (\cos x)y = 2\sin x$$

$$y\sin x = \int 2\sin x \, dx = -2\cos x + C$$

$$y = -2\cot x + C\csc x$$

$(1,1):\ 1 = -2\cot 1 + C\csc 1 \Rightarrow C = \dfrac{1 + 2\cot 1}{\csc 1} = \sin 1 + 2\cos 1$

$$y = -2\cot x + (\sin 1 + 2\cos 1)\csc x$$

$(3,-1):\ -1 = -2\cot 3 + C\csc 3 \Rightarrow C = \dfrac{2\cot 3 - 1}{\csc 3} = 2\cos 3 - \sin 3$

$$y = -2\cot x + (2\cos 3 - \sin 3)\csc x$$

(c)

**51.** $e^{2x+y}\, dx - e^{x-y}\, dy = 0$

Separation of variables:

$$e^{2x}e^{y}\, dx = e^{x}e^{-y}\, dy$$

$$\int e^{x}\, dx = \int e^{-2y}\, dy$$

$$e^{x} = -\tfrac{1}{2}e^{-2y} + C_1$$

$$2e^{x} + e^{-2y} = C$$

**53.** $(y\cos x - \cos x)\, dx + dy = 0$

Separation of variables:

$$\int \cos x \, dx = \int \frac{-1}{y-1}\, dy$$

$$\sin x = -\ln(y-1) + \ln C$$

$$\ln(y-1) = -\sin x + \ln C$$

$$y = Ce^{-\sin x} + 1$$

**55.** $(2y - e^{x})\, dx + x\, dy = 0$

Linear: $y' + \left(\dfrac{2}{x}\right)y = \dfrac{1}{x}e^{x}$

Integrating factor: $e^{\int (2/x)\, dx} = e^{\ln x^2} = x^2$

$$yx^2 = \int x^2 \frac{1}{x}e^{x}\, dx = e^{x}(x-1) + C$$

$$y = \frac{e^{x}}{x^2}(x-1) + \frac{C}{x^2}$$

**57.** $3(y - 4x^2)\, dx = -x\, dy$

$$x\frac{dy}{dx} = -3y + 12x^2$$

$$y' + \frac{3}{x}y = 12x$$

Integrating factor: $e^{\int (3/x)\, dx} = e^{3\ln x} = x^3$

$$y'x^3 + \frac{3}{x}x^3 y = 12x(x^3) = 12x^4$$

$$yx^3 = \int 12x^4\, dx = \frac{12}{5}x^5 + C$$

$$y = \frac{12}{5}x^2 + \frac{C}{x^3}$$

**59.** $y' + 3x^2 y = x^2 y^3$

$n = 3, Q = x^2, P = 3x^2$

$y^{-2} e^{\int (-2)3x^2 \, dx} = \int (-2) x^2 e^{\int (-2)3x^2 \, dx} \, dx$

$y^{-2} e^{-2x^3} = -\int 2x^2 e^{-2x^3} \, dx$

$y^{-2} e^{-2x^3} = \frac{1}{3} e^{-2x^3} + C$

$y^{-2} = \frac{1}{3} + Ce^{2x^3}$

$\frac{1}{y^2} + Ce^{2x^3} + \frac{1}{3}$

**61.** $y' + \left(\frac{1}{x}\right) y = xy^2$

$n = 2, Q = x, P = x^{-1}$

$e^{\int -(1/x) \, dx} = e^{-\ln|x|} = x^{-1}$

$y^{-1} x^{-1} = \int -x(x^{-1}) \, dx = -x + C$

$\frac{1}{y} = -x^2 + Cx$

$y = \frac{1}{Cx - x^2}$

**63.** $xy' + y = xy^3$

$y' + \frac{1}{x} y = y^3$

$n = 3, Q = 1, P = \frac{1}{x}, \qquad e^{\int \frac{-2}{x} \, dx} = e^{-2 \ln x} = x^{-2}$

$y^{-2} x^{-2} = \int -2x^{-2} \, dx + C = 2x^{-1} + C$

$y^{-2} = 2x + Cx^2$

$y^2 = \frac{1}{2x + Cx^2} \qquad \text{or} \qquad \frac{1}{y^2} = 2x + Cx^2$

**65.** $y' - y = e^x \sqrt[3]{y}, n = \frac{1}{3}, Q = e^x, P = -1$

$e^{\int -(2/3) \, dx} = e^{-(2/3)x}$

$y^{2/3} e^{-(2/3)x} = \int \frac{2}{3} e^x e^{-(2/3)x} \, dx = \int \frac{2}{3} e^{(1/3)x} \, dx$

$y^{2/3} e^{-(2/3)x} = 2e^{(1/3)x} + C$

$y^{2/3} = 2e^x + Ce^{2x/3}$

**67.** False. The equation contains $\sqrt{y}$.

# Section 6.6   Predator-Prey Differential Equations

**1.** $\dfrac{dx}{dt} = ax - bxy = 0.9x - 0.05xy$

$\dfrac{dy}{dt} = -my + nxy = -0.6y + 0.008xy$

$\dfrac{dx}{dt} = \dfrac{dy}{dt} = 0 \Rightarrow 0.9x - 0.05xy = x(0.9 - 0.05y) = 0$

$-0.6y + 0.008xy = y(-0.6 + 0.008x) = 0$

If $x = 0$, then $y = 0$.

If $y = \dfrac{0.9}{0.05} = \dfrac{90}{5} = 18$, then $x = \dfrac{0.6}{0.008} = \dfrac{600}{8} = 75$.

Solutions: $(0, 0)$ and $(75, 18)$

**3.** $\dfrac{dx}{dt} = ax - bxy = 0.5x - 0.01xy$

$\dfrac{dy}{dt} = -my + nxy = -0.49y + 0.007xy$

$\dfrac{dx}{dt} = \dfrac{dy}{dt} = 0 \Rightarrow 0.5x - 0.01xy = x(0.5 - 0.01y) = 0$

$-0.49y + 0.007xy = y(-0.49 + 0.007x) = 0$

If $x = 0$, then $y = 0$.

If $y = \dfrac{0.5}{0.01} = \dfrac{50}{1} = 50$, then $x = \dfrac{0.49}{0.007} = \dfrac{490}{7} = 70$.

Solutions: $(0, 0)$ and $(70, 50)$

**5. (a)**

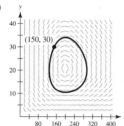

**(b)**

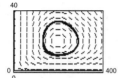

**7. (a)** The initial conditions are $x(0) = 40$ and $y(0) = 20$.

**(b)**

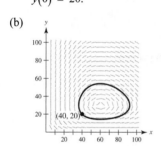

**9.** Critical points are $(x, y) = (0,0)$ and

$$(x, y) = \left(\frac{m}{n}, \frac{a}{b}\right) = \left(\frac{0.3}{0.006}, \frac{0.8}{0.04}\right) = (50, 20).$$

**11.**

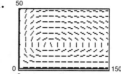

**13.** Critical points are $(x, y) = (0,0)$ and

$$(x, y) = \left(\frac{m}{n}, \frac{a}{b}\right)$$
$$= \left(\frac{0.4}{0.00004}, \frac{0.1}{0.00008}\right)$$
$$= (10{,}000, 1250).$$

**15.**

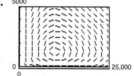

**17.** Using $x(0) = 50$ and $y(0) = 20$, you obtain the constant solutions $x = 50$ and $y = 20$.

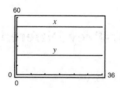

The slope field is the same, but the solution curve reduces to a single point at $(50, 20)$.

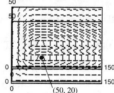

**19.** $\dfrac{dx}{dt} = ax - bx^2 - cxy = 2x - 3x^2 - 2xy$

$\dfrac{dy}{dt} = my - ny^2 - pxy = 2y - 3y^2 - 2xy$

From Example 5, you have $(0, 0)$, $\left(0, \dfrac{m}{n}\right) = \left(0, \dfrac{2}{3}\right)$, $\left(\dfrac{a}{b}, 0\right) = \left(\dfrac{2}{3}, 0\right)$ and

$$\left(\frac{an - mc}{bn - cp}, \frac{bm - ap}{bn - cp}\right) = \left(\frac{6 - 4}{9 - 4}, \frac{6 - 4}{9 - 4}\right) = \left(\frac{2}{5}, \frac{2}{5}\right).$$

**21.** $\dfrac{dx}{dt} = ax - bx^2 - cxy = 0.15x - 0.6x^2 - 0.75xy$

$\dfrac{dy}{dt} = my - ny^2 - pxy = 0.15y - 12y^2 - 0.45xy$

From Example 5, you have $(0, 0)$, $\left(0, \dfrac{m}{n}\right) = \left(0, \dfrac{1}{8}\right)$, $\left(\dfrac{a}{b}, 0\right) = \left(\dfrac{1}{4}, 0\right)$ and

$\left(\dfrac{an - mc}{bn - cp}, \dfrac{bm - ap}{bn - cp}\right) = \left(\dfrac{0.18 - 0.1125}{0.72 - 0.3375}, \dfrac{0.09 - 0.0675}{0.72 - 0.3375}\right) = \left(\dfrac{3}{17}, \dfrac{1}{17}\right) \approx (0.1765, 0.0588)$.

**23.** $a = 0.8$, $b = 0.4$, $c = 0.1$, $m = 0.3$, $n = 0.6$, $p = 0.1$

Four critical points:

$(0, 0)$

$\left(0, \dfrac{m}{n}\right) = \left(0, \dfrac{0.3}{0.6}\right) = \left(0, \dfrac{1}{2}\right)$

$\left(\dfrac{a}{b}, 0\right) = \left(\dfrac{0.8}{0.4}, 0\right) = (2, 0)$

$\left(\dfrac{an - mc}{bn - cp}, \dfrac{bm - ap}{bn - cp}\right) = \left(\dfrac{0.45}{0.23}, \dfrac{0.04}{0.23}\right) = \left(\dfrac{45}{23}, \dfrac{4}{23}\right)$

**25.** $a = 0.8$, $b = 0.4$, $c = 1$, $m = 0.3$, $n = 0.6$, $p = 1$

Four critical points:

$(0, 0)$

$\left(0, \dfrac{m}{n}\right) = \left(0, \dfrac{0.3}{0.6}\right) = \left(0, \dfrac{1}{2}\right)$

$\left(\dfrac{a}{b}, 0\right) = \left(\dfrac{0.8}{0.4}, 0\right) = (2, 0)$

$\left(\dfrac{an - mc}{bn - cp}, \dfrac{bm - ap}{bn - cp}\right) = \left(\dfrac{0.18}{-0.76}, \dfrac{-0.68}{-0.76}\right)$

$= \left(-\dfrac{9}{38}, \dfrac{17}{19}\right)$

**27.** Assuming the initial conditions are the critical points

$\left(x(0), y(0)\right) = \left(\dfrac{45}{23}, \dfrac{4}{23}\right)$

you obtain constant solutions.

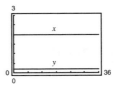

**29.** Yes, they are separable. See bottom of page 437.

**31.** As in Exercise 30, using any of the four critical points as initial conditions will yield constant solutions.

**33.** (a) If $y = 0$, then $\dfrac{dx}{dt} = ax\left(1 - \dfrac{x}{L}\right)$, which is a logistic equation.

(b) $\dfrac{dx}{dt} = 0.4x\left(1 - \dfrac{x}{100}\right) - 0.01xy$

$\dfrac{dy}{dt} = -0.3y + 0.005xy$

$(0, 0)$ is a critical point. If $y = 0$, then $x = 100$ and $(100, 0)$ is a critical point. If $x, y \neq 0$, then

$0.4\left(1 - \dfrac{x}{100}\right) = 0.01y$

$-0.3 + 0.05x = 0$.

So, $x = \dfrac{0.3}{0.005} = 60$ and $0.4\left(1 - \dfrac{60}{100}\right) = 0.01y \Rightarrow y = 16$.

The third critical point is $(60, 16)$.

(c)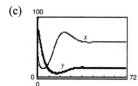

(d)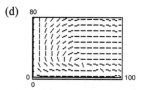

(e)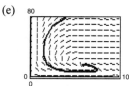

# Review Exercises for Chapter 6

**1.** $y = x^3$, $y' = 3x^2$

$2xy' + 4y = 2x(3x^2) + 4(x^3) = 10x^3$.

Yes, it is a solution.

**3.** $\dfrac{dy}{dx} = 4x^2 + 7$

$y = \int (4x^2 + 7)\, dx = \dfrac{4x^3}{3} + 7x + C$

**5.** $\dfrac{dy}{dx} = \cos 2x$

$y = \int \cos 2x\, dx = \dfrac{1}{2} \sin 2x + C$

**7.** $\dfrac{dy}{dx} = e^{2-x}$

$y = \int e^{2-x}\, dx = -e^{2-x} + C$

**9.** $\dfrac{dy}{dx} = 2x - y$

| $x$ | $-4$ | $-2$ | 0 | 2 | 4 | 8 |
|---|---|---|---|---|---|---|
| $y$ | 2 | 0 | 4 | 4 | 6 | 8 |
| $dy/dx$ | $-10$ | $-4$ | $-4$ | 0 | 2 | 8 |

**11.** $y' = 2x^2 - x$, $\quad (0, 2)$

(a) and (b)

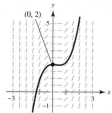

**13.** $y' = x - y$, $y(0) = 4$, $n = 10$, $h = 0.05$

$y_1 = y_0 + hf(x_0, y_0) = 4 + (0.05)(0 - 4) = 3.8$

$y_2 = y_1 + hf(x_1, y_1) = 3.8 + (0.05)(0.05 - 3.8) = 3.6125$, etc.

| $n$ | 0 | 1 | 2 | 3 | 4 | 5 | 6 | 7 | 8 | 9 | 10 |
|---|---|---|---|---|---|---|---|---|---|---|---|
| $x_n$ | 0 | 0.05 | 0.1 | 0.15 | 0.2 | 0.25 | 0.3 | 0.35 | 0.4 | 0.45 | 0.5 |
| $y_n$ | 4 | 3.8 | 3.6125 | 3.437 | 3.273 | 3.119 | 2.975 | 2.842 | 2.717 | 2.601 | 2.494 |

**15.** $\dfrac{dy}{dx} = 2x - 5x^2$

$y = \int (2x - 5x^2)\, dx = x^2 - \dfrac{5}{3}x^3 + C$

**17.** $\dfrac{dy}{dx} = (3 + y)^2$

$\int (3 + y)^{-2}\, dy = \int dx$

$-(3 + y)^{-1} = x + C$

$3 + y = \dfrac{-1}{x + C}$

$y = -3 - \dfrac{1}{x + C}$

**19.** $(2 + x)y' - xy = 0$

$$(2 + x)\frac{dy}{dx} = xy$$

$$\frac{1}{y}\,dy = \frac{x}{2 + x}dx$$

$$\frac{1}{y}\,dy = \left(1 - \frac{2}{2 + x}\right)dx$$

$$\ln|y| = x - 2\ln|2 + x| + C_1$$

$$y = Ce^x(2 + x)^{-2} = \frac{Ce^x}{(2 + x)^2}$$

**21.** $\dfrac{dy}{dt} = \dfrac{k}{t^3}$

$$\int dy = \int kt^{-3}\,dt$$

$$y = -\frac{k}{2t^2} + C$$

**23.** $y = Ce^{kt}$

$\left(0, \frac{3}{4}\right): \frac{3}{4} = C$

$(5, 5): 5 = \frac{3}{4}e^{k(5)}$

$$\frac{20}{3} = e^{5k}$$

$$k = \frac{1}{5}\ln\left(\frac{20}{3}\right)$$

$$y = \frac{3}{4}e^{[\ln(20/3)/5]t} \approx \frac{3}{4}e^{0.379t}$$

**25.** $y = Ce^{kt}$

$\left(2, \frac{3}{2}\right): \frac{3}{2} = Ce^{2k} \Rightarrow C = \frac{3}{2}e^{-2k}$

$(4, 5): 5 = Ce^{4k} = \left(\frac{3}{2}e^{-2k}\right)e^{4k} = \frac{3}{2}e^{2k}$

$$\frac{10}{3} = e^{2k} \Rightarrow k = \frac{1}{2}\ln\left(\frac{10}{3}\right)$$

So, $C = \frac{3}{2}e^{-2(1/2)\ln(10/3)} = \frac{3}{2}\left(\frac{3}{10}\right) = \frac{9}{20}$.

$$y = \frac{9}{20}e^{1/2\ln(10/3)t} \approx \frac{9}{20}e^{0.602t}$$

**27.** $\dfrac{dP}{dh} = kp, \quad P(0) = 30$

$$P(h) = 30e^{kh}$$

$$P(18{,}000) = 30e^{18{,}000k} = 15$$

$$k = \frac{\ln(1/2)}{18{,}000} = \frac{-\ln 2}{18{,}000}$$

$$P(h) = 30e^{-(h\ln 2)/18{,}000}$$

$$P(35{,}000) = 30e^{-(35{,}000\ln 2)/18{,}000} \approx 7.79 \text{ inches}$$

**29.** $P = Ce^{0.0185t}$

$$2C = Ce^{0.0185t}$$

$$2 = e^{0.0185t}$$

$$\ln 2 = 0.0185t$$

$$t = \frac{\ln 2}{0.0185} \approx 37.5 \text{ years}$$

**31.** $S = Ce^{k/t}$

(a) $\qquad S = 5$ when $t = 1$

$$5 = Ce^k$$

$$\lim_{t \to \infty} Ce^{k/t} = C = 30$$

$$5 = 30e^k$$

$$k = \ln\frac{1}{6} \approx -1.7918$$

$$S = 30e^{-1.7918/t}$$

(b) When $t = 5$, $S \approx 20.9646$ which is 20,965 units.

(c)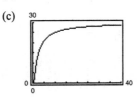

**33.** $\dfrac{dy}{dx} = \dfrac{5x}{y}$

$$\int y\,dy = \int 5x\,dx$$

$$\frac{y^2}{2} = \frac{5x^2}{2} + C_1$$

$$y^2 = 5x^2 + C$$

**35.** $y' - 16xy = 0$

$$\frac{dy}{dx} = 16xy$$

$$\int \frac{1}{y}\,dy = \int 16x\,dx$$

$$\ln|y| = 8x^2 + C_1$$

$$e^{8x^2 + C_1} = y$$

$$y = Ce^{8x^2}$$

**37.** $y^3 y' - 3x = 0$, $y(2) = 2$

$$y^3 \frac{dy}{dx} = 3x$$

$$\int y^3 \, dy = \int 3x \, dx$$

$$\frac{y^4}{4} = \frac{3x^2}{2} + C_1$$

$$y^4 = 6x^2 + C$$

Initial condition: $y(2) = 2 : 16 = 24 + C$

$$C = -8$$

Particular solution: $y^4 = 6x^2 - 8$

**39.** $y^3(x^4 + 1)y' - x^3(y^4 + 1) = 0$, $y(0) = 1$

$$y^3(x^4 + 1)\frac{dy}{dx} = x^3(y^4 + 1)$$

$$\int \frac{y^3}{y^4 + 1} \, dy = \int \frac{x^3}{x^4 + 1} \, dx$$

$$\frac{1}{4} \ln(y^4 + 1) = \frac{1}{4} \ln(x^4 + 1) + \frac{1}{4} \ln C_1$$

$$\ln(y^4 + 1) = \ln\left[C(x^4 + 1)\right]$$

$$y^4 + 1 = C(x^4 + 1)$$

Initial condition: $y(0) = 1 : 1 + 1 = C(0 + 1)$

$$C = 2$$

Particular solution: $y^4 + 1 = 2(x^4 + 1)$

$$y^4 = 2x^4 + 1$$

**41.** $\dfrac{dy}{dx} = \dfrac{-4x}{y}$

$$\int y \, dy = \int -4x \, dx$$

$$\frac{y^2}{2} = -2x^2 + C_1$$

$$4x^2 + y^2 = C \qquad \text{ellipses}$$

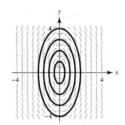

**43.** $P(t) = \dfrac{5250}{1 + 34e^{-0.55t}}$

(a) $k = 0.55$

(b) $L = 5250$

(c) $P(0) = \dfrac{5250}{1 + 34} = 150$

(d) $\qquad 2625 = \dfrac{5250}{1 + 34e^{-0.55t}}$

$$1 + 34e^{-0.55t} = 2$$

$$e^{-0.55t} = \frac{1}{34}$$

$$t = \frac{-1}{0.55} \ln\left(\frac{1}{34}\right) \approx 6.41 \text{ yr}$$

(e) $\dfrac{dP}{dt} = 0.55P\left(1 - \dfrac{P}{5250}\right)$

**45.** $\dfrac{dy}{dt} = y\left(1 - \dfrac{y}{80}\right)$, $\quad (0, 8)$

$$k = 1, L = 80$$

$$y = \frac{L}{1 + be^{-kt}} = \frac{80}{1 + be^{-t}}$$

$$y(0) = 8: \quad 8 = \frac{80}{1 + b} \Rightarrow b = 9$$

Solution: $y = \dfrac{80}{1 + 9e^{-t}}$

**47.** (a) $L = 20{,}400$, $y(0) = 1200$, $y(1) = 2000$

$$y = \frac{20{,}400}{1 + be^{-kt}}$$

$$y(0) = 1200 = \frac{20{,}400}{1 + b} \Rightarrow b = 16$$

$$y(1) = 2000 = \frac{20{,}400}{1 + 16e^{-k}}$$

$$16e^{-k} = \frac{46}{5}$$

$$k = -\ln\frac{23}{40} = \ln\frac{40}{23} \approx 0.553$$

$$y = \frac{20{,}400}{1 + 16e^{-0.553t}}$$

(b) $y(8) \approx 17{,}118$ trout

(c) $10{,}000 = \dfrac{20{,}400}{1 + 16e^{-0.553t}} \Rightarrow t \approx 4.94$ yr

**49.**
$$\frac{dS}{dt} = k(L - S)$$

$$\int \frac{dS}{L - S} = \int k \, dt$$

$$-\ln|L - S| = kt + C_1$$

$$L - S = e^{-kt - C_1}$$

$$S = L + Ce^{-kt}$$

Because $S = 0$ when $t = 0$, you have

$0 = L + C \Rightarrow C = -L$. So, $S = L(1 - e^{-kt})$.

**51.** The differential equation is given by the following.

$$\frac{dP}{dn} = kP(L - P)$$

$$\int \frac{1}{P(L - P)} \, dP = \int k \, dn$$

$$\frac{1}{L}\left[\ln|P| - \ln|L - P|\right] = kn + C_1$$

$$\frac{P}{L - P} = Ce^{Lkn}$$

$$P = \frac{CLe^{Lkn}}{1 + Ce^{Lkn}} = \frac{CL}{e^{-Lkn} + C}$$

**53. (a)**

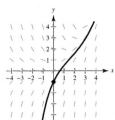

**(b)** $y' = e^{x/2} - y$

$y' + y = e^{x/2}$, Integrating factor: $e^{\int dx} = e^x$

$$ye^x = \int e^{x/2}e^x \, dx = \int e^{(3/2)x} \, dx$$

$$= \frac{2}{3}e^{(3/2)x} + C$$

$$y = \frac{2}{3}e^{x/2} + Ce^{-x}$$

$$y(0) = -1 = \frac{2}{3} + C \Rightarrow C = -\frac{5}{3}$$

$$y = \frac{2}{3}e^{x/2} - \frac{5}{3}e^{-x} = \frac{1}{3}\left[2e^{x/2} - 5e^{-x}\right]$$

**(c)**

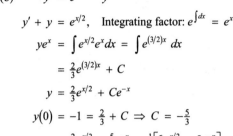

**55. (a)**

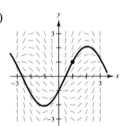

**(b)**
$$\frac{dy}{dx} = \csc x + y \cot x$$

$$\frac{dy}{dx} - (\cot x)y = \csc x$$

Integrating factor: $e^{\int -\cot x \, dx} = e^{-\ln|\sin x|} = \csc x$

$$\csc x \cdot y' - \csc x \cot x \cdot y = \csc^2 x$$

$$(y \csc x)' = \csc^2 x$$

$$y \csc x = \int \csc^2 x \, dx = -\cot x + C$$

$$y = -\cos x + C \sin x$$

$$y(1) = 1 \Rightarrow 1 = -\cos 1 + C \sin 1$$

$$\Rightarrow C = \frac{1 + \cos 1}{\sin 1} \approx 1.8305$$

$$y = -\cos x + 1.8305 \sin x$$

**(c)**

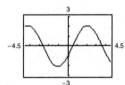

**57.** $y' - y = 10$

$P(x) = -1, Q(x) = 10$

$u(x) = e^{\int -dx} = e^{-x}$

$$y = \frac{1}{e^{-x}} \int 10e^{-x} \, dx$$

$$= e^x\left(-10e^{-x} + C\right)$$

$$= -10 + Ce^x$$

**59.** $4y' = e^{x/y} + y$

$$y' - \frac{1}{4}y = \frac{1}{4}e^{x/4}$$

$$P(x) = -\frac{1}{4}, Q(x) = \frac{1}{4}e^{x/4}$$

$$u(x) = e^{\int -(1/4) \, dx} = e^{-(1/4)x}$$

$$y = \frac{1}{e^{-(1/4)x}} \int \frac{1}{4}e^{x/4}e^{-(1/4)x} \, dx$$

$$= e^{(1/4)x}\left(\frac{1}{4}x + C\right)$$

$$= \frac{1}{4}xe^{x/4} + Ce^{x/4}$$

**61.** $(x - 2)y' + y = 1$

$$\frac{dy}{dx} + \frac{1}{x-2}y = \frac{1}{x-2}$$

$$P(x) = \frac{1}{x-2}, Q(x) = \frac{1}{x-2}$$

$$u(x) = e^{\int (1/x-2)\, dx} = e^{\ln|x-2|} = x - 2$$

$$y = \frac{1}{x-2}\int \left(\frac{1}{x-2}\right)(x-2)\, dx = \frac{1}{x-2}(x + C)$$

**63.** $y' + 5y = e^{5x}$

Integrating factor: $e^{\int 5\, dx} = e^{5x}$

$$ye^{5x} = \int e^{10x}\, dx = \tfrac{1}{10}e^{10x} + C$$

$$y = \tfrac{1}{10}e^{5x} + Ce^{-5x}$$

**65.** $y' + 5y = e^{5x}, y(0) = 3$

$$P(x) = 5, Q(x) = e^{5x}$$

$$u(x) = e^{\int 5\, dx} = e^{5x}$$

$$y = \frac{1}{e^{5x}}\int (e^{5x})(e^{5x})\, dx$$

$$= \frac{1}{e^{5x}}\int e^{10x}\, dx$$

$$= \frac{1}{e^{5x}}\int \left(\frac{1}{10}e^{10x} + C\right)$$

$$= \frac{1}{10}e^{5x} + Ce^{-5x}$$

Initial condition:

$$y(0) = 3 : 3 = \frac{1}{10}e^0 + Ce^0 \Rightarrow C = \frac{29}{10}$$

Particular solution: $y = \dfrac{1}{10}e^{5x} + \dfrac{29}{10}e^{-5x}$

**67.** Answers will vary. *Sample answer:* $(x^2 + 3y^2)\, dx - 2xy\, dy = 0$

Solution: Let $y = vx,\ dy = x\, dv + v\, dx$.

$$(x^2 + 3v^2x^2)\, dx - 2x(vx)(x\, dv + v\, dx) = 0$$

$$(x^2 + v^2x^2)\, dx - 2x^3v\, dv = 0$$

$$(1 + v^2)\, dx = 2\, xv\, dv$$

$$\int \frac{dx}{x} = \int \frac{2v}{1+v^2}\, dv$$

$$\ln|x| = \ln|1 + v^2| + C_1$$

$$x = C(1 + v^2) = C\left(1 + \frac{y^2}{x^2}\right)$$

$$x^3 = C(x^2 + y^2)$$

**69.** Answers will vary.

*Sample answer:* $x^3y' + 2x^2y = 1$

$$y' + \frac{2}{x}y = \frac{1}{x^3}$$

$$u(x) = e^{\int (2/x)\, dx} = x^2$$

$$y = \frac{1}{x^2}\int \frac{1}{x^3}(x^2)\, dx = \frac{1}{x^2}\left[\ln|x| + C\right]$$

**71.** $A_0 = 500{,}000, \quad r = 0.10$

(a) $P = 40{,}000$

$$A = \frac{40{,}000}{0.10} + \left(500{,}000 - \frac{40{,}000}{0.10}\right)e^{0.10t} = 100{,}000(4 + e^{0.10t})$$

The balance continues to increase.

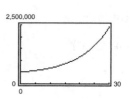

(b) $P = 50,000$

$$A = \frac{50,000}{0.10} + \left(500,000 - \frac{50,000}{0.10}\right)e^{0.10t} = 500,000$$

The balance remains at $500,000.

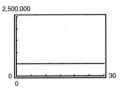

(c) $P = 60,000$

$$A = \frac{60,000}{0.10} + \left(500,000 - \frac{60,000}{0.10}\right)e^{0.10t} = 100,000\left(6 - e^{0.10t}\right)$$

The balance decreases and is depleted in $t = (\ln 6)/0.10 \approx 17.9$ years.

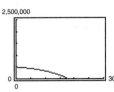

**73.** (a) $\dfrac{dx}{dt} = ax - bxy = 0.3x - 0.02xy$

$\dfrac{dy}{dt} = -my + nxy = -0.4y + 0.01xy$

(b) $x' = y' = 0$ when $(x, y) = (0, 0)$ and

$$(x, y) = \left(\frac{m}{n}, \frac{a}{b}\right) = \left(\frac{0.4}{0.01}, \frac{0.3}{0.02}\right) = (40, 14).$$

(c)

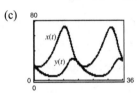

**75.** (a) $\dfrac{dx}{dt} = ax - bx^2 - cxy = 3x - x^2 - xy$

$\dfrac{dy}{dt} = my - ny^2 - pxy = 2y - y^2 - 0.5xy$

(b) $x' = y' = 0$ when $(x, y) = (0, 0)$,

$$(x, y) = \left(0, \frac{m}{n}\right) = (0, 2),$$

$$(x, y) = \left(\frac{a}{b}, 0\right) = (3, 0),$$

$$(x, y) = \left(\frac{an - mc}{bn - cp}, \frac{bm - ap}{bn - cp}\right)$$

$$= \left(\frac{1}{1/2}, \frac{1/2}{1/2}\right) = (2, 1).$$

(c)

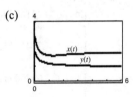

# Problem Solving for Chapter 6

**1.** (a) $\dfrac{dy}{dt} = y^{1.01}$

$$\int y^{-1.01}\, dy = \int dt$$

$$\frac{y^{-0.01}}{-0.01} = t + C_1$$

$$\frac{1}{y^{0.01}} = -0.01t + C$$

$$y^{0.01} = \frac{1}{C - 0.01t}$$

$$y = \frac{1}{(C - 0.01t)^{100}}$$

$y(0) = 1:\ 1 = \dfrac{1}{C^{100}} \Rightarrow C = 1$

So, $y = \dfrac{1}{(1 - 0.01t)^{100}}$.

For $T = 100$, $\lim\limits_{t \to T^-} y = \infty$.

(b) $\displaystyle\int y^{-(1+\varepsilon)}\, dy = \int k\, dt$

$$\frac{y^{-\varepsilon}}{-\varepsilon} = kt + C_1$$

$$y^{-\varepsilon} = -\varepsilon kt + C$$

$$y = \frac{1}{(C - \varepsilon kt)^{1/\varepsilon}}$$

$y(0) = y_0 = \dfrac{1}{C^{1/\varepsilon}} \Rightarrow C^{1/\varepsilon} = \dfrac{1}{y_0} \Rightarrow C = \left(\dfrac{1}{y_0}\right)^{\varepsilon}$

So, $y = \dfrac{1}{\left(\dfrac{1}{y_0^{\varepsilon}} - \varepsilon kt\right)^{1/\varepsilon}}$.

For $t \to \dfrac{1}{y_0^{\varepsilon}\varepsilon k}$, $y \to \infty$.

**3. (a)** $y' = x - y, y(0) = 1, h = 0.1$

Using the modified Euler Method, you obtain:

| $x$ | $y$ |
|---|---|
| 0 | 1.0 |
| 0.1 | 0.91 |
| 0.2 | 0.83805 |
| $\vdots$ | $\vdots$ |
| 1.0 | 0.73708 |

**(b)**

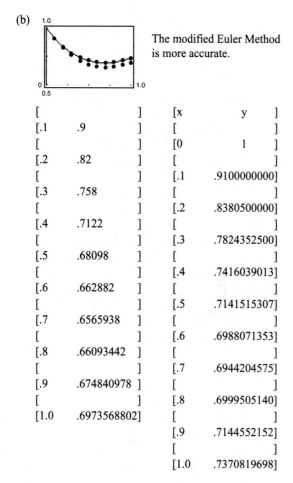

The modified Euler Method is more accurate.

| [ | ] | [x | y | ] |
|---|---|---|---|---|
| [.1 | .9 ] | [ | | ] |
| [ | ] | [0 | 1 | ] |
| [.2 | .82 ] | [ | | ] |
| [ | ] | [.1 | .9100000000] | |
| [.3 | .758 ] | [ | | ] |
| [ | ] | [.2 | .8380500000] | |
| [.4 | .7122 ] | [ | | ] |
| [ | ] | [.3 | .7824352500] | |
| [.5 | .68098 ] | [ | | ] |
| [ | ] | [.4 | .7416039013] | |
| [.6 | .662882 ] | [ | | ] |
| [ | ] | [.5 | .7141515307] | |
| [.7 | .6565938 ] | [ | | ] |
| [ | ] | [.6 | .6988071353] | |
| [.8 | .66093442 ] | [ | | ] |
| [ | ] | [.7 | .6944204575] | |
| [.9 | .674840978 ] | [ | | ] |
| [ | ] | [.8 | .6999505140] | |
| [1.0 | .6973568802] | [ | | ] |
| | | [.9 | .7144552152] | |
| | | [ | | ] |
| | | [1.0 | .7370819698] | |

**5.** $k = \left(\dfrac{1}{12}\right)^2 \pi$

$g = 32$

$x^2 + (y - 6)^2 = 36$   Equation of tank

$x^2 = 36 - (y - 6)^2 = 12y - y^2$

Area of cross section: $A(h) = (12h - h^2)\pi$

$$A(h)\frac{dh}{dt} = -k\sqrt{2gh}$$

$$(12h - h^2)\pi\frac{dh}{dt} = -\frac{1}{144}\pi\sqrt{64h}$$

$$(12h - h^2)\frac{dh}{dt} = -\frac{1}{18}h^{1/2}$$

$$\int(18h^{3/2} - 216h^{1/2})\,dh = \int dt$$

$$\frac{36}{5}h^{5/2} - 144h^{3/2} = t + C$$

$$\frac{h^{3/2}}{5}(36h - 720) = t + C$$

When $h = 6, t = 0$ and $C = \dfrac{6^{3/2}}{5}(-504) \approx -1481.45.$

The tank is completely drained when

$h = 0 \Rightarrow t = 1481.45 \text{ sec} \approx 24 \text{ min, } 41 \text{ sec}$

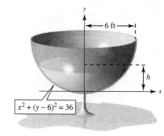

**7.** $A(h)\dfrac{dh}{dt} = -k\sqrt{2gh}$

$$\pi 64\frac{dh}{dt} = \frac{-\pi}{36}8\sqrt{h}$$

$$\int h^{-1/2}\,dh = \int\frac{-1}{288}\,dt$$

$$2\sqrt{h} = \frac{-t}{288} + C$$

$h = 20: 2\sqrt{20} = C = 4\sqrt{5}$

$$2\sqrt{h} = \frac{-t}{288} + 4\sqrt{5}$$

$h = 0 \Rightarrow t = 4\sqrt{5}(288)$

$\approx 2575.95 \text{ sec} \approx 42 \text{ min, } 56 \text{ sec}$

**9.** $\dfrac{ds}{dt} = 3.5 - 0.019s$

(a) $\displaystyle\int \dfrac{-ds}{3.5 - 0.019s} = -\int dt$

$\dfrac{1}{0.019}\ln|3.5 - 0.019s| = -t + C_1$

$\ln|3.5 - 0.019s| = -0.019t + C_2$

$3.5 - 0.019s = C_3 e^{-0.019t}$

$0.019s = 3.5 - C_3 e^{-0.019t}$

$s = 184.21 - Ce^{-0.019t}$

(b)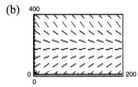

(c) As $t \to \infty$, $Ce^{-0.019t} \to 0$, and $s \to 184.21$.

**11.** (a) $\displaystyle\int \dfrac{dC}{C} = \int -\dfrac{R}{V}\,dt$

$\ln|C| = -\dfrac{R}{V}t + K_1$

$C = Ke^{-Rt/V}$

Since $C = C_0$ when $t = 0$, it follows that $K = C_0$ and the function is $C = C_0 e^{-Rt/V}$.

(b) Finally, as $t \to \infty$, we have

$\displaystyle\lim_{t \to \infty} C = \lim_{t \to \infty} C_0 e^{-Rt/V} = 0$.

**13.** (a) $\displaystyle\int \dfrac{1}{Q - RC}\,dC = \int \dfrac{1}{V}\,dt$

$-\dfrac{1}{R}\ln|Q - RC| = \dfrac{t}{V} + K_1$

$Q - RC = e^{-R[(t/V) + K_1]}$

$C = \dfrac{1}{R}\left(Q - e^{-R[(t/V) + K_1]}\right) = \dfrac{1}{R}\left(Q - Ke^{-Rt/V}\right)$

Because $C = 0$ when $t = 0$, it follows that $K = Q$ and you have $C = \dfrac{Q}{R}\left(1 - e^{-Rt/V}\right)$.

(b) As $t \to \infty$, the limit of $C$ is $Q/R$.

# C H A P T E R  7
## Applications of Integration

# CHAPTER 7
## Applications of Integration

### Section 7.1  Area of a Region Between Two Curves

**1.** $A = \int_0^6 \left[ 0 - \left( x^2 - 6x \right) \right] dx = -\int_0^6 \left( x^2 - 6x \right) dx$

**3.** $A = \int_0^3 \left[ \left( -x^2 + 2x + 3 \right) - \left( x^2 - 4x + 3 \right) \right] dx$

$= \int_0^3 \left( -2x^2 + 6x \right) dx$

**5.** $A = 2\int_{-1}^0 3\left( x^3 - x \right) dx = 6\int_{-1}^0 \left( x^3 - x \right) dx$

or $-6\int_0^1 \left( x^3 - x \right) dx$

**7.** $\int_0^4 \left[ (x + 1) - \frac{x}{2} \right] dx$

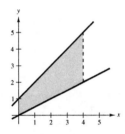

**9.** $\int_2^3 \left[ \left( \frac{x^3}{3} - x \right) - \frac{x}{3} \right] dx$

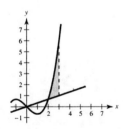

**11.** $\int_{-2}^1 \left[ (2 - y) - y^2 \right] dy$

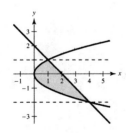

**13.** $f(x) = x + 1$

$g(x) = (x - 1)^2$

$A \approx 4$

Matches (d)

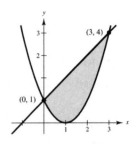

**14.** $f(x) = 2 - \frac{1}{2}x$

$g(x) = 2 - \sqrt{x}$

$A \approx 1$

Matches (a)

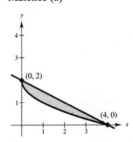

**15.** (a)
$$x = 4 - y^2$$
$$x = y - 2$$
$$4 - y^2 = y - 2$$
$$y^2 + y - 6 = 0$$
$$(y + 3)(y - 2) = 0$$

Intersection points: $(0, 2)$ and $(-5, -3)$

$A = \int_{-5}^0 \left[ (x + 2) + \sqrt{4 - x} \right] dx + \int_0^4 2\sqrt{4 - x}\, dx$

$= \frac{61}{6} + \frac{32}{3} = \frac{125}{6}$

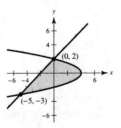

(b) $A = \int_{-3}^2 \left[ \left( 4 - y^2 \right) - (y - 2) \right] dy = \frac{125}{6}$

(c) The second method is simpler. Explanations will vary.

**17.**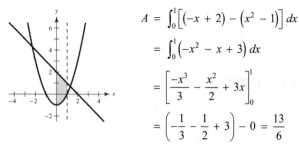

$$A = \int_0^1 \left[ (-x + 2) - \left(x^2 - 1\right) \right] dx$$

$$= \int_0^1 \left(-x^2 - x + 3\right) dx$$

$$= \left[ \frac{-x^3}{3} - \frac{x^2}{2} + 3x \right]_0^1$$

$$= \left( -\frac{1}{3} - \frac{1}{2} + 3 \right) - 0 = \frac{13}{6}$$

**19.** The points of intersection are given by:

$$x^2 + 2x = x + 2$$

$$x^2 + x - 2 = 0$$

$$(x + 2)(x - 1) = 0 \quad \text{when } x = -2, 1$$

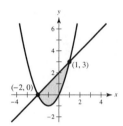

$$A = \int_{-2}^1 \left[ g(x) - f(x) \right] dx$$

$$= \int_{-2}^1 \left[ (x + 2) - \left(x^2 + 2x\right) \right] dx$$

$$= \left[ \frac{-x^3}{3} - \frac{x^2}{2} + 2x \right]_{-2}^1$$

$$= \left( -\frac{1}{3} - \frac{1}{2} + 2 \right) - \left( \frac{8}{3} - 2 - 4 \right) = \frac{9}{2}$$

**21.** The points of intersection are given by:

$$x = 2 - x \quad \text{and} \quad x = 0 \quad \text{and} \quad 2 - x = 0$$

$$x = 1 \qquad\qquad x = 0 \qquad\qquad x = 2$$

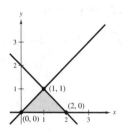

$$A = \int_0^1 \left[ (2 - y) - (y) \right] dy = \left[ 2y - y^2 \right]_0^1 = 1$$

Note that if you integrate with respect to $x$, you need two integrals. Also, note that the region is a triangle.

**23.** The points of intersection are given by:

$$\sqrt{x} + 3 = \frac{1}{2}x + 3$$

$$\sqrt{x} = \frac{1}{2}x$$

$$x = \frac{x^2}{4} \quad \text{when } x = 0, 4$$

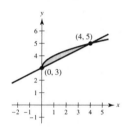

$$A = \int_0^4 \left[ \left(\sqrt{x} + 3\right) - \left( \frac{1}{2}x + 3 \right) \right] dx$$

$$= \left[ \frac{2}{3}x^{3/2} - \frac{x^2}{4} \right]_0^4 = \frac{16}{3} - 4 = \frac{4}{3}$$

**25.** The points of intersection are given by:

$$y^2 = y + 2$$

$$(y - 2)(y + 1) = 0 \quad \text{when } y = -1, 2$$

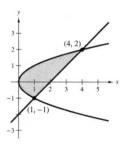

$$A = \int_{-1}^2 \left[ g(y) - f(y) \right] dy$$

$$= \int_{-1}^2 \left[ (y + 2) - y^2 \right] dy$$

$$= \left[ 2y + \frac{y^2}{2} - \frac{y^3}{3} \right]_{-1}^2 = \frac{9}{2}$$

**27.**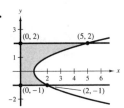

$$A = \int_{-1}^{2} \left[ f(y) - g(y) \right] dy$$

$$= \int_{-1}^{2} \left[ (y^2 + 1) - 0 \right] dy$$

$$= \left[ \frac{y^3}{3} + y \right]_{-1}^{2} = 6$$

**29.** $y = \dfrac{10}{x} \Rightarrow x = \dfrac{10}{y}$

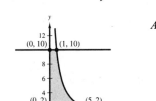

$$A = \int_{2}^{10} \frac{10}{y} \, dy$$

$$= \left[ 10 \ln y \right]_{2}^{10}$$

$$= 10(\ln 10 - \ln 2)$$

$$= 10 \ln 5 \approx 16.0944$$

**31. (a)**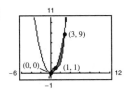

**(b)** The points of intersection are given by:

$$x^3 - 3x^2 + 3x = x^2$$

$$x(x - 1)(x - 3) = 0 \quad \text{when } x = 0, 1, 3$$

$$A = \int_{0}^{1} \left[ f(x) - g(x) \right] dx + \int_{1}^{3} \left[ g(x) - f(x) \right] dx$$

$$= \int_{0}^{1} \left[ (x^3 - 3x^2 + 3x) - x^2 \right] dx + \int_{1}^{3} \left[ x^2 - (x^3 - 3x^2 + 3x) \right] dx$$

$$= \int_{0}^{1} (x^3 - 4x^2 + 3x) \, dx + \int_{1}^{3} (-x^3 + 4x^2 - 3x) \, dx = \left[ \frac{x^4}{4} - \frac{4}{3}x^3 + \frac{3}{2}x^2 \right]_{0}^{1} + \left[ \frac{-x^4}{4} + \frac{4}{3}x^3 - \frac{3}{2}x^2 \right]_{1}^{3} = \frac{5}{12} + \frac{8}{3} = \frac{37}{12}$$

**(c)** Numerical approximation: $0.417 + 2.667 \approx 3.083$

**33. (a)** $f(x) = x^4 - 4x^2, \quad g(x) = x^2 - 4$

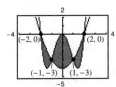

**(b)** The points of intersection are given by:

$$x^4 - 4x^2 = x^2 - 4$$

$$x^4 - 5x^2 + 4 = 0$$

$$(x^2 - 4)(x^2 - 1) = 0 \quad \text{when } x = \pm 2, \pm 1$$

By symmetry:

$$A = 2 \int_{0}^{1} \left[ (x^4 - 4x^2) - (x^2 - 4) \right] dx + 2 \int_{1}^{2} \left[ (x^2 - 4) - (x^4 - 4x^2) \right] dx$$

$$= 2 \int_{0}^{1} (x^4 - 5x^2 + 4) \, dx + 2 \int_{1}^{2} (-x^4 + 5x^2 - 4) \, dx$$

$$= 2 \left[ \frac{x^5}{5} - \frac{5x^3}{3} + 4x \right]_{0}^{1} + 2 \left[ -\frac{x^5}{5} + \frac{5x^3}{3} - 4x \right]_{1}^{2} = 2 \left[ \frac{1}{5} - \frac{5}{3} + 4 \right] + 2 \left[ \left( -\frac{32}{5} + \frac{40}{3} - 8 \right) - \left( -\frac{1}{5} + \frac{5}{3} - 4 \right) \right] = 8$$

**(c)** Numerical approximation: $5.067 + 2.933 = 8.0$

**35.** (a)

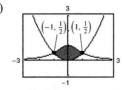

(b) The points of intersection are given by:

$$\frac{1}{1 + x^2} = \frac{x^2}{2}$$

$$x^4 + x^2 - 2 = 0$$

$$(x^2 + 2)(x^2 - 1) = 0 \quad \text{when } x = \pm 1$$

$$A = 2 \int_0^1 [f(x) - g(x)] \, dx$$

$$= 2 \int_0^1 \left[ \frac{1}{1 + x^2} - \frac{x^2}{2} \right] dx$$

$$= 2 \left[ \arctan x - \frac{x^3}{6} \right]_0^1$$

$$= 2\left( \frac{\pi}{4} - \frac{1}{6} \right) = \frac{\pi}{2} - \frac{1}{3} \approx 1.237$$

(c) Numerical approximation: 1.237

**37.** $A = \int_0^{2\pi} \left[ (2 - \cos x) - \cos x \right] dx$

$$= 2 \int_0^{2\pi} (1 - \cos x) \, dx$$

$$= 2[x - \sin x]_0^{2\pi} = 4\pi \approx 12.566$$

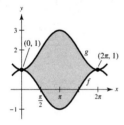

**39.** $A = 2 \int_0^{\pi/3} [f(x) - g(x)] \, dx$

$$= 2 \int_0^{\pi/3} (2 \sin x - \tan x) \, dx$$

$$= 2 \left[ -2 \cos x + \ln |\cos x| \right]_0^{\pi/3} = 2(1 - \ln 2) \approx 0.614$$

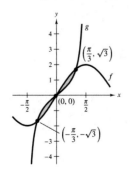

**41.** $A = \int_0^1 \left[ xe^{-x^2} - 0 \right] dx$

$$= \left[ -\frac{1}{2} e^{-x^2} \right]_0^1 = \frac{1}{2}\left( 1 - \frac{1}{e} \right) \approx 0.316$$

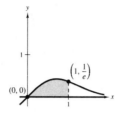

**43.** (a)

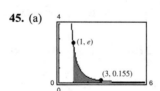

(b) $A = \int_0^{\pi} (2 \sin x + \sin 2x) \, dx$

$$= \left[ -2 \cos x - \tfrac{1}{2} \cos 2x \right]_0^{\pi}$$

$$= \left( 2 - \tfrac{1}{2} \right) - \left( -2 - \tfrac{1}{2} \right) = 4$$

(c) Numerical approximation: 4.0

**45.** (a)

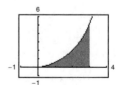

(b) $A = \int_1^3 \frac{1}{x^2} e^{1/x} \, dx$

$$= \left[ -e^{-1/x} \right]_1^3$$

$$= e - e^{1/3}$$

(c) Numerical approximation: 1.323

**47.** (a)

(b) The integral

$$A = \int_0^3 \sqrt{\frac{x^3}{4 - x}} \, dx$$

does not have an elementary antiderivative.

(c) $A \approx 4.7721$

**49.** (a)

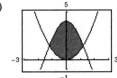

(b) The intersection points are difficult to determine by hand.

(c)  Area $= \int_{-c}^{c} \left[ 4 \cos x - x^2 \right] dx \approx 6.3043$ where $c \approx 1.201538$.

**51.** $F(x) = \int_{0}^{x} \left( \frac{1}{2}t + 1 \right) dt = \left[ \frac{t^2}{4} + t \right]_{0}^{x} = \frac{x^2}{4} + x$

(a) $F(0) = 0$

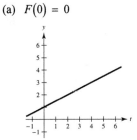

(b) $F(2) = \frac{2^2}{4} + 2 = 3$

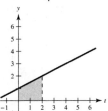

(c) $F(6) = \frac{6^2}{4} + 6 = 15$

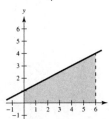

**53.** $F(\alpha) = \int_{-1}^{\alpha} \cos \frac{\pi \theta}{2} \, d\theta = \left[ \frac{2}{\pi} \sin \frac{\pi \theta}{2} \right]_{-1}^{\alpha} = \frac{2}{\pi} \sin \frac{\pi \alpha}{2} + \frac{2}{\pi}$

(a) $F(-1) = 0$

(b) $F(0) = \frac{2}{\pi} \approx 0.6366$

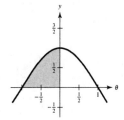

(c) $F\left(\frac{1}{2}\right) = \frac{2 + \sqrt{2}}{\pi} \approx 1.0868$

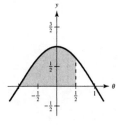

**55.** $A = \int_{2}^{4} \left[ \left( \frac{9}{2}x - 12 \right) - (x - 5) \right] dx + \int_{4}^{6} \left[ \left( -\frac{5}{2}x + 16 \right) - (x - 5) \right] dx$

$= \int_{2}^{4} \left( \frac{7}{2}x - 7 \right) dx + \int_{4}^{6} \left( -\frac{7}{2}x + 21 \right) dx = \left[ \frac{7}{4}x^2 - 7x \right]_{2}^{4} + \left[ -\frac{7}{4}x^2 + 21x \right]_{4}^{6} = 7 + 7 = 14$

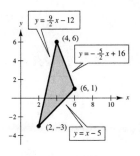

**57.**

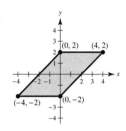

Left boundary line: $y = x + 2 \Leftrightarrow x = y - 2$

Right boundary line: $y = x - 2 \Leftrightarrow x = y + 2$

$$A = \int_{-2}^{2} \left[ (y + 2) - (y - 2) \right] dy$$

$$= \int_{-2}^{2} 4 \, dy = \left[ 4y \right]_{-2}^{2} = 8 - (-8) = 16$$

**59.** Answers will vary. *Sample answer:* If you let $\Delta x = 6$ and $n = 10$, $b - a = 10(6) = 60$.

(a)  Area $\approx \dfrac{60}{2(10)} \Big[ 0 + 2(14) + 2(14) + 2(12) + 2(12) + 2(15) + 2(20) + 2(23) + 2(25) + 2(26) + 0 \Big] = 3[322] = 966 \text{ ft}^2$

(b)  Area $\approx \dfrac{60}{3(10)} \Big[ 0 + 4(14) + 2(14) + 4(12) + 2(12) + 4(15) + 2(20) + 4(23) + 2(25) + 4(26) + 0 \Big] = 2[502] = 1004 \text{ ft}^2$

**61.**  $f(x) = x^3$

$f'(x) = 3x^2$

At $(1, 1)$, $f'(1) = 3$.

Tangent line: $y - 1 = 3(x - 1)$ or $y = 3x - 2$

The tangent line intersects $f(x) = x^3$ at $x = -2$.

$$A = \int_{-2}^{1} \left[ x^3 - (3x - 2) \right] dx = \left[ \frac{x^4}{4} - \frac{3x^2}{2} + 2x \right]_{-2}^{1} = \frac{27}{4}$$

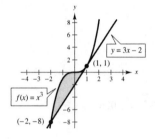
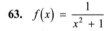

**63.**  $f(x) = \dfrac{1}{x^2 + 1}$

$f'(x) = -\dfrac{2x}{\left(x^2 + 1\right)^2}$

At $\left(1, \dfrac{1}{2}\right)$, $f'(1) = -\dfrac{1}{2}$.

Tangent line: $y - \dfrac{1}{2} = -\dfrac{1}{2}(x - 1)$ or $y = -\dfrac{1}{2}x + 1$

The tangent line intersects $f(x) = \dfrac{1}{x^2 + 1}$ at $x = 0$.

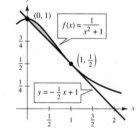

$$A = \int_{0}^{1} \left[ \frac{1}{x^2 + 1} - \left( -\frac{1}{2}x + 1 \right) \right] dx = \left[ \arctan x + \frac{x^2}{4} - x \right]_{0}^{1} = \frac{\pi - 3}{4} \approx 0.0354$$

**65.** $x^4 - 2x^2 + 1 \le 1 - x^2$ on $[-1, 1]$

$$A = \int_{-1}^{1} \left[ (1 - x^2) - (x^4 - 2x^2 + 1) \right] dx$$

$$= \int_{-1}^{1} (x^2 - x^4) \, dx$$

$$= \left[ \frac{x^3}{3} - \frac{x^5}{5} \right]_{-1}^{1} = \frac{4}{15}$$

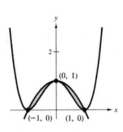

You can use a single integral because $x^4 - 2x^2 + 1 \le 1 - x^2$ on $[-1, 1]$.

**67.** (a) $\int_{0}^{5} \left[ v_1(t) - v_2(t) \right] dt = 10$ means that Car 1 traveled

10 more meters than Car 2 on the interval $0 \le t \le 5$.

$\int_{0}^{10} \left[ v_1(t) - v_2(t) \right] dt = 30$ means that Car 1

traveled 30 more meters than Car 2 on the interval $0 \le t \le 10$.

$\int_{20}^{30} \left[ v_1(t) - v_2(t) \right] dt = -5$ means that Car 2

traveled 5 more meters than Car 1 on the interval $20 \le t \le 30$.

(b) No, it is not possible because you do not know the initial distance between the cars.

(c) At $t = 10$, Car 1 is ahead by 30 meters.

(d) At $t = 20$, Car 1 is ahead of Car 2 by 13 meters. From part (a), at $t = 30$, Car 1 is ahead by $13 - 5 = 8$ meters.

**69.**

$$A = \int_{-3}^{3} (9 - x^2) \, dx = 36$$

$$\int_{-\sqrt{9-b}}^{\sqrt{9-b}} \left[ (9 - x^2) - b \right] dx = 18$$

$$\int_{0}^{\sqrt{9-b}} \left[ (9 - b) - x^2 \right] dx = 9$$

$$\left[ (9 - b)x - \frac{x^3}{3} \right]_{0}^{\sqrt{9-b}} = 9$$

$$\frac{2}{3}(9 - b)^{3/2} = 9$$

$$(9 - b)^{3/2} = \frac{27}{2}$$

$$9 - b = \frac{9}{\sqrt[3]{4}}$$

$$b = 9 - \frac{9}{\sqrt[3]{4}} \approx 3.330$$

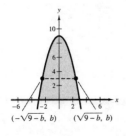

$(-\sqrt{9-b}, b)$   $(\sqrt{9-b}, b)$

**71.** Area of triangle $OAB$ is $\frac{1}{2}(4)(4) = 8$.

$$4 = \int_{0}^{a} (4 - x) \, dx = \left[ 4x - \frac{x^2}{2} \right]_{0}^{a} = 4a - \frac{a^2}{2}$$

$$a^2 - 8a + 8 = 0$$

$$a = 4 \pm 2\sqrt{2}$$

Because $0 < a < 4$, select $a = 4 - 2\sqrt{2} \approx 1.172$.

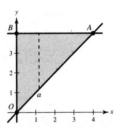

**73.** $\displaystyle \lim_{\|\Delta\| \to 0} \sum_{i=1}^{n} \left( x_i - x_i^2 \right) \Delta x$

where $x_i = \dfrac{i}{n}$ and $\Delta x = \dfrac{1}{n}$ is the same as

$$\int_{0}^{1} (x - x^2) \, dx = \left[ \frac{x^2}{2} - \frac{x^3}{3} \right]_{0}^{1} = \frac{1}{6}.$$

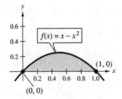

**75.** $R_1$ projects the greater revenue because the area under the curve is greater.

$$\int_{15}^{20} \left[ (7.21 + 0.58t) - (7.21 + 0.45t) \right] dt$$

$$= \int_{15}^{20} 0.13t \, dt = \left[ \frac{0.13t^2}{2} \right]_{15}^{20} = \$11.375 \text{ billion}$$

**77. (a)** $y_1 = 0.0124x^2 - 0.385x + 7.85$

**(b)**

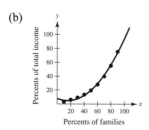

**(c)**

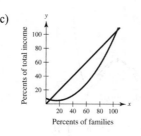

**(d)** Income inequality $= \int_0^{100} [x - y_1] \, dx \approx 2006.7$

**79. (a)** $A = 2\left[ \int_0^5 \left(1 - \frac{1}{3}\sqrt{5 - x}\right) dx + \int_5^{5.5} (1 - 0) \, dx \right]$

$= 2\left( \left[ x + \frac{2}{9}(5 - x)^{3/2} \right]_0^5 + \left[ x \right]_5^{5.5} \right)$

$= 2\left( 5 - \frac{10\sqrt{5}}{9} + 5.5 - 5 \right) \approx 6.031 \text{ m}^2$

**(b)** $V = 2A \approx 2(6.031) \approx 12.062 \text{ m}^3$

**(c)** $5000 \, V \approx 5000(12.062) = 60{,}310 \text{ pounds}$

**81.** Line: $y = \dfrac{-3}{7\pi}x$

$A = \int_0^{7\pi/6} \left[ \sin x + \frac{3x}{7\pi} \right] dx$

$= \left[ -\cos x + \frac{3x^2}{14\pi} \right]_0^{7\pi/6}$

$= \frac{\sqrt{3}}{2} + \frac{7\pi}{24} + 1$

$\approx 2.7823$

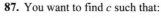

**83.** True. The region has been shifted $C$ units upward (if $C > 0$), or $C$ units downward (if $C < 0$).

**85.** False. Let $f(x) = x$ and $g(x) = 2x - x^2$, $f$ and $g$ intersect at $(1, 1)$, the midpoint of $[0, 2]$, but

$\int_a^b \left[ f(x) - g(x) \right] dx = \int_0^2 \left[ x - (2x - x^2) \right] dx = \frac{2}{3} \neq 0.$

**87.** You want to find $c$ such that:

$\int_0^b \left[ (2x - 3x^3) - c \right] dx = 0$

$\left[ x^2 - \frac{3}{4}x^4 - cx \right]_0^b = 0$

$b^2 - \frac{3}{4}b^4 - cb = 0$

But, $c = 2b - 3b^3$ because $(b, c)$ is on the graph.

$b^2 - \frac{3}{4}b^4 - (2b - 3b^3)b = 0$

$4 - 3b^2 - 8 + 12b^2 = 0$

$9b^2 = 4$

$b = \frac{2}{3}$

$c = \frac{4}{9}$

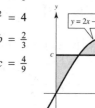

# Section 7.2  Volume: The Disk Method

**1.** $V = \pi \int_0^1 (-x + 1)^2 \, dx = \pi \int_0^1 (x^2 - 2x + 1) \, dx = \pi \left[ \frac{x^3}{3} - x^2 + x \right]_0^1 = \frac{\pi}{3}$

**3.** $V = \pi \int_1^4 (\sqrt{x})^2 \, dx = \pi \int_1^4 x \, dx = \pi \left[ \frac{x^2}{2} \right]_1^4 = \frac{15\pi}{2}$

**5.** $V = \pi \int_0^1 \left[ (x^2)^2 - (x^5)^2 \right] dx$

$= \pi \int_0^1 (x^4 - x^{10}) \, dx$

$= \pi \left[ \dfrac{x^5}{5} - \dfrac{x^{11}}{11} \right]_0^1$

$= \pi \left( \dfrac{1}{5} - \dfrac{1}{11} \right) = \dfrac{6\pi}{55}$

**7.** $y = x^2 \Rightarrow x = \sqrt{y}$

$V = \pi \int_0^4 (\sqrt{y})^2 \, dy = \pi \int_0^4 y \, dy = \pi \left[ \dfrac{y^2}{2} \right]_0^4 = 8\pi$

**9.** $y = x^{2/3} \Rightarrow x = y^{3/2}$

$V = \pi \int_0^1 (y^{3/2})^2 \, dy = \pi \int_0^1 y^3 \, dy = \pi \left[ \dfrac{y^4}{4} \right]_0^1 = \dfrac{\pi}{4}$

**11.** $y = \sqrt{x},\ y = 0,\ x = 3$

(a)  $R(x) = \sqrt{x},\ r(x) = 0$

$V = \pi \int_0^3 (\sqrt{x})^2 \, dx = \pi \int_0^3 x \, dx = \pi \left[ \dfrac{x^2}{2} \right]_0^3 = \dfrac{9\pi}{2}$

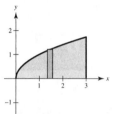

(b)  $R(y) = 3,\ r(y) = y^2$

$V = \pi \int_0^{\sqrt{3}} \left[ 3^2 - (y^2)^2 \right] dy = \pi \int_0^{\sqrt{3}} (9 - y^4) \, dy = \pi \left[ 9y - \dfrac{y^5}{5} \right]_0^{\sqrt{3}} = \pi \left[ 9\sqrt{3} - \dfrac{9}{5}\sqrt{3} \right] = \dfrac{36\sqrt{3}\pi}{5}$

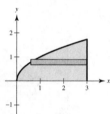

(c)  $R(y) = 3 - y^2,\ r(y) = 0$

$V = \pi \int_0^{\sqrt{3}} (3 - y^2)^2 \, dy = \pi \int_0^{\sqrt{3}} (9 - 6y^2 + y^4) \, dy$

$= \pi \left[ 9y - 2y^3 + \dfrac{y^5}{5} \right]_0^{\sqrt{3}} = \pi \left[ 9\sqrt{3} - 6\sqrt{3} + \dfrac{9\sqrt{3}}{5} \right]$

$= \dfrac{24\sqrt{3}\pi}{5}$

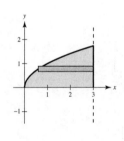

(d)  $R(y) = 3 + (3 - y^2) = 6 - y^2,\ r(y) = 3$

$V = \pi \int_0^{\sqrt{3}} \left[ (6 - y^2)^2 - 3^2 \right] dy = \pi \int_0^{\sqrt{3}} (y^4 - 12y^2 + 27) \, dy$

$= \pi \left[ \dfrac{y^5}{5} - 4y^3 + 27y \right]_0^{\sqrt{3}} = \pi \left[ \dfrac{9\sqrt{3}}{5} - 12\sqrt{3} + 27\sqrt{3} \right]$

$= \dfrac{84\sqrt{3}\pi}{5}$

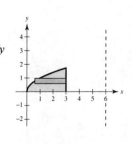

**13.** $y = x^2$, $y = 4x - x^2$ intersect at $(0, 0)$ and $(2, 4)$.

(a) $R(x) = 4x - x^2$, $r(x) = x^2$

$$V = \pi \int_0^2 \left[ \left(4x - x^2\right)^2 - x^4 \right] dx$$

$$= \pi \int_0^2 \left(16x^2 - 8x^3\right) dx$$

$$= \pi \left[ \frac{16}{3}x^3 - 2x^4 \right]_0^2 = \frac{32\pi}{3}$$

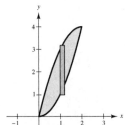

(b) $R(x) = 6 - x^2$, $r(x) = 6 - \left(4x - x^2\right)$

$$V = \pi \int_0^2 \left[ \left(6 - x^2\right)^2 - \left(6 - 4x + x^2\right)^2 \right] dx$$

$$= 8\pi \int_0^2 \left(x^3 - 5x^2 + 6x\right) dx$$

$$= 8\pi \left[ \frac{x^4}{4} - \frac{5}{3}x^3 + 3x^2 \right]_0^2 = \frac{64\pi}{3}$$

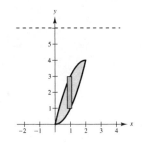

**15.** $R(x) = 4 - x$, $r(x) = 1$

$$V = \pi \int_0^3 \left[ (4 - x)^2 - (1)^2 \right] dx$$

$$= \pi \int_0^3 \left(x^2 - 8x + 15\right) dx$$

$$= \pi \left[ \frac{x^3}{3} - 4x^2 + 15x \right]_0^3$$

$$= 18\pi$$

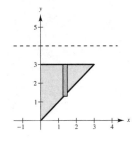

**17.** $R(x) = 4$, $r(x) = 4 - \dfrac{3}{1 + x}$

$$V = \pi \int_0^3 \left[ 4^2 - \left(4 - \frac{3}{1 + x}\right)^2 \right] dx$$

$$= \pi \int_0^3 \left[ \frac{24}{1 + x} - \frac{9}{(1 + x)^2} \right] dx$$

$$= \pi \left[ 24 \ln|1 + x| + \frac{9}{1 + x} \right]_0^3$$

$$= \pi \left[ \left(24 \ln 4 + \frac{9}{4}\right) - 9 \right]$$

$$= \left(48 \ln 2 - \frac{27}{4}\right) \pi \approx 83.318$$

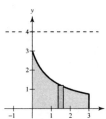

**19.** $R(y) = 5 - y$, $r(y) = 0$

$$V = \pi \int_0^4 (5 - y)^2 \, dy$$

$$= \pi \int_0^4 \left(25 - 10y + y^2\right) dy$$

$$= \pi \left[ 25y - 5y^2 + \frac{y^3}{3} \right]_0^4$$

$$= \pi \left[ 100 - 80 + \frac{64}{3} \right]$$

$$= \frac{124\pi}{3}$$

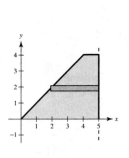

**21.** $R(y) = 5 - y^2$, $r(y) = 1$

$$V = \pi \int_{-2}^2 \left[ \left(5 - y^2\right)^2 - 1 \right] dy$$

$$= 2\pi \int_0^2 \left[ y^4 - 10y^2 + 24 \right] dy$$

$$= 2\pi \left[ \frac{y^5}{5} - \frac{10y^3}{3} + 24y \right]_0^2$$

$$= 2\pi \left[ \frac{32}{5} - \frac{80}{3} + 48 \right] = \frac{832\pi}{15}$$

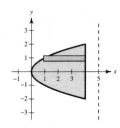

**23.** $R(x) = \dfrac{1}{\sqrt{x+1}}, \ r(x) = 0$

$$V = \pi \int_0^4 \left( \dfrac{1}{\sqrt{x+1}} \right)^2 dx$$

$$= \pi \int_0^4 \dfrac{1}{x+1} dx = \pi \Big[ \ln|x+1| \Big]_0^4 = \pi \ln 5$$

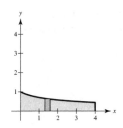

**25.** $R(x) = \dfrac{1}{x}, \ r(x) = 0$

$$V = \pi \int_1^3 \left( \dfrac{1}{x} \right)^2 dx$$

$$= \pi \left[ -\dfrac{1}{x} \right]_1^3$$

$$= \pi \left[ -\dfrac{1}{3} + 1 \right] = \dfrac{2}{3}\pi$$

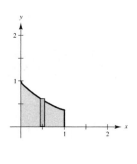

**27.** $R(x) = e^{-x}, \ r(x) = 0$

$$V = \pi \int_0^1 \left( e^{-x} \right)^2 dx$$

$$= \pi \int_0^1 e^{-2x} dx$$

$$= \left[ -\dfrac{\pi}{2} e^{-2x} \right]_0^1$$

$$= \dfrac{\pi}{2}\left( 1 - e^{-2} \right) \approx 1.358$$

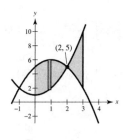

**29.**
$$x^2 + 1 = -x^2 + 2x + 5$$
$$2x^2 - 2x - 4 = 0$$
$$x^2 - x - 2 = 0$$
$$(x - 2)(x + 1) = 0$$

The curves intersect at $(-1, 2)$ and $(2, 5)$.

$$V = \pi \int_0^2 \left[ \left(5 + 2x - x^2\right)^2 - \left(x^2 + 1\right)^2 \right] dx + \pi \int_2^3 \left[ \left(x^2 + 1\right)^2 - \left(5 + 2x - x^2\right)^2 \right] dx$$

$$= \pi \int_0^2 \left( -4x^3 - 8x^2 + 20x + 24 \right) dx + \pi \int_2^3 \left( 4x^3 + 8x^2 - 20x - 24 \right) dx$$

$$= \pi \left[ -x^4 - \dfrac{8}{3}x^3 + 10x^2 + 24x \right]_0^2 + \pi \left[ x^4 + \dfrac{8}{3}x^3 - 10x^2 - 24x \right]_2^3$$

$$= \pi \dfrac{152}{3} + \pi \dfrac{125}{3} = \dfrac{277\pi}{3}$$

**31.** $y = 6 - 3x \Rightarrow x = \dfrac{1}{3}(6 - y)$

$$V = \pi \int_0^6 \left[ \dfrac{1}{3}(6 - y) \right]^2 dy$$

$$= \dfrac{\pi}{9} \int_0^6 \left[ 36 - 12y + y^2 \right] dy$$

$$= \dfrac{\pi}{9} \left[ 36y - 6y^2 + \dfrac{y^3}{3} \right]_0^6$$

$$= \dfrac{\pi}{9} \left[ 216 - 216 + \dfrac{216}{3} \right]$$

$$= 8\pi = \dfrac{1}{3}\pi r^2 h, \ \text{Volume of cone}$$

**33.** $V = \pi \int_0^\pi (\sin x)^2 \, dx$

$= \pi \int_0^\pi \dfrac{1 - \cos 2x}{2} \, dx$

$= \dfrac{\pi}{2} \left[ x - \dfrac{1}{2} \sin 2x \right]_0^\pi = \dfrac{\pi}{2} [\pi] = \dfrac{\pi^2}{2}$

Numerical approximation: 4.9348

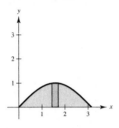

**35.** $V = \pi \int_1^2 \left( e^{x-1} \right)^2 \, dx$

$= \pi \int_1^2 e^{2x-2} \, dx$

$= \dfrac{\pi}{2} e^{2x-2} \Big]_1^2$

$= \dfrac{\pi}{2} \left( e^2 - 1 \right)$

Numerical approximation: 10.0359

**37.** $V = \pi \int_0^2 \left[ e^{-x^2} \right]^2 \, dx \approx 1.9686$

**39.** $V = \pi \int_0^5 \left[ 2 \arctan(0.2x) \right]^2 \, dx$

$\approx 15.4115$

**41.** $V = \pi \int_0^1 y^2 \, dy = \pi \dfrac{y^3}{3} \Big]_0^1 = \dfrac{\pi}{3}$

**43.** $V = \pi \int_0^1 \left( x^2 - x^4 \right) dx$

$= \pi \left[ \dfrac{x^3}{3} - \dfrac{x^5}{5} \right]_0^1$

$= \pi \left( \dfrac{1}{3} - \dfrac{1}{5} \right)$

$= \dfrac{2\pi}{15}$

**45.** $V = \pi \int_0^1 (1 - y) \, dy$

$= \pi \left[ y - \dfrac{y^2}{2} \right]_0^1 = \pi \left( 1 - \dfrac{1}{2} \right) = \dfrac{\pi}{2}$

**47.** $V = \pi \int_0^1 \left( y - y^2 \right) dy$

$= \pi \left[ \dfrac{y^2}{2} - \dfrac{y^3}{3} \right]_0^1 = \pi \left( \dfrac{1}{2} - \dfrac{1}{3} \right) = \dfrac{\pi}{6}$

**49.** $\pi \int_0^{\pi/2} \sin^2 x \, dx$ represents the volume of the solid

generated by revolving the region bounded by
$y = \sin x, \, y = 0, \, x = 0, \, x = \pi/2$ about the $x$-axis.

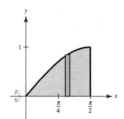

**51.**

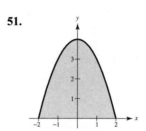

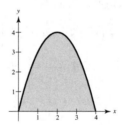

The volumes are the same because the solid has been

translated horizontally. $\left( 4x - x^2 = 4 - (x - 2)^2 \right)$

**53.** (a) True. Answers will vary.

(b) False. Answers will vary.

**55.** $V = \pi \int_0^4 \left( \sqrt{x} \right)^2 \, dx = \pi \int_0^4 x \, dx = \left[ \dfrac{\pi x^2}{2} \right]_0^4 = 8\pi$

Let $0 < c < 4$ and set

$\pi \int_0^c x \, dx = \left[ \dfrac{\pi x^2}{2} \right]_0^c = \dfrac{\pi c^2}{2} = 4\pi.$

$c^2 = 8$

$c = \sqrt{8} = 2\sqrt{2}$

So, when $x = 2\sqrt{2}$, the solid is divided into two parts
of equal volume.

**57.** $V = \pi \int_{-\sqrt{R^2-r^2}}^{\sqrt{R^2-r^2}} \left[ \left( \sqrt{R^2 - x^2} \right)^2 - r^2 \right] dx$

$= 2\pi \int_0^{\sqrt{R^2-r^2}} \left( R^2 - r^2 - x^2 \right) dx$

$= 2\pi \left[ \left( R^2 - r^2 \right)x - \frac{x^3}{3} \right]_0^{\sqrt{R^2-r^2}}$

$= 2\pi \left[ \left( R^2 - r^2 \right)^{3/2} - \frac{\left( R^2 - r^2 \right)^{3/2}}{3} \right] = \frac{4}{3}\pi \left( R^2 - r^2 \right)^{3/2}$

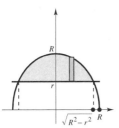

**59.** $R(x) = \dfrac{r}{h}x, \, r(x) = 0$

$V = \pi \int_0^h \frac{r^2}{h^2}x^2 \, dx = \left[ \frac{r^2 \pi}{3h^2}x^3 \right]_0^h = \frac{r^2 \pi}{3h^2}h^3 = \frac{1}{3}\pi r^2 h$

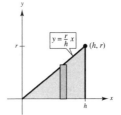

**61.** $x = r - \dfrac{r}{H}y = r\left( 1 - \dfrac{y}{H} \right), \, R(y) = r\left( 1 - \dfrac{y}{H} \right), \, r(y) = 0$

$V = \pi \int_0^h \left[ r\left( 1 - \frac{y}{H} \right) \right]^2 dy = \pi r^2 \int_0^h \left( 1 - \frac{2}{H}y + \frac{1}{H^2}y^2 \right) dy$

$= \pi r^2 \left[ y - \frac{1}{H}y^2 + \frac{1}{3H^2}y^3 \right]_0^h$

$= \pi r^2 \left( h - \frac{h^2}{H} + \frac{h^3}{3H^2} \right) = \pi r^2 h\left( 1 - \frac{h}{H} + \frac{h^2}{3H^2} \right)$

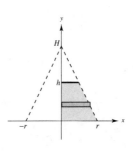

**63.**

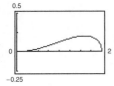

$V = \pi \int_0^2 \left( \frac{1}{8}x^2\sqrt{2 - x} \right)^2 dx = \frac{\pi}{64} \int_0^2 x^4(2 - x) \, dx = \frac{\pi}{64} \left[ \frac{2x^5}{5} - \frac{x^6}{6} \right]_0^2 = \frac{\pi}{30} \, m^3$

**65.** (a) $R(x) = \dfrac{3}{5}\sqrt{25 - x^2}, \, r(x) = 0$

$V = \frac{9\pi}{25} \int_{-5}^5 \left( 25 - x^2 \right) dx = \frac{18\pi}{25} \int_0^5 \left( 25 - x^2 \right) dx = \frac{18\pi}{25} \left[ 25x - \frac{x^3}{3} \right]_0^5 = 60\pi$

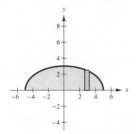

(b)  $R(y) = \dfrac{5}{3}\sqrt{9 - y^2}, r(y) = 0, x \geq 0$

$$V = \frac{25\pi}{9}\int_0^3 \left(9 - y^2\right) dy = \frac{25\pi}{9}\left[9y - \frac{y^3}{3}\right]_0^3 = 50\pi$$

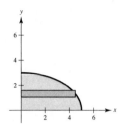

**67.** (a)  First find where  $y = b$  intersects the parabola:

$$b = 4 - \frac{x^2}{4}$$

$$x^2 = 16 - 4b = 4(4 - b)$$

$$x = 2\sqrt{4 - b}$$

$$V = \int_0^{2\sqrt{4-b}} \pi\left[4 - \frac{x^2}{4} - b\right]^2 dx + \int_{2\sqrt{4-b}}^4 \pi\left[b - 4 + \frac{x^2}{4}\right]^2 dx$$

$$= \int_0^4 \pi\left[4 - \frac{x^2}{4} - b\right]^2 dx$$

$$= \pi\int_0^4 \left[\frac{x^4}{16} - 2x^2 + \frac{bx^2}{2} + b^2 - 8b + 16\right] dx$$

$$= \pi\left[\frac{x^5}{80} - \frac{2x^3}{3} + \frac{bx^3}{6} + b^2 x - 8bx + 16x\right]_0^4$$

$$= \pi\left(\frac{64}{5} - \frac{128}{3} + \frac{32}{3}b + 4b^2 - 32b + 64\right) = \pi\left(4b^2 - \frac{64}{3}b + \frac{512}{15}\right)$$

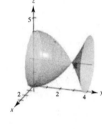

(b)  Graph of  $V(b) = \pi\left(4b^2 - \dfrac{64}{3}b + \dfrac{512}{15}\right)$

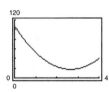

Minimum volume is 17.87 for  $b = 2.67.$

(c)  $V'(b) = \pi\left(8b - \dfrac{64}{3}\right) = 0 \Rightarrow b = \dfrac{64/3}{8} = \dfrac{8}{3} = 2\dfrac{2}{3}$

$V''(b) = 8\pi > 0 \Rightarrow b = \dfrac{8}{3}$  is a relative minimum.

**69.** (a)  $\pi \int_0^h r^2 \, dx$   (ii)

is the volume of a right circular cylinder with radius $r$ and height $h$.

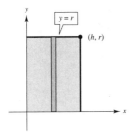

(b)  $\pi \int_{-b}^{b} \left( a\sqrt{1 - \dfrac{x^2}{b^2}} \right)^2 dx$   (iv)

is the volume of an ellipsoid with axes $2a$ and $2b$.

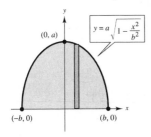

(c)  $\pi \int_{-r}^{r} \left( \sqrt{r^2 - x^2} \right)^2 dx$   (iii)

is the volume of a sphere with radius $r$.

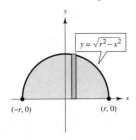

(d)  $\pi \int_0^h \left( \dfrac{rx}{h} \right)^2 dx$   (i)

is the volume of a right circular cone with the radius of the base as $r$ and height $h$.

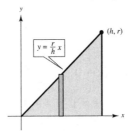

(e)  $\pi \int_{-r}^{r} \left[ \left( R + \sqrt{r^2 - x^2} \right)^2 - \left( R - \sqrt{r^2 - x^2} \right)^2 \right] dx$  (v)

is the volume of a torus with the radius of its circular cross section as $r$ and the distance from the axis of the torus to the center of its cross section as $R$.

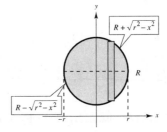

**71.**

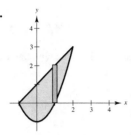

Base of cross section $= (x + 1) - (x^2 - 1) = 2 + x - x^2$

(a) $A(x) = b^2 = (2 + x - x^2)^2 = 4 + 4x - 3x^2 - 2x^3 + x^4$

$$V = \int_{-1}^{2} (4 + 4x - 3x^2 - 2x^3 + x^4)\, dx = \left[ 4x + 2x^2 - x^3 - \frac{1}{2}x^4 + \frac{1}{5}x^5 \right]_{-1}^{2} = \frac{81}{10}$$

$2 + x - x^2$

$\longleftarrow 2 + x - x^2 \longrightarrow$

(b) $A(x) = bh = (2 + x - x^2)1$

$$V = \int_{-1}^{2} (2 + x - x^2)\, dx = \left[ 2x + \frac{x^2}{2} - \frac{x^3}{3} \right]_{-1}^{2} = \frac{9}{2}$$

$\longleftarrow 2 + x - x^2 \longrightarrow$ , $1$

**73.** The cross sections are squares. By symmetry, you can set up an integral for an eighth of the volume and multiply by 8.

$$A(y) = b^2 = \left( \sqrt{r^2 - y^2} \right)^2$$

$$V = 8 \int_{0}^{r} (r^2 - y^2)\, dy$$

$$= 8 \left[ r^2 y - \frac{1}{3}y^3 \right]_{0}^{r}$$

$$= \frac{16}{3} r^3$$

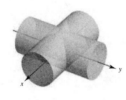

**75.** (a) Because the cross sections are isosceles right triangles:

$$A(x) = \frac{1}{2}bh = \frac{1}{2}\left( \sqrt{r^2 - y^2} \right)\left( \sqrt{r^2 - y^2} \right) = \frac{1}{2}(r^2 - y^2)$$

$$V = \frac{1}{2} \int_{-r}^{r} (r^2 - y^2)\, dy = \int_{0}^{r} (r^2 - y^2)\, dy = \left[ r^2 y - \frac{y^3}{3} \right]_{0}^{r} = \frac{2}{3}r^3$$

(b) $A(x) = \frac{1}{2}bh = \frac{1}{2}\sqrt{r^2 - y^2}\left( \sqrt{r^2 - y^2}\, \tan \theta \right) = \frac{\tan \theta}{2}(r^2 - y^2)$

$$V = \frac{\tan \theta}{2} \int_{-r}^{r} (r^2 - y^2)\, dy = \tan \theta \int_{0}^{r} (r^2 - y^2)\, dy = \tan \theta \left[ r^2 y - \frac{y^3}{3} \right]_{0}^{r} = \frac{2}{3}r^3 \tan \theta$$

As $\theta \to 90°$, $V \to \infty$.

## Section 7.3  Volume: The Shell Method

**1.** $p(x) = x, h(x) = x$

$$V = 2\pi \int_0^2 x(x)\,dx = \left[\frac{2\pi x^3}{3}\right]_0^2 = \frac{16\pi}{3}$$

**3.** $p(x) = x, h(x) = \sqrt{x}$

$$V = 2\pi \int_0^4 x\sqrt{x}\,dx = 2\pi \int_0^4 x^{3/2}\,dx = \left[\frac{4\pi}{5}x^{5/2}\right]_0^4 = \frac{128\pi}{5}$$

**5.** $p(x) = x, h(x) = \frac{1}{4}x^2$

$$V = 2\pi \int_0^4 x\left(\frac{1}{4}x^2\right)dx$$

$$= \frac{\pi}{2}\left[\frac{x^4}{4}\right]_0^4$$

$$= 32\pi$$

**7.** $p(x) = x, h(x) = \left(4x - x^2\right) - x^2 = 4x - 2x^2$

$$V = 2\pi \int_0^2 x\left(4x - 2x^2\right)dx$$

$$= 4\pi \int_0^2 \left(2x^2 - x^3\right)dx$$

$$= 4\pi\left[\frac{2}{3}x^3 - \frac{1}{4}x^4\right]_0^2 = \frac{16\pi}{3}$$

**9.** $p(x) = x$

$h(x) = 4 - \left(4x - x^2\right)$

$\qquad = x^2 - 4x + 4$

$$V = 2\pi \int_0^2 x\left(x^2 - 4x + 4\right)dx$$

$$V = 2\pi \int_0^2 \left(x^3 - 4x^2 + 4x\right)dx$$

$$= 2\pi\left[\frac{x^4}{4} - \frac{4}{3}x^3 + 2x^2\right]_0^2$$

$$= \frac{8\pi}{3}$$

**11.** $p(x) = x, h(x) = \sqrt{x - 2}$

$$V = 2\pi \int_2^4 x\sqrt{x - 2}\,dx$$

Let $u = x - 2, x = u + 2, du = dx$.

When $x = 2, u = 0$.

When $x = 4, u = 2$.

$$V = 2\pi \int_0^2 \left(u + 2\right)u^{1/2}\,du$$

$$= 2\pi\left[\frac{2}{5}u^{5/2} + \frac{4}{3}u^{3/2}\right]_0^2$$

$$= 2\pi\left[\frac{2}{5}(2)^{5/2} + \frac{4}{3}(2)^{3/2}\right]$$

$$= 2\pi\sqrt{2}\left[\frac{2}{5}(4) + \frac{4}{3}(2)\right] = \frac{128\sqrt{2}\pi}{15}$$

**13.** $p(x) = x, h(x) = \frac{1}{\sqrt{2\pi}}e^{-x^2/2}$

$$V = 2\pi \int_0^1 x\left(\frac{1}{\sqrt{2\pi}}e^{-x^2/2}\right)dx$$

$$= \sqrt{2\pi}\int_0^1 e^{-x^2/2}x\,dx$$

$$= \left[-\sqrt{2\pi}e^{-x^2/2}\right]_0^1$$

$$= \sqrt{2\pi}\left(1 - \frac{1}{\sqrt{e}}\right)$$

$$\approx 0.986$$

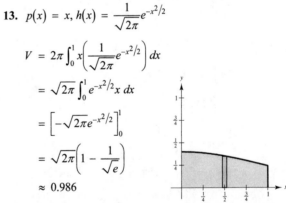

**15.** $p(y) = y, h(y) = 2 - y$

$$V = 2\pi \int_0^2 y(2 - y)\,dy$$

$$= 2\pi \int_0^2 \left(2y - y^2\right)dy$$

$$= 2\pi\left[y^2 - \frac{y^3}{3}\right]_0^2 = \frac{8\pi}{3}$$

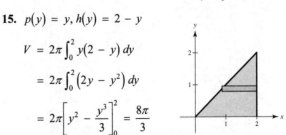

**17.** $p(y) = y$ and $h(y) = 1$ if $0 \le y < \dfrac{1}{2}$.

$p(y) = y$ and $h(y) = \dfrac{1}{y} - 1$ if $\dfrac{1}{2} \le y \le 1$.

$V = 2\pi \displaystyle\int_0^{1/2} y \, dy + 2\pi \displaystyle\int_{1/2}^1 (1 - y) \, dy$

$= 2\pi \left[ \dfrac{y^2}{2} \right]_0^{1/2} + 2\pi \left[ y - \dfrac{y^2}{2} \right]_{1/2}^1 = \dfrac{\pi}{4} + \dfrac{\pi}{4} = \dfrac{\pi}{2}$

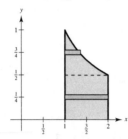

**19.** $p(y) = y$, $h(y) = \sqrt[3]{y}$

$V = 2\pi \displaystyle\int_0^8 y \sqrt[3]{y} \, dy$

$= 2\pi \displaystyle\int_0^8 y^{4/3} \, dy$

$= \left[ 2\pi \left( \dfrac{3}{7} \right) y^{7/3} \right]_0^8$

$= \dfrac{6\pi}{7} (2^7) = \dfrac{768\pi}{7}$

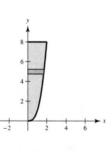

**21.** $p(y) = y$, $h(y) = (4 - y) - (y) = 4 - 2y$

$V = 2\pi \displaystyle\int_0^2 y(4 - 2y) \, dy$

$= 2\pi \displaystyle\int_0^2 (4y - 2y^2) \, dy$

$= 2\pi \left[ 2y^2 - \dfrac{2}{3} y^3 \right]_0^2$

$= 2\pi \left( 8 - \dfrac{16}{3} \right) = \dfrac{16\pi}{3}$

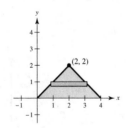

**23.** $p(x) = 4 - x$, $h(x) = 2x - x^2$

$V = 2\pi \displaystyle\int_0^2 (4 - x)(2x - x^2) \, dx$

$= 2\pi \displaystyle\int_0^2 (8x - 6x^2 + x^3) \, dx$

$= 2\pi \left[ 4x^2 - 2x^3 + \dfrac{x^4}{4} \right]_0^2$

$= 2\pi[16 - 16 + 4] = 8\pi$

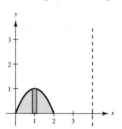

**25.** $p(x) = 4 - x$, $h(x) = 4x - x^2 - x^2 = 4x - 2x^2$

$V = 2\pi \displaystyle\int_0^2 (4 - x)(4x - 2x^2) \, dx$

$= 2\pi(2) \displaystyle\int_0^2 (x^3 - 6x^2 + 8x) \, dx$

$= 4\pi \left[ \dfrac{x^4}{4} - 2x^3 + 4x^2 \right]_0^2 = 16\pi$

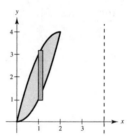

**27.** The shell method would be easier:

$V = 2\pi \displaystyle\int_0^4 \left[ 4 - (y - 2)^2 \right] y \, dy$

Using the disk method:

$V = \pi \displaystyle\int_0^4 \left[ \left( 2 + \sqrt{4 - x} \right)^2 - \left( 2 - \sqrt{4 - x} \right)^2 \right] dx$

$\left[ \textbf{Note:} \ V = \dfrac{128\pi}{3} \right]$

**29. (a) Disk**

$$R(x) = x^3, r(x) = 0$$

$$V = \pi \int_0^2 x^6\, dx = \pi \left[ \frac{x^7}{7} \right]_0^2 = \frac{128\pi}{7}$$

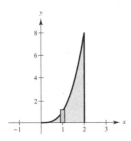

**(b) Shell**

$$p(x) = x, h(x) = x^3$$

$$V = 2\pi \int_0^2 x^4\, dx = 2\pi \left[ \frac{x^5}{5} \right]_0^2 = \frac{64\pi}{5}$$

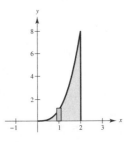

**(c) Shell**

$$p(x) = 4 - x, h(x) = x^3$$

$$V = 2\pi \int_0^2 (4 - x)x^3\, dx$$

$$= 2\pi \int_0^2 (4x^3 - x^4)\, dx$$

$$= 2\pi \left[ x^4 - \frac{1}{5}x^5 \right]_0^2 = \frac{96\pi}{5}$$

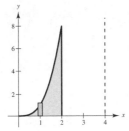

**31. (a) Shell**

$$p(y) = y, h(y) = \left( a^{1/2} - y^{1/2} \right)^2$$

$$V = 2\pi \int_0^a y \left( a - 2a^{1/2}y^{1/2} + y \right) dy$$

$$= 2\pi \int_0^a \left( ay - 2a^{1/2}y^{3/2} + y^2 \right) dy$$

$$= 2\pi \left[ \frac{a}{2}y^2 - \frac{4a^{1/2}}{5}y^{5/2} + \frac{y^3}{3} \right]_0^a$$

$$= 2\pi \left( \frac{a^3}{2} - \frac{4a^3}{5} + \frac{a^3}{3} \right) = \frac{\pi a^3}{15}$$

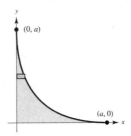

**(b)** Same as part (a) by symmetry

**(c) Shell**

$$p(x) = a - x, h(x) = \left( a^{1/2} - x^{1/2} \right)^2$$

$$V = 2\pi \int_0^a (a - x)\left( a^{1/2} - x^{1/2} \right)^2 dx$$

$$= 2\pi \int_0^a \left( a^2 - 2a^{3/2}x^{1/2} + 2a^{1/2}x^{3/2} - x^2 \right) dx$$

$$= 2\pi \left[ a^2x - \frac{4}{3}a^{3/2}x^{3/2} + \frac{4}{5}a^{1/2}x^{5/2} - \frac{1}{3}x^3 \right]_0^a$$

$$= \frac{4\pi a^3}{15}$$

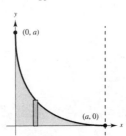

**33. (a)**

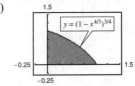

**(b)** $x^{4/3} + y^{4/3} = 1, x = 0, y = 0$

$$y = \left( 1 - x^{4/3} \right)^{3/4}$$

$$V = 2\pi \int_0^1 x\left( 1 - x^{4/3} \right)^{3/4} dx \approx 1.5056$$

**35.** (a)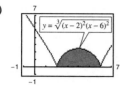

(b) $V = 2\pi \int_2^6 x\sqrt[3]{(x-2)^2(x-6)^2} \, dx \approx 187.249$

**37.** Answers will vary.

(a) The rectangles would be vertical.

(b) The rectangles would be horizontal.

**39.** $\pi \int_1^5 (x - 1) \, dx = \pi \int_1^5 (\sqrt{x-1})^2 \, dx$

This integral represents the volume of the solid generated by revolving the region bounded by $y = \sqrt{x-1}$, $y = 0$, and $x = 5$ about the $x$-axis by using the disk method.

$2\pi \int_0^2 y[5 - (y^2 + 1)] \, dy$

represents this same volume by using the shell method.

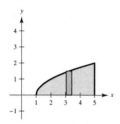

Disk method

**41.**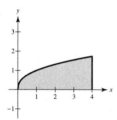

(a) Around $x$-axis: $V = \pi \int_0^4 (x^{2/5})^2 \, dx = \left[\pi \frac{5}{9} x^{9/5}\right]_0^4$

$= \frac{5}{9}\pi(4)^{9/5} \approx 6.7365\pi$

(b) Around $y$-axis: $V = 2\pi \int_0^4 x(x^{2/5}) \, dx$

$= \left[2\pi \frac{5}{12} x^{12/5}\right]_0^4 \approx 23.2147\pi$

(c) Around $x = 4$:

$V = 2\pi \int_0^4 (4 - x)x^{2/5} \, dx \approx 16.5819\pi$

So, $(a) < (c) < (b)$.

**43.** $2\pi \int_0^2 x^3 \, dx = 2\pi \int_0^2 x(x^2) \, dx$

(a) Plane region bounded by

$y = x^2, y = 0, x = 0, x = 2$

(b) Revolved about the $y$-axis

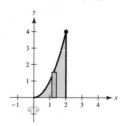

Other answers possible.

**45.** $2\pi \int_0^6 (y + 2)\sqrt{6 - y} \, dy$

(a) Plane region bounded by

$x = \sqrt{6 - y}, x = 0, y = 0$

(b) Revolved around line $y = -2$

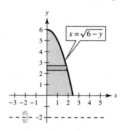

Other answers possible.

**47.** $p(x) = x, h(x) = 2 - \frac{1}{2}x^2$

$$V = 2\pi \int_0^2 x\left(2 - \frac{1}{2}x^2\right) dx$$

$$= 2\pi \int_0^2 \left(2x - \frac{1}{2}x^3\right) dx$$

$$= 2\pi \left[x^2 - \frac{1}{8}x^4\right]_0^2 = 4\pi \quad \text{(total volume)}$$

Now find $x_0$ such that:

$$\pi = 2\pi \int_0^{x_0} \left(2x - \frac{1}{2}x^3\right) dx$$

$$1 = 2\left[x^2 - \frac{1}{8}x^4\right]_0^{x_0}$$

$$1 = 2x_0{}^2 - \frac{1}{4}x_0{}^4$$

$$x_0{}^4 - 8x_0{}^2 + 4 = 0$$

$$x_0{}^2 = 4 \pm 2\sqrt{3} \quad \text{(Quadratic Formula)}$$

Take $x_0 = \sqrt{4 - 2\sqrt{3}} \approx 0.73205$, because the other root is too large.

Diameter: $2\sqrt{4 - 2\sqrt{3}} \approx 1.464$

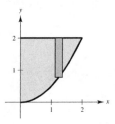

**49.** $V = 4\pi \int_{-1}^1 (2 - x)\sqrt{1 - x^2}\, dx$

$$= 8\pi \int_{-1}^1 \sqrt{1 - x^2}\, dx - 4\pi \int_{-1}^1 x\sqrt{1 - x^2}\, dx$$

$$= 8\pi\left(\frac{\pi}{2}\right) + 2\pi \int_{-1}^1 x\left(1 - x^2\right)^{1/2}(-2)\, dx$$

$$= 4\pi^2 + \left[2\pi\left(\frac{2}{3}\right)\left(1 - x^2\right)^{3/2}\right]_{-1}^1 = 4\pi^2$$

**51. (a)** $\dfrac{d}{dx}[\sin x - x\cos x + C] = \cos x + x\sin x - \cos x$

$$= x\sin x$$

So, $\int x\sin x\, dx = \sin x - x\cos x + C$.

**(b) (i)** $p(x) = x, h(x) = \sin x$

$$V = 2\pi \int_0^{\pi/2} x\sin x\, dx$$

$$= 2\pi\left[\sin x - x\cos x\right]_0^{\pi/2}$$

$$= 2\pi\left[(1 - 0) - 0\right] = 2\pi$$

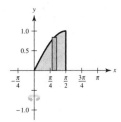

**(ii)** $p(x) = x, h(x) = 2\sin x - (-\sin x) = 3\sin x$

$$V = 2\pi \int_0^{\pi} x(3\sin x)\, dx$$

$$= 6\pi \int_0^{\pi} x\sin x\, dx$$

$$= 6\pi\left[\sin x - x\cos x\right]_0^{\pi}$$

$$= 6\pi(\pi) = 6\pi^2$$

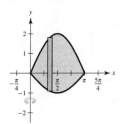

**53. Disk Method**

$$R(y) = \sqrt{r^2 - y^2}$$

$$r(y) = 0$$

$$V = \pi \int_{r-h}^{r} \left(r^2 - y^2\right) dy$$

$$= \pi\left[r^2 y - \frac{y^3}{3}\right]_{r-h}^{r} = \frac{1}{3}\pi h^2(3r - h)$$

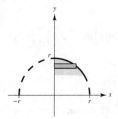

**55.** (a) Area of region $= \int_0^b \left[ ab^n - ax^n \right] dx$

$$= \left[ ab^n x - a\frac{x^{n+1}}{n+1} \right]_0^b$$

$$= ab^{n+1} - a\frac{b^{n+1}}{n+1}$$

$$= ab^{n+1}\left( 1 - \frac{1}{n+1} \right)$$

$$= ab^{n+1}\left( \frac{n}{n+1} \right)$$

$$R_1(n) = \frac{ab^{n+1}\left[ n/(n+1) \right]}{(ab^n)b} = \frac{n}{n+1}$$

(b) $\displaystyle\lim_{n\to\infty} R_1(n) = \lim_{n\to\infty} \frac{n}{n+1} = 1$

$\displaystyle\lim_{n\to\infty}\left( ab^n \right)b = \infty$

(c) **Disk Method:**

$$V = 2\pi \int_0^b x\left( ab^n - ax^n \right) dx$$

$$= 2\pi a \int_0^b \left( xb^n - x^{n+1} \right) dx$$

$$= 2\pi a\left[ \frac{b^n}{2}x^2 - \frac{x^{n+2}}{n+2} \right]_0^b$$

$$= 2\pi a\left[ \frac{b^{n+2}}{2} - \frac{b^{n+2}}{n+2} \right] = \pi ab^{n+2}\left( \frac{n}{n+2} \right)$$

$$R_2(n) = \frac{\pi ab^{n+2}\left[ n/(n+2) \right]}{(\pi b^2)(ab^n)} = \left( \frac{n}{n+2} \right)$$

(d) $\displaystyle\lim_{n\to\infty} R_2(n) = \lim_{n\to\infty}\left( \frac{n}{n+2} \right) = 1$

$\displaystyle\lim_{n\to\infty}\left( \pi b^2 \right)\left( ab^n \right) = \infty$

(e) As $n \to \infty$, the graph approaches the line $x = b$.

**57.** (a) $V = 2\pi \int_0^4 xf(x)\,dx = \dfrac{2\pi(40)}{3(4)}\left[ 0 + 4(10)(45) + 2(20)(40) + 4(30)(20) + 0 \right] = \dfrac{20\pi}{3}(5800) \approx 121{,}475 \text{ ft}^3$

(b) Top line: $y - 50 = \dfrac{40 - 50}{20 - 0}(x - 0) = -\dfrac{1}{2}x \Rightarrow y = -\dfrac{1}{2}x + 50$

Bottom line: $y - 40 = \dfrac{0 - 40}{40 - 20}(x - 20) = -2(x - 20) \Rightarrow y = -2x + 80$

$$V = 2\pi \int_0^{20} x\left( -\frac{1}{2}x + 50 \right) dx + 2\pi \int_{20}^{40} x(-2x + 80)\,dx$$

$$= 2\pi \int_0^{20}\left( -\frac{1}{2}x^2 + 50x \right) dx + 2\pi \int_{20}^{40}\left( -2x^2 + 80x \right) dx$$

$$= 2\pi\left[ -\frac{x^3}{6} + 25x^2 \right]_0^{20} + 2\pi\left[ -\frac{2x^3}{3} + 40x^2 \right]_{20}^{40} = 2\pi\left( \frac{26{,}000}{3} \right) + 2\pi\left( \frac{32{,}000}{3} \right) \approx 121{,}475 \text{ ft}^3$$

(Note that Simpson's Rule is exact for this problem.)

**59.** $V_1 = \pi \int_{1/4}^{c} \frac{1}{x^2}\, dx = \pi \left[ -\frac{1}{x} \right]_{1/4}^{c} = \pi \left[ -\frac{1}{c} + 4 \right] = \frac{4c - 1}{c}\pi$

$V_2 = \left[ 2\pi \int_{1/4}^{c} x\left(\frac{1}{x}\right) dx = 2\pi x \right]_{1/4}^{c} = 2\pi\left( c - \frac{1}{4} \right)$

$V_1 = V_2 \Rightarrow \frac{4c - 1}{c}\pi = 2\pi\left( c - \frac{1}{4} \right)$

$\qquad\qquad\quad 4c - 1 = 2c\left( c - \frac{1}{4} \right)$

$\qquad\qquad 4c^2 - 9c + 2 = 0$

$\qquad\quad (4c - 1)(c - 2) = 0$

$\qquad\qquad\qquad c = 2 \ \left( c = \frac{1}{4} \text{ yields no volume.} \right)$

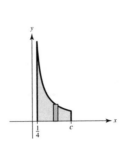

**61.** $y^2 = x(4 - x)^2, \quad 0 \le x \le 4$

$y_1 = \sqrt{x(4 - x)^2} = (4 - x)\sqrt{x}$

$y_2 = -\sqrt{x(4 - x)^2} = -(4 - x)\sqrt{x}$

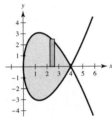

(a) $V = \pi \int_{0}^{4} x(4 - x)^2\, dx = \pi \int_{0}^{4} (x^3 - 8x^2 + 16x)\, dx = \pi\left[ \frac{x^4}{4} - \frac{8x^3}{3} + 8x^2 \right]_{0}^{4} = \frac{64\pi}{3}$

(b) $V = 4\pi \int_{0}^{4} x(4 - x)\sqrt{x}\, dx = 4\pi \int_{0}^{4} \left( 4x^{3/2} - x^{5/2} \right) dx = 4\pi\left[ \frac{8}{5}x^{5/2} - \frac{2}{7}x^{7/2} \right]_{0}^{4} = \frac{2048\pi}{35}$

(c) $V = 4\pi \int_{0}^{4} (4 - x)(4 - x)\sqrt{x}\, dx = 4\pi \int_{0}^{4} \left( 16\sqrt{x} - 8x^{3/2} + x^{5/2} \right) dx = 4\pi\left[ \frac{32}{3}x^{3/2} - \frac{16}{5}x^{5/2} + \frac{2}{7}x^{7/2} \right]_{0}^{4} = \frac{8192\pi}{105}$

# Section 7.4    Arc Length and Surfaces of Revolution

**1.** $(0, 0), (8, 15)$

(a) $d = \sqrt{(8 - 0)^2 + (15 - 0)^2}$

$\quad = \sqrt{64 + 225}$

$\quad = \sqrt{289} = 17$

(b) $y = \frac{15}{8}x$

$y' = \frac{15}{8}$

$s = \int_{0}^{8} \sqrt{1 + \left(\frac{15}{8}\right)^2}\, dx = \int_{0}^{8} \frac{17}{8}\, dx = \left[ \frac{17}{8}x \right]_{0}^{8} = 17$

**3.** $y = \frac{2}{3}(x^2 + 1)^{3/2}$

$y' = (x^2 + 1)^{1/2}(2x), \quad 0 \le x \le 1$

$1 + (y')^2 = 1 + 4x^2(x^2 + 1)$

$\qquad = 4x^4 + 4x^2 + 1 = (2x^2 + 1)^2$

$s = \int_0^1 \sqrt{1 + (y')^2}\, dx$

$\qquad = \int_0^1 (2x^2 + 1)\, dx = \left[\frac{2x^3}{3} + x\right]_0^1 = \frac{5}{3}$

**5.** $y = \frac{2}{3}x^{3/2} + 1$

$y' = x^{1/2}, \quad 0 \le x \le 1$

$s = \int_0^1 \sqrt{1 + x}\, dx$

$\qquad = \left[\frac{2}{3}(1 + x)^{3/2}\right]_0^1 = \frac{2}{3}(\sqrt{8} - 1) \approx 1.219$

**7.** $y = \frac{3}{2}x^{2/3}$

$y' = \frac{1}{x^{1/3}}, \quad 1 \le x \le 8$

$s = \int_1^8 \sqrt{1 + \left(\frac{1}{x^{1/3}}\right)^2}\, dx$

$\qquad = \int_1^8 \sqrt{\frac{x^{2/3} + 1}{x^{2/3}}}\, dx$

$\qquad = \frac{3}{2}\int_1^8 \sqrt{x^{2/3} + 1}\left(\frac{2}{3x^{1/3}}\right) dx$

$\qquad = \frac{3}{2}\left[\frac{2}{3}(x^{2/3} + 1)^{3/2}\right]_1^8$

$\qquad = 5\sqrt{5} - 2\sqrt{2} \approx 8.352$

**9.** $y = \frac{x^5}{10} + \frac{1}{6x^3}, \quad 2 \le x \le 5$

$y' = \frac{x^4}{2} - \frac{1}{2x^4} = \frac{1}{2}\left(x^4 - \frac{1}{x^4}\right)$

$1 + (y')^2 = 1 + \frac{1}{4}\left(x^4 - \frac{1}{x^4}\right)^2 = 1 + \frac{1}{4}\left(x^8 - 2 + \frac{1}{x^8}\right)$

$\qquad = \frac{1}{4}\left(x^8 + 2 + \frac{1}{x^8}\right) = \frac{1}{4}\left(x^4 + \frac{1}{x^4}\right)^2$

$s = \int_2^5 \sqrt{1 + (y')^2}\, dx = \int_2^5 \frac{1}{2}\left(x^4 + \frac{1}{x^4}\right) dx$

$\qquad = \frac{1}{2}\left[\frac{x^5}{5} - \frac{1}{3x^3}\right]_2^5 = \frac{1}{2}\left[\left(625 - \frac{1}{375}\right) - \left(\frac{32}{5} - \frac{1}{24}\right)\right]$

$\qquad = \frac{618639}{2000} \approx 309.320$

**11.** $y = \ln(\sin x), \quad \left[\frac{\pi}{4}, \frac{3\pi}{4}\right]$

$y' = \frac{1}{\sin x}\cos x = \cot x$

$1 + (y')^2 = 1 + \cot^2 x = \csc^2 x$

$s = \int_{\pi/4}^{3\pi/4} \csc x\, dx$

$\qquad = \left[\ln|\csc x - \cot x|\right]_{\pi/4}^{3\pi/4}$

$\qquad = \ln(\sqrt{2} + 1) - \ln(\sqrt{2} - 1) \approx 1.763$

**13.** $y = \frac{1}{2}(e^x + e^{-x})$

$y' = \frac{1}{2}(e^x - e^{-x}), \quad [0, 2]$

$1 + (y')^2 = \left[\frac{1}{2}(e^x + e^{-x})\right]^2, \quad [0, 2]$

$s = \int_0^2 \sqrt{\left[\frac{1}{2}(e^x + e^{-x})\right]^2}\, dx$

$\qquad = \frac{1}{2}\int_0^2 (e^x + e^{-x})\, dx$

$\qquad = \frac{1}{2}\left[e^x - e^{-x}\right]_0^2 = \frac{1}{2}\left(e^2 - \frac{1}{e^2}\right) \approx 3.627$

**15.**  $x = \dfrac{1}{3}\left(y^2 + 2\right)^{3/2}, \quad 0 \le y \le 4$

$\dfrac{dx}{dy} = y\left(y^2 + 2\right)^{1/2}$

$s = \displaystyle\int_0^4 \sqrt{1 + y^2\left(y^2 + 2\right)}\ dy$

$= \displaystyle\int_0^4 \sqrt{y^4 + 2y^2 + 1}\ dy$

$= \displaystyle\int_0^4 \left(y^2 + 1\right) dy$

$= \left[\dfrac{y^3}{3} + y\right]_0^4 = \dfrac{64}{3} + 4 = \dfrac{76}{3}$

**17. (a)**  $y = 4 - x^2, \quad 0 \le x \le 2$

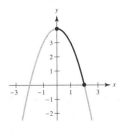

**(b)**  $y' = -2x$

$1 + \left(y'\right)^2 = 1 + 4x^2$

$L = \displaystyle\int_0^2 \sqrt{1 + 4x^2}\ dx$

**(c)**  $L \approx 4.647$

**19. (a)**  $y = \dfrac{1}{x}, \quad 1 \le x \le 3$

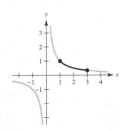

**(b)**  $y' = -\dfrac{1}{x^2}$

$1 + \left(y'\right)^2 = 1 + \dfrac{1}{x^4}$

$L = \displaystyle\int_1^3 \sqrt{1 + \dfrac{1}{x^4}}\ dx$

**(c)**  $L \approx 2.147$

**21. (a)**  $y = \sin x, \quad 0 \le x \le \pi$

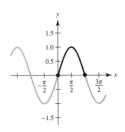

**(b)**  $y' = \cos x$

$1 + \left(y'\right)^2 = 1 + \cos^2 x$

$L = \displaystyle\int_0^\pi \sqrt{1 + \cos^2 x}\ dx$

**(c)**  $L \approx 3.820$

**23. (a)**  $x = e^{-y}, \quad 0 \le y \le 2$

$y = -\ln x$

$1 \ge x \ge e^{-2} \approx 0.135$

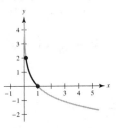

**(b)**  $y' = -\dfrac{1}{x}$

$1 + \left(y'\right)^2 = 1 + \dfrac{1}{x^2}$

$L = \displaystyle\int_{e^{-2}}^1 \sqrt{1 + \dfrac{1}{x^2}}\ dx$

**(c)**  $L \approx 2.221$

Alternatively, you can do all the computations with respect to $y$.

**(a)**  $x = e^{-y}, \quad 0 \le y \le 2$

**(b)**  $\dfrac{dx}{dy} = -e^{-y}$

$1 + \left(\dfrac{dx}{dy}\right)^2 = 1 + e^{-2y}$

$L = \displaystyle\int_0^2 \sqrt{1 + e^{-2y}}\ dy$

**(c)**  $L \approx 2.221$

**25.** (a) $y = 2 \arctan x, \quad 0 \le x \le 1$

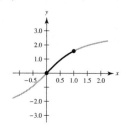

(b) $y' = \dfrac{2}{1 + x^2}$

$$L = \int_0^1 \sqrt{1 + \dfrac{4}{\left(1 + x^2\right)^2}}\, dx$$

(c) $L \approx 1.871$

**27.** $\displaystyle\int_0^2 \sqrt{1 + \left[\dfrac{d}{dx}\left(\dfrac{5}{x^2 + 1}\right)\right]^2}\, dx$

$s \approx 5$

Matches (b)

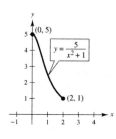

**28.** $\displaystyle\int_0^{\pi/4} \sqrt{1 + \left[\dfrac{d}{dx}(\tan x)\right]^2}\, dx$

$s \approx 1$

Matches (e)

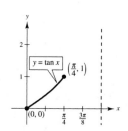

**29.** $y = x^3, \quad [0, 4]$

(a) $d = \sqrt{(4 - 0)^2 + (64 - 0)^2} \approx 64.125$

(b) $d = \sqrt{(1 - 0)^2 + (1 - 0)^2} + \sqrt{(2 - 1)^2 + (8 - 1)^2} + \sqrt{(3 - 2)^2 + (27 - 8)^2} + \sqrt{(4 - 3)^2 + (64 - 27)^2} \approx 64.525$

(c) $s = \displaystyle\int_0^4 \sqrt{1 + \left(3x^2\right)^2}\, dx = \int_0^4 \sqrt{1 + 9x^4}\, dx \approx 64.666 \quad$ (Simpson's Rule, $n = 10$)

(d) $64.672$

**31.** $\qquad y = 20 \cosh \dfrac{x}{20}, \quad -20 \le x \le 20$

$$y' = \sinh \dfrac{x}{20}$$

$$1 + \left(y'\right)^2 = 1 + \sinh^2 \dfrac{x}{20} = \cosh^2 \dfrac{x}{20}$$

$$L = \int_{-20}^{20} \cosh \dfrac{x}{20}\, dx = 2 \int_0^{20} \cosh \dfrac{x}{20}\, dx = \left[2(20) \sinh \dfrac{x}{20}\right]_0^{20} = 40 \sinh(1) \approx 47.008 \text{ m}$$

**33.** $y = 693.8597 - 68.7672 \cosh 0.0100333x$

$y' = -0.6899619478 \sinh 0.0100333x$

$$s = \int_{-299.2239}^{299.2239} \sqrt{1 + \left(-0.6899619478 \sinh 0.0100333x\right)^2}\, dx \approx 1480$$

(Use Simpson's Rule with $n = 100$ or a graphing utility.)

**35.** $\qquad y = \sqrt{9 - x^2}$

$$y' = \dfrac{-x}{\sqrt{9 - x^2}}$$

$$1 + \left(y'\right)^2 = \dfrac{9}{9 - x^2}$$

$$s = \int_0^2 \sqrt{\dfrac{9}{9 - x^2}}\, dx = \int_0^2 \dfrac{3}{\sqrt{9 - x^2}}\, dx$$

$$= \left[3 \arcsin \dfrac{x}{3}\right]_0^2 = 3\left(\arcsin \dfrac{2}{3} - \arcsin 0\right)$$

$$= 3 \arcsin \dfrac{2}{3} \approx 2.1892$$

**37.** $y = \dfrac{x^3}{3}$

$y' = x^2, \quad [0, 3]$

$$S = 2\pi \int_0^3 \dfrac{x^3}{3} \sqrt{1 + x^4}\, dx$$

$$= \dfrac{\pi}{6} \int_0^3 \left(1 + x^4\right)^{1/2}\left(4x^3\right)\, dx$$

$$= \left[\dfrac{\pi}{9}\left(1 + x^4\right)^{3/2}\right]_0^3$$

$$= \dfrac{\pi}{9}\left(82\sqrt{82} - 1\right) \approx 258.85$$

**39.**
$$y = \frac{x^3}{6} + \frac{1}{2x}$$

$$y' = \frac{x^2}{2} - \frac{1}{2x^2}$$

$$1 + (y')^2 = \left(\frac{x^2}{2} + \frac{1}{2x^2}\right)^2, \quad [1, 2]$$

$$S = 2\pi \int_1^2 \left(\frac{x^3}{6} + \frac{1}{2x}\right)\left(\frac{x^2}{2} + \frac{1}{2x^2}\right) dx$$

$$= 2\pi \int_1^2 \left(\frac{x^5}{12} + \frac{x}{3} + \frac{1}{4x^3}\right) dx$$

$$= 2\pi \left[\frac{x^6}{72} + \frac{x^2}{6} - \frac{1}{8x^2}\right]_1^2 = \frac{47\pi}{16}$$

**41.**
$$y = \sqrt{4 - x^2}$$

$$y' = \frac{1}{2}(4 - x^2)^{-1/2}(-2x) = \frac{-x}{\sqrt{4 - x^2}}, \quad -1 \le x \le 1$$

$$1 + (y')^2 = 1 + \frac{x^2}{4 - x^2} = \frac{4}{4 - x^2}$$

$$S = 2\pi \int_{-1}^1 \sqrt{4 - x^2} \cdot \sqrt{\frac{4}{4 - x^2}} \, dx$$

$$= 4\pi \int_{-1}^1 dx = 4\pi [x]_{-1}^1 = 8\pi$$

**43.** $y = \sqrt[3]{x} + 2$

$$y' = \frac{1}{3x^{2/3}}, \quad [1, 8]$$

$$S = 2\pi \int_1^8 x\sqrt{1 + \frac{1}{9x^{4/3}}} \, dx$$

$$= \frac{2\pi}{3} \int_1^8 x^{1/3} \sqrt{9x^{4/3} + 1} \, dx$$

$$= \frac{\pi}{18} \int_1^8 (9x^{4/3} + 1)^{1/2}(12x^{1/3}) \, dx$$

$$= \left[\frac{\pi}{27}(9x^{4/3} + 1)^{3/2}\right]_1^8$$

$$= \frac{\pi}{27}(145\sqrt{145} - 10\sqrt{10}) \approx 199.48$$

**45.**
$$y = 1 - \frac{x^2}{4}$$

$$y' = -\frac{x}{2}, \quad 0 \le x \le 2$$

$$1 + (y')^2 = 1 + \frac{x^2}{4} = \frac{4 + x^2}{4}$$

$$S = 2\pi \int_0^2 x\sqrt{\frac{4 + x^2}{4}} \, dx$$

$$= \pi \int_0^2 x\sqrt{4 + x^2} \, dx$$

$$= \frac{1}{2}\pi \int_0^2 (4 + x^2)^{1/2}(2x) \, dx$$

$$= \frac{1}{2}\pi \left[\frac{2}{3}(4 + x^2)^{3/2}\right]_0^2$$

$$= \frac{\pi}{3}(8^{3/2} - 4^{3/2})$$

$$= \frac{\pi}{3}(16\sqrt{2} - 8) \approx 15.318$$

**47.** $y = \sin x$

$$y' = \cos x, \quad [0, \pi]$$

$$S = 2\pi \int_0^\pi \sin x\sqrt{1 + \cos^2 x} \, dx \approx 14.4236$$

**49.** A rectifiable curve is one that has a finite arc length.

**51.** The precalculus formula is the surface area formula for the lateral surface of the frustum of a right circular cone. The formula is $S = 2\pi r L$, where $r = \frac{1}{2}(r_1 + r_2)$, which is the average radius of the frustum, and $L$ is the length of a line segment on the frustum. The representative element is

$$2\pi f(d_i)\sqrt{\Delta x_i^2 + \Delta y_i^2} = 2\pi f(d_i)\sqrt{1 + \left(\frac{\Delta y_i}{\Delta x_i}\right)^2} \, \Delta x_i.$$

**53. (a)**

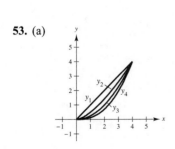

**(b)** $y_1, y_2, y_3, y_4$

**(c)** $y_1' = 1, \quad s_1 = \int_0^4 \sqrt{2} \, dx \approx 5.657$

$$y_2' = \frac{3}{4}x^{1/2}, \quad s_2 = \int_0^4 \sqrt{1 + \frac{9x}{16}} \, dx \approx 5.759$$

$$y_3' = \frac{1}{2}x, \quad s_3 = \int_0^4 \sqrt{1 + \frac{x^2}{4}} \, dx \approx 5.916$$

$$y_4' = \frac{5}{16}x^{3/2}, \quad s_4 = \int_0^4 \sqrt{1 + \frac{25}{256}x^3} \, dx \approx 6.063$$

**55.** $\qquad y = \dfrac{3x}{4}, \quad y' = \dfrac{3}{4}$

$1 + (y')^2 = 1 + \dfrac{9}{16} = 25/16$

$S = 2\pi \displaystyle\int_0^4 x\sqrt{\dfrac{25}{16}}\, dx = \dfrac{5\pi}{2}\left[\dfrac{x^2}{2}\right]_0^4 = 20\pi$

**57.** $\qquad y = \sqrt{9 - x^2}$

$y' = \dfrac{-x}{\sqrt{9 - x^2}}$

$\sqrt{1 + (y')^2} = \dfrac{3}{\sqrt{9 - x^2}}$

$S = 2\pi \displaystyle\int_0^2 \dfrac{3x}{\sqrt{9 - x^2}}\, dx$

$= -3\pi \displaystyle\int_0^2 \dfrac{-2x}{\sqrt{9 - x^2}}\, dx$

$= \left[-6\pi\sqrt{9 - x^2}\right]_0^2$

$= 6\pi\left(3 - \sqrt{5}\right) \approx 14.40$

See figure in Exercise 58.

**59.** (a) Approximate the volume by summing six disks of thickness 3 and circumference $C_i$ equal to the average of the given circumferences:

$V \approx \displaystyle\sum_{i=1}^{6} \pi r_i^2(3) = \sum_{i=1}^{6} \pi\left(\dfrac{C_i}{2\pi}\right)^2(3) = \dfrac{3}{4\pi}\sum_{i=1}^{6} C_i^2$

$= \dfrac{3}{4\pi}\left[\left(\dfrac{50 + 65.5}{2}\right)^2 + \left(\dfrac{65.5 + 70}{2}\right)^2 + \left(\dfrac{70 + 66}{2}\right)^2 + \left(\dfrac{66 + 58}{2}\right)^2 + \left(\dfrac{58 + 51}{2}\right)^2 + \left(\dfrac{51 + 48}{2}\right)^2\right]$

$= \dfrac{3}{4\pi}\left[57.75^2 + 67.75^2 + 68^2 + 62^2 + 54.5^2 + 49.5^2\right] = \dfrac{3}{4\pi}(21813.625) = 5207.62 \text{ in.}^3$

(b) The lateral surface area of a frustum of a right circular cone is $\pi s(R + r)$. For the first frustum:

$S_1 \approx \pi\left[3^2 + \left(\dfrac{65.5 - 50}{2\pi}\right)^2\right]^{1/2}\left[\dfrac{50}{2\pi} + \dfrac{65.5}{2\pi}\right]$

$= \left(\dfrac{50 + 65.5}{2}\right)\left[9 + \left(\dfrac{65.5 - 50}{2\pi}\right)^2\right]^{1/2}.$

Adding the six frustums together:

$S \approx \left(\dfrac{50 + 65.5}{2}\right)\left[9 + \left(\dfrac{15.5}{2\pi}\right)^2\right]^{1/2} + \left(\dfrac{65.5 + 70}{2}\right)\left[9 + \left(\dfrac{4.5}{2\pi}\right)^2\right]^{1/2}$

$\quad + \left(\dfrac{70 + 66}{2}\right)\left[9 + \left(\dfrac{4}{2\pi}\right)^2\right]^{1/2} + \left(\dfrac{66 + 58}{2}\right)\left[9 + \left(\dfrac{8}{2\pi}\right)^2\right]^{1/2}$

$\quad + \left(\dfrac{58 + 51}{2}\right)\left[9 + \left(\dfrac{7}{2\pi}\right)^2\right]^{1/2} + \left(\dfrac{51 + 48}{2}\right)\left[9 + \left(\dfrac{3}{2\pi}\right)^2\right]^{1/2}$

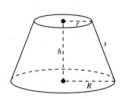

$\approx 224.30 + 208.96 + 208.54 + 202.06 + 174.41 + 150.37 = 1168.64$

(c) $r = 0.00401y^3 - 0.1416y^2 + 1.232y + 7.943$

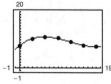

(d) $V = \displaystyle\int_0^{18} \pi r^2\, dy \approx 5275.9 \text{ in.}^3$

$S = \displaystyle\int_0^{18} 2\pi r(y)\sqrt{1 + r'(y)^2}\, dy \approx 1179.5 \text{ in.}^2$

**61.** (a) $V = \pi \int_1^b \frac{1}{x^2}\, dx = \left[ -\frac{\pi}{x} \right]_1^b = \pi\left( 1 - \frac{1}{b} \right)$

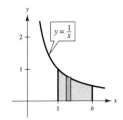

(b) $S = 2\pi \int_1^b \frac{1}{x} \sqrt{1 + \left( -\frac{1}{x^2} \right)^2}\, dx$

$= 2\pi \int_1^b \frac{1}{x} \sqrt{1 + \frac{1}{x^4}}\, dx$

$= 2\pi \int_1^b \frac{\sqrt{x^4 + 1}}{x^3}\, dx$

(c) $\displaystyle\lim_{b \to \infty} V = \lim_{b \to \infty} \pi\left( 1 - \frac{1}{b} \right) = \pi$

(d) Because

$$\frac{\sqrt{x^4 + 1}}{x^3} > \frac{\sqrt{x^4}}{x^3} = \frac{1}{x} > 0 \text{ on } [1, b],$$

you have

$$\int_1^b \frac{\sqrt{x^4 + 1}}{x^3}\, dx > \int_1^b \frac{1}{x}\, dx = \left[ \ln x \right]_1^b = \ln b$$

and $\displaystyle\lim_{b \to \infty} \ln b \to \infty.$ So,

$$\lim_{b \to \infty} 2\pi \int_1^b \frac{\sqrt{x^4 + 1}}{x^3}\, dx = \infty.$$

**65.** $x^{2/3} + y^{2/3} = 4$

$y^{2/3} = 4 - x^{2/3}$

$y = \left( 4 - x^{2/3} \right)^{3/2}, \quad 0 \le x \le 8$

$y' = \frac{3}{2}\left( 4 - x^{2/3} \right)^{1/2}\left( -\frac{2}{3}x^{-1/3} \right) = \frac{-\left( 4 - x^{2/3} \right)^{1/2}}{x^{1/3}}$

$1 + (y')^2 = 1 + \frac{4 - x^{2/3}}{x^{2/3}} = \frac{4}{x^{2/3}}$

$S = 2\pi \int_0^8 \left( 4 - x^{2/3} \right)^{3/2} \sqrt{\frac{4}{x^{2/3}}}\, dx = 4\pi \int_0^8 \frac{\left( 4 - x^{2/3} \right)^{3/2}}{x^{1/3}}\, dx = \left[ -\frac{12\pi}{5}\left( 4 - x^{2/3} \right)^{5/2} \right]_0^8 = \frac{384\pi}{5}$

[Surface area of portion above the *x*-axis]

**63.** $y = \frac{1}{3}\left( x^{3/2} - 3x^{1/2} + 2 \right)$

When $x = 0$, $y = \frac{2}{3}$. So, the fleeing object has traveled

$\frac{2}{3}$ unit when it is caught.

$$y' = \frac{1}{3}\left( \frac{3}{2}x^{1/2} - \frac{3}{2}x^{-1/2} \right) = \left( \frac{1}{2} \right)\frac{x - 1}{x^{1/2}}$$

$$1 + (y')^2 = 1 + \frac{(x - 1)^2}{4x} = \frac{(x + 1)^2}{4x}$$

$$s = \int_0^1 \frac{x + 1}{2x^{1/2}}\, dx = \frac{1}{2}\int_0^1 \left( x^{1/2} + x^{-1/2} \right) dx$$

$$= \frac{1}{2}\left[ \frac{2}{3}x^{3/2} + 2x^{1/2} \right]_0^1 = \frac{4}{3} = 2\left( \frac{2}{3} \right)$$

The pursuer has traveled twice the distance that the fleeing object has traveled when it is caught.

**67.** $y = kx^2$, $y' = 2kx$

$1 + (y')^2 = 1 + 4k^2x^2$

$h = kw^2 \Rightarrow k = \dfrac{h}{w^2} \Rightarrow 1 + (y') = 1 + \dfrac{4h^2}{w^4}x^2$

By symmetry, $C = 2\displaystyle\int_0^w \sqrt{1 + \dfrac{4h^2}{w^4}x^2}\ dx$.

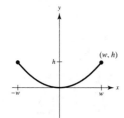

**69.**
$y = f(x) = \cosh x$

$y' = \sinh x$

$1 + (y')^2 = 1 + \sinh^2 x = \cosh^2 x$

Area $= \displaystyle\int_0^t \cosh x\ dx = [\sinh x]_0^t = \sinh t$

Arc length $= \displaystyle\int_0^t \sqrt{1 + (y')^2}\ dx$

$= \displaystyle\int_0^t \cosh x\ dx = \sinh x\ \Big]_0^t$

$= \sinh t.$

Another curve with this property is $g(x) = 1$.

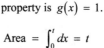

Area $= \displaystyle\int_0^t dx = t$

Arc length $= t$

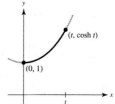

# Section 7.5  Work

**1.** $W = Fd = 1200(40) = 48{,}000$ ft-lb

**3.** $W = Fd = (112)(8) = 896$ joules (Newton-meters)

**5.** $F(x) = kx$

$5 = k(3)$

$k = \dfrac{5}{3}$

$F(x) = \dfrac{5}{3}x$

$W = \displaystyle\int_0^7 F(x)\ dx = \int_0^7 \dfrac{5}{3}x\ dx = \left[\dfrac{5}{6}x^2\right]_0^7 = \dfrac{245}{6}$ in.-lb

$\approx 40.833$ in.-lb $\approx 3.403$ ft-lb

**7.** $F(x) = kx$

$20 = k(9)$

$k = \dfrac{20}{9}$

$W = \displaystyle\int_0^{12} \dfrac{20}{9}x\ dx = \left[\dfrac{10}{9}x^2\right]_0^{12} = 160$ in.-lb $= \dfrac{40}{3}$ ft-lb

**9.** $W = 18 = \displaystyle\int_0^{1/3} kx\ dx = \dfrac{kx^2}{2}\Bigg]_0^{1/3} = \dfrac{k}{18} \Rightarrow k = 324$

$W = \displaystyle\int_{1/3}^{7/12} 324x\ dx = \left[162x^2\right]_{1/3}^{7/12} = 37.125$ ft-lb

$\left[\text{\textbf{Note:} 4 inches} = \dfrac{1}{3}\text{ foot}\right]$

**11.** Assume that Earth has a radius of 4000 miles.

$F(x) = \dfrac{k}{x^2}$

$5 = \dfrac{k}{(4000)^2}$

$k = 80{,}000{,}000$

$F(x) = \dfrac{80{,}000{,}000}{x^2}$

(a) $W = \displaystyle\int_{4000}^{4100} \dfrac{80{,}000{,}000}{x^2}\ dx = \left[\dfrac{-80{,}000{,}000}{x}\right]_{4000}^{4100}$

$\approx 487.8$ mi-tons $\approx 5.15 \times 10^9$ ft-lb

(b) $W = \displaystyle\int_{4000}^{4300} \dfrac{80{,}000{,}000}{x^2}\ dx$

$\approx 1395.3$ mi-ton $\approx 1.47 \times 10^{10}$ ft-ton

**13.** Assume that Earth has a radius of 4000 miles.

$$F(x) = \frac{k}{x^2}$$

$$10 = \frac{k}{(4000)^2}$$

$$k = 160{,}000{,}000$$

$$F(x) = \frac{160{,}000{,}000}{x^2}$$

(a) $W = \displaystyle\int_{4000}^{15{,}000} \frac{160{,}000{,}000}{x^2} \, dx = \left[ -\frac{160{,}000{,}000}{x} \right]_{4000}^{15{,}000} \approx -10{,}666.667 + 40{,}000$

$$= 29{,}333.333 \text{ mi-ton}$$

$$\approx 2.93 \times 10^4 \text{ mi-ton}$$

$$\approx 3.10 \times 10^{11} \text{ ft-lb}$$

(b) $W = \displaystyle\int_{4000}^{26{,}000} \frac{160{,}000{,}000}{x^2} \, dx = \left[ -\frac{160{,}000{,}000}{x} \right]_{4000}^{26{,}000} \approx -6{,}153.846 + 40{,}000$

$$= 33{,}846.154 \text{ mi-ton}$$

$$\approx 3.38 \times 10^4 \text{ mi-ton}$$

$$\approx 3.57 \times 10^{11} \text{ ft-lb}$$

**15.** Weight of each layer: $62.4(20) \, \Delta y$

Distance: $4 - y$

(a) $W = \displaystyle\int_{2}^{4} 62.4(20)(4 - y) \, dy = \left[ 4992y - 624y^2 \right]_{2}^{4} = 2496 \text{ ft-lb}$

(b) $W = \displaystyle\int_{0}^{4} 62.4(20)(4 - y) \, dy = \left[ 4992y - 624y^2 \right]_{0}^{4} = 9984 \text{ ft-lb}$

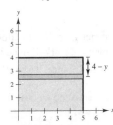

**17.** Volume of disk: $\pi(2)^2 \, \Delta y = 4\pi \, \Delta y$

Weight of disk of water: $9800(4\pi) \, \Delta y$

Distance the disk of water is moved: $5 - y$

$$W = \int_{0}^{4} (5 - y)(9800)4\pi \, dy = 39{,}200\pi \int_{0}^{4} (5 - y) \, dy$$

$$= 39{,}200\pi \left[ 5y - \frac{y^2}{2} \right]_{0}^{4}$$

$$= 39{,}200\pi(12) = 470{,}400\pi \text{ newton–meters}$$

**19.** Volume of disk: $\pi\left(\dfrac{2}{3}y\right)^2 \Delta y$

Weight of disk: $62.4\pi\left(\dfrac{2}{3}y\right)^2 \Delta y$

Distance: $6 - y$

$$W = \frac{4(62.4)\pi}{9} \int_{0}^{6} (6 - y)y^2 \, dy$$

$$= \frac{4}{9}(62.4)\pi \left[ 2y^3 - \frac{1}{4}y^4 \right]_{0}^{6}$$

$$= 2995.2\pi \text{ ft-lb}$$

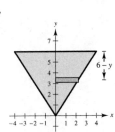

**21.** Volume of disk: $\pi\left(\sqrt{36 - y^2}\right)^2 \Delta y$

Weight of disk: $62.4\pi\left(36 - y^2\right) \Delta y$

Distance: $y$

$$W = 62.4\pi \int_0^6 y\left(36 - y^2\right) dy$$

$$= 62.4\pi \int_0^6 \left(36y - y^3\right) dy = 62.4\pi\left[18y^2 - \tfrac{1}{4}y^4\right]_0^6$$

$$= 20{,}217.6\pi \text{ ft-lb}$$

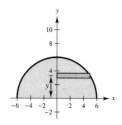

**23.** Volume of layer: $V = lwh = 4(2)\sqrt{(9/4) - y^2}\,\Delta y$

Weight of layer: $W = 42(8)\sqrt{(9/4) - y^2}\,\Delta y$

Distance: $\tfrac{13}{2} - y$

$$W = \int_{-1.5}^{1.5} 42(8)\sqrt{\tfrac{9}{4} - y^2}\left(\tfrac{13}{2} - y\right) dy$$

$$= 336\left[\tfrac{13}{2} \int_{-1.5}^{1.5} \sqrt{\tfrac{9}{4} - y^2}\, dy - \int_{-1.5}^{1.5} \sqrt{\tfrac{9}{4} - y^2}\, y\, dy\right]$$

The second integral is zero because the integrand is odd and the limits of integration are symmetric to the origin. The first integral represents the area of a semicircle of radius $\tfrac{3}{2}$. So, the work is

$$W = 336\left(\tfrac{13}{2}\right)\pi\left(\tfrac{3}{2}\right)^2\left(\tfrac{1}{2}\right) = 2457\pi \text{ ft-lb}.$$

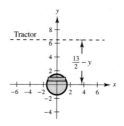

**25.** Weight of section of chain: $3\,\Delta y$

Distance: $20 - y$. $\quad \Delta W = (\text{force increment})(\text{distance}) = (3\,\Delta y)(20 - y)$

$$W = \int_0^{20} (20 - y)3\, dy = 3\left[20y - \frac{y^2}{2}\right]_0^{20} = 3\left[400 - \frac{400}{2}\right] = 600 \text{ ft-lb}$$

**27.** The lower 10 feet of fence are raised 10 feet with a constant force.

$$W_1 = 3(10)(10) = 300 \text{ ft-lb}$$

The top 10 feet are raised with a variable force.

Weight of section: $3\,\Delta y$

Distance: $10 - y$

$$W_2 = \int_0^{10} 3(10 - y)\, dy = 3\left[10y - \frac{y^2}{2}\right]_0^{10} = 150 \text{ ft-lb}$$

$$W = W_1 + W_2 = 300 + 150 = 450 \text{ ft-lb}$$

**29.** Weight of section of chain: $3\,\Delta y$

Distance: $15 - 2y$

$$W = 3\int_0^{7.5} (15 - 2y)\, dy = \left[-\tfrac{3}{4}(15 - 2y)^2\right]_0^{7.5}$$

$$= \tfrac{3}{4}(15)^2 = 168.75 \text{ ft-lb}$$

**31.** If an object is moved a distance $D$ in the direction of an applied constant force $F$, then the work $W$ done by the force is defined as force times distance, $W = FD$.

**33.** (a) requires more work. In part (b) no work is done because the books are not moved:

$$W = \text{force} \times \text{distance}$$

**35.** (a) $W = \int_0^9 6\, dx = 54 \text{ ft-lb}$

(b) $W = \int_0^7 20\, dx + \int_7^9 (-10x + 90)\, dx = 140 + 20$

$$= 160 \text{ ft-lb}$$

(c) $W = \int_0^9 \frac{1}{27}x^2\, dx = \frac{x^3}{81}\Big]_0^9 = 9 \text{ ft-lb}$

(d) $W = \int_0^9 \sqrt{x}\, dx = \frac{2}{3}x^{3/2}\Big]_0^9 = \frac{2}{3}(27) = 18 \text{ ft-lb}$

**37.**  $p = \dfrac{k}{V}$

$1000 = \dfrac{k}{2}$

$k = 2000$

$W = \displaystyle\int_2^3 \dfrac{2000}{V}\, dV$

$= \Big[\, 2000 \ln|V|\, \Big]_2^3 = 2000 \ln\!\left(\dfrac{3}{2}\right) \approx 810.93 \text{ ft-lb}$

**39.**  $W = \displaystyle\int_0^5 1000\big[1.8 - \ln(x + 1)\big]\, dx \approx 3249.44 \text{ ft-lb}$

**41.**  $W = \displaystyle\int_0^5 100x\sqrt{125 - x^3}\, dx \approx 10{,}330.3 \text{ ft-lb}$

**43.** (a)  $W = FD = (8000\pi)(2) = 16{,}000\pi \text{ ft} \cdot \text{lb}$

(b)  $W \approx \dfrac{2 - 0}{3(6)}\big[0 + 4(20{,}000) + 2(22{,}000) + 4(15{,}000) + 2(10{,}000) + 4(5000) + 0\big] \approx 24888.889 \text{ ft-lb}$

(c)  $F(x) = -16{,}261.36x^4 + 85{,}295.45x^3 - 157{,}738.64x^2 + 104{,}386.36x - 32.4675$

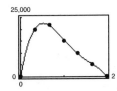

(d)  $F(x)$ is a maximum when $x \approx 0.524$ feet.

(e)  $W = \displaystyle\int_0^2 F(x)\, dx \approx 25{,}180.5 \text{ ft-lb}$

## Section 7.6    Moments, Centers of Mass, and Centroids

**1.**  $\bar{x} = \dfrac{7(-5) + 3(0) + 5(3)}{7 + 3 + 5} = \dfrac{-20}{15} = -\dfrac{4}{3}$

**3.**  $\bar{x} = \dfrac{1(6) + 3(10) + 2(3) + 9(2) + 5(4)}{1 + 3 + 2 + 9 + 5} = \dfrac{80}{20} = 4$

**5.** (a)  Add 4 to each $x$-value because each point is translated to the right 4 units.

$\bar{x} = \dfrac{1(10) + 3(14) + 2(7) + 9(6) + 5(8)}{1 + 3 + 2 + 9 + 5} = \dfrac{160}{20} = 8$

**Note:** From Exercise 3, $4 + 4 = 8$.

(b)  Subtract 2 from each $x$-value because each point is translated 2 units to the left.

$\bar{x} = \dfrac{8(-4) + 5(4) + 5(-2) + 12(1) + 2(-7)}{8 + 5 + 5 + 12 + 2} = \dfrac{-24}{32} = -\dfrac{3}{4}$

**Note:** From Exercise 4, $\dfrac{5}{4} - 2 = -\dfrac{3}{4}$.

**7.**  $48x = 72(L - x) = 72(10 - x)$

$48x = 720 - 72x$

$120x = 720$

$x = 6 \text{ ft}$

**9.**  $\bar{x} = \dfrac{5(2) + 1(-3) + 3(1)}{5 + 1 + 3} = \dfrac{10}{9}$

$\bar{y} = \dfrac{5(2) + 1(1) + 3(-4)}{5 + 1 + 3} = -\dfrac{1}{9}$

$(\bar{x}, \bar{y}) = \left(\dfrac{10}{9}, -\dfrac{1}{9}\right)$

**11.** $\bar{x} = \dfrac{12(2) + 6(-1) + (9/2)(6) + 15(2)}{12 + 6 + (9/2) + 15} = \dfrac{75}{37.5} = 2$

$\bar{y} = \dfrac{12(3) + 6(5) + (9/2)(8) + 15(-2)}{12 + 6 + (9/2) + 15} = \dfrac{72}{37.5} = \dfrac{48}{25}$

$(\bar{x}, \bar{y}) = \left(2, \dfrac{48}{25}\right)$

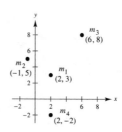

**13.** $m = \rho \displaystyle\int_0^2 \dfrac{x}{2}\, dx = \left[\rho\dfrac{x^2}{4}\right]_0^2 = \rho$

$M_x = \rho \displaystyle\int_0^2 \dfrac{1}{2}\left(\dfrac{x}{2}\right)^2 dx = \dfrac{\rho}{8}\left[\dfrac{x^3}{3}\right]_0^2 = \dfrac{\rho}{3}$

$\bar{y} = \dfrac{M_x}{m} = \dfrac{\rho/3}{\rho} = \dfrac{1}{3}$

$M_y = \rho \displaystyle\int_0^2 x\left(\dfrac{x}{2}\right) dx = \dfrac{\rho}{2}\left[\dfrac{x^3}{3}\right]_0^2 = \dfrac{4}{3}\rho$

$\bar{x} = \dfrac{M_y}{m} = \dfrac{4/3\rho}{\rho} = \dfrac{4}{3}$

$(\bar{x}, \bar{y}) = \left(\dfrac{4}{3}, \dfrac{1}{3}\right)$

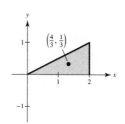

**15.** $m = \rho \displaystyle\int_0^4 \sqrt{x}\, dx = \left[\dfrac{2\rho}{3}x^{3/2}\right]_0^4 = \dfrac{16\rho}{3}$

$M_x = \rho \displaystyle\int_0^4 \dfrac{\sqrt{x}}{2}\left(\sqrt{x}\right) dx = \left[\rho\dfrac{x^2}{4}\right]_0^4 = 4\rho$

$\bar{y} = \dfrac{M_x}{m} = 4\rho\left(\dfrac{3}{16\rho}\right) = \dfrac{3}{4}$

$M_y = \rho \displaystyle\int_0^4 x\sqrt{x}\, dx = \left[\rho\dfrac{2}{5}x^{5/2}\right]_0^4 = \dfrac{64\rho}{5}$

$\bar{x} = \dfrac{M_y}{m} = \dfrac{64\rho}{5}\left(\dfrac{3}{16\rho}\right) = \dfrac{12}{5}$

$(\bar{x}, \bar{y}) = \left(\dfrac{12}{5}, \dfrac{3}{4}\right)$

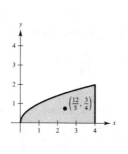

**17.** $m = \rho \displaystyle\int_0^1 \left(x^2 - x^3\right) dx = \rho\left[\dfrac{x^3}{3} - \dfrac{x^4}{4}\right]_0^1 = \dfrac{\rho}{12}$

$M_x = \rho \displaystyle\int_0^1 \dfrac{\left(x^2 + x^3\right)}{2}\left(x^2 - x^3\right) dx = \dfrac{\rho}{2}\displaystyle\int_0^1 \left(x^4 - x^6\right) dx = \dfrac{\rho}{2}\left[\dfrac{x^5}{5} - \dfrac{x^7}{7}\right]_0^1 = \dfrac{\rho}{35}$

$\bar{y} = \dfrac{M_x}{m} = \dfrac{\rho}{35}\left(\dfrac{12}{\rho}\right) = \dfrac{12}{35}$

$M_y = \rho \displaystyle\int_0^1 x\left(x^2 - x^3\right) dx = \rho\displaystyle\int_0^1 \left(x^3 - x^4\right) dx = \rho\left[\dfrac{x^4}{4} - \dfrac{x^5}{5}\right]_0^1 = \dfrac{\rho}{20}$

$\bar{x} = \dfrac{M_y}{m} = \dfrac{\rho}{20}\left(\dfrac{12}{\rho}\right) = \dfrac{3}{5}$

$(\bar{x}, \bar{y}) = \left(\dfrac{3}{5}, \dfrac{12}{35}\right)$

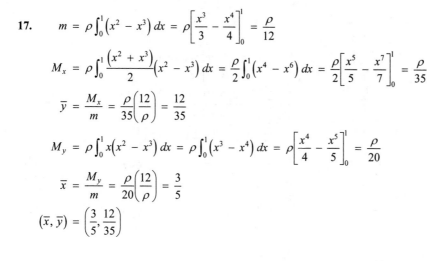

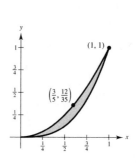

**19.**   $m = \rho \int_0^3 \left[ (-x^2 + 4x + 2) - (x + 2) \right] dx = -\rho \left[ \dfrac{x^3}{3} + \dfrac{3x^2}{2} \right]_0^3 = \dfrac{9\rho}{2}$

$M_x = \rho \int_0^3 \left[ \dfrac{(-x^2 + 4x + 2) + (x + 2)}{2} \right] \left[ (-x^2 + 4x + 2) - (x + 2) \right] dx$

$\quad = \dfrac{\rho}{2} \int_0^3 (-x^2 + 5x + 4)(-x^2 + 3x) \, dx = \dfrac{\rho}{2} \int_0^3 (x^4 - 8x^3 + 11x^2 + 12x) \, dx = \dfrac{\rho}{2} \left[ \dfrac{x^5}{5} - 2x^4 + \dfrac{11x^3}{3} + 6x^2 \right]_0^3 = \dfrac{99\rho}{5}$

$\bar{y} = \dfrac{M_x}{m} = \dfrac{99\rho}{5} \left( \dfrac{2}{9\rho} \right) = \dfrac{22}{5}$

$M_y = \rho \int_0^3 x \left[ (-x^2 + 4x - 2) - (x + 2) \right] dx = \rho \int_0^3 (-x^3 + 3x^2) \, dx = \rho \left[ -\dfrac{x^4}{4} + x^3 \right]_0^3 = \dfrac{27\rho}{4}$

$\bar{x} = \dfrac{M_y}{m} = \dfrac{27\rho}{4} \left( \dfrac{2}{9\rho} \right) = \dfrac{3}{2}$

$(\bar{x}, \bar{y}) = \left( \dfrac{3}{2}, \dfrac{22}{5} \right)$

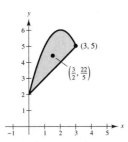

**21.**   $m = \rho \int_0^8 x^{2/3} \, dx = \rho \left[ \dfrac{3}{5} x^{5/3} \right]_0^8 = \dfrac{96\rho}{5}$

$M_x = \rho \int_0^8 \dfrac{x^{2/3}}{2} \left( x^{2/3} \right) dx = \dfrac{\rho}{2} \left[ \dfrac{3}{7} x^{7/3} \right]_0^8 = \dfrac{192\rho}{7}$

$\bar{y} = \dfrac{M_x}{m} = \dfrac{192\rho}{7} \left( \dfrac{5}{96\rho} \right) = \dfrac{10}{7}$

$M_y = \rho \int_0^8 x \left( x^{2/3} \right) dx = \rho \left[ \dfrac{3}{8} x^{8/3} \right]_0^8 = 96\rho$

$\bar{x} = \dfrac{M_y}{m} = 96\rho \left( \dfrac{5}{96\rho} \right) = 5$

$(\bar{x}, \bar{y}) = \left( 5, \dfrac{10}{7} \right)$

**23.**   $m = 2\rho \int_0^2 (4 - y^2) \, dy = 2\rho \left[ 4y - \dfrac{y^3}{3} \right]_0^2 = \dfrac{32\rho}{3}$

$M_y = 2\rho \int_0^2 \left( \dfrac{4 - y^2}{2} \right) (4 - y^2) \, dy = \rho \left[ 16y - \dfrac{8}{3} y^3 + \dfrac{y^5}{5} \right]_0^2 = \dfrac{256\rho}{15}$

$\bar{x} = \dfrac{M_y}{m} = \dfrac{256\rho}{15} \left( \dfrac{3}{32\rho} \right) = \dfrac{8}{5}$

By symmetry, $M_x$ and $\bar{y} = 0$.

$(\bar{x}, \bar{y}) = \left( \dfrac{8}{5}, 0 \right)$

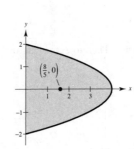

**25.**
$$m = \rho \int_0^3 \left[ (2y - y^2) - (-y) \right] dy = \rho \left[ \frac{3y^2}{2} - \frac{y^3}{3} \right]_0^3 = \frac{9\rho}{2}$$

$$M_y = \rho \int_0^3 \frac{\left[ (2y - y^2) + (-y) \right]}{2} \left[ (2y - y^2) - (-y) \right] dy = \frac{\rho}{2} \int_0^3 (y - y^2)(3y - y^2) \, dy$$

$$= \frac{\rho}{2} \int_0^3 (y^4 - 4y^3 + 3y^2) \, dy = \frac{\rho}{2} \left[ \frac{y^5}{5} - y^4 + y^3 \right]_0^3 = -\frac{27\rho}{10}$$

$$\bar{x} = \frac{M_y}{m} = -\frac{27\rho}{10}\left( \frac{2}{9\rho} \right) = -\frac{3}{5}$$

$$M_x = \rho \int_0^3 y \left[ (2y - y^2) - (-y) \right] dy = \rho \int_0^3 (3y^2 - y^3) \, dy = \rho \left[ y^3 - \frac{y^4}{4} \right]_0^3 = \frac{27\rho}{4}$$

$$\bar{y} = \frac{M_x}{m} = \frac{27\rho}{4}\left( \frac{2}{9\rho} \right) = \frac{3}{2}$$

$$(\bar{x}, \bar{y}) = \left( -\frac{3}{5}, \frac{3}{2} \right)$$

**27.**  $m = \rho \int_0^5 10x\sqrt{125 - x^3} \, dx \approx 1033.0\rho$

$$M_x = \rho \int_0^5 \left( \frac{10x\sqrt{125 - x^3}}{2} \right)\left( 10x\sqrt{125 - x^3} \right) dx = 50\rho \int_0^5 x^2(125 - x^3) \, dx = \frac{3{,}124{,}375\rho}{24} \approx 130{,}208\rho$$

$$M_y = \rho \int_0^5 10x^2\sqrt{125 - x^3} \, dx = -\frac{10\rho}{3} \int_0^5 \sqrt{125 - x^3}\,(-3x^2) \, dx = \frac{12{,}500\sqrt{5}\rho}{9} \approx 3105.6\rho$$

$$\bar{x} = \frac{M_y}{m} \approx 3.0$$

$$\bar{y} = \frac{M_x}{m} \approx 126.0$$

Therefore, the centroid is $(3.0, 126.0)$.

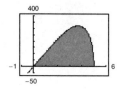

**29.**  $m = \rho \int_{-20}^{20} 5\sqrt[3]{400 - x^2} \, dx \approx 1239.76\rho$

$$M_x = \rho \int_{-20}^{20} \frac{5\sqrt[3]{400 - x^2}}{2}\left( 5\sqrt[3]{400 - x^2} \right) dx$$

$$= \frac{25\rho}{2} \int_{-20}^{20} (400 - x^2)^{2/3} \, dx \approx 20064.27$$

$$\bar{y} = \frac{M_x}{m} \approx 16.18$$

$\bar{x} = 0$ by symmetry. Therefore, the centroid is $(0, 16.2)$.

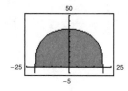

**31.** Centroids of the given regions: $(1, 0)$ and $(3, 0)$

Area: $A = 4 + \pi$

$$\bar{x} = \frac{4(1) + \pi(3)}{4 + \pi} = \frac{4 + 3\pi}{4 + \pi}$$

$$\bar{y} = \frac{4(0) + \pi(0)}{4 + \pi} = 0$$

$$(\bar{x}, \bar{y}) = \left( \frac{4 + 3\pi}{4 + \pi}, 0 \right) \approx (1.88, 0)$$

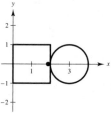

**33.** Centroids of the given regions: $\left(0, \dfrac{3}{2}\right)$, $(0, 5)$, and

$\left(0, \dfrac{15}{2}\right)$

Area: $A = 15 + 12 + 7 = 34$

$$\bar{x} = \frac{15(0) + 12(0) + 7(0)}{34} = 0$$

$$\bar{y} = \frac{15(3/2) + 12(5) + 7(15/2)}{34} = \frac{135}{34}$$

$$(\bar{x}, \bar{y}) = \left(0, \frac{135}{34}\right) \approx (0, 3.97)$$

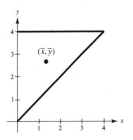

**35.** Centroids of the given regions: $(1, 0)$ and $(3, 0)$

Mass: $4 + 2\pi$

$$\bar{x} = \frac{4(1) + 2\pi(3)}{4 + 2\pi} = \frac{2 + 3\pi}{2 + \pi}$$

$$\bar{y} = 0$$

$$(\bar{x}, \bar{y}) = \left(\frac{2 + 3\pi}{2 + \pi}, 0\right) \approx (2.22, 0)$$

**37.** $r = 5$ is distance between center of circle and $y$-axis.

$A \approx \pi(4)^2 = 16\pi$ is the area of circle. So,

$V = 2\pi r A = 2\pi(5)(16\pi) = 160\pi^2 \approx 1579.14.$

**39.** $A = \dfrac{1}{2}(4)(4) = 8$

$$\bar{y} = \left(\frac{1}{8}\right)\frac{1}{2}\int_0^4 (4 + x)(4 - x)\, dx = \frac{1}{16}\left[16x - \frac{x^3}{3}\right]_0^4 = \frac{8}{3}$$

$$r = \bar{y} = \frac{8}{3}$$

$$V = 2\pi r A = 2\pi\left(\frac{8}{3}\right)(8) = \frac{128\pi}{3} \approx 134.04$$

**41.** The center of mass $(\bar{x}, \bar{y})$ is $\bar{x} = M_y/m$ and
$\bar{y} = M_x/m$, where:

1.    $m = m_1 + m_2 + \cdots + m_n$ is the total mass of the system.

2.    $M_y = m_1 x_1 + m_2 x_2 + \cdots + m_n x_n$ is the moment about the $y$-axis.

3.    $M_x = m_1 y_1 + m_2 y_2 + \cdots + m_n y_n$ is the moment about the $x$-axis.

**43.** Let $R$ be a region in a plane and let $L$ be a line such that $L$ does not intersect the interior of $R$. If $r$ is the distance between the centroid of $R$ and $L$, then the volume $V$ of the solid of revolution formed by revolving $R$ about $L$ is $V = 2\pi r A$ where $A$ is the area of $R$.

**45.**

$$A = \frac{1}{2}(2a)c = ac$$

$$\frac{1}{A} = \frac{1}{ac}$$

$$\bar{x} = \left(\frac{1}{ac}\right)\frac{1}{2}\int_0^c \left[\left(\frac{b - a}{c}y + a\right)^2 - \left(\frac{b + a}{c}y - a\right)^2\right]dy$$

$$= \frac{1}{2ac}\int_0^c \left[\frac{4ab}{c}y - \frac{4ab}{c^2}y^2\right]dy = \frac{1}{2ac}\left[\frac{2ab}{c}y^2 - \frac{4ab}{3c^2}y^3\right]_0^c = \frac{1}{2ac}\left(\frac{2}{3}abc\right) = \frac{b}{3}$$

$$\bar{y} = \frac{1}{ac}\int_0^c y\left[\left(\frac{b - a}{c}y + a\right) - \left(\frac{b + a}{c}y - a\right)\right]dy$$

$$= \frac{1}{ac}\int_0^c y\left(-\frac{2a}{c}y + 2a\right)dy = \frac{2}{c}\int_0^c \left(y - \frac{y^2}{c}\right)dy = \frac{2}{c}\left[\frac{y^2}{2} - \frac{y^3}{3c}\right]_0^c = \frac{c}{3}$$

$$(\bar{x}, \bar{y}) = \left(\frac{b}{3}, \frac{c}{3}\right)$$

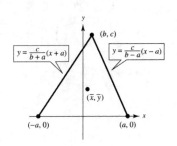

From elementary geometry, $(b/3, c/3)$ is the point of intersection of the medians.

**47.** $A = \dfrac{c}{2}(a + b)$

$$\frac{1}{A} = \frac{2}{c(a + b)}$$

$$\bar{x} = \frac{2}{c(a + b)} \int_0^c x\left(\frac{b - a}{c}x + a\right) dx = \frac{2}{c(a + b)} \int_0^c \left(\frac{b - a}{c}x^2 + ax\right) dx = \frac{2}{c(a + b)}\left[\frac{b - a}{c}\frac{x^3}{3} + \frac{ax^2}{2}\right]_0^c$$

$$= \frac{2}{c(a + b)}\left[\frac{(b - a)c^2}{3} + \frac{ac^2}{2}\right] = \frac{2}{c(a + b)}\left[\frac{2bc^2 - 2ac^2 + 3ac^2}{6}\right] = \frac{c(2b + a)}{3(a + b)} = \frac{(a + 2b)c}{3(a + b)}$$

$$\bar{y} = \frac{2}{c(a + b)}\frac{1}{2}\int_0^c \left(\frac{b - a}{c}x + a\right)^2 dx = \frac{1}{c(a + b)}\int_0^c \left[\left(\frac{b - a}{c}\right)^2 x^2 + \frac{2a(b - a)}{c}x + a^2\right] dx$$

$$= \frac{1}{c(a + b)}\left[\left(\frac{b - a}{c}\right)^2 \frac{x^3}{3} + \frac{2a(b - a)}{c}\frac{x^2}{2} + a^2 x\right]_0^c = \frac{1}{c(a + b)}\left[\frac{(b - a)^2 c}{3} + ac(b - a) + a^2 c\right]$$

$$= \frac{1}{3c(a + b)}\left[(b^2 - 2ab + a^2)c + 3ac(b - a) + 3a^2 c\right]$$

$$= \frac{1}{3(a + b)}\left[b^2 - 2ab + a^2 + 3ab - 3a^2 + 3a^2\right] = \frac{a^2 + ab + b^2}{3(a + b)}$$

So, $(\bar{x}, \bar{y}) = \left(\dfrac{(a + 2b)c}{3(a + b)}, \dfrac{a^2 + ab + b^2}{3(a + b)}\right)$.

The one line passes through $\left(0, \dfrac{a}{2}\right)$ and $\left(c, \dfrac{b}{2}\right)$. Its equation is $y = \dfrac{b - a}{2c}x + \dfrac{a}{2}$. The other line passes through

$(0, -b)$ and $(c, a + b)$. Its equation is $y = \dfrac{a + 2b}{c}x - b$. $(\bar{x}, \bar{y})$ is the point of intersection of these two lines.

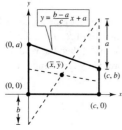

**49.** $\bar{x} = 0$ by symmetry.

$$A = \frac{1}{2}\pi ab$$

$$\frac{1}{A} = \frac{2}{\pi ab}$$

$$\bar{y} = \frac{2}{\pi ab}\frac{1}{2}\int_{-a}^a \left(\frac{b}{a}\sqrt{a^2 - x^2}\right)^2 dx = \frac{1}{\pi ab}\left(\frac{b^2}{a^2}\right)\left[a^2 x - \frac{x^3}{3}\right]_{-a}^a = \frac{b}{\pi a^3}\left(\frac{4a^3}{3}\right) = \frac{4b}{3\pi}$$

$$(\bar{x}, \bar{y}) = \left(0, \frac{4b}{3\pi}\right)$$

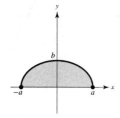

**51.** (a)

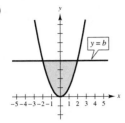

(b) $\bar{x} = 0$ by symmetry.

(c) $M_y = \int_{-\sqrt{b}}^{\sqrt{b}} x(b - x^2)\, dx = 0$ because $bx - x^3$ is odd.

(d) $\bar{y} > \dfrac{b}{2}$ because there is more area above $y = \dfrac{b}{2}$ than below.

(e) $M_x = \int_{-\sqrt{b}}^{\sqrt{b}} \dfrac{(b + x^2)(b - x^2)}{2}\, dx = \int_{-\sqrt{b}}^{\sqrt{b}} \dfrac{b^2 - x^4}{2}\, dx = \dfrac{1}{2}\left[ b^2 x - \dfrac{x^5}{5} \right]_{-\sqrt{b}}^{\sqrt{b}} = b^2\sqrt{b} - \dfrac{b^2\sqrt{b}}{5} = \dfrac{4b^2\sqrt{b}}{5}$

$A = \int_{-\sqrt{b}}^{\sqrt{b}} (b - x^2)\, dx = \left[ bx - \dfrac{x^3}{3} \right]_{-\sqrt{b}}^{\sqrt{b}} = \left( b\sqrt{b} - \dfrac{b\sqrt{b}}{3} \right)2 = 4\dfrac{b\sqrt{b}}{3}$

$\bar{y} = \dfrac{M_x}{A} = \dfrac{4b^2\sqrt{b}/5}{4b\sqrt{b}/3} = \dfrac{3}{5}b$

**53.** (a) $\bar{x} = 0$ by symmetry.

$A = 2\int_{0}^{40} f(x)\, dx = \dfrac{2(40)}{3(4)}\left[ 30 + 4(29) + 2(26) + 4(20) + 0 \right] = \dfrac{20}{3}(278) = \dfrac{5560}{3}$

$M_x = \int_{-40}^{40} \dfrac{f(x)^2}{2}\, dx = \dfrac{40}{3(4)}\left[ 30^2 + 4(29)^2 + 2(26)^2 + 4(20)^2 + 0 \right] = \dfrac{10}{3}(7216) = \dfrac{72{,}160}{3}$

$\bar{y} = \dfrac{M_x}{A} = \dfrac{72{,}160/3}{5560/3} = \dfrac{72{,}160}{5560} \approx 12.98$

$(\bar{x}, \bar{y}) = (0, 12.98)$

(b) $y = \left(-1.02 \times 10^{-5}\right)x^4 - 0.0019x^2 + 29.28$   (Use nine data points.)

(c) $\bar{y} = \dfrac{M_x}{A} \approx \dfrac{23{,}697.68}{1843.54} \approx 12.85$

$(\bar{x}, \bar{y}) = (0, 12.85)$

**55.** The surface area of the sphere is $S = 4\pi r^2$. The arc length of $C$ is $s = \pi r$. The distance traveled by the centroid is

$d = \dfrac{S}{s} = \dfrac{4\pi r^2}{\pi r} = 4r.$

This distance is also the circumference of the circle of radius $y$.

$d = 2\pi y$

So, $2\pi y = 4r$ and you have $y = 2r/\pi$. Therefore, the centroid of the semicircle $y = \sqrt{r^2 - x^2}$ is $(0, 2r/\pi)$.

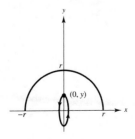

**57.**  $A = \int_0^1 x^n \, dx = \left[ \dfrac{x^{n+1}}{n+1} \right]_0^1 = \dfrac{1}{n+1}$

$m = \rho A = \dfrac{\rho}{n+1}$

$M_x = \dfrac{\rho}{2} \int_0^1 \left( x^n \right)^2 dx = \left[ \dfrac{\rho}{2} \cdot \dfrac{x^{2n+1}}{2n+1} \right]_0^1 = \dfrac{\rho}{2(2n+1)}$

$M_y = \rho \int_0^1 x \left( x^n \right) dx = \left[ \rho \cdot \dfrac{x^{n+2}}{n+2} \right]_0^1 = \dfrac{\rho}{n+2}$

$\bar{x} = \dfrac{M_y}{m} = \dfrac{n+1}{n+2}$

$\bar{y} = \dfrac{M_x}{m} = \dfrac{n+1}{2(2n+1)} = \dfrac{n+1}{4n+2}$

Centroid:  $\left( \dfrac{n+1}{n+2}, \dfrac{n+1}{4n+2} \right)$

As $n \to \infty, (\bar{x}, \bar{y}) \to \left( 1, \dfrac{1}{4} \right)$. The graph approaches the $x$-axis and the line $x = 1$ as $n \to \infty$.

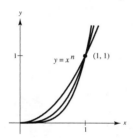

# Section 7.7  Fluid Pressure and Fluid Force

**1.**  $F = PA = \left[ 62.4 \, (8) \right] 3 = 1497.6 \text{ lb}$

**3.**  $F = PA = \left[ 62.4 \, (8) \right] 10 = 4992 \text{ lb}$

**5.**  $F = 62.4(h + 2)(6) - (62.4)(h)(6)$
$= 62.4(2)(6) = 748.8 \text{ lb}$

**7.**  $h(y) = 3 - y$
$L(y) = 4$
$F = 62.4 \int_0^3 (3 - y)(4) \, dy$
$= 249.6 \int_0^3 (3 - y) \, dy$
$= 249.6 \left[ 3y - \dfrac{y^2}{2} \right]_0^3 = 1123.2 \text{ lb}$

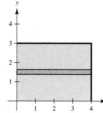

**9.**  $h(y) = 3 - y$
$L(y) = 2 \left( \dfrac{y}{3} + 1 \right)$
$F = 2(62.4) \int_0^3 (3 - y) \left( \dfrac{y}{3} + 1 \right) dy$
$= 124.8 \int_0^3 \left( 3 - \dfrac{y^2}{3} \right) dy$
$= 124.8 \left[ 3y - \dfrac{y^3}{9} \right]_0^3 = 748.8 \text{ lb}$

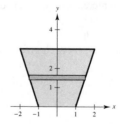

**11.** $h(y) = 4 - y$

$L(y) = 2\sqrt{y}$

$F = 2(62.4) \int_0^4 (4 - y)\sqrt{y} \, dy$

$\quad = 124.8 \int_0^4 \left(4y^{1/2} - y^{3/2}\right) dy$

$\quad = 124.8 \left[\dfrac{8y^{3/2}}{3} - \dfrac{2y^{5/2}}{5}\right]_0^4 = 1064.96 \text{ lb}$

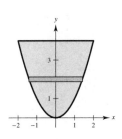

**13.** $h(y) = 4 - y$

$L(y) = 2$

$F = 9800 \int_0^2 2(4 - y) \, dy$

$\quad = 9800 \left[8y - y^2\right]_0^2 = 117{,}600 \text{ newtons}$

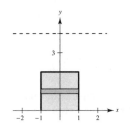

**15.** $h(y) = 12 - y$

$L(y) = 6 - \dfrac{2y}{3}$

$F = 9800 \int_0^9 (12 - y)\left(6 - \dfrac{2y}{3}\right) dy = 9800 \left[72y - 7y^2 + \dfrac{2y^3}{9}\right]_0^9 = 2{,}381{,}400 \text{ newtons}$

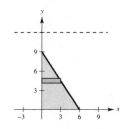

**17.** $h(y) = 2 - y$

$L(y) = 10$

$F = 140.7 \int_0^2 (2 - y)(10) \, dy$

$\quad = 1407 \int_0^2 (2 - y) \, dy$

$\quad = 1407 \left[2y - \dfrac{y^2}{2}\right]_0^2 = 2814 \text{ lb}$

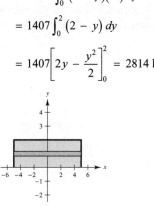

**19.** $h(y) = 4 - y$

$L(y) = 6$

$F = 140.7 \int_0^4 (4 - y)(6) \, dy$

$\quad = 844.2 \int_0^4 (4 - y) \, dy$

$\quad = 844.2 \left[4y - \dfrac{y^2}{2}\right]_0^4 = 6753.6 \text{ lb}$

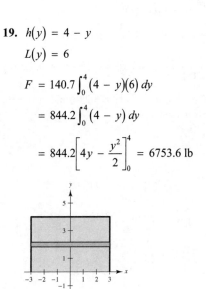

**21.** $h(y) = -y$

$L(y) = 2(\frac{1}{2})\sqrt{9 - 4y^2}$

$F = 42 \int_{-3/2}^{0} (-y)\sqrt{9 - 4y^2} \, dy$

$\quad = \frac{42}{8} \int_{-3/2}^{0} (9 - 4y^2)^{1/2} (-8y) \, dy$

$\quad = \left[ (\frac{21}{4})(\frac{2}{3})(9 - 4y^2)^{3/2} \right]_{-3/2}^{0} = 94.5 \text{ lb}$

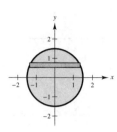

**23.** $h(y) = k - y$

$L(y) = 2\sqrt{r^2 - y^2}$

$F = w \int_{-r}^{r} (k - y)\sqrt{r^2 - y^2} \,(2)\, dy = w\left[ 2k \int_{-r}^{r} \sqrt{r^2 - y^2} \, dy + \int_{-r}^{r} \sqrt{r^2 - y^2} \,(-2y)\, dy \right]$

The second integral is zero because its integrand is odd and the limits of integration are symmetric to the origin. The first integral is the area of a semicircle with radius $r$.

$F = w\left[ (2k)\frac{\pi r^2}{2} + 0 \right] = wk\pi r^2$

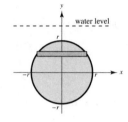

**25.** $h(y) = k - y$

$L(y) = b$

$F = w \int_{-h/2}^{h/2} (k - y)b \, dy$

$\quad = wb\left[ ky - \frac{y^2}{2} \right]_{-h/2}^{h/2} = wb(hk) = wkhb$

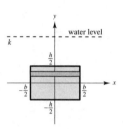

**27.** From Exercise 25: $F = 64(15)(1)(1) = 960 \text{ lb}$

**29.** $h(y) = 4 - y$

$F = 62.4 \int_{0}^{4} (4 - y)L(y) \, dy$

Using Simpson's Rule with $n = 8$ you have:

$F \approx 62.4\left(\frac{4 - 0}{3(8)}\right)\left[ 0 + 4(3.5)(3) + 2(3)(5) + 4(2.5)(8) + 2(2)(9) + 4(1.5)(10) + 2(1)(10.25) + 4(0.5)(10.5) + 0 \right] = 3010.8 \text{ lb}$

**31.** If the fluid force is one-half of 1123.2 lb, and the height of the water is $b$, then

$h(y) = b - y$

$L(y) = 4$

$F = 62.4 \int_{0}^{b} (b - y)(4) \, dy = \frac{1}{2}(1123.2)$

$\int_{0}^{b} (b - y) \, dy = 2.25$

$\left[ by - \frac{y^2}{2} \right]_{0}^{b} = 2.25$

$b^2 - \frac{b^2}{2} = 2.25$

$b^2 = 4.5 \Rightarrow b \approx 2.12 \text{ ft.}$

The pressure increases with increasing depth.

**33.** You use horizontal representative rectangles because you are measuring total force against a region between two depths.

# Review Exercises for Chapter 7

**1.** $A = \int_{-2}^{2} \left[ \left( 6 - \dfrac{x^2}{2} \right) - \dfrac{3}{4}x \right] dx$

$= \left[ 6x - \dfrac{x^3}{6} - \dfrac{3x^2}{8} \right]_{-2}^{2}$

$= \left( 12 - \dfrac{4}{3} - \dfrac{3}{2} \right) - \left( -12 + \dfrac{4}{3} - \dfrac{3}{2} \right) = \dfrac{64}{3}$

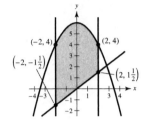

**3.** $A = \int_{-1}^{1} \dfrac{1}{x^2 + 1}\, dx = [\arctan x]_{-1}^{1} = \dfrac{\pi}{4} - \left( -\dfrac{\pi}{4} \right) = \dfrac{\pi}{2}$

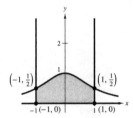

**5.** $A = 2\int_{0}^{1} \left( x - x^3 \right) dx = 2\left[ \dfrac{1}{2}x^2 - \dfrac{1}{4}x^4 \right]_{0}^{1} = \dfrac{1}{2}$

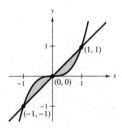

**7.** $A = \int_{0}^{2} \left( e^2 - e^x \right) dx = \left[ xe^2 - e^x \right]_{0}^{2} = e^2 + 1$

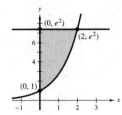

**9.** $A = \int_{\pi/4}^{5\pi/4} \left( \sin x - \cos x \right) dx$

$= \left[ -\cos x - \sin x \right]_{\pi/4}^{5\pi/4}$

$= \left( \dfrac{1}{\sqrt{2}} + \dfrac{1}{\sqrt{2}} \right) - \left( -\dfrac{1}{\sqrt{2}} - \dfrac{1}{\sqrt{2}} \right)$

$= \dfrac{4}{\sqrt{2}} = 2\sqrt{2}$

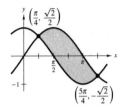

**11.** Points of intersection:

$x^2 - 8x + 3 = 3 + 8x - x^2$

$2x^2 - 16x = 0 \quad \text{when} \quad x = 0, 8$

$A = \int_{0}^{8} \left[ \left( 3 + 8x - x^2 \right) - \left( x^2 - 8x + 3 \right) \right] dx$

$= \int_{0}^{8} \left( 16x - 2x^2 \right) dx$

$= \left[ 8x^2 - \dfrac{2}{3}x^3 \right]_{0}^{8} = \dfrac{512}{3} \approx 170.667$

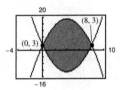

**13.** $y = \left( 1 - \sqrt{x} \right)^2$

$A = \int_{0}^{1} \left( 1 - \sqrt{x} \right)^2 dx$

$= \int_{0}^{1} \left( 1 - 2x^{1/2} + x \right) dx$

$= \left[ x - \dfrac{4}{3}x^{3/2} + \dfrac{1}{2}x^2 \right]_{0}^{1} = \dfrac{1}{6} \approx 0.1667$

**15.** (a) Trapezoidal: Area $\approx \dfrac{160}{2(8)}\Big[0 + 2(50) + 2(54) + 2(82) + 2(82) + 2(73) + 2(75) + 2(80) + 0\Big] = 9920 \text{ ft}^2$

(b) Simpson's: Area $\approx \dfrac{160}{3(8)}\Big[0 + 4(50) + 2(54) + 4(82) + 2(82) + 4(73) + 2(75) + 4(80) + 0\Big] = 10{,}413\tfrac{1}{3} \text{ ft}^2$

**17.** (a) **Disk**

$$V = \pi \int_0^3 x^2 \, dx = \left[\frac{\pi x^3}{3}\right]_0^3 = 9\pi$$

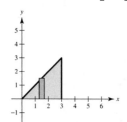

(b) **Shell**

$$V = 2\pi \int_0^3 x(x) \, dx = 2\pi \left[\frac{x^3}{3}\right]_0^3 = 18\pi$$

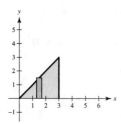

(c) **Shell**

$$V = 2\pi \int_0^3 (3 - x)x \, dx = 2\pi \left[\frac{3x^2}{2} - \frac{x^3}{3}\right]_0^3 = 9\pi$$

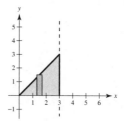

(d) **Shell**

$$V = 2\pi \int_0^3 (6 - x)x \, dx = 2\pi \left[3x^2 - \frac{x^3}{3}\right]_0^3 = 36\pi$$

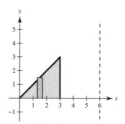

**19. Shell**

$$V = 2\pi \int_0^1 \frac{x}{x^4 + 1} \, dx = \pi \int_0^1 \frac{(2x)}{\left(x^2\right)^2 + 1} \, dx = \Big[\pi \arctan\left(x^2\right)\Big]_0^1 = \pi\left(\frac{\pi}{4} - 0\right) = \frac{\pi^2}{4}$$

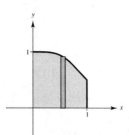

**21. Shell**

$$V = 2\pi \int_2^5 x\left(\frac{1}{x^2}\right) dx$$

$$= 2\pi \int_2^5 \frac{1}{x} \, dx$$

$$= \Big[2\pi \ln|x|\Big]_2^5$$

$$= 2\pi(\ln 5 - \ln 2)$$

$$= 2\pi \ln\left(\frac{5}{2}\right)$$

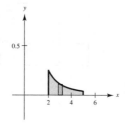

**23.** The volume of the spheroid is given by:

$$V = 4\pi \int_0^4 x\left(\tfrac{3}{4}\right)\sqrt{16 - x^2}\, dx$$

$$= \left[3\pi\left(-\tfrac{1}{2}\right)\left(\tfrac{2}{3}\right)(16 - x^2)^{3/2}\right]_0^4$$

$$= 64\pi$$

$$\tfrac{1}{4}V = 16\pi$$

**Disk:**  $$\pi\int_{-3}^{y_0} \tfrac{16}{9}(9 - y^2)\, dy = 16\pi$$

$$\tfrac{1}{9}\int_{-3}^{y_0}(9 - y^2)\, dy = 1$$

$$\left[9y - \tfrac{1}{3}y^3\right]_{-3}^{y_0} = 9$$

$$\left(9y_0 - \tfrac{1}{3}y_0^3\right) - (-27 + 9) = 9$$

$$y_0^3 - 27y_0 - 27 = 0$$

By Newton's Method, $y_0 \approx -1.042$ and the depth of the gasoline is $3 - 1.042 = 1.958$ feet.

**25.**  $$f(x) = \tfrac{4}{5}x^{5/4}$$

$$f'(x) = x^{1/4}$$

$$1 + \left[f'(x)\right]^2 = 1 + \sqrt{x}$$

$$u = 1 + \sqrt{x}$$

$$x = (u - 1)^2$$

$$dx = 2(u - 1)\, du$$

$$s = \int_0^4 \sqrt{1 + \sqrt{x}}\, dx = 2\int_1^3 \sqrt{u}(u - 1)\, du$$

$$= 2\int_1^3 \left(u^{3/2} - u^{1/2}\right) du$$

$$= 2\left[\tfrac{2}{5}u^{5/2} - \tfrac{2}{3}u^{3/2}\right]_1^3 = \tfrac{4}{15}\left[u^{3/2}(3u - 5)\right]_1^3$$

$$= \tfrac{8}{15}(1 + 6\sqrt{3}) = 6.076$$

**27.**  $$y = 300\cosh\left(\tfrac{x}{2000}\right) - 280,\ -2000 \le x \le 2000$$

$$y' = \tfrac{3}{20}\sinh\left(\tfrac{x}{2000}\right)$$

$$s = \int_{-2000}^{2000}\sqrt{1 + \left[\tfrac{3}{20}\sinh\left(\tfrac{x}{2000}\right)\right]^2}\, dx$$

$$= \tfrac{1}{20}\int_{-2000}^{2000}\sqrt{400 + 9\sinh^2\left(\tfrac{x}{2000}\right)}\, dx$$

$$= 4018.2\ \text{ft (by Simpson's Rule or graphing utility)}$$

**29.**  $$y = \tfrac{3}{4}x$$

$$y' = \tfrac{3}{4}$$

$$1 + (y')^2 = \tfrac{25}{16}$$

$$S = 2\pi\int_0^4\left(\tfrac{3}{4}x\right)\sqrt{\tfrac{25}{16}}\, dx = \left[\left(\tfrac{15\pi}{8}\right)\tfrac{x^2}{2}\right]_0^4 = 15\pi$$

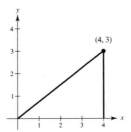

**31.**  $$F = kx$$

$$5 = k(1)$$

$$F = 5x$$

$$W = \int_0^5 5x\, dx = \tfrac{5x^2}{2}\Big]_0^5 = \tfrac{125}{2}\ \text{in.-lb} \approx 5.21\ \text{ft-lb}$$

**33.** Volume of disk: $\pi\left(\tfrac{1}{3}\right)^2 \Delta y$   $\left[\text{diameter} = \tfrac{2}{3}\ \text{ft}\right]$

Weight of disk: $62.4\pi\left(\tfrac{1}{3}\right)^2 \Delta y$

Distance: $190 - y$

$$W = \tfrac{62.4\pi}{9}\int_0^{165}(190 - y)\, dy$$

$$= \tfrac{62.4\pi}{9}\left[190y - \tfrac{y^2}{2}\right]_0^{165}$$

$$= \tfrac{62.4\pi}{9}\left[\tfrac{35,475}{2}\right] = 122,980\pi\ \text{ft-lb}$$

$$\approx 193.2\ \text{foot-tons}$$

**35.** Weight of section of chain: $4\,\Delta x$

Distance moved: $10 - x$

$$W = 4\int_0^{10}(10 - x)\, dx = 4\left[10x - \tfrac{x^2}{2}\right]_0^{10}$$

$$= 200\ \text{ft-lb}$$

**37.** $W = \int_a^b F(x)\, dx$

$$80 = \int_0^4 ax^2\, dx = \left[\frac{ax^3}{3}\right]_0^4 = \frac{64}{3}a$$

$$a = \frac{3(80)}{64} = \frac{15}{4} = 3.75$$

**39.** $\bar{x} = \dfrac{8(-1) + 12(2) + 6(5) + 14(7)}{8 + 12 + 6 + 14} = \dfrac{144}{40} = \dfrac{18}{5} = 3.6$

**41.** $A = \int_{-1}^3 \left[(2x + 3) - x^2\right] dx = \left[x^2 + 3x - \frac{1}{3}x^3\right]_{-1}^3 = \frac{32}{3}$

$$\frac{1}{A} = \frac{3}{32}$$

$$\bar{x} = \frac{3}{32}\int_{-1}^3 x\left(2x + 3 - x^2\right) dx = \frac{3}{32}\int_{-1}^3 \left(3x + 2x^2 - x^3\right) dx = \frac{3}{32}\left[\frac{3}{2}x^2 + \frac{2}{3}x^3 - \frac{1}{4}x^4\right]_{-1}^3 = 1$$

$$\bar{y} = \left(\frac{3}{32}\right)\frac{1}{2}\int_{-1}^3 \left[(2x + 3)^2 - x^4\right] dx = \frac{3}{64}\int_{-1}^3 \left(9 + 12x + 4x^2 - x^4\right) dx$$

$$= \frac{3}{64}\left[9x + 6x^2 + \frac{4}{3}x^3 - \frac{1}{5}x^5\right]_{-1}^3 = \frac{17}{5}$$

$$(\bar{x}, \bar{y}) = \left(1, \frac{17}{5}\right)$$

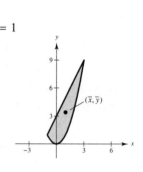

**43.** $\bar{y} = 0$ by symmetry.

For the trapezoid:

$$m = \left[(4)(6) - (1)(6)\right]\rho = 18\rho$$

$$M_y = \rho\int_0^6 x\left[\left(\frac{1}{6}x + 1\right) - \left(-\frac{1}{6}x - 1\right)\right] dx = \rho\int_0^6 \left(\frac{1}{3}x^2 + 2x\right) dx = \rho\left[\frac{x^3}{9} + x^2\right]_0^6 = 60\rho$$

For the semicircle:

$$m = \left(\frac{1}{2}\right)(\pi)(2)^2\rho = 2\pi\rho$$

$$M_y = \rho\int_6^8 x\left[\sqrt{4 - (x - 6)^2} - \left(-\sqrt{4 - (x - 6)^2}\right)\right] dx = 2\rho\int_6^8 x\sqrt{4 - (x - 6)^2}\, dx$$

Let $u = x - 6$, then $x = u + 6$ and $dx = du$. When $x = 6, u = 0$. When $x = 8, u = 2$.

$$M_y = 2\rho\int_0^2 (u + 6)\sqrt{4 - u^2}\, du = 2\rho\int_0^2 u\sqrt{4 - u^2}\, du + 12\rho\int_0^2 \sqrt{4 - u^2}\, du$$

$$= 2\rho\left[\left(-\frac{1}{2}\right)\left(\frac{2}{3}\right)(4 - u^2)^{3/2}\right]_0^2 + 12\rho\left[\frac{\pi(2)^2}{4}\right] = \frac{16\rho}{3} + 12\pi\rho = \frac{4\rho(4 + 9\pi)}{3}$$

So, you have: $\bar{x}(18\rho + 2\pi\rho) = 60\rho + \dfrac{4\rho(4 + 9\pi)}{3}$

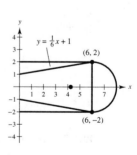

$$\bar{x} = \frac{180\rho + 4\rho(4 + 9\pi)}{3} \cdot \frac{1}{2\rho(9 + \pi)} = \frac{2(9\pi + 49)}{3(\pi + 9)}$$

The centroid of the blade is $\left(\dfrac{2(9\pi + 49)}{3(\pi + 9)}, 0\right)$.

**45.** $h(y) = 9 - y$

$$L(y) = 4 - \frac{4}{3}y$$

$$F = 64\int_0^3 (9 - y)\left(4 - \frac{4}{3}y\right) dy$$

$$= 64\int_0^3 \left(36 - 16y + \frac{4}{3}y^2\right) dy$$

$$= 64\left[36y - 8y^2 + \frac{4}{9}y^3\right]_0^3$$

$$= 64\left[36(3) - 8(9) + 4(3)\right] = 64(48)$$

$$= 3072 \text{ lb}$$

**47.** Wall at shallow end:

$$F = 62.4\int_0^5 y(20) \, dy = \left[(1248)\frac{y^2}{2}\right]_0^5 = 15,600 \text{ lb}$$

Wall at deep end:

$$F = 62.4\int_0^{10} y(20) \, dy = \left[(624)y^2\right]_0^{10} = 62,400 \text{ lb}$$

Side wall:

$$F_1 = 62.4\int_0^5 y(40) \, dy = \left[(1248)y^2\right]_0^5 = 31,200 \text{ lb}$$

$$F_2 = 62.4\int_0^5 (10 - y)8y \, dy = 62.4\int_0^5 (80y - 8y^2) \, dy$$

$$= 62.4(8)\left[5y^2 - \frac{y^3}{3}\right]_0^5 = 41,600 \text{ lb}$$

$$F = F_1 + F_2 = 72,800 \text{ lb}$$

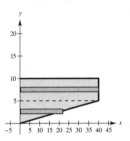

# Problem Solving for Chapter 7

**1.** $T = \frac{1}{2}c(c^2) = \frac{1}{2}c^3$

$$R = \int_0^c (cx - x^2) \, dx = \left[\frac{cx^2}{2} - \frac{x^3}{3}\right]_0^c = \frac{c^3}{2} - \frac{c^3}{3} = \frac{c^3}{6}$$

$$\lim_{c \to 0^+} \frac{T}{R} = \lim_{c \to 0^+} \frac{\frac{1}{2}c^3}{\frac{1}{6}c^3} = 3$$

**3.** $R = \int_0^1 x(1 - x) \, dx = \left[\frac{x^2}{2} - \frac{x^3}{3}\right]_0^1 = \frac{1}{2} - \frac{1}{3} = \frac{1}{6}$

Let $(c, mc)$ be the intersection of the line and the parabola.

Then, $mc = c(1 - c) \Rightarrow m = 1 - c$ or $c = 1 - m$.

$$\frac{1}{2}\left(\frac{1}{6}\right) = \int_0^{1-m} (x - x^2 - mx) \, dx$$

$$\frac{1}{12} = \left[\frac{x^2}{2} - \frac{x^3}{3} - m\frac{x^2}{2}\right]_0^{1-m}$$

$$= \frac{(1-m)^2}{2} - \frac{(1-m)^3}{3} - m\frac{(1-m)^2}{2}$$

$$1 = 6(1-m)^2 - 4(1-m)^3 - 6m(1-m)^2$$

$$= (1-m)^2\left(6 - 4(1-m) - 6m\right)$$

$$= (1-m)^2(2 - 2m)$$

$$\frac{1}{2} = (1-m)^3$$

$$\left(\frac{1}{2}\right)^{1/3} = 1 - m$$

$$m = 1 - \left(\frac{1}{2}\right)^{1/3} \approx 0.2063$$

So, $y = 0.2063x$.

**5.** $8y^2 = x^2(1 - x^2)$

$$y = \pm \frac{|x|\sqrt{1 - x^2}}{2\sqrt{2}}$$

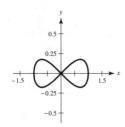

For $x > 0$, $y' = \dfrac{1 - 2x^2}{2\sqrt{2}\sqrt{1 - x^2}}$.

$$S = 2(2\pi)\int_0^1 x\sqrt{1 + \left(\frac{1 - 2x^2}{2\sqrt{2}\sqrt{1 - x^2}}\right)^2}\, dx$$

$$= \frac{5\sqrt{2}\pi}{3}$$

**7.** By the Theorem of Pappus,

$$V = 2\pi r A$$

$$= 2\pi\left[d + \tfrac{1}{2}\sqrt{w^2 + l^2}\right]lw.$$

**13.** (a) $\bar{y} = 0$ by symmetry

$$M_y = 2\int_1^6 x\frac{1}{x^4}\, dx = 2\int_1^6 \frac{1}{x^3}\, dx = \frac{35}{36}$$

$$m = 2\int_1^6 \frac{1}{x^4}\, dx = \frac{215}{324}$$

$$\bar{x} = \frac{35/36}{215/324} = \frac{63}{43} \qquad (\bar{x}, \bar{y}) = \left(\frac{63}{43}, 0\right)$$

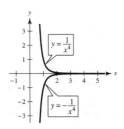

(b) $\displaystyle M_y = 2\int_1^b \frac{1}{x^3}\, dx = \frac{b^2 - 1}{b^2}$

$\displaystyle m = 2\int_1^b \frac{1}{x^4}\, dx = \frac{2(b^3 - 1)}{3b^3}$

$\displaystyle \bar{x} = \frac{(b^2 - 1)/b^2}{2(b^3 - 1)/3b^3} = \frac{3b(b + 1)}{2(b^2 + b + 1)} \qquad (\bar{x}, \bar{y}) = \left(\frac{3b(b + 1)}{2(b^2 + b + 1)}, 0\right)$

(c) $\displaystyle \lim_{b\to\infty} \bar{x} = \frac{3b(b + 1)}{2(b^2 + b + 1)} = \frac{3}{2} \qquad (\bar{x}, \bar{y}) = \left(\frac{3}{2}, 0\right)$

**9.** $f'(x)^2 = e^x$

$f'(x) = e^{x/2}$

$f(x) = 2e^{x/2} + C$

$f(0) = 0 \Rightarrow C = -2$

$f(x) = 2e^{x/2} - 2$

**11.** Let $\rho_f$ be the density of the fluid and $\rho_0$ the density of the iceberg. The buoyant force is

$$F = \rho_f g \int_{-h}^{0} A(y)\, dy$$

where $A(y)$ is a typical cross section and $g$ is the acceleration due to gravity. The weight of the object is

$$W = \rho_0 g \int_{-h}^{L-h} A(y)\, dy.$$

$$F = W$$

$$\rho_f g \int_{-h}^{0} A(y)\, dy = \rho_0 g \int_{-h}^{L-h} A(y)\, dy$$

$$\frac{\rho_0}{\rho_f} = \frac{\text{submerged volume}}{\text{total volume}}$$

$$= \frac{0.92 \times 10^3}{1.03 \times 10^3} = 0.893 \text{ or } 89.3\%$$

**15.** Point of equilibrium: $50 - 0.5x = 0.125x$

$$x = 80, \ p = 10$$

$(P_0, x_0) = (10, 80)$

Consumer surplus $= \int_0^{80} \big[(50 - 0.5x) - 10\big] \, dx = 1600$

Producer surplus $= \int_0^{80} (10 - 0.125x) \, dx = 400$

**17.** Use Exercise 25, Section 7.7, which gives $F = wkhb$ for a rectangle plate.

Wall at shallow end

From Exercise 25: $F = 62.4(2)(4)(20) = 9984$ lb

Wall at deep end

From Exercise 25: $F = 62.4(4)(8)(20) = 39{,}936$ lb

Side wall

From Exercise 25: $F_1 = 62.4(2)(4)(40) = 19{,}968$ lb

$$F_2 = 62.4 \int_0^4 (8 - y)(10y) \, dy$$

$$= 624 \int_0^4 (8y - y^2) \, dy = 624\left[ 4y^2 - \frac{y^3}{3} \right]_0^4$$

$$= 26{,}624 \text{ lb}$$

Total force: $F_1 + F_2 = 46{,}592$ lb

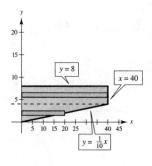

# CHAPTER 8
# Integration Techniques, L'Hôpital's Rule, and Improper Integrals

# CHAPTER 8
## Integration Techniques, L'Hôpital's Rule, and Improper Integrals

### Section 8.1  Basic Integration Rules

**1. (a)** $\dfrac{d}{dx}\left[2\sqrt{x^2+1}+C\right] = 2\left(\dfrac{1}{2}\right)\left(x^2+1\right)^{-1/2}(2x) = \dfrac{2x}{\sqrt{x^2+1}}$

**(b)** $\dfrac{d}{dx}\left[\sqrt{x^2+1}+C\right] = \dfrac{1}{2}\left(x^2+1\right)^{-1/2}(2x) = \dfrac{x}{\sqrt{x^2+1}}$

**(c)** $\dfrac{d}{dx}\left[\dfrac{1}{2}\sqrt{x^2+1}+C\right] = \dfrac{1}{2}\left(\dfrac{1}{2}\right)\left(x^2+1\right)^{-1/2}(2x) = \dfrac{x}{2\sqrt{x^2+1}}$

**(d)** $\dfrac{d}{dx}\left[\ln\left(x^2+1\right)+C\right] = \dfrac{2x}{x^2+1}$

$\displaystyle\int \dfrac{x}{\sqrt{x^2+1}}\,dx$ matches (b).

**2. (a)** $\dfrac{d}{dx}\left[\ln\sqrt{x^2+1}+C\right] = \dfrac{1}{2}\left(\dfrac{2x}{x^2+1}\right) = \dfrac{x}{x^2+1}$

**(b)** $\dfrac{d}{dx}\left[\dfrac{2x}{\left(x^2+1\right)^2}+C\right] = \dfrac{\left(x^2+1\right)^2(2)-(2x)(2)\left(x^2+1\right)(2x)}{\left(x^2+1\right)^4} = \dfrac{2\left(1-3x^2\right)}{\left(x^2+1\right)^3}$

**(c)** $\dfrac{d}{dx}\left[\arctan x + C\right] = \dfrac{1}{1+x^2}$

**(d)** $\dfrac{d}{dx}\left[\ln\left(x^2+1\right)+C\right] = \dfrac{2x}{x^2+1}$

$\displaystyle\int \dfrac{x}{x^2+1}\,dx$ matches (a).

**3. (a)** $\dfrac{d}{dx}\left[\ln\sqrt{x^2+1}+C\right] = \dfrac{1}{2}\left(\dfrac{2x}{x^2+1}\right) = \dfrac{x}{x^2+1}$

**(b)** $\dfrac{d}{dx}\left[\dfrac{2x}{\left(x^2+1\right)^2}+C\right] = \dfrac{\left(x^2+1\right)^2(2)-(2x)(2)\left(x^2+1\right)(2x)}{\left(x^2+1\right)^4} = \dfrac{2\left(1-3x^2\right)}{\left(x^2+1\right)^3}$

**(c)** $\dfrac{d}{dx}\left[\arctan x + C\right] = \dfrac{1}{1+x^2}$

**(d)** $\dfrac{d}{dx}\left[\ln\left(x^2+1\right)+C\right] = \dfrac{2x}{x^2+1}$

$\displaystyle\int \dfrac{1}{x^2+1}\,dx$ matches (c).

**4.** (a) $\dfrac{d}{dx}\left[2x\sin\left(x^2+1\right)+C\right)\right] = 2x\left[\cos\left(x^2+1\right)(2x)\right] + 2\sin\left(x^2+1\right) = 2\left[2x^2\cos\left(x^2+1\right)+\sin\left(x^2+1\right)\right]$

(b) $\dfrac{d}{dx}\left[-\dfrac{1}{2}\sin\left(x^2+1\right)+C\right] = -\dfrac{1}{2}\cos\left(x^2+1\right)(2x) = -x\cos\left(x^2+1\right)$

(c) $\dfrac{d}{dx}\left[\dfrac{1}{2}\sin\left(x^2+1\right)+C\right] = \dfrac{1}{2}\cos\left(x^2+1\right)(2x) = x\cos\left(x^2+1\right)$

(d) $\dfrac{d}{dx}\left[-2x\sin\left(x^2+1\right)+C\right] = -2x\left[\cos\left(x^2+1\right)(2x)\right] - 2\sin\left(x^2+1\right) = -2\left[2x^2\cos\left(x^2+1\right)+\sin\left(x^2+1\right)\right]$

$\int x\cos\left(x^2+1\right)dx$ matches (c).

**5.** $\int(5x-3)^4\,dx$

$u = 5x - 3,\ du = 5\,dx,\ n = 4$

Use $\int u^n\,du$.

**7.** $\int\dfrac{1}{\sqrt{x}\left(1-2\sqrt{x}\right)}\,dx$

$u = 1 - 2\sqrt{x},\ du = -\dfrac{1}{\sqrt{x}}\,dx$

Use $\int\dfrac{du}{u}$.

**9.** $\int\dfrac{3}{\sqrt{1-t^2}}\,dt$

$u = t,\ du = dt,\ a = 1$

Use $\int\dfrac{du}{\sqrt{a^2-u^2}}$.

**11.** $\int t\sin t^2\,dt$

$u = t^2,\ du = 2t\,dt$

Use $\int\sin u\,du$.

**13.** $\int(\cos x)e^{\sin x}\,dx$

$u = \sin x,\ du = \cos x\,dx$

Use $\int e^u\,du$.

**15.** Let $u = x - 5,\ du = dx$.

$\int 14(x-5)^6\,dx = 14\int(x-5)^6\,dx = 2(x-5)^7 + C$

**17.** Let $u = z - 10,\ du = dz$.

$\int\dfrac{7}{(z-10)^7}\,dz = 7\int(z-10)^{-7}\,dz = -\dfrac{7}{6(z-10)^6} + C$

**19.** $\int\left[v+\dfrac{1}{(3v-1)^3}\right]dv = \int v\,dv + \dfrac{1}{3}\int(3v-1)^{-3}(3)\,dv$

$\qquad = \dfrac{1}{2}v^2 - \dfrac{1}{6(3v-1)^2} + C$

**21.** Let $u = -t^3 + 9t + 1$,

$du = \left(-3t^2+9\right)dt = -3\left(t^2-3\right)dt.$

$\int\dfrac{t^2-3}{-t^3+9t+1}\,dt = -\dfrac{1}{3}\int\dfrac{-3\left(t^2-3\right)}{-t^3+9t+1}\,dt$

$\qquad = -\dfrac{1}{3}\ln\left|-t^3+9t+1\right| + C$

**23.** $\int\dfrac{x^2}{x-1}\,dx = \int(x+1)\,dx + \int\dfrac{1}{x-1}\,dx$

$\qquad = \dfrac{1}{2}x^2 + x + \ln|x-1| + C$

**25.** Let $u = 1 + e^x,\ du = e^x\,dx$.

$\int\dfrac{e^x}{1+e^x}\,dx = \ln\left(1+e^x\right) + C$

**27.** $\int\left(5+4x^2\right)^2\,dx = \int\left(25+40x^2+16x^4\right)dx$

$\qquad = 25x + \dfrac{40}{3}x^3 + \dfrac{16}{5}x^5 + C$

$\qquad = \dfrac{x}{15}\left(48x^5+200x^3+375\right) + C$

**29.** Let $u = 2\pi x^2,\ du = 4\pi x\,dx$.

$\int x\left(\cos 2\pi x^2\right)dx = \dfrac{1}{4\pi}\int\left(\cos 2\pi x^2\right)(4\pi x)\,dx$

$\qquad = \dfrac{1}{4\pi}\sin 2\pi x^2 + C$

**31.** Let $u = \cos x,\ du = -\sin x\,dx$.

$\int\dfrac{\sin x}{\sqrt{\cos x}}\,dx = -\int(\cos x)^{-1/2}(-\sin x)\,dx$

$\qquad = -2\sqrt{\cos x} + C$

**33.** Let $u = 1 + e^x$, $du = e^x \, dx$.

$$\int \frac{2}{e^{-x} + 1} \, dx = 2 \int \left( \frac{2}{e^{-x} + 1} \right) \left( \frac{e^x}{e^x} \right) dx$$

$$= 2 \int \frac{e^x}{1 + e^x} \, dx = 2 \ln \left( 1 + e^x \right) + C$$

**35.** $\int \dfrac{\ln x^2}{x} \, dx = 2 \int (\ln x) \dfrac{1}{x} \, dx$

$$= 2 \frac{(\ln x)^2}{2} + C = (\ln x)^2 + C$$

**37.** $\int \dfrac{1 + \cos \alpha}{\sin \alpha} \, d\alpha = \int \csc \alpha \, d\alpha + \int \cot \alpha \, d\alpha$

$$= -\ln \left| \csc \alpha + \cot \alpha \right| + \ln \left| \sin \alpha \right| + C$$

**39.** Let $u = 4t + 1$, $du = 4 \, dt$.

$$\int \frac{-1}{\sqrt{1 - (4t + 1)^2}} \, dt = -\frac{1}{4} \int \frac{4}{\sqrt{1 - (4t + 1)^2}} \, dt$$

$$= -\frac{1}{4} \arcsin(4t + 1) + C$$

**41.** Let $u = \cos \left( \dfrac{2}{t} \right)$, $du = \dfrac{2 \sin(2/t)}{t^2} \, dt$.

$$\int \frac{\tan(2/t)}{t^2} \, dt = \frac{1}{2} \int \frac{1}{\cos(2/t)} \left[ \frac{2 \sin(2/t)}{t^2} \right] dt$$

$$= \frac{1}{2} \ln \left| \cos \left( \frac{2}{t} \right) \right| + C$$

**43.** Note: $10x - x^2 = 25 - \left( 25 - 10x + x^2 \right)$

$$= 25 - (5 - x)^2$$

$$\int \frac{6}{\sqrt{10x - x^2}} \, dx = 6 \int \frac{1}{\sqrt{25 - (5 - x)^2}} \, dx$$

$$= -6 \int \frac{-1}{\sqrt{5^2 - (5 - x)^2}} \, dx$$

$$= -6 \arcsin \frac{(5 - x)}{5} + C$$

$$= 6 \arcsin \left( \frac{x - 5}{5} \right) + C$$

**45.** $\int \dfrac{4}{4x^2 + 4x + 65} \, dx = \int \dfrac{1}{\left[ x + (1/2) \right]^2 + 16} \, dx$

$$= \frac{1}{4} \arctan \left[ \frac{x + (1/2)}{4} \right] + C$$

$$= \frac{1}{4} \arctan \left( \frac{2x + 1}{8} \right) + C$$

**47.** $\dfrac{ds}{dt} = \dfrac{t}{\sqrt{1 - t^4}}$, $\left( 0, -\dfrac{1}{2} \right)$

(a)

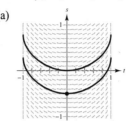

(b) $u = t^2$, $du = 2t \, dt$

$$\int \frac{t}{\sqrt{1 - t^4}} \, dt = \frac{1}{2} \int \frac{2t}{\sqrt{1 - \left( t^2 \right)^2}} \, dt$$

$$= \frac{1}{2} \arcsin t^2 + C$$

$$\left( 0, -\frac{1}{2} \right): \; -\frac{1}{2} = \frac{1}{2} \arcsin 0 + C \Rightarrow C = -\frac{1}{2}$$

$$s = \frac{1}{2} \arcsin t^2 - \frac{1}{2}$$

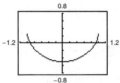

**49.**

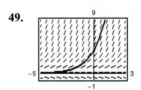

$$y = 4e^{0.8x}$$

**51.** $\dfrac{dy}{dx} = \left( e^x + 5 \right)^2 = e^{2x} + 10e^x + 25$

$$y = \int \left( e^{2x} + 10e^x + 25 \right) dx$$

$$= \frac{1}{2} e^{2x} + 10e^x + 25x + C$$

**53.** $\dfrac{dr}{dt} = \dfrac{10e^t}{\sqrt{1 - e^{2t}}}$

$$r = \int \frac{10e^t}{\sqrt{1 - \left( e^t \right)^2}} \, dt$$

$$= 10 \arcsin \left( e^t \right) + C$$

**55.** $\dfrac{dy}{dx} = \dfrac{\sec^2 x}{4 + \tan^2 x}$

Let $u = \tan x$, $du = \sec^2 x \, dx$.

$y = \displaystyle\int \dfrac{\sec^2 x}{4 + \tan^2 x} \, dx = \dfrac{1}{2} \arctan\left(\dfrac{\tan x}{2}\right) + C$

**57.** Let $u = 2x$, $du = 2 \, dx$.

$\displaystyle\int_0^{\pi/4} \cos 2x \, dx = \dfrac{1}{2} \int_0^{\pi/4} \cos 2x(2) \, dx$

$\qquad\qquad = \left[\dfrac{1}{2} \sin 2x\right]_0^{\pi/4} = \dfrac{1}{2}$

**59.** Let $u = -x^2$, $du = -2x \, dx$.

$\displaystyle\int_0^1 xe^{-x^2} \, dx = -\dfrac{1}{2} \int_0^1 e^{-x^2}(-2x) \, dx = \left[-\dfrac{1}{2}e^{-x^2}\right]_0^1$

$\qquad\qquad = \dfrac{1}{2}\left(1 - e^{-1}\right) \approx 0.316$

**61.** Let $u = x^2 + 36$, $du = 2x \, dx$.

$\displaystyle\int_0^8 \dfrac{2x}{\sqrt{x^2 + 36}} \, dx = \int_0^8 \left(x^2 + 36\right)^{-1/2}(2x) \, dx$

$\qquad\qquad = 2\left[\left(x^2 + 36\right)^{1/2}\right]_0^8 = 8$

**63.** Let $u = 3x$, $du = 3 \, dx$.

$\displaystyle\int_0^{2/\sqrt{3}} \dfrac{1}{4 + 9x^2} \, dx = \dfrac{1}{3} \int_0^{2/\sqrt{3}} \dfrac{3}{4 + (3x)^2} \, dx$

$\qquad\qquad = \left[\dfrac{1}{6} \arctan\left(\dfrac{3x}{2}\right)\right]_0^{2/\sqrt{3}}$

$\qquad\qquad = \dfrac{\pi}{18} \approx 0.175$

**69.** $\displaystyle\int \dfrac{1}{x^2 + 4x + 13} \, dx = \dfrac{1}{3} \arctan\left(\dfrac{x + 2}{3}\right) + C$

The antiderivatives are vertical translations of each other.

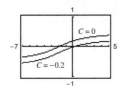

**71.** $\displaystyle\int \dfrac{1}{1 + \sin \theta} \, d\theta = \tan \theta - \sec \theta + C \quad \left(\text{or } \dfrac{-2}{1 + \tan(\theta/2)}\right)$

The antiderivatives are vertical translations of each other.

**65.** $A = \displaystyle\int_0^{3/2} \left(-4x + 6\right)^{3/2} \, dx$

$\quad = -\dfrac{1}{4} \displaystyle\int_0^{3/2} \left(6 - 4x\right)^{3/2}(-4) \, dx$

$\quad = -\dfrac{1}{4}\left[\dfrac{2}{5}(6 - 4x)^{5/2}\right]_0^{3/2}$

$\quad = -\dfrac{1}{10}\left(0 - 6^{5/2}\right)$

$\quad = \dfrac{18}{5}\sqrt{6} \approx 8.8182$

**67.** $y^2 = x^2\left(1 - x^2\right)$

$\quad y = \pm\sqrt{x^2\left(1 - x^2\right)}$

$\quad A = 4\displaystyle\int_0^1 x\sqrt{1 - x^2} \, dx$

$\quad = -2\displaystyle\int_0^1 \left(1 - x^2\right)^{1/2}(-2x) \, dx$

$\quad = -\dfrac{4}{3}\left[\left(1 - x\right)^{3/2}\right]_0^1$

$\quad = -\dfrac{4}{3}(0 - 1) = \dfrac{4}{3}$

**73.** Power Rule: $\int u^n \, du = \dfrac{u^{n+1}}{n+1} + C, \quad n \ne -1$

$$u = x^2 + 1, \, n = 3$$

**75.** Log Rule: $\int \dfrac{du}{u} = \ln|u| + C, \quad u = x^2 + 1$

**77.** $\sin x + \cos x = a \sin(x + b)$

$\sin x + \cos x = a \sin x \cos b + a \cos x \sin b$

$\sin x + \cos x = (a \cos b) \sin x + (a \sin b) \cos x$

Equate coefficients of like terms to obtain the following.

$1 = a \cos b \quad \text{and} \quad 1 = a \sin b$

So, $a = 1/\cos b$. Now, substitute for $a$ in $1 = a \sin b$.

$$1 = \left(\frac{1}{\cos b}\right) \sin b$$

$$1 = \tan b \implies b = \frac{\pi}{4}$$

Because $b = \dfrac{\pi}{4}$, $a = \dfrac{1}{\cos(\pi/4)} = \sqrt{2}$. So,

$$\sin x + \cos x = \sqrt{2} \sin\left(x + \frac{\pi}{4}\right).$$

$$\int \frac{dx}{\sin x + \cos x} = \int \frac{dx}{\sqrt{2} \sin(x + (\pi/4))}$$

$$= \frac{1}{\sqrt{2}} \int \csc\left(x + \frac{\pi}{4}\right) dx$$

$$= -\frac{1}{\sqrt{2}} \ln\left|\csc\left(x + \frac{\pi}{4}\right) + \cot\left(x + \frac{\pi}{4}\right)\right| + C$$

**79.** $\int_0^{1/a} \left(x - ax^2\right) dx = \left[\frac{1}{2}x^2 - \frac{a}{3}x^3\right]_0^{1/a} = \frac{1}{6a^2}$

Let $\dfrac{1}{6a^2} = \dfrac{2}{3}$, $12a^2 = 3$, $a = \dfrac{1}{2}$.

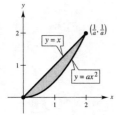

**81.** (a) They are equivalent because

$$e^{x + C_1} = e^x \cdot e^{C_1} = Ce^x, \, C = e^{C_1}.$$

(b) They differ by a constant.

$$\sec^2 x + C_1 = \left(\tan^2 x + 1\right) + C_1 = \tan^2 x + C$$

**83.** $\int_0^2 \dfrac{4x}{x^2 + 1} \, dx \approx 3$

Matches (a).

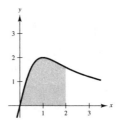

**84.** $\int_0^2 \dfrac{4}{x^2 + 1} \, dx \approx 4$

Matches (d).

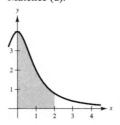

**85.** (a) $y = 2\pi x^2, \quad 0 \le x \le 2$

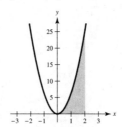

(b) $y = \sqrt{2}x, \quad 0 \le x \le 2$

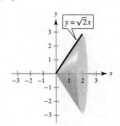

(c) $y = x, \quad 0 \le x \le 2$

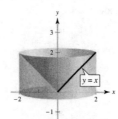

**87. (a) Shell Method:**

Let $u = -x^2$, $du = -2x\ dx$.

$$V = 2\pi \int_0^1 xe^{-x^2}\ dx$$

$$= -\pi \int_0^1 e^{-x^2}(-2x)\ dx$$

$$= \left[-\pi e^{-x^2}\right]_0^1$$

$$= \pi\left(1 - e^{-1}\right) \approx 1.986$$

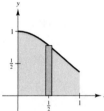

**(b) Shell Method:**

$$V = 2\pi \int_0^b xe^{-x^2}\ dx$$

$$= \left[-\pi e^{-x^2}\right]_0^b$$

$$= \pi\left(1 - e^{-b^2}\right) = \frac{4}{3}$$

$$e^{-b^2} = \frac{3\pi - 4}{3\pi}$$

$$b = \sqrt{\ln\left(\frac{3\pi}{3\pi - 4}\right)} \approx 0.743$$

**89.**  $y = f(x) = \ln(\sin x)$

$$f'(x) = \frac{\cos x}{\sin x}$$

$$s = \int_{\pi/4}^{\pi/2} \sqrt{1 + \frac{\cos^2 x}{\sin^2 x}}\ dx = \int_{\pi/4}^{\pi/2} \sqrt{\frac{\sin^2 x + \cos^2 x}{\sin^2 x}}\ dx$$

$$= \int_{\pi/4}^{\pi/2} \frac{1}{\sin x}\ dx = \int_{\pi/4}^{\pi/2} \csc x\ dx$$

$$= \left[-\ln\left|\csc x + \cot x\right|\right]_{\pi/4}^{\pi/2}$$

$$= -\ln(1) + \ln\left(\sqrt{2} + 1\right)$$

$$= \ln\left(\sqrt{2} + 1\right) \approx 0.8814$$

**91.**  $y = 2\sqrt{x}$

$$y' = \frac{1}{\sqrt{x}}$$

$$1 + (y')^2 = 1 + \frac{1}{x} = \frac{x + 1}{x}$$

$$S = 2\pi \int_0^9 2\sqrt{x}\sqrt{\frac{x + 1}{x}}\ dx$$

$$= 2\pi \int_0^9 2\sqrt{x + 1}\ dx$$

$$= \left[4\pi\left(\frac{2}{3}\right)(x + 1)^{3/2}\right]_0^9 = \frac{8\pi}{3}\left(10\sqrt{10} - 1\right) \approx 256.545$$

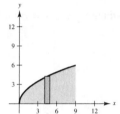

**93.**  Average value $= \dfrac{1}{b - a} \displaystyle\int_a^b f(x)\ dx$

$$= \frac{1}{3 - (-3)} \int_{-3}^3 \frac{1}{1 + x^2}\ dx$$

$$= \frac{1}{6}\left[\arctan(x)\right]_{-3}^3$$

$$= \frac{1}{6}\left[\arctan(3) - \arctan(-3)\right]$$

$$= \frac{1}{3}\arctan(3) \approx 0.4163$$

**95.**  $y = \tan(\pi x)$

$$y' = \pi \sec^2(\pi x)$$

$$1 + (y')^2 = 1 + \pi^2 \sec^4(\pi x)$$

$$s = \int_0^{1/4} \sqrt{1 + \pi^2 \sec^4(\pi x)}\ dx \approx 1.0320$$

**97.** (a) $\int \cos^3 x \, dx = \int (1 - \sin^2 x) \cos x \, dx = \sin x - \dfrac{\sin^3 x}{3} + C = \dfrac{1}{3} \sin x (\cos^2 x + 2) + C$

(b) $\int \cos^5 x \, dx = \int (1 - \sin^2 x)^2 \cos x \, dx = \int (1 - 2 \sin^2 x + \sin^4 x) \cos x \, dx$

$$= \sin x - \dfrac{2}{3} \sin^3 x + \dfrac{\sin^5 x}{5} + C = \dfrac{1}{15} \sin x (3 \cos^4 x + 4 \cos^2 x + 8) + C$$

(c) $\int \cos^7 x \, dx = \int (1 - \sin^2 x)^3 \cos x \, dx$

$$= \int (1 - 3 \sin^2 x + 3 \sin^4 x - \sin^6 x) \cos x \, dx$$

$$= \sin x - \sin^3 x + \dfrac{3}{5} \sin^5 x - \dfrac{1}{7} \sin^7 x + C$$

$$= \dfrac{1}{35} \sin x (5 \cos^6 x + 6 \cos^4 x + 8 \cos^2 x + 16) + C$$

(d) $\int \cos^{15} x \, dx = \int (1 - \sin^2 x)^7 \cos x \, dx$

You would expand $(1 - \sin^2 x)^7$.

**99.** Let $f(x) = \dfrac{1}{2} \left( x \sqrt{x^2 + 1} + \ln \left| x + \sqrt{x^2 + 1} \right| \right) + C.$

$$f'(x) = \dfrac{1}{2} \left( x \dfrac{1}{2} (x^2 + 1)^{-1/2} (2x) + \sqrt{x^2 + 1} + \dfrac{1}{x + \sqrt{x^2 + 1}} \left( 1 + \dfrac{1}{2} (x^2 + 1)^{-1/2} (2x) \right) \right)$$

$$= \dfrac{1}{2} \left( \dfrac{x^2}{\sqrt{x^2 + 1}} + \sqrt{x^2 + 1} + \dfrac{1}{x + \sqrt{x^2 + 1}} \left( 1 + \dfrac{x}{\sqrt{x^2 + 1}} \right) \right)$$

$$= \dfrac{1}{2} \left( \dfrac{x^2 + (x^2 + 1)}{\sqrt{x^2 + 1}} + \dfrac{1}{x + \sqrt{x^2 + 1}} \left( \dfrac{\sqrt{x^2 + 1} + x}{\sqrt{x^2 + 1}} \right) \right)$$

$$= \dfrac{1}{2} \left( \dfrac{2x^2 + 1}{\sqrt{x^2 + 1}} + \dfrac{1}{\sqrt{x^2 + 1}} \right) = \dfrac{1}{2} \left( \dfrac{2(x^2 + 1)}{\sqrt{x^2 + 1}} \right) = \sqrt{x^2 + 1}$$

So, $\int \sqrt{x^2 + 1} \, dx = \dfrac{1}{2} \left( x \sqrt{x^2 + 1} + \ln \left| x + \sqrt{x^2 + 1} \right| \right) + C.$

Let $g(x) = \dfrac{1}{2} \left( x \sqrt{x^2 + 1} + \operatorname{arcsinh}(x) \right).$

$$g'(x) = \dfrac{1}{2} \left( x \dfrac{1}{2} (x^2 + 1)^{-1/2} (2x) + \sqrt{x^2 + 1} + \dfrac{1}{\sqrt{x^2 + 1}} \right)$$

$$= \dfrac{1}{2} \left( \dfrac{x^2}{\sqrt{x^2 + 1}} + \sqrt{x^2 + 1} + \dfrac{1}{\sqrt{x^2 + 1}} \right)$$

$$= \dfrac{1}{2} \left( \dfrac{x^2 + (x^2 + 1) + 1}{\sqrt{x^2 + 1}} \right)$$

$$= \dfrac{1}{2} \left( \dfrac{2(x^2 + 1)}{\sqrt{x^2 + 1}} \right) = \sqrt{x^2 + 1}$$

So, $\int \sqrt{x^2 + 1} \, dx = \dfrac{1}{2} \left( x \sqrt{x^2 + 1} + \operatorname{arcsinh}(x) \right) + C.$

## Section 8.2 Integration by Parts

**1.** $\int xe^{2x}\,dx$

$u = x,\ dv = e^{2x}\,dx$

**3.** $\int (\ln x)^2\,dx$

$u = (\ln x)^2,\ dv = dx$

**5.** $\int x \sec^2 x\,dx$

$u = x,\ dv = \sec^2 x\,dx$

**7.** $dv = x^3\,dx \Rightarrow v = \int x^3\,dx = \dfrac{x^4}{4}$

$u = \ln x \Rightarrow du = \dfrac{1}{x}\,dx$

$\int x^3 \ln x\,dx = uv - \int v\,du$

$\qquad = (\ln x)\dfrac{x^4}{4} - \int \left(\dfrac{x^4}{4}\right)\dfrac{1}{x}\,dx$

$\qquad = \dfrac{x^4}{4}\ln x - \dfrac{1}{4}\int x^3\,dx$

$\qquad = \dfrac{x^4}{4}\ln x - \dfrac{1}{16}x^4 + C$

$\qquad = \dfrac{1}{16}x^4(4 \ln x - 1) + C$

**9.** $dv = \sin 3x\,dx \Rightarrow v = \int \sin 3x\,dx = -\dfrac{1}{3}\cos 3x$

$u = x \qquad\quad \Rightarrow du = dx$

$\int x \sin 3x\,dx = uv - \int v\,du$

$\qquad = x\left(-\dfrac{1}{3}\cos 3x\right) - \int -\dfrac{1}{3}\cos 3x\,dx$

$\qquad = -\dfrac{x}{3}\cos 3x + \dfrac{1}{9}\sin 3x + C$

**11.** $dv = e^{-4x}\,dx \Rightarrow v = \int e^{-4x}\,dx = -\dfrac{1}{4}e^{-4x}$

$u = x \qquad\quad \Rightarrow du = dx$

$\int xe^{-4x}\,dx = x\left(-\dfrac{1}{4}e^{-4x}\right) - \int -\dfrac{1}{4}e^{-4x}\,dx$

$\qquad = -\dfrac{x}{4}e^{-4x} - \dfrac{1}{16}e^{-4x} + C$

$\qquad = -\dfrac{1}{16e^{4x}}(1 + 4x) + C$

**13.** Use integration by parts three times.

(1) $dv = e^x\,dx \Rightarrow\ v = \int e^x\,dx = e^x$

$\quad u = x^3 \qquad \Rightarrow du = 3x^2\,dx$

(2) $dv = e^x\,dx \Rightarrow\ v = \int e^x\,dx = e^x$

$\quad u = x^2 \qquad \Rightarrow du = 2x\,dx$

(3) $dv = e^x\,dx \Rightarrow\ v = \int e^x\,dx = e^x$

$\quad u = x \qquad \Rightarrow du = dx$

$\int x^3 e^x\,dx = x^3 e^x - 3\int x^2 e^x\,dx = x^3 e^x - 3x^2 e^x + 6\int xe^x\,dx$

$\qquad = x^3 e^x - 3x^2 e^x + 6xe^x - 6e^x + C = e^x(x^3 - 3x^2 + 6x - 6) + C$

**15.** $dv = t\,dt \qquad \Rightarrow\ v = \int t\,dt = \dfrac{t^2}{2}$

$u = \ln(t + 1) \Rightarrow du = \dfrac{1}{t + 1}\,dt$

$\int t \ln(t + 1)\,dt = \dfrac{t^2}{2}\ln(t + 1) - \dfrac{1}{2}\int \dfrac{t^2}{t + 1}\,dt$

$\qquad = \dfrac{t^2}{2}\ln(t + 1) - \dfrac{1}{2}\int \left(t - 1 + \dfrac{1}{t + 1}\right)dt$

$\qquad = \dfrac{t^2}{2}\ln(t + 1) - \dfrac{1}{2}\left[\dfrac{t^2}{2} - t + \ln(t + 1)\right] + C$

$\qquad = \dfrac{1}{4}\left[2(t^2 - 1)\ln|t + 1| - t^2 + 2t\right] + C$

**17.** Let $u = \ln x,\ du = \dfrac{1}{x}\,dx$.

$\int \dfrac{(\ln x)^2}{x}\,dx = \int (\ln x)^2\left(\dfrac{1}{x}\right)dx = \dfrac{(\ln x)^3}{3} + C$

**19.** $dv = \dfrac{1}{(2x+1)^2}\, dx \;\Rightarrow\; v = \int (2x+1)^{-2}\, dx$

$$= -\dfrac{1}{2(2x+1)}$$

$u = xe^{2x} \qquad\qquad \Rightarrow\; du = \left(2xe^{2x} + e^{2x}\right) dx$

$$= e^{2x}(2x+1)\, dx$$

$\displaystyle\int \dfrac{xe^{2x}}{(2x+1)^2}\, dx = -\dfrac{xe^{2x}}{2(2x+1)} + \int \dfrac{e^{2x}}{2}\, dx$

$$= \dfrac{-xe^{2x}}{2(2x+1)} + \dfrac{e^{2x}}{4} + C = \dfrac{e^{2x}}{4(2x+1)} + C$$

**21.** $dv = \sqrt{x-5}\, dx \;\Rightarrow\; v = \int (x-5)^{1/2}\, dx = \tfrac{2}{3}(x-5)^{3/2}$

$u = x \qquad\qquad \Rightarrow\; du = dx$

$\displaystyle\int x\sqrt{x-5}\, dx = x\tfrac{2}{3}(x-5)^{3/2} - \int \tfrac{2}{3}(x-5)^{3/2}\, dx$

$$= \tfrac{2}{3}x(x-5)^{3/2} - \tfrac{4}{15}(x-5)^{5/2} + C$$

$$= \tfrac{2}{15}(x-5)^{3/2}\big(5x - 2(x-5)\big) + C$$

$$= \tfrac{2}{15}(x-5)^{3/2}(3x+10) + C$$

**23.** $dv = \cos x\, dx \;\Rightarrow\; v = \int \cos x\, dx = \sin x$

$u = x \qquad\quad \Rightarrow\; du = dx$

$\displaystyle\int x\cos x\, dx = x\sin x - \int \sin x\, dx = x\sin x + \cos x + C$

**25.** Use integration by parts three times.

(1) $u = x^3,\, du = 3x^2\, dx,\, dv = \sin x\, dx,\, v = -\cos x$

$\displaystyle\int x^3 \sin dx = -x^3 \cos x + 3\int x^2 \cos x\, dx$

(2) $u = x^2,\, du = 2x\, dx,\, dv = \cos x\, dx,\, v = \sin x$

$\displaystyle\int x^3 \sin x\, dx = -x^3 \cos x + 3\left(x^2 \sin x - 2\int x\sin x\, dx\right) = -x^3 \cos x + 3x^2 \sin x - 6\int x\sin x\, dx$

(3) $u = x,\, du = dx,\, dv = \sin x\, dx,\, v = -\cos x$

$\displaystyle\int x^3 \sin x\, dx = -x^3 \cos x + 3x^2 \sin x - 6\left(-x\cos x + \int \cos x\, dx\right)$

$$= -x^3 \cos x + 3x^2 \sin x + 6x\cos x - 6\sin x + C$$

$$= \left(6x - x^3\right)\cos x + \left(3x^2 - 6\right)\sin x + C$$

**27.** $dv = dx \qquad\quad \Rightarrow\; v = \int dx = x$

$u = \arctan x \;\Rightarrow\; du = \dfrac{1}{1+x^2}\, dx$

$\displaystyle\int \arctan x\, dx = x\arctan x - \int \dfrac{x}{1+x^2}\, dx$

$$= x\arctan x - \dfrac{1}{2}\ln\!\left(1+x^2\right) + C$$

**29.** Use integration by parts twice.

(1) $dv = e^{-3x}\, dx \quad \Rightarrow \quad v = \int e^{-3x}\, dx = -\frac{1}{3}e^{-3x}$

$u = \sin 5x \quad \Rightarrow \quad du = 5\cos 5x\, dx$

$\int e^{-3x}\sin 5x\, dx = \sin 5x\left(-\frac{1}{3}e^{-3x}\right) - \int\left(-\frac{1}{3}e^{-3x}\right)5\cos x\, dx = -\frac{1}{3}e^{-3x}\sin 5x + \frac{5}{3}\int e^{-3x}\cos 5x\, dx$

(2) $dv = e^{-3x}\, dx \quad \Rightarrow \quad v = \int e^{-3x}\, dx = -\frac{1}{3}e^{-3x}$

$u = \cos 5x \quad \Rightarrow \quad du = -5\sin 5x\, dx$

$\int e^{-3x}\sin 5x\, dx = -\frac{1}{3}e^{-3x}\sin 5x + \frac{5}{3}\left[\left(-\frac{1}{3}e^{-3x}\cos 5x - \int\left(-\frac{1}{3}e^{-3x}\right)(-5\sin 5x)\right)dx\right]$

$\qquad\qquad\qquad\quad = -\frac{1}{3}e^{-3x}\sin 5x - \frac{5}{9}e^{-3x}\cos 5x - \frac{25}{9}\int e^{-3x}\sin 5x\, dx$

$\left(1 + \frac{25}{9}\right)\int e^{-3x}\sin 5x\, dx = -\frac{1}{3}e^{-3x}\sin 5x - \frac{5}{9}e^{-3x}\cos 5x$

$\int e^{-3x}\sin 5x\, dx = \frac{9}{34}\left(-\frac{1}{3}e^{-3x}\sin 5x - \frac{5}{9}e^{-3x}\cos 5x\right) + C = -\frac{3}{34}e^{-3x}\sin 5x - \frac{5}{34}e^{-3x}\cos 5x + C$

**31.** $dv = dx \quad \Rightarrow \quad v = x$

$u = \ln x \quad \Rightarrow \quad du = \frac{1}{x}\, dx$

$y' = \ln x$

$y = \int \ln x\, dx = x\ln x - \int x\left(\frac{1}{x}\right)dx = x\ln x - x + C = x(-1 + \ln x) + C$

**33.** Use integration by parts twice.

(1) $dv = \frac{1}{\sqrt{3 + 5t}}\, dt \quad \Rightarrow \quad v = \int (3 + 5t)^{-1/2}\, dt = \frac{2}{5}(3 + 5t)^{1/2}$

$u = t^2 \qquad\qquad \Rightarrow \quad du = 2t\, dt$

$\int \frac{t^2}{\sqrt{3 + 5t}}\, dt = \frac{2}{5}t^2(3 + 5t)^{1/2} - \int \frac{2}{5}(3 + 5t)^{1/2}\,2t\, dt$

$\qquad\qquad\qquad = \frac{2}{5}t^2(3 + 5t)^{1/2} - \frac{4}{5}\int t(3 + 5t)^{1/2}\, dt$

(2) $dv = (3 + 5t)^{1/2}\, dt \quad \Rightarrow \quad v = \int (3 + 5t)^{1/2}\, dt = \frac{2}{15}(3 + 5t)^{3/2}$

$u = t \qquad\qquad\qquad \Rightarrow \quad du = dt$

$\int \frac{t^2}{\sqrt{3 + 5t}}\, dt = \frac{2}{5}t^2(3 + 5t)^{1/2} - \frac{4}{5}\left[\frac{2}{15}t(3 + 5t)^{3/2} - \int \frac{2}{15}(3 + 5t)^{3/2}\, dt\right]$

$\qquad\qquad\qquad = \frac{2}{5}t^2(3 + 5t)^{1/2} - \frac{8}{75}t(3 + 5t)^{3/2} + \frac{8}{75}\int (3 + 5t)^{3/2}\, dt$

$\qquad\qquad\qquad = \frac{2}{5}t^2(3 + 5t)^{1/2} - \frac{8}{75}t(3 + 5t)^{3/2} + \frac{16}{1875}(3 + 5t)^{5/2} + C$

$\qquad\qquad\qquad = \frac{2}{1875}\sqrt{3 + 5t}\left(375t^2 - 100t(3 + 5t) + 8(3 + 5t)^2\right) + C$

$\qquad\qquad\qquad = \frac{2}{625}\sqrt{3 + 5t}\left(25t^2 - 20t + 24\right) + C$

**35.** (a)

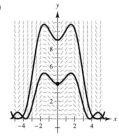

(b)  $\dfrac{dy}{dx} = x\sqrt{y}\,\cos x, \quad (0, 4)$

$\displaystyle\int\dfrac{dy}{\sqrt{y}} = \int x \cos x \, dx$

$\displaystyle\int y^{-1/2}\,dy = \int x \cos x \, dx \qquad (u = x, du = dx, dv = \cos x \, dx, v = \sin x)$

$2y^{1/2} = x \sin x - \displaystyle\int \sin x \, dx = x \sin x + \cos x + C$

$(0, 4): \ 2(4)^{1/2} = 0 + 1 + C \Rightarrow C = 3$

$2\sqrt{y} = x \sin x + \cos x + 3$

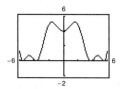

**37.** $\dfrac{dy}{dx} = \dfrac{x}{y}\,e^{x/8}, \ y(0) = 2$

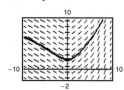

**39.** $u = x, du = dx, dv = e^{x/2}\,dx, v = 2e^{x/2}$

$\displaystyle\int xe^{x/2}\,dx = 2xe^{x/2} - \int 2e^{x/2}\,dx$

$= 2xe^{x/2} - 4e^{x/2} + C$

So,

$\displaystyle\int_0^3 xe^{x/2}\,dx = \left[2xe^{x/2} - 4e^{x/2}\right]_0^3$

$= \left(6e^{3/2} - 4e^{3/2}\right) - (-4)$

$= 4 + 2e^{3/2} \approx 12.963$

**41.** $u = x, du = dx, dv = \cos 2x \, dx, v = \dfrac{1}{2}\sin 2x$

$\displaystyle\int x \cos 2x \, dx = \dfrac{1}{2}x \sin 2x - \int\dfrac{1}{2}\sin 2x \, dx$

$= \dfrac{1}{2}x \sin 2x + \dfrac{1}{4}\cos 2x + C$

So,

$\displaystyle\int_0^{\pi/4} x \cos 2x \, dx = \left[\dfrac{1}{2}x \sin 2x + \dfrac{1}{4}\cos 2x\right]_0^{\pi/4}$

$= \left(\dfrac{\pi}{8}(1) + 0\right) - \left(0 + \dfrac{1}{4}\right)$

$= \dfrac{\pi}{8} - \dfrac{1}{4} \approx 0.143$

**43.** $u = \arccos x, du = -\dfrac{1}{\sqrt{1 - x^2}}\,dx, dv = dx, v = x$

$\displaystyle\int \arccos x \, dx = x \arccos x + \int\dfrac{x}{\sqrt{1 - x^2}}\,dx$

$= x \arccos x - \sqrt{1 - x^2} + C$

So,

$\displaystyle\int_0^{1/2} \arccos x = \left[x \arccos x - \sqrt{1 - x^2}\right]_0^{1/2}$

$= \dfrac{1}{2}\arccos\left(\dfrac{1}{2}\right) - \sqrt{\dfrac{3}{4}} + 1$

$= \dfrac{\pi}{6} - \dfrac{\sqrt{3}}{2} + 1 \approx 0.658.$

**45.** Use integration by parts twice.

$$(1) \quad dv = e^x \, dx \implies v = \int e^x \, dx = e^x \qquad (2) \qquad dv = e^x \, dx \implies v = \int e^x \, dx = e^x$$

$$u = \sin x \implies du = \cos x \, dx \qquad\qquad u = \cos x \implies du = -\sin x \, dx$$

$$\int e^x \sin x \, dx = e^x \sin x - \int e^x \cos x \, dx = e^x \sin x - e^x \cos x - \int e^x \sin x \, dx$$

$$2 \int e^x \sin x \, dx = e^x (\sin x - \cos x)$$

$$\int e^x \sin x \, dx = \frac{e^x}{2}(\sin x - \cos x) + C$$

So, $\displaystyle \int_0^1 e^x \sin x \, dx = \left[ \frac{e^x}{2}(\sin x - \cos x) \right]_0^1 = \frac{e}{2}(\sin 1 - \cos 1) + \frac{1}{2} = \frac{e(\sin 1 - \cos 1) + 1}{2} \approx 0.909.$

**47.** $\displaystyle dv = x \, dx, \, v = \frac{x^2}{2}, \, u = \text{arcsec } x, \, du = \frac{1}{x\sqrt{x^2 - 1}} \, dx$

$$\int x \, \text{arcsec } x \, dx = \frac{x^2}{2} \text{arcsec } x - \int \frac{x^2/2}{x\sqrt{x^2 - 1}} \, dx = \frac{x^2}{2} \text{arcsec } x - \frac{1}{4} \int \frac{2x}{\sqrt{x^2 - 1}} \, dx = \frac{x^2}{2} \text{arcsec } x - \frac{1}{2}\sqrt{x^2 - 1} + C$$

So,

$$\int_2^4 x \, \text{arcsec } x \, dx = \left[ \frac{x^2}{2} \text{arcsec } x - \frac{1}{2}\sqrt{x^2 - 1} \right]_2^4 = \left( 8 \, \text{arcsec } 4 - \frac{\sqrt{15}}{2} \right) - \left( \frac{2\pi}{3} - \frac{\sqrt{3}}{2} \right) = 8 \, \text{arcsec } 4 - \frac{\sqrt{15}}{2} + \frac{\sqrt{3}}{2} - \frac{2\pi}{3} \approx 7.380.$$

**49.** $\displaystyle \int x^2 e^{2x} \, dx = x^2 \left( \frac{1}{2} e^{2x} \right) - (2x)\left( \frac{1}{4} e^{2x} \right) + 2 \left( \frac{1}{8} e^{2x} \right) + C$

$$= \frac{1}{2} x^2 e^{2x} - \frac{1}{2} x e^{2x} + \frac{1}{4} e^{2x} + C$$

$$= \frac{1}{4} e^{2x} \left( 2x^2 - 2x + 1 \right) + C$$

| Alternate signs | $u$ and its derivatives | $v'$ and its antiderivatives |
|---|---|---|
| + | $x^2$ | $e^{2x}$ |
| − | $2x$ | $\frac{1}{2} e^{2x}$ |
| + | $2$ | $\frac{1}{4} e^{2x}$ |
| − | $0$ | $\frac{1}{8} e^{2x}$ |

**51.** $\displaystyle \int x^3 \sin x \, dx = x^3 (-\cos x) - 3x^2 (-\sin x) + 6x \cos x - 6 \sin x + C$

$$= -x^3 \cos x + 3x^2 \sin x + 6x \cos x - 6 \sin x + C$$

$$= \left( 3x^2 - 6 \right) \sin x - \left( x^3 - 6x \right) \cos x + C$$

| Alternate signs | $u$ and its derivatives | $v'$ and its antiderivatives |
|---|---|---|
| + | $x^3$ | $\sin x$ |
| − | $3x^2$ | $-\cos x$ |
| + | $6x$ | $-\sin x$ |
| − | $6$ | $\cos x$ |
| + | $0$ | $\sin x$ |

**53.** $\int x \sec^2 x \, dx = x \tan x + \ln|\cos x| + C$

| Alternate signs | $u$ and its derivatives | $v'$ and its antiderivatives |
|---|---|---|
| + | $x$ | $\sec^2 x$ |
| − | $1$ | $\tan x$ |
| + | $0$ | $-\ln|\cos x|$ |

**55.** $u = \sqrt{x} \implies u^2 = x \implies 2u \, du = dx$

$$\int \sin \sqrt{x} \, dx = \int \sin u (2u \, du) = 2 \int u \sin u \, du$$

Integration by parts:

$$w = u, dw = du, dv = \sin u \, du, v = -\cos u$$

$$2 \int u \sin u \, du = 2\left(-u \cos u + \int \cos u \, du\right)$$
$$= 2(-u \cos u + \sin u) + C$$
$$= 2\left(-\sqrt{x} \cos \sqrt{x} + \sin \sqrt{x}\right) + C$$

**57.** $u = x^2, du = 2x \, dx$

$$\int x^5 e^{x^2} \, dx = \frac{1}{2} \int e^{x^2} x^4 \, 2x \, dx = \frac{1}{2} \int e^u u^2 \, du$$

Integration by parts twice.

(1) $w = u^2, dw = 2u \, du, dv = e^u \, du, v = e^u$

$$\frac{1}{2} \int e^u u^2 \, du = \frac{1}{2}\left[u^2 e^u - \int 2u e^u \, du\right]$$
$$= \frac{1}{2} u^2 e^u - \int u e^u \, du$$

(2) $w = u, dw = du, dv = e^u \, du, v = e^u$

$$\frac{1}{2} \int e^u u^2 \, du = \frac{1}{2} u^2 e^u - \left(u e^u - \int e^u du\right)$$
$$= \frac{1}{2} u^2 e^u - u e^u + e^u + C$$
$$= \frac{1}{2} x^4 e^{x^2} - x^2 e^{x^2} + e^{x^2} + C$$
$$= \frac{e^{x^2}}{2}\left(x^4 - 2x^2 + 2\right) + C$$

**59.** (a) Integration by parts is based on the Product Rule.

(b) Answers will vary. *Sample answer:* You want $dv$ to be the most complicated portion of the integrand.

**61.** (a) No

Substitution

(b) Yes

$u = \ln x, dv = x \, dx$

(c) Yes

$u = x^2, dv = e^{-3x} \, dx$

(d) No

Substitution

(e) Yes. Let $u = x$ and

$$dv = \frac{1}{\sqrt{x+1}} \, dx.$$

(Substitution also works. Let $u = \sqrt{x+1}$.)

(f) No

Substitution

**63. (a)** $dv = \dfrac{x}{\sqrt{4 + x^2}}\, dx \quad \Rightarrow \quad v = \int (4 + x^2)^{-1/2} x\, dx = \sqrt{4 + x^2}$

$u = x^2 \qquad\qquad \Rightarrow \quad du = 2x\, dx$

$\displaystyle \int \dfrac{x^3}{\sqrt{4 + x^2}}\, dx = x^2 \sqrt{4 + x^2} - 2 \int x \sqrt{4 + x^2}\, dx$

$\qquad\qquad\qquad = x^2 \sqrt{4 + x^2} - \dfrac{2}{3}(4 + x^2)^{3/2} + C = \dfrac{1}{3}\sqrt{4 + x^2}(x^2 - 8) + C$

**(b)** $u = 4 + x^2 \Rightarrow x^2 = u - 4$ and $2x\, dx = du \Rightarrow x\, dx = \dfrac{1}{2}du$

$\displaystyle \int \dfrac{x^3}{\sqrt{4 + x^2}}\, dx = \int \dfrac{x^2}{\sqrt{4 + x^2}} x\, dx = \int \left( \dfrac{u - 4}{\sqrt{u}} \right) \dfrac{1}{2}\, du$

$\qquad\qquad\qquad = \dfrac{1}{2} \int (u^{1/2} - 4u^{-1/2})\, du = \dfrac{1}{2}\left( \dfrac{2}{3} u^{3/2} - 8u^{1/2} \right) + C$

$\qquad\qquad\qquad\qquad = \dfrac{1}{3} u^{1/2}(u - 12) + C$

$\qquad\qquad\qquad\qquad = \dfrac{1}{3}\sqrt{4 + x^2}\left[ (4 + x^2) - 12 \right] + C = \dfrac{1}{3}\sqrt{4 + x^2}(x^2 - 8) + C$

**65.** $n = 0: \displaystyle \int \ln x\, dx = x(\ln x - 1) + C$

$n = 1: \displaystyle \int x \ln x\, dx = \dfrac{x^2}{4}(2 \ln x - 1) + C$

$n = 2: \displaystyle \int x^2 \ln x\, dx = \dfrac{x^3}{9}(3 \ln x - 1) + C$

$n = 3: \displaystyle \int x^3 \ln x\, dx = \dfrac{x^4}{16}(4 \ln x - 1) + C$

$n = 4: \displaystyle \int x^4 \ln x\, dx = \dfrac{x^5}{25}(5 \ln x - 1) + C$

In general, $\displaystyle \int x^n \ln x\, dx = \dfrac{x^{n+1}}{(n + 1)^2}\left[ (n + 1)\ln x - 1 \right] + C.$

**67.** $dv = \sin x\, dx \quad \Rightarrow \quad v = -\cos x$

$u = x^n \qquad\qquad \Rightarrow \quad du = nx^{n-1}\, dx$

$\displaystyle \int x^n \sin x\, dx = -x^n \cos x + n \int x^{n-1} \cos x\, dx$

**69.** $dv = x^n\, dx \quad \Rightarrow \quad v = \dfrac{x^{n+1}}{n + 1}$

$u = \ln x \quad \Rightarrow \quad du = \dfrac{1}{x}\, dx$

$\displaystyle \int x^n \ln x\, dx = \dfrac{x^{n+1}}{n + 1} \ln x - \int \dfrac{x^n}{n + 1}\, dx$

$\qquad\qquad\qquad = \dfrac{x^{n+1}}{n + 1} \ln x - \dfrac{x^{n+1}}{(n + 1)^2} + C$

$\qquad\qquad\qquad = \dfrac{x^{n+1}}{(n + 1)^2}\left[ (n + 1)\ln x - 1 \right] + C$

**71.** Use integration by parts twice.

(1)  $dv = e^{ax}\, dx \;\Rightarrow\; v = \dfrac{1}{a}\, e^{ax}$          (2)  $dv = e^{ax}\, dx \;\Rightarrow\; v = \dfrac{1}{a}\, e^{ax}$

$\quad\; u = \sin bx \;\Rightarrow\; du = b \cos bx\, dx$          $\quad\; u = \cos bx \;\Rightarrow\; du = -b \sin bx\, dx$

$$\int e^{ax} \sin bx\, dx = \frac{e^{ax} \sin bx}{a} - \frac{b}{a}\int e^{ax} \cos bx\, dx$$

$$= \frac{e^{ax} \sin bx}{a} - \frac{b}{a}\left(\frac{e^{ax}\cos bx}{a} + \frac{b}{a}\int e^{ax}\sin bx\, dx\right) = \frac{e^{ax}\sin bx}{a} - \frac{b}{a^2}\, e^{ax}\cos bx - \frac{b^2}{a^2}\int e^{ax}\sin bx\, dx$$

Therefore,  $\left(1 + \dfrac{b^2}{a^2}\right)\displaystyle\int e^{ax}\sin bx\, dx = \dfrac{e^{ax}\left(a\sin bx - b\cos bx\right)}{a^2}$

$$\int e^{ax}\sin bx\, dx = \frac{e^{ax}\left(a\sin bx - b\cos bx\right)}{a^2 + b^2} + C.$$

**73.**  $n = 2$          (Use formula in Exercise 67.)

$$\int x^2 \sin x\, dx = -x^2 \cos x + 2\int x \cos x\, dx$$

$$= -x^2 \cos x + 2\left[x \sin x - \int \sin x\, dx\right] \text{(Use formula in Exercise 68; } (n = 1).\text{)}$$

$$= -x^2 \cos x + 2x \sin x + 2\cos x + C$$

**75.**  $n = 5$   (Use formula in Exercise 69.)

$$\int x^5 \ln x\, dx = \frac{x^6}{6^2}(-1 + 6\ln x) + C = \frac{x^6}{36}(-1 + 6\ln x) + C$$

**77.**  $a = -3, b = 4$          (Use formula in Exercise 71.)

$$\int e^{-3x}\sin 4x\, dx = \frac{e^{-3x}\left(-3\sin 4x - 4\cos 4x\right)}{(-3)^2 + (4)^2} + C$$

$$= \frac{-e^{-3x}\left(3\sin 4x + 4\cos 4x\right)}{25} + C$$

**79.**

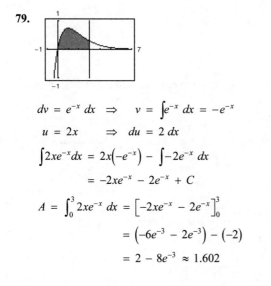

$dv = e^{-x}\, dx \;\Rightarrow\; v = \displaystyle\int e^{-x}\, dx = -e^{-x}$

$u = 2x \qquad\Rightarrow\; du = 2\, dx$

$$\int 2xe^{-x}dx = 2x(-e^{-x}) - \int -2e^{-x}\, dx$$

$$= -2xe^{-x} - 2e^{-x} + C$$

$$A = \int_0^3 2xe^{-x}\, dx = \left[-2xe^{-x} - 2e^{-x}\right]_0^3$$

$$= \left(-6e^{-3} - 2e^{-3}\right) - (-2)$$

$$= 2 - 8e^{-3} \approx 1.602$$

**81.**

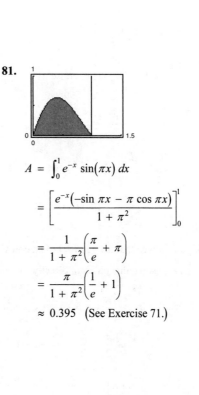

$$A = \int_0^1 e^{-x}\sin(\pi x)\, dx$$

$$= \left[\frac{e^{-x}\left(-\sin \pi x - \pi \cos \pi x\right)}{1 + \pi^2}\right]_0^1$$

$$= \frac{1}{1 + \pi^2}\left(\frac{\pi}{e} + \pi\right)$$

$$= \frac{\pi}{1 + \pi^2}\left(\frac{1}{e} + 1\right)$$

$$\approx 0.395 \quad \text{(See Exercise 71.)}$$

**83.** (a) $dv = dx \quad \Rightarrow \quad v = x$

$u = \ln x \quad \Rightarrow \quad du = \dfrac{1}{x}\,dx$

$A = \displaystyle\int_1^e \ln x\,dx = \left[x \ln x - x\right]_1^e = 1$    (Use integration by parts once.)

(b) $R(x) = \ln x,\, r(x) = 0$

$V = \pi \displaystyle\int_1^e (\ln x)^2\,dx$

$\quad = \pi\left[x(\ln x)^2 - 2x \ln x + 2x\right]_1^e$    (Use integration by parts twice, see Exercise 3.)

$\quad = \pi(e - 2) \approx 2.257$

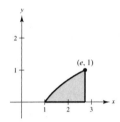

(c) $p(x) = x,\, h(x) = \ln x$

$V = 2\pi \displaystyle\int_1^e x \ln x\,dx = 2\pi\left[\dfrac{x^2}{4}(-1 + 2 \ln x)\right]_1^e$

$\quad = \dfrac{(e^2 + 1)\pi}{2} \approx 13.177$  (See Exercise 91.)

(d) $\bar{x} = \dfrac{\displaystyle\int_1^e x \ln x\,dx}{1} = \dfrac{e^2 + 1}{4} \approx 2.097$

$\bar{y} = \dfrac{\dfrac{1}{2}\displaystyle\int_1^e (\ln x)^2\,dx}{1} = \dfrac{e - 2}{2} \approx 0.359$

$(\bar{x}, \bar{y}) = \left(\dfrac{e^2 + 1}{4}, \dfrac{e - 2}{2}\right) \approx (2.097, 0.359)$

**85.** In Example 6, you showed that the centroid of an equivalent region was $(1, \pi/8)$. By symmetry, the centroid of this region is $(\pi/8, 1)$. You can also solve this problem directly.

$A = \displaystyle\int_0^1 \left(\dfrac{\pi}{2} - \arcsin x\right) dx = \left[\dfrac{\pi}{2}x - x \arcsin x - \sqrt{1 - x^2}\right]_0^1$    (Example 3)

$\quad = \left(\dfrac{\pi}{2} - \dfrac{\pi}{2} - 0\right) - (-1) = 1$

$\bar{x} = \dfrac{M_y}{A} = \displaystyle\int_0^1 x\left(\dfrac{\pi}{2} - \arcsin x\right) dx = \dfrac{\pi}{8}, \quad \bar{y} = \dfrac{M_x}{A} = \displaystyle\int_0^1 \dfrac{(\pi/2) + \arcsin x}{2}\left(\dfrac{\pi}{2} - \arcsin x\right) dx = 1$

**87.** Average value $= \dfrac{1}{\pi}\displaystyle\int_0^\pi e^{-4t}(\cos 2t + 5 \sin 2t)\,dt$

$\quad = \dfrac{1}{\pi}\left[e^{-4t}\left(\dfrac{-4 \cos 2t + 2 \sin 2t}{20}\right) + 5e^{-4t}\left(\dfrac{-4 \sin 2t - 2 \cos 2t}{20}\right)\right]_0^\pi$    (From Exercises 71 and 72)

$\quad = \dfrac{7}{10\pi}\left(1 - e^{-4\pi}\right) \approx 0.223$

**89.** $c(t) = 100,000 + 4000t, r = 5\%, t_1 = 10$

$$P = \int_0^{10} (100,000 + 4000t)e^{-0.05t}\, dt = 4000 \int_0^{10} (25 + t)e^{-0.05t}\, dt$$

Let $u = 25 + t, dv = e^{-0.05t}\, dt, du = dt, v = -\dfrac{100}{5}e^{-0.05t}$.

$$P = 4000\left\{ \left[ (25+t)\left( -\frac{100}{5}e^{-0.05t} \right) \right]_0^{10} + \frac{100}{5} \int_0^{10} e^{-0.05t}\, dt \right\} = 4000\left\{ \left[ (25+t)\left( -\frac{100}{5}e^{-0.05t} \right) \right]_0^{10} - \left[ \frac{10,000}{25}e^{-0.05t} \right]_0^{10} \right\} \approx \$931,265$$

**91.** $\displaystyle \int_{-\pi}^{\pi} x \sin nx\, dx = \left[ -\frac{x}{n}\cos nx + \frac{1}{n^2}\sin nx \right]_{-\pi}^{\pi} = -\frac{\pi}{n}\cos \pi n - \frac{\pi}{n}\cos(-\pi n) = -\frac{2\pi}{n}\cos \pi n = \begin{cases} -(2\pi/n), & \text{if } n \text{ is even} \\ (2\pi/n), & \text{if } n \text{ is odd} \end{cases}$

**93.** Let $u = x, dv = \sin\left( \dfrac{n\pi}{2}x \right) dx, du = dx, v = -\dfrac{2}{n\pi}\cos\left( \dfrac{n\pi}{2}x \right)$.

$$I_1 = \int_0^1 x \sin\left( \frac{n\pi}{2}x \right) dx = \left[ \frac{-2x}{n\pi}\cos\left( \frac{n\pi}{2}x \right) \right]_0^1 + \frac{2}{n\pi}\int_0^1 \cos\left( \frac{n\pi}{2}x \right) dx$$

$$= -\frac{2}{n\pi}\cos\left( \frac{n\pi}{2} \right) + \left[ \left( \frac{2}{n\pi} \right)^2 \sin\left( \frac{n\pi}{2}x \right) \right]_0^1$$

$$= -\frac{2}{n\pi}\cos\left( \frac{n\pi}{2} \right) + \left( \frac{2}{n\pi} \right)^2 \sin\left( \frac{n\pi}{2} \right)$$

Let $u = (-x + 2), dv = \sin\left( \dfrac{n\pi}{2}x \right) dx, du = -dx, v = -\dfrac{2}{n\pi}\cos\left( \dfrac{n\pi}{2}x \right)$.

$$I_2 = \int_1^2 (-x + 2) \sin\left( \frac{n\pi}{2}x \right) dx = \left[ \frac{-2(-x+2)}{n\pi}\cos\left( \frac{n\pi}{2}x \right) \right]_1^2 - \frac{2}{n\pi}\int_1^2 \cos\left( \frac{n\pi}{2}x \right) dx$$

$$= \frac{2}{n\pi}\cos\left( \frac{n\pi}{2} \right) - \left[ \left( \frac{2}{n\pi} \right)^2 \sin\left( \frac{n\pi}{2}x \right) \right]_1^2$$

$$= \frac{2}{n\pi}\cos\left( \frac{n\pi}{2} \right) + \left( \frac{2}{n\pi} \right)^2 \sin\left( \frac{n\pi}{2} \right)$$

$$h(I_1 + I_2) = b_n = h\left[ \left( \frac{2}{n\pi} \right)^2 \sin\left( \frac{n\pi}{2} \right) + \left( \frac{2}{n\pi} \right)^2 \sin\left( \frac{n\pi}{2} \right) \right] = \frac{8h}{(n\pi)^2}\sin\left( \frac{n\pi}{2} \right)$$

**95.** $f'(x) = 3x \sin(2x)$, $f(0) = 0$

(a) $f(x) = \int 3x \sin 2x \, dx$

$$= -\tfrac{3}{4}(2x \cos 2x - \sin 2x) + C$$

(Parts: $u = 3x$, $dv = \sin 2x \, dx$)

$$f(0) = 0 = -\tfrac{3}{4}(0) + C \Rightarrow C = 0$$

$$f(x) = -\tfrac{3}{4}(2x \cos 2x - \sin 2x)$$

(b)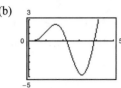

(d) Using $h = 0.1$, you obtain the points:

| $n$ | $x_n$ | $y_n$ |
|-----|-------|-------|
| 0 | 0 | 0 |
| 1 | 0.1 | 0 |
| 2 | 0.2 | 0.0060 |
| 3 | 0.3 | 0.0293 |
| 4 | 0.4 | 0.0801 |
| ⋮ | ⋮ | ⋮ |
| 40 | 4.0 | 1.0210 |

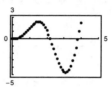

(c) Using $h = 0.05$, you obtain the points:

| $n$ | $x_n$ | $y_n$ |
|-----|-------|-------|
| 0 | 0 | 0 |
| 1 | 0.05 | 0 |
| 2 | 0.10 | $7.4875 \times 10^{-4}$ |
| 3 | 0.15 | 0.0037 |
| 4 | 0.20 | 0.0104 |
| ⋮ | ⋮ | ⋮ |
| 80 | 4.0 | 1.3181 |

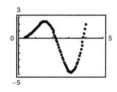

**97.** On $\left[0, \dfrac{\pi}{2}\right]$, $\sin x \le 1 \Rightarrow x \sin x \le x \Rightarrow \displaystyle\int_0^{\pi/2} x \sin x \, dx \le \int_0^{\pi/2} x \, dx$.

**99.** For any integrable function, $\displaystyle\int f(x)\,dx = C + \int f(x)\,dx$, but this cannot be used to imply that $C = 0$.

## Section 8.3   Trigonometric Integrals

**1.** Let $u = \cos x$, $du = -\sin x \, dx$.

$$\int \cos^5 x \sin x \, dx = -\int \cos^5 x \, (-\sin x) \, dx = -\frac{\cos^6 x}{6} + C$$

**3.** Let $u = \sin 2x$, $du = 2 \cos 2x \, dx$.

$$\int \sin^7 2x \cos 2x \, dx = \frac{1}{2}\int \sin^7 2x (2 \cos 2x) \, dx$$

$$= \frac{1}{2}\left(\frac{\sin^8 2x}{8}\right) + C$$

$$= \frac{1}{16}\sin^8 2x + C$$

**5.** $\displaystyle\int \sin^3 x \cos^2 x \, dx = \int (1 - \cos^2 x) \cos^2 x \sin x \, dx$

$$= \int (\cos^2 x - \cos^4 x) \sin x \, dx$$

$$= -\int (\cos^2 x - \cos^4 x)(-\sin x) \, dx$$

$$= -\frac{\cos^3 x}{3} + \frac{\cos^5 x}{5} + C$$

**7.** $\int \sin^3 2\theta \sqrt{\cos 2\theta}\, d\theta = \int (1 - \cos^2 2\theta) \sqrt{\cos 2\theta} \sin 2\theta\, d\theta$

$$= \int \left[ (\cos 2\theta)^{1/2} - (\cos 2\theta)^{5/2} \right] \sin 2\theta\, d\theta$$

$$= -\frac{1}{2} \int \left[ (\cos 2\theta)^{1/2} - (\cos 2\theta)^{5/2} \right] (-2 \sin 2\theta)\, d\theta$$

$$= -\frac{1}{2} \left[ \frac{2}{3}(\cos 2\theta)^{3/2} - \frac{2}{7}(\cos 2\theta)^{7/2} \right] + C$$

$$= -\frac{1}{3}(\cos 2\theta)^{3/2} + \frac{1}{7}(\cos 2\theta)^{7/2} + C$$

**9.** $\int \cos^2 3x\, dx = \int \frac{1 + \cos 6x}{2}\, dx = \frac{1}{2}\left( x + \frac{1}{6} \sin 6x \right) + C = \frac{1}{12}(6x + \sin 6x) + C$

**11.** Integration by parts:

$$dv = \sin^2 x\, dx = \frac{1 - \cos 2x}{2} \Rightarrow v = \frac{x}{2} - \frac{\sin 2x}{4} = \frac{1}{4}(2x - \sin 2x)$$

$$u = x \Rightarrow du = dx$$

$$\int x \sin^2 x\, dx = \frac{1}{4}x(2x - \sin 2x) - \frac{1}{4}\int(2x - \sin 2x)\, dx$$

$$= \frac{1}{4}x(2x - \sin 2x) - \frac{1}{4}\left( x^2 + \frac{1}{2}\cos 2x \right) + C = \frac{1}{8}(2x^2 - 2x \sin 2x - \cos 2x) + C$$

**13.** $\int_0^{\pi/2} \cos^7 x\, dx = \left( \frac{2}{3} \right)\left( \frac{4}{5} \right)\left( \frac{6}{7} \right) = \frac{16}{35},\ (n = 7)$

**17.** $\int_0^{\pi/2} \sin^6 x\, dx = \left( \frac{1}{2} \right)\left( \frac{3}{4} \right)\left( \frac{5}{6} \right)\frac{\pi}{2} = \frac{5\pi}{32},\ (n = 6)$

**15.** $\int_0^{\pi/2} \cos^{10} x\, dx = \left( \frac{1}{2} \right)\left( \frac{3}{4} \right)\left( \frac{5}{6} \right)\left( \frac{7}{8} \right)\left( \frac{9}{10} \right)\left( \frac{\pi}{2} \right)$

$$= \frac{63}{512}\pi,\ (n = 10)$$

**19.** $\int \sec 4x\, dx = \frac{1}{4}\int \sec 4x(4\, dx)$

$$= \frac{1}{4}\ln|\sec 4x + \tan 4x| + C$$

**21.** $dv = \sec^2 \pi x\, dx \Rightarrow v = \frac{1}{\pi}\tan \pi x$

$$u = \sec \pi x \Rightarrow du = \pi \sec \pi x \tan \pi x\, dx$$

$$\int \sec^3 \pi x\, dx = \frac{1}{\pi}\sec \pi x \tan \pi x - \int \sec \pi x \tan^2 \pi x\, dx = \frac{1}{\pi}\sec \pi x \tan \pi x - \int \sec \pi x(\sec^2 \pi x - 1)\, dx$$

$$2\int \sec^3 \pi x\, dx = \frac{1}{\pi}\left( \sec \pi x \tan \pi x + \ln|\sec \pi x + \tan \pi x| \right) + C_1$$

$$\int \sec^3 \pi x\, dx = \frac{1}{2\pi}\left( \sec \pi x \tan \pi x + \ln|\sec \pi x + \tan \pi x| \right) + C$$

**23.** $\int \tan^5 \frac{x}{2}\, dx = \int \left( \sec^2 \frac{x}{2} - 1 \right)\tan^3 \frac{x}{2}\, dx$

$$= \int \tan^3 \frac{x}{2} \sec^2 \frac{x}{2}\, dx - \int \tan^3 \frac{x}{2}\, dx$$

$$= \frac{\tan^4 \frac{x}{2}}{2} - \int \left( \sec^2 \frac{x}{2} - 1 \right)\tan \frac{x}{2}\, dx$$

$$= \frac{1}{2}\tan^4 \frac{x}{2} - \tan^2 \frac{x}{2} - 2\ln\left| \cos \frac{x}{2} \right| + C$$

**25.** Let $u = \sec 2t$, $du = 2 \sec 2t \tan 2t$.

$$\int \tan^3 2t \cdot \sec^3 2t \, dt = \int \left(\sec^2 2t - 1\right) \sec^3 2t \cdot \tan 2t \, dt$$

$$= \int \left(\sec^4 2t - \sec^2 2t\right)\left(\sec 2t \tan 2t\right) dt = \frac{\sec^5 2t}{10} - \frac{\sec^3 2t}{6} + C$$

**27.** $\int \sec^6 4x \tan 4x \, dx = \dfrac{1}{4} \int \sec^5 4x\left(4 \sec 4x \tan 4x\right) dx$

$$= \frac{\sec^6 4x}{24} + C$$

**29.** $\int \sec^5 x \tan^3 x \, dx = \int \sec^4 x \tan^2 x \, (\sec x \tan x) \, dx$

$$= \int \sec^4 x\left(\sec^2 x - 1\right)(\sec x \tan x) \, dx$$

$$= \int \left(\sec^6 x - \sec^4 x\right)(\sec x \tan x) \, dx$$

$$= \frac{\sec^7 x}{7} - \frac{\sec^5 x}{5} + C$$

**31.** $\int \dfrac{\tan^2 x}{\sec x} \, dx = \int \dfrac{\left(\sec^2 x - 1\right)}{\sec x} \, dx$

$$= \int (\sec x - \cos x) \, dx$$

$$= \ln\left|\sec x + \tan x\right| - \sin x + C$$

**33.** $r = \int \sin^4(\pi\theta) \, d\theta = \dfrac{1}{4} \int \left[1 - \cos(2\pi\theta)\right]^2 d\theta$

$$= \frac{1}{4} \int \left[1 - 2\cos(2\pi\theta) + \cos^2(2\pi\theta)\right] d\theta$$

$$= \frac{1}{4} \int \left[1 - 2\cos(2\pi\theta) + \frac{1 + \cos(4\pi\theta)}{2}\right] d\theta$$

$$= \frac{1}{4}\left[\theta - \frac{1}{\pi}\sin(2\pi\theta) + \frac{\theta}{2} + \frac{1}{8\pi}\sin(4\pi\theta)\right] + C$$

$$= \frac{1}{32\pi}\left[12\pi\theta - 8\sin(2\pi\theta) + \sin(4\pi\theta)\right] + C$$

**35.** $y = \int \tan^3 3x \sec 3x \, dx$

$$= \int \left(\sec^2 3x - 1\right) \sec 3x \tan 3x \, dx$$

$$= \frac{1}{3} \int \sec^2 3x\left(3 \sec 3x \tan 3x\right) dx - \frac{1}{3} \int 3 \sec 3x \tan 3x \, dx$$

$$= \frac{1}{9} \sec^3 3x - \frac{1}{3} \sec 3x + C$$

**37. (a)**

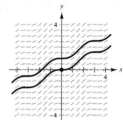

**(b)** $\dfrac{dy}{dx} = \sin^2 x, \quad (0, 0)$

$$y = \int \sin^2 x \, dx = \int \frac{1 - \cos 2x}{2} \, dx$$

$$= \frac{1}{2}x - \frac{\sin 2x}{4} + C$$

$(0, 0)$: $0 = C$, $y = \dfrac{1}{2}x - \dfrac{\sin 2x}{4}$

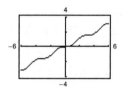

**39.** $\dfrac{dy}{dx} = \dfrac{3 \sin x}{y}, \; y(0) = 2$

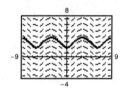

**41.** $\displaystyle\int \cos 2x \cos 6x \, dx = \frac{1}{2}\int\Big[\cos\big((2-6)x\big) + \cos\big((2+6)x\big)\Big]\,dx$

$\displaystyle\qquad\qquad\qquad\qquad = \frac{1}{2}\int\Big[\cos(-4x) + \cos 8x\Big]\,dx$

$\displaystyle\qquad\qquad\qquad\qquad = \frac{1}{2}\int(\cos 4x + \cos 8x)\,dx$

$\displaystyle\qquad\qquad\qquad\qquad = \frac{1}{2}\left[\frac{\sin 4x}{4} + \frac{\sin 8x}{8}\right] + C$

$\displaystyle\qquad\qquad\qquad\qquad = \frac{\sin 4x}{8} + \frac{\sin 8x}{16} + C$

$\displaystyle\qquad\qquad\qquad\qquad = \frac{1}{16}(2\sin 4x + \sin 8x) + C$

**43.** $\displaystyle\int \sin 2x \cos 4x \, dx = \frac{1}{2}\int\Big[\sin\big((2-4)x\big) + \sin\big((2+4)x\big)\Big]\,dx$

$\displaystyle\qquad\qquad\qquad\qquad = \frac{1}{2}\int\big(\sin(-2x) + \sin 6x\big)\,dx$

$\displaystyle\qquad\qquad\qquad\qquad = \frac{1}{2}\int(-\sin 2x + \sin 6x)\,dx$

$\displaystyle\qquad\qquad\qquad\qquad = \frac{1}{2}\left[\frac{\cos 2x}{2} - \frac{\cos 6x}{6}\right] + C$

$\displaystyle\qquad\qquad\qquad\qquad = \frac{1}{4}\cos 2x - \frac{1}{12}\cos 6x + C$

$\displaystyle\qquad\qquad\qquad\qquad = \frac{1}{12}(3\cos 2x - \cos 6x) + C$

**45.** $\displaystyle\int \sin\theta \sin 3\theta \, d\theta = \frac{1}{2}\int(\cos 2\theta - \cos 4\theta)\,d\theta$

$\displaystyle\qquad\qquad\qquad\qquad = \tfrac{1}{2}\big(\tfrac{1}{2}\sin 2\theta - \tfrac{1}{4}\sin 4\theta\big) + C$

$\displaystyle\qquad\qquad\qquad\qquad = \tfrac{1}{8}(2\sin 2\theta - \sin 4\theta) + C$

**47.** $\displaystyle\int \cot^3 2x \, dx = \int(\csc^2 2x - 1)\cot 2x \, dx$

$\displaystyle\qquad\qquad\quad = -\frac{1}{2}\int \cot 2x(-2\csc^2 2x)\,dx - \frac{1}{2}\int\frac{2\cos 2x}{\sin 2x}\,dx$

$\displaystyle\qquad\qquad\quad = -\frac{1}{4}\cot^2 2x - \frac{1}{2}\ln|\sin 2x| + C$

$\displaystyle\qquad\qquad\quad = \frac{1}{4}\big(\ln|\csc^2 2x| - \cot^2 2x\big) + C$

**49.** $\displaystyle\int \csc^4 3x \, dx = \int \csc^2 3x(1 + \cot^2 3x)\,dx$

$\displaystyle\qquad\qquad\quad = \int \csc^2 3x \, dx + \int \cot^2 3x \csc^2 3x \, dx$

$\displaystyle\qquad\qquad\quad = -\frac{1}{3}\cot 3x - \frac{1}{9}\cot^3 3x + C$

**53.** $\displaystyle\int \frac{1}{\sec x \tan x}\,dx = \int \frac{\cos^2 x}{\sin x}\,dx = \int\frac{1 - \sin^2 x}{\sin x}\,dx$

$\displaystyle\qquad\qquad\qquad\qquad = \int(\csc x - \sin x)\,dx$

$\displaystyle\qquad\qquad\qquad\qquad = \ln|\csc x - \cot x| + \cos x + C$

**51.** $\displaystyle\int \frac{\cot^2 t}{\csc t}\,dt = \int\frac{\csc^2 t - 1}{\csc t}\,dt$

$\displaystyle\qquad\qquad\quad = \int(\csc t - \sin t)\,dt$

$\displaystyle\qquad\qquad\quad = \ln|\csc t - \cot t| + \cos t + C$

**55.** $\int \left( \tan^4 t - \sec^4 t \right) dt = \int \left( \tan^2 t + \sec^2 t \right)\left( \tan^2 t - \sec^2 t \right) dt, \qquad \left( \tan^2 t - \sec^2 t = -1 \right)$

$\qquad\qquad\qquad\qquad\quad = -\int \left( \tan^2 t + \sec^2 t \right) dt = -\int \left( 2 \sec^2 t - 1 \right) dt = -2 \tan t + t + C$

**57.** $\int_{-\pi}^{\pi} \sin^2 x \, dx = 2 \int_0^{\pi} \dfrac{1 - \cos 2x}{2} \, dx$

$\qquad\qquad\qquad = \left[ x - \dfrac{1}{2} \sin 2x \right]_0^{\pi} = \pi$

**59.** $\int_0^{\pi/4} 6 \tan^3 x \, dx = 6 \int_0^{\pi/4} \left( \sec^2 x - 1 \right) \tan x \, dx$

$\qquad\qquad\qquad\quad = 6 \int_0^{\pi/4} \left[ \tan x \sec^2 x - \tan x \right] dx$

$\qquad\qquad\qquad\quad = 6 \left[ \dfrac{\tan^2 x}{2} + \ln|\cos x| \right]_0^{\pi/4}$

$\qquad\qquad\qquad\quad = 6 \left[ \dfrac{1}{2} + \ln\left( \dfrac{\sqrt{2}}{2} \right) \right] = 6 \left( \dfrac{1}{2} - \ln \sqrt{2} \right)$

$\qquad\qquad\qquad\quad = 3(1 - \ln 2)$

**61.** Let $u = 1 + \sin t, \, du = \cos t \, dt.$

$\qquad \int_0^{\pi/2} \dfrac{\cos t}{1 + \sin t} \, dt = \left[ \ln|1 + \sin t| \right]_0^{\pi/2} = \ln 2$

**63.** $\int_{-\pi/2}^{\pi/2} 3 \cos^3 x \, dx = 3 \int_{-\pi/2}^{\pi/2} \left( 1 - \sin^2 x \right) \cos x \, dx$

$\qquad\qquad\qquad\qquad = 3 \left[ \sin x - \dfrac{\sin^3 x}{3} \right]_{-\pi/2}^{\pi/2}$

$\qquad\qquad\qquad\qquad = 3 \left[ \left( 1 - \dfrac{1}{3} \right) - \left( -1 + \dfrac{1}{3} \right) \right] = 4$

**65.** (a) Save one sine factor and convert the remaining factors to cosines. Then expand and integrate.

(b) Save one cosine factor and convert the remaining factors to sines. Then expand and integrate.

(c) Make repeated use of the power reducing formulas to convert the integrand to odd powers of the cosine. Then proceed as in part (b).

**67.** (a) $\int \sin x \cos x \, dx = \dfrac{\sin^2 x}{2} + C$

(b) $-\int \cos x \left( -\sin x \right) dx = -\dfrac{\cos^2 x}{2} + C$

(c) $dv = \cos x \, dx \Rightarrow v = \sin x$

$\qquad u = \sin x \quad\;\; \Rightarrow du = \cos x \, dx$

$\qquad \int \sin x \cos x \, dx = \sin^2 x - \int \sin x \cos x \, dx$

$\qquad 2 \int \sin x \cos x \, dx = \sin^2 x$

$\qquad \int \sin x \cos x \, dx = \dfrac{\sin^2 x}{2} + C$

(Answers will vary.)

(d) $\int \sin x \cos x \, dx = \int \dfrac{1}{2} \sin 2x \, dx = -\dfrac{1}{4} \cos 2x + C$

The answers all differ by a constant.

**69. (a)** Let $u = \tan 3x$, $du = 3 \sec^2 3x \, dx$.

$$\int \sec^4 3x \tan^3 3x \, dx = \int \sec^2 3x \tan^3 3x \sec^2 3x \, dx = \frac{1}{3} \int (\tan^2 3x + 1) \tan^3 3x (3 \sec^2 3x) \, dx$$

$$= \frac{1}{3} \int (\tan^5 3x + \tan^3 3x)(3 \sec^2 3x) \, dx = \frac{\tan^6 3x}{18} + \frac{\tan^4 3x}{12} + C_1$$

Or let $u = \sec 3x$, $du = 3 \sec 3x \tan 3x \, dx$.

$$\int \sec^4 3x \tan^3 3x \, dx = \int \sec^3 3x \tan^2 3x \sec 3x \tan 3x \, dx$$

$$= \frac{1}{3} \int \sec^3 3x (\sec^2 3x - 1)(3 \sec 3x \tan 3x) \, dx = \frac{\sec^6 3x}{18} - \frac{\sec^4 3x}{12} + C$$

**(b)**

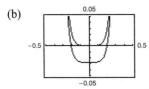

**(c)** $\dfrac{\sec^6 3x}{18} - \dfrac{\sec^4 3x}{12} + C = \dfrac{\left(1 + \tan^2 3x\right)^3}{18} - \dfrac{\left(1 + \tan^2 3x\right)^2}{12} + C$

$$= \frac{1}{18} \tan^6 3x + \frac{1}{6} \tan^4 3x + \frac{1}{6} \tan^2 3x + \frac{1}{18} - \frac{1}{12} \tan^4 3x - \frac{1}{6} \tan^2 3x - \frac{1}{12} + C$$

$$= \frac{\tan^6 3x}{18} + \frac{\tan^4 3x}{12} + \left(\frac{1}{18} - \frac{1}{12}\right) + C$$

$$= \frac{\tan^6 3x}{18} + \frac{\tan^4 3x}{12} + C_2$$

**71.** $A = \displaystyle\int_0^{\pi/2} \left(\sin x - \sin^3 x\right) dx$

$$= \int_0^{\pi/2} \sin x \, dx - \int_0^{\pi/2} \sin^3 x \, dx$$

$$= \left[-\cos x\right]_0^{\pi/2} - \frac{2}{3} \qquad \text{(Wallis's Formula)}$$

$$= 1 - \frac{2}{3} = \frac{1}{3}$$

**73.** $A = \displaystyle\int_{-\pi/4}^{\pi/4} \left[\cos^2 x - \sin^2 x\right] dx$

$$= \int_{-\pi/4}^{\pi/4} \cos 2x \, dx$$

$$= \left[\frac{\sin 2x}{2}\right]_{-\pi/4}^{\pi/4} = \frac{1}{2} + \frac{1}{2} = 1$$

**75. Disks**

$$R(x) = \tan x, \, r(x) = 0$$

$$V = 2\pi \int_0^{\pi/4} \tan^2 x \, dx$$

$$= 2\pi \int_0^{\pi/4} \left(\sec^2 x - 1\right) dx$$

$$= 2\pi \left[\tan x - x\right]_0^{\pi/4}$$

$$= 2\pi \left(1 - \frac{\pi}{4}\right) \approx 1.348$$

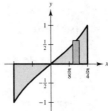

**77.** (a) $V = \pi \int_0^\pi \sin^2 x \, dx = \frac{\pi}{2} \int_0^\pi (1 - \cos 2x) \, dx = \frac{\pi}{2}\left[ x - \frac{1}{2} \sin 2x \right]_0^\pi = \frac{\pi^2}{2}$

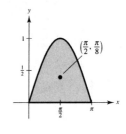

(b) $A = \int_0^\pi \sin x \, dx = \left[ -\cos x \right]_0^\pi = 1 + 1 = 2$

Let $u = x, dv = \sin x \, dx, du = dx, v = -\cos x$.

$\bar{x} = \frac{1}{A} \int_0^\pi x \sin x \, dx = \frac{1}{2}\left[ \left[ -x \cos x \right]_0^\pi + \int_0^\pi \cos x \, dx \right] = \frac{1}{2}\left[ -x \cos x + \sin x \right]_0^\pi = \frac{\pi}{2}$

$\bar{y} = \frac{1}{2A} \int_0^\pi \sin^2 x \, dx = \frac{1}{8} \int_0^\pi (1 - \cos 2x) \, dx = \frac{1}{8}\left[ x - \frac{1}{2} \sin 2x \right]_0^\pi = \frac{\pi}{8}$

$(\bar{x}, \bar{y}) = \left( \dfrac{\pi}{2}, \dfrac{\pi}{8} \right)$

**79.** $dv = \sin x \, dx \Rightarrow \quad v = -\cos x$

$u = \sin^{n-1} x \Rightarrow du = (n - 1) \sin^{n-2} x \cos x \, dx$

$\int \sin^n x \, dx = -\sin^{n-1} x \cos x + (n - 1) \int \sin^{n-2} x \cos^2 x \, dx = -\sin^{n-1} x \cos x + (n - 1) \int \sin^{n-2} x (1 - \sin^2 x) \, dx$

$\qquad = -\sin^{n-1} x \cos x + (n - 1) \int \sin^{n-2} x \, dx - (n - 1) \int \sin^n x \, dx$

Therefore, $n \int \sin^n x \, dx = -\sin^{n-1} x \cos x + (n - 1) \int \sin^{n-2} x \, dx$

$\qquad \int \sin^n x \, dx = \dfrac{-\sin^{n-1} x \cos x}{n} + \dfrac{n-1}{n} \int \sin^{n-2} x \, dx.$

**81.** Let $u = \sin^{n-1} x, du = (n - 1) \sin^{n-2} x \cos x \, dx, dv = \cos^m x \sin x \, dx, v = \dfrac{-\cos^{m+1} x}{m+1}.$

$\int \cos^m x \sin^n x \, dx = \dfrac{-\sin^{n-1} x \cos^{m+1} x}{m+1} + \dfrac{n-1}{m+1} \int \sin^{n-2} x \cos^{m+2} x \, dx$

$\qquad = \dfrac{-\sin^{n-1} x \cos^{m+1} x}{m+1} + \dfrac{n-1}{m+1} \int \sin^{n-2} x \cos^m x (1 - \sin^2 x) \, dx$

$\qquad = \dfrac{-\sin^{n-1} x \cos^{m+1} x}{m+1} + \dfrac{n-1}{m+1} \int \sin^{n-2} x \cos^m x \, dx - \dfrac{n-1}{m+1} \int \sin^n x \cos^m x \, dx$

$\dfrac{m+n}{m+1} \int \cos^m x \sin^n x \, dx = \dfrac{-\sin^{n-1} x \cos^{m+1} x}{m+1} + \dfrac{n-1}{m+1} \int \sin^{n-2} x \cos^m x \, dx$

$\int \cos^m x \sin^n x \, dx = \dfrac{-\cos^{m+1} x \sin^{n-1} x}{m+n} + \dfrac{n-1}{m+n} \int \cos^m x \sin^{n-2} x \, dx$

**83.** $\int \sin^5 x \, dx = -\dfrac{\sin^4 x \cos x}{5} + \dfrac{4}{5} \int \sin^3 x \, dx$

$\qquad = -\dfrac{\sin^4 x \cos x}{5} + \dfrac{4}{5}\left( -\dfrac{\sin^2 x \cos x}{3} + \dfrac{2}{3} \int \sin x \, dx \right)$

$\qquad = -\dfrac{1}{5} \sin^4 x \cos x - \dfrac{4}{15} \sin^2 x \cos x - \dfrac{8}{15} \cos x + C$

$\qquad = -\dfrac{\cos x}{15}\left( 3 \sin^4 x + 4 \sin^2 x + 8 \right) + C$

**85.** $\displaystyle\int \sec^4 \frac{2\pi x}{5}\, dx = \frac{5}{2\pi}\int \sec^4\left(\frac{2\pi x}{5}\right)\frac{2\pi}{5}\, dx$

$$= \frac{5}{2\pi}\left[\frac{1}{3}\sec^2\left(\frac{2\pi x}{5}\right)\tan\left(\frac{2\pi x}{5}\right) + \frac{2}{3}\int \sec^2\left(\frac{2\pi x}{5}\right)\frac{2\pi}{5}\, dx\right]$$

$$= \frac{5}{6\pi}\left[\sec^2\left(\frac{2\pi x}{5}\right)\tan\left(\frac{2\pi x}{5}\right) + 2\tan\left(\frac{2\pi x}{5}\right)\right] + C$$

$$= \frac{5}{6\pi}\tan\left(\frac{2\pi x}{5}\right)\left[\sec^2\left(\frac{2\pi x}{5}\right) + 2\right] + C$$

**87.** $\displaystyle f(t) = a_0 + a_1 \cos\frac{\pi t}{6} + b_1 \sin\frac{\pi t}{6}$

$$a_0 = \frac{1}{12}\int_0^{12} f(t)\, dt, \; a_1 = \frac{1}{6}\int_0^{12} f(t)\cos\frac{\pi t}{6}\, dt, \; b_1 = \frac{1}{6}\int_0^{12} f(t)\sin\frac{\pi t}{6}\, dt$$

(a)   $a_0 \approx \dfrac{1}{12}\cdot\dfrac{(12-0)}{3(12)}[33.5 + 4(35.4) + 2(44.7) + 4(55.6) + 2(67.4) + 4(76.2) + 2(80.4) + 4(79.0) + 2(72.0)$

$$+ 4(61.0) + 2(49.3) + 4(38.6) + 33.5]$$

$$\approx 57.72$$

$$a_1 \approx -23.36$$

$$b_1 \approx -2.75 \qquad \text{(Answers will vary.)}$$

$$H(t) \approx 57.72 - 23.36\cos\left(\frac{\pi t}{6}\right) - 2.75\sin\left(\frac{\pi t}{6}\right)$$

(b)   $L(t) \approx 42.04 - 20.91\cos\left(\dfrac{\pi t}{6}\right) - 4.33\sin\left(\dfrac{\pi t}{6}\right)$

(c)

Temperature difference is greatest in the summer $(t \approx 4.9$ or end of May$)$.

**89.** $\displaystyle\int_{-\pi}^{\pi} \cos(mx)\cos(nx)\, dx = \frac{1}{2}\left[\frac{\sin(m+n)x}{m+n} + \frac{\sin(m-n)x}{m-n}\right]_{-\pi}^{\pi} = 0, \; (m \neq n)$

$$\int_{-\pi}^{\pi} \sin(mx)\sin(nx)\, dx = \frac{1}{2}\int_{-\pi}^{\pi}\left[\cos(m-n)x - \cos(m+n)x\right] dx$$

$$= \frac{1}{2}\left[\frac{\sin(m-n)x}{m-n} - \frac{\sin(m+n)x}{m+n}\right]_{-\pi}^{\pi} = 0, \; (m \neq n)$$

$$\int_{-\pi}^{\pi} \sin(mx)\cos(nx)\, dx = \frac{1}{2}\int_{-\pi}^{\pi}\left[\sin(m+n)x + \sin(m-n)x\right] dx$$

$$= -\frac{1}{2}\left[\frac{\cos(m+n)x}{m+n} + \frac{\cos(m-n)x}{m-n}\right]_{-\pi}^{\pi}, \quad (m \neq n)$$

$$= -\frac{1}{2}\left[\left(\frac{\cos(m+n)\pi}{m+n} + \frac{\cos(m-n)\pi}{m-n}\right) - \left(\frac{\cos(m+n)(-\pi)}{m+n} + \frac{\cos(m-n)(-\pi)}{m-n}\right)\right]$$

$$= 0, \text{ because } \cos(-\theta) = \cos\theta.$$

$$\int_{-\pi}^{\pi} \sin(mx)\cos(mx)\, dx = \frac{1}{m}\left[\frac{\sin^2(mx)}{2}\right]_{-\pi}^{\pi} = 0$$

# Section 8.4  Trigonometric Substitution

**1.** Use $x = 3 \tan \theta$.

**3.** Use $x = 5 \sin \theta$.

**5.** Let $x = 4 \sin \theta, dx = 4 \cos \theta \, d\theta, \sqrt{16 - x^2} = 4 \cos \theta.$

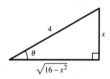

$$\int \frac{1}{\left(16 - x^2\right)^{3/2}} \, dx = \int \frac{4 \cos \theta}{\left(4 \cos \theta\right)^3} \, d\theta = \frac{1}{16} \int \sec^2 \theta \, d\theta = \frac{1}{16} \tan \theta + C = \frac{1}{16} \left( \frac{x}{\sqrt{16 - x^2}} \right) + C$$

**7.** Same substitution as in Exercise 5

$$\int \frac{\sqrt{16 - x^2}}{x} \, dx = \int \frac{4 \cos \theta}{4 \sin \theta} \, 4 \cos \theta \, d\theta$$

$$= 4 \int \frac{\cos^2 \theta}{\sin \theta} \, d\theta$$

$$= 4 \int \frac{1 - \sin^2 \theta}{\sin \theta} \, d\theta$$

$$= 4 \int \left( \csc \theta - \sin \theta \right) d\theta$$

$$= -4 \ln \left| \csc \theta + \cot \theta \right| + 4 \cos \theta + C$$

$$= -4 \ln \left| \frac{4}{x} + \frac{\sqrt{16 - x^2}}{x} \right| + 4 \frac{\sqrt{16 - x^2}}{4} + C$$

$$= -4 \ln \left| \frac{4 + \sqrt{16 - x^2}}{x} \right| + \sqrt{16 - x^2} + C$$

$$= 4 \ln \left| \frac{4 - \sqrt{16 - x^2}}{x} \right| + \sqrt{16 - x^2} + C$$

**9.** Let $x = 5 \sec \theta, dx = 5 \sec \theta \tan \theta \, d\theta,$

$\sqrt{x^2 - 25} = 5 \tan \theta.$

$$\int \frac{1}{\sqrt{x^2 - 25}} \, dx = \int \frac{5 \sec \theta \tan \theta}{5 \tan \theta} \, d\theta$$

$$= \int \sec \theta \, d\theta$$

$$= \ln \left| \sec \theta + \tan \theta \right| + C$$

$$= \ln \left| \frac{x}{5} + \frac{\sqrt{x^2 - 25}}{5} \right| + C$$

$$= \ln \left| x + \sqrt{x^2 - 25} \right| + C$$

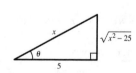

**11.** Same substitution as in Exercise 9

$$\int x^3 \sqrt{x^2 - 25}\, dx = \int (5 \sec \theta)^3 (5 \tan \theta)(5 \sec \theta \tan \theta)\, d\theta$$

$$= 3125 \int \sec^4 \theta \tan^2 \theta\, d\theta$$

$$= 3125 \int (1 + \tan^2 \theta) \tan^2 \theta \sec^2 \theta\, d\theta$$

$$= 3125 \int (\tan^2 \theta + \tan^4 \theta) \sec^2 \theta\, d\theta$$

$$= 3125 \left[ \frac{\tan^3 \theta}{3} + \frac{\tan^5 \theta}{5} \right] + C$$

$$= 3125 \left[ \frac{(x^2 - 25)^{3/2}}{125(3)} + \frac{(x^2 - 25)^{5/2}}{5^5(5)} \right] + C$$

$$= \frac{1}{15}(x^2 - 25)^{3/2} \left[ 125 + 3(x^2 - 25) \right] + C$$

$$= \frac{1}{15}(x^2 - 25)^{3/2}(50 + 3x^2) + C$$

**13.** Let $x = \tan \theta$, $dx = \sec^2 \theta\, d\theta$, $\sqrt{1 + x^2} = \sec \theta$.

$$\int x\sqrt{1 + x^2}\, dx = \int \tan \theta(\sec \theta) \sec^2 \theta\, d\theta = \frac{\sec^3 \theta}{3} + C = \frac{1}{3}(1 + x^2)^{3/2} + C$$

**Note:** This integral could have been evaluated with the Power Rule.

**15.** Same substitution as in Exercise 13

$$\int \frac{1}{(1 + x^2)^2}\, dx = \int \frac{1}{(\sqrt{1 + x^2})^4}\, dx = \int \frac{\sec^2 \theta\, d\theta}{\sec^4 \theta}$$

$$= \int \cos^2 \theta\, d\theta = \frac{1}{2} \int (1 + \cos 2\theta)\, d\theta$$

$$= \frac{1}{2} \left[ \theta + \frac{\sin 2\theta}{2} \right]$$

$$= \frac{1}{2} [\theta + \sin \theta \cos \theta] + C$$

$$= \frac{1}{2} \left[ \arctan x + \left( \frac{x}{\sqrt{1 + x^2}} \right)\left( \frac{1}{\sqrt{1 + x^2}} \right) \right] + C$$

$$= \frac{1}{2} \left( \arctan x + \frac{x}{1 + x^2} \right) + C$$

**17.** Let $u = 4x$, $a = 3$, $du = 4\, dx$.

$$\int \sqrt{9 + 16x^2}\, dx = \frac{1}{4} \int \sqrt{(4x)^2 + 3^2}\, (4)\, dx$$

$$= \frac{1}{4} \cdot \frac{1}{2} \left[ 4x\sqrt{16x^2 + 9} + 9 \ln \left| 4x + \sqrt{16x^2 + 9} \right| \right] + C$$

$$= \frac{1}{2} x\sqrt{16x^2 + 9} + \frac{9}{8} \ln \left| 4x + \sqrt{16x^2 + 9} \right| + C$$

**19.** $\int \sqrt{25 - 4x^2} \, dx = \int 2\sqrt{\dfrac{25}{4} - x^2} \, dx, \quad a = \dfrac{5}{2}$

$\qquad = 2\left(\dfrac{1}{2}\right)\left[\dfrac{25}{4} \arcsin\left(\dfrac{2x}{5}\right) + x\sqrt{\dfrac{25}{4} - x^2}\right] + C$

$\qquad = \dfrac{25}{4} \arcsin\left(\dfrac{2x}{5}\right) + \dfrac{x}{2}\sqrt{25 - 4x^2} + C$

**21.** $\int \dfrac{1}{\sqrt{16 - x^2}} \, dx = \arcsin\left(\dfrac{x}{4}\right) + C$

**23.** Let $x = 2 \sin \theta, \, dx = 2 \cos \theta \, d\theta,$
$\quad \sqrt{4 - x^2} = 2 \cos \theta.$

$\int \sqrt{16 - 4x^2} \, dx = 2 \int \sqrt{4 - x^2} \, dx$

$\qquad = 2 \int 2 \cos \theta (2 \cos \theta \, d\theta)$

$\qquad = 8 \int \cos^2 \theta \, d\theta$

$\qquad = 4 \int (1 + \cos 2\theta) \, d\theta$

$\qquad = 4\left(\theta + \dfrac{1}{2} \sin 2\theta\right) + C$

$\qquad = 4\theta + 4 \sin \theta \cos \theta + C$

$\qquad = 4 \arcsin\left(\dfrac{x}{2}\right) + x\sqrt{4 - x^2} + C$

**25.** Let $x = \sin \theta, \, dx = \cos \theta \, d\theta, \, \sqrt{1 - x^2} = \cos \theta.$

$\int \dfrac{\sqrt{1 - x^2}}{x^4} \, dx = \int \dfrac{\cos \theta (\cos \theta \, d\theta)}{\sin^4 \theta}$

$\qquad = \int \cot^2 \theta \csc^2 \theta \, d\theta$

$\qquad = -\dfrac{1}{3} \cot^3 \theta + C$

$\qquad = -\dfrac{\left(1 - x^2\right)^{3/2}}{3x^3} + C$

**27.** Let $2x = 3 \tan \theta \Rightarrow x = \dfrac{3}{2} \tan \theta, \, dx = \dfrac{3}{2} \sec^2 \theta \, d\theta, \, \sqrt{4x^2 + 9} = 3 \sec \theta.$

$\int \dfrac{1}{x\sqrt{4x^2 + 9}} \, dx = \int \dfrac{(3/2) \sec^2 \theta \, d\theta}{(3/2) \tan \theta \, 3 \sec \theta}$

$\qquad = \dfrac{1}{3} \int \csc \theta \, d\theta$

$\qquad = -\dfrac{1}{3} \ln|\csc \theta + \cot \theta| + C$

$\qquad = -\dfrac{1}{3} \ln\left|\dfrac{\sqrt{4x^2 + 9} + 3}{2x}\right| + C$

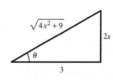

**29.** Let $u = x^2 + 3, \, du = 2x \, dx.$

$\int \dfrac{-3x}{\left(x^2 + 3\right)^{3/2}} \, dx = -\dfrac{3}{2} \int \left(x^2 + 3\right)^{-3/2} (2x) \, dx$

$\qquad = -\dfrac{3}{2} \dfrac{\left(x^2 + 3\right)^{-1/2}}{(-1/2)} + C$

$\qquad = \dfrac{3}{\sqrt{x^2 + 3}} + C$

**31.** Let $e^x = \sin \theta$, $e^x \, dx = \cos \theta \, d\theta$, $\sqrt{1 - e^{2x}} = \cos \theta$.

$$\int e^x \sqrt{1 - e^{2x}} \, dx = \int \cos^2 \theta \, d\theta$$

$$= \frac{1}{2} \int (1 + \cos 2\theta) \, d\theta$$

$$= \frac{1}{2} \left( \theta + \frac{\sin 2\theta}{2} \right)$$

$$= \frac{1}{2} (\theta + \sin \theta \cos \theta) + C$$

$$= \frac{1}{2} \left( \arcsin e^x + e^x \sqrt{1 - e^{2x}} \right) + C$$

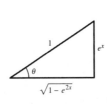

**33.** Let $x = \sqrt{2} \tan \theta$, $dx = \sqrt{2} \sec^2 \theta \, d\theta$, $x^2 + 2 = 2 \sec^2 \theta$.

$$\int \frac{1}{4 + 4x^2 + x^4} \, dx = \int \frac{1}{\left(x^2 + 2\right)^2} \, dx = \int \frac{\sqrt{2} \sec^2 \theta \, d\theta}{4 \sec^4 \theta}$$

$$= \frac{\sqrt{2}}{4} \int \cos^2 \theta \, d\theta$$

$$= \frac{\sqrt{2}}{4} \left( \frac{1}{2} \right) \int (1 + \cos 2\theta) \, d\theta$$

$$= \frac{\sqrt{2}}{8} \left( \theta + \frac{1}{2} \sin 2\theta \right) + C$$

$$= \frac{\sqrt{2}}{8} (\theta + \sin \theta \cos \theta) + C$$

$$= \frac{\sqrt{2}}{8} \left( \arctan \frac{x}{\sqrt{2}} + \frac{x}{\sqrt{x^2 + 2}} \cdot \frac{\sqrt{2}}{\sqrt{x^2 + 2}} \right) = \frac{1}{4} \left( \frac{x}{x^2 + 2} + \frac{1}{\sqrt{2}} \arctan \frac{x}{\sqrt{2}} \right) + C$$

**35.** Use integration by parts. Because $x > \frac{1}{2}$,

$$u = \text{arcsec } 2x \Rightarrow du = \frac{1}{x\sqrt{4x^2 - 1}} \, dx, \, dv = dx \Rightarrow v = x$$

$$\int \text{arcsec } 2x \, dx = x \, \text{arcsec } 2x - \int \frac{1}{\sqrt{4x^2 - 1}} \, dx$$

$$2x = \sec \theta, \, dx = \frac{1}{2} \sec \theta \tan \theta \, d\theta, \, \sqrt{4x^2 - 1} = \tan \theta$$

$$\int \text{arcsec } 2x \, dx = x \, \text{arcsec } 2x - \int \frac{(1/2) \sec \theta \tan \theta \, d\theta}{\tan \theta} = x \, \text{arcsec } 2x - \frac{1}{2} \int \sec \theta \, d\theta$$

$$= x \, \text{arcsec } 2x - \frac{1}{2} \ln |\sec \theta + \tan \theta| + C = x \, \text{arcsec } 2x - \frac{1}{2} \ln \left| 2x + \sqrt{4x^2 - 1} \right| + C.$$

**37.** $\int \frac{1}{\sqrt{4x - x^2}} \, dx = \int \frac{1}{\sqrt{4 - (x - 2)^2}} \, dx = \arcsin \left( \frac{x - 2}{2} \right) + C$

**39.** $x^2 + 6x + 12 = x^2 + 6x + 9 + 3 = (x + 3)^2 + \left(\sqrt{3}\right)^2$

Let $x + 3 = \sqrt{3}\tan\theta$, $dx = \sqrt{3}\sec^2\theta\,d\theta$.

$$\sqrt{x^2 + 6x + 12} = \sqrt{(x + 3)^2 + \left(\sqrt{3}\right)^2} = \sqrt{3}\sec\theta$$

```
       √x²+6x+12 ╱|
               ╱  | x+3
              ╱ θ |
             ‾‾‾‾‾
               √3
```

$$\int \frac{x}{\sqrt{x^2 + 6x + 12}}\,dx = \int \frac{\sqrt{3}\tan\theta - 3}{\sqrt{3}\sec\theta}\sqrt{3}\sec^2\theta\,d\theta$$

$$= \int \sqrt{3}\sec\theta\tan\theta\,d\theta - 3\int\sec\theta\,d\theta$$

$$= \sqrt{3}\sec\theta - 3\ln\left|\sec\theta + \tan\theta\right| + C$$

$$= \sqrt{3}\left(\frac{\sqrt{x^2 + 6x + 12}}{\sqrt{3}}\right) - 3\ln\left|\frac{\sqrt{x^2 + 6x + 12}}{\sqrt{3}} + \frac{x + 3}{\sqrt{3}}\right| + C$$

$$= \sqrt{x^2 + 6x + 12} - 3\ln\left|\sqrt{x^2 + 6x + 12} + (x + 3)\right| + C$$

**41.** Let $t = \sin\theta$, $dt = \cos\theta\,d\theta$, $1 - t^2 = \cos^2\theta$.

(a) $\int \dfrac{t^2}{\left(1 - t^2\right)^{3/2}}\,dt = \int \dfrac{\sin^2\theta\cos\theta\,d\theta}{\cos^3\theta} = \int\tan^2\theta\,d\theta = \int\left(\sec^2\theta - 1\right)d\theta = \tan\theta - \theta + C = \dfrac{t}{\sqrt{1 - t^2}} - \arcsin t + C$

So, $\displaystyle\int_0^{\sqrt{3}/2} \frac{t^2}{\left(1 - t^2\right)^{3/2}}\,dt = \left[\frac{t}{\sqrt{1 - t^2}} - \arcsin t\right]_0^{\sqrt{3}/2} = \frac{\sqrt{3}/2}{\sqrt{1/4}} - \arcsin\frac{\sqrt{3}}{2} = \sqrt{3} - \frac{\pi}{3} \approx 0.685.$

(b) When $t = 0$, $\theta = 0$. When $t = \sqrt{3}/2$, $\theta = \pi/3$. So,

$$\int_0^{\sqrt{3}/2} \frac{t^2}{\left(1 - t^2\right)^{3/2}}\,dt = \left[\tan\theta - \theta\right]_0^{\pi/3} = \sqrt{3} - \frac{\pi}{3} \approx 0.685.$$

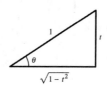

**43.** (a) Let $x = 3\tan\theta$, $dx = 3\sec^2\theta\,d\theta$, $\sqrt{x^2 + 9} = 3\sec\theta$.

$$\int \frac{x^3}{\sqrt{x^2 + 9}}\,dx = \int \frac{\left(27\tan^3\theta\right)\left(3\sec^2\theta\,d\theta\right)}{3\sec\theta}$$

$$= 27\int\left(\sec^2\theta - 1\right)\sec\theta\tan\theta\,d\theta$$

$$= 27\left[\frac{1}{3}\sec^3\theta - \sec\theta\right] + C = 9\left[\sec^3\theta - 3\sec\theta\right] + C$$

$$= 9\left[\left(\frac{\sqrt{x^2 + 9}}{3}\right)^3 - 3\left(\frac{\sqrt{x^2 + 9}}{3}\right)\right] + C = \frac{1}{3}\left(x^2 + 9\right)^{3/2} - 9\sqrt{x^2 + 9} + C$$

So, $\displaystyle\int_0^3 \frac{x^3}{\sqrt{x^2 + 9}}\,dx = \left[\frac{1}{3}\left(x^2 + 9\right)^{3/2} - 9\sqrt{x^2 + 9}\right]_0^3$

$$= \left(\frac{1}{3}\left(54\sqrt{2}\right) - 27\sqrt{2}\right) - (9 - 27) = 18 - 9\sqrt{2} = 9\left(2 - \sqrt{2}\right) \approx 5.272.$$

(b) When $x = 0$, $\theta = 0$. When $x = 3$, $\theta = \pi/4$. So,

$$\int_0^3 \frac{x^3}{\sqrt{x^2 + 9}}\,dx = 9\left[\sec^3\theta - 3\sec\theta\right]_0^{\pi/4} = 9\left(2\sqrt{2} - 3\sqrt{2}\right) - 9(1 - 3) = 9\left(2 - \sqrt{2}\right) \approx 5.272.$$

**45.** (a) Let $x = 3\sec\theta$, $dx = 3\sec\theta\tan\theta\,d\theta$, $\sqrt{x^2 - 9} = 3\tan\theta$.

$$\int \frac{x^2}{\sqrt{x^2 - 9}}\,dx = \int \frac{9\sec^2\theta}{3\tan\theta}\,3\sec\theta\tan\theta\,d\theta$$

$$= 9\int \sec^3\theta\,d\theta$$

$$= 9\left(\frac{1}{2}\sec\theta\tan\theta + \frac{1}{2}\int \sec\theta\,d\theta\right) \quad \text{(8.3 Exercise 102 or Example 5, Section 8.2)}$$

$$= \frac{9}{2}\left(\sec\theta\tan\theta + \ln|\sec\theta + \tan\theta|\right)$$

$$= \frac{9}{2}\left(\frac{x}{3} \cdot \frac{\sqrt{x^2 - 9}}{3} + \ln\left|\frac{x}{3} + \frac{\sqrt{x^2 - 9}}{3}\right|\right)$$

So,

$$\int_4^6 \frac{x^2}{\sqrt{x^2 - 9}}\,dx = \frac{9}{2}\left[\frac{x\sqrt{x^2 - 9}}{9} + \ln\left|\frac{x}{3} + \frac{\sqrt{x^2 - 9}}{3}\right|\right]_4^6$$

$$= \frac{9}{2}\left[\left(\frac{6\sqrt{27}}{9} + \ln\left|2 + \frac{\sqrt{27}}{3}\right|\right) - \left(\frac{4\sqrt{7}}{9} + \ln\left|\frac{4}{3} + \frac{\sqrt{7}}{3}\right|\right)\right]$$

$$= 9\sqrt{3} - 2\sqrt{7} + \frac{9}{2}\left[\ln\left(\frac{6 + \sqrt{27}}{3}\right) - \ln\left(\frac{4 + \sqrt{7}}{3}\right)\right]$$

$$= 9\sqrt{3} - 2\sqrt{7} + \frac{9}{2}\ln\left(\frac{6 + 3\sqrt{3}}{4 + \sqrt{7}}\right) \approx 12.644.$$

(b) When $x = 4$, $\theta = \operatorname{arcsec}\left(\frac{4}{3}\right)$. When $x = 6$, $\theta = \operatorname{arcsec}(2) = \frac{\pi}{3}$.

$$\int_4^6 \frac{x^2}{\sqrt{x^2 - 9}}\,dx = \frac{9}{2}\left[\sec\theta\tan\theta + \ln|\sec\theta + \tan\theta|\right]_{\operatorname{arcsec}(4/3)}^{\pi/3}$$

$$= \frac{9}{2}\left(2 \cdot \sqrt{3} + \ln\left|2 + \sqrt{3}\right|\right) - \frac{9}{2}\left(\frac{4}{3}\left(\frac{\sqrt{7}}{3}\right) + \ln\left|\frac{4}{3} + \frac{\sqrt{7}}{3}\right|\right)$$

$$= 9\sqrt{3} - 2\sqrt{7} + \frac{9}{2}\ln\left(\frac{6 + 3\sqrt{3}}{4 + \sqrt{7}}\right) \approx 12.644$$

**47.** (a) Let $u = a\sin\theta$, $\sqrt{a^2 - u^2} = a\cos\theta$, where $-\pi/2 \le \theta \le \pi/2$.

(b) Let $u = a\tan\theta$, $\sqrt{a^2 + u^2} = a\sec\theta$, where $-\pi/2 < \theta < \pi/2$.

(c) Let $u = a\sec\theta$, $\sqrt{u^2 - a^2} = \tan\theta$ if $u > a$ and $\sqrt{u^2 - a^2} = -\tan\theta$ if $u < -a$, where $0 \le \theta < \pi/2$ or $\pi/2 < \theta \le \pi$.

**49. (a)** $u = x^2 + 9, du = 2x\,dx$

$$\int \frac{x}{x^2 + 9}\,dx = \frac{1}{2}\int \frac{du}{u} = \frac{1}{2}\ln|u| + C = \frac{1}{2}\ln(x^2 + 9) + C$$

Let $x = 3\tan\theta$, $x^2 + 9 = 9\sec^2\theta$, $dx = 3\sec^2\theta\,d\theta$.

$$\int \frac{x}{x^2 + 9}\,dx = \int \frac{3\tan\theta}{9\sec^2\theta}3\sec^2\theta\,d\theta = \int \tan\theta\,d\theta$$

$$= -\ln|\cos\theta| + C_1$$

$$= -\ln\left|\frac{3}{\sqrt{x^2 + 9}}\right| + C_1$$

$$= -\ln 3 + \ln\sqrt{x^2 + 9} + C_1 = \frac{1}{2}\ln(x^2 + 9) + C_2$$

The answers are equivalent.

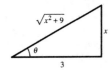

**(b)** $\int \frac{x^2}{x^2 + 9}\,dx = \int \frac{x^2 + 9 - 9}{x^2 + 9}\,dx = \int\left(1 - \frac{9}{x^2 + 9}\right)dx = x - 3\arctan\left(\frac{x}{3}\right) + C$

Let $x = 3\tan\theta$, $x^2 + 9 = 9\sec^2\theta$, $dx = 3\sec^2\theta\,d\theta$.

$$\int \frac{x^2}{x^2 + 9}\,dx = \int \frac{9\tan^2\theta}{9\sec^2\theta}3\sec^2\theta\,d\theta$$

$$= 3\int \tan^2\theta\,d\theta = 3\int(\sec^2\theta - 1)\,d\theta$$

$$= 3\tan\theta - 3\theta + C_1$$

$$= x - 3\arctan\left(\frac{x}{3}\right) + C_1$$

The answers are equivalent.

**51. True**

$$\int \frac{dx}{\sqrt{1 - x^2}} = \int \frac{\cos\theta\,d\theta}{\cos\theta} = \int d\theta$$

**53. False**

$$\int_0^{\sqrt{3}} \frac{dx}{\left(\sqrt{1 + x^2}\right)^3} = \int_0^{\pi/3} \frac{\sec^2\theta\,d\theta}{\sec^3\theta} = \int_0^{\pi/3} \cos\theta\,d\theta$$

**55.** $A = 4\int_0^a \frac{b}{a}\sqrt{a^2 - x^2}\,dx$

$$= \frac{4b}{a}\int_0^a \sqrt{a^2 - x^2}\,dx$$

$$= \left[\frac{4b}{a}\left(\frac{1}{2}\right)\left(a^2 \arcsin\frac{x}{a} + x\sqrt{a^2 - x^2}\right)\right]_0^a$$

$$= \frac{2b}{a}\left(a^2\left(\frac{\pi}{2}\right)\right) = \pi ab$$

**Note:** See Theorem 8.2 for $\int \sqrt{a^2 - x^2}\,dx$.

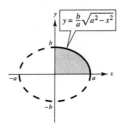

**57.** (a)  $x^2 + (y - k)^2 = 25$

Radius of circle $= 5$

$k^2 = 5^2 + 5^2 = 50$

$k = 5\sqrt{2}$

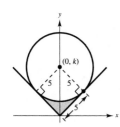

(b)  Area $=$ square $- \dfrac{1}{4}$(circle)

$$= 25 - \frac{1}{4}\pi(5)^2 = 25\left(1 - \frac{\pi}{4}\right)$$

(c)  Area $= r^2 - \dfrac{1}{4}\pi r^2 = r^2\left(1 - \dfrac{\pi}{4}\right)$

**59.** Let $x - 3 = \sin\theta$, $dx = \cos\theta\, d\theta$, $\sqrt{1 - (x-3)^2} = \cos\theta$.

**Shell Method:**

$$V = 4\pi \int_2^4 x\sqrt{1 - (x-3)^2}\, dx$$

$$= 4\pi \int_{-\pi/2}^{\pi/2} (3 + \sin\theta)\cos^2\theta\, d\theta$$

$$= 4\pi\left[\frac{3}{2}\int_{-\pi/2}^{\pi/2} (1 + \cos 2\theta)\, d\theta + \int_{-\pi/2}^{\pi/2} \cos^2\theta \sin\theta\, d\theta\right]$$

$$= 4\pi\left[\frac{3}{2}\left(\theta + \frac{1}{2}\sin 2\theta\right) - \frac{1}{3}\cos^3\theta\right]_{-\pi/2}^{\pi/2} = 6\pi^2$$

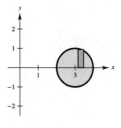

**61.**  $y = \ln x$, $y' = \dfrac{1}{x}$, $1 + (y')^2 = 1 + \dfrac{1}{x^2} = \dfrac{x^2 + 1}{x^2}$

Let $x = \tan\theta$, $dx = \sec^2\theta\, d\theta$, $\sqrt{x^2 + 1} = \sec\theta$.

$$s = \int_1^5 \sqrt{\frac{x^2 + 1}{x^2}}\, dx = \int_1^5 \frac{\sqrt{x^2 + 1}}{x}\, dx$$

$$= \int_a^b \frac{\sec\theta}{\tan\theta}\sec^2\theta\, d\theta = \int_a^b \frac{\sec\theta}{\tan\theta}(1 + \tan^2\theta)\, d\theta$$

$$= \int_a^b (\csc\theta + \sec\theta\tan\theta)\, d\theta = \left[-\ln|\csc\theta + \cot\theta| + \sec\theta\right]_a^b$$

$$= \left[-\ln\left|\frac{\sqrt{x^2 + 1}}{x} + \frac{1}{x}\right| + \sqrt{x^2 + 1}\right]_1^5$$

$$= \left[-\ln\left(\frac{\sqrt{26} + 1}{5}\right) + \sqrt{26}\right] - \left[-\ln\left(\sqrt{2} + 1\right) + \sqrt{2}\right]$$

$$= \ln\left[\frac{5\left(\sqrt{2} + 1\right)}{\sqrt{26} + 1}\right] + \sqrt{26} - \sqrt{2} \approx 4.367 \text{ or } \ln\left[\frac{\sqrt{26} - 1}{5\left(\sqrt{2} - 1\right)}\right] + \sqrt{26} - \sqrt{2}$$

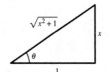

**63.** Length of one arch of sine curve: $y = \sin x$, $y' = \cos x$

$$L_1 = \int_0^\pi \sqrt{1 + \cos^2 x}\ dx$$

Length of one arch of cosine curve: $y = \cos x$, $y' = -\sin x$

$$L_2 = \int_{-\pi/2}^{\pi/2} \sqrt{1 + \sin^2 x}\ dx$$

$$= \int_{-\pi/2}^{\pi/2} \sqrt{1 + \cos^2\left(x - \frac{\pi}{2}\right)}\ dx, \quad u = x - \frac{\pi}{2},\ du = dx$$

$$= \int_{-\pi}^{0} \sqrt{1 + \cos^2 u}\ du$$

$$= \int_{0}^{\pi} \sqrt{1 + \cos^2 u}\ du = L_1$$

**65.** Let $x = 3\tan\theta$, $dx = 3\sec^2\theta\ d\theta$, $\sqrt{x^2 + 9} = 3\sec\theta$.

$$A = 2\int_0^4 \frac{3}{\sqrt{x^2 + 9}}\ dx = 6\int_0^4 \frac{dx}{\sqrt{x^2 + 9}} = 6\int_a^b \frac{3\sec^2\theta\ d\theta}{3\sec\theta}$$

$$= 6\int_a^b \sec\theta\ d\theta = \left[6\ln|\sec\theta + \tan\theta|\right]_a^b = \left[6\ln\left|\frac{\sqrt{x^2 + 9} + x}{3}\right|\right]_0^4 = 6\ln 3$$

$\bar{x} = 0$ (by symmetry)

$$\bar{y} = \frac{1}{2}\left(\frac{1}{A}\right)\int_{-4}^4 \left(\frac{3}{\sqrt{x^2 + 9}}\right)^2 dx = \frac{9}{12\ln 3}\int_{-4}^4 \frac{1}{x^2 + 9}\ dx = \frac{3}{4\ln 3}\left[\frac{1}{3}\arctan\frac{x}{3}\right]_{-4}^4 = \frac{2}{4\ln 3}\arctan\frac{4}{3} \approx 0.422$$

$$(\bar{x}, \bar{y}) = \left(0, \frac{1}{2\ln 3}\arctan\frac{4}{3}\right) \approx (0, 0.422)$$

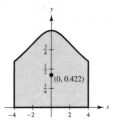

**67.** $y = x^2, y' = 2x, 1 + (y')^2 = 1 + 4x^2$

$2x = \tan\theta, dx = \dfrac{1}{2}\sec^2\theta\,d\theta, \sqrt{1 + 4x^2} = \sec\theta$

(For $\int \sec^5\theta\,d\theta$ and $\int \sec^3\theta\,d\theta$, see Exercise 82 in Section 8.3.)

$$S = 2\pi\int_0^{\sqrt{2}} x^2\sqrt{1 + 4x^2}\,dx = 2\pi\int_a^b \left(\frac{\tan\theta}{2}\right)^2(\sec\theta)\left(\frac{1}{2}\sec^2\theta\right)d\theta$$

$$= \frac{\pi}{4}\int_a^b \sec^3\theta\tan^2\theta\,d\theta = \frac{\pi}{4}\left[\int_a^b \sec^5\theta\,d\theta - \int_a^b \sec^3\theta\,d\theta\right]$$

$$= \frac{\pi}{4}\left\{\frac{1}{4}\left[\sec^3\theta\tan\theta + \frac{3}{2}\left(\sec\theta\tan\theta + \ln\left|\sec\theta + \tan\theta\right|\right)\right] - \frac{1}{2}\left(\sec\theta\tan\theta + \ln\left|\sec\theta + \tan\theta\right|\right)\right\}\Bigg]_a^b$$

$$= \frac{\pi}{4}\left[\frac{1}{4}\left[\left(1 + 4x^2\right)^{3/2}(2x)\right] - \frac{1}{8}\left[\left(1 + 4x^2\right)^{1/2}(2x) + \ln\left|\sqrt{1 + 4x^2} + 2x\right|\right]\right]_0^{\sqrt{2}}$$

$$= \frac{\pi}{4}\left[\frac{54\sqrt{2}}{4} - \frac{6\sqrt{2}}{8} - \frac{1}{8}\ln\left(3 + 2\sqrt{2}\right)\right]$$

$$= \frac{\pi}{4}\left(\frac{51\sqrt{2}}{4} - \frac{\ln\left(3 + 2\sqrt{2}\right)}{8}\right) = \frac{\pi}{32}\left[102\sqrt{2} - \ln\left(3 + 2\sqrt{2}\right)\right] \approx 13.989$$

**69.** (a)  Area of representative rectangle: $2\sqrt{1 - y^2}\,\Delta y$

Force: $2(62.4)(3 - y)\sqrt{1 - y^2}\,\Delta y$

$$F = 124.8\int_{-1}^{1} (3 - y)\sqrt{1 - y^2}\,dy$$

$$= 124.8\left[3\int_{-1}^{1} \sqrt{1 - y^2}\,dy - \int_{-1}^{1} y\sqrt{1 - y^2}\,dy\right]$$

$$= 124.8\left[\frac{3}{2}\left(\arcsin y + y\sqrt{1 - y^2}\right) + \frac{1}{2}\left(\frac{2}{3}\right)\left(1 - y^2\right)^{3/2}\right]_{-1}^{1} = (62.4)3\left[\arcsin 1 - \arcsin(-1)\right] = 187.2\pi \text{ lb}$$

(b)  $F = 124.8\int_{-1}^{1} (d - y)\sqrt{1 - y^2}\,dy = 124.8d\int_{-1}^{1} \sqrt{1 - y^2}\,dy - 124.8\int_{-1}^{1} y\sqrt{1 - y^2}\,dy$

$$= 124.8\left(\frac{d}{2}\right)\left[\arcsin y + y\sqrt{1 - y^2}\right]_{-1}^{1} - 124.8(0) = 62.4\pi d \text{ lb}$$

**71.** Let $u = a \sin\theta$, $du = a \cos\theta\, d\theta$, $\sqrt{a^2 - u^2} = a\cos\theta$.

$$\int \sqrt{a^2 - u^2}\, du = \int a^2 \cos^2\theta\, d\theta = a^2 \int \frac{1 + \cos 2\theta}{2}\, d\theta$$

$$= \frac{a^2}{2}\left(\theta + \frac{1}{2}\sin 2\theta\right) + C = \frac{a^2}{2}(\theta + \sin\theta\cos\theta) + C$$

$$= \frac{a^2}{2}\left[\arcsin\frac{u}{a} + \left(\frac{u}{a}\right)\left(\frac{\sqrt{a^2 + u^2}}{a}\right)\right] + C = \frac{1}{2}\left(a^2 \arcsin\frac{u}{a} + u\sqrt{a^2 - u^2}\right) + C$$

Let $u = a \sec\theta$, $du = a \sec\theta \tan\theta\, d\theta$, $\sqrt{u^2 - a^2} = a\tan\theta$.

$$\int \sqrt{u^2 - a^2}\, du = \int a\tan\theta(a\sec\theta\tan\theta)\, d\theta = a^2 \int \tan^2\theta \sec\theta\, d\theta$$

$$= a^2 \int (\sec^2\theta - 1)\sec\theta\, d\theta = a^2 \int (\sec^3\theta - \sec\theta)\, d\theta$$

$$= a^2\left[\frac{1}{2}\sec\theta\tan\theta + \frac{1}{2}\int\sec\theta\, d\theta\right] - a^2\int\sec\theta\, d\theta = a^2\left[\frac{1}{2}\sec\theta\tan\theta - \frac{1}{2}\ln|\sec\theta + \tan\theta|\right]$$

$$= \frac{a^2}{2}\left[\frac{u}{a}\cdot\frac{\sqrt{u^2 - a^2}}{a} - \ln\left|\frac{u}{a} + \frac{\sqrt{u^2 - a^2}}{a}\right|\right] + C_1 = \frac{1}{2}\left[u\sqrt{u^2 - a^2} - a^2\ln\left|u + \sqrt{u^2 - a^2}\right|\right] + C$$

Let $u = a\tan\theta$, $du = a\sec^2\theta\, d\theta$, $\sqrt{u^2 + a^2} = a\sec\theta$.

$$\int \sqrt{u^2 + a^2}\, du = \int (a\sec\theta)(a\sec^2\theta)\, d\theta$$

$$= a^2 \int \sec^3\theta\, d\theta = a^2\left[\frac{1}{2}\sec\theta\tan\theta + \frac{1}{2}\ln|\sec\theta + \tan\theta|\right] + C_1$$

$$= \frac{a^2}{2}\left[\frac{\sqrt{u^2 + a^2}}{a}\cdot\frac{u}{a} + \ln\left|\frac{\sqrt{u^2 + a^2}}{a} + \frac{u}{a}\right|\right] + C_1 = \frac{1}{2}\left[u\sqrt{u^2 + a^2} + a^2\ln\left|u + \sqrt{u^2 + a^2}\right|\right] + C$$

**73.** Large circle: $x^2 + y^2 = 25$

$$y = \sqrt{25 - x^2}, \quad \text{upper half}$$

From the right triangle, the center of the small circle is $(0, 4)$.

$$x^2 + (y - 4)^2 = 9$$

$$y = 4 + \sqrt{9 - x^2}, \quad \text{upper half}$$

$$A = 2\int_0^3\left[\left(4 + \sqrt{9 - x^2}\right) - \sqrt{25 - x^2}\right]dx$$

$$= 2\left[4x + \frac{1}{2}\left[9\arcsin\left(\frac{x}{3}\right) + x\sqrt{9 - x^2}\right] - \frac{1}{2}\left[25\arcsin\left(\frac{x}{5}\right) + x\sqrt{25 - x^2}\right]\right]_0^3$$

$$= 2\left[12 + \frac{9}{2}\arcsin(1) - \frac{25}{2}\arcsin\frac{3}{5} - 6\right]$$

$$= 12 + \frac{9\pi}{2} - 25\arcsin\frac{3}{5} \approx 10.050$$

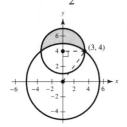

**75.** Let $I = \int_0^1 \dfrac{\ln(x + 1)}{x^2 + 1}\, dx$

Let $x = \dfrac{1 - u}{1 + u}, \quad dx = \dfrac{-2}{(1 + u)^2}\, du$

$x + 1 = \dfrac{2}{1 + u}, \quad x^2 + 1 = \dfrac{2 + 2u^2}{(1 + u)^2}$

$I = \displaystyle\int_1^0 \dfrac{\ln\left(\dfrac{2}{1 + u}\right)}{\dfrac{2 + 2u^2}{(1 + u)^2}}\left(\dfrac{-2}{(1 + u)^2}\right) du$

$= \displaystyle\int_1^0 \dfrac{-\ln\left(\dfrac{2}{1 + u}\right)}{1 + u^2}\, du = \int_0^1 \dfrac{\ln\left(\dfrac{2}{1 + u}\right)}{1 + u^2}\, du = \int_0^1 \dfrac{\ln 2}{1 + u^2}\, du - \int_0^1 \dfrac{\ln(1 + u)}{1 + u^2}\, du = (\ln 2)[\arctan u]_0^1 - I$

$\Rightarrow 2I = \ln 2\left(\dfrac{\pi}{4}\right)$

$I = \dfrac{\pi}{8}\ln 2 \approx 0.272198$

# Section 8.5   Partial Fractions

**1.** $\dfrac{4}{x^2 - 8x} = \dfrac{4}{x(x - 8)} = \dfrac{A}{x} + \dfrac{B}{x - 8}$

**3.** $\dfrac{2x - 3}{x^3 + 10x} = \dfrac{2x - 3}{x(x^2 + 10)} = \dfrac{A}{x} + \dfrac{Bx + C}{x^2 + 10}$

**5.** $\dfrac{1}{x^2 - 9} = \dfrac{1}{(x - 3)(x + 3)} = \dfrac{A}{x + 3} + \dfrac{B}{x - 3}$

$1 = A(x - 3) + B(x + 3)$

When $x = 3, \quad 1 = 6B \Rightarrow B = \dfrac{1}{6}$.

When $x = -3, \ 1 = -6A \Rightarrow A = -\dfrac{1}{6}$.

$\displaystyle\int \dfrac{1}{x^2 - 9}\, dx = -\dfrac{1}{6}\int \dfrac{1}{x + 3}\, dx + \dfrac{1}{6}\int \dfrac{1}{x - 3}\, dx$

$= -\dfrac{1}{6}\ln|x + 3| + \dfrac{1}{6}\ln|x - 3| + C$

$= \dfrac{1}{6}\ln\left|\dfrac{x - 3}{x + 3}\right| + C$

**7.** $\dfrac{5}{x^2 + 3x - 4} = \dfrac{5}{(x + 4)(x - 1)} = \dfrac{A}{x + 4} + \dfrac{B}{x - 1}$

$5 = A(x - 1) + B(x + 4)$

When $x = 1, \quad 5 = 5B \Rightarrow B = 1$.

When $x = -4, \quad 5 = -5A \Rightarrow A = -1$.

$\displaystyle\int \dfrac{5}{x^2 + 3x - 4}\, dx = \int \dfrac{-1}{x + 4}\, dx + \int \dfrac{1}{x - 1}\, dx$

$= -\ln|x + 4| + \ln|x - 1| + C$

$= \ln\left|\dfrac{x - 1}{x + 4}\right| + C$

**9.** $\dfrac{x^2 + 12x + 12}{x(x + 2)(x - 2)} = \dfrac{A}{x} + \dfrac{B}{x + 2} + \dfrac{C}{x - 2}$

$x^2 + 12x + 12 = A(x + 2)(x - 2) + Bx(x - 2) + Cx(x + 2)$

When $x = 0, 12 = -4A \Rightarrow A = -3$.

When $x = -2, -8 = 8B \Rightarrow B = -1$.

When $x = 2, 40 = 8C \Rightarrow C = 5$.

$\displaystyle\int \dfrac{x^2 + 12x + 12}{x^3 - 4x}\, dx = 5\int \dfrac{1}{x - 2}\, dx - \int \dfrac{1}{x + 2}\, dx - 3\int \dfrac{1}{x}\, dx = 5\ln|x - 2| - \ln|x + 2| - 3\ln|x| + C$

**11.** $\dfrac{2x^3 - 4x^2 - 15x + 5}{x^2 - 2x - 8} = 2x + \dfrac{x + 5}{(x - 4)(x + 2)} = 2x + \dfrac{A}{x - 4} + \dfrac{B}{x + 2}$

$$x + 5 = A(x + 2) + B(x - 4)$$

When $x = 4, 9 = 6A \Rightarrow A = \dfrac{3}{2}$.

When $x = -2, 3 = -6B \Rightarrow B = -\dfrac{1}{2}$.

$\displaystyle\int \dfrac{2x^3 - 4x^2 - 15x + 5}{x^2 - 2x - 8}\,dx = \int \left(2x + \dfrac{3/2}{x - 4} - \dfrac{1/2}{x + 2}\right) dx = x^2 + \dfrac{3}{2}\ln|x - 4| - \dfrac{1}{2}\ln|x + 2| + C$

**13.** $\dfrac{4x^2 + 2x - 1}{x^2(x + 1)} = \dfrac{A}{x} + \dfrac{B}{x^2} + \dfrac{C}{x + 1}$

$$4x^2 + 2x - 1 = Ax(x + 1) + B(x + 1) + Cx^2$$

When $x = 0, B = -1$.

When $x = -1, C = 1$.

When $x = 1, A = 3$.

$\displaystyle\int \dfrac{4x^2 + 2x - 1}{x^3 + x^2}\,dx = \int \left(\dfrac{3}{x} - \dfrac{1}{x^2} + \dfrac{1}{x + 1}\right) dx$

$$= 3\ln|x| + \dfrac{1}{x} + \ln|x + 1| + C$$

$$= \dfrac{1}{x} + \ln\left|x^4 + x^3\right| + C$$

**15.** $\dfrac{x^2 + 3x - 4}{x^3 - 4x^2 + 4x} = \dfrac{x^2 + 3x - 4}{x(x - 2)^2} = \dfrac{A}{x} + \dfrac{B}{(x - 2)} + \dfrac{C}{(x - 2)^2}$

$$x^2 + 3x - 4 = A(x - 2)^2 + Bx(x - 2) + Cx$$

When $x = 0, -4 = 4A \Rightarrow A = -1$.

When $x = 2, 6 = 2C \Rightarrow C = 3$.

When $x = 1, 0 = -1 - B + 3 \Rightarrow B = 2$.

$\displaystyle\int \dfrac{x^2 + 3x - 4}{x^3 - 4x^2 + 4x}\,dx = \int \dfrac{-1}{x}\,dx + \int \dfrac{2}{(x - 2)}\,dx + \int \dfrac{3}{(x - 2)^2}\,dx = -\ln|x| + 2\ln|x - 2| - \dfrac{3}{(x - 2)} + C$

**17.** $\dfrac{x^2 - 1}{x(x^2 + 1)} = \dfrac{A}{x} + \dfrac{Bx + C}{x^2 + 1}$

$$x^2 - 1 = A(x^2 + 1) + (Bx + C)x$$

When $x = 0, A = -1$.

When $x = 1, 0 = -2 + B + C$.

When $x = -1, 0 = -2 + B - C$.

Solving these equations you have $A = -1, B = 2, C = 0$.

$\displaystyle\int \dfrac{x^2 - 1}{x^3 + x}\,dx = -\int \dfrac{1}{x}\,dx + \int \dfrac{2x}{x^2 + 1}\,dx = -\ln|x| + \ln\left|x^2 + 1\right| + C = \ln\left|\dfrac{x^2 + 1}{x}\right| + C$

**19.** $\dfrac{x^2}{x^4 - 2x^2 - 8} = \dfrac{A}{x - 2} + \dfrac{B}{x + 2} + \dfrac{Cx + D}{x^2 + 2}$

$$x^2 = A(x + 2)(x^2 + 2) + B(x - 2)(x^2 + 2) + (Cx + D)(x + 2)(x - 2)$$

When $x = 2, 4 = 24A$.

When $x = -2, 4 = -24B$.

When $x = 0, 0 = 4A - 4B - 4D$.

When $x = 1, 1 = 9A - 3B - 3C - 3D$.

Solving these equations you have $A = \dfrac{1}{6}, B = -\dfrac{1}{6}, C = 0, D = \dfrac{1}{3}$.

$$\int \dfrac{x^2}{x^4 - 2x^2 - 8}\,dx = \dfrac{1}{6}\left(\int \dfrac{1}{x - 2}\,dx - \int \dfrac{1}{x + 2}\,dx + 2\int \dfrac{1}{x^2 + 2}\,dx\right) = \dfrac{1}{6}\left(\ln\left|\dfrac{x - 2}{x + 2}\right| + \sqrt{2}\arctan\dfrac{x}{\sqrt{2}}\right) + C$$

**21.** $\dfrac{x^2 + 5}{(x + 1)(x^2 - 2x + 3)} = \dfrac{A}{x + 1} + \dfrac{Bx + C}{x^2 - 2x + 3}$

$$x^2 + 5 = A(x^2 - 2x + 3) + (Bx + C)(x + 1)$$
$$= (A + B)x^2 + (-2A + B + C)x + (3A + C)$$

When $x = -1, A = 1$.

By equating coefficients of like terms, you have $A + B = 1, -2A + B + C = 0, 3A + C = 5$.

Solving these equations you have $A = 1, B = 0, C = 2$.

$$\int \dfrac{x^2 + 5}{x^3 - x^2 + x + 3}\,dx = \int \dfrac{1}{x + 1}\,dx + 2\int \dfrac{1}{(x - 1)^2 + 2}\,dx = \ln|x + 1| + \sqrt{2}\arctan\left(\dfrac{x - 1}{\sqrt{2}}\right) + C$$

**23.** $\dfrac{3}{4x^2 + 5x + 1} = \dfrac{3}{(4x + 1)(x + 1)} = \dfrac{A}{4x + 1} + \dfrac{B}{x + 1}$

$$3 = A(x + 1) + B(4x + 1)$$

When $x = -1, 3 = -3B \Rightarrow B = -1$.

When $-\dfrac{1}{4}, 3 = \dfrac{3}{4}A \Rightarrow A = 4$.

$$\int_0^2 \dfrac{3}{4x^2 + 5x + 1}\,dx = \int_0^2 \dfrac{4}{4x + 1}\,dx + \int_0^2 \dfrac{-1}{x + 1}\,dx$$
$$= \Big[\ln|4x + 1| - \ln|x + 1|\Big]_0^2$$
$$= \ln 9 - \ln 3$$
$$= 2\ln 3 - \ln 3 = \ln 3$$

**25.** $\dfrac{x + 1}{x(x^2 + 1)} = \dfrac{A}{x} + \dfrac{Bx + C}{x^2 + 1}$

$$x + 1 = A(x^2 + 1) + (Bx + C)x$$

When $x = 0, A = 1$.

When $x = 1, 2 = 2A + B + C$.

When $x = -1, 0 = 2A + B - C$.

Solving these equations we have
$A = 1, B = -1, C = 1$.

$$\int_1^2 \dfrac{x + 1}{x(x^2 + 1)}\,dx = \int_1^2 \dfrac{1}{x}\,dx - \int_1^2 \dfrac{x}{x^2 + 1}\,dx + \int_1^2 \dfrac{1}{x^2 + 1}\,dx$$

$$= \left[\ln|x| - \dfrac{1}{2}\ln(x^2 + 1) + \arctan x\right]_1^2$$

$$= \dfrac{1}{2}\ln\dfrac{8}{5} - \dfrac{\pi}{4} + \arctan 2$$

$$\approx 0.557$$

**27.** Let $u = \cos x$, $du = -\sin x\, dx$.

$$\frac{1}{u(u + 1)} = \frac{A}{u} + \frac{B}{u + 1}$$

$$1 = A(u + 1) + Bu$$

When $u = 0$, $A = 1$.

When $u = -1$, $B = -1$.

$$\int \frac{\sin x}{\cos x + \cos^2 x}\, dx = -\int \frac{1}{u(u + 1)}\, du$$

$$= \int \frac{1}{u + 1}\, du - \int \frac{1}{u}\, du$$

$$= \ln|u + 1| - \ln|u| + C$$

$$= \ln\left|\frac{u + 1}{u}\right| + C$$

$$= \ln\left|\frac{\cos x + 1}{\cos x}\right| + C$$

$$= \ln|1 + \sec x| + C$$

**29.** Let $u = \tan x$, $du = \sec^2 x\, dx$.

$$\frac{1}{u^2 + 5u + 6} = \frac{1}{(u + 3)(u + 2)} = \frac{A}{u + 3} + \frac{B}{u + 2}$$

$$1 = A(u + 2) + B(u + 3)$$

When $u = -2$, $1 = B$.

When $u = -3$, $1 = -A \Rightarrow A = -1$.

$$\int \frac{\sec^2 x}{\tan^2 x + 5\tan x + 6}\, dx = \int \frac{1}{u^2 + 5u + 6}\, du$$

$$= \int \frac{-1}{u + 3}\, du + \int \frac{1}{u + 2}\, du$$

$$= -\ln|u + 3| + \ln|u + 2| + C$$

$$= \ln\left|\frac{\tan x + 2}{\tan x + 3}\right| + C$$

**31.** Let $u = e^x$, $du = e^x\, dx$.

$$\frac{1}{(u - 1)(u + 4)} = \frac{A}{u - 1} + \frac{B}{u + 4}$$

$$1 = A(u + 4) + B(u - 1)$$

When $u = 1$, $A = \frac{1}{5}$.

When $u = -4$, $B = -\frac{1}{5}$.

$$\int \frac{e^x}{(e^x - 1)(e^x + 4)}\, dx = \int \frac{1}{(u - 1)(u + 4)}\, du$$

$$= \frac{1}{5}\left(\int \frac{1}{u - 1}\, du - \int \frac{1}{u + 4}\, du\right)$$

$$= \frac{1}{5} \ln\left|\frac{u - 1}{u + 4}\right| + C$$

$$= \frac{1}{5} \ln\left|\frac{e^x - 1}{e^x + 4}\right| + C$$

**33.** Let $u = \sqrt{x}$, $u^2 = x$, $2u\, du = dx$.

$$\int \frac{\sqrt{x}}{x - 4}\, dx = \int \frac{u(2u)du}{u^2 - 4} = \int \left(\frac{2u^2 - 8}{u^2 - 4} + \frac{8}{u^2 - 4}\right) du = \int \left(2 + \frac{8}{u^2 - 4}\right) du$$

$$\frac{8}{u^2 - 4} = \frac{8}{(u - 2)(u + 2)} = \frac{A}{u - 2} + \frac{B}{u + 2}$$

$$8 = A(u + 2) + B(u - 2)$$

When $u = -2$, $8 = -4B \Rightarrow B = -2$.

When $u = 2$, $8 = 4A \Rightarrow A = 2$.

$$\int \left(2 + \frac{8}{u^2 - 4}\right) du = 2u + \int \left(\frac{2}{u - 2} - \frac{2}{u + 2}\right) du$$

$$= 2u + 2\ln|u - 2| - 2\ln|u + 2| + C$$

$$= 2\sqrt{x} + 2\ln\left|\frac{\sqrt{x} - 2}{\sqrt{x} + 2}\right| + C$$

**35.** $\dfrac{1}{x(a + bx)} = \dfrac{A}{x} + \dfrac{B}{a + bx}$

$\qquad\qquad 1 = A(a + bx) + Bx$

When $x = 0, 1 = aA \Rightarrow A = 1/a$.

When $x = -a/b, 1 = -(a/b)B \Rightarrow B = -b/a$.

$\displaystyle\int \frac{1}{x(a + bx)}\, dx = \frac{1}{a}\int \left(\frac{1}{x} - \frac{b}{a + bx}\right) dx$

$\qquad\qquad\qquad = \frac{1}{a}\big(\ln|x| - \ln|a + bx|\big) + C$

$\qquad\qquad\qquad = \frac{1}{a}\ln\left|\frac{x}{a + bx}\right| + C$

**37.** $\dfrac{x}{(a + bx)^2} = \dfrac{A}{a + bx} + \dfrac{B}{(a + bx)^2}$

$\qquad\qquad x = A(a + bx) + B$

When $x = -a/b, B = -a/b$.

When $x = 0, 0 = aA + B \Rightarrow A = 1/b$.

$\displaystyle\int \frac{x}{(a + bx)^2}\, dx = \int \left(\frac{1/b}{a + bx} + \frac{-a/b}{(a + bx)^2}\right) dx$

$\qquad\qquad\qquad = \frac{1}{b}\int \frac{1}{a + bx}\, dx - \frac{a}{b}\int \frac{1}{(a + bx)^2}\, dx$

$\qquad\qquad\qquad = \frac{1}{b^2}\ln|a + bx| + \frac{a}{b^2}\left(\frac{1}{a + bx}\right) + C$

$\qquad\qquad\qquad = \frac{1}{b^2}\left(\frac{a}{a + bx} + \ln|a + bx|\right) + C$

**45.** Average cost $= \dfrac{1}{80 - 75}\displaystyle\int_{75}^{80} \frac{124p}{(10 + p)(100 - p)}\, dp$

$\qquad\qquad = \dfrac{1}{5}\displaystyle\int_{75}^{80} \left(\frac{-124}{(10 + p)11} + \frac{1240}{(100 - p)11}\right) dp$

$\qquad\qquad = \dfrac{1}{5}\left[\frac{-124}{11}\ln(10 + p) - \frac{1240}{11}\ln(100 - p)\right]_{75}^{80}$

$\qquad\qquad \approx \dfrac{1}{5}(24.51) = 4.9$

Approximately \$490,000

**39.** Dividing $x^3$ by $x - 5$

**41.** (a) Substitution: $u = x^2 + 2x - 8$

(b) Partial fractions

(c) Trigonometric substitution (tan) or inverse tangent rule

**43.** $\qquad\qquad A = \displaystyle\int_0^1 \frac{12}{x^2 + 5x + 6}\, dx$

$\dfrac{12}{x^2 + 5x + 6} = \dfrac{12}{(x + 2)(x + 3)} = \dfrac{A}{x + 2} + \dfrac{B}{x + 3}$

$\qquad\qquad 12 = A(x + 3) + B(x + 2)$

Let $x = -3$: $12 = B(-1) \Rightarrow B = -12$

Let $x = -2$: $12 = A(1) \Rightarrow A = 12$

$A = \displaystyle\int_0^1 \left(\frac{12}{x + 2} - \frac{12}{x + 3}\right) dx$

$\qquad = \big[12\ln|x + 2| - 12\ln|x + 3|\big]_0^1$

$\qquad = 12(\ln 3 - \ln 4 - \ln 2 + \ln 3)$

$\qquad = 12\ln\left(\dfrac{9}{8}\right) \approx 1.4134$

**47.** $V = \pi \int_0^3 \left(\dfrac{2x}{x^2+1}\right)^2 dx = 4\pi \int_0^3 \dfrac{x^2}{\left(x^2+1\right)^2} dx$

$$= 4\pi \int_0^3 \left(\dfrac{1}{x^2+1} - \dfrac{1}{\left(x^2+1\right)^2}\right) dx \qquad \text{(partial fractions)}$$

$$= 4\pi \left[\arctan x - \dfrac{1}{2}\left(\arctan x + \dfrac{x}{x^2+1}\right)\right]_0^3 \qquad \text{(trigonometric substitution)}$$

$$= 2\pi \left[\arctan x - \dfrac{x}{x^2+1}\right]_0^3 = 2\pi \left(\arctan 3 - \dfrac{3}{10}\right) \approx 5.963$$

$A = \int_0^3 \dfrac{2x}{x^2+1} dx = \left[\ln\left(x^2+1\right)\right]_0^3 = \ln 10$

$\bar{x} = \dfrac{1}{A}\int_0^3 \dfrac{2x^2}{x^2+1} dx = \dfrac{1}{\ln 10}\int_0^3 \left(2 - \dfrac{2}{x^2+1}\right) dx = \dfrac{1}{\ln 10}\left[2x - 2\arctan x\right]_0^3 = \dfrac{2}{\ln 10}(3 - \arctan 3) \approx 1.521$

$\bar{y} = \dfrac{1}{A}\left(\dfrac{1}{2}\right)\int_0^3 \left(\dfrac{2x}{x^2+1}\right)^2 dx = \dfrac{2}{\ln 10}\int_0^3 \dfrac{x^2}{\left(x^2+1\right)^2} dx$

$$= \dfrac{2}{\ln 10}\int_0^3 \left(\dfrac{1}{x^2+1} - \dfrac{1}{\left(x^2+1\right)^2}\right) dx \qquad \text{(partial fractions)}$$

$$= \dfrac{2}{\ln 10}\left[\arctan x - \dfrac{1}{2}\left(\arctan x + \dfrac{x}{x^2+1}\right)\right]_0^3 \qquad \text{(trigonometric substitution)}$$

$$= \dfrac{2}{\ln 10}\left[\dfrac{1}{2}\arctan x - \dfrac{x}{2\left(x^2+1\right)}\right]_0^3 = \dfrac{1}{\ln 10}\left[\arctan x - \dfrac{x}{x^2+1}\right]_0^3 = \dfrac{1}{\ln 10}\left(\arctan 3 - \dfrac{3}{10}\right) \approx 0.412$$

$\left(\bar{x}, \bar{y}\right) \approx \left(1.521, 0.412\right)$

**49.**
$$\frac{1}{(x+1)(n-x)} = \frac{A}{x+1} + \frac{B}{n-x}, A = B = \frac{1}{n+1}$$

$$\frac{1}{n+1}\int\left(\frac{1}{x+1} + \frac{1}{n-x}\right)dx = kt + C$$

$$\frac{1}{n+1}\ln\left|\frac{x+1}{n-x}\right| = kt + C$$

When $t = 0, x = 0, C = \frac{1}{n+1}\ln\frac{1}{n}$.

$$\frac{1}{n+1}\ln\left|\frac{x+1}{n-x}\right| = kt + \frac{1}{n+1}\ln\frac{1}{n}$$

$$\frac{1}{n+1}\left[\ln\left|\frac{x+1}{n-x}\right| - \ln\frac{1}{n}\right] = kt$$

$$\ln\frac{nx+n}{n-x} = (n+1)kt$$

$$\frac{nx+n}{n-x} = e^{(n+1)kt}$$

$$x = \frac{n\left[e^{(n+1)kt} - 1\right]}{n + e^{(n+1)kt}} \qquad \textbf{Note: } \lim_{t\to\infty} x = n$$

**51.** $\dfrac{x}{1+x^4} = \dfrac{Ax+B}{x^2+\sqrt{2}x+1} + \dfrac{Cx+D}{x^2-\sqrt{2}x+1}$

$x = (Ax+B)(x^2 - \sqrt{2}x + 1) + (Cx+D)(x^2 + \sqrt{2}x + 1)$

$= (A+C)x^3 + (B+D - \sqrt{2}A + \sqrt{2}C)x^2 + (A+C - \sqrt{2}B + \sqrt{2}D)x + (B+D)$

$0 = A + C \Rightarrow C = -A$

$0 = B + D - \sqrt{2}A + \sqrt{2}C \qquad -2\sqrt{2}A = 0 \Rightarrow A = 0 \text{ and } C = 0$

$1 = A + C - \sqrt{2}B + \sqrt{2}D \qquad -2\sqrt{2}B = 1 \Rightarrow B = -\dfrac{\sqrt{2}}{4} \text{ and } D = \dfrac{\sqrt{2}}{4}$

$0 = B + D \Rightarrow D = -B$

So,

$$\int_0^1 \frac{x}{1+x^4}\,dx = \int_0^1\left(\frac{-\sqrt{2}/4}{x^2+\sqrt{2}x+1} + \frac{\sqrt{2}/4}{x^2-\sqrt{2}x+1}\right)dx$$

$$= \frac{\sqrt{2}}{4}\int_0^1\left[\frac{-1}{\left[x+\left(\sqrt{2}/2\right)\right]^2 + (1/2)} + \frac{1}{\left[x-\left(\sqrt{2}/2\right)\right]^2 + (1/2)}\right]dx$$

$$= \frac{\sqrt{2}}{4}\cdot\frac{1}{1/\sqrt{2}}\left[-\arctan\left(\frac{x+\left(\sqrt{2}/2\right)}{1/\sqrt{2}}\right) + \arctan\left(\frac{x-\left(\sqrt{2}/2\right)}{1/\sqrt{2}}\right)\right]_0^1$$

$$= \frac{1}{2}\left[-\arctan\left(\sqrt{2}x+1\right) + \arctan\left(\sqrt{2}x-1\right)\right]_0^1$$

$$= \frac{1}{2}\left[\left(-\arctan\left(\sqrt{2}+1\right) + \arctan\left(\sqrt{2}-1\right)\right) - \left(-\arctan 1 + \arctan(-1)\right)\right]$$

$$= \frac{1}{2}\left[\arctan\left(\sqrt{2}-1\right) - \arctan\left(\sqrt{2}+1\right) + \frac{\pi}{4} + \frac{\pi}{4}\right].$$

Because $\arctan x - \arctan y = \arctan\left[(x-y)/(1+xy)\right]$, you have:

$$\int_0^1 \frac{x}{1+x^4}\,dx = \frac{1}{2}\left[\arctan\left(\frac{\left(\sqrt{2}-1\right) - \left(\sqrt{2}+1\right)}{1 + \left(\sqrt{2}-1\right)\left(\sqrt{2}+1\right)}\right) + \frac{\pi}{2}\right] = \frac{1}{2}\left[\arctan\left(\frac{-2}{2}\right) + \frac{\pi}{2}\right] = \frac{1}{2}\left(-\frac{\pi}{4} + \frac{\pi}{2}\right) = \frac{\pi}{8}$$

## Section 8.6   Integration by Tables and Other Integration Techniques

**1.** By Formula 6: $(a = 5, b = 1)$

$$\int \frac{x^2}{5 + x}\, dx = \left[ -\frac{x}{2}(10 - x) + 25 \ln|5 + x| \right] + C$$

**3.** By Formula 44: $\displaystyle\int \frac{1}{x^2\sqrt{1 - x^2}}\, dx = -\frac{\sqrt{1 - x^2}}{x} + C$

**5.** By Formulas 51 and 49:

$$\int \cos^4 3x\, dx = \frac{1}{3}\int \cos^4 3x\,(3)\, dx$$

$$= \frac{1}{3}\left[ \frac{\cos^3 3x \sin 3x}{4} + \frac{3}{4}\int \cos^2 3x\, dx \right]$$

$$= \frac{1}{12}\cos^3 3x \sin 3x + \frac{1}{4}\cdot\frac{1}{3}\int \cos^2 3x\,(3)\, dx$$

$$= \frac{1}{12}\cos^3 3x \sin 3x + \frac{1}{12}\cdot\frac{1}{2}(3x + \sin 3x \cos 3x) + C$$

$$= \frac{1}{24}\left( 2\cos^3 3x \sin 3x + 3x + \sin 3x \cos 3x \right) + C$$

**7.** By Formula 57: $\displaystyle\int \frac{1}{\sqrt{x}\left(1 - \cos \sqrt{x}\right)}\, dx = 2\int \frac{1}{1 - \cos \sqrt{x}}\left( \frac{1}{2\sqrt{x}} \right) dx = -2\left( \cot \sqrt{x} + \csc \sqrt{x} \right) + C$

$$u = \sqrt{x},\, du = \frac{1}{2\sqrt{x}}\, dx$$

**9.** By Formula 84:

$$\int \frac{1}{1 + e^{2x}}\, dx = 2x - \frac{1}{2}\ln\left(1 + e^{2x}\right) + C$$

**11.** By Formula 89: $(n = 7)$

$$\int x^7 \ln x\, dx = \frac{x^8}{64}\left[-1 + 8\ln x\right] + C = \frac{1}{64}x^8\left(8\ln x - 1\right) + C$$

**13.** (a)  Let $u = 3x$, $x = \dfrac{u}{3}$, $du = 3\, dx$.

$$\int x^2 e^{3x}\, dx = \int \left( \frac{u}{3} \right)^2 e^u \frac{1}{3}\, du = \frac{1}{27}\int u^2 e^u\, du$$

By Formulas 83 and 82:

$$\int x^2 e^{3x}\, dx = \frac{1}{27}\left[ u^2 e^u - 2\int u e^u\, du \right] = \frac{1}{27}\left[ u^2 e^u - 2\left((u - 1)e^u\right) \right] + C = \frac{1}{27}e^{3x}\left(9x^2 - 6x + 2\right) + C$$

(b)  Integration by parts: $u = x^2$, $du = 2x\, dx$, $dv = e^{3x}\, dx$, $v = \dfrac{1}{3}e^{3x}$

$$\int x^2 e^{3x}\, dx = x^2 \frac{1}{3}e^{3x} - \int \frac{2}{3}x e^{3x}\, dx$$

Parts again: $u = x$, $du = dx$, $dv = e^{3x}$, $v = \dfrac{1}{3}e^{3x}$

$$\int x^2 e^{3x}\, dx = \frac{1}{3}x^2 e^{3x} - \frac{2}{3}\left[ \frac{x}{3}e^{3x} - \int \frac{1}{3}e^{3x}\, dx \right] = \frac{1}{3}x^2 e^{3x} - \frac{2}{9}x e^{3x} + \frac{2}{27}e^{3x} + C = \frac{1}{27}e^{3x}\left[9x^2 - 6x + 2\right] + C$$

**15. (a)** By Formula 12: $(a = b = 1, u = x)$

$$\int \frac{1}{x^2(x+1)}\, dx = \frac{-1}{1}\left(\frac{1}{x} + \frac{1}{1}\ln\left|\frac{x}{1+x}\right|\right) + C$$

$$= \frac{-1}{x} - \ln\left|\frac{x}{1+x}\right| + C$$

$$= \frac{-1}{x} + \ln\left|\frac{x+1}{x}\right| + C$$

**(b)** Partial fractions:

$$\frac{1}{x^2(x+1)} = \frac{A}{x} + \frac{B}{x^2} + \frac{C}{x+1}$$

$$1 = Ax(x+1) + B(x+1) + Cx^2$$

$$x = 0: 1 = B$$

$$x = -1: 1 = C$$

$$x = 1: 1 = 2A + 2 + 1 \Rightarrow A = -1$$

$$\int \frac{1}{x^2(x+1)}\, dx = \int\left[\frac{-1}{x} + \frac{1}{x^2} + \frac{1}{x+1}\right] dx$$

$$= -\ln|x| - \frac{1}{x} + \ln|x+1| + C$$

$$= -\frac{1}{x} - \ln\left|\frac{x}{x+1}\right| + C$$

**17.** By Formula 80:

$$\int x\, \text{arccsc}(x^2 + 1)\, dx = \frac{1}{2}\int \text{arccsc}(x^2+1)(2x)\, dx$$

$$= \frac{1}{2}\left[(x^2+1)\text{arccsc}(x^2+1) + \ln\left|x^2 + 1 + \sqrt{(x^2+1)^2 - 1}\right|\right] + C$$

$$= \frac{1}{2}(x^2+1)\text{arccsc}(x^2+1) + \frac{1}{2}\ln\left(x^2 + 1 + \sqrt{x^4 + 2x^2}\right) + C$$

**19.** By Formula 35: $\int \dfrac{1}{x^2\sqrt{x^2-4}}\, dx = \dfrac{\sqrt{x^2-4}}{4x} + C$

**21.** By Formula 4: $(a = 2, b = -5)$

$$\int \frac{4x}{(2-5x)^2}\, dx = 4\left[\frac{1}{25}\left(\frac{2}{2-5x} + \ln|2-5x|\right)\right] + C$$

$$= \frac{4}{25}\left(\frac{2}{2-5x} + \ln|2-5x|\right) + C$$

**23.** By Formula 76:

$$\int e^x \arccos e^x\, dx = e^x \arccos e^x - \sqrt{1 - e^{2x}} + C$$

$$u = e^x,\, du = e^x\, dx$$

**25.** By Formula 73:

$$\int \frac{x}{1 - \sec x^2}\, dx = \frac{1}{2}\int \frac{2x}{1 - \sec x^2}\, dx$$

$$= \frac{1}{2}(x^2 + \cot x^2 + \csc x^2) + C$$

**27.** By Formula 14: $\int \dfrac{\cos\theta}{3 + 2\sin\theta + \sin^2\theta}\, d\theta = \dfrac{\sqrt{2}}{2}\arctan\left(\dfrac{1+\sin\theta}{\sqrt{2}}\right) + C$   $(b^2 = 4 < 12 = 4ac)$

$$u = \sin\theta,\, du = \cos\theta\, d\theta$$

**29.** By Formula 35: $\int \dfrac{1}{x^2\sqrt{2 + 9x^2}}\, dx = 3\int \dfrac{3}{(3x)^2\sqrt{(\sqrt{2})^2 + (3x)^2}}\, dx = -\dfrac{3\sqrt{2 + 9x^2}}{6x} + C = -\dfrac{\sqrt{2 + 9x^2}}{2x} + C$

**31.** By Formula 3: $\int \dfrac{\ln x}{x(3 + 2\ln x)}\,dx = \dfrac{1}{4}\Big(2\ln|x| - 3\ln\big|3 + 2\ln|x|\big|\Big) + C$

$$u = \ln x, \, du = \frac{1}{x}\,dx$$

**33.** By Formulas 1, 23, and 35: $\int \dfrac{x}{\left(x^2 - 6x + 10\right)^2}\,dx = \dfrac{1}{2}\int \dfrac{2x - 6 + 6}{\left(x^2 - 6x + 10\right)^2}\,dx$

$$= \frac{1}{2}\int \left(x^2 - 6x + 10\right)^{-2}(2x - 6)\,dx + 3\int \frac{1}{\left[(x - 3)^2 + 1\right]^2}\,dx$$

$$= -\frac{1}{2\left(x^2 - 6x + 10\right)} + \frac{3}{2}\left[\frac{x - 3}{x^2 - 6x + 10} + \arctan(x - 3)\right] + C$$

$$= \frac{3x - 10}{2\left(x^2 - 6x + 10\right)} + \frac{3}{2}\arctan(x - 3) + C$$

**35.** By Formula 31: $\int \dfrac{x}{\sqrt{x^4 - 6x^2 + 5}}\,dx = \dfrac{1}{2}\int \dfrac{2x}{\sqrt{\left(x^2 - 3\right)^2 - 4}}\,dx = \dfrac{1}{2}\ln\left|x^2 - 3 + \sqrt{x^4 - 6x^2 + 5}\right| + C$

$$u = x^2 - 3, \, du = 2x\,dx$$

**37.** By Formula 8:

$$\int \frac{e^{3x}}{\left(1 + e^x\right)^3}\,dx = \int \frac{\left(e^x\right)^2}{\left(1 + e^x\right)^3}\left(e^x\right)\,dx$$

$$= \frac{2}{1 + e^x} - \frac{1}{2\left(1 + e^x\right)^2} + \ln\left|1 + e^x\right| + C$$

$$u = e^x, \, du = e^x\,dx$$

**39.** By Formula 81:

$$\int_0^1 xe^{x^2}\,dx = \left[\tfrac{1}{2}e^{x^2}\right]_0^1 = \tfrac{1}{2}(e - 1) \approx 0.8591$$

**41.** By Formula 89: $(n = 4)$

$$\int_1^2 x^4 \ln x\,dx = \left[\frac{x^5}{25}(-1 + 5\ln x)\right]_1^2$$

$$= \frac{32}{25}[-1 + 5\ln 2] - \frac{1}{25}[-1 + 0]$$

$$= -\frac{31}{25} + \frac{32}{5}\ln 2 \approx 3.1961$$

**43.** By Formula 23, and letting $u = \sin x$:

$$\int_{-\pi/2}^{\pi/2} \frac{\cos x}{1 + \sin^2 x}\,dx = \left[\arctan(\sin x)\right]_{-\pi/2}^{\pi/2}$$

$$= \arctan(1) - \arctan(-1) = \frac{\pi}{2}$$

**45.** By Formulas 54 and 55:

$$\int t^3 \cos t\,dt = t^3 \sin t - 3\int t^2 \sin t\,dt$$

$$= t^3 \sin t - 3\left(-t^2 \cos t + 2\int t \cos t\,dt\right)$$

$$= t^3 \sin t + 3t^2 \cos t - 6\left(t \sin t - \int \sin t\,dt\right)$$

$$= t^3 \sin t + 3t^2 \cos t - 6t \sin t - 6\cos t + C$$

So,

$$\int_0^{\pi/2} t^3 \cos t\,dt = \left[t^3 \sin t + 3t^2 \cos t - 6t \sin t - 6\cos t\right]_0^{\pi/2}$$

$$= \left(\frac{\pi^3}{8} - 3\pi\right) + 6 = \frac{\pi^3}{8} + 6 - 3\pi \approx 0.4510.$$

**47.** $\dfrac{u^2}{(a+bu)^2} = \dfrac{1}{b^2} - \dfrac{(2a/b)u + (a^2/b^2)}{(a+bu)^2} = \dfrac{1}{b^2} + \dfrac{A}{a+bu} + \dfrac{B}{(a+bu)^2}$

$-\dfrac{2a}{b}u - \dfrac{a^2}{b^2} = A(a+bu) + B = (aA+B) + bAu$

Equating the coefficients of like terms you have $aA + B = -a^2/b^2$ and $bA = -2a/b$. Solving these equations you have $A = -2a/b^2$ and $B = a^2/b^2$.

$\displaystyle\int \dfrac{u^2}{(a+bu)^2}\,du = \dfrac{1}{b^2}\int du - \dfrac{2a}{b^2}\left(\dfrac{1}{b}\right)\int \dfrac{1}{a+bu}b\,du + \dfrac{a^2}{b^2}\left(\dfrac{1}{b}\right)\int \dfrac{1}{(a+bu)^2}b\,du = \dfrac{1}{b^2}u - \dfrac{2a}{b^3}\ln|a+bu| - \dfrac{a^2}{b^3}\left(\dfrac{1}{a+bu}\right) + C$

$= \dfrac{1}{b^3}\left(bu - \dfrac{a^2}{a+bu} - 2a\ln|a+bu|\right) + C$

**49.** When you have $u^2 + a^2$:

$u = a\tan\theta$

$du = a\sec^2\theta\,d\theta$

$u^2 + a^2 = a^2\sec^2\theta$

$\displaystyle\int \dfrac{1}{(u^2+a^2)^{3/2}}\,du = \int \dfrac{a\sec^2\theta\,d\theta}{a^3\sec^3\theta} = \dfrac{1}{a^2}\int\cos\theta\,d\theta = \dfrac{1}{a^2}\sin\theta + C = \dfrac{u}{a^2\sqrt{u^2+a^2}} + C$

When you have $u^2 - a^2$:

$u = a\sec\theta$

$du = a\sec\theta\tan\theta\,d\theta$

$u^2 - a^2 = a^2\tan^2\theta$

$\displaystyle\int \dfrac{1}{(u^2-a^2)^{3/2}}\,du = \int \dfrac{a\sec\theta\tan\theta\,d\theta}{a^3\tan^3\theta} = \dfrac{1}{a^2}\int \dfrac{\cos\theta}{\sin^2\theta}\,d\theta = \dfrac{1}{a^2}\int\csc\theta\cot\theta\,d\theta = -\dfrac{1}{a^2}\csc\theta + C = \dfrac{-u}{a^2\sqrt{u^2-a^2}} + C$

**51.** $\displaystyle\int(\arctan u)\,du = u\arctan u - \dfrac{1}{2}\int \dfrac{2u}{1+u^2}\,du$

$= u\arctan u - \dfrac{1}{2}\ln(1+u^2) + C$

$= u\arctan u - \ln\sqrt{1+u^2} + C$

$w = \arctan u,\ dv = du,\ dw = \dfrac{du}{1+u^2},\ v = u$

**53.** $\displaystyle\int \dfrac{1}{2 - 3\sin\theta}\,d\theta = \int\left[\dfrac{\dfrac{2\,du}{1+u^2}}{2 - 3\left(\dfrac{2u}{1+u^2}\right)}\right],\ u = \tan\dfrac{\theta}{2}$

$= \displaystyle\int \dfrac{2}{2(1+u^2) - 6u}\,du$

$= \displaystyle\int \dfrac{1}{u^2 - 3u + 1}\,du$

$= \displaystyle\int \dfrac{1}{\left(u - \dfrac{3}{2}\right)^2 - \dfrac{5}{4}}\,du$

$= \dfrac{1}{\sqrt{5}}\ln\left|\dfrac{\left(u - \dfrac{3}{2}\right) - \dfrac{\sqrt{5}}{2}}{\left(u - \dfrac{3}{2}\right) + \dfrac{\sqrt{5}}{2}}\right| + C$

$= \dfrac{1}{\sqrt{5}}\ln\left|\dfrac{2u - 3 - \sqrt{5}}{2u - 3 + \sqrt{5}}\right| + C$

$= \dfrac{1}{\sqrt{5}}\ln\left|\dfrac{2\tan\left(\dfrac{\theta}{2}\right) - 3 - \sqrt{5}}{2\tan\left(\dfrac{\theta}{2}\right) - 3 + \sqrt{5}}\right| + C$

**55.** $\displaystyle\int_0^{\pi/2} \frac{1}{1 + \sin\theta + \cos\theta}\,d\theta = \int_0^1 \left[ \dfrac{\dfrac{2\,du}{1 + u^2}}{1 + \dfrac{2u}{1 + u^2} + \dfrac{1 - u^2}{1 + u^2}} \right]$

$$= \int_0^1 \frac{1}{1 + u}\,du$$

$$= \Big[\ln|1 + u|\Big]_0^1$$

$$= \ln 2$$

$$u = \tan\frac{\theta}{2}$$

**57.** $\displaystyle\int \frac{\sin\theta}{3 - 2\cos\theta}\,d\theta = \frac{1}{2}\int \frac{2\sin\theta}{3 - 2\cos\theta}\,d\theta$

$$= \frac{1}{2}\ln|u| + C$$

$$= \frac{1}{2}\ln(3 - 2\cos\theta) + C$$

$$u = 3 - 2\cos\theta,\ du = 2\sin\theta\,d\theta$$

**59.** $\displaystyle\int \frac{\sin\sqrt{\theta}}{\sqrt{\theta}}\,d\theta = 2\int \sin\sqrt{\theta}\left(\frac{1}{2\sqrt{\theta}}\right)d\theta$

$$= -2\cos\sqrt{\theta} + C$$

$$u = \sqrt{\theta},\ du = \frac{1}{2\sqrt{\theta}}\,d\theta$$

**61.** By Formula 21: $(a = 3,\ b = 1)$

$$A = \int_0^6 \frac{x}{\sqrt{x + 3}}\,dx = \left[ \frac{-2(6 - x)}{3}\sqrt{x + 3} \right]_0^6$$

$$= 4\sqrt{3} \approx 6.928 \text{ square units}$$

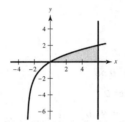

**63.** (a) $n = 1$: $u = \ln x,\ du = \dfrac{1}{x}\,dx,\ dv = x\,dx,\ v = \dfrac{x^2}{2}$

$$\int x\ln x\,dx = \frac{x^2}{2}\ln x - \int \left(\frac{x^2}{2}\right)\frac{1}{x}\,dx = \frac{x^2}{2}\ln x - \frac{x^2}{4} + C$$

$n = 2$: $u = \ln x,\ du = \dfrac{1}{x}\,dx,\ dv = x^2\,dx,\ v = \dfrac{x^3}{3}$

$$\int x^2\ln x\,dx = \frac{x^3}{3}\ln x - \int \left(\frac{x^3}{3}\right)\frac{1}{x}\,dx = \frac{x^3}{3}\ln x - \frac{x^3}{9} + C$$

$n = 3$: $u = \ln x,\ du = \dfrac{1}{x}\,dx,\ dv = x^3\,dx,\ v = \dfrac{x^4}{4}$

$$\int x^3\ln x\,dx = \frac{x^4}{4}\ln x - \int \left(\frac{x^4}{4}\right)\frac{1}{x}\,dx = \frac{x^4}{4}\ln x - \frac{x^4}{16} + C$$

(b) $\displaystyle\int x^n\ln x\,dx = \frac{x^{n+1}}{n + 1}\ln x - \frac{x^{n+1}}{(n + 1)^2} + C$

**65.** (a) Arctangent Formula, Formula 23,

$$\int \frac{1}{u^2 + 1}\,du,\ u = e^x$$

(b) Log Rule: $\displaystyle\int \frac{1}{u}\,du,\ u = e^x + 1$

(c) Substitution: $u = x^2,\ du = 2x\,dx$, then Formula 81

(d) Integration by parts

(e) Cannot be integrated.

(f) Formula 16 with $u = e^{2x}$

**67.** False. You might need to convert your integral using substitution or algebra.

**69.** $W = \displaystyle\int_0^5 2000xe^{-x}\,dx$

$$= -2000\int_0^5 -xe^{-x}\,dx$$

$$= 2000\int_0^5 (-x)e^{-x}(-1)\,dx$$

$$= 2000\Big[(-x)e^{-x} - e^{-x}\Big]_0^5$$

$$= 2000\left(-\frac{6}{e^5} + 1\right)$$

$$\approx 1919.145 \text{ ft-lb}$$

**71.**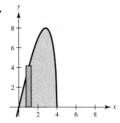

$$V = 2\pi \int_0^4 x\left(x\sqrt{16 - x^2}\right) dx$$

$$= 2\pi \int_0^4 x^2\sqrt{16 - x^2}\, dx$$

By Formula 38: $(a = 4)$

$$V = 2\pi\left[\frac{1}{8}\left(x\left(2x^2 - 16\right)\sqrt{16 - x^2} + 256\arcsin\left(\frac{x}{4}\right)\right)\right]_0^4$$

$$= 2\pi\left[32\left(\frac{\pi}{2}\right)\right] = 32\pi^2$$

**73.** $\dfrac{1}{2 - 0}\displaystyle\int_0^2 \dfrac{5000}{1 + e^{4.8-1.9t}}\, dt = \dfrac{2500}{-1.9}\displaystyle\int_0^2 \dfrac{-1.9\, dt}{1 + e^{4.8-1.9t}}$

$$= -\frac{2500}{1.9}\left[(4.8 - 1.9t) - \ln\left(1 + e^{4.8-1.9t}\right)\right]_0^2$$

$$= -\frac{2500}{1.9}\left[\left(1 - \ln(1 + e)\right) - \left(4.8 - \ln\left(1 + e^{4.8}\right)\right)\right]$$

$$= \frac{2500}{1.9}\left[3.8 + \ln\left(\frac{1 + e}{1 + e^{4.8}}\right)\right] \approx 401.4$$

## Section 8.7   Indeterminate Forms and L'Hôpital's Rule

**1.** $\displaystyle\lim_{x\to 0}\frac{\sin 4x}{\sin 3x} \approx 1.3333\ \left(\text{exact: }\frac{4}{3}\right)$

| $x$ | $-0.1$ | $-0.01$ | $-0.001$ | $0.001$ | $0.01$ | $0.1$ |
|---|---|---|---|---|---|---|
| $f(x)$ | 1.3177 | 1.3332 | 1.3333 | 1.3333 | 1.3332 | 1.3177 |

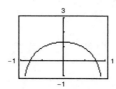

**3.** $\displaystyle\lim_{x\to\infty} x^5 e^{-x/100} \approx 0$

| $x$ | $1$ | $10$ | $10^2$ | $10^3$ | $10^4$ | $10^5$ |
|---|---|---|---|---|---|---|
| $f(x)$ | 0.9900 | 90,484 | $3.7 \times 10^9$ | $4.5 \times 10^{10}$ | 0 | 0 |

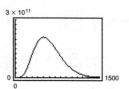

**5.** (a) $\displaystyle\lim_{x\to 4}\frac{3(x - 4)}{x^2 - 16} = \lim_{x\to 4}\frac{3(x - 4)}{(x - 4)(x + 4)} = \lim_{x\to 4}\frac{3}{x + 4} = \frac{3}{8}$

    (b) $\displaystyle\lim_{x\to 4}\frac{3(x - 4)}{x^2 - 16} = \lim_{x\to 4}\frac{d/dx\left[3(x - 4)\right]}{d/dx\left[x^2 - 16\right]} = \lim_{x\to 4}\frac{3}{2x} = \frac{3}{8}$

7. (a) $\lim\limits_{x \to 6} \dfrac{\sqrt{x + 10} - 4}{x - 6} = \lim\limits_{x \to 6} \dfrac{\sqrt{x + 10} - 4}{x - 6} \cdot \dfrac{\sqrt{x + 10} + 4}{\sqrt{x + 10} + 4} = \lim\limits_{x \to 6} \dfrac{(x + 10) - 16}{(x - 6)\left(\sqrt{x + 10} + 4\right)} = \lim\limits_{x \to 6} \dfrac{1}{\sqrt{x + 10} + 4} = \dfrac{1}{8}$

(b) $\lim\limits_{x \to 6} \dfrac{\sqrt{x + 10} - 4}{x - 6} = \lim\limits_{x \to 6} \dfrac{d/dx\left[\sqrt{x + 10} - 4\right]}{d/dx[x - 6]} = \lim\limits_{x \to 6} \dfrac{\frac{1}{2}(x + 10)^{-1/2}}{1} = 1/8$

9. (a) $\lim\limits_{x \to \infty} \dfrac{5x^2 - 3x + 1}{3x^2 - 5} = \lim\limits_{x \to \infty} \dfrac{5 - (3/x) + \left(1/x^2\right)}{3 - \left(5/x^2\right)} = \dfrac{5}{3}$

(b) $\lim\limits_{x \to \infty} \dfrac{5x^2 - 3x + 1}{3x^2 - 5} = \lim\limits_{x \to \infty} \dfrac{(d/dx)\left[5x^2 - 3x + 1\right]}{(d/dx)\left[3x^2 - 5\right]} = \lim\limits_{x \to \infty} \dfrac{10x - 3}{6x} = \lim\limits_{x \to \infty} \dfrac{(d/dx)[10x - 3]}{(d/dx)[6x]} = \lim\limits_{x \to \infty} \dfrac{10}{6} = \dfrac{5}{3}$

11. $\lim\limits_{x \to 3} \dfrac{x^2 - 2x - 3}{x - 3} = \lim\limits_{x \to 3} \dfrac{2x - 2}{1} = 4$

13. $\lim\limits_{x \to 0} \dfrac{\sqrt{25 - x^2} - 5}{x} = \lim\limits_{x \to 0} \dfrac{\frac{1}{2}\left(25 - x^2\right)^{-1/2}(-2x)}{1}$

$= \lim\limits_{x \to 0} \dfrac{-x}{\sqrt{25 - x^2}} = 0$

15. $\lim\limits_{x \to 0^+} \dfrac{e^x - (1 + x)}{x^3} = \lim\limits_{x \to 0^+} \dfrac{e^x - 1}{3x^2} = \lim\limits_{x \to 0^+} \dfrac{e^x}{6x} = \infty$

17. $\lim\limits_{x \to 1} \dfrac{x^{11} - 1}{x^4 - 1} = \lim\limits_{x \to 1} \dfrac{11x^{10}}{4x^3} = \dfrac{11}{4}$

19. $\lim\limits_{x \to 0} \dfrac{\sin 3x}{\sin 5x} = \lim\limits_{x \to 0} \dfrac{3 \cos 3x}{5 \cos 5x} = \dfrac{3}{5}$

21. $\lim\limits_{x \to 0} \dfrac{\arcsin x}{x} = \lim\limits_{x \to 0} \dfrac{1/\sqrt{1 - x^2}}{1} = 1$

23. $\lim\limits_{x \to \infty} \dfrac{5x^2 + 3x - 1}{4x^2 + 5} = \lim\limits_{x \to \infty} \dfrac{10x + 3}{8x} = \lim\limits_{x \to \infty} \dfrac{10}{8} = \dfrac{5}{4}$

25. $\lim\limits_{x \to \infty} \dfrac{x^2 + 4x + 7}{x - 6} = \lim\limits_{x \to \infty} \dfrac{2x + 4}{1} = \infty$

27. $\lim\limits_{x \to \infty} \dfrac{x^3}{e^{x/2}} = \lim\limits_{x \to \infty} \dfrac{3x^2}{(1/2)e^{x/2}}$

$= \lim\limits_{x \to \infty} \dfrac{6x}{(1/4)e^{x/2}} = \lim\limits_{x \to \infty} \dfrac{6}{(1/8)e^{x/2}} = 0$

29. $\lim\limits_{x \to \infty} \dfrac{x}{\sqrt{x^2 + 1}} = \lim\limits_{x \to \infty} \dfrac{1}{\sqrt{1 + \left(1/x^2\right)}} = 1$

**Note:** L'Hôpital's Rule does not work on this limit. See Exercise 83.

31. $\lim\limits_{x \to \infty} \dfrac{\cos x}{x} = 0$ by Squeeze Theorem

$\left(\dfrac{\cos x}{x} \le \dfrac{1}{x}, \text{ for } x > 0\right)$

33. $\lim\limits_{x \to \infty} \dfrac{\ln x}{x^2} = \lim\limits_{x \to \infty} \dfrac{1/x}{2x} = \lim\limits_{x \to \infty} \dfrac{1}{2x^2} = 0$

35. $\lim\limits_{x \to \infty} \dfrac{e^x}{x^4} = \lim\limits_{x \to \infty} \dfrac{e^x}{4x^3}$

$= \lim\limits_{x \to \infty} \dfrac{e^x}{12x^2}$

$= \lim\limits_{x \to \infty} \dfrac{e^x}{24x}$

$= \lim\limits_{x \to \infty} \dfrac{e^x}{24} = \infty$

37. $\lim\limits_{x \to 0} \dfrac{\sin 5x}{\tan 9x} = \lim\limits_{x \to 0} \dfrac{5 \cos 5x}{9 \sec^2 9x} = \dfrac{5}{9}$

39. $\lim\limits_{x \to 0} \dfrac{\arctan x}{\sin x} = \lim\limits_{x \to 0} \dfrac{1/\left(1 + x^2\right)}{\cos x} = 1$

41. $\lim\limits_{x \to \infty} \dfrac{\int_1^x \ln\left(e^{4t - 1}\right) dt}{x}$

$= \lim\limits_{x \to \infty} \dfrac{\int_1^x (4t - 1)\, dt}{x}$

$= \lim\limits_{x \to \infty} \dfrac{4x - 1}{1} = \infty$

43. (a) $\lim\limits_{x \to \infty} x \ln x$, not indeterminate

(b) $\lim\limits_{x \to \infty} x \ln x = (\infty)(\infty) = \infty$

(c)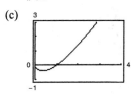

**45.** (a) $\displaystyle\lim_{x\to\infty}\left(x\sin\frac{1}{x}\right)=(\infty)(0)$

(b) $\displaystyle\lim_{x\to\infty}x\sin\frac{1}{x}=\lim_{x\to\infty}\frac{\sin(1/x)}{1/x}$

$$=\lim_{x\to\infty}\frac{\left(-1/x^2\right)\cos(1/x)}{-1/x^2}$$

$$=\lim_{x\to\infty}\cos\left(\frac{1}{x}\right)=1$$

(c)
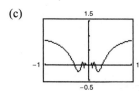

**47.** (a) $\displaystyle\lim_{x\to0^+}x^{1/x}=0^\infty=0$, not indeterminate

(See Exercise 108).

(b) Let $\quad y=x^{1/x}$

$$\ln y=\ln x^{1/x}=\frac{1}{x}\ln x.$$

Because $x\to0^+,\dfrac{1}{x}\ln x\to(\infty)(-\infty)=-\infty$. So,

$$\ln y\to-\infty\Rightarrow y\to0^+.$$

Therefore, $\displaystyle\lim_{x\to0^+}x^{1/x}=0.$

(c)
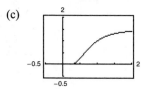

**49.** (a) $\displaystyle\lim_{x\to\infty}x^{1/x}=\infty^0$

(b) Let $y=\displaystyle\lim_{x\to\infty}x^{1/x}.$

$$\ln y=\lim_{x\to\infty}\frac{\ln x}{x}=\lim_{x\to\infty}\left(\frac{1/x}{1}\right)=0$$

So, $\ln y=0\Rightarrow y=e^0=1.$ Therefore,

$$\lim_{x\to\infty}x^{1/x}=1.$$

(c)
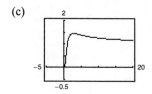

**51.** (a) $\displaystyle\lim_{x\to0^+}(1+x)^{1/x}=1^\infty$

(b) Let $y=\displaystyle\lim_{x\to0^+}(1+x)^{1/x}.$

$$\ln y=\lim_{x\to0^+}\frac{\ln(1+x)}{x}$$

$$=\lim_{x\to0^+}\left(\frac{1/(1+x)}{1}\right)=1$$

So, $\ln y=1\Rightarrow y=e^1=e.$

Therefore, $\displaystyle\lim_{x\to0^+}(1+x)^{1/x}=e.$

(c)
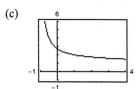

**53.** (a) $\displaystyle\lim_{x\to0^+}\left[3(x)^{x/2}\right]=0^0$

(b) Let $y=\displaystyle\lim_{x\to0^+}3(x)^{x/2}.$

$$\ln y=\lim_{x\to0^+}\left[\ln 3+\frac{x}{2}\ln x\right]$$

$$=\lim_{x\to0^+}\left[\ln 3+\frac{\ln x}{2/x}\right]$$

$$=\lim_{x\to0^+}\ln 3+\lim_{x\to0^+}\frac{1/x}{-2/x^2}$$

$$=\lim_{x\to0^+}\ln 3-\lim_{x\to0^+}\frac{x}{2}$$

$$=\ln 3$$

So, $\displaystyle\lim_{x\to0^+}3(x)^{x/2}=3.$

(c)
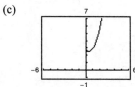

**55.** (a) $\lim\limits_{x\to1^+}(\ln x)^{x-1} = 0^0$

(b) Let $y = (\ln x)^{x-1}$.

$$\ln y = \ln\left[(\ln x)^{x-1}\right] = (x-1)\ln(\ln x)$$

$$= \frac{\ln(\ln x)}{(x-1)^{-1}}$$

$$\lim\limits_{x\to1^+}\ln y = \lim\limits_{x\to1^+}\frac{\ln(\ln x)}{(x-1)^{-1}}$$

$$= \lim\limits_{x\to1^+}\frac{1/(x\ln x)}{-(x-1)^{-2}}$$

$$= \lim\limits_{x\to1^+}\frac{-(x-1)^2}{x\ln x}$$

$$= \lim\limits_{x\to1^+}\frac{-2(x-1)}{1+\ln x} = 0$$

Because $\lim\limits_{x\to1^+}\ln y = 0$, $\lim\limits_{x\to1^+}y = 1$.

(c)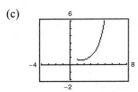

**57.** (a) $\lim\limits_{x\to2^+}\left(\frac{8}{x^2-4}-\frac{x}{x-2}\right) = \infty - \infty$

(b) $\lim\limits_{x\to2^+}\left(\frac{8}{x^2-4}-\frac{x}{x-2}\right) = \lim\limits_{x\to2^+}\frac{8-x(x+2)}{x^2-4}$

$$= \lim\limits_{x\to2^+}\frac{(2-x)(4+x)}{(x+2)(x-2)}$$

$$= \lim\limits_{x\to2^+}\frac{-(x+4)}{x+2} = \frac{-3}{2}$$

(c)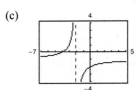

**59.** (a) $\lim\limits_{x\to1^+}\left(\frac{3}{\ln x}-\frac{2}{x-1}\right) = \infty - \infty$

(b) $\lim\limits_{x\to1^+}\left(\frac{3}{\ln x}-\frac{2}{x-1}\right) = \lim\limits_{x\to1^+}\frac{3x-3-2\ln x}{(x-1)\ln x}$

$$= \lim\limits_{x\to1^+}\frac{3-(2/x)}{\left[(x-1)/x\right]+\ln x} = \infty$$

(c)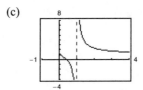

**61.** $\dfrac{0}{0}, \dfrac{\infty}{\infty}, 0\cdot\infty, 1^\infty, 0^0, \infty-\infty, \infty^0$

**63.** (a) Let $f(x) = x^2 - 25$ and $g(x) = x - 5$.

(b) Let $f(x) = (x-5)^2$ and $g(x) = x^2 - 25$.

(c) Let $f(x) = x^2 - 25$ and $g(x) = (x-5)^3$.

(Answers will vary.)

**65.** (a) Yes: $\dfrac{0}{0}$

(b) No: $\dfrac{0}{-1}$

(c) Yes: $\dfrac{\infty}{\infty}$

(d) Yes: $\dfrac{0}{0}$

(e) No: $\dfrac{-1}{0}$

(f) Yes: $\dfrac{0}{0}$

**67.**

| $x$ | 10 | $10^2$ | $10^4$ | $10^6$ | $10^8$ | $10^{10}$ |
|---|---|---|---|---|---|---|
| $\dfrac{(\ln x)^4}{x}$ | 2.811 | 4.498 | 0.720 | 0.036 | 0.001 | 0.000 |

**69.** $\lim\limits_{x\to\infty}\dfrac{x^2}{e^{5x}} = \lim\limits_{x\to\infty}\dfrac{2x}{5e^{5x}} = \lim\limits_{x\to\infty}\dfrac{2}{25e^{5x}} = 0$

**71.** $\lim\limits_{x\to\infty} \dfrac{(\ln x)^3}{x} = \lim\limits_{x\to\infty} \dfrac{3(\ln x)^2(1/x)}{1}$

$\qquad = \lim\limits_{x\to\infty} \dfrac{3(\ln x)^2}{x}$

$\qquad = \lim\limits_{x\to\infty} \dfrac{6(\ln x)(1/x)}{1}$

$\qquad = \lim\limits_{x\to\infty} \dfrac{6(\ln x)}{x} = \lim\limits_{x\to\infty} \dfrac{6}{x} = 0$

**73.** $\lim\limits_{x\to\infty} \dfrac{(\ln x)^n}{x^m} = \lim\limits_{x\to\infty} \dfrac{n(\ln x)^{n-1}/x}{mx^{m-1}}$

$\qquad = \lim\limits_{x\to\infty} \dfrac{n(\ln x)^{n-1}}{mx^m}$

$\qquad = \lim\limits_{x\to\infty} \dfrac{n(n-1)(\ln x)^{n-2}}{m^2 x^m}$

$\qquad = \cdots = \lim\limits_{x\to\infty} \dfrac{n!}{m^n x^m} = 0$

**75.** $y = x^{1/x},\ x > 0$

Horizontal asymptote: $y = 1$ (See Exercise 49.)

$\ln y = \dfrac{1}{x}\ln x$

$\left(\dfrac{1}{y}\right)\dfrac{dy}{dx} = \dfrac{1}{x}\left(\dfrac{1}{x}\right) + (\ln x)\left(-\dfrac{1}{x^2}\right)$

$\dfrac{dy}{dx} = x^{1/x}\left(\dfrac{1}{x^2}\right)(1 - \ln x) = x^{(1/x)-2}(1 - \ln x) = 0$

Critical number:       $x = e$

Intervals:       $(0, e)$       $(e, \infty)$

Sign of $dy/dx$:       $+$       $-$

$y = f(x)$:   Increasing   Decreasing

Relative maximum:   $\left(e, e^{1/e}\right)$

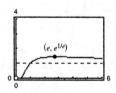

**77.** $y = 2xe^{-x}$

$\lim\limits_{x\to\infty} \dfrac{2x}{e^x} = \lim\limits_{x\to\infty} \dfrac{2}{e^x} = 0$

Horizontal asymptote: $y = 0$

$\dfrac{dy}{dx} = 2x\left(-e^{-x}\right) + 2e^{-x}$

$\qquad = 2e^{-x}(1 - x) = 0$

Critical number:       $x = 1$

Intervals:       $(-\infty, 1)$       $(1, \infty)$

Sign of $dy/dx$:       $+$       $-$

$y = f(x)$:   Increasing   Decreasing

Relative maximum: $\left(1, \dfrac{2}{e}\right)$

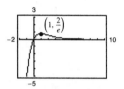

**79.** $\lim\limits_{x\to 2} \dfrac{3x^2 + 4x + 1}{x^2 - x - 2} = \dfrac{21}{0}$

Limit is not of the form $\dfrac{0}{0}$ or $\dfrac{\infty}{\infty}$.

L'Hôpital's Rule does not apply.

**81.** $\lim\limits_{x\to\infty} \dfrac{e^{-x}}{1 + e^{-x}} = \dfrac{0}{1 + 0} = 0$

Limit is not of the form $0/0$ or $\infty/\infty$.

L'Hôpital's Rule does not apply.

**83. (a)** Applying L'Hôpital's Rule twice results in the original limit, so L'Hôpital's Rule fails:

$\lim\limits_{x\to\infty} \dfrac{x}{\sqrt{x^2 + 1}} = \lim\limits_{x\to\infty} \dfrac{1}{x/\sqrt{x^2 + 1}}$

$\qquad = \lim\limits_{x\to\infty} \dfrac{\sqrt{x^2 + 1}}{x}$

$\qquad = \lim\limits_{x\to\infty} \dfrac{x/\sqrt{x^2 + 1}}{1}$

$\qquad = \lim\limits_{x\to\infty} \dfrac{x}{\sqrt{x^2 + 1}}$

**(b)** $\lim\limits_{x\to\infty} \dfrac{x}{\sqrt{x^2 + 1}} = \lim\limits_{x\to\infty} \dfrac{x/x}{\sqrt{x^2 + 1}/x}$

$\qquad = \lim\limits_{x\to\infty} \dfrac{1}{\sqrt{1 + 1/x^2}} = \dfrac{1}{\sqrt{1 + 0}} = 1$

**(c)**

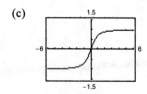

**85.** $f(x) = \sin(3x)$, $g(x) = \sin(4x)$

$f'(x) = 3\cos(3x)$, $g'(x) = 4\cos(4x)$

$y_1 = \dfrac{f(x)}{g(x)} = \dfrac{\sin 3x}{\sin 4x}$,

$y_2 = \dfrac{f'(x)}{g'(x)} = \dfrac{3\cos 3x}{4\cos 4x}$

As $x \to 0$, $y_1 \to 0.75$ and $y_2 \to 0.75$

By L'Hôpital's Rule,

$$\lim_{x \to 0} \frac{\sin 3x}{\sin 4x} = \lim_{x \to 0} \frac{3\cos 3x}{4\cos 4x} = \frac{3}{4}$$

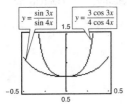

**87.** $\displaystyle\lim_{k \to 0} \frac{32\left(1 - e^{-kt} + \dfrac{v_0 k e^{-kt}}{32}\right)}{k} = \lim_{k \to 0} \frac{32(1 - e^{-kt})}{k} + \lim_{k \to 0} \left(v_0 e^{-kt}\right) = \lim_{k \to 0} \frac{32(0 + te^{-kt})}{1} + \lim_{k \to 0} \left(\frac{v_0}{e^{kt}}\right) = 32t + v_0$

**89.** Let $N$ be a fixed value for $n$. Then

$$\lim_{x \to \infty} \frac{x^{N-1}}{e^x} = \lim_{x \to \infty} \frac{(N-1)x^{N-2}}{e^x} = \lim_{x \to \infty} \frac{(N-1)(N-2)x^{N-3}}{e^x} = \cdots = \lim_{x \to \infty} \left[\frac{(N-1)!}{e^x}\right] = 0. \quad \text{(See Exercise 74.)}$$

**91.** $f(x) = x^3$, $g(x) = x^2 + 1$, $[0, 1]$

$\dfrac{f(b) - f(a)}{g(b) - g(a)} = \dfrac{f'(c)}{g'(c)}$

$\dfrac{f(1) - f(0)}{g(1) - g(0)} = \dfrac{3c^2}{2c}$

$\dfrac{1}{1} = \dfrac{3c}{2}$

$c = \dfrac{2}{3}$

**93.** $f(x) = \sin x$, $g(x) = \cos x$, $\left[0, \dfrac{\pi}{2}\right]$

$\dfrac{f(\pi/2) - f(0)}{g(\pi/2) - g(0)} = \dfrac{f'(c)}{g'(c)}$

$\dfrac{1}{-1} = \dfrac{\cos c}{-\sin c}$

$-1 = -\cot c$

$c = \dfrac{\pi}{4}$

**95.** False. L'Hôpital's Rule does not apply because

$$\lim_{x \to 0} \left(x^2 + x + 1\right) \neq 0.$$

$$\lim_{x \to 0^+} \frac{x^2 + x + 1}{x} = \lim_{x \to 0^+} \left(x + 1 + \frac{1}{x}\right) = 1 + \infty = \infty$$

**97.** True

**99.** Area of triangle: $\dfrac{1}{2}(2x)(1 - \cos x) = x - x \cos x$

Shaded area:  Area of rectangle $-$ Area under curve

$$2x(1 - \cos x) - 2\int_0^x (1 - \cos t)\,dt = 2x(1 - \cos x) - 2[t - \sin t]_0^x$$
$$= 2x(1 - \cos x) - 2(x - \sin x)$$
$$= 2 \sin x - 2x \cos x$$

Ratio: $\displaystyle\lim_{x\to 0} \frac{x - x \cos x}{2 \sin x - 2x \cos x} = \lim_{x\to 0} \frac{1 + x \sin x - \cos x}{2 \cos x + 2x \sin x - 2 \cos x}$

$$= \lim_{x\to 0} \frac{1 + x \sin x - \cos x}{2x \sin x}$$

$$= \lim_{x\to 0} \frac{x \cos x + \sin x + \sin x}{2x \cos x + 2 \sin x}$$

$$= \lim_{x\to 0} \frac{x \cos x + 2 \sin x}{2x \cos x + 2 \sin x} \cdot \frac{1/\cos x}{1/\cos x} = \lim_{x\to 0} \frac{x + 2 \tan x}{2x + 2 \tan x} = \lim_{x\to 0} \frac{1 + 2 \sec^2 x}{2 + 2 \sec^2 x} = \frac{3}{4}$$

**101.** $\displaystyle\lim_{x\to 0} \frac{4x - 2 \sin 2x}{2x^3} = \lim_{x\to 0} \frac{4 - 4 \cos 2x}{6x^2} = \lim_{x\to 0} \frac{8 \sin 2x}{12x} = \lim_{x\to 0} \frac{16 \cos 2x}{12} = \frac{16}{12} = \frac{4}{3}$

Let $c = \dfrac{4}{3}$.

**103.** $\displaystyle\lim_{x\to 0} \frac{a - \cos bx}{x^2} = 2$

Near $x = 0,\ \cos bx \approx 1$ and $x^2 \approx 0 \Rightarrow a = 1$.

Using L'Hôpital's Rule,

$$\lim_{x\to 0} \frac{1 - \cos bx}{x^2} = \lim_{x\to 0} \frac{b \sin bx}{2x} = \lim_{x\to 0} \frac{b^2 \cos bx}{2} = 2.$$

So, $b^2 = 4$ and $b = \pm 2$.

Answer: $a = 1,\ b = \pm 2$

**105. (a)** $\displaystyle\lim_{h\to 0} \frac{f(x + h) - f(x - h)}{2h} = \lim_{h\to 0} \frac{f'(x + h)(1) - f'(x - h)(-1)}{2} = \lim_{h\to 0}\left[\frac{f'(x + h) + f'(x - h)}{2}\right] = \frac{f'(x) + f'(x)}{2} = f'(x)$

**(b)**

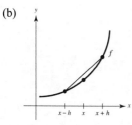

Graphically, the slope of the line joining $\big(x - h, f(x - h)\big)$ and $\big(x + h, f(x + h)\big)$ is approximately $f'(x)$.

So, $\displaystyle\lim_{h\to 0} \frac{f(x + h) - f(x - h)}{2h} = f'(x)$.

**107. (a)** $\lim\limits_{x \to 0^+} (-x \ln x)$ is the form $0 \cdot \infty$.

**(b)** $\lim\limits_{x \to 0^+} \dfrac{-\ln x}{1/x} = \lim\limits_{x \to 0^+} \dfrac{-1/x}{-1/x^2} = \lim\limits_{x \to 0^+} (x) = 0$

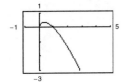

**109.** $\lim\limits_{x \to a} f(x)^{g(x)}$

$y = f(x)^{g(x)}$

$\ln y = g(x) \ln f(x)$

$\lim\limits_{x \to a} g(x) \ln f(x) = (-\infty)(-\infty) = \infty$

As $x \to a$, $\ln y \Rightarrow \infty$, and therefore $y = \infty$. So, $\lim\limits_{x \to a} f(x)^{g(x)} = \infty$.

**111. (a)** $\lim\limits_{x \to 0^+} x^{(\ln 2)/(1 + \ln x)}$ is of form $0^0$.

Let $y = x^{(\ln 2)/(1 + \ln x)}$

$\ln y = \dfrac{\ln 2}{1 + \ln x} \ln x$

$\lim\limits_{x \to 0^+} \ln y = \dfrac{\ln 2 (1/x)}{1/x} = \ln 2$.

So, $\lim\limits_{x \to 0^+} x^{(\ln 2)/(1 + \ln x)} = 2$.

**(b)** $\lim\limits_{x \to \infty} x^{(\ln 2)/(1 + \ln x)}$ is of form $\infty^0$.

Let $y = x^{(\ln 2)/(1 + \ln x)}$

$\ln y = \dfrac{\ln 2}{1 + \ln x} \ln x$

$\lim\limits_{x \to \infty} \ln y = \dfrac{\ln 2 (1/x)}{1/x} = \ln 2$.

So, $\lim\limits_{x \to \infty} x^{(\ln 2)/(1 + \ln x)} = 2$.

**(c)** $\lim\limits_{x \to 0} (x + 1)^{(\ln 2)/(x)}$ is of form $1^\infty$.

Let $y = (x + 1)^{(\ln 2)/(x)}$

$\ln y = \dfrac{\ln 2}{x} \ln(x + 1)$

$\lim\limits_{x \to 0} \ln y = \lim\limits_{x \to 0} \dfrac{(\ln 2)1/(x + 1)}{1} = \ln 2$.

So, $\lim\limits_{x \to 0} (x + 1)^{(\ln 2)/(x)} = 2$.

**113.** (a) $h(x) = \dfrac{x + \sin x}{x}$

$\displaystyle\lim_{x \to \infty} h(x) = 1$

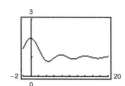

(b) $h(x) = \dfrac{x + \sin x}{x} = \dfrac{x}{x} + \dfrac{\sin x}{x} = 1 + \dfrac{\sin x}{x}, \; x > 0$

So, $\displaystyle\lim_{x \to \infty} h(x) = \lim_{x \to \infty}\left[1 + \dfrac{\sin x}{x}\right] = 1 + 0 = 1.$

(c) No. $h(x)$ is not an indeterminate form.

**115.** Let $f(x) = \left[\dfrac{1}{x} \cdot \dfrac{a^x - 1}{a - 1}\right]^{1/x}.$

For $a > 1$ and $x > 0,$

$\ln f(x) = \dfrac{1}{x}\left[\ln\dfrac{1}{x} + \ln\left(a^x - 1\right) - \ln(a - 1)\right] = -\dfrac{\ln x}{x} + \dfrac{\ln\left(a^x - 1\right)}{x} - \dfrac{\ln(a - 1)}{x}.$

As $x \to \infty, \dfrac{\ln x}{x} \to 0, \dfrac{\ln(a - 1)}{x} \to 0,$ and $\dfrac{\ln\left(a^x - 1\right)}{x} = \dfrac{\ln\left[\left(1 - a^{-x}\right)a^x\right]}{x} = \dfrac{\ln\left(1 - a^{-x}\right)}{x} + \ln a \to \ln a.$

So, $\ln f(x) \to \ln a.$

For $0 < a < 1$ and $x > 0,$

$\ln f(x) = \dfrac{-\ln x}{x} + \dfrac{\ln\left(1 - a^x\right)}{x} - \dfrac{\ln(1 - a)}{x} \to 0$ as $x \to \infty.$

Combining these results, $\displaystyle\lim_{x \to \infty} f(x) = \begin{cases} a & \text{if} \quad a > 1 \\ 1 & \text{if} \quad 0 < a < 1 \end{cases}.$

# Section 8.8   Improper Integrals

**1.** $\displaystyle\int_0^1 \dfrac{dx}{5x - 3}$ is improper because $5x - 3 = 0$ when

$x = \dfrac{3}{5},$ and $0 \le \dfrac{3}{5} \le 1.$

**3.** $\displaystyle\int_0^1 \dfrac{2x - 5}{x^2 - 5x + 6}\, dx = \int_0^1 \dfrac{2x - 5}{(x - 2)(x - 3)}\, dx$ is not

improper because

$\dfrac{2x - 5}{(x - 2)(x - 3)}$ is continuous on $[0, 1].$

**5.** $\displaystyle\int_0^2 e^{-x}\, dx$ is not improper because $f(x) = e^{-x}$ is

continuous on $[0, 2].$

**7.** $\displaystyle\int_{-\infty}^{\infty} \dfrac{\sin x}{4 + x^2}\, dx$ is improper because the limits of

integration are $-\infty$ and $\infty.$

**9.** Infinite discontinuity at $x = 0.$

$\displaystyle\int_0^4 \dfrac{1}{\sqrt{x}}\, dx = \lim_{b \to 0^+} \int_b^4 \dfrac{1}{\sqrt{x}}\, dx$

$= \displaystyle\lim_{b \to 0^+} \left[2\sqrt{x}\,\right]_b^4$

$= \displaystyle\lim_{b \to 0^+} \left(4 - 2\sqrt{b}\,\right) = 4$

Converges

**11.** Infinite discontinuity at $x = 1$.

$$\int_0^2 \frac{1}{(x-1)^2}\,dx = \int_0^1 \frac{1}{(x-1)^2}\,dx + \int_1^2 \frac{1}{(x-1)^2}\,dx$$

$$= \lim_{b \to 1^-} \int_0^b \frac{1}{(x-1)^2}\,dx + \lim_{c \to 1^+} \int_c^2 \frac{1}{(x-1)^2}\,dx$$

$$= \lim_{b \to 1^-} \left[ -\frac{1}{x-1} \right]_0^b + \lim_{c \to 1^+} \left[ -\frac{1}{x-1} \right]_c^2$$

$$= (\infty - 1) + (-1 + \infty)$$

Diverges

**13.** $\int_{-1}^1 \frac{1}{x^2}\,dx \neq -2$

because the integrand is not defined at $x = 0$.
The integral diverges.

**15.** $\int_0^\infty e^{-x}\,dx \neq 0$. You need to evaluate the limit.

$$\lim_{b \to \infty} \int_0^b e^{-x}\,dx = \lim_{b \to \infty} \left[ -e^{-x} \right]_0^b$$

$$= \lim_{b \to \infty} \left[ -e^{-b} + 1 \right] = 1$$

**17.** $\int_1^\infty \frac{1}{x^3}\,dx = \lim_{b \to \infty} \int_1^b x^{-3}\,dx$

$$= \lim_{b \to \infty} \left[ \frac{x^{-2}}{-2} \right]_1^b$$

$$= \lim_{b \to \infty} \left[ \frac{-1}{2b^2} + \frac{1}{2} \right] = \frac{1}{2}$$

**19.** $\int_1^\infty \frac{3}{\sqrt[3]{x}}\,dx = \lim_{b \to \infty} \int_1^b 3x^{-1/3}\,dx$

$$= \lim_{b \to \infty} \left[ \frac{9}{2} x^{2/3} \right]_1^b = \infty$$

Diverges

**21.** $\int_{-\infty}^0 xe^{-4x}\,dx = \lim_{b \to -\infty} \int_b^0 xe^{-4x}\,dx$

$$= \lim_{b \to -\infty} \left[ \left( \frac{-x}{4} - \frac{1}{16} \right) e^{-4x} \right]_b^0 \qquad \text{(Integration by parts)}$$

$$= \lim_{b \to -\infty} \left[ -\frac{1}{16} + \frac{b}{4} + \frac{1}{16} e^{-4b} \right] = -\infty$$

Diverges

**23.** $\int_0^\infty x^2 e^{-x}\,dx = \lim_{b \to \infty} \int_0^b x^2 e^{-x}\,dx$

$$= \lim_{b \to \infty} \left[ -e^{-x}(x^2 + 2x + 2) \right]_0^b$$

$$= \lim_{b \to \infty} \left( -\frac{b^2 + 2b + 2}{e^b} + 2 \right) = 2$$

Because $\lim_{b \to \infty} \left( -\frac{b^2 + 2b + 2}{e^b} \right) = 0$ by L'Hôpital's Rule.

**25.** $\int_4^\infty \frac{1}{x(\ln x)^3}\,dx = \lim_{b \to \infty} \int_4^b (\ln x)^{-3} \frac{1}{x}\,dx$

$$= \lim_{b \to \infty} \left[ -\frac{1}{2} (\ln x)^{-2} \right]_4^b$$

$$= \lim_{b \to \infty} \left[ -\frac{1}{2} (\ln b)^{-2} + \frac{1}{2} (\ln 4)^{-2} \right]$$

$$= \frac{1}{2} \frac{1}{(2 \ln 2)^2} = \frac{1}{2(\ln 4)^2}$$

**27.** $\int_{-\infty}^\infty \frac{4}{16 + x^2}\,dx = \int_{-\infty}^0 \frac{4}{16 + x^2}\,dx + \int_0^\infty \frac{4}{16 + x^2}\,dx$

$$= \lim_{b \to -\infty} \int_b^0 \frac{4}{16 + x^2}\,dx + \lim_{c \to \infty} \int_0^c \frac{4}{16 + x^2}\,dx$$

$$= \lim_{b \to -\infty} \left[ \arctan\left( \frac{x}{4} \right) \right]_b^0 + \lim_{c \to \infty} \left[ \arctan\left( \frac{x}{4} \right) \right]_0^c$$

$$= \lim_{b \to -\infty} \left[ 0 - \arctan\left( \frac{b}{4} \right) \right] + \lim_{c \to \infty} \left[ \arctan\left( \frac{c}{4} \right) - 0 \right]$$

$$= -\left( -\frac{\pi}{2} \right) + \frac{\pi}{2} = \pi$$

**29.** $\displaystyle\int_0^\infty \frac{1}{e^x + e^{-x}}\,dx = \lim_{b\to\infty}\int_0^b \frac{e^x}{1 + e^{2x}}\,dx$

$\qquad\qquad\qquad = \lim_{b\to\infty}\Big[\arctan\!\big(e^x\big)\Big]_0^b$

$\qquad\qquad\qquad = \dfrac{\pi}{2} - \dfrac{\pi}{4} = \dfrac{\pi}{4}$

**31.** $\displaystyle\int_0^\infty \cos \pi x\,dx = \lim_{b\to\infty}\Big[\frac{1}{\pi}\sin \pi x\Big]_0^b$

Diverges because $\sin \pi b$ does not approach a limit as $b \to \infty$.

**33.** $\displaystyle\int_0^1 \frac{1}{x^2}\,dx = \lim_{b\to 0^+}\Big[\frac{-1}{x}\Big]_b^1 = \lim_{b\to 0^+}\Big(-1 + \frac{1}{b}\Big) = -1 + \infty$

Diverges

**35.** $\displaystyle\int_0^2 \frac{1}{\sqrt[3]{x-1}}\,dx = \int_0^1 \frac{1}{\sqrt[3]{x-1}}\,dx + \int_1^2 \frac{1}{\sqrt[3]{x-1}}\,dx = \lim_{b\to 1^-}\Big[\frac{3}{2}(x-1)^{2/3}\Big]_0^b + \lim_{c\to 1^+}\Big[\frac{3}{2}(x-1)^{2/3}\Big]_c^2 = \frac{-3}{2} + \frac{3}{2} = 0$

**37.** $\displaystyle\int_0^1 x \ln x\,dx = \lim_{b\to 0^+}\Big[\frac{x^2}{2}\ln|x| - \frac{x^2}{4}\Big]_b^1$

$\qquad\qquad\qquad = \lim_{b\to 0^+}\Big(\frac{-1}{4} - \frac{b^2 \ln b}{2} + \frac{b^2}{4}\Big) = \frac{-1}{4}$

because $\displaystyle\lim_{b\to 0^+}\big(b^2 \ln b\big) = 0$ by L'Hôpital's Rule.

**39.** $\displaystyle\int_0^{\pi/2} \tan \theta\,d\theta = \lim_{b\to(\pi/2)^-}\Big[\ln|\sec \theta|\Big]_0^b = \infty$

Diverges

**41.** $\displaystyle\int_2^4 \frac{2}{x\sqrt{x^2 - 4}}\,dx = \lim_{b\to 2^+}\int_b^4 \frac{2}{x\sqrt{x^2 - 4}}\,dx$

$\qquad\qquad\qquad = \lim_{b\to 2^+}\Big[\arcsec\Big|\frac{x}{2}\Big|\Big]_b^4$

$\qquad\qquad\qquad = \lim_{b\to 2^+}\Big(\arcsec 2 - \arcsec\Big(\frac{b}{2}\Big)\Big)$

$\qquad\qquad\qquad = \dfrac{\pi}{3} - 0 = \dfrac{\pi}{3}$

**43.** $\displaystyle\int_3^5 \frac{1}{\sqrt{x^2 - 9}}\,dx = \lim_{b\to 3^+}\Big[\ln\big|x + \sqrt{x^2 - 9}\big|\Big]_b^5$

$\qquad\qquad\qquad = \lim_{b\to 3^+}\Big[\ln 9 - \ln\big(b + \sqrt{b^2 - 9}\big)\Big]$

$\qquad\qquad\qquad = \ln 9 - \ln 3$

$\qquad\qquad\qquad = \ln\dfrac{9}{3} = \ln 3$

**45.** $\displaystyle\int_3^\infty \frac{1}{x\sqrt{x^2 - 9}}\,dx = \lim_{b\to 3^+}\int_b^5 \frac{1}{x\sqrt{x^2 - 9}}\,dx + \lim_{c\to\infty}\int_5^\infty \frac{1}{x\sqrt{x^2 - 9}}\,dx$

$\qquad\qquad\qquad = \lim_{b\to 3^+}\Big[\frac{1}{3}\arcsec \frac{x}{3}\Big]_b^5 + \lim_{c\to\infty}\Big[\frac{1}{3}\arcsec\Big(\frac{x}{3}\Big)\Big]_5^\infty$

$\qquad\qquad\qquad = \lim_{b\to 3^+}\Big[\frac{1}{3}\arcsec\Big(\frac{5}{3}\Big) - \frac{1}{3}\arcsec\Big(\frac{b}{3}\Big)\Big] + \lim_{c\to\infty}\Big[\frac{1}{3}\arcsec\Big(\frac{c}{3}\Big) - \frac{1}{3}\arcsec\Big(\frac{5}{3}\Big)\Big] = -0 + \frac{1}{3}\Big(\frac{\pi}{2}\Big) = \frac{\pi}{6}$

**47.** $\displaystyle\int_0^\infty \frac{4}{\sqrt{x}(x + 6)}\,dx = \int_0^1 \frac{4}{\sqrt{x}(x + 6)}\,dx + \int_1^\infty \frac{4}{\sqrt{x}(x + 6)}\,dx$

Let $u = \sqrt{x},\ u^2 = x,\ 2u\,du = dx$.

$\displaystyle\int \frac{4}{\sqrt{x}(x + 6)}\,dx = \int \frac{4(2u\,du)}{u(u^2 + 6)} = 8\int \frac{du}{u^2 + 6} = \frac{8}{\sqrt{6}}\arctan\Big(\frac{u}{\sqrt{6}}\Big) + C = \frac{8}{\sqrt{6}}\arctan\Big(\frac{\sqrt{x}}{\sqrt{6}}\Big) + C$

So, $\displaystyle\int_0^\infty \frac{4}{\sqrt{x}(x + 6)}\,dx = \lim_{b\to 0^+}\Big[\frac{8}{\sqrt{6}}\arctan\Big(\frac{\sqrt{x}}{\sqrt{6}}\Big)\Big]_b^1 + \lim_{c\to\infty}\Big[\frac{8}{\sqrt{6}}\arctan\Big(\frac{\sqrt{x}}{\sqrt{6}}\Big)\Big]_1^c$

$\qquad\qquad\qquad = \Big[\frac{8}{\sqrt{6}}\arctan\Big(\frac{1}{\sqrt{6}}\Big) - \frac{8}{\sqrt{6}}(0)\Big] + \Big[\frac{8}{\sqrt{6}}\Big(\frac{\pi}{2}\Big) - \frac{8}{\sqrt{6}}\arctan\Big(\frac{1}{\sqrt{6}}\Big)\Big] = \frac{8\pi}{2\sqrt{6}} = \frac{2\pi\sqrt{6}}{3}.$

**49.** If $p = 1$, $\displaystyle\int_1^\infty \frac{1}{x}\,dx = \lim_{b\to\infty}\int_1^b \frac{1}{x}\,dx = \lim_{b\to\infty}\left[\ln x\right]_1^b$

$$= \lim_{b\to\infty}(\ln b) = \infty.$$

Diverges. For $p \neq 1$,

$$\int_1^\infty \frac{1}{x^p}\,dx = \lim_{b\to\infty}\left[\frac{x^{1-p}}{1-p}\right]_1^b = \lim_{b\to\infty}\left(\frac{b^{1-p}}{1-p} - \frac{1}{1-p}\right).$$

This converges to $\dfrac{1}{p-1}$ if $1 - p < 0$ or $p > 1$.

**51.** For $n = 1$:

$$\int_0^\infty xe^{-x}\,dx = \lim_{b\to\infty}\int_0^b xe^{-x}\,dx$$

$$= \lim_{b\to\infty}\left[-e^{-x}x - e^{-x}\right]_0^b \qquad \left(\text{Parts: } u = x,\, dv = e^{-x}\,dx\right)$$

$$= \lim_{b\to\infty}\left(-e^{-b}b - e^{-b} + 1\right)$$

$$= \lim_{b\to\infty}\left(\frac{-b}{e^b} - \frac{1}{e^b} + 1\right) = 1 \quad \text{(L'Hôpital's Rule)}$$

Assume that $\displaystyle\int_0^\infty x^n e^{-x}\,dx$ converges. Then for $n + 1$ you have

$$\int x^{n+1}e^{-x}\,dx = -x^{n+1}e^{-x} + (n+1)\int x^n e^{-x}\,dx$$

by parts $\left(u = x^{n+1},\, du = (n+1)x^n\,dx,\, dv = e^{-x}\,dx,\, v = -e^{-x}\right)$.

So,

$$\int_0^\infty x^{n+1}e^{-x}\,dx = \lim_{b\to\infty}\left[-x^{n+1}e^{-x}\right]_0^b + (n+1)\int_0^\infty x^n e^{-x}\,dx = 0 + (n+1)\int_0^\infty x^n e^{-x}\,dx, \text{ which converges.}$$

**53.** $\displaystyle\int_0^1 \frac{1}{x^5}\,dx$ diverges by Exercise 50. $(p = 5)$

**55.** $\displaystyle\int_1^\infty \frac{1}{x^5}\,dx$ converges by Exercise 49. $(p = 5)$

**57.** Because $\dfrac{1}{x^2 + 5} \le \dfrac{1}{x^2}$ on $[1, \infty)$ and

$\displaystyle\int_1^\infty \frac{1}{x^2}\,dx$ converges by Exercise 49,

$\displaystyle\int_1^\infty \frac{1}{x^2 + 5}\,dx$ converges.

**59.** Because $\dfrac{1}{\sqrt[3]{x(x-1)}} \ge \dfrac{1}{\sqrt[3]{x^2}}$ on $[2, \infty)$ and

$\displaystyle\int_2^\infty \frac{1}{\sqrt[3]{x^2}}\,dx$ diverges by Exercise 49,

$\displaystyle\int_2^\infty \frac{1}{\sqrt[3]{x(x-1)}}\,dx$ diverges.

**61.** $\displaystyle\int_1^\infty \frac{2}{x^2}\,dx$ converges, and $\dfrac{1 - \sin x}{x^2} \le \dfrac{2}{x^2}$ on $[1, \infty)$, so

$\displaystyle\int_1^\infty \frac{1 - \sin x}{x^2}\,dx$ converges.

**63.** Answers will vary. *Sample answer:*

An integral with infinite integration limits or an integral with an infinite discontinuity at or between the integration limits

**65.** $\displaystyle\int_{-1}^1 \frac{1}{x^3}\,dx = \int_{-1}^0 \frac{1}{x^3}\,dx + \int_0^1 \frac{1}{x^3}\,dx$

These two integrals diverge by Exercise 50.

**67.** $A = \displaystyle\int_{-\infty}^1 e^x\,dx$

$$= \lim_{b\to-\infty}\int_b^1 e^x\,dx$$

$$= \lim_{b\to-\infty}\left[e^x\right]_b^1$$

$$= \lim_{b\to-\infty}\left(e - e^b\right) = e$$

**69.** $A = \displaystyle\int_{-\infty}^{\infty} \dfrac{1}{x^2+1}\, dx$

$= \displaystyle\lim_{b \to -\infty} \int_{b}^{0} \dfrac{1}{x^2+1}\, dx + \lim_{b \to \infty} \int_{0}^{b} \dfrac{1}{x^2+1}\, dx$

$= \displaystyle\lim_{b \to -\infty} \Big[\arctan(x)\Big]_{b}^{0} + \lim_{b \to \infty}\Big[\arctan(x)\Big]_{0}^{b}$

$= \displaystyle\lim_{b \to -\infty} \Big[0 - \arctan(b)\Big] + \lim_{b \to \infty}\Big[\arctan(b) - 0\Big]$

$= -\left(-\dfrac{\pi}{2}\right) + \dfrac{\pi}{2} = \pi$

**71. (a)** $A = \displaystyle\int_{0}^{\infty} e^{-x}\, dx$

$= \displaystyle\lim_{b \to \infty}\Big[-e^{-x}\Big]_{0}^{b} = 0 - (-1) = 1$

**(b) Disk:**

$V = \pi \displaystyle\int_{0}^{\infty} \left(e^{-x}\right)^2 dx$

$= \displaystyle\lim_{b \to \infty} \pi\left[-\dfrac{1}{2}e^{-2x}\right]_{0}^{b} = \dfrac{\pi}{2}$

**(c) Shell:**

$V = 2\pi \displaystyle\int_{0}^{\infty} xe^{-x}\, dx$

$= \displaystyle\lim_{b \to \infty} 2\pi\Big[-e^{-x}(x+1)\Big]_{0}^{b} = 2\pi$

**73.**   $x^{2/3} + y^{2/3} = 4$

$\dfrac{2}{3}x^{-1/3} + \dfrac{2}{3}y^{-1/3}y' = 0$

$y' = \dfrac{-y^{1/3}}{x^{1/3}}$

$\sqrt{1 + (y')^2} = \sqrt{1 + \dfrac{y^{2/3}}{x^{2/3}}} = \sqrt{\dfrac{x^{2/3}+y^{2/3}}{x^{2/3}}} = \sqrt{\dfrac{4}{x^{2/3}}} = \dfrac{2}{x^{1/3}}, \quad (x > 0)$

$s = 4\displaystyle\int_{0}^{8} \dfrac{2}{x^{1/3}}\, dx = \lim_{b \to 0^+}\left[8 \cdot \dfrac{3}{2}x^{2/3}\right]_{b}^{8} = 48$

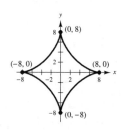

**75.** $(x - 2)^2 + y^2 = 1$

$2(x - 2) + 2yy' = 0$

$y' = \dfrac{-(x-2)}{y}$

$\sqrt{1 + (y')^2} = \sqrt{1 + \left[(x-2)^2/y^2\right]} = \dfrac{1}{y} \text{ (Assume } y > 0.)$

$S = 4\pi\displaystyle\int_{1}^{3} \dfrac{x}{y}\, dx = 4\pi\int_{1}^{3} \dfrac{x}{\sqrt{1-(x-2)^2}}\, dx = 4\pi\int_{1}^{3}\left[\dfrac{x-2}{\sqrt{1-(x-2)^2}} + \dfrac{2}{\sqrt{1-(x-2)^2}}\right] dx$

$= \displaystyle\lim_{\substack{a \to 1^+ \\ b \to 3^-}} 4\pi\left[-\sqrt{1-(x-2)^2} + 2\arcsin(x-2)\right]_{a}^{b} = 4\pi\Big[0 + 2\arcsin(1) - 2\arcsin(-1)\Big] = 8\pi^2$

**77. (a)** $F(x) = \dfrac{K}{x^2},\ 5 = \dfrac{K}{(4000)^2},\ K = 80{,}000{,}000$

$W = \displaystyle\int_{4000}^{\infty} \dfrac{80{,}000{,}000}{x^2}\, dx = \lim_{b \to \infty}\left[\dfrac{-80{,}000{,}000}{x}\right]_{4000}^{b} = 20{,}000 \text{ mi-ton}$

**(b)** $\dfrac{W}{2} = 10{,}000 = \left[\dfrac{-80{,}000{,}000}{x}\right]_{4000}^{b} = \dfrac{-80{,}000{,}000}{b} + 20{,}000$

$\dfrac{80{,}000{,}000}{b} = 10{,}000$

$b = 8000$

Therefore, the rocket has traveled 4000 miles above the earth's surface.

**79.** (a) $\int_{-\infty}^{\infty} \frac{1}{7}e^{-t/7}\,dt = \int_{0}^{\infty} \frac{1}{7}e^{-t/7}\,dt = \lim_{b\to\infty}\left[-e^{-t/7}\right]_{0}^{b} = 1$

   (b) $\int_{0}^{4} \frac{1}{7}e^{-t/7}\,dt = \left[-e^{-t/7}\right]_{0}^{4} = -e^{-4/7} + 1$

   $\qquad\qquad \approx 0.4353 = 43.53\%$

   (c) $\int_{0}^{\infty} t\left[\frac{1}{7}e^{-t/7}\right]dt = \lim_{b\to\infty}\left[-te^{-t/7} - 7e^{-t/7}\right]_{0}^{b}$

   $\qquad\qquad = 0 + 7 = 7$

**81.** (a) $C = 650,000 + \int_{0}^{5} 25,000\,e^{-0.06t}\,dt = 650,000 - \left[\frac{25,000}{0.06}e^{-0.06t}\right]_{0}^{5} \approx \$757,992.41$

   (b) $C = 650,000 + \int_{0}^{10} 25,000 e^{-0.06t}\,dt \approx \$837,995.15$

   (c) $C = 650,000 + \int_{0}^{\infty} 25,000 e^{-0.06t}\,dt = 650,000 - \lim_{b\to\infty}\left[\frac{25,000}{0.06}e^{-0.06t}\right]_{0}^{b} \approx \$1,066,666.67$

**83.** Let $K = \dfrac{2\pi NI\,r}{k}$. Then

$P = K\displaystyle\int_{c}^{\infty} \frac{1}{\left(r^2 + x^2\right)^{3/2}}\,dx.$

Let $x = r\tan\theta,\ dx = r\sec^2\theta\,d\theta,\ \sqrt{r^2 + x^2} = r\sec\theta.$

$\displaystyle\int\frac{1}{\left(r^2 + x^2\right)^{3/2}}\,dx = \int\frac{r\sec^2\theta\,d\theta}{r^3\sec^3\theta} = \frac{1}{r^2}\int\cos\theta\,d\theta$

$\qquad\qquad\qquad = \frac{1}{r^2}\sin\theta + C = \frac{1}{r^2}\frac{x}{\sqrt{r^2 + x^2}} + C$

So,

$P = K\dfrac{1}{r^2}\displaystyle\lim_{b\to\infty}\left[\frac{x}{\sqrt{r^2 + x^2}}\right]_{c}^{b}$

$\quad = \dfrac{K}{r^2}\left[1 - \dfrac{c}{\sqrt{r^2 + c^2}}\right]$

$\quad = \dfrac{K\left(\sqrt{r^2 + c^2} - c\right)}{r^2\sqrt{r^2 + c^2}}$

$\quad = \dfrac{2\pi NI\left(\sqrt{r^2 + c^2} - c\right)}{kr\sqrt{r^2 + c^2}}.$

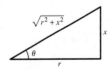

**85.** False. $f(x) = 1/(x + 1)$ is continuous on

$[0, \infty),\ \displaystyle\lim_{x\to\infty} 1/(x + 1) = 0$, but

$\displaystyle\int_{0}^{\infty} \frac{1}{x + 1}\,dx = \lim_{b\to\infty}\left[\ln|x + 1|\right]_{0}^{b} = \infty.$

Diverges

**87.** True

**89.** (a) $\displaystyle\int_{-\infty}^{\infty} \sin x \, dx = \int_{-\infty}^{0} \sin x \, dx + \int_{0}^{\infty} \sin x \, dx$

$\displaystyle = \lim_{b \to -\infty} \int_{b}^{0} \sin x \, dx + \lim_{c \to \infty} \int_{0}^{c} \sin x \, dx$

$\displaystyle = \lim_{b \to -\infty} \left[-\cos x\right]_{b}^{0} + \lim_{c \to \infty} \left[-\cos x\right]_{0}^{c}$

Because $\displaystyle\lim_{b \to -\infty} \left[-\cos b\right]$ diverges, as does

$\displaystyle\lim_{c \to \infty} \left[-\cos c\right]$,

$\displaystyle\int_{-\infty}^{\infty} \sin x \, dx$ diverges.

(b) $\displaystyle\lim_{a \to \infty} \int_{-a}^{a} \sin x \, dx = \lim_{a \to \infty} \left[-\cos x\right]_{-a}^{a}$

$\displaystyle = \lim_{a \to \infty} \left[-\cos(a) + \cos(-a)\right] = 0$

(c) The definition of $\displaystyle\int_{-\infty}^{\infty} f(x) \, dx$ is not

$\displaystyle\lim_{a \to \infty} \int_{-a}^{a} f(x) \, dx.$

**91.** (a) $\displaystyle\int_{1}^{\infty} \frac{1}{x} \, dx = \lim_{b \to \infty} \left[\ln|x|\right]_{1}^{b} = \infty$

$\displaystyle\int_{1}^{\infty} \frac{1}{x^2} \, dx = \lim_{b \to \infty} \left[-\frac{1}{x}\right]_{1}^{b} = 1$

$\displaystyle\int_{1}^{\infty} \frac{1}{x^n} \, dx$ will converge if $n > 1$ and will diverge if

$n \le 1$.

(b) It would appear to converge.

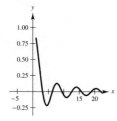

(c) Let $dv = \sin x \, dx \quad \Rightarrow \quad v = -\cos x$

$\displaystyle u = \frac{1}{x} \quad\quad \Rightarrow \quad du = -\frac{1}{x^2} \, dx.$

$\displaystyle\int_{1}^{\infty} \frac{\sin x}{x} \, dx = \lim_{b \to 0} \left[-\frac{\cos x}{x}\right]_{1}^{b} - \int_{1}^{\infty} \frac{\cos x}{x^2} \, dx$

$\displaystyle = \cos 1 - \int_{1}^{\infty} \frac{\cos x}{x^2} \, dx$

Converges

**93.** (a) $\displaystyle f(x) = \frac{1}{2.85\sqrt{2\pi}} e^{-(x-70)^2/16.245}$

$\displaystyle\int_{50}^{90} f(x) \, dx \approx 1.0$

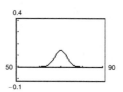

(b) $P(72 \le x < \infty) \approx 0.2414$

(c) $0.5 - P(70 \le x \le 72) \approx 0.5 - 0.2586 = 0.2414$

These are the same answers because of symmetry,

$P(70 \le x < \infty) = 0.5$

and

$0.5 = P(70 \le x < \infty)$

$\quad = P(70 \le x \le 72) + P(72 \le x < \infty).$

**95.** $f(t) = 1$

$\displaystyle F(s) = \int_{0}^{\infty} e^{-st} \, dt = \lim_{b \to \infty} \left[-\frac{1}{s} e^{-st}\right]_{0}^{b} = \frac{1}{s}, \; s > 0$

**97.** $f(t) = t^2$

$\displaystyle F(s) = \int_{0}^{\infty} t^2 e^{-st} \, dt = \lim_{b \to \infty} \left[\frac{1}{s^3}\left(-s^2 t^2 - 2st - 2\right)e^{-st}\right]_{0}^{b}$

$\displaystyle = \frac{2}{s^3}, \; s > 0$

**99.** $f(t) = \cos at$

$\displaystyle F(s) = \int_{0}^{\infty} e^{-st} \cos at \, dt$

$\displaystyle = \lim_{b \to \infty} \left[\frac{e^{-st}}{s^2 + a^2}\left(-s \cos at + a \sin at\right)\right]_{0}^{b}$

$\displaystyle = 0 + \frac{s}{s^2 + a^2} = \frac{s}{s^2 + a^2}, \; s > 0$

**101.** $f(t) = \cosh at$

$$F(s) = \int_0^\infty e^{-st} \cosh at \; dt = \int_0^\infty e^{-st}\left(\frac{e^{at} + e^{-at}}{2}\right) dt = \frac{1}{2}\int_0^\infty \left[e^{t(-s+a)} + e^{t(-s-a)}\right] dt$$

$$= \lim_{b\to\infty}\frac{1}{2}\left[\frac{1}{(-s+a)}e^{t(-s+a)} + \frac{1}{(-s-a)}e^{t(-s-a)}\right]_0^b = 0 - \frac{1}{2}\left[\frac{1}{(-s+a)} + \frac{1}{(-s-a)}\right]$$

$$= \frac{-1}{2}\left[\frac{1}{(-s+a)} + \frac{1}{(-s-a)}\right] = \frac{s}{s^2 - a^2}, \; s > |a|$$

**103.** $\Gamma(n) = \int_0^\infty x^{n-1}e^{-x} \; dx$

(a)  $\Gamma(1) = \int_0^\infty e^{-x} \; dx = \lim_{b\to\infty}\left[-e^{-x}\right]_0^b = 1$

  $\Gamma(2) = \int_0^\infty xe^{-x} \; dx = \lim_{b\to\infty}\left[-e^{-x}(x+1)\right]_0^b = 1$

  $\Gamma(3) = \int_0^\infty x^2 e^{-x} \; dx = \lim_{b\to\infty}\left[-x^2 e^{-x} - 2xe^{-x} - 2e^{-x}\right]_0^b = 2$

(b)  $\Gamma(n+1) = \int_0^\infty x^n e^{-x} \; dx = \lim_{b\to\infty}\left[-x^n e^{-x}\right]_0^b + \lim_{b\to\infty} n\int_0^b x^{n-1} e^{-x} \; dx = 0 + n\Gamma(n) \quad (u = x^n, \; dv = e^{-x} \; dx)$

(c)  $\Gamma(n) = (n-1)!$

**105.** $\int_0^\infty \left(\frac{1}{\sqrt{x^2+1}} - \frac{c}{x+1}\right) dx = \lim_{b\to\infty}\int_0^b \left(\frac{1}{\sqrt{x^2+1}} - \frac{c}{x+1}\right) dx$

$$= \lim_{b\to\infty}\left[\ln\left|x + \sqrt{x^2+1}\right| - c\ln\left|x+1\right|\right]_0^b$$

$$= \lim_{b\to\infty}\left[\ln\left(b + \sqrt{b^2+1}\right) - \ln(b+1)^c\right] = \lim_{b\to\infty}\ln\left[\frac{b + \sqrt{b^2+1}}{(b+1)^c}\right]$$

This limit exists for $c = 1$, and you have

$$\lim_{b\to\infty}\ln\left[\frac{b + \sqrt{b^2+1}}{(b+1)}\right] = \ln 2.$$

**107.** $f(x) = \begin{cases} x \ln x, & 0 < x \le 2 \\ 0, & x = 0 \end{cases}$

$V = \pi\int_0^2 (x \ln x)^2 \; dx$

Let $u = \ln x, e^u = x, e^u \; du = dx$.

$V = \pi\int_{-\infty}^{\ln 2} e^{2u}u^2\left(e^4 \; du\right)$

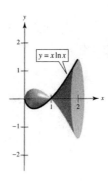

$y = x \ln x$

$\quad = \pi\int_{-\infty}^{\ln 2} e^{3u}u^2 \; du$

$\quad = \lim_{b\to-\infty}\left[\pi\left[\frac{u^2}{3} - \frac{2u}{9} + \frac{2}{27}\right]e^{3u}\right]_b^{\ln 2}$

$\quad = 8\pi\left[\frac{(\ln 2)^2}{3} - \frac{2\ln 2}{9} + \frac{2}{27}\right] \approx 2.0155$

**109.** $u = \sqrt{x}, u^2 = x, 2u\,du = dx$

$$\int_0^1 \frac{\sin x}{\sqrt{x}}\,dx = \int_0^1 \frac{\sin\left(u^2\right)}{u}(2u\,du) = \int_0^1 2\sin\left(u^2\right)du$$

Trapezoidal Rule $(n = 5)$: $0.6278$

**111.** Assume $a < b.$ The proof is similar if $a > b.$

$$\int_{-\infty}^{a} f(x)\,dx + \int_a^{\infty} f(x)\,dx = \lim_{c \to -\infty} \int_c^a f(x)\,dx + \lim_{d \to \infty} \int_a^d f(x)\,dx$$

$$= \lim_{c \to -\infty} \int_c^a f(x)\,dx + \lim_{d \to \infty}\left[\int_a^b f(x)\,dx + \int_b^d f(x)\,dx\right]$$

$$= \lim_{c \to -\infty} \int_c^a f(x)\,dx + \int_a^b f(x)\,dx + \lim_{d \to \infty} \int_b^d f(x)\,dx$$

$$= \lim_{c \to -\infty}\left[\int_c^a f(x)\,dx + \int_a^b f(x)\,dx\right] + \lim_{d \to \infty} \int_b^d f(x)\,dx$$

$$= \lim_{c \to -\infty} \int_c^b f(x)\,dx + \lim_{d \to \infty} \int_b^d f(x)\,dx$$

$$= \int_{-\infty}^{b} f(x)\,dx + \int_b^{\infty} f(x)\,dx$$

# Review Exercises for Chapter 8

**1.** $\int x\sqrt{x^2 - 36}\,dx = \dfrac{1}{2}\int\left(x^2 - 36\right)^{1/2}(2x)\,dx$

$$= \frac{1}{2}\left[\frac{\left(x^2 - 36\right)^{3/2}}{3/2}\right] + C$$

$$= \frac{1}{3}\left(x^2 - 36\right)^{3/2} + C$$

**3.** $\int \dfrac{x}{x^2 - 49}\,dx = \dfrac{1}{2}\int \dfrac{2x}{x^2 - 49}\,dx = \dfrac{1}{2}\ln\left|x^2 - 49\right| + C$

**5.** Let $u = \ln(2x),\ du = \dfrac{1}{x}\,dx.$

$$\int_1^e \frac{\ln(2x)}{x}\,dx = \int_{\ln 2}^{1 + \ln 2} u\,du$$

$$= \frac{u^2}{2}\bigg]_{\ln 2}^{1 + \ln 2}$$

$$= \frac{1}{2}\left[1 + 2\ln 2 + (\ln 2)^2 - (\ln 2)^2\right]$$

$$= \frac{1}{2} + \ln 2 \approx 1.1931$$

**7.** $\int \dfrac{100}{\sqrt{100 - x^2}}\,dx = 100\arcsin\left(\dfrac{x}{10}\right) + C$

**9.** $\int xe^{3x}\,dx = \dfrac{x}{3}e^{3x} - \int \dfrac{1}{3}e^{3x}\,dx$

$$= \frac{x}{3}e^{3x} - \frac{1}{9}e^{3x} + C$$

$$= \frac{1}{9}e^{3x}(3x - 1) + C$$

$$dv = e^{3x}\,dx \quad \Rightarrow \quad v = \frac{1}{3}e^{3x}$$

$$u = x \quad\quad\ \Rightarrow \quad du = dx$$

**11.** $\displaystyle\int e^{2x} \sin 3x\, dx = -\frac{1}{3}e^{2x}\cos 3x + \frac{2}{3}\int e^{2x}\cos 3x\, dx$

$\displaystyle\qquad\qquad\qquad\quad = -\frac{1}{3}e^{2x}\cos 3x + \frac{2}{3}\left(\frac{1}{3}e^{2x}\sin 3x - \frac{2}{3}\int e^{2x}\sin 3x\, dx\right)$

$\displaystyle\frac{13}{9}\int e^{2x}\sin 3x\, dx = -\frac{1}{3}e^{2x}\cos 3x + \frac{2}{9}e^{2x}\sin 3x$

$\displaystyle\int e^{2x}\sin 3x\, dx = \frac{e^{2x}}{13}(2\sin 3x - 3\cos 3x) + C$

(1) $dv = \sin 3x\, dx \quad\Rightarrow\quad v = -\frac{1}{3}\cos 3x$

$\qquad u = e^{2x} \qquad\qquad\Rightarrow\quad du = 2e^{2x}\, dx$

(2) $dv = \cos 3x\, dx \quad\Rightarrow\quad v = \frac{1}{3}\sin 3x$

$\qquad u = e^{2x} \qquad\qquad\Rightarrow\quad du = 2e^{2x}\, dx$

**13.** $\displaystyle\int x^2 \sin 2x\, dx = -\frac{1}{2}x^2\cos 2x + \int x\cos 2x\, dx$

$\displaystyle\qquad\qquad\qquad = -\frac{1}{2}x^2\cos 2x + \frac{1}{2}x\sin 2x - \frac{1}{2}\int \sin 2x\, dx$

$\displaystyle\qquad\qquad\qquad = -\frac{1}{2}x^2\cos 2x + \frac{x}{2}\sin 2x + \frac{1}{4}\cos 2x + C$

(1) $dv = \sin 2x\, dx \quad\Rightarrow\quad v = -\frac{1}{2}\cos 2x$

$\qquad u = x^2 \qquad\qquad\Rightarrow\quad du = 2x\, dx$

(2) $dv = \cos 2x\, dx \quad\Rightarrow\quad v = \frac{1}{2}\sin 2x$

$\qquad u = x \qquad\qquad\Rightarrow\quad du = dx$

**15.** $\displaystyle\int x\arcsin 2x\, dx = \frac{x^2}{2}\arcsin 2x - \int \frac{x^2}{\sqrt{1 - 4x^2}}\, dx$

$\displaystyle\qquad\qquad\qquad = \frac{x^2}{2}\arcsin 2x - \frac{1}{4}\int \frac{(2x)^2}{\sqrt{1 - (2x)^2}}\, dx$

$\displaystyle\qquad\qquad\qquad = \frac{x^2}{2}\arcsin 2x - \frac{1}{4}\left(\frac{1}{2}\right)\left[-(2x)\sqrt{1 - 4x^2} + \arcsin 2x\right] + C \text{ (by Formula 43 of Integration Tables)}$

$\displaystyle\qquad\qquad\qquad = \frac{1}{8}\left[(4x^2 - 1)\arcsin 2x + 2x\sqrt{1 - 4x^2}\right] + C$

$dv = x\, dx \qquad\Rightarrow\quad v = \frac{x^2}{2}$

$u = \arcsin 2x \quad\Rightarrow\quad du = \frac{2}{\sqrt{1 - 4x^2}}\, dx$

**17.** $\displaystyle\int \cos^3(\pi x - 1)\, dx = \int \left[1 - \sin^2(\pi x - 1)\right]\cos(\pi x - 1)\, dx$

$\displaystyle\qquad\qquad\qquad\quad = \frac{1}{\pi}\left[\sin(\pi x - 1) - \frac{1}{3}\sin^3(\pi x - 1)\right] + C$

$\displaystyle\qquad\qquad\qquad\quad = \frac{1}{3\pi}\sin(\pi x - 1)\left[3 - \sin^2(\pi x - 1)\right] + C$

$\displaystyle\qquad\qquad\qquad\quad = \frac{1}{3\pi}\sin(\pi x - 1)\left[3 - \left(1 - \cos^2(\pi x - 1)\right)\right] + C$

$\displaystyle\qquad\qquad\qquad\quad = \frac{1}{3\pi}\sin(\pi x - 1)\left[2 + \cos^2(\pi x - 1)\right] + C$

**19.** $\displaystyle\int \sec^4\left(\frac{x}{2}\right) dx = \int\left[\tan^2\left(\frac{x}{2}\right) + 1\right]\sec^2\left(\frac{x}{2}\right) dx$

$$= \int \tan^2\left(\frac{x}{2}\right)\sec^2\left(\frac{x}{2}\right) dx + \int \sec^2\left(\frac{x}{2}\right) dx$$

$$= \frac{2}{3}\tan^3\left(\frac{x}{2}\right) + 2\tan\left(\frac{x}{2}\right) + C = \frac{2}{3}\left[\tan^3\left(\frac{x}{2}\right) + 3\tan\left(\frac{x}{2}\right)\right] + C$$

**21.** $\displaystyle\int \frac{1}{1 - \sin\theta}\, d\theta = \int \frac{1}{1 - \sin\theta} \cdot \frac{1 + \sin\theta}{1 + \sin\theta}\, d\theta = \int \frac{1 + \sin\theta}{\cos^2\theta}\, d\theta = \int\left(\sec^2\theta + \sec\theta\tan\theta\right) d\theta = \tan\theta + \sec\theta + C$

**23.** $\displaystyle A = \int_{\pi/4}^{3\pi/4} \sin^4 x\, dx.$ Using the Table of Integrals,

$$\int \sin^4 x\, dx = -\frac{\sin^3 x \cos x}{4} + \frac{3}{4}\int \sin^2 x\, dx = \frac{-\sin^3 x \cos x}{4} + \frac{3}{4}\left[\frac{1}{2}(x - \sin x \cos x)\right] + C$$

$$\int_{\pi/4}^{3\pi/4} \sin^4 x\, dx = \left[\frac{-\sin^3 x \cos x}{4} + \frac{3}{8}x - \frac{3}{8}\sin x \cos x\right]_{\pi/4}^{3\pi/4} = \left(\frac{1}{16} + \frac{9\pi}{32} + \frac{3}{16}\right) - \left(\frac{-1}{16} + \frac{3\pi}{32} - \frac{3}{16}\right) = \frac{3\pi}{16} + \frac{1}{2} \approx 1.0890$$

**25.** $\displaystyle\int \frac{-12}{x^2\sqrt{4 - x^2}}\, dx = \int \frac{-24\cos\theta\, d\theta}{\left(4\sin^2\theta\right)(2\cos\theta)}$

$$= -3\int \csc^2\theta\, d\theta$$

$$= 3\cot\theta + C$$

$$= \frac{3\sqrt{4 - x^2}}{x} + C$$

$x = 2\sin\theta,\ dx = 2\cos\theta\, d\theta,\ \sqrt{4 - x^2} = 2\cos\theta$

**27.** $\quad x = 2\tan\theta$

$dx = 2\sec^2\theta\, d\theta$

$4 + x^2 = 4\sec^2\theta$

$$\int \frac{x^3}{\sqrt{4 + x^2}}\, dx = \int \frac{8\tan^3\theta}{2\sec\theta}\, 2\sec^2\theta\, d\theta$$

$$= 8\int \tan^3\theta\sec\theta\, d\theta$$

$$= 8\int\left(\sec^2\theta - 1\right)\tan\theta\sec\theta\, d\theta$$

$$= 8\left[\frac{\sec^3\theta}{3} - \sec\theta\right] + C$$

$$= 8\left[\frac{(x^2 + 4)^{3/2}}{24} - \frac{\sqrt{x^2 + 4}}{2}\right] + C$$

$$= \sqrt{x^2 + 4}\left[\frac{1}{3}(x^2 + 4) - 4\right] + C$$

$$= \frac{1}{3}x^2\sqrt{x^2 + 4} - \frac{8}{3}\sqrt{x^2 + 4} + C$$

$$= \frac{1}{3}(x^2 + 4)^{1/2}(x^2 - 8) + C$$

**29.** $x = 4 \tan \theta,\ dx = 4 \sec^2 \theta\, d\theta,\ \sqrt{16 + x^2} = 4 \sec \theta$

$$\int \frac{6x^3}{\sqrt{16 + x^2}}\, dx = \int \frac{6(4 \tan \theta)^3}{4 \sec \theta} 4 \sec^2 \theta\, d\theta$$

$$= 384 \int \tan^3 \theta \sec \theta\, d\theta$$

$$= 384 \int (\sec^2 \theta - 1) \sec \theta \tan \theta\, d\theta$$

$$= 384 \left[ \frac{\sec^3 \theta}{3} - \sec \theta \right] + C$$

$$= \frac{384}{3} \cdot \frac{(16 + x^2)^{3/2}}{64} - \frac{384 \sqrt{16 + x^2}}{4} + C$$

$$= 2\sqrt{x^2 + 16}(16 + x^2 - 48) + C$$

$$= 2\sqrt{x^2 + 16}(x^2 - 32) + C$$

$$\int_0^1 \frac{6x^3}{\sqrt{16 + x^2}}\, dx = \left[ 2\sqrt{x^2 + 16}\,(x^2 - 32) \right]_0^1$$

$$= 2\sqrt{17}(-31) - 2(4)(-32)$$

$$= 256 - 62\sqrt{17}$$

**31. (a)** Let $x = 2 \tan \theta,\ dx = 2 \sec^2 \theta\, d\theta.$

$$\int \frac{x^3}{\sqrt{4 + x^2}}\, dx = \int \frac{8 \tan^3 \theta}{2 \sec \theta} 2 \sec^2 \theta\, d\theta$$

$$= 8 \int \tan^3 \theta \sec \theta\, d\theta$$

$$= 8 \int \frac{\sin^3 \theta}{\cos^4 \theta}\, d\theta$$

$$= 8 \int (1 - \cos^2 \theta) \cos^{-4} \theta \sin \theta\, d\theta$$

$$= 8 \int (\cos^{-4} \theta - \cos^{-2} \theta) \sin \theta\, d\theta$$

$$= 8 \left[ \frac{\cos^{-3} \theta}{3} - \frac{\cos^{-1} \theta}{-1} \right] + C$$

$$= \frac{8}{3} \sec \theta (\sec^2 \theta - 3) + C$$

$$= \frac{8}{3} \left( \frac{\sqrt{4 + x^2}}{2} \right) \left( \frac{4 + x^2}{4} - 3 \right) + C$$

$$= \frac{1}{3} \sqrt{4 + x^2}(x^2 - 8) + C$$

**(b)** $$\int \frac{x^3}{\sqrt{4 + x^2}}\, dx = \int \frac{x^2}{\sqrt{4 + x^2}} x\, dx$$

$$= \int \frac{(u^2 - 4)u\, du}{u}$$

$$= \int (u^2 - 4)\, du$$

$$= \frac{1}{3}u^3 - 4u + C$$

$$= \frac{u}{3}(u^2 - 12) + C$$

$$= \frac{\sqrt{4 + x^2}}{3}(x^2 - 8) + C$$

$$u^2 = 4 + x^2,\ 2u\, du = 2x\, dx$$

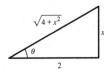

**(c)** $$\int \frac{x^3}{\sqrt{4 + x^2}}\, dx = x^2 \sqrt{4 + x^2} - \int 2x \sqrt{4 + x^2}\, dx$$

$$= x^2 \sqrt{4 + x^2} - \frac{2}{3}(4 + x^2)^{3/2} + C = \frac{\sqrt{4 + x^2}}{3}(x^2 - 8) + C$$

$$dv = \frac{x}{\sqrt{4 + x^2}}\, dx \quad \Rightarrow \quad v = \sqrt{4 + x^2}$$

$$u = x^2 \qquad \qquad \Rightarrow \quad du = 2x\, dx$$

**33.** $\dfrac{x-39}{x^2-x-12} = \dfrac{x-39}{(x-4)(x+3)} = \dfrac{A}{x-4} + \dfrac{B}{x+3}$

$$x - 39 = A(x+3) + B(x-4)$$

When $x = -3$, $-42 = -7B \Rightarrow B = 6$.

When $x = 4$, $-35 = 7A \Rightarrow A = -5$.

$$\int \dfrac{x-39}{x^2-x-12}\,dx = \int \dfrac{-5}{x-4}\,dx + \int \dfrac{6}{x+3}\,dx$$
$$= -5\ln|x-4| + 6\ln|x+3| + C$$

**35.** $\dfrac{x^2+2x}{(x-1)(x^2+1)} = \dfrac{A}{x-1} + \dfrac{Bx+C}{x^2+1}$

$$x^2 + 2x = A(x^2+1) + (Bx+C)(x-1)$$

When $x = 1$, $3 = 2A \Rightarrow A = \dfrac{3}{2}$.

When $x = 0$, $0 = A - C \Rightarrow C = \dfrac{3}{2}$.

When $x = 2$, $8 = 5A + 2B + C \Rightarrow B = -\dfrac{1}{2}$.

$$\int \dfrac{x^2+2x}{x^3-x^2+x-1}\,dx = \dfrac{3}{2}\int \dfrac{1}{x-1}\,dx - \dfrac{1}{2}\int \dfrac{x-3}{x^2+1}\,dx$$
$$= \dfrac{3}{2}\int \dfrac{1}{x-1}\,dx - \dfrac{1}{4}\int \dfrac{2x}{x^2+1}\,dx + \dfrac{3}{2}\int \dfrac{1}{x^2+1}\,dx$$
$$= \dfrac{3}{2}\ln|x-1| - \dfrac{1}{4}\ln|x^2+1| + \dfrac{3}{2}\arctan x + C$$
$$= \dfrac{1}{4}\Big[6\ln|x-1| - \ln(x^2+1) + 6\arctan x\Big] + C$$

**37.** $\dfrac{x^2}{x^2+5x-24} = 1 - \dfrac{5x-24}{x^2+5x-24} = 1 - \dfrac{5x-24}{(x+8)(x-3)}$

$$\dfrac{5x-24}{(x+8)(x-3)} = \dfrac{A}{x+8} + \dfrac{B}{x-3}$$

$$5x - 24 = A(x-3) + B(x+8)$$

When $x = 3$, $-9 = 11B \Rightarrow B = -\dfrac{9}{11}$.

When $x = -8$, $-64 = -11A \Rightarrow A = \dfrac{64}{11}$.

$$\int \dfrac{x^2}{x^2+5x-24}\,dx = \int \left[1 - \dfrac{64/11}{x+8} + \dfrac{9/11}{x-3}\right]dx$$
$$= x - \dfrac{64}{11}\ln|x+8| + \dfrac{9}{11}\ln|x-3| + C$$

**39.** Using Formula 4: $(a = 4, b = 5)$

$$\int \dfrac{x}{(4+5x)^2}\,dx = \dfrac{1}{25}\left(\dfrac{4}{4+5x} + \ln|4+5x|\right) + C$$

**41.** Let $u = x^2$, $du = 2x\,dx$.

$$\int_0^{\sqrt{\pi/2}} \dfrac{x}{1+\sin x^2}\,dx = \dfrac{1}{2}\int_0^{\pi/4} \dfrac{1}{1+\sin u}\,du$$
$$= \dfrac{1}{2}\big[\tan u - \sec u\big]_0^{\pi/4}$$
$$= \dfrac{1}{2}\Big[(1-\sqrt{2}) - (0-1)\Big]$$
$$= 1 - \dfrac{\sqrt{2}}{2}$$

**43.** $\int \dfrac{x}{x^2 + 4x + 8}\, dx = \dfrac{1}{2}\left[\ln\left|x^2 + 4x + 8\right| - 4\int \dfrac{1}{x^2 + 4x + 8}\, dx\right]$     (Formula 15)

$\qquad = \dfrac{1}{2}\left[\ln\left|x^2 + 4x + 8\right|\right] - 2\left[\dfrac{2}{\sqrt{32 - 16}}\arctan\left(\dfrac{2x + 4}{\sqrt{32 - 16}}\right)\right] + C$     (Formula 14)

$\qquad = \dfrac{1}{2}\ln\left|x^2 + 4x + 8\right| - \arctan\left(1 + \dfrac{x}{2}\right) + C$

**45.** $\int \dfrac{1}{\sin \pi x \cos \pi x}\, dx = \dfrac{1}{\pi}\int \dfrac{1}{\sin \pi x \cos \pi x}\, (\pi)\, dx \quad (u = \pi x)$

$\qquad = \dfrac{1}{\pi}\ln\left|\tan \pi x\right| + C$     (Formula 58)

**47.** $dv = dx \quad \Rightarrow \quad v = x$

$\qquad u = (\ln x)^n \quad \Rightarrow \quad du = n(\ln x)^{n-1}\dfrac{1}{x}\, dx$

$\qquad \int (\ln x)^n\, dx = x(\ln x)^n - n\int (\ln x)^{n-1}\, dx$

**49.** $\int \theta \sin \theta \cos \theta\, d\theta = \dfrac{1}{2}\int \theta \sin 2\theta\, d\theta$

$\qquad = -\dfrac{1}{4}\theta \cos 2\theta + \dfrac{1}{4}\int \cos 2\theta\, d\theta = -\dfrac{1}{4}\theta \cos 2\theta + \dfrac{1}{8}\sin 2\theta + C = \dfrac{1}{8}(\sin 2\theta - 2\theta \cos 2\theta) + C$

$\qquad dv = \sin 2\theta\, d\theta \quad \Rightarrow \quad v = -\dfrac{1}{2}\cos 2\theta$

$\qquad u = \theta \qquad \qquad \Rightarrow \quad du = d\theta$

**51.** $\int \dfrac{x^{1/4}}{1 + x^{1/2}}\, dx = 4\int \dfrac{u(u^3)}{1 + u^2}\, du$

$\qquad = 4\int \left(u^2 - 1 + \dfrac{1}{u^2 + 1}\right) du$

$\qquad = 4\left(\dfrac{1}{3}u^3 - u + \arctan u\right) + C$

$\qquad = \dfrac{4}{3}\left[x^{3/4} - 3x^{1/4} + 3\arctan\left(x^{1/4}\right)\right] + C$

$\qquad u = \sqrt[4]{x},\, x = u^4,\, dx = 4u^3\, du$

**53.** $\int \sqrt{1 + \cos x}\, dx = \int \dfrac{\sqrt{1 + \cos x}}{1} \cdot \dfrac{\sqrt{1 - \cos x}}{\sqrt{1 - \cos x}}\, dx$

$\qquad = \int \dfrac{\sin x}{\sqrt{1 - \cos x}}\, dx$

$\qquad = \int (1 - \cos x)^{-1/2}(\sin x)\, dx$

$\qquad = 2\sqrt{1 - \cos x} + C$

$\qquad u = 1 - \cos x,\, du = \sin x\, dx$

**55.** $\int \cos x \ln(\sin x)\, dx = \sin x \ln(\sin x) - \int \cos x\, dx = \sin x \ln(\sin x) - \sin x + C$

$\qquad dv = \cos x\, dx \quad \Rightarrow \quad v = \sin x$

$\qquad u = \ln(\sin x) \quad \Rightarrow \quad du = \dfrac{\cos x}{\sin x}\, dx$

**57.** $y = \int \dfrac{25}{x^2 - 25}\, dx = 25\left(\dfrac{1}{10}\right)\ln\left|\dfrac{x - 5}{x + 5}\right| + C$

$\qquad = \dfrac{5}{2}\ln\left|\dfrac{x - 5}{x + 5}\right| + C$

(Formula 24)

**59.** $y = \int \ln(x^2 + x)\, dx = x \ln\left|x^2 + x\right| - \int \dfrac{2x^2 + x}{x^2 + x}\, dx$

$\qquad = x \ln\left|x^2 + x\right| - \int \dfrac{2x + 1}{x + 1}\, dx$

$\qquad = x \ln\left|x^2 + x\right| - \int 2\, dx + \int \dfrac{1}{x + 1}\, dx$

$\qquad = x \ln\left|x^2 + x\right| - 2x + \ln\left|x + 1\right| + C$

$dv = dx \qquad\qquad \Rightarrow \quad v = x$

$u = \ln(x^2 + x) \quad \Rightarrow \quad du = \dfrac{2x + 1}{x^2 + x}\, dx$

**61.** $\displaystyle\int_2^{\sqrt{5}} x(x^2 - 4)^{3/2}\, dx = \left[\dfrac{1}{5}(x^2 - 4)^{5/2}\right]_2^{\sqrt{5}} = \dfrac{1}{5}$

**63.** $\displaystyle\int_1^4 \dfrac{\ln x}{x}\, dx = \left[\dfrac{1}{2}(\ln x)^2\right]_1^4 = \dfrac{1}{2}(\ln 4)^2 \approx 0.961$

**65.** $\displaystyle\int_0^{\pi} x \sin x\, dx = \left[-x \cos x + \sin x\right]_0^{\pi} = \pi$

**67.** $A = \displaystyle\int_0^4 x\sqrt{4 - x}\, dx = \int_2^0 (4 - u^2)\, u(-2u)\, du$

$\qquad = \displaystyle\int_2^0 2(u^4 - 4u^2)\, du$

$\qquad = \left[2\left(\dfrac{u^5}{5} - \dfrac{4u^3}{3}\right)\right]_2^0 = \dfrac{128}{15}$

$u = \sqrt{4 - x},\ x = 4 - u^2,\ dx = -2u\, du$

**69.** By symmetry, $\bar{x} = 0$, $A = \dfrac{1}{2}\pi$.

$\bar{y} = \dfrac{2}{\pi}\left(\dfrac{1}{2}\right)\displaystyle\int_{-1}^1 \left(\sqrt{1 - x^2}\right)^2 dx = \dfrac{1}{\pi}\left[x - \dfrac{1}{3}x^3\right]_{-1}^1 = \dfrac{4}{3\pi}$

$(\bar{x}, \bar{y}) = \left(0, \dfrac{4}{3\pi}\right)$

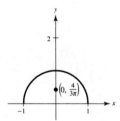

**71.** $s = \displaystyle\int_0^{\pi} \sqrt{1 + \cos^2 x}\, dx \approx 3.82$

**73.** $\displaystyle\lim_{x \to 1}\left[\dfrac{(\ln x)^2}{x - 1}\right] = \lim_{x \to 1}\left[\dfrac{2(1/x)\ln x}{1}\right] = 0$

**75.** $\displaystyle\lim_{x \to \infty} \dfrac{e^{2x}}{x^2} = \lim_{x \to \infty} \dfrac{2e^{2x}}{2x} = \lim_{x \to \infty} \dfrac{4e^{2x}}{2} = \infty$

**77.** $y = \displaystyle\lim_{x \to \infty} (\ln x)^{2/x}$

$\ln y = \displaystyle\lim_{x \to \infty} \dfrac{2 \ln(\ln x)}{x} = \lim_{x \to \infty}\left[\dfrac{2/(x \ln x)}{1}\right] = 0$

Because $\ln y = 0$, $y = 1$.

**79.** $\lim\limits_{n \to \infty} 1000\left(1 + \dfrac{0.09}{n}\right)^n = 1000 \lim\limits_{n \to \infty}\left(1 + \dfrac{0.09}{n}\right)^n$

Let $y = \lim\limits_{n \to \infty}\left(1 + \dfrac{0.09}{n}\right)^n$.

$\ln y = \lim\limits_{n \to \infty} n \ln\left(1 + \dfrac{0.09}{n}\right) = \lim\limits_{n \to \infty} \dfrac{\ln\left(1 + \dfrac{0.09}{n}\right)}{\dfrac{1}{n}} = \lim\limits_{n \to \infty}\left(\dfrac{\dfrac{-0.09/n^2}{1 + (0.09/n)}}{-\dfrac{1}{n^2}}\right) = \lim\limits_{n \to \infty} \dfrac{0.09}{1 + \left(\dfrac{0.09}{n}\right)} = 0.09$

So, $\ln y = 0.09 \Rightarrow y = e^{0.09}$ and $\lim\limits_{n \to \infty} 1000\left(1 + \dfrac{0.09}{n}\right)^n = 1000e^{0.09} \approx 1094.17$.

**81.** $\displaystyle\int_0^{16} \dfrac{1}{\sqrt[4]{x}}\, dx = \lim\limits_{b \to 0^+}\left[\dfrac{4}{3}x^{3/4}\right]_b^{16} = \dfrac{32}{3}$

**83.** $\displaystyle\int_1^{\infty} x^2 \ln x\, dx = \lim\limits_{b \to \infty}\left[\dfrac{x^3}{9}(-1 + 3\ln x)\right]_1^b = \infty$

Diverges

**85.** Let $u = \ln x$, $du = \dfrac{1}{x}\, dx$, $dv = x^{-2}\, dx$, $v = -x^{-1}$.

$\displaystyle\int \dfrac{\ln x}{x^2}\, dx = \dfrac{-\ln x}{x} + \int \dfrac{1}{x^2}\, dx = \dfrac{-\ln x}{x} - \dfrac{1}{x} + C$

$\displaystyle\int_1^{\infty} \dfrac{\ln x}{x^2}\, dx = \lim\limits_{b \to \infty}\left[\dfrac{-\ln x}{x} - \dfrac{1}{x}\right]_1^b$

$= \lim\limits_{b \to \infty}\left(\dfrac{-\ln b}{b} - \dfrac{1}{b}\right) - (-1)$

$= 0 + 1 = 1$

**87.** $\displaystyle\int_2^{\infty} \dfrac{1}{x\sqrt{x^2 - 4}}\, dx = \int_2^3 \dfrac{1}{x\sqrt{x^2 - 4}}\, dx + \int_3^{\infty} \dfrac{1}{x\sqrt{x^2 - 4}}\, dx$

$= \lim\limits_{b \to 2^+}\left[\dfrac{1}{2}\,\text{arcsec}\left(\dfrac{x}{2}\right)\right]_b^3 + \lim\limits_{c \to \infty}\left[\dfrac{1}{2}\,\text{arcsec}\left(\dfrac{x}{2}\right)\right]_3^c$

$= \dfrac{1}{2}\,\text{arcsec}\left(\dfrac{3}{2}\right) - \dfrac{1}{2}(0) + \dfrac{1}{2}\left(\dfrac{\pi}{2}\right) - \dfrac{1}{2}\,\text{arcsec}\left(\dfrac{3}{2}\right)$

$= \dfrac{\pi}{4}$

**89.** $\displaystyle\int_0^{t_0} 500{,}000e^{-0.05t}\, dt = \left[\dfrac{500{,}000}{-0.05}e^{-0.05t}\right]_0^{t_0}$

$= \dfrac{-500{,}000}{0.05}\left(e^{-0.05t_0} - 1\right)$

$= 10{,}000{,}000\left(1 - e^{-0.05t_0}\right)$

(a) $t_0 = 20$: \$6,321,205.59

(b) $t_0 \to \infty$: \$10,000,000

**91.** (a) $P(13 \le x < \infty) = \dfrac{1}{0.95\sqrt{2\pi}}\displaystyle\int_{13}^{\infty} e^{-(x-12.9)^2/2(0.95)^2}\, dx \approx 0.4581$

(b) $P(15 \le x < \infty) = \dfrac{1}{0.95\sqrt{2\pi}}\displaystyle\int_{15}^{\infty} e^{-(x-12.9)^2/2(0.95)^2}\, dx \approx 0.0135$

# Problem Solving for Chapter 8

**1. (a)** $\displaystyle\int_{-1}^{1}\left(1-x^2\right)dx = \left[x - \frac{x^3}{3}\right]_{-1}^{1} = 2\left(1 - \frac{1}{3}\right) = \frac{4}{3}$

$\displaystyle\int_{-1}^{1}\left(1-x^2\right)^2 dx = \int_{-1}^{1}\left(1 - 2x^2 + x^4\right)dx = \left[x - \frac{2x^3}{3} + \frac{x^5}{5}\right]_{-1}^{1} = 2\left(1 - \frac{2}{3} + \frac{1}{5}\right) = \frac{16}{15}$

**(b)** Let $x = \sin u, \; dx = \cos u \, du, \; 1 - x^2 = 1 - \sin^2 u = \cos^2 u.$

$$\int_{-1}^{1}\left(1-x^2\right)^n dx = \int_{-\pi/2}^{\pi/2}\left(\cos^2 u\right)^n \cos u \, du$$

$$= \int_{-\pi/2}^{\pi/2}\cos^{2n+1} u \, du$$

$$= 2\left[\frac{2}{3}\cdot\frac{4}{5}\cdot\frac{6}{7}\cdots\frac{(2n)}{(2n+1)}\right] \qquad \text{(Wallis's Formula)}$$

$$= 2\left[\frac{2^2\cdot 4^2\cdot 6^2\cdots(2n)^2}{2\cdot 3\cdot 4\cdot 5\cdots(2n)(2n+1)}\right]$$

$$= \frac{2(2^{2n})(n!)^2}{(2n+1)!} = \frac{2^{2n+1}(n!)^2}{(2n+1)!}$$

**3.**
$$\lim_{x\to\infty}\left(\frac{x+c}{x-c}\right)^x = 9$$

$$\lim_{x\to\infty} x\ln\left(\frac{x+c}{x-c}\right) = \ln 9$$

$$\lim_{x\to\infty}\frac{\ln(x+c) - \ln(x-c)}{1/x} = \ln 9$$

$$\lim_{x\to\infty}\frac{\dfrac{1}{x+c} - \dfrac{1}{x-c}}{-\dfrac{1}{x^2}} = \ln 9$$

$$\lim_{x\to\infty}\frac{-2c}{(x+c)(x-c)}(-x^2) = \ln 9$$

$$\lim_{x\to\infty}\left(\frac{2cx^2}{x^2 - c^2}\right) = \ln 9$$

$$2c = \ln 9$$

$$2c = 2\ln 3$$

$$c = \ln 3$$

**5.** $\sin\theta = \dfrac{PB}{OP} = PB, \; \cos\theta = OB$

$AQ = \overset{\frown}{AP} = \theta$

$BR = OR + OB = OR + \cos\theta$

The triangles $\triangle AQR$ and $\triangle BPR$ are similar:

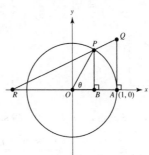

$$\frac{AR}{AQ} = \frac{BR}{BP} \Rightarrow \frac{OR+1}{\theta} = \frac{OR+\cos\theta}{\sin\theta}$$

$$\sin\theta(OR) + \sin\theta = (OR)\theta + \theta\cos\theta$$

$$OR = \frac{\theta\cos\theta - \sin\theta}{\sin\theta - \theta}$$

$$\lim_{\theta\to 0^+} OR = \lim_{\theta\to 0^+}\frac{\theta\cos\theta - \sin\theta}{\sin\theta - \theta}$$

$$= \lim_{\theta\to 0^+}\frac{-\theta\sin\theta + \cos\theta - \cos\theta}{\cos\theta - 1}$$

$$= \lim_{\theta\to 0^+}\frac{-\theta\sin\theta}{\cos\theta - 1}$$

$$= \lim_{\theta\to 0^+}\frac{-\sin\theta - \theta\cos\theta}{-\sin\theta}$$

$$= \lim_{\theta\to 0^+}\frac{\cos\theta + \cos\theta - \theta\sin\theta}{\cos\theta}$$

$$= 2$$

**7. (a)**

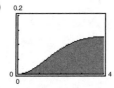

Area ≈ 0.2986

(b) Let $x = 3 \tan \theta$, $dx = 3 \sec^2 \theta \, d\theta$, $x^2 + 9 = 9 \sec^2 \theta$.

$$\int \frac{x^2}{(x^2 + 9)^{3/2}} \, dx = \int \frac{9 \tan^2 \theta}{(9 \sec^2 \theta)^{3/2}} (3 \sec^2 \theta \, d\theta)$$

$$= \int \frac{\tan^2 \theta}{\sec \theta} \, d\theta$$

$$= \int \frac{\sin^2 \theta}{\cos \theta} \, d\theta$$

$$= \int \frac{1 - \cos^2 \theta}{\cos \theta} \, d\theta$$

$$= \ln|\sec \theta + \tan \theta| - \sin \theta + C$$

$$\text{Area} = \int_0^4 \frac{x^2}{(x^2 + 9)^{3/2}} \, dx = \Big[\ln|\sec \theta + \tan \theta| - \sin \theta\Big]_0^{\tan^{-1}(4/3)}$$

$$= \left[\ln\left(\frac{\sqrt{x^2 + 9}}{3} + \frac{x}{3}\right) - \frac{x}{\sqrt{x^2 + 9}}\right]_0^4$$

$$= \ln\left(\frac{5}{3} + \frac{4}{3}\right) - \frac{4}{5} = \ln 3 - \frac{4}{5}$$

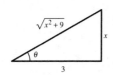

(c) $x = 3 \sinh u$, $dx = 3 \cosh u \, du$, $x^2 + 9 = 9 \sinh^2 u + 9 = 9 \cosh^2 u$

$$A = \int_0^4 \frac{x^2}{(x^2 + 9)^{3/2}} \, dx = \int_0^{\sinh^{-1}(4/3)} \frac{9 \sinh^2 u}{(9 \cosh^2 u)^{3/2}} (3 \cosh u \, du) = \int_0^{\sinh^{-1}(4/3)} \tanh^2 u \, du$$

$$= \int_0^{\sinh^{-1}(4/3)} \left(1 - \text{sech}^2 u\right) du = \Big[u - \tanh u\Big]_0^{\sinh^{-1}(4/3)}$$

$$= \sinh^{-1}\left(\frac{4}{3}\right) - \tanh\left(\sinh^{-1}\left(\frac{4}{3}\right)\right) = \ln\left(\frac{4}{3} + \sqrt{\frac{16}{9} + 1}\right) - \tanh\left[\ln\left(\frac{4}{3} + \sqrt{\frac{16}{9} + 1}\right)\right]$$

$$= \ln\left(\frac{4}{3} + \frac{5}{3}\right) - \tanh\left(\ln\left(\frac{4}{3} + \frac{5}{3}\right)\right) = \ln 3 - \tanh(\ln 3)$$

$$= \ln 3 - \frac{3 - (1/3)}{3 + (1/3)} = \ln 3 - \frac{4}{5}$$

**9.** $y = \ln(1 - x^2), \; y' = \dfrac{-2x}{1 - x^2}$

$$1 + (y')^2 = 1 + \dfrac{4x^2}{(1 - x^2)^2}$$

$$= \dfrac{1 - 2x^2 + x^4 + 4x^2}{(1 - x^2)^2}$$

$$= \left(\dfrac{1 + x^2}{1 - x^2}\right)^2$$

Arc length $= \displaystyle\int_0^{1/2} \sqrt{1 + (y')^2} \, dx$

$$= \int_0^{1/2} \left(\dfrac{1 + x^2}{1 - x^2}\right) dx$$

$$= \int_0^{1/2} \left(-1 + \dfrac{2}{1 - x^2}\right) dx$$

$$= \int_0^{1/2} \left(-1 + \dfrac{1}{x + 1} + \dfrac{1}{1 - x}\right) dx$$

$$= \left[-x + \ln(1 + x) - \ln(1 - x)\right]_0^{1/2}$$

$$= \left(-\dfrac{1}{2} + \ln\dfrac{3}{2} - \ln\dfrac{1}{2}\right)$$

$$= -\dfrac{1}{2} + \ln 3 - \ln 2 + \ln 2$$

$$= \ln 3 - \dfrac{1}{2} \approx 0.5986$$

**11.** Using a graphing utility,

(a) $\displaystyle\lim_{x \to 0^+} \left(\cot x + \dfrac{1}{x}\right) = \infty.$

(b) $\displaystyle\lim_{x \to 0^+} \left(\cot x - \dfrac{1}{x}\right) = 0.$

(c) $\displaystyle\lim_{x \to 0^+} \left(\cot x + \dfrac{1}{x}\right)\left(\cot x - \dfrac{1}{x}\right) \approx -\dfrac{2}{3}.$

Analytically,

(a) $\displaystyle\lim_{x \to 0^+} \left(\cot x + \dfrac{1}{x}\right) = \infty + \infty = \infty.$

(b) $\displaystyle\lim_{x \to 0^+} \left(\cot x - \dfrac{1}{x}\right) = \lim_{x \to 0^+} \dfrac{x \cot x - 1}{x} = \lim_{x \to 0^+} \dfrac{x \cos x - \sin x}{x \sin x}$

$$= \lim_{x \to 0^+} \dfrac{\cos x - x \sin x - \cos x}{\sin x + x \cos x}$$

$$= \lim_{x \to 0^+} \dfrac{-x \sin x}{\sin x + x \cos x}$$

$$= \lim_{x \to 0^+} \dfrac{-\sin x - x \cos x}{\cos x + \cos x - x \sin x} = 0.$$

(c) $\left(\cot x + \dfrac{1}{x}\right)\left(\cot x - \dfrac{1}{x}\right) = \cot^2 x - \dfrac{1}{x^2}$

$$= \dfrac{x^2 \cot^2 x - 1}{x^2}$$

$$\lim_{x \to 0^+} \dfrac{x^2 \cot^2 x - 1}{x^2} = \lim_{x \to 0^+} \dfrac{2x \cot^2 x - 2x^2 \cot x \csc^2 x}{2x}$$

$$= \lim_{x \to 0^+} \dfrac{\cot^2 x - x \cot x \csc^2 x}{1}$$

$$= \lim_{x \to 0^+} \dfrac{\cos^2 x \sin x - x \cos x}{\sin^3 x}$$

$$= \lim_{x \to 0^+} \dfrac{(1 - \sin^2 x)\sin x - x \cos x}{\sin^3 x}$$

$$= \lim_{x \to 0^+} \dfrac{\sin x - x \cos x}{\sin^3 x} - 1.$$

Now, $\displaystyle\lim_{x \to 0^+} \dfrac{\sin x - x \cos x}{\sin^3 x} = \lim_{x \to 0^+} \dfrac{\cos x - \cos x + x \sin x}{3 \sin^2 x \cos x}$

$$= \lim_{x \to 0^+} \dfrac{x}{3 \sin x \cdot \cos x}$$

$$= \lim_{x \to 0^+} \left(\dfrac{x}{\sin x}\right)\dfrac{1}{3 \cos x} = \dfrac{1}{3}.$$

So, $\displaystyle\lim_{x \to 0^+} \left(\cot x + \dfrac{1}{x}\right)\left(\cot x - \dfrac{1}{x}\right) = \dfrac{1}{3} - 1 = -\dfrac{2}{3}.$

The form $0 \cdot \infty$ is indeterminant.

**13.** $x^4 + 1 = \left(x^2 + ax + b\right)\left(x^2 + cx + d\right)$

$$= x^4 + (a + c)x^3 + (ac + b + d)x^2 + (ad + bc)x + bd$$

$a = -c, b = d = 1, a = \sqrt{2}$

$$x^4 + 1 = \left(x^2 + \sqrt{2}x + 1\right)\left(x^2 - \sqrt{2}x + 1\right)$$

$$\int_0^1 \dfrac{1}{x^4 + 1}\, dx = \int_0^1 \dfrac{Ax + B}{x^2 + \sqrt{2}x + 1}\, dx + \int_0^1 \dfrac{Cx + D}{x^2 - \sqrt{2}x + 1}\, dx$$

$$= \int_0^1 \dfrac{\dfrac{1}{2} + \dfrac{\sqrt{2}}{4}x}{x^2 + \sqrt{2}x + 1}\, dx - \int_0^1 \dfrac{-\dfrac{1}{2} + \dfrac{\sqrt{2}}{4}x}{x^2 + \sqrt{2}x + 1}\, dx$$

$$= \dfrac{\sqrt{2}}{4}\left[\arctan\left(\sqrt{2}x + 1\right) + \arctan\left(\sqrt{2}x - 1\right)\right]_0^1 + \dfrac{\sqrt{2}}{8}\left[\ln\left(x^2 + \sqrt{2}x + 1\right) - \ln\left(x^2 - \sqrt{2}x + 1\right)\right]_0^1$$

$$= \dfrac{\sqrt{2}}{4}\left[\arctan\left(\sqrt{2} + 1\right) + \arctan\left(\sqrt{2} - 1\right)\right] + \dfrac{\sqrt{2}}{8}\left[\ln\left(2 + \sqrt{2}\right) - \ln\left(2 - \sqrt{2}\right)\right] - \dfrac{\sqrt{2}}{4}\left[\dfrac{\pi}{4} - \dfrac{\pi}{4}\right] - \dfrac{\sqrt{2}}{8}[0]$$

$$\approx 0.5554 + 0.3116$$

$$\approx 0.8670$$

**15.** $\dfrac{x^3 - 3x^2 + 1}{x^4 - 13x^2 + 12x} = \dfrac{P_1}{x} + \dfrac{P_2}{x - 1} + \dfrac{P_3}{x + 4} + \dfrac{P_4}{x - 3} \Rightarrow c_1 = 0, c_2 = 1, c_3 = -4, c_4 = 3$

$N(x) = x^3 - 3x^2 + 1$

$D'(x) = 4x^3 - 26x + 12$

$P_1 = \dfrac{N(0)}{D'(0)} = \dfrac{1}{12}$

$P_2 = \dfrac{N(1)}{D'(1)} = \dfrac{-1}{-10} = \dfrac{1}{10}$

$P_3 = \dfrac{N(-4)}{D'(-4)} = \dfrac{-111}{-140} = \dfrac{111}{140}$

$P_4 = \dfrac{N(3)}{D'(3)} = \dfrac{1}{42}$

So, $\dfrac{x^3 - 3x^2 + 1}{x^4 - 13x^2 + 12x} = \dfrac{1/12}{x} + \dfrac{1/10}{x - 1} + \dfrac{111/140}{x + 4} + \dfrac{1/42}{x - 3}$.

**17.** Consider $\displaystyle\int \dfrac{1}{\ln x}\, dx$.

Let $u = \ln x$, $du = \dfrac{1}{x}\, dx$, $x = e^u$. Then $\displaystyle\int \dfrac{1}{\ln x}\, dx = \int \dfrac{1}{u} e^u\, du = \int \dfrac{e^u}{u}\, du$.

If $\displaystyle\int \dfrac{1}{\ln x}\, dx$ were elementary, then $\displaystyle\int \dfrac{e^u}{u}\, du$ would be too, which is false.

So, $\displaystyle\int \dfrac{1}{\ln x}\, dx$ is not elementary.

**19.** By parts,

$$\int_a^b f(x)g''(x)\, dx = \left[ f(x)g'(x) \right]_a^b - \int_a^b f'(x)g'(x)\, dx \qquad \left[ u = f(x), dv = g''(x)\, dx \right]$$

$$= -\int_a^b f'(x)g'(x)\, dx$$

$$= \left[ -f'(x)g(x) \right]_a^b + \int_a^b g(x)f''(x)\, dx \qquad \left[ u = f'(x), dv = g'(x)\, dx \right]$$

$$= \int_a^b f''(x)g(x)\, dx.$$

**21.** $\displaystyle\int_2^\infty \left[ \dfrac{1}{x^5} + \dfrac{1}{x^{10}} + \dfrac{1}{x^{15}} \right] dx < \int_2^\infty \dfrac{1}{x^5 - 1}\, dx < \int_2^\infty \left[ \dfrac{1}{x^5} + \dfrac{1}{x^{10}} + \dfrac{2}{x^{15}} \right] dx$

$\displaystyle\lim_{b \to \infty}\left[ -\dfrac{1}{4x^4} - \dfrac{1}{9x^9} - \dfrac{1}{14x^{14}} \right]_2^b < \int_2^\infty \dfrac{1}{x^5 - 1}\, dx < \lim_{b \to \infty}\left[ -\dfrac{1}{4x^4} - \dfrac{1}{9x^9} - \dfrac{1}{7x^{14}} \right]_2^b$

$0.015846 < \displaystyle\int_2^\infty \dfrac{2}{x^5 - 1}\, dx < 0.015851$

# CHAPTER 9
# Infinite Series

# C H A P T E R   9
# Infinite Series

## Section 9.1   Sequences

**1.** $a_n = 3^n$

$a_1 = 3^1 = 3$

$a_2 = 3^2 = 9$

$a_3 = 3^3 = 27$

$a_4 = 3^4 = 81$

$a_5 = 3^5 = 243$

**3.** $a_n = \sin \dfrac{n\pi}{2}$

$a_1 = \sin \dfrac{\pi}{2} = 1$

$a_2 = \sin \pi = 0$

$a_3 = \sin \dfrac{3\pi}{2} = -1$

$a_4 = \sin 2\pi = 0$

$a_5 = \sin \dfrac{5\pi}{2} = 1$

**5.** $a_n = (-1)^{n+1} \left( \dfrac{2}{n} \right)$

$a_1 = \dfrac{2}{1} = 2$

$a_2 = -\dfrac{2}{2} = -1$

$a_3 = \dfrac{2}{3}$

$a_4 = -\dfrac{2}{4} = -\dfrac{1}{2}$

$a_5 = \dfrac{2}{5}$

**7.** $a_1 = 3, \, a_{k+1} = 2(a_k - 1)$

$a_2 = 2(a_1 - 1)$

$\quad = 2(3 - 1) = 4$

$a_3 = 2(a_2 - 1)$

$\quad = 2(4 - 1) = 6$

$a_4 = 2(a_3 - 1)$

$\quad = 2(6 - 1) = 10$

$a_5 = 2(a_4 - 1)$

$\quad = 2(10 - 1) = 18$

**9.** $a_n = \dfrac{10}{n + 1}, \, a_1 = \dfrac{10}{1 + 1} = 5, \, a_2 = \dfrac{10}{3}$

Matches (c).

**10.** $a_n = \dfrac{10n}{n + 1}, \, a_1 = \dfrac{10}{2} = 5, \, a_2 = \dfrac{20}{3}$

Matches (a).

**11.** $a_n = (-1)^n, \, a_1 = -1, \, a_2 = 1, \, a_3 = -1, \ldots$

Matches (d).

**12.** $a_n = \dfrac{(-1)^n}{n}, \, a_1 = \dfrac{-1}{1} = -1, \, a_2 = \dfrac{1}{2}.$

Matches (b).

**13.** $a_n = 3n - 1$

$a_5 = 3(5) - 1 = 14$

$a_6 = 3(6) - 1 = 17$

Add 3 to preceding term.

**15.** $a_{n+1} = 2a_n, \, a_1 = 5$

$a_5 = 2(40) = 80$

$a_6 = 2(80) = 160$

Multiply the preceding term by 2.

**17.** $\dfrac{(n + 1)!}{n!} = \dfrac{n!(n + 1)}{n!} = n + 1$

**19.** $\dfrac{(2n - 1)!}{(2n + 1)!} = \dfrac{(2n - 1)!}{(2n - 1)!(2n)(2n + 1)} = \dfrac{1}{2n(2n + 1)}$

**21.** $\displaystyle\lim_{n \to \infty} \dfrac{5n^2}{n^2 + 2} = 5$

**23.** $\displaystyle\lim_{n \to \infty} \dfrac{2n}{\sqrt{n^2 + 1}} = \lim_{n \to \infty} \dfrac{2}{\sqrt{1 + (1/n^2)}} = \dfrac{2}{1} = 2$

**25.**

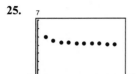

The graph seems to indicate that the sequence converges to 4. Analytically,

$$\lim_{n\to\infty} a_n = \lim_{n\to\infty} \frac{4n+1}{n} = \lim_{x\to\infty} \frac{4x+1}{x} = 4.$$

**27.**

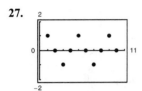

The graph seems to indicate that the sequence diverges. Analytically, the sequence is

$$\{a_n\} = \{1, 0, -1, 0, 1, \dots\}.$$

So, $\lim_{n\to\infty} a_n$ does not exist.

**29.** $\lim_{n\to\infty} \dfrac{5}{n+2} = 0$, converges

**31.** $\lim_{n\to\infty} (-1)^n \left(\dfrac{n}{n+1}\right)$

does not exist (oscillates between $-1$ and $1$), diverges.

**33.** $\lim_{n\to\infty} \dfrac{10n^2 + 3n + 7}{2n^2 - 6} = \lim_{n\to\infty} \dfrac{10 + 3/n + 7/n^2}{2 - 6/n^2}$

$$= \frac{10}{2} = 5, \text{ converges}$$

**35.** $\lim_{n\to\infty} \dfrac{\ln(n^3)}{2n} = \lim_{n\to\infty} \dfrac{3}{2} \dfrac{\ln(n)}{n}$

$$= \lim_{n\to\infty} \frac{3}{2}\left(\frac{1}{n}\right) = 0, \text{ converges}$$

(L'Hôpital's Rule)

**37.** $\lim_{n\to\infty} \dfrac{(n+1)!}{n!} = \lim_{n\to\infty} (n+1) = \infty$, diverges

**39.** $\lim_{n\to\infty} \dfrac{n^p}{e^n} = 0$, converges

$$\left(p > 0, n \geq 2\right)$$

**41.** $\lim_{n\to\infty} 2^{1/n} = 2^0 = 1$, converges

**43.** $\lim_{n\to\infty} \dfrac{\sin n}{n} = \lim_{n\to\infty} (\sin n)\dfrac{1}{n} = 0$,

converges (because $(\sin n)$ is bounded)

**45.** $a_n = -4 + 6n$

**47.** $a_n = n^2 - 3$

**49.** $a_n = \dfrac{n+1}{n+2}$

**51.** $a_n = 1 + \dfrac{1}{n} = \dfrac{n+1}{n}$

**53.** $a_n = 4 - \dfrac{1}{n} < 4 - \dfrac{1}{n+1} = a_{n+1}$,

Monotonic; $|a_n| < 4$, bounded

**55.** $a_n = ne^{-n/2}$

$a_1 = 0.6065$

$a_2 = 0.7358$

$a_3 = 0.6694$

Not monotonic; $|a_n| \leq 0.7358$, bounded

**57.** $a_n = \left(\dfrac{2}{3}\right)^n > \left(\dfrac{2}{3}\right)^{n+1} = a_{n+1}$

Monotonic; $|a_n| \leq \dfrac{2}{3}$, bounded

**59.** $a_n = \sin\left(\dfrac{n\pi}{6}\right)$

$a_1 = 0.500$

$a_2 = 0.8660$

$a_3 = 1.000$

$a_4 = 0.8660$

Not monotonic; $|a_n| \leq 1$, bounded

**61.** (a) $a_n = 7 + \dfrac{1}{n}$

$$\left|7 + \frac{1}{n}\right| \leq 8 \Rightarrow \{a_n\}, \text{ bounded}$$

$$a_n = 7 + \frac{1}{n} > 7 + \frac{1}{n+1} = a_{n+1} \Rightarrow \{a_n\}, \text{ monotonic}$$

Therefore, $\{a_n\}$ converges.

(b)

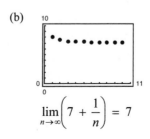

$$\lim_{n\to\infty}\left(7 + \frac{1}{n}\right) = 7$$

**63.** (a) $a_n = \frac{1}{3}\left(1 - \frac{1}{3^n}\right)$

(b)

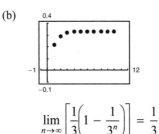

$$\left|\frac{1}{3}\left(1 - \frac{1}{3^n}\right)\right| < \frac{1}{3} \Rightarrow \{a_n\}, \text{ bounded}$$

$$a_n = \frac{1}{3}\left(1 - \frac{1}{3^n}\right) < \frac{1}{3}\left(1 - \frac{1}{3^{n+1}}\right)$$

$$= a_{n+1} \Rightarrow \{a_n\}, \text{ monotonic}$$

$$\lim_{n \to \infty}\left[\frac{1}{3}\left(1 - \frac{1}{3^n}\right)\right] = \frac{1}{3}$$

Therefore, $\{a_n\}$ converges.

**65.** $\{a_n\}$ has a limit because it is a bounded, monotonic sequence. The limit is less than or equal to 4, and greater than or equal to 2.

$$2 \le \lim_{n \to \infty} a_n \le 4$$

**67.** $A_n = P\left(1 + \frac{r}{12}\right)^n$

(a) Because $P > 0$ and $\left(1 + \frac{r}{12}\right) > 1$, the sequence

diverges. $\lim_{n \to \infty} A_n = \infty$

(b) $P = 10{,}000, r = 0.055, A_n = 10{,}000\left(1 + \frac{0.055}{12}\right)^n$

$A_0 = 10{,}000$

$A_1 = 10{,}045.83$

$A_2 = 10{,}091.88$

$A_3 = 10{,}138.13$

$A_4 = 10{,}184.60$

$A_5 = 10{,}231.28$

$A_6 = 10{,}278.17$

$A_7 = 10{,}325.28$

$A_8 = 10{,}372.60$

$A_9 = 10{,}420.14$

$A_{10} = 10{,}467.90$

**69.** No, it is not possible. See the "Definition of the Limit of a sequence". The number $L$ is unique.

**71.** (a) $a_n = 10 - \frac{1}{n}$

(b) Impossible. The sequence converges by Theorem 9.5.

(c) $a_n = \frac{3n}{4n + 1}$

(d) Impossible. An unbounded sequence diverges.

**73.** (a) $A_n = (0.8)^n \, 4{,}500{,}000{,}000$

(b) $A_1 = \$3{,}600{,}000{,}000$

$A_2 = \$2{,}880{,}000{,}000$

$A_3 = \$2{,}304{,}000{,}000$

$A_4 = \$1{,}843{,}200{,}000$

(c) $\lim_{n \to \infty} A_n = \lim_{n \to \infty} (0.8)^n (4.5) = 0$, converges

**75.** $a_n = \sqrt[n]{n} = n^{1/n}$

$a_1 = 1^{1/1} = 1$

$a_2 = \sqrt{2} \approx 1.4142$

$a_3 = \sqrt[3]{3} \approx 1.4422$

$a_4 = \sqrt[4]{4} \approx 1.4142$

$a_5 = \sqrt[5]{5} \approx 1.3797$

$a_6 = \sqrt[6]{6} \approx 1.3480$

Let $y = \lim_{n \to \infty} n^{1/n}$.

$$\ln y = \lim_{n \to \infty}\left(\frac{1}{n} \ln n\right) = \lim_{n \to \infty} \frac{\ln n}{n} = \lim_{n \to \infty} \frac{1/n}{n} = 0$$

Because $\ln y = 0$, you have $y = e^0 = 1$. Therefore, $\lim_{n \to \infty} \sqrt[n]{n} = 1$.

**77.** Because

$$\lim_{n \to \infty} s_n = L > 0,$$

there exists for each $\varepsilon > 0$,

an integer $N$ such that $|s_n - L| < \varepsilon$ for every $n > N$.

Let $\varepsilon = L > 0$ and you have,

$|s_n - L| < L, -L < s_n - L < L$, or $0 < s_n < 2L$ for each $n > N$.

**79.** True

**81.** True

**83.** $a_{n+2} = a_n + a_{n+1}$

(a)
$$a_1 = 1 \qquad\qquad a_7 = 8 + 5 = 13$$
$$a_2 = 1 \qquad\qquad a_8 = 13 + 8 = 21$$
$$a_3 = 1 + 1 = 2 \qquad a_9 = 21 + 13 = 34$$
$$a_4 = 2 + 1 = 3 \qquad a_{10} = 34 + 21 = 55$$
$$a_5 = 3 + 2 = 5 \qquad a_{11} = 55 + 34 = 89$$
$$a_6 = 5 + 3 = 8 \qquad a_{12} = 89 + 55 = 144$$

(b) $b_n = \dfrac{a_{n+1}}{a_n},\ n \ge 1$

$$b_1 = \frac{1}{1} = 1 \qquad\qquad b_6 = \frac{13}{8} = 1.625$$
$$b_2 = \frac{2}{1} = 2 \qquad\qquad b_7 = \frac{21}{13} \approx 1.6154$$
$$b_3 = \frac{3}{2} = 1.5 \qquad\qquad b_8 = \frac{34}{21} \approx 1.6190$$
$$b_4 = \frac{5}{3} \approx 1.6667 \qquad b_9 = \frac{55}{34} \approx 1.6176$$
$$b_5 = \frac{8}{5} = 1.6 \qquad\qquad b_{10} = \frac{89}{55} \approx 1.6182$$

(c)
$$1 + \frac{1}{b_{n-1}} = 1 + \frac{1}{a_n / a_{n-1}}$$
$$= 1 + \frac{a_{n-1}}{a_n} = \frac{a_n + a_{n-1}}{a_n} = \frac{a_{n+1}}{a_n} = b_n$$

(d) If $\lim\limits_{n \to \infty} b_n = \rho$, then $\lim\limits_{n \to \infty} \left(1 + \dfrac{1}{b_{n-1}}\right) = \rho$.

Because $\lim\limits_{n \to \infty} b_n = \lim\limits_{n \to \infty} b_{n-1}$, you have

$$1 + (1/\rho) = \rho.$$
$$\rho + 1 = \rho^2$$
$$0 = \rho^2 - \rho - 1$$
$$\rho = \frac{1 \pm \sqrt{1 + 4}}{2} = \frac{1 \pm \sqrt{5}}{2}$$

Because $a_n$, and therefore $b_n$, is positive,

$$\rho = \frac{1 + \sqrt{5}}{2} \approx 1.6180.$$

**85.** (a) $a_1 = \sqrt{2} \approx 1.4142$

$$a_2 = \sqrt{2 + \sqrt{2}} \approx 1.8478$$
$$a_3 = \sqrt{2 + \sqrt{2 + \sqrt{2}}} \approx 1.9616$$
$$a_4 = \sqrt{2 + \sqrt{2 + \sqrt{2 + \sqrt{2}}}} \approx 1.9904$$
$$a_5 = \sqrt{2 + \sqrt{2 + \sqrt{2 + \sqrt{2 + \sqrt{2}}}}} \approx 1.9976$$

(b) $a_n = \sqrt{2 + a_{n-1}}, \qquad n \ge 2, a_1 = \sqrt{2}$

(c) First use mathematical induction to show that $a_n \le 2$; clearly $a_1 \le 2$. So assume $a_k \le 2$. Then

$$a_k + 2 \le 4$$
$$\sqrt{a_k + 2} \le 2$$
$$a_{k+1} \le 2.$$

Now show that $\{a_n\}$ is an increasing sequence. Because $a_n \ge 0$ and $a_n \le 2$,

$$(a_n - 2)(a_n + 1) \le 0$$
$$a_n^2 - a_n - 2 \le 0$$
$$a_n^2 \le a_n + 2$$
$$a_n \le \sqrt{a_n + 2}$$
$$a_n \le a_{n+1}.$$

Because $\{a_n\}$ is a bounding increasing sequence, it converges to some number $L$, by Theorem 9.5.

$$\lim_{n \to \infty} a_n = L \Rightarrow \sqrt{2 + L} = L \Rightarrow 2 + L = L^2 \Rightarrow L^2 - L - 2 = 0$$
$$\Rightarrow (L - 2)(L + 1) = 0 \Rightarrow L = 2 \quad (L \ne -1)$$

**87. (a)**

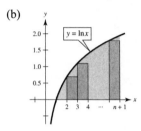

$$\int_1^n \ln x \, dx < \ln 2 + \ln 3 + \cdots + \ln n$$

$$= \ln(1 \cdot 2 \cdot 3 \cdots n) = \ln(n!)$$

**(b)**

$$\int_1^{n+1} \ln x \, dx > \ln 2 + \ln 3 + \cdots + \ln n = \ln(n!)$$

**(c)** $\int \ln x \, dx = x \ln x - x + C$

$$\int_1^n \ln x \, dx = n \ln n - n + 1 = \ln n^n - n + 1$$

From part (a): $\ln n^n - n + 1 < \ln(n!)$

$$e^{\ln n^n - n + 1} < n!$$

$$\frac{n^n}{e^{n-1}} < n!$$

$$\int_1^{n+1} \ln x \, dx = (n+1) \ln(n+1) - (n+1) + 1$$

$$= \ln(n+1)^{n+1} - n$$

From part (b): $\ln(n+1)^{n+1} - n > \ln(n!)$

$$e^{\ln(n+1)^{n+1} - n} > n!$$

$$\frac{(n+1)^{n+1}}{e^n} > n!$$

**(d)** $\dfrac{n^n}{e^{n-1}} < n! < \dfrac{(n+1)^{n+1}}{e^n}$

$$\frac{n}{e^{1-(1/n)}} < \sqrt[n]{n!} < \frac{(n+1)^{(n+1)/n}}{e}$$

$$\frac{1}{e^{1-(1/n)}} < \frac{\sqrt[n]{n!}}{n} < \frac{(n+1)^{1+(1/n)}}{ne}$$

$$\lim_{n\to\infty} \frac{1}{e^{1-(1/n)}} = \frac{1}{e}$$

$$\lim_{n\to\infty} \frac{(n+1)^{1+(1/n)}}{ne} = \lim_{n\to\infty} \frac{(n+1)}{n} \frac{(n+1)^{1/n}}{e}$$

$$= (1)\frac{1}{e}$$

$$= \frac{1}{e}$$

By the Squeeze Theorem, $\displaystyle\lim_{n\to\infty} \frac{\sqrt[n]{n!}}{n} = \frac{1}{e}$.

**(e)** $n = 20$: $\dfrac{\sqrt[20]{20!}}{20} \approx 0.4152$

$n = 50$: $\dfrac{\sqrt[50]{50!}}{50} \approx 0.3897$

$n = 100$: $\dfrac{\sqrt[100]{100!}}{100} \approx 0.3799$

$\dfrac{1}{e} \approx 0.3679$

**89.** For a given $\varepsilon > 0$, you must find $M > 0$ such that

$$|a_n - L| = |r^n| \, \varepsilon \text{ whenever } n > M. \text{ That is,}$$

$$n \ln|r| < \ln(\varepsilon) \text{ or}$$

$$n > \frac{\ln(\varepsilon)}{\ln|r|} \text{ (because } \ln|r| < 0 \text{ for } |r| < 1 \text{).}$$

So, let $\varepsilon > 0$ be given. Let $M$ be an integer satisfying

$$M > \frac{\ln(\varepsilon)}{\ln|r|}.$$

For $n > M$, you have

$$n > \frac{\ln(\varepsilon)}{\ln|r|}$$

$$n \ln|r| < \ln(\varepsilon)$$

$$\ln|r|^n < \ln(\varepsilon)$$

$$|r|^n < \varepsilon$$

$$|r^n - 0| < \varepsilon.$$

So, $\displaystyle\lim_{n\to\infty} r^n = 0$ for $-1 < r < 1$.

**91.** If $\{a_n\}$ is bounded, monotonic and nonincreasing, then $a_1 \geq a_2 \geq a_3 \geq \cdots \geq a_n \geq \cdots$. Then

$-a_1 \leq -a_2 \leq -a_3 \leq \cdots \leq -a_n \leq \cdots$ is a bounded, monotonic, nondecreasing sequence which converges by the first half of the theorem. Because $\{-a_n\}$ converges, then so does $\{a_n\}$.

**93.** $T_n = n! + 2^n$

Use mathematical induction to verify the formula.

$T_0 = 1 + 1 = 2$

$T_1 = 1 + 2 = 3$

$T_2 = 2 + 4 = 6$

Assume $T_k = k! + 2^k$. Then

$$T_{k+1} = (k + 1 + 4)T_k - 4(k + 1)T_{k-1} + (4(k + 1) - 8)T_{k-2}$$
$$= (k + 5)\left[k! + 2^k\right] - 4(k + 1)\left((k - 1)! + 2^{k-1}\right) + (4k - 4)\left((k - 2)! + 2^{k-2}\right)$$
$$= \left[(k + 5)(k)(k - 1) - 4(k + 1)(k - 1) + 4(k - 1)\right](k - 2)! + \left[(k + 5)4 - 8(k + 1) + 4(k - 1)\right]2^{k-2}$$
$$= \left[k^2 + 5k - 4k - 4 + 4\right](k - 1)! + 8 \cdot 2^{k-2}$$
$$= (k + 1)! + 2^{k+1}.$$

By mathematical induction, the formula is valid for all $n$.

## Section 9.2   Series and Convergence

**1.** $S_1 = 1$

$S_2 = 1 + \frac{1}{4} = 1.2500$

$S_3 = 1 + \frac{1}{4} + \frac{1}{9} \approx 1.3611$

$S_4 = 1 + \frac{1}{4} + \frac{1}{9} + \frac{1}{16} \approx 1.4236$

$S_5 = 1 + \frac{1}{4} + \frac{1}{9} + \frac{1}{16} + \frac{1}{25} \approx 1.4636$

**3.** $S_1 = 3$

$S_2 = 3 - \frac{9}{2} = -1.5$

$S_3 = 3 - \frac{9}{2} + \frac{27}{4} = 5.25$

$S_4 = 3 - \frac{9}{2} + \frac{27}{4} - \frac{81}{8} = -4.875$

$S_5 = 3 - \frac{9}{2} + \frac{27}{4} - \frac{81}{8} + \frac{243}{16} = 10.3125$

**5.** $S_1 = 3$

$S_2 = 3 + \frac{3}{2} = 4.5$

$S_3 = 3 + \frac{3}{2} + \frac{3}{4} = 5.250$

$S_4 = 3 + \frac{3}{2} + \frac{3}{4} + \frac{3}{8} = 5.625$

$S_5 = 3 + \frac{3}{2} + \frac{3}{4} + \frac{3}{8} + \frac{3}{16} = 5.8125$

**7.** $\displaystyle\sum_{n=0}^{\infty} \left(\frac{7}{6}\right)^n$

Geometric series

$r = \frac{7}{6} > 1$

Diverges by Theorem 9.6

**9.** $\displaystyle\sum_{n=1}^{\infty} \frac{n}{n + 1}$

$\displaystyle\lim_{n\to\infty} \frac{n}{n + 1} = 1 \neq 0$

Diverges by Theorem 9.9

**11.** $\displaystyle\sum_{n=1}^{\infty} \frac{n^2}{n^2 + 1}$

$\displaystyle\lim_{n\to\infty} \frac{n^2}{n^2 + 1} = 1 \neq 0$

Diverges by Theorem 9.9

**13.** $\displaystyle\sum_{n=1}^{\infty} \frac{2^n + 1}{2^{n+1}}$

$\displaystyle\lim_{n\to\infty} \frac{2^n + 1}{2^{n+1}} = \lim_{n\to\infty} \frac{1 + 2^{-n}}{2} = \frac{1}{2} \neq 0$

Diverges by Theorem 9.9

**15.** $\displaystyle\sum_{n=0}^{\infty} \left(\frac{5}{6}\right)^n$

Geometric series with $r = \frac{5}{6} < 1$

Converges by Theorem 9.6

**17.** $\displaystyle\sum_{n=0}^{\infty} (0.9)^n$

Geometric series with $r = 0.9 < 1$

Converges by Theorem 9.6

**19.** $\displaystyle\sum_{n=1}^{\infty} \frac{1}{n(n+1)} = \sum_{n=1}^{\infty}\left(\frac{1}{n} - \frac{1}{n+1}\right) = \left(1 - \frac{1}{2}\right) + \left(\frac{1}{2} - \frac{1}{3}\right) + \left(\frac{1}{3} - \frac{1}{4}\right) + \left(\frac{1}{4} - \frac{1}{5}\right) + \cdots,\quad S_n = 1 - \frac{1}{n+1}$

$\displaystyle\sum_{n=1}^{\infty} \frac{1}{n(n+1)} = \lim_{n\to\infty} S_n = \lim_{n\to\infty}\left(1 - \frac{1}{n+1}\right) = 1$

**21.** (a) $\displaystyle\sum_{n=1}^{\infty} \frac{6}{n(n+3)} = 2\sum_{n=1}^{\infty}\left(\frac{1}{n} - \frac{1}{n+3}\right) = 2\left[\left(1 - \frac{1}{4}\right) + \left(\frac{1}{2} - \frac{1}{5}\right) + \left(\frac{1}{3} - \frac{1}{6}\right) + \left(\frac{1}{4} - \frac{1}{7}\right) + \cdots\right]$

$\left(S_n = 2\left[1 + \frac{1}{2} + \frac{1}{3} - \left(\frac{1}{n+1} + \frac{1}{n+2} + \frac{1}{n+3}\right)\right]\right) = 2\left(1 + \frac{1}{2} + \frac{1}{3}\right) = \frac{11}{3} \approx 3.667$

(b)

| $n$ | 5 | 10 | 20 | 50 | 100 |
|---|---|---|---|---|---|
| $S_n$ | 2.7976 | 3.1643 | 3.3936 | 3.5513 | 3.6078 |

(c)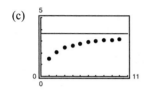

(d) The terms of the series decrease in magnitude slowly. So, the sequence of partial sums approaches the sum slowly.

**23.** (a) $\displaystyle\sum_{n=1}^{\infty} 2(0.9)^{n-1} = \sum_{n=0}^{\infty} 2(0.9)^n = \frac{2}{1 - 0.9} = 20$

(b)

| $n$ | 5 | 10 | 20 | 50 | 100 |
|---|---|---|---|---|---|
| $S_n$ | 8.1902 | 13.0264 | 17.5685 | 19.8969 | 19.9995 |

(c)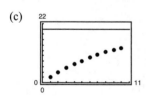

(d) The terms of the series decrease in magnitude slowly. So, the sequence of partial sums approaches the sum slowly.

**25.** $\displaystyle\sum_{n=0}^{\infty} 5\left(\frac{2}{3}\right)^n = \frac{5}{1 - (2/3)} = 15$

**27.** $\displaystyle\sum_{n=1}^{\infty} \frac{4}{n(n+2)} = 2\sum_{n=1}^{\infty}\left(\frac{1}{n} - \frac{1}{n+2}\right)$

$S_n = 2\left[\left(1 - \frac{1}{3}\right) + \left(\frac{1}{2} - \frac{1}{4}\right) + \left(\frac{1}{3} - \frac{1}{5}\right) + \cdots + \left(\frac{1}{n-1} - \frac{1}{n+1}\right) + \left(\frac{1}{n} - \frac{1}{n+2}\right)\right] = 2\left(1 + \frac{1}{2} - \frac{1}{n+1} - \frac{1}{n+2}\right)$

$\displaystyle\sum_{n=1}^{\infty} \frac{4}{n(n+2)} = \lim_{n\to\infty} S_n = \lim_{n\to\infty} 2\left(1 + \frac{1}{2} - \frac{1}{n+1} - \frac{1}{n+2}\right) = 3$

**29.** $\displaystyle\sum_{n=0}^{\infty} 8\left(\frac{3}{4}\right)^n = \frac{8}{1 - (3/4)} = 32$

**31.** $\displaystyle\sum_{n=0}^{\infty}\left(\frac{1}{2^n} - \frac{1}{3^n}\right) = \sum_{n=0}^{\infty}\left(\frac{1}{2}\right)^n - \sum_{n=0}^{\infty}\left(\frac{1}{3}\right)^n$

$= \frac{1}{1 - (1/2)} - \frac{1}{1 - (1/3)} = 2 - \frac{3}{2} = \frac{1}{2}$

**33.** Note that $\sin(1) \approx 0.8415 < 1$. The series $\sum\limits_{n=1}^{\infty} \left[\sin(1)\right]^n$ is geometric with $r = \sin(1) < 1$. So,

$$\sum_{n=1}^{\infty} \left[\sin(1)\right]^n = \sin(1) \sum_{n=0}^{\infty} \left[\sin(1)\right]^n = \frac{\sin(1)}{1 - \sin(1)} \approx 5.3080.$$

**35.** (a) $0.\overline{4} = \sum\limits_{n=0}^{\infty} \frac{4}{10}\left(\frac{1}{10}\right)^n$

(b) Geometric series with $a = \dfrac{4}{10}$ and $r = \dfrac{1}{10}$

$$S = \frac{a}{1 - r} = \frac{4/10}{1 - (1/10)} = \frac{4}{9}$$

**37.** (a) $0.\overline{81} = \sum\limits_{n=0}^{\infty} \frac{81}{100}\left(\frac{1}{100}\right)^n$

(b) Geometric series with $a = \dfrac{81}{100}$ and $r = \dfrac{1}{100}$

$$S = \frac{a}{1 - r} = \frac{81/100}{1 - (1/100)} = \frac{81}{99} = \frac{9}{11}$$

**39.** (a) $0.0\overline{75} = \sum\limits_{n=0}^{\infty} \frac{3}{40}\left(\frac{1}{100}\right)^n$

(b) Geometric series with $a = \dfrac{3}{40}$ and $r = \dfrac{1}{100}$

$$S = \frac{a}{1 - r} = \frac{3/40}{99/100} = \frac{5}{66}$$

**41.** $\sum\limits_{n=0}^{\infty} (1.075)^n$

Geometric series with $r = 1.075$

Diverges by Theorem 9.6

**43.** $\sum\limits_{n=1}^{\infty} \dfrac{n + 10}{10n + 1}$

$$\lim_{n \to \infty} \frac{n + 10}{10n + 1} = \frac{1}{10} \neq 0$$

Diverges by Theorem 9.9

**45.** $\sum\limits_{n=1}^{\infty} \left(\dfrac{1}{n} - \dfrac{1}{n + 2}\right)$

$$S_n = \left(1 - \frac{1}{3}\right) + \left(\frac{1}{2} - \frac{1}{4}\right) + \left(\frac{1}{3} - \frac{1}{5}\right) + \cdots + \left(\frac{1}{n-1} - \frac{1}{n+1}\right) + \left(\frac{1}{n} - \frac{1}{n+2}\right) = 1 + \frac{1}{2} - \frac{1}{n+1} - \frac{1}{n+2}$$

$$\sum_{n=1}^{\infty} \left(\frac{1}{n} - \frac{1}{n+2}\right) = \lim_{n \to \infty} S_n = \lim_{n \to \infty}\left(1 + \frac{1}{2} - \frac{1}{n+1} - \frac{1}{n+2}\right) = \frac{3}{2}, \text{ converges}$$

**47.** $\sum\limits_{n=1}^{\infty} \dfrac{3^n}{n^3}$

$$\lim_{n \to \infty} \frac{3^n}{n^3} = \lim_{n \to \infty} \frac{(\ln 2)3^n}{3n^2}$$

$$= \lim_{n \to \infty} \frac{(\ln 2)^2 3^n}{6n} = \lim_{n \to \infty} \frac{(\ln n)^3 3^n}{6} = \infty$$

(by L'Hôpital's Rule); diverges by Theorem 9.9

**49.** Because $n > \ln(n)$, the terms $a_n = \dfrac{n}{\ln(n)}$ do not approach 0 as $n \to \infty$. So, the series $\sum\limits_{n=2}^{\infty} \dfrac{n}{\ln(n)}$ diverges.

**51.** For $k \neq 0$,

$$\lim_{n \to \infty}\left(1 + \frac{k}{n}\right)^n = \lim_{n \to \infty}\left[\left(1 + \frac{k}{n}\right)^{n/k}\right]^k$$

$$= e^k \neq 0.$$

For $k = 0, \lim\limits_{n \to \infty} (1 + 0)^n = 1 \neq 0$.

So, $\sum\limits_{n=1}^{\infty} \left[1 + \dfrac{k}{n}\right]^n$ diverges.

**53.** $\lim\limits_{n \to \infty} \arctan n = \dfrac{\pi}{2} \neq 0$

So, $\sum\limits_{n=1}^{\infty} \arctan n$ diverges.

**55.** See definitions on page 595.

**57.** The series given by

$$\sum_{n=0}^{\infty} ar^n = a + ar + ar^2 + \cdots + ar^n + \cdots, a \neq 0$$

is a geometric series with ratio $r$. When $0 < |r| < 1$, the series converges to $a/(1 - r)$. The series diverges if $|r| \geq 1$.

**59. (a)** $\displaystyle\sum_{n=1}^{\infty} a_n = a_1 + a_2 + a_3 + \cdots$

**(b)** $\displaystyle\sum_{k=1}^{\infty} a_k = a_1 + a_2 + a_3 + \cdots$

These are the same. The third series is different, unless $a_1 = a_2 = \cdots = a$ is constant.

**(c)** $\displaystyle\sum_{n=1}^{\infty} a_k = a_k + a_k + \cdots$

**63.** $\displaystyle\sum_{n=1}^{\infty} (x - 1)^n = (x - 1)\sum_{n=0}^{\infty} (x - 1)^n$

Geometric series: converges for $|x - 1| < 1 \Rightarrow 0 < x < 2$

$$f(x) = (x - 1)\sum_{n=0}^{\infty} (x - 1)^n$$

$$= (x - 1)\frac{1}{1 - (x - 1)} = \frac{x - 1}{2 - x}, \quad 0 < x < 2$$

**65.** $\displaystyle\sum_{n=0}^{\infty} (-1)^n x^n = \sum_{n=0}^{\infty} (-x)^n$

Geometric series: converges for

$$|-x| < 1 \Rightarrow |x| < 1 \Rightarrow -1 < x < 1$$

$$f(x) = \sum_{n=0}^{\infty} (-x)^n = \frac{1}{1 + x}, \quad -1 < x < 1$$

**67. (a)** $x$ is the common ratio.

**(b)** $1 + x + x^2 + \cdots = \displaystyle\sum_{n=0}^{\infty} x^n = \frac{1}{1 - x}, \quad |x| < 1$

**(c)** $y_1 = \dfrac{1}{1 - x}$

$y_2 = S_3 = 1 + x + x^2$

$y_3 = S_5 = 1 + x + x^2 + x^3 + x^4$

Answers will vary.

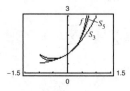

**61.** $\displaystyle\sum_{n=1}^{\infty} (3x)^n = (3x)\sum_{n=0}^{\infty} (3x)^n$

Geometric series: converges for $|3x| < 1 \Rightarrow |x| < \dfrac{1}{3}$

$$f(x) = (3x)\sum_{n=0}^{\infty} (3x)^n = (3x)\frac{1}{1 - 3x} = \frac{3x}{1 - 3x}, \quad |x| < \frac{1}{3}$$

**69.** $\dfrac{1}{n(n + 1)} < 0.0001$

$$10,000 < n^2 + n$$

$$0 < n^2 + n - 10,000$$

$$n = \frac{-1 \pm \sqrt{1^2 - 4(1)(-10,000)}}{2}$$

Choosing the positive value for $n$ you have $n \approx 99.5012$. The first *term* that is less than 0.0001 is $n = 100$.

$$\left(\frac{1}{8}\right)^n < 0.0001$$

$$10,000 < 8^n$$

This inequality is true when $n = 5$. This series converges at a faster rate.

**71.** $\displaystyle\sum_{i=0}^{n-1} 8000(0.95)^i = \frac{8000\left[1 - 0.95^n\right]}{1 - 0.95}$

$$= 160,000\left[1 - 0.95^n\right], \quad n > 0$$

**73.** $\displaystyle\sum_{i=0}^{\infty} 200(0.75)^i = 800$ million dollars

**75.** $D_1 = 16$

$D_2 = \underbrace{0.81(16)}_{up} + \underbrace{0.81(16)}_{down} = 32(0.81)$

$D_3 = 16(0.81)^2 + 16(0.81)^2 = 32(0.81)^2$

$\vdots$

$D = 16 + 32(0.81) + 32(0.81)^2 + \cdots$

$= -16 + \sum_{n=0}^{\infty} 32(0.81)^n = -16 + \dfrac{32}{1-0.81}$

$\approx 152.42$ feet

**77.** $P(n) = \dfrac{1}{2}\left(\dfrac{1}{2}\right)^n$

$P(2) = \dfrac{1}{2}\left(\dfrac{1}{2}\right)^2 = \dfrac{1}{8}$

$\sum_{n=0}^{\infty} \dfrac{1}{2}\left(\dfrac{1}{2}\right)^n = \dfrac{1/2}{1-(1/2)} = 1$

**79. (a)** $\sum_{n=1}^{\infty}\left(\dfrac{1}{2}\right)^n = \sum_{n=0}^{\infty}\dfrac{1}{2}\left(\dfrac{1}{2}\right)^n = \dfrac{1}{2}\dfrac{1}{(1-(1/2))} = 1$

**(b)** No, the series is not geometric.

**(c)** $\sum_{n=1}^{\infty} n\left(\dfrac{1}{2}\right)^n = 2$

**81. (a)** $64 + 32 + 16 + 8 + 4 + 2 = 126$ in.$^2$

**(b)** $\sum_{n=0}^{\infty} 64\left(\dfrac{1}{2}\right)^n = \dfrac{64}{1-(1/2)} = 128$ in.$^2$

**Note:** This is one-half of the area of the original square

**83.** $\sum_{n=1}^{20} 100{,}000\left(\frac{1}{1.06}\right)^n = \dfrac{100{,}000}{1.06}\sum_{i=0}^{19}\left(\dfrac{1}{1.06}\right)^i = \dfrac{100{,}000}{1.06}\left[\dfrac{1-1.06^{-20}}{1-1.06^{-1}}\right]$ $\left(n=20, r=1.06^{-1}\right) \approx \$1{,}146{,}992.12$

The \$2,000,000 sweepstakes has a present value of \$1,146,992.12. After accruing interest over the 20-year period, it attains its full value.

**85.** $w = \sum_{i=0}^{n-1} 0.01(2)^i = \dfrac{0.01(1-2^n)}{1-2} = 0.01(2^n - 1)$

**(a)** When $n = 29$: $w = \$5{,}368{,}709.11$

**(b)** When $n = 30$: $w = \$10{,}737{,}418.23$

**(c)** When $n = 31$: $w = \$21{,}474{,}836.47$

**87.** $P = 45, \quad r = 0.03, \quad t = 20$

**(a)** $A = 45\left(\dfrac{12}{0.03}\right)\left[\left(1+\dfrac{0.03}{12}\right)^{12(20)} - 1\right] \approx \$14{,}773.59$

**(b)** $A = \dfrac{45\left(e^{0.03(20)} - 1\right)}{e^{0.03/12} - 1} \approx \$14{,}779.65$

**89.** $P = 100, \quad r = 0.04, \quad t = 35$

**(a)** $A = 100\left(\dfrac{12}{0.04}\right)\left[\left(1+\dfrac{0.04}{12}\right)^{12(35)} - 1\right] \approx \$91{,}373.09$

**(b)** $A = \dfrac{100\left(e^{0.04(35)} - 1\right)}{e^{0.04/12} - 1} \approx \$91{,}503.32$

**91.** False. $\lim_{n\to\infty}\dfrac{1}{n} = 0$, but $\sum_{n=1}^{\infty}\dfrac{1}{n}$ diverges.

**93.** False; $\sum_{n=1}^{\infty} ar^n = \left(\dfrac{a}{1-r}\right) - a$

The formula requires that the geometric series begins with $n = 0$.

**95.** True

$0.74999\ldots = 0.74 + \dfrac{9}{10^3} + \dfrac{9}{10^4} + \cdots$

$= 0.74 + \dfrac{9}{10^3}\sum_{n=0}^{\infty}\left(\dfrac{1}{10}\right)^n$

$= 0.74 + \dfrac{9}{10^3}\cdot\dfrac{1}{1-(1/10)}$

$= 0.74 + \dfrac{9}{10^3}\cdot\dfrac{10}{9}$

$= 0.74 + \dfrac{1}{100} = 0.75$

**97.** Let $\sum a_n = \sum_{n=0}^{\infty} 1$ and $\sum b_n = \sum_{n=0}^{\infty} (-1)$.

Both are divergent series.

$\sum(a_n + b_n) = \sum_{n=0}^{\infty}\left[1 + (-1)\right] = \sum_{n=0}^{\infty}[1-1] = 0$

**99. (a)** $\dfrac{1}{a_{n+1}a_{n+2}} - \dfrac{1}{a_{n+2}a_{n+3}} = \dfrac{a_{n+3} - a_{n+1}}{a_{n+1}a_{n+2}a_{n+3}} = \dfrac{a_{n+2}}{a_{n+1}a_{n+2}a_{n+3}} = \dfrac{1}{a_{n+1}a_{n+3}}$

**(b)** $S_n = \displaystyle\sum_{k=0}^{n} \dfrac{1}{a_{k+1}a_{k+3}}$

$= \displaystyle\sum_{k=0}^{n} \left[ \dfrac{1}{a_{k+1}a_{k+2}} - \dfrac{1}{a_{k+2}a_{k+3}} \right]$

$= \left[ \dfrac{1}{a_1 a_2} - \dfrac{1}{a_2 a_3} \right] + \left[ \dfrac{1}{a_2 a_3} - \dfrac{1}{a_3 a_4} \right] + \cdots + \left[ \dfrac{1}{a_{n+1}a_{n+2}} - \dfrac{1}{a_{n+2}a_{n+3}} \right] = \dfrac{1}{a_1 a_2} - \dfrac{1}{a_{n+2}a_{n+3}} = 1 - \dfrac{1}{a_{n+2}a_{n+3}}$

$\displaystyle\sum_{n=0}^{\infty} \dfrac{1}{a_{n+1}a_{n+3}} = \lim_{n\to\infty} S_n = \lim_{n\to\infty} \left[ 1 - \dfrac{1}{a_{n+2}a_{n+3}} \right] = 1$

**101.** $\dfrac{1}{r} + \dfrac{1}{r^2} + \dfrac{1}{r^3} + \cdots = \displaystyle\sum_{n=0}^{\infty} \dfrac{1}{r}\left(\dfrac{1}{r}\right)^n = \dfrac{1/r}{1 - (1/r)} = \dfrac{1}{r-1}$ $\quad \left( \text{since } \left| \dfrac{1}{r} \right| < 1 \right)$

This is a geometric series which converges if

$\left| \dfrac{1}{r} \right| < 1 \Leftrightarrow |r| > 1.$

**103.** The series is telescoping:

$S_n = \displaystyle\sum_{k=1}^{n} \dfrac{6^k}{\left(3^{k+1} - 2^{k+1}\right)\left(3^k - 2^k\right)}$

$= \displaystyle\sum_{k=1}^{n} \left[ \dfrac{3^k}{3^k - 2^k} - \dfrac{3^{k+1}}{3^{k+1} - 2^{k+1}} \right]$

$= 3 - \dfrac{3^{n+1}}{3^{n+1} - 2^{n+1}}$

$\lim_{n\to\infty} S_n = 3 - 1 = 2$

# Section 9.3 The Integral Test and *p*-Series

**1.** $\displaystyle\sum_{n=1}^{\infty} \dfrac{1}{n + 3}$

Let

$f(x) = \dfrac{1}{x + 3}, \quad f'(x) = -\dfrac{1}{(x + 3)^2} < 0 \text{ for } x \geq 1.$

$f$ is positive, continuous, and decreasing for $x \geq 1$.

$\displaystyle\int_1^{\infty} \dfrac{1}{x + 3} \, dx = \left[ \ln(x + 3) \right]_1^{\infty} = \infty$

So, the series diverges by Theorem 9.10.

**3.** $\displaystyle\sum_{n=1}^{\infty} \dfrac{1}{2^n}$

Let $f(x) = \dfrac{1}{2^x}, \quad f'(x) = -(\ln 2)2^{-x} < 0 \text{ for } x \geq 1.$

$f$ is positive, continuous, and decreasing for $x \geq 1$.

$\displaystyle\int_1^{\infty} \dfrac{1}{2^x} \, dx = \left[ \dfrac{-1}{(\ln 2)\, 2^x} \right]_1^{\infty} = \dfrac{1}{2 \ln 2}$

So, the series converges by Theorem 9.10.

**5.** $\displaystyle\sum_{n=1}^{\infty} e^{-n}$

Let $f(x) = e^{-x}, \quad f'(x) = -e^{-x} < 0 \text{ for } x \geq 1.$

$f$ is positive, continuous, and decreasing for $x \geq 1$.

$\displaystyle\int_1^{\infty} e^{-x} \, dx = \left[ -e^{-x} \right]_1^{\infty} = \dfrac{1}{e}$

So, the series converges by Theorem 9.10.

**7.** $\displaystyle\sum_{n=1}^{\infty} \dfrac{1}{n^2 + 1}$

Let

$f(x) = \dfrac{1}{x^2 + 1}, \quad f'(x) = -\dfrac{2x}{\left(x^2 + 1\right)^2} < 0 \text{ for } x \geq 1.$

$f$ is positive, continuous, and decreasing for $x \geq 1$.

$\displaystyle\int_1^{\infty} \dfrac{1}{x^2 + 1} \, dx = \left[ \arctan x \right]_1^{\infty} = \dfrac{\pi}{4}$

So, the series converges by Theorem 9.10.

**9.** $\displaystyle\sum_{n=1}^{\infty} \frac{\ln(n + 1)}{n + 1}$

Let $f(x) = \dfrac{\ln(x + 1)}{x + 1}$, $f'(x) = \dfrac{1 - \ln(x + 1)}{(x + 1)^2} < 0$ for $x \geq 2$.

$f$ is positive, continuous, and decreasing for $x \geq 2$.

$$\int_1^\infty \frac{\ln(x + 1)}{x + 1}\, dx = \left[ \frac{[\ln(x + 1)]^2}{2} \right]_1^\infty = \infty$$

So, the series diverges by Theorem 9.10.

**11.** $\displaystyle\sum_{n=1}^{\infty} \frac{1}{\sqrt{n}\left(\sqrt{n} + 1\right)}$

Let $f(x) = \dfrac{1}{\sqrt{x}\left(\sqrt{x} + 1\right)}$,

$f'(x) = -\dfrac{1 + 2\sqrt{x}}{2x^{3/2}\left(\sqrt{x} + 1\right)^2} < 0$.

$f$ is positive, continuous, and decreasing for $x \geq 1$.

$$\int_1^\infty \frac{1}{\sqrt{x}\left(\sqrt{x} + 1\right)}\, dx = \left[ 2 \ln\left(\sqrt{x} + 1\right) \right]_1^\infty = \infty$$

So, the series diverges by Theorem 9.10.

**13.** $\displaystyle\sum_{n=1}^{\infty} \frac{\arctan n}{n^2 + 1}$

Let $f(x) = \dfrac{\arctan x}{x^2 + 1}$,

$f'(x) = \dfrac{1 - 2x \arctan x}{(x^2 + 1)^2} < 0$ for $x \geq 1$.

$f$ is positive, continuous, and decreasing for $x \geq 1$.

$$\int_1^\infty \frac{\arctan x}{x^2 + 1}\, dx = \left[ \frac{(\arctan x)^2}{2} \right]_1^\infty = \frac{3\pi^2}{32}$$

So, the series converges by Theorem 9.10.

**15.** $\displaystyle\sum_{n=1}^{\infty} \frac{\ln n}{n^2}$

Let $f(x) = \dfrac{\ln x}{x^2}$, $f'(x) = \dfrac{1 - 2 \ln x}{x^3}$.

$f$ is positive, continuous, and decreasing for $x > e^{1/2} \approx 1.6$.

$$\int_1^\infty \frac{\ln x}{x^2}\, dx = \left[ \frac{-(\ln x + 1)}{x} \right]_1^\infty = 1$$

So, the series converges by Theorem 9.10.

**17.** $\displaystyle\sum_{n=1}^{\infty} \frac{1}{(2n + 3)^3}$

Let $f(x) = (2x + 3)^{-3}$, $f'(x) = \dfrac{-6}{(2x + 3)^4} < 0$

$f$ is positive, continuous, and decreasing for $x \geq 1$.

$$\int_1^\infty (2x + 3)^{-3}\, dx = \left[ \frac{-1}{4(2x + 3)^2} \right]_1^\infty = \frac{1}{100}$$

So, the series converges by Theorem 9.10.

**19.** $\displaystyle\sum_{n=1}^{\infty} \frac{4n}{2n^2 + 1}$

Let $f(x) = \dfrac{4x}{2x^2 + 1}$, $f'(x) = \dfrac{-4(2x^2 - 1)}{(2x^2 + 1)^2} < 0$

for $x \geq 1$.

$f$ is positive, continuous, and decreasing for $x \geq 1$.

$$\int_1^\infty \frac{4x}{2x^2 + 1}\, dx = \left[ \ln(2x^2 + 1) \right]_1^\infty = \infty$$

So, the series diverges by Theorem 9.10.

**21.** $\displaystyle\sum_{n=1}^{\infty} \frac{n}{n^4 + 1}$

Let $f(x) = \dfrac{x}{x^4 + 1}$, $f'(x) = \dfrac{1 - 3x^4}{(x^4 + 1)^2} < 0$ for $x > 1$.

$f$ is positive, continuous, and decreasing for $x > 1$.

$$\int_1^\infty \frac{x}{x^4 + 1}\, dx = \left[ \frac{1}{2} \arctan(x^2) \right]_1^\infty = \frac{\pi}{8}$$

So, the series converges by Theorem 9.10.

**23.** $\displaystyle\sum_{n=1}^{\infty} \frac{n^{k-1}}{n^k + c}$

Let

$$f(x) = \frac{x^{k-1}}{x^k + c}, \quad f'(x) = \frac{x^{k-2}\big[c(k-1) - x^k\big]}{\big(x^k + c\big)^2} < 0$$

for $x > \sqrt[k]{c(k-1)}$.

$f$ is positive, continuous, and decreasing for $x > \sqrt[k]{c(k-1)}$.

$$\int_1^{\infty} \frac{x^{k-1}}{x^k + c}\, dx = \left[\frac{1}{k}\ln\big(x^k + c\big)\right]_1^{\infty} = \infty$$

So, the series diverges by Theorem 9.10.

**25.** Let $f(x) = \dfrac{(-1)^x}{x}$, $f(n) = a_n$.

The function $f$ is not positive for $x \geq 1$.

**27.** Let $f(x) = \dfrac{2 + \sin x}{x}$, $f(n) = a_n$.

The function $f$ is not decreasing for $x \geq 1$.

**29.** $\displaystyle\sum_{n=1}^{\infty} \frac{1}{n^3}$

Let $f(x) = \dfrac{1}{x^3}$.

$f$ is positive, continuous, and decreasing for $x \geq 1$.

$$\int_1^{\infty} \frac{1}{x^3}\, dx = \left[-\frac{1}{2x^2}\right]_1^{\infty} = \frac{1}{2}$$

Converges by Theorem 9.10

**31.** $\displaystyle\sum_{n=1}^{\infty} \frac{1}{n^{1/4}}$

Let $f(x) = \dfrac{1}{x^{1/4}}$, $f'(x) = \dfrac{-1}{4x^{5/4}} < 0$ for $x \geq 1$

$f$ is positive, continuous, and decreasing for $x \geq 1$.

$$\int_1^{\infty} \frac{1}{x^{1/4}}\, dx = \left[\frac{4x^{3/4}}{3}\right]_1^{\infty} = \infty$$

Diverges by Theorem 9.10

**33.** $\displaystyle\sum_{n=1}^{\infty} \frac{1}{\sqrt[5]{n}} = \sum_{n=1}^{\infty} \frac{1}{n^{1/5}}$

Divergent $p$-series with $p = \dfrac{1}{5} < 1$

**35.** $\displaystyle\sum_{n=1}^{\infty} \frac{1}{n^{3/2}}$

Convergent $p$-series with $p = \dfrac{3}{2} > 1$

**37.** $\displaystyle\sum_{n=1}^{\infty} \frac{1}{n^{1.04}}$

Convergent $p$-series with $p = 1.04 > 1$

**39.** (a)

| $n$ | 5 | 10 | 20 | 50 | 100 |
|---|---|---|---|---|---|
| $S_n$ | 3.7488 | 3.75 | 3.75 | 3.75 | 3.75 |

The partial sums approach the sum 3.75 very rapidly.

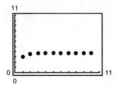

(b)

| $n$ | 5 | 10 | 20 | 50 | 100 |
|---|---|---|---|---|---|
| $S_n$ | 1.4636 | 1.5498 | 1.5962 | 1.6251 | 1.635 |

The partial sums approach the sum $\pi^2/6 \approx 1.6449$ slower than the series in part (a).

**41.** Let $f$ be positive, continuous, and decreasing for $x \geq 1$ and $a_n = f(n)$. Then,

$$\sum_{n=1}^{\infty} a_n \quad \text{and} \quad \int_1^{\infty} f(x)\,dx$$

either both converge or both diverge (Theorem 9.10). See Example 1, page 620.

**43.** Your friend is not correct. The series

$$\sum_{n=10,000}^{\infty} \frac{1}{n} = \frac{1}{10,000} + \frac{1}{10,001} + \cdots$$

is the harmonic series, starting with the $10,000^{\text{th}}$ term, and therefore diverges.

**45. (a)**

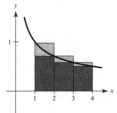

$$\sum_{n=1}^{\infty} \frac{1}{\sqrt{n}} > \int_1^{\infty} \frac{1}{\sqrt{x}}\,dx$$

The area under the rectangle is greater than the area under the curve.

Because $\int_1^{\infty} \dfrac{1}{\sqrt{x}}\,dx = \left[2\sqrt{x}\right]_1^{\infty} = \infty$, diverges,

$$\sum_{n=1}^{\infty} \frac{1}{\sqrt{n}} \text{ diverges.}$$

**(b)**

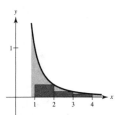

$$\sum_{n=2}^{\infty} \frac{1}{n^2} < \int_1^{\infty} \frac{1}{x^2}\,dx$$

The area under the rectangles is less than the area under the curve.

Because $\int_1^{\infty} \dfrac{1}{x^2}\,dx = \left[-\dfrac{1}{x}\right]_1^{\infty} = 1$, converges,

$$\sum_{n=2}^{\infty} \frac{1}{n^2} \text{ converges} \left(\text{and so does } \sum_{n=1}^{\infty} \frac{1}{n^2}\right).$$

**47.** $\displaystyle\sum_{n=2}^{\infty} \frac{1}{n(\ln n)^p}$

If $p = 1$, then the series diverges by the Integral Test. If $p \neq 1$,

$$\int_2^{\infty} \frac{1}{x(\ln x)^p}\,dx = \int_2^{\infty} (\ln x)^{-p} \frac{1}{x}\,dx = \left[\frac{(\ln x)^{-p+1}}{-p+1}\right]_2^{\infty}.$$

Converges for $-p + 1 < 0$ or $p > 1$

**49.** $\displaystyle\sum_{n=1}^{\infty} \frac{n}{\left(1 + n^2\right)^p}$

If $p = 1$, $\displaystyle\sum_{n=1}^{\infty} \frac{n}{1 + n^2}$ diverges (see Example 1). Let

$$f(x) = \frac{x}{\left(1 + x^2\right)^p}, \quad p \neq 1$$

$$f'(x) = \frac{1 - (2p - 1)x^2}{\left(1 + x^2\right)^{p+1}}.$$

For a fixed $p > 0$, $p \neq 1$, $f'(x)$ is eventually negative. $f$ is positive, continuous, and eventually decreasing.

$$\int_1^{\infty} \frac{x}{\left(1 + x^2\right)^p}\,dx = \left[\frac{1}{\left(x^2 + 1\right)^{p-1}(2 - 2p)}\right]_1^{\infty}$$

For $p > 1$, this integral converges. For $0 < p < 1$, it diverges.

**51.** $\displaystyle\sum_{n=1}^{\infty} \left(\frac{3}{p}\right)^n$, Geometric series.

Converges for $\left|\dfrac{3}{p}\right| < 1 \Rightarrow |p| > 3 \Rightarrow p > 3$

**53.**

$f(1) = a_1$
$f(2) = a_2$
$f(N+1) = a_{N+1}$
$f(N) = a_N$

$$S_N = \sum_{n=1}^{N} a_n = a_1 + a_2 + \cdots + a_N$$

$$R_N = S - S_N = \sum_{n=N+1}^{\infty} a_n > 0$$

$$R_N = S - S_N = \sum_{n=N+1}^{\infty} a_n = a_{N+1} + a_{N+2} + \cdots$$

$$\leq \int_N^{\infty} f(x)\,dx$$

So, $0 \leq R_n \leq \displaystyle\int_N^{\infty} f(x)\,dx$

**55.** $S_5 = 1 + \dfrac{1}{2^2} + \dfrac{1}{3^2} + \dfrac{1}{4^2} + \dfrac{1}{5^2} \approx 1.4636$

$0 \le R_5 \le \displaystyle\int_5^\infty \dfrac{1}{x^2}\,dx = \left[-\dfrac{1}{x}\right]_5^\infty = \dfrac{1}{5} = 0.2$

$1.4636 \le \displaystyle\sum_{n=1}^\infty \dfrac{1}{n^2} \le 1.4636 + 0.2 = 1.6636$

**57.** $S_{10} = \dfrac{1}{2} + \dfrac{1}{5} + \dfrac{1}{10} + \dfrac{1}{17} + \dfrac{1}{26} + \dfrac{1}{37} + \dfrac{1}{50} + \dfrac{1}{65} + \dfrac{1}{82} + \dfrac{1}{101} \approx 0.9818$

$0 \le R_{10} \le \displaystyle\int_{10}^\infty \dfrac{1}{x^2 + 1}\,dx = [\arctan x]_{10}^\infty = \dfrac{\pi}{2} - \arctan 10 \approx 0.0997$

$0.9818 \le \displaystyle\sum_{n=1}^\infty \dfrac{1}{n^2 + 1} \le 0.9818 + 0.0997 = 1.0815$

**59.** $S_4 = \dfrac{1}{e} + \dfrac{2}{e^4} + \dfrac{3}{e^9} + \dfrac{4}{e^{16}} \approx 0.4049$

$0 \le R_4 \le \displaystyle\int_4^\infty xe^{-x^2}\,dx = \left[-\dfrac{1}{2}e^{-x^2}\right]_4^\infty = \dfrac{e^{-16}}{2} \approx 5.6 \times 10^{-8}$

$0.4049 \le \displaystyle\sum_{n=1}^\infty ne^{-n^2} \le 0.4049 + 5.6 \times 10^{-8}$

**61.** $0 \le R_N \le \displaystyle\int_N^\infty \dfrac{1}{x^4}\,dx = \left[-\dfrac{1}{3x^3}\right]_N^\infty = \dfrac{1}{3N^3} < 0.001$

$\dfrac{1}{N^3} < 0.003$

$N^3 > 333.33$

$N > 6.93$

$N \ge 7$

**63.** $R_N \le \displaystyle\int_N^\infty e^{-x/2}\,dx = \left[-2e^{-x/2}\right]_N^\infty = \dfrac{2}{e^{N/2}} < 0.001$

$\dfrac{2}{e^{N/2}} < 0.001$

$e^{N/2} > 2000$

$\dfrac{N}{2} > \ln 2000$

$N > 2\ln 2000 \approx 15.2$

$N \ge 16$

**65. (a)** $\displaystyle\sum_{n=2}^\infty \dfrac{1}{n^{1.1}}$. This is a convergent p-series with $p = 1.1 > 1$. $\displaystyle\sum_{n=2}^\infty \dfrac{1}{n \ln n}$ is a divergent series. Use the Integral Test.

$f(x) = \dfrac{1}{x \ln x}$ is positive, continuous, and decreasing for $x \ge 2$.

$\displaystyle\int_2^\infty \dfrac{1}{x \ln x}\,dx = \Big[\ln|\ln x|\Big]_2^\infty = \infty$

**(b)** $\displaystyle\sum_{n=2}^6 \dfrac{1}{n^{1.1}} = \dfrac{1}{2^{1.1}} + \dfrac{1}{3^{1.1}} + \dfrac{1}{4^{1.1}} + \dfrac{1}{5^{1.1}} + \dfrac{1}{6^{1.1}} \approx 0.4665 + 0.2987 + 0.2176 + 0.1703 + 0.1393$

$\displaystyle\sum_{n=2}^6 \dfrac{1}{n \ln n} = \dfrac{1}{2 \ln 2} + \dfrac{1}{3 \ln 3} + \dfrac{1}{4 \ln 4} + \dfrac{1}{5 \ln 5} + \dfrac{1}{6 \ln 6} \approx 0.7213 + 0.3034 + 0.1803 + 0.1243 + 0.0930$

For $n \ge 4$, the terms of the convergent series **seem** to be larger than those of the divergent series.

**(c)** $\dfrac{1}{n^{1.1}} < \dfrac{1}{n \ln n}$

$n \ln n < n^{1.1}$

$\ln n < n^{0.1}$

This inequality holds when $n \ge 3.5 \times 10^{15}$. Or, $n > e^{40}$. Then $\ln e^{40} = 40 < \left(e^{40}\right)^{0.1} = e^4 \approx 55$.

**67.** (a) Let $f(x) = 1/x$. $f$ is positive, continuous, and decreasing on $[1, \infty)$.

$$S_n - 1 \le \int_1^n \frac{1}{x} \, dx$$

$$S_n - 1 \le \ln n$$

So, $S_n \le 1 + \ln n$. Similarly,

$$S_n \ge \int_1^{n+1} \frac{1}{x} \, dx = \ln(n + 1).$$

So, $\ln(n + 1) \le S_n \le 1 + \ln n$.

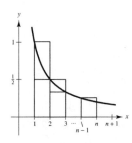

(b) Because $\ln(n + 1) \le S_n \le 1 + \ln n$, you have $\ln(n + 1) - \ln n \le S_n - \ln n \le 1$. Also, because $\ln x$ is an increasing function, $\ln(n + 1) - \ln n > 0$ for $n \ge 1$. So, $0 \le S_n - \ln n \le 1$ and the sequence $\{a_n\}$ is bounded.

(c) $a_n - a_{n+1} = [S_n - \ln n] - [S_{n+1} - \ln(n + 1)] = \int_n^{n+1} \frac{1}{x} \, dx - \frac{1}{n + 1} \ge 0$

So, $a_n \ge a_{n+1}$ and the sequence is decreasing.

(d) Because the sequence is bounded and monotonic, it converges to a limit, $\gamma$.

(e) $a_{100} = S_{100} - \ln 100 \approx 0.5822$ (Actually $\gamma \approx 0.577216$.)

**69.** $\displaystyle\sum_{n=2}^{\infty} x^{\ln n}$

(a) $x = 1$: $\displaystyle\sum_{n=2}^{\infty} 1^{\ln n} = \sum_{n=2}^{\infty} 1$, diverges

(b) $x = \dfrac{1}{e}$: $\displaystyle\sum_{n=2}^{\infty} \left(\frac{1}{e}\right)^{\ln n} = \sum_{n=2}^{\infty} e^{-\ln n} = \sum_{n=2}^{\infty} \frac{1}{n}$, diverges

(c) Let $x$ be given, $x > 0$. Put $x = e^{-p} \Leftrightarrow \ln x = -p$.

$$\sum_{n=2}^{\infty} x^{\ln n} = \sum_{n=2}^{\infty} e^{-p \ln n} = \sum_{n=2}^{\infty} n^{-p} = \sum_{n=2}^{\infty} \frac{1}{n^p}$$

This series converges for $p > 1 \Rightarrow x < \dfrac{1}{e}$.

**71.** Let $f(x) = \dfrac{1}{3x - 2}$, $f'(x) = \dfrac{-3}{(3x - 2)^2} < 0$ for $x \ge 1$.

$f$ is positive, continuous, and decreasing for $x \ge 1$.

$$\int_1^{\infty} \frac{1}{3x - 2} \, dx = \left[ \frac{1}{3} \ln |3x - 2| \right]_1^{\infty} = \infty$$

So, the series $\displaystyle\sum_{n=1}^{\infty} \frac{1}{3n - 2}$

diverges by Theorem 9.10.

**73.** $\displaystyle\sum_{n=1}^{\infty} \frac{1}{n \sqrt[4]{n}} = \sum_{n=1}^{\infty} \frac{1}{n^{5/4}}$

$p$-series with $p = \dfrac{5}{4}$

Converges by Theorem 9.11

**75.** $\displaystyle\sum_{n=0}^{\infty} \left(\frac{2}{3}\right)^n$

Geometric series with $r = \dfrac{2}{3}$

Converges by Theorem 9.6

**77.** $\displaystyle\sum_{n=1}^{\infty} \frac{n}{\sqrt{n^2 + 1}}$

$$\lim_{n \to \infty} \frac{n}{\sqrt{n^2 + 1}} = \lim_{n \to \infty} \frac{1}{\sqrt{1 + (1/n^2)}} = 1 \ne 0$$

Diverges by Theorem 9.9

**79.** $\displaystyle\sum_{n=1}^{\infty}\left(1+\frac{1}{n}\right)^{n}$

$$\lim_{n\to\infty}\left(1+\frac{1}{n}\right)^{n}=e\neq 0$$

Fails *n*th-Term Test

Diverges by Theorem 9.9

**81.** $\displaystyle\sum_{n=2}^{\infty}\frac{1}{n(\ln n)^{3}}$

Let $f(x)=\dfrac{1}{x(\ln x)^{3}}$.

*f* is positive, continuous, and decreasing for $x\geq 2$.

$$\int_{2}^{\infty}\frac{1}{x(\ln x)^{3}}\,dx=\int_{2}^{\infty}(\ln x)^{-3}\frac{1}{x}\,dx$$

$$=\left[\frac{(\ln x)^{-2}}{-2}\right]_{2}^{\infty}$$

$$=\left[-\frac{1}{2(\ln x)^{2}}\right]_{2}^{\infty}=\frac{1}{2(\ln 2)^{2}}$$

Converges by Theorem 9.10. See Exercise 47.

# Section 9.4   Comparisons of Series

**1. (a)** $\displaystyle\sum_{n=1}^{\infty}\frac{6}{n^{3/2}}=\frac{6}{1}+\frac{6}{2^{3/2}}+\cdots;\ S_{1}=6$

$\displaystyle\sum_{n=1}^{\infty}\frac{6}{n^{3/2}+3}=\frac{6}{4}+\frac{6}{2^{3/2}+3}+\cdots;\ S_{1}=\frac{3}{2}$

$\displaystyle\sum_{n=1}^{\infty}\frac{6}{n\sqrt{n^{2}+0.5}}=\frac{6}{1\sqrt{1.5}}+\frac{6}{2\sqrt{4.5}}+\cdots;\ S_{1}=\frac{6}{\sqrt{1.5}}\approx 4.9$

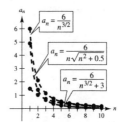

**(b)** The first series is a *p*-series. It converges $\left(p=\dfrac{3}{2}>1\right)$.

**(c)** The magnitude of the terms of the other two series are less than the corresponding terms at the convergent *p*-series. So, the other two series converge.

**(d)** The smaller the magnitude of the terms, the smaller the magnitude of the terms of the sequence of partial sums.

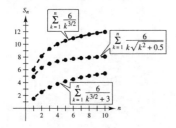

**3.** $\dfrac{1}{2n-1}>\dfrac{1}{2n}>0$ for $n\geq 1$

Therefore,

$$\sum_{n=1}^{\infty}\frac{1}{2n-1}$$

diverges by comparison with the divergent *p*-series

$$\frac{1}{2}\sum_{n=1}^{\infty}\frac{1}{n}.$$

**5.** $\dfrac{1}{\sqrt{n}-1}>\dfrac{1}{\sqrt{n}}$ for $n\geq 2$

Therefore,

$$\sum_{n=2}^{\infty}\frac{1}{\sqrt{n}-1}$$

diverges by comparison with the divergent *p*-series

$$\sum_{n=2}^{\infty}\frac{1}{\sqrt{n}}.$$

**7.** For $n \geq 3$, $\dfrac{\ln n}{n + 1} > \dfrac{1}{n + 1} > 0$.

Therefore,

$$\sum_{n=1}^{\infty} \frac{\ln n}{n + 1}$$

diverges by comparison with the divergent series

$$\sum_{n=1}^{\infty} \frac{1}{n + 1}.$$

**Note:** $\displaystyle\sum_{n=1}^{\infty} \frac{1}{n + 1}$ diverges by the Integral Test.

**9.** For $n > 3$, $\dfrac{1}{n^2} > \dfrac{1}{n!} > 0$.

Therefore,

$$\sum_{n=0}^{\infty} \frac{1}{n!}$$

converges by comparison with the convergent $p$-series

$$\sum_{n=1}^{\infty} \frac{1}{n^2}.$$

**11.** $0 < \dfrac{1}{e^{n^2}} \leq \dfrac{1}{e^n}$

Therefore,

$$\sum_{n=0}^{\infty} \frac{1}{e^{n^2}}$$

converges by comparison with the convergent geometric series

$$\sum_{n=0}^{\infty} \left(\frac{1}{e}\right)^n.$$

**13.** $\displaystyle\lim_{n \to \infty} \frac{n/(n^2 + 1)}{1/n} = \lim_{n \to \infty} \frac{n^2}{n^2 + 1} = 1$

Therefore,

$$\sum_{n=1}^{\infty} \frac{n}{n^2 + 1}$$

diverges by a limit comparison with the divergent $p$-series

$$\sum_{n=1}^{\infty} \frac{1}{n}.$$

**15.** $\displaystyle\lim_{n \to \infty} \frac{1/\sqrt{n^2 + 1}}{1/n} = \lim_{n \to \infty} \frac{n}{\sqrt{n^2 + 1}} = 1$

Therefore,

$$\sum_{n=0}^{\infty} \frac{1}{\sqrt{n^2 + 1}}$$

diverges by a limit comparison with the divergent $p$-series

$$\sum_{n=1}^{\infty} \frac{1}{n}.$$

**17.** $\displaystyle\lim_{n \to \infty} \frac{\dfrac{2n^2 - 1}{3n^5 + 2n + 1}}{1/n^3} = \lim_{n \to \infty} \frac{2n^5 - n^3}{3n^5 + 2n + 1} = \frac{2}{3}$

Therefore,

$$\sum_{n=1}^{\infty} \frac{2n^2 - 1}{3n^5 + 2n + 1}$$

converges by a limit comparison with the convergent $p$-series

$$\sum_{n=1}^{\infty} \frac{1}{n^3}.$$

**19.** $\displaystyle\lim_{n \to \infty} \frac{1/\left(n\sqrt{n^2 + 1}\right)}{1/n^2} = \lim_{n \to \infty} \frac{n^2}{n\sqrt{n^2 + 1}} = 1$

Therefore,

$$\sum_{n=1}^{\infty} \frac{1}{n\sqrt{n^2 + 1}}$$

converges by a limit comparison with the convergent $p$-series

$$\sum_{n=1}^{\infty} \frac{1}{n^2}.$$

**21.** $\displaystyle\lim_{n \to \infty} \frac{\left(n^{k-1}\right)/\left(n^k + 1\right)}{1/n} = \lim_{n \to \infty} \frac{n^k}{n^k + 1} = 1$

Therefore,

$$\sum_{n=1}^{\infty} \frac{n^{k-1}}{n^k + 1}$$

diverges by a limit comparison with the divergent $p$-series

$$\sum_{n=1}^{\infty} \frac{1}{n}.$$

**23.** $\displaystyle\sum_{n=1}^{\infty} \frac{\sqrt[3]{n}}{n} = \sum_{n=1}^{\infty} \frac{1}{n^{2/3}}$

Diverges;

$p$-series with $p = \dfrac{2}{3}$

**25.** $\displaystyle\sum_{n=1}^{\infty} \frac{1}{5^n + 1}$

Converges;

Direct comparison with convergent geometric series

$\displaystyle\sum_{n=1}^{\infty} \left(\frac{1}{5}\right)^n$

**27.** $\displaystyle\sum_{n=1}^{\infty} \frac{2n}{3n - 2}$

Diverges; $n^{\text{th}}$-Term Test

$\displaystyle\lim_{n\to\infty} \frac{2n}{3n - 2} = \frac{2}{3} \neq 0$

**29.** $\displaystyle\sum_{n=1}^{\infty} \frac{n}{\left(n^2 + 1\right)^2}$

Converges; Integral Test

**31.** $\displaystyle\lim_{n\to\infty} \frac{a_n}{1/n} = \lim_{n\to\infty} na_n$. By given conditions $\displaystyle\lim_{n\to\infty} na_n$ is

finite and nonzero. Therefore,

$\displaystyle\sum_{n=1}^{\infty} a_n$

diverges by a limit comparison with the $p$-series

$\displaystyle\sum_{n=1}^{\infty} \frac{1}{n}.$

**33.** $\dfrac{1}{2} + \dfrac{2}{5} + \dfrac{3}{10} + \dfrac{4}{17} + \dfrac{5}{26} + \cdots = \displaystyle\sum_{n=1}^{\infty} \frac{n}{n^2 + 1},$

which diverges because the degree of the numerator is only one less than the degree of the denominator.

**35.** $\displaystyle\sum_{n=1}^{\infty} \frac{1}{n^3 + 1}$

converges because the degree of the numerator is three less than the degree of the denominator.

**37.** $\displaystyle\lim_{n\to\infty} n\left(\frac{n^3}{5n^4 + 3}\right) = \lim_{n\to\infty} \frac{n^4}{5n^4 + 3} = \frac{1}{5} \neq 0$

Therefore, $\displaystyle\sum_{n=1}^{\infty} \frac{n^3}{5n^4 + 3}$ diverges.

**39.** $\dfrac{1}{200} + \dfrac{1}{400} + \dfrac{1}{600} + \cdots = \displaystyle\sum_{n=1}^{\infty} \frac{1}{200n}$

diverges, (harmonic)

**41.** $\dfrac{1}{201} + \dfrac{1}{204} + \dfrac{1}{209} + \dfrac{1}{216} = \displaystyle\sum_{n=1}^{\infty} \frac{1}{200 + n^2}$

converges

**43.** Some series diverge or converge very slowly. You cannot decide convergence or divergence of a series by comparing the first few terms.

**45.** See Theorem 9.13, page 614. One example is

$\displaystyle\sum_{n=2}^{\infty} \frac{1}{\sqrt{n} - 1}$ diverges because $\displaystyle\lim_{n\to\infty} \frac{1/\sqrt{n} - 1}{1/\sqrt{n}} = 1$ and

$\displaystyle\sum_{n=2}^{\infty} \frac{1}{\sqrt{n}}$ diverges ($p$-series).

**47.** (a) $\displaystyle\sum_{n=1}^{\infty} \frac{1}{(2n - 1)^2} = \sum_{n=1}^{\infty} \frac{1}{4n^2 - 4n + 1}$

converges because the degree of the numerator is two less than the degree of the denominator. (See Exercise 32.)

(b)

| $n$ | 5 | 10 | 20 | 50 | 100 |
|-----|------|------|------|------|------|
| $S_n$ | 1.1839 | 1.2087 | 1.2212 | 1.2287 | 1.2312 |

(c) $\displaystyle\sum_{n=3}^{\infty} \frac{1}{(2n - 1)^2} = \frac{\pi^2}{8} - S_2 \approx 0.1226$

(d) $\displaystyle\sum_{n=10}^{\infty} \frac{1}{(2n - 1)^2} = \frac{\pi^2}{8} - S_9 \approx 0.0277$

**49.** False. Let $a_n = \dfrac{1}{n^3}$ and $b_n - \dfrac{1}{n^2}$. $0 < a_n \le b_n$ and both

$$\sum_{n=1}^{\infty} \frac{1}{n^3} \text{ and } \sum_{n=1}^{\infty} \frac{1}{n^2} \text{ converge.}$$

**51.** True

**53.** True

**55.** Because $\displaystyle\sum_{n=1}^{\infty} b_n$ converges, $\displaystyle\lim_{n \to \infty} b_n = 0$. There exists $N$

such that $b_n < 1$ for $n > N$. So, $a_n b_n < a_n$ for

$n > N$ and $\displaystyle\sum_{n=1}^{\infty} a_n b_n$ converges by comparison to the

convergent series $\displaystyle\sum_{i=1}^{\infty} a_n$.

**57.** $\displaystyle\sum \frac{1}{n^2}$ and $\displaystyle\sum \frac{1}{n^3}$ both converge, and therefore, so does

$$\sum \left( \frac{1}{n^2} \right)\left( \frac{1}{n^3} \right) = \sum \frac{1}{n^5}.$$

**59.** Suppose $\displaystyle\lim_{n \to \infty} \frac{a_n}{b_n} = 0$ and $\Sigma b_n$ converges.

From the definition of limit of a sequence, there exists
$M > 0$ such that

$$\left| \frac{a_n}{b_n} - 0 \right| < 1$$

whenever $n > M$. So, $a_n < b_n$ for $n > M$. From the
Comparison Test, $\Sigma a_n$ converges.

**61.** (a) Let $\displaystyle\sum a_n = \sum \frac{1}{(n+1)^3}$, and $\displaystyle\sum b_n = \sum \frac{1}{n^2}$,

converges.

$$\lim_{n \to \infty} \frac{a_n}{b_n} = \lim_{n \to \infty} \frac{1/\left[(n+1)^3\right]}{1/(n^2)} = \lim_{n \to \infty} \frac{n^2}{(n+1)^3} = 0$$

By Exercise 59, $\displaystyle\sum_{n=1}^{\infty} \frac{1}{(n+1)^3}$ converges.

(b) Let $\displaystyle\sum a_n = \sum \frac{1}{\sqrt{n}\pi^n}$, and $\displaystyle\sum b_n = \sum \frac{1}{\pi^n}$,

converges.

$$\lim_{n \to \infty} \frac{a_n}{b_n} = \lim_{n \to \infty} \frac{1/\left(\sqrt{n}\pi^n\right)}{1/(\pi^n)} = \lim_{n \to \infty} \frac{1}{\sqrt{n}} = 0$$

By Exercise 59, $\displaystyle\sum_{n=1}^{\infty} \frac{1}{\sqrt{n}\pi^n}$ converges.

**63.** Because $\displaystyle\lim_{n \to \infty} a_n = 0$, the terms of $\Sigma \sin(a_n)$ are positive
for sufficiently large $n$. Because

$$\lim_{n \to \infty} \frac{\sin(a_n)}{a_n} = 1 \text{ and } \sum a_n$$

converges, so does $\Sigma \sin(a_n)$.

**65.** First note that $f(x) = \ln x - x^{1/4} = 0$ when
$x \approx 5503.66$. That is,

$\ln n < n^{1/4}$ for $n > 5504$

which implies that

$\dfrac{\ln n}{n^{3/2}} < \dfrac{1}{n^{5/4}}$ for $n > 5504$.

Because $\displaystyle\sum_{n=1}^{\infty} \frac{1}{n^{5/4}}$ is a convergent $p$-series,

$$\sum_{n=1}^{\infty} \frac{\ln n}{n^{3/2}}$$

converges by direct comparison.

**67.** Consider two cases:

If $a_n \ge \dfrac{1}{2^{n+1}}$, then $a_n^{1/(n+1)} \ge \left( \dfrac{1}{2^{n+1}} \right)^{1/(n+1)} = \dfrac{1}{2}$, and

$$a_n^{n/(n+1)} = \frac{a_n}{a_n^{1/(n+1)}} \le 2a_n.$$

If $a_n \le \dfrac{1}{2^{n+1}}$, then $a_n^{n/(n+1)} \le \left( \dfrac{1}{2^{n+1}} \right)^{n/(n+1)} = \dfrac{1}{2^n}$, and

combining, $a_n^{n/(n+1)} \le 2a_n + \dfrac{1}{2^n}$.

Because $\displaystyle\sum_{n=1}^{\infty} \left( 2a_n + \frac{1}{2^n} \right)$ converges, so does $\displaystyle\sum_{n=1}^{\infty} a_n^{n/(n+1)}$

by the Comparison Test.

## Section 9.5   Alternating Series

**1.** $\displaystyle\sum_{n=1}^{\infty}\frac{(-1)^{n-1}}{2n-1}=\frac{\pi}{4}\approx 0.7854$

(a)

| $n$ | 1 | 2 | 3 | 4 | 5 | 6 | 7 | 8 | 9 | 10 |
|---|---|---|---|---|---|---|---|---|---|---|
| $S_n$ | 1 | 0.6667 | 0.8667 | 0.7238 | 0.8349 | 0.7440 | 0.8209 | 0.7543 | 0.8131 | 0.7605 |

(b)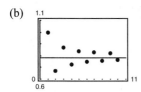

(c) The points alternate sides of the horizontal line $y=\dfrac{\pi}{4}$ that represents the sum of the series.

   The distance between successive points and the line decreases.

(d) The distance in part (c) is always less than the magnitude of the next term of the series.

**3.** $\displaystyle\sum_{n=1}^{\infty}\frac{(-1)^{n-1}}{n^2}=\frac{\pi^2}{12}\approx 0.8225$

(a)

| $n$ | 1 | 2 | 3 | 4 | 5 | 6 | 7 | 8 | 9 | 10 |
|---|---|---|---|---|---|---|---|---|---|---|
| $S_n$ | 1 | 0.75 | 0.8611 | 0.7986 | 0.8386 | 0.8108 | 0.8312 | 0.8156 | 0.8280 | 0.8180 |

(b)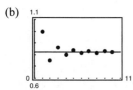

(c) The points alternate sides of the horizontal line $y=\dfrac{\pi^2}{12}$ that represents the sum of the series.

   The distance between successive points and the line decreases.

(d) The distance in part (c) is always less than the magnitude of the next term in the series.

**5.** $\displaystyle\sum_{n=1}^{\infty}\frac{(-1)^{n+1}}{n+1}$

$a_{n+1}=\dfrac{1}{n+2}<\dfrac{1}{n+1}=a_n$

$\displaystyle\lim_{n\to\infty}a_n=\lim_{n\to\infty}\frac{1}{n+1}=0$

Converges by Theorem 9.14

**7.** $\displaystyle\sum_{n=1}^{\infty}\frac{(-1)^{n}}{3^n}$

$a_{n+1}=\dfrac{1}{3^{n+1}}<\dfrac{1}{3^n}=a_n$

$\displaystyle\lim_{n\to 0}\frac{1}{3^n}=0$

Converges by Theorem 9.14

(**Note:** $\displaystyle\sum_{n=1}^{\infty}\left(\frac{-1}{3}\right)^n$ is a convergent geometric series)

**9.** $\displaystyle\sum_{n=1}^{\infty}\frac{(-1)^n(5n-1)}{4n+1}$

$\displaystyle\lim_{n\to\infty}\frac{5n-1}{4n+1}=\frac{5}{4}$

Diverges by $n$th-Term test

**11.** $\displaystyle\sum_{n=1}^{\infty}\frac{(-1)^n n}{\ln(n+1)}$

$\displaystyle\lim_{n\to\infty}\frac{n}{\ln(n+1)}=\infty$

Diverges by $n$th-Term test

**13.** $\displaystyle\sum_{n=1}^{\infty}\frac{(-1)^n}{\sqrt{n}}$

$\displaystyle a_{n+1}=\frac{1}{\sqrt{n+1}}<\frac{1}{\sqrt{n}}=a_n$

$\displaystyle\lim_{n\to\infty}\frac{1}{\sqrt{n}}=0$

Converges by Theorem 9.14

**15.** $\displaystyle\sum_{n=1}^{\infty}\frac{(-1)^{n+1}(n+1)}{\ln(n+1)}$

$\displaystyle\lim_{n\to\infty}\frac{n+1}{\ln(n+1)}=\lim_{n\to\infty}\frac{1}{1/(n+1)}=\lim_{n\to\infty}(n+1)=\infty$

Diverges by the $n$th-Term Test

**17.** $\displaystyle\sum_{n=1}^{\infty}\sin\left[\frac{(2n-1)\pi}{2}\right]=\sum_{n=1}^{\infty}(-1)^{n+1}$

Diverges by the $n$th-Term Test

**19.** $\displaystyle\sum_{n=0}^{\infty}\frac{(-1)^n}{n!}$

$\displaystyle a_{n+1}=\frac{1}{(n+1)!}<\frac{1}{n!}=a_n$

$\displaystyle\lim_{n\to\infty}\frac{1}{n!}=0$

Converges by Theorem 9.14

**21.** $\displaystyle\sum_{n=1}^{\infty}\frac{(-1)^{n+1}\sqrt{n}}{n+2}$

$\displaystyle a_{n+1}=\frac{\sqrt{n+1}}{(n+1)+2}<\frac{\sqrt{n}}{n+2}\text{ for }n\geq 2$

$\displaystyle\lim_{n\to\infty}\frac{\sqrt{n}}{n+2}=0$

Converges by Theorem 9.14

**23.** $\displaystyle\sum_{n=1}^{\infty}\frac{(-1)^{n+1}n!}{1\cdot 3\cdot 5\cdots(2n-1)}$

$\displaystyle a_{n+1}=\frac{(n+1)!}{1\cdot 3\cdot 5\cdots(2n-1)(2n+1)}=\frac{n!}{1\cdot 3\cdot 5\cdots(2n-1)}\cdot\frac{n+1}{2n+1}=a_n\left(\frac{n+1}{2n+1}\right)<a_n$

$\displaystyle\lim_{n\to\infty}a_n=\lim_{n\to\infty}\frac{n!}{1\cdot 3\cdot 5\cdots(2n-1)}=\lim_{n\to\infty}\frac{1\cdot 2\cdot 3\cdots n}{1\cdot 3\cdot 5\cdots(2n-1)}=\lim_{n\to\infty}2\left[\frac{3}{3}\cdot\frac{4}{5}\cdot\frac{5}{7}\cdots\frac{n}{2n-3}\right]\cdot\frac{1}{2n-1}=0$

Converges by Theorem 9.14

**25.** $\displaystyle\sum_{n=1}^{\infty}\frac{(-1)^{n+1}(2)}{e^n-e^{-n}}=\sum_{n=1}^{\infty}\frac{(-1)^{n+1}(2e^n)}{e^{2n}-1}$

Let $f(x)=\dfrac{2e^x}{e^{2x}-1}$. Then

$\displaystyle f'(x)=\frac{-2e^x(e^{2x}+1)}{(e^{2x}-1)^2}<0.$

So, $f(x)$ is decreasing. Therefore, $a_{n+1}<a_n$, and

$\displaystyle\lim_{n\to\infty}\frac{2e^n}{e^{2n}-1}=\lim_{n\to\infty}\frac{2e^n}{2e^{2n}}=\lim_{n\to\infty}\frac{1}{e^n}=0.$

The series converges by Theorem 9.14.

**27.** $\displaystyle S_6=\sum_{n=0}^{5}\frac{(-1)^n 5}{n!}=\frac{11}{6}$

$\displaystyle |R_6|=|S-S_6|\leq a_7=\frac{5}{720}=\frac{1}{144}$

$\displaystyle\frac{11}{6}-\frac{1}{144}\leq S\leq\frac{11}{6}+\frac{1}{144}$

$1.8264\leq S\leq 1.8403$

**29.** $\displaystyle S_6=\sum_{n=1}^{6}\frac{(-1)^{n+1}2}{n^3}\approx 1.7996$

$\displaystyle |R_6|=|S-S_6|\leq a_7=\frac{2}{7^3}\approx 0.0058$

$1.7796-0.0058\leq S\leq 1.7796+0.0058$

$1.7938\leq S\leq 1.8054$

**31.** $\displaystyle\sum_{n=1}^{\infty} \frac{(-1)^{n+1}}{n^3}$

By Theorem 9.15,

$$|R_N| \le a_{N+1} = \frac{1}{(N+1)^3} < 0.001$$

$$\Rightarrow (N+1)^3 > 1000 \Rightarrow N+1 > 10.$$

Use 10 terms.

**33.** $\displaystyle\sum_{n=1}^{\infty} \frac{(-1)^{n+1}}{2n^3 - 1}$

By Theorem 9.15,

$$|R_N| \le a_{N+1} = \frac{1}{2(N+1)^3 - 1} < 0.001$$

$$\Rightarrow 2(N+1)^3 - 1 > 1000.$$

By trial and error, this inequality is valid when

$N = 7\left[2(8^3) - 1 = 1024\right]$.

Use 7 terms.

**35.** $\displaystyle\sum_{n=0}^{\infty} \frac{(-1)^n}{n!}$

By Theorem 9.15,

$$|R_N| \le a_{N+1} = \frac{1}{(N+1)!} < 0.001$$

$$\Rightarrow (N+1)! > 1000.$$

By trial and error, this inequality is valid when
$N = 6(7! = 5040)$. Use 7 terms since the sum begins
with $n = 0$.

**37.** $\displaystyle\sum_{n=1}^{\infty} \frac{(-1)^n}{2^n}$

$\displaystyle\sum_{n=1}^{\infty} \frac{1}{2^n}$ is a convergent geometric series.

Therefore, $\displaystyle\sum_{n=1}^{\infty} \frac{(-1)^n}{2^n}$ converges absolutely.

**39.** $\displaystyle\sum_{n=1}^{\infty} \frac{(-1)^n}{n!}$

$$\frac{1}{n!} < \frac{1}{n^2} \text{ for } n \ge 4$$

and $\displaystyle\sum_{n=1}^{\infty} \frac{1}{n^2}$ is a convergent *p*-series.

So, $\displaystyle\sum_{n=1}^{\infty} \frac{1}{n!}$ converges, and

$\displaystyle\sum_{n=1}^{\infty} \frac{(-1)^n}{n!}$ converges absolutely.

**41.** $\displaystyle\sum_{n=1}^{\infty} \frac{(-1)^{n+1}}{\sqrt{n}}$

The given series converges by the Alternating Series
Test, but does not converge absolutely because

$$\sum_{n=1}^{\infty} \frac{1}{\sqrt{n}}$$

is a divergent *p*-series. Therefore, the series converges
conditionally.

**43.** $\displaystyle\sum_{n=1}^{\infty} \frac{(-1)^{n+1} n^2}{(n+1)^2}$

$$\lim_{n \to \infty} \frac{n^2}{(n+1)^2} = 1$$

Therefore, the series diverges by the *n*th-Term Test.

**45.** $\displaystyle\sum_{n=2}^{\infty} \frac{(-1)^n}{n \ln n}$

The series converges by the Alternating Series Test.

Let $f(x) = \dfrac{1}{x \ln x}$.

$$\int_2^{\infty} \frac{1}{x \ln x}\, dx = \left[\ln(\ln x)\right]_2^{\infty} = \infty$$

By the Integral Test, $\displaystyle\sum_{n=2}^{\infty} \frac{1}{n \ln n}$ diverges.

So, the series $\displaystyle\sum_{n=2}^{\infty} \frac{(-1)^n}{n \ln n}$ converges conditionally.

**47.** $\displaystyle\sum_{n=2}^{\infty}\frac{(-1)^n n}{n^3-5}$

$\displaystyle\sum_{n=2}^{\infty}\frac{n}{n^3-5}$ converges by a limit comparison to the $p$-series $\displaystyle\sum_{n=2}^{\infty}\frac{1}{n^2}$. Therefore, the given series converges absolutely.

**49.** $\displaystyle\sum_{n=0}^{\infty}\frac{(-1)^n}{(2n+1)!}$

$\displaystyle\sum_{n=0}^{\infty}\frac{1}{(2n+1)!}$

is convergent by comparison to the convergent geometric series

$\displaystyle\sum_{n=0}^{\infty}\left(\frac{1}{2}\right)^n$

because

$$\frac{1}{(2n+1)!} < \frac{1}{2^n} \text{ for } n > 0.$$

Therefore, the given series converges absolutely.

**51.** $\displaystyle\sum_{n=0}^{\infty}\frac{\cos n\pi}{n+1} = \sum_{n=0}^{\infty}\frac{(-1)^n}{n+1}$

The given series converges by the Alternating Series Test, but

$$\sum_{n=0}^{\infty}\frac{|\cos n\pi|}{n+1} = \sum_{n=0}^{\infty}\frac{1}{n+1}$$

diverges by a limit comparison to the divergent harmonic series,

$\displaystyle\sum_{n=1}^{\infty}\frac{1}{n}.$

$$\lim_{n\to\infty}\frac{|\cos n\pi|/(n+1)}{1/n} = 1, \text{ therefore, the series}$$

converges conditionally.

**53.** $\displaystyle\sum_{n=1}^{\infty}\frac{\cos n\pi}{n^2} = \sum_{n=1}^{\infty}\frac{(-1)^n}{n^2}$

$\displaystyle\sum_{n=1}^{\infty}\frac{1}{n^2}$ is a convergent $p$-series.

Therefore, the given series converges absolutely.

**55.** An alternating series is a series whose terms alternate in sign.

**57.** $|S - S_N| = |R_N| \le a_{N+1}$ (Theorem 9.15)

**59.** (a) False. For example, let $a_n = \dfrac{(-1)^n}{n}$.

Then $\displaystyle\sum a_n = \sum\frac{(-1)^n}{n}$ converges

and $\displaystyle\sum(-a_n) = \sum\frac{(-1)^{n+1}}{n}$ converges.

But, $\displaystyle\sum|a_n| = \sum\frac{1}{n}$ diverges.

(b) True. For if $\displaystyle\sum|a_n|$ converged, then so would $\displaystyle\sum a_n$ by Theorem 9.16.

**61.** True. $S_{100} = -1 + \frac{1}{2} - \frac{1}{3} + \cdots + \frac{1}{100}$

Because the next term $-\frac{1}{101}$ is negative, $S_{100}$ is an overestimate of the sum.

**63.** $\displaystyle\sum_{n=1}^{\infty}(-1)^n\frac{1}{n^p}$

If $p = 0$, then $\displaystyle\sum_{n=1}^{\infty}(-1)^n$ diverges.

If $p < 0$, then $\displaystyle\sum_{n=1}^{\infty}(-1)^n n^{-p}$ diverges.

If $p > 0$, then $\displaystyle\lim_{n\to\infty}\frac{1}{n^p} = 0$ and

$$a_{n+1} = \frac{1}{(n+1)^p} < \frac{1}{n^p} = a_n.$$

Therefore, the series converges for $p > 0$.

**65.** Because

$\displaystyle\sum_{n=1}^{\infty}|a_n|$

converges you have $\displaystyle\lim_{n\to\infty}|a_n| = 0$. So, there must exist an $N > 0$ such that $|a_N| < 1$ for all $n > N$ and it follows that $a_n^2 \le |a_n|$ for all $n > N$. So, by the Comparison Test,

$\displaystyle\sum_{n=1}^{\infty}a_n^2$

converges. Let $a_n = 1/n$ to see that the converse is false.

**67.** $\displaystyle\sum_{n=1}^{\infty}\frac{1}{n^2}$ converges, and so does $\displaystyle\sum_{n=1}^{\infty}\frac{1}{n^4}$.

**69.** (a) No, the series does not satisfy $a_{n+1} \le a_n$ for all $n$.

For example, $\dfrac{1}{9} < \dfrac{1}{8}$.

(b) Yes, the series converges.

$$S_{2n} = \frac{1}{2} - \frac{1}{3} + \cdots + \frac{1}{2^n} - \frac{1}{3^n}$$

$$= \left( \frac{1}{2} + \cdots + \frac{1}{2^n} \right) - \left( \frac{1}{3} + \cdots + \frac{1}{3^n} \right)$$

$$= \left( 1 + \frac{1}{2} + \cdots + \frac{1}{2^n} \right) - \left( 1 + \frac{1}{3} + \cdots + \frac{1}{3^n} \right)$$

As $n \to \infty$,

$$S_{2n} \to \frac{1}{1 - (1/2)} - \frac{1}{1 - (1/3)} = 2 - \frac{3}{2} = \frac{1}{2}.$$

**71.** $\displaystyle\sum_{n=1}^{\infty} \frac{10}{n^{3/2}} = 10 \sum_{n=1}^{\infty} \frac{1}{n^{3/2}}$,

convergent $p$-series

**73.** Diverges by $n$th-Term Test

$$\lim_{n \to \infty} a_n = \infty$$

**75.** Convergent geometric series

$$\left( r = \frac{7}{8} < 1 \right)$$

**77.** Convergent geometric series $\left( r = 1/\sqrt{e} \right)$ or Integral Test

**79.** Converges (absolutely) by Alternating Series Test

**81.** The first term of the series is zero, not one. You cannot regroup series terms arbitrarily.

## Section 9.6   The Ratio and Root Tests

**1.** $\dfrac{(n+1)!}{(n-2)!} = \dfrac{(n+1)(n)(n-1)(n-2)!}{(n-2)!} = (n+1)(n)(n-1)$

**3.** Use the Principle of Mathematical Induction. When $k = 1$, the formula is valid because $1 = \dfrac{(2(1))!}{2^1 \cdot 1!}$. Assume that

$$1 \cdot 3 \cdot 5 \cdots (2n-1) = \frac{(2n)!}{2^n n!}$$

and show that

$$1 \cdot 3 \cdot 5 \cdots (2n-1)(2n+1) = \frac{(2n+2)!}{2^{n+1}(n+1)!}.$$

To do this, note that:

$$1 \cdot 3 \cdot 5 \cdots (2n-1)(2n+1) = \left[ 1 \cdot 3 \cdot 5 \cdots (2n-1) \right](2n+1)$$

$$= \frac{(2n)!}{2^n n!} \cdot (2n+1) \quad \text{(Induction hypothesis)}$$

$$= \frac{(2n)!(2n+1)}{2^n n!} \cdot \frac{(2n+2)}{2(n+1)}$$

$$= \frac{(2n)!(2n+1)(2n+2)}{2^{n+1} n!(n+1)}$$

$$= \frac{(2n+2)!}{2^{n+1}(n+1)!}$$

The formula is valid for all $n \ge 1$.

**5.** $\displaystyle\sum_{n=1}^{\infty} n\left(\frac{3}{4}\right)^n = 1\left(\frac{3}{4}\right) + 2\left(\frac{9}{16}\right) + \cdots$

$S_1 = \frac{3}{4}$, $S_2 \approx 1.875$

Matches (d).

**6.** $\displaystyle\sum_{n=1}^{\infty} \left(\frac{3}{4}\right)^n \left(\frac{1}{n!}\right) = \frac{3}{4} + \frac{9}{16}\left(\frac{1}{2}\right) + \cdots$

$S_1 = \frac{3}{4}$, $S_2 = 1.03$

Matches (c).

**7.** $\displaystyle\sum_{n=1}^{\infty} \frac{(-3)^{n+1}}{n!} = 9 - \frac{3^3}{2} + \cdots$

$S_1 = 9$

Matches (f).

**8.** $\displaystyle\sum_{n=1}^{\infty} \frac{(-1)^{n-1} 4}{(2n)!} = \frac{4}{2} - \frac{4}{24} + \cdots$

$S_1 = 2$

Matches (b).

**9.** $\displaystyle\sum_{n=1}^{\infty} \left(\frac{4n}{5n-3}\right)^n = \frac{4}{2} + \left(\frac{8}{7}\right)^2 + \cdots$

$S_1 = 2,\ S_2 = 3.31$

Matches (a).

**10.** $\displaystyle\sum_{n=0}^{\infty} 4e^{-n} = 4 + \frac{4}{e} + \cdots$

$S_1 = 4$

Matches (e).

**11. (a)** Ratio Test: $\displaystyle\lim_{n\to\infty}\left|\frac{a_{n+1}}{a_n}\right| = \lim_{n\to\infty} \frac{(n+1)^3 (1/2)^{n+1}}{n^3 (1/2)^n}$

$$= \lim_{n\to\infty}\left(\frac{n+1}{n}\right)^3 \frac{1}{2} = \frac{1}{2} < 1,\ \text{converges}$$

**(b)**

| $n$ | 5 | 10 | 15 | 20 | 25 |
|---|---|---|---|---|---|
| $S_n$ | 13.7813 | 24.2363 | 25.8468 | 25.9897 | 25.9994 |

**(c)**

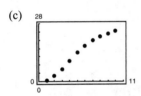

**(d)** The sum is approximately 26.

**(e)** The more rapidly the terms of the series approach 0, the more rapidly the sequence of partial sums approaches the sum of the series.

**13.** $\displaystyle\sum_{n=1}^{\infty} \frac{1}{5^n}$

$$\lim_{n\to\infty}\left|\frac{a_{n+1}}{a_n}\right| = \lim_{n\to\infty}\left|\frac{1/5^{(n+1)}}{1/5^n}\right| = \lim_{n\to\infty} \frac{5^n}{5^{n+1}} = \frac{1}{5} < 1$$

Therefore, the series converges by the Ratio Test.

**15.** $\displaystyle\sum_{n=0}^{\infty} \frac{n!}{3^n}$

$$\lim_{n\to\infty}\left|\frac{a_{n+1}}{a_n}\right| = \lim_{n\to\infty}\left|\frac{(n+1)!}{3^{n+1}} \cdot \frac{3^n}{n!}\right| = \lim_{n\to\infty} \frac{n+1}{3} = \infty$$

Therefore, by the Ratio Test, the series diverges.

**17.** $\displaystyle\sum_{n=1}^{\infty} n\left(\frac{6}{5}\right)^n$

$$\lim_{n\to\infty}\left|\frac{a_{n+1}}{a_n}\right| = \lim_{n\to\infty} \frac{(n+1)(6/5)^{n+1}}{n(6/5)^n}$$

$$= \lim_{n\to\infty} \frac{n+1}{n}\left(\frac{6}{5}\right) = \frac{6}{5} > 1$$

Therefore, the series diverges by the Ratio Test.

**19.** $\displaystyle\sum_{n=1}^{\infty} \frac{n}{4^n}$

$$\lim_{n\to\infty}\left|\frac{a_{n+1}}{a_n}\right| = \lim_{n\to\infty} \frac{(n+1)/4^{n+1}}{n/4^n} = \lim_{n\to\infty} \frac{n+1}{4n} = 1/4 < 1$$

Therefore, the series converges by the Ratio Test.

**21.** $\displaystyle\sum_{n=1}^{\infty} \frac{n^3}{3^n}$

$$\lim_{n\to\infty}\left|\frac{a_{n+1}}{a_n}\right| = \lim_{n\to\infty}\left|\frac{(n+1)^3 / 3^{(n+1)}}{n^3 / 3^n}\right|$$

$$= \lim_{n\to\infty}\left(\frac{n+1}{n}\right)^3 \frac{1}{3} = \frac{1}{3} < 1$$

Therefore, the series converges by the Ratio Test.

**23.** $\displaystyle\sum_{n=0}^{\infty} \frac{(-1)^n 2^n}{n!}$

$$\lim_{n\to\infty}\left|\frac{a_{n+1}}{a_n}\right| = \lim_{n\to\infty}\left|\frac{2^{n+1}}{(n+1)!}\cdot\frac{n!}{2^n}\right|$$

$$= \lim_{n\to\infty}\frac{2}{n+1} = 0$$

Therefore, by the Ratio Test, the series converges.

**25.** $\displaystyle\sum_{n=1}^{\infty} \frac{n!}{n3^n}$

$$\lim_{n\to\infty}\left|\frac{a_{n+1}}{a_n}\right| = \lim_{n\to\infty}\left|\frac{(n+1)!}{(n+1)3^{n+1}}\cdot\frac{n3^n}{n!}\right| = \lim_{n\to\infty}\frac{n}{3} = \infty$$

Therefore, by the Ratio Test, the series diverges.

**29.** $\displaystyle\sum_{n=0}^{\infty} \frac{6^n}{(n+1)^n}$

$$\lim_{n\to\infty}\left|\frac{a_{n+1}}{a_n}\right| = \lim_{n\to\infty}\frac{6^{n+1}/(n+2)^{n+1}}{6^n/(n+1)^n} = \lim_{n\to\infty}\frac{6}{n+2}\left(\frac{n+1}{n+2}\right)^n = 0\left(\frac{1}{e}\right) = 0.$$

To find $\displaystyle\lim_{n\to\infty}\left(\frac{n+1}{n+2}\right)^n$ : Let $y = \left(\frac{n+1}{n+2}\right)^n$

$$\ln y = n\ln\left(\frac{n+1}{n+2}\right) = \frac{\ln(n+1)-\ln(n+2)}{1/n}$$

$$\lim_{n\to\infty}[\ln y] = \lim_{n\to\infty}\left[\frac{1/(n+1)-1/(n+2)}{-1/n^2}\right] = \lim_{n\to\infty}\left[\frac{-n^2[(n+2)-(n+1)]}{(n+1)(n+2)}\right] = -1$$

by L'Hôpital's Rule. So, $y \to \dfrac{1}{e}$.

Therefore, the series converges by the Ratio Test.

**31.** $\displaystyle\sum_{n=0}^{\infty} \frac{5^n}{2^n+1}$

$$\lim_{n\to\infty}\left|\frac{a_{n+1}}{a_n}\right| = \lim_{n\to\infty}\frac{5^{n+1}/(2^{n+1}+1)}{5^n/(2^n+1)} = \lim_{n\to\infty}\frac{5(2^n+1)}{(2^{n+1}+1)} = \lim_{n\to\infty}\frac{5(1+1/2^n)}{2+1/2^n} = \frac{5}{2} > 1$$

Therefore, the series diverges by the Ratio Test.

**33.** $\displaystyle\sum_{n=0}^{\infty} \frac{(-1)^{n+1}n!}{1\cdot 3\cdot 5\cdots(2n+1)}$

$$\lim_{n\to\infty}\left|\frac{a_{n+1}}{a_n}\right| = \lim_{n\to\infty}\left|\frac{(n+1)!}{1\cdot 3\cdot 5\cdots(2n+1)(2n+3)}\cdot\frac{1\cdot 3\cdot 5\cdots(2n+1)}{n!}\right| = \lim_{n\to\infty}\frac{n+1}{2n+3} = \frac{1}{2}$$

Therefore, by the Ratio Test, the series converges.

**Note:** The first few terms of this series are $-1 + \dfrac{1}{1\cdot 3} - \dfrac{2!}{1\cdot 3\cdot 5} + \dfrac{3!}{1\cdot 3\cdot 5\cdot 7} - \cdots$.

**27.** $\displaystyle\sum_{n=0}^{\infty} \frac{e^n}{n!}$

$$\lim_{n\to\infty}\left|\frac{a_{n+1}}{a_n}\right| = \lim_{n\to\infty}\frac{e^{n+1}/(n+1)!}{e^n/n!}$$

$$= \lim_{n\to\infty}e\left(\frac{n!}{(n+1)!}\right) = \lim_{n\to\infty}\frac{e}{n+1} = 0$$

Therefore, the series converges by the Ratio Test.

**35.** $\displaystyle\sum_{n=1}^{\infty} \frac{1}{5^n}$

$$\lim_{n\to\infty} \sqrt[n]{|a_n|} = \lim_{n\to\infty} \left[\frac{1}{5^n}\right]^{1/n} = \frac{1}{5} < 1$$

Therefore, by the Root Test, the series converges.

**37.** $\displaystyle\sum_{n=1}^{\infty} \left(\frac{n}{2n+1}\right)^n$

$$\lim_{n\to\infty} \sqrt[n]{|a_n|} = \lim_{n\to\infty} \sqrt[n]{\left(\frac{n}{2n+1}\right)^n} = \lim_{n\to\infty} \frac{n}{2n+1} = \frac{1}{2}$$

Therefore, by the Root Test, the series converges.

**39.** $\displaystyle\sum_{n=1}^{\infty} \left(\frac{3n+2}{n+3}\right)^n$

$$\lim_{n\to\infty} \sqrt[n]{|a_n|} = \lim_{n\to\infty} \sqrt[n]{\left(\frac{3n+2}{n+3}\right)^n}$$

$$= \lim_{n\to\infty} \frac{3n+2}{n+3} = 3 > 1$$

Therefore, the series diverges by the Root Test.

**41.** $\displaystyle\sum_{n=2}^{\infty} \frac{(-1)^n}{(\ln n)^n}$

$$\lim_{n\to\infty} \sqrt[n]{|a_n|} = \lim_{n\to\infty} \sqrt[n]{\left|\frac{(-1)^n}{(\ln n)^n}\right|} = \lim_{n\to\infty} \frac{1}{|\ln n|} = 0$$

Therefore, by the Root Test, the series converges.

**43.** $\displaystyle\sum_{n=1}^{\infty} \left(2\sqrt[n]{n} + 1\right)^n$

$$\lim_{n\to\infty} \sqrt[n]{|a_n|} = \lim_{n\to\infty} \sqrt[n]{\left(2\sqrt[n]{n} + 1\right)^n} = \lim_{n\to\infty} \left(2\sqrt[n]{n} + 1\right)$$

To find $\displaystyle\lim_{n\to\infty} \sqrt[n]{n}$, let $y = \displaystyle\lim_{n\to\infty} \sqrt[n]{n}$. Then

$$\ln y = \lim_{n\to\infty} \left(\ln \sqrt[n]{n}\right)$$

$$= \lim_{n\to\infty} \frac{1}{n} \ln n = \lim_{n\to\infty} \frac{\ln n}{n} = \lim_{n\to\infty} \frac{1/n}{1} = 0.$$

So, $\ln y = 0$, so $y = e^0 = 1$ and

$$\lim_{n\to\infty} \left(2\sqrt[n]{n} + 1\right) = 2(1) + 1 = 3.$$

Therefore, by the Root Test, the series diverges.

**45.** $\displaystyle\sum_{n=1}^{\infty} \frac{n}{3^n}$

$$\lim_{n\to\infty} \sqrt[n]{|a_n|} = \lim_{n\to\infty} \left(\frac{n}{3^n}\right)^{1/n} = \lim_{n\to\infty} \frac{n^{1/n}}{3} = \frac{1}{3}$$

Therefore, the series converges by the Root Test.

**Note:** You can use L'Hôpital's Rule to show

$$\lim_{n\to\infty} n^{1/n} = 1:$$

Let $y = n^{1/n}$, $\ln y = \dfrac{1}{n} \ln n = \dfrac{\ln n}{n}$

$$\lim_{n\to\infty} \frac{\ln n}{n} = \lim_{n\to\infty} \frac{1/n}{1} = 0 \Rightarrow y \to 1$$

**47.** $\displaystyle\sum_{n=1}^{\infty} \left(\frac{1}{n} - \frac{1}{n^2}\right)^n$

$$\lim_{n\to\infty} \sqrt[n]{|a_n|} = \lim_{n\to\infty} \sqrt[n]{\left(\frac{1}{n} - \frac{1}{n^2}\right)^n}$$

$$= \lim_{n\to\infty} \left(\frac{1}{n} - \frac{1}{n^2}\right) = 0 - 0 = 0 < 1$$

Therefore, by the Root Test, the series converges.

**49.** $\displaystyle\sum_{n=2}^{\infty} \frac{n}{(\ln n)^n}$

$$\lim_{n\to\infty} \sqrt[n]{|a_n|} = \lim_{n\to\infty} \sqrt[n]{\frac{n}{(\ln n)^n}} = \lim_{n\to\infty} \frac{n^{1/n}}{\ln n} = 0$$

Therefore, by the Root Test, the series converges.

**51.** $\displaystyle\sum_{n=1}^{\infty} \frac{(-1)^{n+1} 5}{n}$

$$a_{n+1} = \frac{5}{n+1} < \frac{5}{n} = a_n$$

$$\lim_{n\to\infty} \frac{5}{n} = 0$$

Therefore, by the Alternating Series Test, the series converges (conditional convergence).

**53.** $\displaystyle\sum_{n=1}^{\infty} \frac{3}{n\sqrt{n}} = 3\sum_{n=1}^{\infty} \frac{1}{n^{3/2}}$

This is a convergent $p$-series.

**55.** $\displaystyle\sum_{n=1}^{\infty} \frac{5n}{2n-1}$

$$\lim_{n\to\infty} \frac{5n}{2n-1} = \frac{5}{2}$$

Therefore, the series diverges by the $n$th-Term Test

**57.** $\displaystyle\sum_{n=1}^{\infty} \frac{(-1)^n 3^{n-2}}{2^n} = \sum_{n=1}^{\infty} \frac{(-1)^n 3^n 3^{-2}}{2^n} = \sum_{n=1}^{\infty} \frac{1}{9}\left(-\frac{3}{2}\right)^n$

Because $|r| = \dfrac{3}{2} > 1$, this is a divergent geometric series.

**59.** $\displaystyle\sum_{n=1}^{\infty} \frac{10n + 3}{n2^n}$

$\displaystyle\lim_{n\to\infty} \frac{(10n + 3)/n2^n}{1/2^n} = \lim_{n\to\infty} \frac{10n + 3}{n} = 10$

Therefore, the series converges by a Limit Comparison Test with the geometric series

$\displaystyle\sum_{n=0}^{\infty} \left(\frac{1}{2}\right)^n.$

**61.** $\left|\dfrac{\cos n}{3^n}\right| \le \dfrac{1}{3^n}$

Therefore the series $\displaystyle\sum_{n=1}^{\infty} \left|\frac{\cos n}{3^n}\right|$ converges

by Direct comparison with the convergent geometric

series $\displaystyle\sum_{n=1}^{\infty} \frac{1}{3^n}$. So, $\displaystyle\sum \frac{\cos n}{3^n}$ converges.

**63.** $\displaystyle\sum_{n=1}^{\infty} \frac{n!}{n7^n}$

$\displaystyle\lim_{n\to\infty} \left|\frac{a_{n+1}}{a_n}\right| = \lim_{n\to\infty} \left|\frac{(n + 1)!/(n + 1)7^{n+1}}{n!/n7^n}\right|$

$\displaystyle = \lim_{n\to\infty} \frac{(n + 1)! \, n}{(n + 1) \, n!} \frac{1}{7}$

$\displaystyle = \lim_{n\to\infty} 7n = \infty$

Therefore, the series diverges by the Ratio Test.

**65.** $\displaystyle\sum_{n=1}^{\infty} \frac{(-1)^n 3^{n-1}}{n!}$

$\displaystyle\lim_{n\to\infty} \left|\frac{a_{n+1}}{a_n}\right| = \lim_{n\to\infty} \left|\frac{3^n}{(n + 1)!} \cdot \frac{n!}{3^{n-1}}\right| = \lim_{n\to\infty} \frac{3}{n + 1} = 0$

Therefore, by the Ratio Test, the series converges. (Absolutely)

**67.** $\displaystyle\sum_{n=1}^{\infty} \frac{(-3)^n}{3 \cdot 5 \cdot 7 \cdots (2n + 1)}$

$\displaystyle\lim_{n\to\infty} \left|\frac{a_{n+1}}{a_n}\right| = \lim_{n\to\infty} \left|\frac{(-3)^{n+1}}{3 \cdot 5 \cdot 7 \cdots (2n + 1)(2n + 3)} \cdot \frac{3 \cdot 5 \cdot 7 \cdots (2n + 1)}{(-3)^n}\right| = \lim_{n\to\infty} \frac{3}{2n + 3} = 0$

Therefore, by the Ratio Test, the series converges.

**69.** (a) and (c) are the same.

$\displaystyle\sum_{n=1}^{\infty} \frac{n5^n}{n!} = \sum_{n=0}^{\infty} \frac{(n + 1)5^{n+1}}{(n + 1)!}$

$\displaystyle = 5 + \frac{(2)(5)^2}{2!} + \frac{(3)(5)^3}{3!} + \frac{(4)(5)^4}{4!} + \cdots$

**71.** (a) and (b) are the same.

$\displaystyle\sum_{n=0}^{\infty} \frac{(-1)^n}{(2n + 1)!} = \sum_{n=1}^{\infty} \frac{(-1)^{n-1}}{(2n - 1)!}$

$\displaystyle = 1 - \frac{1}{3!} + \frac{1}{5!} - \cdots$

**73.** Replace $n$ with $n + 1$.

$\displaystyle\sum_{n=1}^{\infty} \frac{n}{7^n} = \sum_{n=0}^{\infty} \frac{n + 1}{7^{n+1}}$

**75.** (a) Because

$\dfrac{3^{10}}{2^{10}10!} \approx 1.59 \times 10^{-5},$

use 9 terms.

(b) $\displaystyle\sum_{k=1}^{9} \frac{(-3)^k}{2^k k!} \approx -0.7769$

**77.** $\displaystyle\lim_{n\to\infty} \left|\frac{a_{n+1}}{a_n}\right| = \lim_{n\to\infty} \left|\frac{(4n - 1)/(3n + 2)a_n}{a_n}\right|$

$\displaystyle = \lim_{n\to\infty} \frac{4n - 1}{3n + 2} = \frac{4}{3} > 1$

The series diverges by the Ratio Test.

**79.** $\displaystyle\lim_{n\to\infty} \left|\frac{a_{n+1}}{a_n}\right| = \lim_{n\to\infty} \left|\frac{(\sin n + 1)/(\sqrt{n})a_n}{a_n}\right|$

$\displaystyle = \lim_{n\to\infty} \frac{\sin n + 1}{\sqrt{n}} = 0 < 1$

The series converges by the Ratio Test.

**81.** $\lim\limits_{n\to\infty}\left|\dfrac{a_{n+1}}{a_n}\right| = \lim\limits_{n\to\infty}\left|\dfrac{(1 + (1)/(n))a_n}{a_n}\right| = \lim\limits_{n\to\infty}\left(1 + \dfrac{1}{n}\right) = 1$

The Ratio Test is inconclusive.

But, $\lim\limits_{n\to\infty} a_n \neq 0$, so the series diverges.

**83.** $\lim\limits_{n\to\infty}\left|\dfrac{a_{n+1}}{a_n}\right| = \lim\limits_{n\to\infty}\left|\dfrac{\dfrac{1\cdot 2\cdots n(n+1)}{1\cdot 3\cdots(2n-1)(2n+1)}}{\dfrac{1\cdot 2\cdots n}{1\cdot 3\cdots(2n-1)}}\right|$

$= \lim\limits_{n\to\infty}\dfrac{n+1}{2n+1} = \dfrac{1}{2} < 1$

The series converges by the Ratio Test.

**85.** $\displaystyle\sum_{n=3}^{\infty}\dfrac{1}{(\ln n)^n}$

$\lim\limits_{n\to\infty}\sqrt[n]{|a_n|} = \lim\limits_{n\to\infty}\sqrt[n]{\dfrac{1}{(\ln n)^n}} = \lim\limits_{n\to\infty}\dfrac{1}{\ln n} = 0$

Therefore, by the Root Test, the series converges.

**87.** $\displaystyle\sum_{n=0}^{\infty} 2\left(\dfrac{x}{3}\right)^n$

$\lim\limits_{n\to\infty}\left|\dfrac{a_{n+1}}{a_n}\right| = \lim\limits_{n\to\infty}\left|\dfrac{2(x/3)^{n+1}}{2(x/3)^n}\right| = \lim\limits_{n\to\infty}\left|\dfrac{x}{3}\right| = \left|\dfrac{x}{3}\right|$

For the series to converge, $\left|\dfrac{x}{3}\right| < 1 \Rightarrow -3 < x < 3$.

For $x = 3$, $\displaystyle\sum_{n=0}^{\infty} 2(1)^n$ diverges.

For $x = -3$, $\displaystyle\sum_{n=0}^{\infty} 2(-1)^n$ diverges.

**89.** $\displaystyle\sum_{n=1}^{\infty}\dfrac{(-1)^n(x+1)^n}{n}$

$\lim\limits_{n\to\infty}\left|\dfrac{a_{n+1}}{a_n}\right| = \lim\limits_{n\to\infty}\left|\dfrac{(x+1)^{n+1}/(n+1)}{x^n/n}\right|$

$= \lim\limits_{n\to\infty}\left|\dfrac{n}{n+1}(x+1)\right| = |x+1|$

For the series to converge,

$|x+1| < 1 \Rightarrow -1 < x + 1 < 1$

$\Rightarrow -2 < x < 0.$

For $x = 0$, $\displaystyle\sum_{n=1}^{\infty}\dfrac{(-1)^n}{n}$ converges.

For $x = -2$, $\displaystyle\sum_{n=1}^{\infty}\dfrac{(-1)^n(-1)^n}{n} = \sum_{n=1}^{\infty}\dfrac{1}{n}$ diverges.

**91.** $\displaystyle\sum_{n=0}^{\infty} n!\left(\dfrac{x}{2}\right)^n$

$\lim\limits_{n\to\infty}\left|\dfrac{a_{n+1}}{a_n}\right| = \lim\limits_{n\to\infty}\dfrac{(n+1)!\left|\dfrac{x}{2}\right|^{n+1}}{n!\left|\dfrac{x}{2}\right|^n}$

$= \lim\limits_{n\to\infty}(n+1)\left|\dfrac{x}{2}\right| = \infty$

The series converges only at $x = 0$.

**93.** See Theorem 9.17, page 627.

**95.** No. Let $a_n = \dfrac{1}{n + 10,000}$.

The series $\displaystyle\sum_{n=1}^{\infty}\dfrac{1}{n + 10,000}$ diverges.

**97.** The series converges absolutely. See Theorem 9.17.

**99.** Assume that

$\lim\limits_{n\to\infty}\left|a_{n+1}/a_n\right| = L > 1$ or that $\lim\limits_{n\to\infty}\left|a_{n+1}/a_n\right| = \infty.$

Then there exists $N > 0$ such that $\left|a_{n+1}/a_n\right| > 1$ for all $n > N$. Therefore,

$\left|a_{n+1}\right| > \left|a_n\right|,\ n > N \Rightarrow \lim\limits_{n\to\infty} a_n \neq 0 \Rightarrow \sum a_n$ diverges.

**101.** $\displaystyle\sum_{n=1}^{\infty}\dfrac{1}{n^{3/2}}$

$\lim\limits_{n\to\infty}\left|\dfrac{a_{n+1}}{a_n}\right| = \lim\limits_{n\to\infty}\left|\dfrac{1}{(n+1)^{3/2}}\cdot\dfrac{n^{3/2}}{1}\right| = \lim\limits_{n\to\infty}\left(\dfrac{n}{n+1}\right)^{3/2} = 1$

**103.** $\displaystyle\sum_{n=1}^{\infty}\dfrac{1}{n^4}$

$\lim\limits_{n\to\infty}\left|\dfrac{a_{n+1}}{a_n}\right| = \lim\limits_{n\to\infty}\left|\dfrac{1}{(n+1)^4}\cdot\dfrac{n^4}{1}\right| = \lim\limits_{n\to\infty}\left(\dfrac{n}{n+1}\right)^4 = 1$

**105.** $\displaystyle\sum_{n=1}^{\infty} \frac{1}{n^p}$, *p*-series

$$\lim_{n\to\infty} \sqrt[n]{|a_n|} = \lim_{n\to\infty} \sqrt[n]{\frac{1}{n^p}} = \lim_{n\to\infty} \frac{1}{n^{p/n}} = 1$$

So, the Root Test is inconclusive.

**Note:** $\displaystyle\lim_{n\to\infty} n^{p/n} = 1$ because if $y = n^{p/n}$, then

$\ln y = \dfrac{p}{n} \ln n$ and $\dfrac{p}{n} \ln n \to 0$ as $n \to \infty$.

So $y \to 1$ as $n \to \infty$.

**107.** $\displaystyle\sum_{n=1}^{\infty} \frac{(n!)^2}{(xn)!}$, *x* positive integer

(a)  $x = 1$: $\displaystyle\sum \frac{(n!)^2}{n!} = \sum n!$, diverges

(b)  $x = 2$: $\displaystyle\sum \frac{(n!)^2}{(2n)!}$ converges by the Ratio Test:

$$\lim_{n\to\infty} \frac{[(n+1)!]^2}{(2n+2)!} \bigg/ \frac{(n!)^2}{(2n)!} = \lim_{n\to\infty} \frac{(n+1)^2}{(2n+2)(2n+1)} = \frac{1}{4} < 1$$

(c)  $x = 3$: $\displaystyle\sum \frac{(n!)^2}{(3n)!}$ converges by the Ratio Test:

$$\lim_{n\to\infty} \frac{[(n+1)!]^2}{(3n+3)!} \bigg/ \frac{(n!)^2}{(3n)!} = \lim_{n\to\infty} \frac{(n+1)^2}{(3n+3)(3n+2)(3n+1)} = 0 < 1$$

(d)  Use the Ratio Test:

$$\lim_{n\to\infty} \frac{[(n+1)!]^2}{[x(n+1)]!} \bigg/ \frac{(n!)^2}{(xn)!} = \lim_{n\to\infty} (n+1)^2 \frac{(xn)!}{(xn+x)!}$$

The cases $x = 1, 2, 3$ were solved above. For $x > 3$, the limit is 0. So, the series converges for all integers $x \geq 2$.

**109.** First prove Abel's Summation Theorem:

If the partial sums of $\sum a_n$ are bounded and if $\{b_n\}$ decreases to zero, then $\sum a_n b_n$ converges.

Let $S_k = \displaystyle\sum_{i=1}^{k} a_i$. Let $M$ be a bound for $\{|S_k|\}$.

$$a_1 b_1 + a_2 b_2 + \cdots + a_n b_n = S_1 b_1 + (S_2 - S_1)b_2 + \cdots + (S_n - S_{n-1})b_n$$

$$= S_1(b_1 - b_2) + S_2(b_2 - b_3) + \cdots + S_{n-1}(b_{n-1} - b_n) + S_n b_n$$

$$= \sum_{i=1}^{n-1} S_i(b_i - b_{i+1}) + S_n b_n$$

The series $\displaystyle\sum_{i=1}^{\infty} S_i(b_i - b_{i+1})$ is absolutely convergent because $\left|S_i(b_i - b_{i+1})\right| \leq M(b_i - b_{i+1})$ and $\displaystyle\sum_{i=1}^{\infty} (b_i - b_{i+1})$ converges to $b_1$.

Also, $\displaystyle\lim_{n\to\infty} S_n b_n = 0$ because $\{S_n\}$ bounded and $b_n \to 0$. Thus, $\displaystyle\sum_{n=1}^{\infty} a_n b_n = \lim_{n\to\infty} \sum_{i=1}^{n} a_i b_i$ converges.

Now let $b_n = \dfrac{1}{n}$ to finish the problem.

## Section 9.7 Taylor Polynomials and Approximations

**1.** $y = -\frac{1}{2}x^2 + 1$

Parabola

Matches (d)

**2.** $y = \frac{1}{8}x^4 - \frac{1}{2}x^2 + 1$

$y$-axis symmetry

Three relative extrema

Matches (c)

**3.** $y = e^{-1/2}\left[(x + 1) + 1\right]$

Linear

Matches (a)

**4.** $y = e^{-1/2}\left[\frac{1}{3}(x - 1)^3 - (x - 1) + 1\right]$

Cubic

Matches (b)

**5.** $f(x) = \dfrac{\sqrt{x}}{4}, \ C = 4, \ f(4) = \dfrac{1}{2}$

$f'(x) = \dfrac{1}{8\sqrt{x}}, \ f'(4) = \dfrac{1}{16}$

$P_1(x) = f(4) + f'(4)(x - 4)$

$\qquad = \dfrac{1}{2} + \dfrac{1}{16}(x - 4)$

$\qquad = \dfrac{1}{16}x + \dfrac{1}{4}$

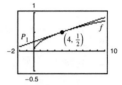

$P_1$ is the first-degree Taylor polynomial for $f$ at 4.

**7.** $f(x) = \sec x \qquad f\left(\dfrac{\pi}{4}\right) = \sqrt{2}$

$f'(x) = \sec x \tan x \qquad f'\left(\dfrac{\pi}{4}\right) = \sqrt{2}$

$P_1(x) = f\left(\dfrac{\pi}{4}\right) + f'\left(\dfrac{\pi}{4}\right)\left(x - \dfrac{\pi}{4}\right)$

$P_1(x) = \sqrt{2} + \sqrt{2}\left(x - \dfrac{\pi}{4}\right)$

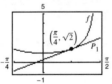

$P_1$ is called the first degree Taylor polynomial for $f$ at $\dfrac{\pi}{4}$.

**9.** $f(x) = \dfrac{4}{\sqrt{x}} = 4x^{-1/2} \qquad f(1) = 4$

$f'(x) = -2x^{-3/2} \qquad f'(1) = -2$

$f''(x) = 3x^{-5/2} \qquad f''(1) = 3$

$P_2 = f(1) + f'(1)(x - 1) + \dfrac{f''(1)}{2}(x - 1)^2$

$\quad = 4 - 2(x - 1) + \dfrac{3}{2}(x - 1)^2$

| $x$ | 0 | 0.8 | 0.9 | 1.0 | 1.1 | 1.2 | 2 |
|---|---|---|---|---|---|---|---|
| $f(x)$ | Error | 4.4721 | 4.2164 | 4.0 | 3.8139 | 3.6515 | 2.8284 |
| $P_2(x)$ | 7.5 | 4.46 | 4.215 | 4.0 | 3.815 | 3.66 | 3.5 |

**11.** $f(x) = \cos x$

$P_2(x) = 1 - \frac{1}{2}x^2$

$P_4(x) = 1 - \frac{1}{2}x^2 + \frac{1}{24}x^4$

$P_6(x) = 1 - \frac{1}{2}x^2 + \frac{1}{24}x^4 - \frac{1}{720}x^6$

(a)

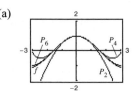

(b)   $f'(x) = -\sin x$        $P_2'(x) = -x$

$f''(x) = -\cos x$        $P_2''(x) = -1$

$f''(0) = P_2''(0) = -1$

$f'''(x) = \sin x$        $P_4'''(x) = x$

$f^{(4)}(x) = \cos x$        $P_4^{(4)}(x) = 1$

$f^{(4)}(0) = 1 = P_4^{(4)}(0)$

$f^{(5)}(x) = -\sin x$        $P_6^{(5)}(x) = -x$

$f^{(6)}(x) = -\cos x$        $P^{(6)}(x) = -1$

$f^{(6)}(0) = -1 = P_6^{(6)}(0)$

(c)  In general, $f^{(n)}(0) = P_n^{(n)}(0)$ for all $n$.

**13.**   $f(x) = e^{4x}$        $f(0) = 1$

$f'(x) = 4e^{4x}$        $f'(0) = 4$

$f''(x) = 16e^{4x}$        $f''(0) = 16$

$f'''(x) = 64e^{4x}$        $f'''(0) = 64$

$f^{(4)}(x) = 256e^{4x}$        $f^{(4)}(0) = 256$

$P_4(x) = f(0) + f'(0)x + \frac{f''(0)}{2!}x^2 + \frac{f'''(0)}{3!}x^3 + \frac{f^{(4)}(0)}{4!}x^4$

$= 1 + 4x + 8x^2 + \frac{32}{3}x^3 + \frac{32}{3}x^4$

**15.**   $f(x) = e^{-x/2}$        $f(0) = 1$

$f'(x) = -\frac{1}{2}e^{-x/2}$        $f'(0) = -\frac{1}{2}$

$f''(x) = \frac{1}{4}e^{-x/2}$        $f''(0) = \frac{1}{4}$

$f'''(x) = -\frac{1}{8}e^{-x/2}$        $f'''(0) = -\frac{1}{8}$

$f^{(4)}(x) = \frac{1}{16}e^{-x/2}$        $f^{(4)}(0) = \frac{1}{16}$

$P_4(x) = f(0) + f'(0)x + \frac{f''(0)}{2!}x^2 + \frac{f'''(0)}{3!}x^3 + \frac{f^{(4)}(0)}{4!}x^4$

$= 1 - \frac{1}{2}x + \frac{1}{8}x^2 - \frac{1}{48}x^3 + \frac{1}{384}x^4$

**17.**   $f(x) = \sin x$        $f(0) = 0$

$f'(x) = \cos x$        $f'(0) = 1$

$f''(x) = -\sin x$        $f''(0) = 0$

$f'''(x) = -\cos x$        $f'''(0) = -1$

$f^{(4)}(x) = \sin x$        $f^{(4)}(0) = 0$

$f^{(5)}(x) = \cos x$        $f^{(5)}(0) = 1$

$P_5(x) = 0 + (1)x + \frac{0}{2!}x^2 + \frac{-1}{3!}x^3 + \frac{0}{4!}x^4 + \frac{1}{5!}x^5$

$= x - \frac{1}{6}x^3 + \frac{1}{120}x^5$

**19.**   $f(x) = xe^x$        $f(0) = 0$

$f'(x) = xe^x + e^x$        $f'(0) = 1$

$f''(x) = xe^x + 2e^x$        $f''(0) = 2$

$f'''(x) = xe^x + 3e^x$        $f'''(0) = 3$

$f^{(4)}(x) = xe^x + 4e^x$        $f^{(4)}(0) = 4$

$P_4(x) = 0 + x + \frac{2}{2!}x^2 + \frac{3}{3!}x^3 + \frac{4}{4!}x^4$

$= x + x^2 + \frac{1}{2}x^3 + \frac{1}{6}x^4$

**21.** $f(x) = \dfrac{1}{x+1} = (x+1)^{-1}$ $\qquad f(0) = 1$

$\qquad f'(x) = -(x+1)^{-2}$ $\qquad\qquad f'(0) = -1$

$\qquad f''(x) = 2(x+1)^{-3}$ $\qquad\qquad f''(0) = 2$

$\qquad f'''(x) = -6(x+1)^{-4}$ $\qquad\qquad f'''(0) = -6$

$\qquad f^{(4)}(x) = 24(x+1)^{-5}$ $\qquad\qquad f^{(4)}(0) = 24$

$\qquad f^{(5)}(x) = -120(x+1)^{-6}$ $\qquad\quad f^{(5)}(0) = -120$

$\qquad P_5(x) = 1 - x + \dfrac{2x^2}{2!} - \dfrac{6x^3}{3!} + \dfrac{24x^4}{4!} - \dfrac{120x^5}{5!}$

$\qquad\qquad = 1 - x + x^2 - x^3 + x^4 - x^5$

**23.** $f(x) = \sec x$ $\qquad\qquad\qquad f(0) = 1$

$\qquad f'(x) = \sec x \tan x$ $\qquad\qquad f'(0) = 0$

$\qquad f''(x) = \sec^3 x + \sec x \tan^2 x$ $\qquad f''(0) = 1$

$\qquad P_2(x) = 1 + 0x + \dfrac{1}{2!}x^2 = 1 + \dfrac{1}{2}x^2$

**27.** $f(x) = \sqrt{x} = x^{1/2}$ $\qquad f(4) = 2$

$\qquad f'(x) = \dfrac{1}{2}x^{-1/2}$ $\qquad\qquad f'(4) = \dfrac{1}{4}$

$\qquad f''(x) = -\dfrac{1}{4}x^{-3/2}$ $\qquad\quad f''(4) = -\dfrac{1}{32}$

$\qquad f'''(x) = \dfrac{3}{8}x^{-5/2}$ $\qquad\quad f'''(4) = \dfrac{3}{256}$

$\qquad P_3(x) = 2 + \dfrac{1}{4}(x-4) - \dfrac{1/32}{2!}(x-4)^2 + \dfrac{3/256}{3!}(x-4)^3$

$\qquad\qquad = 2 + \dfrac{1}{4}(x-4) - \dfrac{1}{64}(x-4)^2 + \dfrac{1}{512}(x-4)^3$

**29.** $f(x) = \ln x$ $\qquad\qquad f(2) = \ln 2$

$\qquad f'(x) = \dfrac{1}{x} = x^{-1}$ $\qquad f'(2) = 1/2$

$\qquad f''(x) = -x^{-2}$ $\qquad\qquad f''(2) = -1/4$

$\qquad f'''(x) = 2x^{-3}$ $\qquad\qquad f'''(2) = 1/4$

$\qquad f^{(4)}(x) = -6x^{-4}$ $\qquad\quad f^{(4)}(2) = -3/8$

$\qquad P_4(x) = \ln 2 + \dfrac{1}{2}(x-2) - \dfrac{1/4}{2!}(x-2)^2 + \dfrac{1/4}{3!}(x-2)^3 - \dfrac{3/8}{4!}(x-2)^4$

$\qquad\qquad = \ln 2 + \dfrac{1}{2}(x-2) - \dfrac{1}{8}(x-2)^2 + \dfrac{1}{24}(x-2)^3 - \dfrac{1}{64}(x-2)^4$

**25.** $f(x) = \dfrac{2}{x} = 2x^{-1}$ $\qquad f(1) = 2$

$\qquad f'(x) = -2x^{-2}$ $\qquad\qquad f'(1) = -2$

$\qquad f''(x) = 4x^{-3}$ $\qquad\qquad f''(1) = 4$

$\qquad f'''(x) = -12x^{-4}$ $\qquad\quad f'''(1) = -12$

$\qquad P_3(x) = 2 - 2(x-1) + \dfrac{4}{2!}(x-1)^2 - \dfrac{12}{3!}(x-1)^3$

$\qquad\qquad = 2 - 2(x-1) + 2(x-1)^2 - 2(x-1)^3$

**31.** (a) $P_3(x) = \pi x + \dfrac{\pi^3}{3}x^3$

(b) $Q_3(x) = 1 + 2\pi\left(x - \dfrac{1}{4}\right) + 2\pi^2\left(x - \dfrac{1}{4}\right)^2 + \dfrac{8}{3}\pi^3\left(x - \dfrac{1}{4}\right)^3$

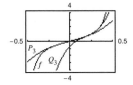

**33.** $f(x) = \sin x$

$P_1(x) = x$

$P_3(x) = x - \dfrac{1}{6}x^3$

$P_5(x) = x - \dfrac{1}{6}x^3 + \dfrac{1}{120}x^5$

(a)

| $x$ | 0.00 | 0.25 | 0.50 | 0.75 | 1.00 |
|---|---|---|---|---|---|
| $\sin x$ | 0.0000 | 0.2474 | 0.4794 | 0.6816 | 0.8415 |
| $P_1(x)$ | 0.0000 | 0.2500 | 0.5000 | 0.7500 | 1.0000 |
| $P_3(x)$ | 0.0000 | 0.2474 | 0.4792 | 0.6797 | 0.8333 |
| $P_5(x)$ | 0.0000 | 0.2474 | 0.4794 | 0.6817 | 0.8417 |

(b)

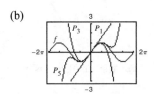

(c) As the distance increases, the accuracy decreases.

**35.** $f(x) = \arcsin x$

(a) $P_3(x) = x + \dfrac{x^3}{6}$

(b)

| $x$ | −0.75 | −0.50 | −0.25 | 0 | 0.25 | 0.50 | 0.75 |
|---|---|---|---|---|---|---|---|
| $f(x)$ | −0.848 | −0.524 | −0.253 | 0 | 0.253 | 0.524 | 0.848 |
| $P_3(x)$ | −0.820 | −0.521 | −0.253 | 0 | 0.253 | 0.521 | 0.820 |

(c)

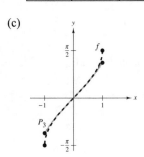

**37.** $f(x) = \cos x$

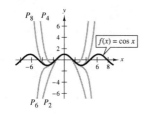

**39.** $f(x) = \ln(x^2 + 1)$

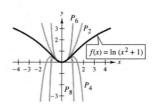

**41.** $f(x) = e^{3x} \approx 1 + 3x + \frac{9}{2}x^2 + \frac{9}{2}x^3 + \frac{27}{8}x^4$

$f\left(\frac{1}{2}\right) \approx 4.3984$

**43.** $f(x) = \ln x \approx \ln(2) + \frac{1}{2}(x - 2) - \frac{1}{8}(x - 2)^2 + \frac{1}{24}(x - 2)^3 - \frac{1}{64}(x - 2)^4$

$f(2.1) \approx 0.7419$

**45.** $f(x) = \cos x; \ f^{(5)}(x) = -\sin x \Rightarrow$ Max on $[0, 0.3]$ is 1.

$R_4(x) \leq \frac{1}{5!}(0.3)^5 = 2.025 \times 10^{-5}$

Note: you could use $R_5(x)$: $f^{(6)}(x) = -\cos x$, max on $[0, 0.3]$ is 1.

$R_5(x) \leq \frac{1}{6!}(0.3)^6 = 1.0125 \times 10^{-6}$

Exact error: $0.000001 = 1.0 \times 10^{-6}$

**47.** $f(x) = \arcsin x; \ f^{(4)}(x) = \frac{x(6x^2 + 9)}{(1 - x^2)^{7/2}} \Rightarrow$ Max on

$[0, 0.4]$ is $f^{(4)}(0.4) \approx 7.3340$.

$R_3(x) \leq \frac{7.3340}{4!}(0.4)^4 \approx 0.00782 = 7.82 \times 10^{-3}$. The

exact error is $8.5 \times 10^{-4}$. [Note: You could use $R_4$.]

**49.** $g(x) = \sin x$

$\left| g^{(n+1)}(x) \right| \leq 1$ for all $x$.

$R_n(x) \leq \frac{1}{(n + 1)!}(0.3)^{n+1} < 0.001$

By trial and error, $n = 3$.

**51.**    $f(x) = e^x$

$f^{(n+1)}(x) = e^x$

Max on $[0, 0.6]$ is $e^{0.6} \approx 1.8221$.

$R_n \leq \frac{1.8221}{(n + 1)!}(0.6)^{n+1} < 0.001$

By trial and error, $n = 5$.

**53.** $f(x) = \ln(x + 1)$

$f^{(n+1)}(x) = \frac{(-1)^n n!}{(x + 1)^{n+1}} \Rightarrow$ Max on $[0, 0.5]$ is $n!$.

$R_n \leq \frac{n!}{(n + 1)!}(0.5)^{n+1} = \frac{(0.5)^{n+1}}{n + 1} < 0.0001$

By trial and error, $n = 9$. (See Example 9.) Using 9 terms, $\ln(1.5) \approx 0.4055$.

**55.**    $f(x) = e^x \approx 1 + x + \frac{x^2}{2} + \frac{x^3}{6}, \ x < 0$

$R_3(x) = \frac{e^z}{4!}x^4 < 0.001$

$e^z x^4 < 0.024$

$\left| xe^{z/4} \right| < 0.3936$

$|x| < \frac{0.3936}{e^{z/4}} < 0.3936, \ z < 0$

$-0.3936 < x < 0$

**57.** $f(x) = \cos x \approx 1 - \dfrac{x^2}{2!} + \dfrac{x^4}{4!}$, fifth degree polynomial

$\left| f^{(n+1)}(x) \right| \le 1$ for all $x$ and all $n$.

$\left| R_5(x) \right| \le \dfrac{1}{6!} |x|^6 < 0.001$

$|x|^6 < 0.72$

$|x| < 0.9467$

$-0.9467 < x < 0.9467$

**Note:** Use a graphing utility to graph

$y = \cos x - \left(1 - x^2/2 + x^4/24\right)$ in the viewing

window $[-0.9467, 0.9467] \times [-0.001, 0.001]$ to verify

the answer.

**59.** The graph of the approximating polynomial $P$ and the elementary function $f$ both pass through the point $(c, f(c))$ and the slopes of $P$ and $f$ agree at $(c, f(c))$. Depending on the degree of $P$, the $n$th derivatives of $P$ and $f$ agree at $(c, f(c))$.

**61.** See definition on page 638.

**63.** As the degree of the polynomial increases, the graph of the Taylor polynomial becomes a better and better approximation of the function within the interval of convergence. Therefore, the accuracy is increased.

**65.** (a) $f(x) = e^x$

$P_4(x) = 1 + x + \dfrac{1}{2}x^2 + \dfrac{1}{6}x^3 + \dfrac{1}{24}x^4$

$g(x) = xe^x$

$Q_5(x) = x + x^2 + \dfrac{1}{2}x^3 + \dfrac{1}{6}x^4 + \dfrac{1}{24}x^5$

$Q_5(x) = x\, P_4(x)$

(b) $f(x) = \sin x$

$P_5(x) = x - \dfrac{x^3}{3!} + \dfrac{x^5}{5!}$

$g(x) = x \sin x$

$Q_6(x) = x\, P_5(x) = x^2 - \dfrac{x^4}{3!} + \dfrac{x^6}{5!}$

(c) $g(x) = \dfrac{\sin x}{x} = \dfrac{1}{x} P_5(x) = 1 - \dfrac{x^2}{3!} + \dfrac{x^4}{5!}$

**67.** (a) $Q_2(x) = -1 + \dfrac{\pi^2 (x + 2)^2}{32}$

(b) $R_2(x) = -1 + \dfrac{\pi^2 (x - 6)^2}{32}$

(c) No. The polynomial will be linear. Horizontal translations of the result in part (a) are possible only at $x = -2 + 8n$ (where $n$ is an integer) because the period of $f$ is 8.

**69.** Let $f$ be an even function and $P_n$ be the $n$th Maclaurin polynomial for $f$. Because $f$ is even, $f'$ is odd, $f''$ is even, $f'''$ is odd, etc. All of the odd derivatives of $f$ are odd and so, all of the odd powers of $x$ will have coefficients of zero. $P_n$ will only have terms with even powers of $x$.

**71.** As you move away from $x = c$, the Taylor Polynomial becomes less and less accurate.

## Section 9.8   Power Series

**1.** Centered at 0

**3.** Centered at 2

**5.** $\displaystyle\sum_{n=0}^{\infty} (-1)^n \dfrac{x^n}{n+1}$

$L = \lim_{n \to \infty} \left| \dfrac{u_{n+1}}{u_n} \right| = \lim_{n \to \infty} \left| \dfrac{(-1)^{n+1} x^{n+1}}{n+2} \cdot \dfrac{n+1}{(-1)^n x^n} \right|$

$= \lim_{n \to \infty} \left| \dfrac{n+1}{n+2} \right| |x| = |x|$

$|x| < 1 \Rightarrow R = 1$

**7.** $\displaystyle\sum_{n=1}^{\infty} \dfrac{(4x)^n}{n^2}$

$L = \lim_{n \to \infty} \left| \dfrac{u_{n+1}}{u_n} \right|$

$= \lim_{n \to \infty} \left| \dfrac{(4x)^{n+1}/(n+1)^2}{(4x)^n/n^2} \right| = \lim_{n \to \infty} \left| \dfrac{n^2}{(n+1)^2}(4x) \right| = 4|x|$

$4|x| < 1 \Rightarrow R = \dfrac{1}{4}$

**9.** $\sum_{n=0}^{\infty} \dfrac{x^{2n}}{(2n)!}$

$$L = \lim_{n \to \infty} \left| \dfrac{u_{n+1}}{u_n} \right|$$

$$= \lim_{n \to \infty} \left| \dfrac{x^{(2n+2)}/(2n + 2)!}{x^{2n}/(2n)!} \right|$$

$$= \lim_{n \to \infty} \left| \dfrac{x^2}{(2n + 2)(2n + 1)} \right| = 0$$

So, the series converges for all $x \Rightarrow R = \infty$.

**11.** $\sum_{n=0}^{\infty} \left( \dfrac{x}{4} \right)^n$

Because the series is geometric, it converges only if

$$\left| \dfrac{x}{4} \right| < 1, \text{ or } -4 < x < 4.$$

**13.** $\sum_{n=1}^{\infty} \dfrac{(-1)^n x^n}{n}$

$$\lim_{n \to \infty} \left| \dfrac{u_{n+1}}{u_n} \right| = \lim_{n \to \infty} \left| \dfrac{(-1)^{n+1} x^{n+1}}{n + 1} \cdot \dfrac{n}{(-1)^n x^n} \right|$$

$$= \lim_{n \to \infty} \left| \dfrac{nx}{n + 1} \right| = |x|$$

Interval: $-1 < x < 1$

When $x = 1$, the alternating series $\sum_{n=1}^{\infty} \dfrac{(-1)^n}{n}$ converges.

When $x = -1$, the $p$-series $\sum_{n=1}^{\infty} \dfrac{1}{n}$ diverges.

Therefore, the interval of convergence is $(-1, 1]$.

**15.** $\sum_{n=0}^{\infty} \dfrac{x^{5n}}{n!}$

$$\lim_{n \to \infty} \left| \dfrac{u_{n+1}}{u_n} \right| = \lim_{n \to \infty} \left| \dfrac{x^{5(n+1)}/(n + 1)!}{5^n/n!} \right| = \lim_{n \to \infty} \left| \dfrac{x^5}{n + 1} \right| = 0$$

The series converges for all $x$. The interval of convergence is $(-\infty, \infty)$.

**17.** $\sum_{n=0}^{\infty} (2n)! \left( \dfrac{x}{3} \right)^n$

$$\lim_{n \to \infty} \left| \dfrac{u_{n+1}}{u_n} \right| = \lim_{n \to \infty} \left| \dfrac{(2n + 2)!(x/3)^{n+1}}{(2n)!(x/3)^n} \right|$$

$$= \left| \dfrac{(2n + 2)(2n + 1)x}{3} \right| = \infty$$

The series converges only for $x = 0$.

**19.** $\sum_{n=1}^{\infty} \dfrac{(-1)^{n+1} x^n}{6^n}$

Because the series is geometric, it converges only if

$$\left| \dfrac{x}{6} \right| < 1 \Rightarrow |x| < 6 \text{ or } -6 < x < 6.$$

**21.** $\sum_{n=1}^{\infty} \dfrac{(-1)^{n+1}(x - 4)^n}{n9^n}$

$$\lim_{n \to \infty} \left| \dfrac{u_{n+1}}{u_n} \right| = \lim_{n \to \infty} \left| \dfrac{(-1)^{n+2}(x - 4)^{n+1}/((n + 1)9^{n+1})}{(-1)^n(x - 4)^n/(n9^n)} \right|$$

$$= \lim_{n \to \infty} \left| \dfrac{n}{n + 1} \dfrac{(x - 4)}{9} \right| = \dfrac{1}{9}|x - 4|$$

$R = 9$

Interval: $-5 < x < 13$

When $x = 13$, $\sum_{n=1}^{\infty} \dfrac{(-1)^{n+1}9^n}{n9^n} = \sum_{n=1}^{\infty} \dfrac{(-1)^{n+1}}{n}$ converges.

When $x = -5$, $\sum_{n=1}^{\infty} \dfrac{(-1)^{n+1}(-9)^n}{n9^n} = \sum_{n=1}^{\infty} \dfrac{-1}{n}$ diverges.

Therefore, the interval of convergence is $(-5, 13]$.

**23.** $\displaystyle\sum_{n=0}^{\infty} \frac{(-1)^{n+1}(x-1)^{n+1}}{n+1}$

$$\lim_{n\to\infty}\left|\frac{u_{n+1}}{u_n}\right| = \lim_{n\to\infty}\left|\frac{(-1)^{n+2}(x-1)^{n+2}}{n+2} \cdot \frac{n+1}{(-1)^{n+1}(x-1)^{n+1}}\right| = \lim_{n\to\infty}\left|\frac{(n+1)(x-1)}{n+2}\right| = |x-1|$$

$R = 1$

Center: $x = 1$

Interval: $-1 < x - 1 < 1$ or $0 < x < 2$

When $x = 0$, the series $\displaystyle\sum_{n=0}^{\infty}\frac{1}{n+1}$ diverges by the integral test.

When $x = 2$, the alternating series $\displaystyle\sum_{n=0}^{\infty}\frac{(-1)^{n+1}}{n+1}$ converges.

Therefore, the interval of convergence is $(0, 2]$.

**25.** $\displaystyle\sum_{n=1}^{\infty}\left(\frac{x-3}{3}\right)^{n-1}$ is geometric. It converges if

$$\left|\frac{x-3}{3}\right| < 1 \Rightarrow |x-3| < 3 \Rightarrow 0 < x < 6.$$

Therefore, the interval of convergence is $(0, 6)$.

**27.** $\displaystyle\sum_{n=1}^{\infty}\frac{n}{n+1}(-2x)^{n-1}$

$$\lim_{n\to\infty}\left|\frac{u_{n+1}}{u_n}\right| = \lim_{n\to\infty}\left|\frac{(n+1)(-2x)^n}{n+2} \cdot \frac{n+1}{n(-2x)^{n-1}}\right|$$

$$= \lim_{n\to\infty}\left|\frac{(-2x)(n+1)^2}{n(n+2)}\right| = 2|x|$$

$R = \dfrac{1}{2}$

Interval: $-\dfrac{1}{2} < x < \dfrac{1}{2}$

When $x = -\dfrac{1}{2}$, the series $\displaystyle\sum_{n=1}^{\infty}\frac{n}{n+1}$ diverges by the $n$th Term Test.

When $x = \dfrac{1}{2}$, the alternating series

$\displaystyle\sum_{n=1}^{\infty}\frac{(-1)^{n-1}n}{n+1}$ diverges.

Therefore, the interval of convergence is $\left(-\dfrac{1}{2}, \dfrac{1}{2}\right)$.

**29.** $\displaystyle\sum_{n=0}^{\infty}\frac{x^{3n+1}}{(3n+1)!}$

$$\lim_{n\to\infty}\left|\frac{u_{n+1}}{u_n}\right| = \lim_{n\to\infty}\left|\frac{x^{3n+4}/(3n+4)!}{x^{3n+1}/(3n+1)!}\right|$$

$$= \lim_{n\to\infty}\left|\frac{x^3}{(3n+4)(3n+3)(3n+2)}\right| = 0$$

Therefore, the interval of convergence is $(-\infty, \infty)$.

**31.** $\displaystyle\sum_{n=1}^{\infty}\frac{2 \cdot 3 \cdot 4 \cdots (n+1)x^n}{n!} = \sum_{n=1}^{\infty}(n+1)x^n$

$$\lim_{n\to\infty}\left|\frac{a_{n+1}}{a_n}\right| = \lim_{n\to\infty}\left|\frac{(n+2)x^{n+1}}{(n+1)x^n}\right| = \lim_{n\to\infty}\left|\frac{n+2}{n+1}x\right| = |x|$$

Converges if $|x| < 1 \Rightarrow -1 < x < 1$.

At $x = \pm 1$, diverges.

Therefore the interval of convergence is $(-1, 1)$.

**33.** $\displaystyle\sum_{n=1}^{\infty} \frac{(-1)^{n+1} 3 \cdot 7 \cdot 11 \cdots (4n-1)(x-3)^n}{4^n}$

$$\lim_{n\to\infty}\left|\frac{u_{n+1}}{u_n}\right| = \lim_{n\to\infty}\left|\frac{(-1)^{n+2} \cdot 3 \cdot 7 \cdot 11 \cdots (4n-1)(4n+3)(x-3)^{n+1}}{4^{n+1}} \cdot \frac{4^n}{(-1)^{n+1} \cdot 3 \cdot 7 \cdot 11 \cdots (4n-1)(x-3)^n}\right|$$

$$= \lim_{n\to\infty}\left|\frac{(4n+3)(x-3)}{4}\right| = \infty$$

$R = 0$

Center: $x = 3$

Therefore, the series converges only for $x = 3$.

**35.** $\displaystyle\sum_{n=1}^{\infty} \frac{(x-c)^{n-1}}{c^{n-1}}$

$$\lim_{n\to\infty}\left|\frac{u_{n+1}}{u_n}\right| = \lim_{n\to\infty}\left|\frac{(x-c)^n}{c^n} \cdot \frac{c^{n-1}}{(x-c)^{n-1}}\right| = \frac{1}{c}|x-c|$$

$R = c$

Center: $x = c$

Interval: $-c < x - c < c$ or $0 < x < 2c$

When $x = 0$, the series $\displaystyle\sum_{n=1}^{\infty}(-1)^{n-1}$ diverges.

When $x = 2c$, the series $\displaystyle\sum_{n=1}^{\infty} 1$ diverges.

Therefore, the interval of convergence is $(0, 2c)$.

**37.** $\displaystyle\sum_{n=0}^{\infty}\left(\frac{x}{k}\right)^n$

Because the series is geometric, it converges only if $|x/k| < 1$ or $-k < x < k$.

Therefore, the interval of convergence is $(-k, k)$.

**39.** $\displaystyle\sum_{n=1}^{\infty} \frac{k(k+1)\cdots(k+n-1)x^n}{n!}$

$$\lim_{n\to\infty}\left|\frac{u_{n+1}}{u_n}\right| = \lim_{n\to\infty}\left|\frac{k(k+1)\cdots(k+n-1)(k+n)x^{n+1}}{(n+1)!} \cdot \frac{n!}{k(k+1)\cdots(k+n-1)x^n}\right| = \lim_{n\to\infty}\left|\frac{(k+n)x}{n+1}\right| = |x|$$

$R = 1$

When $x = \pm 1$, the series diverges and the interval of convergence is $(-1, 1)$.

$$\left[\frac{k(k+1)\cdots(k+n-1)}{1\cdot 2 \cdots n} \geq 1\right]$$

**41.** $\displaystyle\sum_{n=0}^{\infty} \frac{x^n}{n!} = 1 + \frac{x}{1} + \frac{x^2}{2} + \cdots = \sum_{n=1}^{\infty} \frac{x^{n-1}}{(n-1)!}$

**43.** $\displaystyle\sum_{n=0}^{\infty} \frac{x^{2n+1}}{(2n+1)!} = \sum_{n=1}^{\infty} \frac{x^{2n-1}}{(2n-1)!}$

Replace $n$ with $n - 1$.

**45. (a)** $f(x) = \sum_{n=0}^{\infty} \left(\frac{x}{3}\right)^n, (-3, 3)$    (Geometric)

**(b)** $f'(x) = \sum_{n=1}^{\infty} \frac{n}{3}\left(\frac{x}{3}\right)^{n-1}, (-3, 3)$

**(c)** $f''(x) = \sum_{n=2}^{\infty} \frac{n(n-1)}{9}\left(\frac{x}{3}\right)^{n-2}, (-3, 3)$

**(d)** $\int f(x)\, dx = \sum_{n=0}^{\infty} \frac{3}{n+1}\left(\frac{x}{3}\right)^{n+1}, [-3, 3)$

$$\left[\sum \frac{3}{n+1}\left(\frac{-3}{3}\right)^{n+1} = \sum \frac{(-1)^{n+1}3}{n+1}, \text{converges}\right]$$

**47. (a)** $f(x) = \sum_{n=0}^{\infty} \frac{(-1)^{n+1}(x-1)^{n+1}}{n+1}, (0, 2]$

**(b)** $f'(x) = \sum_{n=0}^{\infty} (-1)^{n+1}(x-1)^n, (0, 2)$

**(c)** $f''(x) = \sum_{n=1}^{\infty} (-1)^{n+1}n(x-1)^{n-1}, (0, 2)$

**(d)** $\int f(x)\, dx = \sum_{n=1}^{\infty} \frac{(-1)^{n+1}(x-1)^{n+2}}{(n+1)(n+2)}, [0, 2]$

**49.** A series of the form

$$\sum_{n=0}^{\infty} a_n(x-c)^n = a_0 + a_1(x-c) + a_2(x-c)^2 + \cdots$$
$$+ a_n(x-c)^n + \cdots$$

is called a power series centered at $c$, where $c$ is constant.

**51.** The interval of convergence of a power series is the set of all values of $x$ for which the power series converges.

**53.** You differentiate and integrate the power series term by term. The radius of convergence remains the same. However, the interval of convergence might change.

**55.** Many answers possible.

**(a)** $\sum_{n=1}^{\infty} \left(\frac{x}{2}\right)^n$ Geometric: $\left|\frac{x}{2}\right| < 1 \Rightarrow |x| < 2$

**(b)** $\sum_{n=1}^{\infty} \frac{(-1)^n x^n}{n}$ converges for $-1 < x \le 1$

**(c)** $\sum_{n=1}^{\infty} (2x+1)^n$ Geometric:

$|2x+1| < 1 \Rightarrow -1 < x < 0$

**(d)** $\sum_{n=1}^{\infty} \frac{(x-2)^n}{n4^n}$ converges for $-2 \le x < 6$

**57. (a)** $f(x) = \sum_{n=0}^{\infty} \frac{x^{2n+1}}{(2n+1)!}$

$$\lim_{n\to\infty}\left|\frac{u_{n+1}}{u_n}\right| = \lim_{n\to\infty}\left|\frac{x^{2n+3}}{(2n+3)!} \cdot \frac{(2n+1)!}{x^{2n+1}}\right|$$
$$= \lim_{n\to\infty}\left|\frac{x^2}{(2n+2)(2n+3)}\right| = 0$$

Therefore, the interval of convergence is $(-\infty, \infty)$.

$$g(x) = \sum_{n=0}^{\infty} \frac{(-1)^n x^{2n}}{(2n)!}$$

$$\lim_{n\to\infty}\left|\frac{u_{n+1}}{u_n}\right| = \lim_{n\to\infty}\left|\frac{(-1)^{n+1}x^{2n+2}}{(2n+2)!} \cdot \frac{(2n)!}{(-1)^n x^{2n}}\right|$$
$$= \lim_{n\to\infty}\frac{1}{2n+2} = 0$$

Therefore, the interval of convergence is $(-\infty, \infty)$.

**(b)** $f'(x) = \sum_{n=0}^{\infty} \frac{(-1)^n x^{2n}}{(2n)!} = g(x)$

**(c)** $g'(x) = \sum_{n=1}^{\infty} \frac{(-1)^n x^{2n-1}}{(2n-1)!} = \sum_{n=0}^{\infty} \frac{(-1)^{n+1}x^{2n+1}}{(2n+1)!}$
$$= -\sum_{n=0}^{\infty} \frac{(-1)^n x^{2n+1}}{(2n+1)!} = -f(x)$$

**(d)** $f(x) = \sin x$ and

$g(x) = \cos x$

**59.** $y = \sum_{n=0}^{\infty} \frac{(-1)^n x^{2n+1}}{(2n+1)!} = \sum_{n=1}^{\infty} \frac{(-1)^{n-1}x^{2n-1}}{(2n-1)!}$

$y' = \sum_{n=0}^{\infty} \frac{(-1)^n(2n+1)x^{2n}}{(2n+1)!} = \sum_{n=0}^{\infty} \frac{(-1)^n x^{2n}}{(2n)!}$

$y'' = \sum_{n=1}^{\infty} \frac{(-1)^n(2n)x^{2n-1}}{(2n)!} = \sum_{n=1}^{\infty} \frac{(-1)^n x^{2n-1}}{(2n-1)!}$

$y'' + y = \sum_{n=1}^{\infty} \frac{(-1)^n x^{2n-1}}{(2n-1)!} + \sum_{n=1}^{\infty} \frac{(-1)^{n-1}x^{2n-1}}{(2n-1)!} = 0$

**61.** $y = \sum_{n=0}^{\infty} \frac{x^{2n+1}}{(2n+1)!} = \sum_{n=1}^{\infty} \frac{x^{2n-1}}{(2n-1)!}$

$y' = \sum_{n=0}^{\infty} \frac{(2n+1)x^{2n}}{(2n+1)!} = \sum_{n=0}^{\infty} \frac{x^{2n}}{(2n)!}$

$y'' = \sum_{n=1}^{\infty} \frac{(2n)x^{2n-1}}{(2n)!} = \sum_{n=1}^{\infty} \frac{x^{2n-1}}{(2n-1)!} = y$

$y'' - y = 0$

**63.** $y = \sum_{n=0}^{\infty} \frac{x^{2n}}{2^n n!}$ $\quad y' = \sum_{n=1}^{\infty} \frac{2nx^{2n-1}}{2^n n!}$ $\quad y'' = \sum_{n=1}^{\infty} \frac{2n(2n-1)x^{2n-2}}{2^n n!}$

$$y'' - xy' - y = \sum_{n=1}^{\infty} \frac{2n(2n-1)x^{2n-2}}{2^n n!} - \sum_{n=1}^{\infty} \frac{2nx^{2n}}{2^n n!} - \sum_{n=0}^{\infty} \frac{x^{2n}}{2^n n!}$$

$$= \sum_{n=1}^{\infty} \frac{2n(2n-1)x^{2n-2}}{2^n n!} - \sum_{n=0}^{\infty} \frac{(2n+1)x^{2n}}{2^n n!}$$

$$= \sum_{n=0}^{\infty} \left[ \frac{(2n+2)(2n+1)x^{2n}}{2^{n+1}(n+1)!} - \frac{(2n+1)x^{2n}}{2^n n!} \cdot \frac{2(n+1)}{2(n+1)} \right]$$

$$= \sum_{n=0}^{\infty} \frac{2(n+1)x^{2n}\left[(2n+1) - (2n+1)\right]}{2^{n+1}(n+1)!} = 0$$

**65.** $J_0(x) = \sum_{k=0}^{\infty} \frac{(-1)^k x^{2k}}{2^{2k}(k!)^2}$

(a) $\lim\limits_{k \to \infty} \left| \frac{u_{k+1}}{u_k} \right| = \lim\limits_{k \to \infty} \left| \frac{(-1)^{k+1}x^{2k+2}}{2^{2k+2}\left[(k+1)!\right]^2} \cdot \frac{2^{2k}(k!)^2}{(-1)^k x^{2k}} \right| = \lim\limits_{k \to \infty} \left| \frac{(-1)x^2}{2^2(k+1)^2} \right| = 0$

Therefore, the interval of convergence is $-\infty < x < \infty$.

(b) $\qquad J_0 = \sum_{k=0}^{\infty} (-1)^k \frac{x^{2k}}{4^k(k!)^2}$

$$J_0' = \sum_{k=1}^{\infty} (-1)^k \frac{2kx^{2k-1}}{4^k(k!)^2} = \sum_{k=0}^{\infty} (-1)^{k+1} \frac{(2k+2)x^{2k+1}}{4^{k+1}\left[(k+1)!\right]^2}$$

$$J_0'' = \sum_{k=1}^{\infty} (-1)^k \frac{2k(2k-1)x^{2k-2}}{4^k(k!)^2} = \sum_{k=0}^{\infty} (-1)^{k+1} \frac{(2k+2)(2k+1)x^{2k}}{4^{k+1}\left[(k+1)!\right]^2}$$

$$x^2 J_0'' + x J_0' + x^2 J_0 = \sum_{k=0}^{\infty} (-1)^{k+1} \frac{2(2k+1)x^{2k+2}}{4^{k+1}(k+1)!k!} + \sum_{k=0}^{\infty} (-1)^{k+1} \frac{2x^{2k+2}}{4^{k+1}(k+1)!k!} + \sum_{k=0}^{\infty} (-1)^k \frac{x^{2k+2}}{4^k(k!)^2}$$

$$= \sum_{k=0}^{\infty} \frac{(-1)^k x^{2k+2}}{4^k(k!)^2} \left[ (-1)\frac{2(2k+1)}{4(k+1)} + (-1)\frac{2}{4(k+1)} + 1 \right]$$

$$= \sum_{k=0}^{\infty} \frac{(-1)^k x^{2k+2}}{4^k(k!)^2} \left[ \frac{-4k-2}{4k+4} - \frac{2}{4k+4} + \frac{4k+4}{4k+4} \right] = 0$$

(c) $P_6(x) = 1 - \dfrac{x^2}{4} + \dfrac{x^4}{64} - \dfrac{x^6}{2304}$

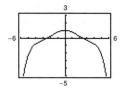

(d) $\displaystyle\int_0^1 J_0\,dx = \int_0^1 \sum_{k=0}^{\infty} \frac{(-1)^k x^{2k}}{4^k(k!)^2}\,dx = \left[ \sum_{k=0}^{\infty} \frac{(-1)^k x^{2k+1}}{4^k(k!)^2(2k+1)} \right]_0^1 = \sum_{k=0}^{\infty} \frac{(-1)^k}{4^k(k!)^2(2k+1)} = 1 - \frac{1}{12} + \frac{1}{320} \approx 0.92$

(integral is approximately 0.9197304101)

**67.** $\displaystyle\sum_{n=0}^{\infty}\left(\frac{x}{4}\right)^{n}$, $(-4, 4)$

(a) $\displaystyle\sum_{n=0}^{\infty}\left(\frac{(5/2)}{4}\right)^{n} = \sum_{n=0}^{\infty}\left(\frac{5}{8}\right)^{n} = \frac{1}{1 - 5/8} = \frac{8}{3}$

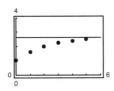

(b) $\displaystyle\sum_{n=0}^{\infty}\left(\frac{(-5/2)}{4}\right)^{n} = \sum_{n=0}^{\infty}\left(-\frac{5}{8}\right)^{n} = \frac{1}{1 + 5/8} = \frac{8}{13}$

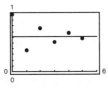

(c) The alternating series converges more rapidly. The partial sums of the series of positive terms approaches the sum from below. The partial sums of the alternating series alternate sides of the horizontal line representing the sum.

(d)

| $M$ | 10 | 100 | 1000 | 10,000 |
|---|---|---|---|---|
| $N$ | 5 | 14 | 24 | 35 |

**69.** $\displaystyle f(x) = \sum_{n=0}^{\infty}(-1)^{n}\frac{x^{2n}}{(2n)!} = \cos x$

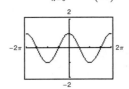

**71.** $\displaystyle f(x) = \sum_{n=0}^{\infty}(-1)^{n}x^{n} = \sum_{n=0}^{\infty}(-x)^{n}$  Geometric

$= \dfrac{1}{1 - (-x)} = \dfrac{1}{1 + x}$ for $-1 < x < 1$

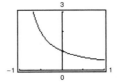

**73.** False;

$$\sum_{n=1}^{\infty}\frac{(-1)^{n}x^{n}}{n2^{n}}$$

converges for $x = 2$ but diverges for $x = -2$.

**75.** True; the radius of convergence is $R = 1$ for both series.

**77.** $\displaystyle\lim_{n\to\infty}\left|\frac{a_{n+1}}{a_{n}}\right| = \lim_{n\to\infty}\left|\frac{(n+1+p)!}{(n+1)!(n+1+q)!}x^{n+1} \Big/ \frac{(n+p)!}{n!(n+q)!}x^{n}\right| = \lim_{n\to\infty}\left|\frac{(n+1+p)x}{(n+1)(n+1+q)}\right| = 0$

So, the series converges for all $x$: $R = \infty$.

**79.** (a) $\displaystyle f(x) = \sum_{n=0}^{\infty}c_{n}x^{n}, c_{n+3} = c_{n}$

$= c_{0} + c_{1}x + c_{2}x^{2} + c_{0}x^{3} + c_{1}x^{4} + c_{2}x^{5} + c_{0}x^{6} + \cdots$

$S_{3n} = c_{0}\left(1 + x^{3} + \cdots + x^{3n}\right) + c_{1}x\left(1 + x^{3} + \cdots + x^{3n}\right) + c_{2}x^{2}\left(1 + x^{3} + \cdots + x^{3n}\right)$

$\displaystyle\lim_{n\to\infty}S_{3n} = c_{0}\sum_{n=0}^{\infty}x^{3n} + c_{1}x\sum_{n=0}^{\infty}x^{3n} + c_{2}x^{2}\sum_{n=0}^{\infty}x^{3n}$

Each series is geometric, $R = 1$, and the interval of convergence is $(-1, 1)$.

(b) For $|x| < 1$, $\displaystyle f(x) = c_{0}\frac{1}{1 - x^{3}} + c_{1}x\frac{1}{1 - x^{3}} + c_{2}x^{2}\frac{1}{1 - x^{3}} = \frac{c_{0} + c_{1}x + c_{2}x^{2}}{1 - x^{3}}$.

**81.** At $x = x_{0} + R$, $\displaystyle\sum_{n=0}^{\infty}c_{n}(x - x_{0})^{n} = \sum_{n=0}^{\infty}c_{n}R^{n}$, diverges.

At $x = x_{0} - R$, $\displaystyle\sum_{n=0}^{\infty}c_{n}(x - x_{0})^{n} = \sum_{n=0}^{\infty}c_{n}(-R)^{n}$, converges.

Furthermore, at $x = x_{0} - R$,

$$\sum_{n=0}^{\infty}\left|c_{n}(x - x_{0})^{n}\right| = \sum_{n=0}^{\infty}C_{n}R^{n}, \text{ diverges.}$$

So, the series converges conditionally at $x_{0} - R$.

## Section 9.9   Representation of Functions by Power Series

**1. (a)** $\dfrac{1}{4 - x} = \dfrac{1/4}{1 - (x/4)}$

$$= \dfrac{a}{1 - r} = \sum_{n=0}^{\infty} \left(\dfrac{1}{4}\right)\left(\dfrac{x}{4}\right)^n = \sum_{n=0}^{\infty} \dfrac{x^n}{4^{n+1}}$$

This series converges on $(-4, 4)$.

**(b)**
$$\begin{array}{r}
\dfrac{1}{4} + \dfrac{x}{16} + \dfrac{x^2}{64} + \cdots \\[4pt]
4 - x \,\overline{)\, 1 \phantom{xxxxxxxxx}} \\[4pt]
1 - \dfrac{x}{4} \phantom{xxxxx} \\[2pt]
\overline{\phantom{xx}\dfrac{x}{4}\phantom{xxxxx}} \\[2pt]
\dfrac{x}{4} - \dfrac{x^2}{16} \\[2pt]
\overline{\phantom{xx}\dfrac{x^2}{16}\phantom{xx}} \\[2pt]
\dfrac{x^2}{16} - \dfrac{x^3}{64} \\[2pt]
\overline{\phantom{xxxxxxxxx}} \\
\vdots
\end{array}$$

**3. (a)** $\dfrac{4}{3 + x} = \dfrac{4/3}{1 - (-x/3)} = \dfrac{a}{1 - r}$

$$= \sum_{n=0}^{\infty} \dfrac{4}{3}\left(\dfrac{-x}{3}\right)^n = \sum_{n=0}^{\infty} \dfrac{4(-1)^n\, x^n}{3^{n+1}}$$

The series converges on $(-3, 3)$.

**(b)**
$$\begin{array}{r}
\dfrac{4}{3} - \dfrac{4}{9}x + \dfrac{4x^2}{27} - \cdots \\[4pt]
3 + x \,\overline{)\, 4 \phantom{xxxxxxxxx}} \\[4pt]
4 + \dfrac{4}{3}x \phantom{xxxxx} \\[2pt]
\overline{\phantom{xx}-\dfrac{4}{3}x\phantom{xxxxx}} \\[2pt]
-\dfrac{4}{3}x - \dfrac{4x^2}{9} \\[2pt]
\overline{\phantom{xxx}\dfrac{4x^2}{9}\phantom{xx}} \\[2pt]
\dfrac{4x^2}{9} + \dfrac{4x^3}{27} \\[2pt]
\overline{\phantom{xxxxxxxxx}} \\
-\dfrac{4x^3}{27} \\
\vdots
\end{array}$$

**5.** $\dfrac{1}{3 - x} = \dfrac{1}{2 - (x - 1)} = \dfrac{1/2}{1 - \left(\dfrac{x - 1}{2}\right)} = \dfrac{a}{1 - r}$

$$= \sum_{n=0}^{\infty} \dfrac{1}{2}\left(\dfrac{x - 1}{2}\right)^n = \sum_{n=0}^{\infty} \dfrac{(x - 1)^n}{2^{n+1}}$$

Interval of convergence:

$$\left|\dfrac{x - 1}{2}\right| < 1 \Rightarrow |x - 1| < 2 \Rightarrow (-1, 3)$$

**7.** $\dfrac{1}{1 - 3x} = \dfrac{a}{1 - r} = \sum_{n=0}^{\infty} (3x)^n$

Interval of convergence: $|3x| < 1 \Rightarrow \left(\dfrac{1}{3}, \dfrac{1}{3}\right)$

**9.** $\dfrac{5}{2x - 3} = \dfrac{5}{-9 + 2(x + 3)} = \dfrac{-5/9}{1 - \dfrac{2}{9}(x + 3)} = \dfrac{a}{1 - r}$

$$= -\dfrac{5}{9}\sum_{n=0}^{\infty} \left(\dfrac{2}{9}(x + 3)\right)^n, \quad \left|\dfrac{2}{9}(x + 3)\right| < 1$$

$$= -5\sum_{n=0}^{\infty} \dfrac{2^n}{9^{n+1}}(x + 3)^n$$

Interval of convergence: $\left|\dfrac{2}{9}(x + 3)\right| < 1 \Rightarrow \left(-\dfrac{15}{2}, \dfrac{3}{2}\right)$

**11.** $\dfrac{3}{3x + 4} = \dfrac{3/4}{1 + \dfrac{3}{4}x} = \dfrac{3/4}{1 - \left(-\dfrac{3}{4}x\right)} = \dfrac{a}{1 - r}$

$$= \sum_{n=0}^{\infty} \dfrac{3}{4}\left(-\dfrac{3}{4}x\right)^n = \sum_{n=0}^{\infty} \dfrac{(-1)^n\, 3^{n+1}\, x^n}{4^{n+1}}$$

Interval of convergence:

$$\left|-\dfrac{3}{4}x\right| < 1 \Rightarrow |3x| < 4 \Rightarrow |x| < \dfrac{4}{3} \Rightarrow \left(-\dfrac{4}{3}, \dfrac{4}{3}\right)$$

**13.** $\dfrac{4x}{x^2 + 2x - 3} = \dfrac{3}{x + 3} + \dfrac{1}{x - 1}$

$$= \dfrac{1}{1 - (-x/3)} + \dfrac{-1}{1 - x}$$

$$= \sum_{n=0}^{\infty} \left(-\dfrac{x}{3}\right)^n - \sum_{n=0}^{\infty} x^n = \sum_{n=0}^{\infty} \left[\dfrac{1}{(-3)^n} - 1\right]x^n$$

Interval of convergence: $\left|\dfrac{x}{3}\right| < 1$ and $|x| < 1 \Rightarrow (-1, 1)$

**15.** $\dfrac{2}{1 - x^2} = \dfrac{1}{1 - x} + \dfrac{1}{1 + x}$

$\qquad\qquad = \displaystyle\sum_{n=0}^{\infty} \left(1 + (-1)^n\right) x^n = 2 \sum_{n=0}^{\infty} x^{2n}$

Interval of convergence: $\left| x^2 \right| < 1$ or $(-1, 1)$  because

$\displaystyle\lim_{n \to \infty} \left| \dfrac{u_{n+1}}{u_n} \right| = \lim_{n \to \infty} \left| \dfrac{2x^{2n+2}}{2x^{2n}} \right| = \left| x^2 \right|$

**17.** $\dfrac{1}{1 + x} = \displaystyle\sum_{n=0}^{\infty} (-1)^n \, x^n$

$\qquad \dfrac{1}{1 - x} = \displaystyle\sum_{n=0}^{\infty} (-1)^n (-x)^n = \sum_{n=0}^{\infty} (-1)^{2n} x^n = \sum_{n=0}^{\infty} x^n$

$\qquad h(x) = \dfrac{-2}{x^2 - 1} = \dfrac{1}{1 + x} + \dfrac{1}{1 - x} = \displaystyle\sum_{n=0}^{\infty} (-1)^n x^n + \sum_{n=0}^{\infty} x^n = \sum_{n=0}^{\infty} \left[ (-1)^n + 1 \right] x^n$

$\qquad\qquad = 2 + 0x + 2x^2 + 0x^3 + 2x^4 + 0x^5 + 2x^6 + \cdots = 2 \displaystyle\sum_{n=0}^{\infty} x^{2n}, \ (-1, 1) \ \text{(See Exercise 15.)}$

**19.** By taking the first derivative, you have $\dfrac{d}{dx} \left[ \dfrac{1}{x + 1} \right] = \dfrac{-1}{(x + 1)^2}$. Therefore,

$\qquad \dfrac{-1}{(x + 1)^2} = \dfrac{d}{dx} \left[ \displaystyle\sum_{n=0}^{\infty} (-1)^n x^n \right] = \sum_{n=1}^{\infty} (-1)^n n x^{n-1} = \sum_{n=0}^{\infty} (-1)^{n+1} (n + 1) x^n, \ (-1, 1).$

**21.** By integrating, you have $\displaystyle\int \dfrac{1}{x + 1} \, dx = \ln(x + 1)$. Therefore,

$\qquad \ln(x + 1) = \displaystyle\int \left[ \sum_{n=0}^{\infty} (-1)^n x^n \right] dx = C + \sum_{n=0}^{\infty} \dfrac{(-1)^n x^{n+1}}{n + 1}, \ -1 < x \le 1.$

To solve for $C$, let $x = 0$ and conclude that $C = 0$. Therefore,

$\qquad \ln(x + 1) = \displaystyle\sum_{n=0}^{\infty} \dfrac{(-1)^n x^{n+1}}{n + 1}, \ (-1, 1].$

**23.** $\dfrac{1}{x^2 + 1} = \displaystyle\sum_{n=0}^{\infty} (-1)^n \left(x^2\right)^n = \sum_{n=0}^{\infty} (-1)^n x^{2n}, \ (-1, 1)$

**25.** Because, $\dfrac{1}{x + 1} = \displaystyle\sum_{n=0}^{\infty} (-1)^n x^n$, you have $\dfrac{1}{4x^2 + 1} = \displaystyle\sum_{n=0}^{\infty} (-1)^n \left(4x^2\right)^n = \sum_{n=0}^{\infty} (-1)^n 4^n x^{2n} = \sum_{n=0}^{\infty} (-1)^n (2x)^{2n}, \ \left( -\dfrac{1}{2}, \dfrac{1}{2} \right).$

**27.** $x - \dfrac{x^2}{2} \leq \ln(x+1) \leq x - \dfrac{x^2}{2} + \dfrac{x^3}{3}$

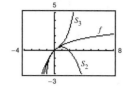

| $x$ | 0.0 | 0.2 | 0.4 | 0.6 | 0.8 | 1.0 |
|---|---|---|---|---|---|---|
| $S_2 = x - \dfrac{x^2}{2}$ | 0.000 | 0.180 | 0.320 | 0.420 | 0.480 | 0.500 |
| $\ln(x+1)$ | 0.000 | 0.182 | 0.336 | 0.470 | 0.588 | 0.693 |
| $S_3 = x - \dfrac{x^2}{2} + \dfrac{x^3}{3}$ | 0.000 | 0.183 | 0.341 | 0.492 | 0.651 | 0.833 |

**29.** $\displaystyle\sum_{n=1}^{\infty} \frac{(-1)^{n+1}(x-1)^n}{n} = \frac{(x-1)}{1} - \frac{(x-1)^2}{2} + \frac{(x-1)^3}{3} - \cdots$

(a)

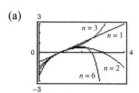

(b) From Example 4,

$$\sum_{n=1}^{\infty} \frac{(-1)^{n+1}(x-1)^n}{n} = \sum_{n=0}^{\infty} \frac{(-1)^n(x-1)^{n+1}}{n+1} = \ln x, \ 0 < x \leq 2, \ R = 1.$$

(c) $x = 0.5$:

$$\sum_{n=1}^{\infty} \frac{(-1)^{n+1}(-1/2)^n}{n} = \sum_{n=1}^{\infty} \frac{-(1/2)^n}{n} \approx -0.693147$$

(d) This is an approximation of $\ln\left(\dfrac{1}{2}\right)$. The error is approximately 0. [The error is less than the first omitted term,

$1\big/(51 \cdot 2^{51}) \approx 8.7 \times 10^{-18}$.]

**In Exercises 31–33, arctan** $x = \displaystyle\sum_{n=0}^{\infty} (-1)^n \frac{x^{2n+1}}{2n+1}$.

**31.** $\arctan \dfrac{1}{4} = \displaystyle\sum_{n=0}^{\infty} (-1)^n \frac{(1/4)^{2n+1}}{2n+1} = \sum_{n=0}^{\infty} \frac{(-1)^n}{(2n+1)4^{2n+1}} = \frac{1}{4} - \frac{1}{192} + \frac{1}{5120} + \cdots$

Because $\dfrac{1}{5120} < 0.001$, you can approximate the series by its first two terms: $\arctan \dfrac{1}{4} \approx \dfrac{1}{4} - \dfrac{1}{192} \approx 0.245$.

**33.**
$$\frac{\arctan x^2}{x} = \frac{1}{x}\sum_{n=0}^{\infty}(-1)^n\frac{\left(x^2\right)^{2n+1}}{2n+1} = \sum_{n=0}^{\infty}(-1)^2\frac{x^{4n+1}}{2n+1}$$

$$\int\frac{\arctan x^2}{x}\,dx = \sum_{n=0}^{\infty}(-1)^n\frac{x^{4n+2}}{(4n+2)(2n+1)} + C \;(\text{Note: } C = 0)$$

$$\int_0^{1/2}\frac{\arctan x^2}{x}\,dx = \sum_{n=0}^{\infty}(-1)^n\frac{1}{(4n+2)(2n+1)2^{4n+2}} = \frac{1}{8} - \frac{1}{1152} + \cdots$$

Because $\dfrac{1}{1152} < 0.001$, you can approximate the series by its first term: $\displaystyle\int_0^{1/2}\frac{\arctan x^2}{x}\,dx \approx 0.125$.

**In Exercises 35–37, use** $\dfrac{1}{1-x} = \sum_{n=0}^{\infty}x^n, |x| < 1.$

**35.** $\dfrac{1}{(1-x)^2} = \dfrac{d}{dx}\left[\dfrac{1}{1-x}\right] = \dfrac{d}{dx}\left[\sum_{n=0}^{\infty}x^n\right] = \sum_{n=1}^{\infty}nx^{n-1}, |x| < 1$

**37.** $\dfrac{1+x}{(1-x)^2} = \dfrac{1}{(1-x)^2} + \dfrac{x}{(1-x)^2}$

$$= \sum_{n=1}^{\infty}n\left(x^{n-1} + x^n\right), \quad |x| < 1$$

$$= \sum_{n=0}^{\infty}(2n+1)x^n, \quad |x| < 1$$

**39.** $P(n) = \left(\dfrac{1}{2}\right)^n$

$$E(n) = \sum_{n=1}^{\infty}nP(n) = \sum_{n=1}^{\infty}n\left(\frac{1}{2}\right)^n = \frac{1}{2}\sum_{n=1}^{\infty}n\left(\frac{1}{2}\right)^{n-1}$$

$$= \frac{1}{2}\frac{1}{\left[1-(1/2)\right]^2} = 2$$

Because the probability of obtaining a head on a single toss is $\dfrac{1}{2}$, it is expected that, on average, a head will be obtained in two tosses.

**41.** Because $\dfrac{1}{1+x} = \dfrac{1}{1-(-x)}$, substitute $(-x)$ into the geometric series.

**43.** Because $\dfrac{1}{1+x} = 5\left(\dfrac{1}{1-(-x)}\right)$, substitute $(-x)$ into the geometric series and then multiply the series by 5.

**45.** Let $\arctan x + \arctan y = \theta$. Then,

$$\tan(\arctan x + \arctan y) = \tan\theta$$

$$\frac{\tan(\arctan x) + \tan(\arctan y)}{1 - \tan(\arctan x)\tan(\arctan y)} = \tan\theta$$

$$\frac{x+y}{1-xy} = \tan\theta$$

$$\arctan\left(\frac{x+y}{1-xy}\right) = \theta.$$

Therefore,

$$\arctan x + \arctan y = \arctan\left(\frac{x+y}{1-xy}\right) \text{ for } xy \neq 1.$$

**47.** (a) $2\arctan\dfrac{1}{2} = \arctan\dfrac{1}{2} + \arctan\dfrac{1}{2} = \arctan\left[\dfrac{\frac{1}{2}+\frac{1}{2}}{1-(1/2)^2}\right] = \arctan\dfrac{4}{3}$

$2\arctan\dfrac{1}{2} - \arctan\dfrac{1}{7} = \arctan\dfrac{4}{3} + \arctan\left(-\dfrac{1}{7}\right) = \arctan\left[\dfrac{(4/3)-(1/7)}{1+(4/3)(1/7)}\right] = \arctan\dfrac{25}{25} = \arctan 1 = \dfrac{\pi}{4}$

(b) $\pi = 8\arctan\dfrac{1}{2} - 4\arctan\dfrac{1}{7} \approx 8\left[\dfrac{1}{2} - \dfrac{(0.5)^3}{3} + \dfrac{(0.5)^5}{5} - \dfrac{(0.5)^7}{7}\right] - 4\left[\dfrac{1}{7} - \dfrac{(1/7)^3}{3} + \dfrac{(1/7)^5}{5} - \dfrac{(1/7)^7}{7}\right] \approx 3.14$

**49.** From Exercise 21, you have

$$\ln(x + 1) = \sum_{n=0}^{\infty} \frac{(-1)^n x^{n+1}}{n + 1} = \sum_{n=1}^{\infty} \frac{(-1)^{n-1} x^n}{n}$$

$$= \sum_{n=1}^{\infty} \frac{(-1)^{n+1} x^n}{n}.$$

So, $\displaystyle\sum_{n=1}^{\infty} (-1)^{n+1} \frac{1}{2^n n} = \sum_{n=1}^{\infty} \frac{(-1)^{n+1} (1/2)^n}{n}$

$$= \ln\left(\frac{1}{2} + 1\right) = \ln\frac{3}{2} \approx 0.4055.$$

**51.** From Exercise 49, you have

$$\sum_{n=1}^{\infty} (-1)^{n+1} \frac{2^n}{5^n n} = \sum_{n=1}^{\infty} \frac{(-1)^{n+1} (2/5)^n}{n}$$

$$= \ln\left(\frac{2}{5} + 1\right) = \ln\frac{7}{5} \approx 0.3365.$$

**53.** From Exercise 52, you have

$$\sum_{n=0}^{\infty} (-1)^n \frac{1}{2^{2n+1}(2n + 1)} = \sum_{n=0}^{\infty} (-1)^n \frac{(1/2)^{2n+1}}{2n + 1}$$

$$= \arctan\frac{1}{2} \approx 0.4636.$$

**55.** The series in Exercise 52 converges to its sum at a slower rate because its terms approach 0 at a much slower rate.

## Section 9.10   Taylor and Maclaurin Series

**1.** For $c = 0$, you have:

$$f(x) = e^{2x}$$

$$f^{(n)}(x) = 2^n e^{2x} \Rightarrow f^{(n)}(0) = 2^n$$

$$e^{2x} = 1 + 2x + \frac{4x^2}{2!} + \frac{8x^3}{3!} + \frac{16x^4}{4!} + \cdots = \sum_{n=0}^{\infty} \frac{(2x)^n}{n!}.$$

**57.** Because the first series is the derivative of the second series, the second series converges for $|x + 1| < 4$ (and perhaps at the endpoints, $x = 3$ and $x = -5$.)

**59.** $\displaystyle\sum_{n=0}^{\infty} \frac{(-1)^n}{3^n(2n + 1)}$

From Example 5 you have $\arctan x = \displaystyle\sum_{n=0}^{\infty} (-1)^n \frac{x^{2n+1}}{2n + 1}$.

$$\sum_{n=0}^{\infty} \frac{(-1)^n}{3^n(2n + 1)} = \sum_{n=0}^{\infty} \frac{(-1)^n}{(\sqrt{3})^{2n}(2n + 1)} \frac{\sqrt{3}}{\sqrt{3}}$$

$$= \sqrt{3} \sum_{n=0}^{\infty} \frac{(-1)^n (1/\sqrt{3})^{2n+1}}{2n + 1}$$

$$= \sqrt{3} \arctan\left(\frac{1}{\sqrt{3}}\right)$$

$$= \sqrt{3}\left(\frac{\pi}{6}\right) \approx 0.9068997$$

**61.** Using a graphing utility, you obtain the following partial sums for the left hand side. Note that $1/\pi = 0.3183098862$.

$n = 0$: $S_0 \approx 0.3183098784$

$n = 1$: $S_1 = 0.3183098862$

**3.** For $c = \pi/4$, you have:

$$f(x) = \cos(x) \qquad f\left(\frac{\pi}{4}\right) = \frac{\sqrt{2}}{2}$$

$$f'(x) = -\sin(x) \qquad f'\left(\frac{\pi}{4}\right) = -\frac{\sqrt{2}}{2}$$

$$f''(x) = -\cos(x) \qquad f''\left(\frac{\pi}{4}\right) = -\frac{\sqrt{2}}{2}$$

$$f'''(x) = \sin(x) \qquad f'''\left(\frac{\pi}{4}\right) = \frac{\sqrt{2}}{2}$$

$$f^{(4)}(x) = \cos(x) \qquad f^{(4)}\left(\frac{\pi}{4}\right) = \frac{\sqrt{2}}{2}$$

and so on. Therefore, you have:

$$\cos x = \sum_{n=0}^{\infty} \frac{f^{(n)}(\pi/4)\left[x - (\pi/4)\right]^n}{n!}$$

$$= \frac{\sqrt{2}}{2}\left[1 - \left(x - \frac{\pi}{4}\right) - \frac{\left[x - (\pi/4)\right]^2}{2!} + \frac{\left[x - (\pi/4)\right]^3}{3!} + \frac{\left[x - (\pi/4)\right]^4}{4!} - \cdots\right]$$

$$= \frac{\sqrt{2}}{2}\sum_{n=0}^{\infty} \frac{(-1)^{n(n+1)/2}\left[x - (\pi/4)\right]^n}{n!}.$$

[**Note:** $(-1)^{n(n+1)/2} = 1, -1, -1, 1, 1, -1, -1, 1, \ldots$]

**5.** For $c = 1$, you have

$$f(x) = \frac{1}{x} = x^{-1} \qquad f(1) = 1$$

$$f'(x) = -x^{-2} \qquad f'(1) = -1$$

$$f''(x) = 2x^{-3} \qquad f''(1) = 2$$

$$f'''(x) = -6x^{-4} \qquad f'''(1) = -6$$

and so on. Therefore, you have

$$\frac{1}{x} = \sum_{n=0}^{\infty} \frac{f^{(n)}(1)(x - 1)^n}{n!}$$

$$= 1 - (x - 1) + \frac{2(x - 1)^2}{2!} - \frac{6(x - 1)^3}{3!} + \cdots$$

$$= 1 - (x - 1) + (x - 1)^2 - (x - 1)^3 + \cdots$$

$$= \sum_{n=0}^{\infty} (-1)^n (x - 1)^n$$

**7.** For $c = 1$, you have,

$$f(x) = \ln x \qquad f(1) = 0$$

$$f'(x) = \frac{1}{x} \qquad f'(1) = 1$$

$$f''(x) = -\frac{1}{x^2} \qquad f''(1) = -1$$

$$f'''(x) = \frac{2}{x^3} \qquad f'''(1) = 2$$

$$f^{(4)}(x) = -\frac{6}{x^4} \qquad f^{(4)}(1) = -6$$

$$f^{(5)}(x) = \frac{24}{x^5} \qquad f^{(5)}(1) = 24$$

and so on. Therefore, you have:

$$\ln x = \sum_{n=0}^{\infty} \frac{f^{(n)}(1)(x-1)^n}{n!}$$

$$= 0 + (x-1) - \frac{(x-1)^2}{2!} + \frac{2(x-1)^3}{3!} - \frac{6(x-1)^4}{4!} + \frac{24(x-1)^5}{5!} - \cdots$$

$$= (x-1) - \frac{(x-1)^2}{2} + \frac{(x-1)^3}{3} - \frac{(x-1)^4}{4} + \frac{(x-1)^5}{5} - \cdots$$

$$= \sum_{n=0}^{\infty} (-1)^n \frac{(x-1)^{n+1}}{n+1}.$$

**9.** For $c = 0$, you have

$$f(x) = \sin 3x \qquad f(0) = 0$$

$$f'(x) = 3 \cos 3x \qquad f'(0) = 3$$

$$f''(x) = -9 \sin 3x \qquad f''(0) = 0$$

$$f'''(x) = -27 \cos 3x \qquad f'''(0) = -27$$

$$f^{(4)}(x) = 81 \sin 3x \qquad f^{(4)}(0) = 0$$

and so on. Therefore you have

$$\sin 3x = \sum_{n=0}^{\infty} \frac{f^{(n)}(0)x^n}{n!} = 0 + 3x + 0 - \frac{27x^3}{3!} + 0 + \cdots = \sum_{n=0}^{\infty} \frac{(-1)^n (3x)^{2n+1}}{(2n+1)!}$$

**11.** For $c = 0$, you have:

$$f(x) = \sec(x) \qquad\qquad f(0) = 1$$

$$f'(x) = \sec(x) \tan(x) \qquad\qquad f'(0) = 0$$

$$f''(x) = \sec^3(x) + \sec(x) \tan^2(x) \qquad\qquad f''(0) = 1$$

$$f'''(x) = 5 \sec^3(x) \tan(x) + \sec(x) \tan^3(x) \qquad\qquad f'''(0) = 0$$

$$f^{(4)}(x) = 5 \sec^5(x) + 18 \sec^3(x) \tan^2(x) + \sec(x) \tan^4(x) \qquad f^{(4)}(0) = 5$$

$$\sec(x) = \sum_{n=0}^{\infty} \frac{f^{(n)}(0)x^n}{n!} = 1 + \frac{x^2}{2!} + \frac{5x^4}{4!} + \cdots.$$

**13.** The Maclaurin series for $f(x) = \cos x$ is $\displaystyle\sum_{n=0}^{\infty} \frac{(-1)^n x^{2n}}{(2n)!}$.

Because $f^{(n+1)}(x) = \pm\sin x$ or $\pm\cos x$, you have $\left| f^{(n+1)}(z) \right| \le 1$ for all $z$. So by Taylor's Theorem,

$$0 \le \left| R_n(x) \right| = \left| \frac{f^{(n+1)}(z)}{(n+1)!} x^{n+1} \right| \le \frac{|x|^{n+1}}{(n+1)!}.$$

Because $\displaystyle\lim_{n \to \infty} \frac{|x|^{n+1}}{(n+1)!} = 0$, it follows that $R_n(x) \to 0$ as $n \to \infty$. So, the Maclaurin series for $\cos x$ converges to

$\cos x$ for all $x$.

**15.** The Maclaurin series for $f(x) = \sinh x$ is $\displaystyle\sum_{n=0}^{\infty} \frac{x^{2n+1}}{(2n+1)!}$.

$f^{(n+1)}(x) = \sinh x$ (or $\cosh x$). For fixed $x$,

$$0 \le \left| R_n(x) \right| = \left| \frac{f^{(n+1)}(z)}{(n+1)!} x^{n+1} \right| = \left| \frac{\sinh(z)}{(n+1)!} x^{n+1} \right| \to 0 \text{ as } n \to \infty.$$

(The argument is the same if $f^{(n+1)}(x) = \cosh x$ ). So, the Maclaurin series for $\sinh x$ converges to $\sinh x$ for all $x$.

**17.** Because $(1+x)^{-k} = 1 - kx + \dfrac{k(k+1)x^2}{2!} - \dfrac{k(k+1)(k+2)x^3}{3!} + \cdots$, you have

$$(1+x)^{-2} = 1 - 2x + \frac{2(3)x^2}{2!} - \frac{2(3)(4)x^3}{3!} + \frac{2(3)(4)(5)x^4}{4!} - \cdots = 1 - 2x + 3x^2 - 4x^3 + 5x^4 - \cdots$$

$$= \sum_{n=0}^{\infty} (-1)^n (n+1) x^n.$$

**19.** Because $(1+x)^{-k} = 1 - kx + \dfrac{k(k+1)x^2}{2!} - \dfrac{k(k+1)(k+2)x^3}{3!} + \cdots$, you have

$$\left[1 + (-x)\right]^{-1/2} = 1 + \left(\frac{1}{2}\right)x + \frac{(1/2)(3/2)x^2}{2!} + \frac{(1/2)(3/2)(5/2)x^3}{3!} + \cdots$$

$$= 1 + \frac{x}{2} + \frac{(1)(3)x^2}{2^2 \, 2!} + \frac{(1)(3)(5)x^3}{2^3 \, 3!} + \cdots$$

$$= 1 + \sum_{n=1}^{\infty} \frac{1 \cdot 3 \cdot 5 \cdots (2n-1)x^n}{2^n \, n!}.$$

**21.** $\dfrac{1}{\sqrt{4+x^2}} = \left(\dfrac{1}{2}\right)\left[1 + \left(\dfrac{x}{2}\right)^2\right]^{-1/2}$ and because $(1+x)^{-1/2} = 1 + \displaystyle\sum_{n=1}^{\infty} \frac{(-1)^n 1 \cdot 3 \cdot 5 \cdots (2n-1)x^n}{2^n \, n!}$, you have

$$\frac{1}{\sqrt{4+x^2}} = \frac{1}{2}\left[1 + \sum_{n=1}^{\infty} \frac{(-1)^n 1 \cdot 3 \cdot 5 \cdots (2n-1)(x/2)^{2n}}{2^n \, n!}\right] = \frac{1}{2} + \sum_{n=1}^{\infty} \frac{(-1)^n 1 \cdot 3 \cdot 5 \cdots (2n-1)x^{2n}}{2^{3n+1} \, n!}.$$

**23.** $\sqrt{1+x} = (1+x)^{1/2}, \qquad k = 1/2$

$$\sqrt{1+x} = 1 + \frac{1}{2}x + \frac{1/2(-1/2)}{2!}x^2 + \frac{1/2(-1/2)(-3/2)}{3!}x^3 + \cdots = 1 + \frac{1}{2}x + \sum_{n=2}^{\infty} (-1)^{n+1} \frac{1 \cdot 3 \cdot 5 \cdots (2n-3)}{2^n \, n!} x^n$$

**25.** Because $(1+x)^{1/2} = 1 + \dfrac{x}{2} + \displaystyle\sum_{n=2}^{\infty} \frac{(-1)^{n+1} 1 \cdot 3 \cdot 5 \cdots (2n-3)x^n}{2^n \, n!}$

you have $(1+x^2)^{1/2} = 1 + \dfrac{x^2}{2} + \displaystyle\sum_{n=2}^{\infty} \frac{(-1)^{n+1} 1 \cdot 3 \cdot 5 \cdots (2n-3)x^{2n}}{2^n \, n!}$.

**27.**   $e^x = \displaystyle\sum_{n=0}^{\infty} \frac{x^n}{n!} = 1 + x + \frac{x^2}{2!} + \frac{x^3}{3!} + \frac{x^4}{4!} + \frac{x^5}{5!} + \cdots$

$e^{x^2/2} = \displaystyle\sum_{n=0}^{\infty} \frac{\left(x^2/2\right)^n}{n!} = \sum_{n=0}^{\infty} \frac{x^{2n}}{2^n n!} = 1 + \frac{x^2}{2} + \frac{x^4}{2^2 2!} + \frac{x^6}{2^3 3!} + \frac{x^8}{2^4 4!} + \cdots$

**29.**   $\ln x = \displaystyle\sum_{n=1}^{\infty} (-1)^{n-1} \frac{(x-1)^n}{n}, \; 0 < x \le 2$

$\ln (x+1) = \displaystyle\sum_{n=1}^{\infty} \frac{(-1)^{n-1} x^n}{n}, \; -1 < x \le 1$

**31.**   $\sin x = \displaystyle\sum_{n=0}^{\infty} \frac{(-1)^n x^{2n+1}}{(2n+1)!}$

$\sin 3x = \displaystyle\sum_{n=0}^{\infty} \frac{(-1)^n (3x)^{2n+1}}{(2n+1)!}$

**33.**   $\cos x = \displaystyle\sum_{n=0}^{\infty} \frac{(-1)^n x^{2n}}{(2n)!} = 1 - \frac{x^2}{2!} + \frac{x^4}{4!} - \frac{x^6}{6!} + \cdots$

$\cos 4x = \displaystyle\sum_{n=0}^{\infty} \frac{(-1)^n (4x)^{2n}}{(2n)!} = \sum_{n=0}^{\infty} \frac{(-1)^n 4^{2n} x^{2n}}{(2n)!}$

$= 1 - \frac{16x^2}{2!} + \frac{256x^4}{4!} - \cdots$

**35.**   $\cos x = \displaystyle\sum_{n=0}^{\infty} \frac{(-1)^n x^{2n}}{(2n)!} = 1 - \frac{x^2}{2!} + \frac{x^4}{4!} - \cdots$

$\cos x^{3/2} = \displaystyle\sum_{n=0}^{\infty} \frac{(-1)^n \left(x^{3/2}\right)^{2n}}{(2n)!}$

$= \displaystyle\sum_{n=0}^{\infty} \frac{(-1)^n x^{3n}}{(2n)!}$

$= 1 - \frac{x^3}{2!} + \frac{x^6}{4!} - \cdots$

**37.**   $e^x = 1 + x + \frac{x^2}{2!} + \frac{x^3}{3!} + \frac{x^4}{4!} + \frac{x^5}{5!} + \cdots$

$e^{-x} = 1 - x + \frac{x^2}{2!} - \frac{x^3}{3!} + \frac{x^4}{4!} - \frac{x^5}{5!} + \cdots$

$e^x - e^{-x} = 2x + \frac{2x^3}{3!} + \frac{2x^5}{5!} + \frac{2x^7}{7!} + \cdots$

$\sinh(x) = \frac{1}{2}\left(e^x - e^{-x}\right)$

$= x + \frac{x^3}{3!} + \frac{x^5}{5!} + \frac{x^7}{7!} + \cdots = \displaystyle\sum_{n=0}^{\infty} \frac{x^{2n+1}}{(2n+1)!}$

**39.**   $\cos^2(x) = \frac{1}{2}\left[1 + \cos(2x)\right]$

$= \frac{1}{2}\left[1 + 1 - \frac{(2x)^2}{2!} + \frac{(2x)^4}{4!} - \frac{(2x)^6}{6!} - \cdots\right] = \frac{1}{2}\left[1 + \displaystyle\sum_{n=0}^{\infty} \frac{(-1)^n (2x)^{2n}}{(2n)!}\right]$

**41.**   $x \sin x = x\left(x - \frac{x^3}{3!} + \frac{x^5}{5!} - \cdots\right) = x^2 - \frac{x^4}{3!} + \frac{x^6}{5!} - \cdots = \displaystyle\sum_{n=0}^{\infty} \frac{(-1)^n x^{2n+2}}{(2n+1)!}$

**43.**   $\dfrac{\sin x}{x} = \dfrac{x - \left(x^3/3!\right) + \left(x^5/5!\right) - \cdots}{x} = 1 - \frac{x^2}{2!} + \frac{x^4}{4!} - \cdots = \displaystyle\sum_{n=0}^{\infty} \frac{(-1)^n x^{2n}}{(2n+1)!}, \; x \ne 0$

$= 1, \; x = 0$

**45.**   $e^{ix} = 1 + ix + \frac{(ix)^2}{2!} + \frac{(ix)^3}{3!} + \frac{(ix)^4}{4!} + \cdots = 1 + ix - \frac{x^2}{2!} - \frac{ix^3}{3!} + \frac{x^4}{4!} + \frac{ix^5}{5!} - \frac{x^6}{6!} - \cdots$

$e^{-ix} = 1 - ix + \frac{(-ix)^2}{2!} + \frac{(-ix)^3}{3!} + \frac{(-ix)^4}{4!} + \cdots = 1 - ix - \frac{x^2}{2!} + \frac{ix^3}{3!} + \frac{x^4}{4!} - \frac{ix^5}{5!} - \frac{x^6}{6!} + \cdots$

$e^{ix} - e^{-ix} = 2ix - \frac{2ix^3}{3!} + \frac{2ix^5}{5!} - \frac{2ix^7}{7!} + \cdots$

$\dfrac{e^{ix} - e^{-ix}}{2i} = x - \frac{x^3}{3!} + \frac{x^5}{5!} - \frac{x^7}{7!} + \cdots = \displaystyle\sum_{n=0}^{\infty} \frac{(-1)^n x^{2n+1}}{(2n+1)!} = \sin(x)$

**47.** $f(x) = e^x \sin x$

$$= \left(1 + x + \frac{x^2}{2} + \frac{x^3}{6} + \frac{x^4}{24} + \cdots\right)\left(x - \frac{x^3}{6} + \frac{x^5}{120} - \cdots\right)$$

$$= x + x^2 + \left(\frac{x^3}{2} - \frac{x^3}{6}\right) + \left(\frac{x^4}{6} - \frac{x^4}{6}\right) + \left(\frac{x^5}{120} - \frac{x^5}{12} + \frac{x^5}{24}\right) + \cdots$$

$$= x + x^2 + \frac{x^3}{3} - \frac{x^5}{30} + \cdots$$

$$P_5(x) = x + x^2 + \frac{x^3}{3} - \frac{x^5}{30}$$

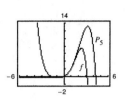

**49.** $h(x) = \cos x \ln(1 + x)$

$$= \left(1 - \frac{x^2}{2} + \frac{x^4}{24} + \cdots\right)\left(x - \frac{x^2}{2} + \frac{x^3}{3} - \frac{x^4}{4} + \frac{x^5}{5} - \cdots\right)$$

$$= x - \frac{x^2}{2} + \left(\frac{x^3}{3} - \frac{x^3}{2}\right) + \left(\frac{x^4}{4} - \frac{x^4}{4}\right) + \left(\frac{x^5}{5} - \frac{x^5}{6} + \frac{x^5}{24}\right) + \cdots$$

$$= x - \frac{x^2}{2} - \frac{x^3}{6} + \frac{3x^5}{40} + \cdots$$

$$P_5(x) = x - \frac{x^2}{2} - \frac{x^3}{6} + \frac{3x^5}{40}$$

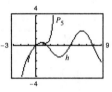

**51.** $g(x) = \dfrac{\sin x}{1 + x}$. Divide the series for $\sin x$ by $(1 + x)$.

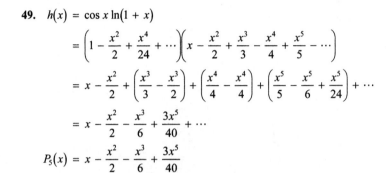

$$g(x) = x - x^2 + \frac{5x^3}{6} - \frac{5x^4}{6} + \cdots$$

$$P_4(x) = x - x^2 + \frac{5x^3}{6} - \frac{5x^4}{6}$$

**53.** 
$$\int_0^x \left(e^{-t^2} - 1\right) dt = \int_0^x \left[\left(\sum_{n=0}^{\infty} \frac{(-1)^n t^{2n}}{n!}\right) - 1\right] dt$$

$$= \int_0^x \left[\sum_{n=0}^{\infty} \frac{(-1)^{n+1} t^{2n+2}}{(n+1)!}\right] dt$$

$$= \left[\sum_{n=0}^{\infty} \frac{(-1)^{n+1} t^{2n+3}}{(2n+3)(n+1)!}\right]_0^x$$

$$= \sum_{n=0}^{\infty} \frac{(-1)^{n+1} x^{2n+3}}{(2n+3)(n+1)!}$$

**55.** Because $\ln x = \displaystyle\sum_{n=0}^{\infty} \frac{(-1)^n (x-1)^{n+1}}{n+1} = (x-1) - \frac{(x-1)^2}{2} + \frac{(x-1)^3}{3} - \frac{(x-1)^4}{4} + \cdots, \quad (0 < x \le 2)$

you have $\ln 2 = 1 - \dfrac{1}{2} + \dfrac{1}{3} - \dfrac{1}{4} + \cdots = \displaystyle\sum_{n=1}^{\infty} (-1)^{n+1} \frac{1}{n} \approx 0.6931.$   (10,001 terms)

**57.** Because $e^x = \sum_{n=0}^{\infty} \dfrac{x^n}{n!} = 1 + x + \dfrac{x^2}{2!} + \dfrac{x^3}{3!} + \cdots,$

you have $e^2 = 1 + 2 + \dfrac{2^2}{2!} + \dfrac{2^3}{3!} + \cdots = \sum_{n=0}^{\infty} \dfrac{2^n}{n!} \approx 7.3891.$  (12 terms)

**59.** Because

$$\cos x = \sum_{n=0}^{\infty} \dfrac{(-1)^n x^{2n}}{(2n)!} = 1 - \dfrac{x^2}{2!} + \dfrac{x^4}{4!} - \dfrac{x^6}{6!} + \dfrac{x^8}{8!} - \cdots$$

$$1 - \cos x = \dfrac{x^2}{2!} - \dfrac{x^4}{4!} + \dfrac{x^6}{6!} - \dfrac{x^8}{8!} + \cdots = \sum_{n=0}^{\infty} \dfrac{(-1)^n x^{2n+2}}{(2n+2)!}$$

$$\dfrac{1 - \cos}{x} = \dfrac{x}{2!} - \dfrac{x^3}{4!} + \dfrac{x^5}{6!} - \dfrac{x^7}{8!} + \cdots = \sum_{n=0}^{\infty} \dfrac{(-1)^n x^{2n+1}}{(2n+2)!}$$

you have $\displaystyle\lim_{x \to 0} \dfrac{1 - \cos x}{x} = \lim_{x \to 0} \sum_{n=0}^{\infty} \dfrac{(-1)x^{2n+1}}{(2n+2)!} = 0.$

**61.** Because $e^x = 1 + x + \dfrac{x^2}{2!} + \dfrac{x^3}{3!} + \cdots$

$e^x - 1 = x + \dfrac{x^2}{2!} + \dfrac{x^3}{3!} + \cdots \sum_{n=0}^{\infty} \dfrac{x^{n+1}}{(n+1)!}$

and $\dfrac{e^x - 1}{x} = 1 + \dfrac{x}{2!} + \dfrac{x^2}{3!} + \cdots \sum_{n=0}^{\infty} \dfrac{x^n}{(n+1)!}$

you have $\displaystyle\lim_{x \to 0} \dfrac{e^x - 1}{x} = \lim_{x \to 0} \sum \dfrac{x^n}{(n+1)!} = 1.$

**63.** $\displaystyle\int_0^1 e^{-x^3} \, dx = \int_0^1 \left[ \sum_{n=0}^{\infty} \dfrac{(-x^3)^n}{n!} \right] dx$

$\qquad = \displaystyle\int_0^1 \left[ \sum_{n=0}^{\infty} \dfrac{(-1)^n x^{3n}}{n!} \right] dx$

$\qquad = \left[ \displaystyle\sum_{n=0}^{\infty} \dfrac{(-1)^n x^{3n+1}}{(3n+1)n!} \right]_0^1$

$\qquad = \displaystyle\sum_{n=0}^{\infty} \dfrac{(-1)^n}{(3n+1)n!}$

$\qquad = 1 - \dfrac{1}{4} + \dfrac{1}{14} - \cdots + (-1)^n \dfrac{1}{(3n+1)n!} + \cdots$

Because $\dfrac{1}{[3(6)+1]6!} < 0.0001,$ you need 6 terms.

$\displaystyle\int_0^1 e^{-x^2} \, dx \approx \sum_{n=0}^{5} \dfrac{(-1)^n}{(3n+1)n!} \approx 0.8075$

**65.** $\displaystyle\int_0^1 \dfrac{\sin x}{x} \, dx = \int_0^1 \left[ \sum_{n=0}^{\infty} \dfrac{(-1)^n x^{2n}}{(2n+1)!} \right] dx = \left[ \sum_{n=0}^{\infty} \dfrac{(-1)^n x^{2n+1}}{(2n+1)(2n+1)!} \right]_0^1 = \sum_{n=0}^{\infty} \dfrac{(-1)^n}{(2n+1)(2n+1)!}$

Because $1/(7 \cdot 7!) < 0.0001,$ you need three terms:

$\displaystyle\int_0^1 \dfrac{\sin x}{x} \, dx = 1 - \dfrac{1}{3 \cdot 3!} + \dfrac{1}{5 \cdot 5!} - \cdots \approx 0.9461.$  (using three nonzero terms)

**Note:** You are using $\displaystyle\lim_{x \to 0^+} \dfrac{\sin x}{x} = 1.$

**67.** $\displaystyle\int_0^{1/2} \frac{\arctan x}{x}\, dx = \int_0^{1/2}\left(1 - \frac{x^2}{3} + \frac{x^4}{5} - \frac{x^6}{7} + \cdots\right) dx$

$$= \left[x - \frac{x^3}{3^2} + \frac{x^5}{5^2} - \frac{x^7}{7^2} + \cdots\right]_0^{1/2}$$

Because $1/(9^2 2^9) < 0.0001$, you have

$$\int_0^{1/2} \frac{\arctan x}{x}\, dx \approx \left(\frac{1}{2} - \frac{1}{3^2 2^3} + \frac{1}{5^2 2^5} - \frac{1}{7^2 2^7} + \frac{1}{9^2 2^9}\right)$$

$$\approx 0.4872.$$

**Note:** You are using $\displaystyle\lim_{x\to 0^+} \frac{\arctan x}{x} = 1$.

**69.** $\displaystyle\int_{0.1}^{0.3} \sqrt{1 + x^3}\, dx = \int_{0.1}^{0.3}\left(1 + \frac{x^3}{2} - \frac{x^6}{8} + \frac{x^9}{16} - \frac{5x^{12}}{128} + \cdots\right) dx = \left[x + \frac{x^4}{8} - \frac{x^7}{56} + \frac{x^{10}}{160} - \frac{5x^{13}}{1664} + \cdots\right]_{0.1}^{0.3}$

Because $\dfrac{1}{56}(0.3^7 - 0.1^7) < 0.0001$, you need two terms.

$$\int_{0.1}^{0.3} \sqrt{1 + x^3}\, dx = \left[(0.3 - 0.1) + \frac{1}{8}(0.3^4 - 0.1^4)\right] \approx 0.201.$$

**71.** $\displaystyle\int_0^{\pi/2} \sqrt{x}\,\cos x\, dx = \int_0^{\pi/2}\left[\sum_{n=0}^{\infty} \frac{(-1)^n x^{(4n+1)/2}}{(2n)!}\right] dx = \left[\sum_{n=0}^{\infty} \frac{(-1)^n x^{(4n+3)/2}}{\left(\frac{4n+3}{2}\right)(2n)!}\right]_0^{\pi/2} = \left[\sum_{n=0}^{\infty} \frac{(-1)^n 2x^{(4n+3)/2}}{(4n+3)(2n)!}\right]_0^{\pi/2}$

Because $2(\pi/2)^{23/2}\big/(23 \cdot 10!) < 0.0001$, you need five terms.

$$\int_0^1 \sqrt{x}\,\cos x\, dx = 2\left[\frac{(\pi/2)^{3/2}}{3} - \frac{(\pi/2)^{7/2}}{14} + \frac{(\pi/2)^{11/2}}{264} - \frac{(\pi/2)^{15/2}}{10{,}800} + \frac{(\pi/2)^{19/2}}{766{,}080}\right] \approx 0.7040.$$

**73.** From Exercise 27, you have

$$\frac{1}{\sqrt{2\pi}}\int_0^1 e^{-x^2/2}\, dx = \frac{1}{\sqrt{2\pi}}\int_0^1 \sum_{n=0}^{\infty} \frac{(-1)^n x^{2n}}{2^n n!}\, dx = \frac{1}{\sqrt{2\pi}}\left[\sum_{n=0}^{\infty} \frac{(-1)^n x^{2n+1}}{2^n n!(2n+1)}\right]_0^1 = \frac{1}{\sqrt{2\pi}}\sum_{n=0}^{\infty} \frac{(-1)^n}{2^n n!(2n+1)}$$

$$\approx \frac{1}{\sqrt{2\pi}}\left(1 - \frac{1}{2 \cdot 1 \cdot 3} + \frac{1}{2^2 \cdot 2! \cdot 5} - \frac{1}{2^3 \cdot 3! \cdot 7}\right) \approx 0.3412.$$

**75.** $f(x) = x\cos 2x = \displaystyle\sum_{n=0}^{\infty} \frac{(-1)^n 4^n x^{2n+1}}{(2n)!}$

$P_5(x) = x - 2x^3 + \dfrac{2x^5}{3}$

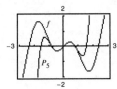

The polynomial is a reasonable approximation on the interval $\left[-\dfrac{3}{4}, \dfrac{3}{4}\right]$.

**77.** $f(x) = \sqrt{x} \ln x, c = 1$

$$P_5(x) = (x - 1) - \frac{(x - 1)^3}{24} + \frac{(x - 1)^4}{24} - \frac{71(x - 1)^5}{1920}$$

The polynomial is a reasonable approximation on the interval $\left[\frac{1}{4}, 2\right]$.

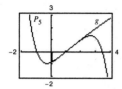

**79.** See Guidelines, page 668.

**81.** (a) Replace $x$ with $(-x)$.

   (b) Replace $x$ with $3x$.

   (c) Multiply series by $x$.

   (d) Replace $x$ with $2x$, then replace $x$ with $-2x$, and add the two together.

**83.** $y = \left(\tan\theta - \dfrac{g}{kv_0\cos\theta}\right)x - \dfrac{g}{k^2}\ln\left(1 - \dfrac{kx}{v_0\cos\theta}\right)$

$$= (\tan\theta)x - \frac{gx}{kv_0\cos\theta} - \frac{g}{k^2}\left[-\frac{kx}{v_0\cos\theta} - \frac{1}{2}\left(\frac{kx}{v_0\cos\theta}\right)^2 - \frac{1}{3}\left(\frac{kx}{v_0\cos\theta}\right)^3 - \frac{1}{4}\left(\frac{kx}{v_0\cos\theta}\right)^4 - \cdots\right]$$

$$= (\tan\theta)x - \frac{gx}{kv_0\cos\theta} + \frac{gx}{kv_0\cos\theta} + \frac{gx^2}{2v_0^2\cos^2\theta} + \frac{gkx^3}{3v_0^3\cos^3\theta} + \frac{gk^2x^4}{4v_0^4\cos^4\theta} + \cdots$$

$$= (\tan\theta)x + \frac{gx^2}{2v_0^2\cos^2\theta} + \frac{kgx^3}{3v_0^3\cos^3\theta} + \frac{k^2gx^4}{4v_0^4\cos^4\theta} + \cdots$$

**85.** $f(x) = \begin{cases} e^{-1/x^2}, & x \neq 0 \\ 0, & x = 0 \end{cases}$

   (a)

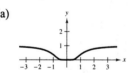

   (b) $f'(0) = \displaystyle\lim_{x\to 0}\frac{f(x) - f(0)}{x - 0} = \lim_{x\to 0}\frac{e^{-1/x^2} - 0}{x}$

   Let $y = \displaystyle\lim_{x\to 0}\frac{e^{-1/x^2}}{x}$. Then

   $\ln y = \displaystyle\lim_{x\to 0}\ln\left(\frac{e^{-1/x^2}}{x}\right) = \lim_{x\to 0^+}\left[-\frac{1}{x^2} - \ln x\right] = \lim_{x\to 0^+}\left[\frac{-1 - x^2\ln x}{x^2}\right] = -\infty.$

   So, $y = e^{-\infty} = 0$ and you have $f'(0) = 0$.

   (c) $\displaystyle\sum_{n=0}^{\infty}\frac{f^{(n)}(0)}{n!}x^n = f(0) + \frac{f'(0)x}{1!} + \frac{f''(0)x^2}{2!} + \cdots = 0 \neq f(x)$ This series converges to $f$ at $x = 0$ only.

**87.** By the Ratio Test: $\lim\limits_{n\to\infty}\left|\dfrac{x^{n+1}}{(n+1)!}\cdot\dfrac{n!}{x^n}\right| = \lim\limits_{n\to\infty}\dfrac{|x|}{n+1} = 0$ which shows that $\sum\limits_{n=0}^{\infty}\dfrac{x^n}{n!}$ converges for all $x$.

**89.** $\dbinom{5}{3} = \dfrac{5\cdot 4\cdot 3}{3!} = \dfrac{60}{6} = 10$

**93.** $(1+x)^k = \sum\limits_{n=0}^{\infty}\dbinom{k}{n}x^n$

**91.** $\dbinom{0.5}{4} = \dfrac{(0.5)(-0.5)(-1.5)(-2.5)}{4!} = -0.0390625 = -\dfrac{5}{128}$

Example: $(1+x)^2 = \sum\limits_{n=0}^{\infty}\dbinom{2}{n}x^n = 1 + 2x + x^2$

**95.** $g(x) = \dfrac{x}{1-x-x^2} = a_0 + a_1 x + a_2 x^2 + \cdots$

$x = (1 - x - x^2)(a_0 + a_1 x + a_2 x^2 + \cdots)$

$x = a_0 + (a_1 - a_0)x + (a_2 - a_1 - a_0)x^2 + (a_3 - a_2 - a_1)x^3 + \cdots$

Equating coefficients,

$a_0 = 0$

$a_1 - a_0 = 1 \Rightarrow a_1 = 1$

$a_2 - a_1 - a_0 = 0 \Rightarrow a_2 = 1$

$a_3 - a_2 - a_1 = 0 \Rightarrow a_3 = 2$

$a_4 = a_3 + a_2 = 3$, etc.

In general, $a_n = a_{n-1} + a_{n-2}$. The coefficients are the Fibonacci numbers.

# Review Exercises for Chapter 9

**1.** $a_n = 5^n$

$a_1 = 5^1 = 5$

$a_2 = 5^2 = 25$

$a_3 = 5^3 = 125$

$a_4 = 5^4 = 625$

$a_5 = 5^5 = 3125$

**3.** $a_n = \left(-\dfrac{1}{4}\right)^n$

$a_1 = \left(-\dfrac{1}{4}\right)^1 = -\dfrac{1}{4}$

$a_2 = \left(-\dfrac{1}{4}\right)^2 = \dfrac{1}{16}$

$a_3 = \left(-\dfrac{1}{4}\right)^3 = -\dfrac{1}{64}$

$a_4 = \left(-\dfrac{1}{4}\right)^4 = \dfrac{1}{256}$

$a_5 = \left(-\dfrac{1}{4}\right)^5 = -\dfrac{1}{1024}$

**5.** $a_n = 4 + \dfrac{2}{n}$: $6, 5, 4.67, \ldots$

Matches (a).

**6.** $a_n = 4 - \dfrac{n}{2}$: $3.5, 3, \ldots$

Matches (c).

**7.** $a_n = 10(0.3)^{n-1}$: $10, 3, \ldots$

Matches (d).

**8.** $a_n = 6\left(-\dfrac{2}{3}\right)^{n-1}$: $6, -4, \ldots$

Matches (b).

**9.** $a_n = \dfrac{5n+2}{n}$

The sequence seems to converge to 5.

$\lim\limits_{n\to\infty} a_n = \lim\limits_{n\to\infty}\dfrac{5n+2}{n}$

$\qquad = \lim\limits_{n\to\infty}\left(5 + \dfrac{2}{n}\right) = 5$

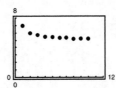

**11.** $\lim\limits_{n\to\infty}\left[\left(\dfrac{2}{5}\right)^n + 5\right] = 0 + 5 = 5$

Converges

**13.** $\lim\limits_{n\to\infty}\dfrac{n^3+1}{n^2}=\infty$

Diverges

**15.** $\lim\limits_{n\to\infty}\dfrac{n}{n^2+1}=0$

Converges

**17.** $\lim\limits_{n\to\infty}\left(\sqrt{n+1}-\sqrt{n}\right)=\lim\limits_{n\to\infty}\left(\sqrt{n+1}-\sqrt{n}\right)\dfrac{\sqrt{n+1}+\sqrt{n}}{\sqrt{n+1}+\sqrt{n}}$

$$=\lim\limits_{n\to\infty}\dfrac{1}{\sqrt{n+1}+\sqrt{n}}=0$$

Converges

**19.** $a_n=5n-2$

**21.** $a_n=\dfrac{1}{n!+1}$

**23.** (a) $A_n=8000\left(1+\dfrac{0.05}{4}\right)^n,\quad n=1,2,3,\dots$

$$A_1=8000\left(1+\dfrac{0.05}{4}\right)^1=\$8100.00$$

$A_2=\$8201.25$

$A_3=\$8303.77$

$A_4=\$8407.56$

$A_5=\$8512.66$

$A_6=\$8619.07$

$A_7=\$8726.80$

$A_8=\$8835.89$

(b) $A_{40}=\$13{,}148.96$

**25.** $S_1=3$

$S_2=3+\dfrac{3}{2}=\dfrac{9}{2}=4.5$

$S_3=3+\dfrac{3}{2}+1=\dfrac{11}{2}=5.5$

$S_4=3+\dfrac{3}{2}+1+\dfrac{3}{4}=\dfrac{25}{4}=6.25$

$S_5=3+\dfrac{3}{2}+1+\dfrac{3}{4}+\dfrac{3}{5}=\dfrac{137}{20}=6.85$

**27.** (a)

| $n$ | 5 | 10 | 15 | 20 | 25 |
|---|---|---|---|---|---|
| $S_n$ | 13.2 | 113.3 | 873.8 | 6648.5 | 50,500.3 |

The series diverges $\left(\text{geometric }r=\dfrac{3}{2}>1\right)$.

(b)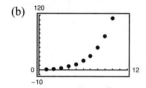

**29.** (a)

| $n$ | 5 | 10 | 15 | 20 | 25 |
|---|---|---|---|---|---|
| $S_n$ | 0.4597 | 0.4597 | 0.4597 | 0.4597 | 0.4597 |

The series converges by the Alternating Series Test.

(b)

**31.** $\sum\limits_{n=0}^{\infty}\left(\dfrac{2}{5}\right)^n=\dfrac{1}{1-(2/5)}=\dfrac{5}{3}$  (Geometric series)

**33.** $\sum\limits_{n=1}^{\infty}\left[(0.6)^n+(0.8)^n\right]=\sum\limits_{n=0}^{\infty}0.6(0.6)^n+\sum\limits_{n=0}^{\infty}0.8(0.8)^n=(0.6)\dfrac{1}{1-0.6}+(0.8)\dfrac{1}{1-0.8}=\dfrac{6}{10}\cdot\dfrac{10}{4}+\dfrac{8}{10}\cdot\dfrac{10}{2}=\dfrac{11}{2}=5.5$

**35.** (a) $0.\overline{09} = 0.09 + 0.0009 + 0.000009 + \cdots = 0.09(1 + 0.01 + 0.0001 + \cdots) = \sum\limits_{n=0}^{\infty} (0.09)(0.01)^n$

(b) $0.\overline{09} = \dfrac{0.09}{1 - 0.01} = \dfrac{1}{11}$

**37.** Diverges. Geometric series with $a = 1$ and $|r| = 1.67 > 1$.

**39.** Diverges. $n$th-Term Test. $\lim\limits_{n\to\infty} a_n \ne 0$.

**41.** $D_1 = 8$

$D_2 = 0.7(8) + 0.7(8) = 16(0.7)$

$\vdots$

$D = 8 + 16(0.7) + 16(0.7)^2 + \cdots + 16(0.7)^n + \cdots$

$= -8 + \sum\limits_{n=0}^{\infty} 16(0.7)^n = -8 + \dfrac{16}{1 - 0.7} = 45\dfrac{1}{3}$ meters

**43.** $\sum\limits_{n=1}^{\infty} \dfrac{2}{6n + 1}$

Let $f(x) = \dfrac{2}{6x + 1}$, $f'(x) = \dfrac{-12}{(6x + 1)^2} < 0$ for $x \ge 1$

$f$ is positive, continuous, and decreasing for $x \ge 1$.

$\int_1^{\infty} \dfrac{2}{6x + 1}\, dx = \left[\dfrac{1}{3}\ln(6x + 1)\right]_1^{\infty} , = \infty$, diverges.

So, the series diverges by Theorem 9.10.

**45.** $\sum\limits_{n=1}^{\infty} \dfrac{1}{n^{5/2}}$ is a $p$-series with $p = \dfrac{5}{2} > 1$.

So, the series converges.

**47.** $\sum\limits_{n=1}^{\infty}\left(\dfrac{1}{n^2} - \dfrac{1}{n}\right) = \sum\limits_{n=1}^{\infty}\dfrac{1}{n^2} - \sum\limits_{n=1}^{\infty}\dfrac{1}{n}$

Because the second series is a divergent $p$-series while the first series is a convergent $p$-series, the difference diverges.

**49.** $\sum\limits_{n=2}^{\infty} \dfrac{1}{\sqrt[3]{n} - 1}$

$\dfrac{1}{\sqrt[3]{n} - 1} > \dfrac{1}{\sqrt[3]{n}}$

Therefore, the series diverges by comparison with the divergent $p$-series

$\sum\limits_{n=2}^{\infty} \dfrac{1}{\sqrt[3]{n}} = \sum\limits_{n=2}^{\infty} \dfrac{1}{n^{1/3}}.$

**51.** $\sum\limits_{n=1}^{\infty} \dfrac{1}{\sqrt{n^3 + 2n}}$

$\lim\limits_{n\to\infty} \dfrac{1/\sqrt{n^3 + 2n}}{1/(n^{3/2})} = \lim\limits_{n\to\infty} \dfrac{n^{3/2}}{\sqrt{n^3 + 2n}} = 1$

By a limit comparison test with the convergent $p$-series

$\sum\limits_{n=1}^{\infty} \dfrac{1}{n^{3/2}}$, the series converges.

**53.** $\sum\limits_{n=1}^{\infty} \dfrac{1 \cdot 3 \cdot 5 \cdots (2n - 1)}{2 \cdot 4 \cdot 6 \cdots (2n)}$

$a_n = \dfrac{1 \cdot 3 \cdot 5 \cdots (2n - 1)}{2 \cdot 4 \cdot 6 \cdots (2n)} = \left(\dfrac{3}{2} \cdot \dfrac{5}{4} \cdots \dfrac{2n - 1}{2n - 2}\right)\dfrac{1}{2n} > \dfrac{1}{2n}$

Because $\sum\limits_{n=1}^{\infty} \dfrac{1}{2n} = \dfrac{1}{2}\sum\limits_{n=1}^{\infty}\dfrac{1}{n}$ diverges (harmonic series), so does the original series.

**55.** $\sum\limits_{n=1}^{\infty} \dfrac{(-1)^n}{n^5}$ converges by the Alternating Series Test.

$\lim\limits_{n\to\infty} \dfrac{1}{n^5} = 0$ and $a_{n+1} = \dfrac{1}{(n + 1)^5} < \dfrac{1}{n^5} = a_n$.

**57.** $\sum\limits_{n=2}^{\infty} \dfrac{(-1)^n - n}{n^2 - 3}$ converges by the Alternating Series Test.

$\lim\limits_{n\to\infty} \dfrac{n}{n^2 - 3} = 0$ and if

$f(x) = \dfrac{n}{n^2 - 3}$, $f'(x) = \dfrac{-(n^2 + 3)}{(n^2 - 3)^2} < 0 \Rightarrow$ terms are decreasing. So, $a_{n+1} < a_n$.

**59.** Diverges by the $n$th-Term Test.

$\lim\limits_{n\to\infty} \dfrac{n}{n - 3} = 1 \ne 0$

**61.** $\lim\limits_{n\to\infty} \sqrt[n]{\left(\dfrac{3n - 1}{2n + 5}\right)^n} = \lim\limits_{n\to\infty}\left(\dfrac{3n - 1}{2n + 5}\right) = \dfrac{3}{2} > 1$

Diverges by Root Test.

**63.** $\displaystyle\sum_{n=1}^{\infty} \frac{n}{e^{n^2}}$

$$\lim_{n\to\infty} \left| \frac{a_{n+1}}{a_n} \right| = \lim_{n\to\infty} \left| \frac{n+1}{e^{(n+1)^2}} \cdot \frac{e^{n^2}}{n} \right|$$

$$= \lim_{n\to\infty} \left| \frac{e^{n^2}(n+1)}{e^{n^2+2n+1}n} \right|$$

$$= \lim_{n\to\infty} \left( \frac{1}{e^{2n+1}} \right)\left( \frac{n+1}{n} \right)$$

$$= (0)(1) = 0 < 1$$

By the Ratio Test, the series converges.

**67.** (a) Ratio Test: $\displaystyle\lim_{n\to\infty} \left| \frac{a_{n+1}}{a_n} \right| = \lim_{n\to\infty} \frac{(n+1)(3/5)^{n+1}}{n(3/5)^n} = \lim_{n\to\infty} \left( \frac{n+1}{n} \right)\left( \frac{3}{5} \right) = \frac{3}{5} < 1$, converges

(b)

| $n$ | 5 | 10 | 15 | 20 | 25 |
|-----|-----|-----|-----|-----|-----|
| $S_n$ | 2.8752 | 3.6366 | 3.7377 | 3.7488 | 3.7499 |

(c)

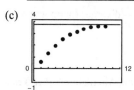

(d) The sum is approximately 3.75.

**69.**  $f(x) = e^{-2x}$,  $\quad f(0) = 1$

$f'(x) = -2e^{-2x}$,  $\quad f'(0) = -2$

$f''(x) = 4e^{-2x}$,  $\quad f''(0) = 4$

$f'''(x) = -8e^{-2x}$,  $\quad f'''(0) = -8$

$P_3(x) = f(0) + f'(0)x + \dfrac{f''(0)}{2!}x^2 + \dfrac{f'''(0)}{3!}x^3$

$\qquad = 1 - 2x + 2x^2 - \dfrac{4}{3}x^3$

**71.**  $f(x) = e^{-3x}$  $\qquad f(0) = 1$

$f'(x) = -3e^{-3x}$  $\qquad f'(0) = -3$

$f''(x) = 9e^{-3x}$  $\qquad f''(0) = 9$

$f'''(x) = -27e^{-3x}$  $\quad f'''(0) = -27$

$P_3(x) = f(0) + f'(0)x + \dfrac{f''(0)}{2!}x^2 + \dfrac{f'''(0)}{3!}x^3$

$\qquad = 1 - 3x + \dfrac{9}{2}x^2 - \dfrac{9}{2}x^3$

**65.** $\displaystyle\sum_{n=1}^{\infty} \frac{2^n}{n^3}$

$$\lim_{n\to\infty} \left| \frac{a_{n+1}}{a_n} \right| = \lim_{n\to\infty} \frac{2^{n+1}}{(n+1)^3} \cdot \frac{n^3}{2^n} = \lim_{n\to\infty} \frac{2n^3}{(n+1)^3} = 2$$

By the Ratio Test, the series diverges.

**73.**  $\quad f(x) = \cos x$

$\left| f^{(n+1)}(x) \right| \le 1$ for all $x$ and all $n$.

$$\left| R_n(x) \right| = \left| \frac{f^{(n+1)}(z)\, x^{n+1}}{(n+1)!} \right| \le \frac{(0.75)^{n+1}}{(n+1)!} < 0.001$$

By trial and error, $n = 5$.   (3 terms)

**75.** $\displaystyle\sum_{n=0}^{\infty} \left( \frac{x}{10} \right)^n$

Geometric series which converges only if $\left| x/10 \right| < 1$ or $-10 < x < 10$.

**77.** $\displaystyle\sum_{n=0}^{\infty} \frac{(-1)^n (x-2)^n}{(n+1)^2}$

$$\lim_{n\to\infty} \left| \frac{u_{n+1}}{u_n} \right| = \lim_{n\to\infty} \left| \frac{(-1)^{n+1}(x-2)^{n+1}}{(n+2)^2} \cdot \frac{(n+1)^2}{(-1)^n(x-2)^n} \right|$$

$$= \left| x - 2 \right|$$

$R = 1$

Center: 2

Because the series converges when $x = 1$ and when $x = 3$, the interval of convergence is $[1, 3]$.

**79.** $\displaystyle\sum_{n=0}^{\infty} n!(x-2)^n$

$$\lim_{n\to\infty}\left|\frac{u_{n+1}}{u_n}\right| = \lim_{n\to\infty}\left|\frac{(n+1)!(x-2)^{n+1}}{n!(x-2)^n}\right| = \infty$$

which implies that the series converges only at the center $x=2$.

**81.** (a) $\displaystyle f(x) = \sum_{n=0}^{\infty}\left(\frac{x}{5}\right)^n, (-5,5)$    (Geometric)

(b) $\displaystyle f'(x) = \sum_{n=1}^{\infty}\frac{n}{5}\left(\frac{x}{5}\right)^{n-1}, (-5,5)$

(c) $\displaystyle f''(x) = \sum_{n=2}^{\infty}\frac{n(n-1)}{25}\left(\frac{x}{5}\right)^{n-2}, (-5,5)$

(d) $\displaystyle \int f(x)\,dx = \sum_{n=0}^{\infty}\frac{5}{n+1}\left(\frac{x}{5}\right)^{n+1}, (-5,5)$

$$\left[\sum_{n=0}^{\infty}\frac{5}{n+1}\left(\frac{-5}{5}\right)^{n+1} = \sum_{n=0}^{\infty}\frac{(-1)^{n+1}\,5}{n+1},\quad \text{converges}\right]$$

**83.**

$$y = \sum_{n=0}^{\infty}(-1)^n\frac{x^{2n}}{4^n(n!)^2}$$

$$y' = \sum_{n=1}^{\infty}\frac{(-1)^n(2n)x^{2n-1}}{4^n(n!)^2} = \sum_{n=0}^{\infty}\frac{(-1)^{n+1}(2n+2)x^{2n+1}}{4^{n+1}\big[(n+1)!\big]^2}$$

$$y'' = \sum_{n=0}^{\infty}\frac{(-1)^{n+1}(2n+2)(2n+1)x^{2n}}{4^{n+1}\big[(n+1)!\big]^2}$$

$$x^2y'' + xy' + x^2y = \sum_{n=0}^{\infty}\frac{(-1)^{n+1}(2n+2)(2n+1)x^{2n+2}}{4^{n+1}\big[(n+1)!\big]^2} + \sum_{n=0}^{\infty}\frac{(-1)^{n+1}(2n+2)x^{2n+2}}{4^{n+1}\big[(n+1)!\big]^2} + \sum_{n=0}^{\infty}(-1)^n\frac{x^{2n+2}}{4^n(n!)^2}$$

$$= \sum_{n=0}^{\infty}\left[\frac{(-1)^{n+1}(2n+2)(2n+1)}{4^{n+1}\big[(n+1)!\big]^2} + \frac{(-1)^{n+1}(2n+2)}{4^{n+1}\big[(n+1)!\big]^2} + \frac{(-1)^n}{4^n(n!)^2}\right]x^{2n+2}$$

$$= \sum_{n=0}^{\infty}\left[\frac{(-1)^{n+1}(2n+2)(2n+1+1)}{4^{n+1}\big[(n+1)!\big]^2} + (-1)^n\frac{1}{4^n(n!)^2}\right]x^{2n+2}$$

$$= \sum_{n=0}^{\infty}\left[\frac{(-1)^{n+1}4(n+1)^2}{4^{n+1}\big[(n+1)!\big]^2} + (-1)^n\frac{1}{4^n(n!)^2}\right]x^{2n+2}$$

$$= \sum_{n=0}^{\infty}\left[\frac{(-1)^{n+1}1}{4^n(n!)^2} + (-1)^n\frac{1}{4^n(n!)^2}\right]x^{2n+2} = 0$$

**85.** $\displaystyle\frac{2}{3-x} = \frac{2/3}{1-(x/3)} = \frac{a}{1-r}$

$$\sum_{n=0}^{\infty}\frac{2}{3}\left(\frac{x}{3}\right)^n = \sum_{n=0}^{\infty}\frac{2x^n}{3^{n+1}}$$

**87.** $\displaystyle\frac{6}{4-x} = \frac{6}{3-(x-1)} = \frac{2}{1-\left(\dfrac{x-1}{3}\right)} = \frac{a}{1-r}$

$$= \sum_{n=0}^{\infty}2\left(\frac{x-1}{3}\right)^n = 2\sum_{n=0}^{\infty}\frac{(x-1)^n}{3^n}$$

Interval of convergence:

$$\left|\frac{x-1}{3}\right| < 1 \Rightarrow |x-1| < 3 \Rightarrow (-2,4)$$

**89.** $\displaystyle \ln x = \sum_{n=1}^{\infty}(-1)^{n+1}\frac{(x-1)^n}{n}, \quad 0 < x \le 2$

$$\ln\left(\frac{5}{4}\right) = \sum_{n=1}^{\infty}(-1)^{n+1}\left(\frac{(5/4)-1}{n}\right)^n = \sum_{n=1}^{\infty}(-1)^{n+1}\frac{1}{4^n n} \approx 0.2231$$

**91.** $\displaystyle e^x = \sum_{n=0}^{\infty}\frac{x^n}{n!}, \quad -\infty < x < \infty$

$$e^{1/2} = \sum_{n=0}^{\infty}\frac{(1/2)^n}{n!} = \sum_{n=0}^{\infty}\frac{1}{2^n n!} \approx 1.6487$$

**93.** $\displaystyle \cos x = \sum_{n=0}^{\infty}(-1)^n\frac{x^{2n}}{(2n)!}, \quad -\infty < x < \infty$

$$\cos\left(\frac{2}{3}\right) = \sum_{n=0}^{\infty}(-1)^n\frac{2^{2n}}{3^{2n}(2n)!} = 0.7859$$

**95.** $f(x) = \sin x$

$f'(x) = \cos x$

$f''(x) = -\sin x$

$f'''(x) = -\cos x, \cdots$

$$\sin(x) = \sum_{n=0}^{\infty} \frac{f^{(n)}(x)\left[x - (3\pi/4)\right]^n}{n!}$$

$$= \frac{\sqrt{2}}{2} - \frac{\sqrt{2}}{2}\left(x - \frac{3\pi}{4}\right) - \frac{\sqrt{2}}{2 \cdot 2!}\left(x - \frac{3\pi}{4}\right)^2 + \cdots = \frac{\sqrt{2}}{2}\sum_{n=0}^{\infty} \frac{(-1)^{n(n+1)/2}\left[x - (3\pi/4)\right]^n}{n!}$$

**97.** $3^x = \left(e^{\ln(3)}\right)^x = e^{x\ln(3)}$ and because $e^x = \sum_{n=0}^{\infty} \dfrac{x^n}{n!}$, you have

$$3^x = \sum_{n=0}^{\infty} \frac{(x\ln 3)^n}{n!} = 1 + x\ln 3 + \frac{x^2\left[\ln 3\right]^2}{2!} + \frac{x^3\left[\ln 3\right]^3}{3!} + \frac{x^4\left[\ln 3\right]^4}{4!} + \cdots.$$

**99.** $f(x) = \dfrac{1}{x}$

$f'(x) = -\dfrac{1}{x^2}$

$f''(x) = \dfrac{2}{x^3}$

$f'''(x) = -\dfrac{6}{x^4}, \cdots$

$$\frac{1}{x} = \sum_{n=0}^{\infty} \frac{f^{(n)}(-1)(x+1)^n}{n!} = \sum_{n=0}^{\infty} \frac{-n!(x+1)^n}{n!} = -\sum_{n=0}^{\infty}(x+1)^n, \; -2 < x < 0$$

**101.** $(1+x)^k = 1 + kx + \dfrac{k(k-1)x^2}{2!} + \dfrac{k(k-1)(k-2)x^3}{3!} + \cdots$

$(1+x)^{1/5} = 1 + \dfrac{x}{5} + \dfrac{(1/5)(-4/5)x^2}{2!} + \dfrac{1/5(-4/5)(-9/5)x^3}{3!} + \cdots$

$$= 1 + \frac{1}{5}x - \frac{1 \cdot 4x^2}{5^2 2!} + \frac{1 \cdot 4 \cdot 9 x^3}{5^3 3!} - \cdots = 1 + \frac{x}{5} + \sum_{n=2}^{\infty} \frac{(-1)^{n+1} 4 \cdot 9 \cdot 14 \cdots (5n-6)x^n}{5^n n!} = 1 + \frac{x}{5} - \frac{2}{25}x^2 + \frac{6}{125}x^3 - \cdots$$

**103. (a)** $f(x) = e^{2x} \qquad f(0) = 1$

$f'(x) = 2e^{2x} \qquad f'(0) = 2$

$f''(x) = 4e^{2x} \qquad f''(0) = 4$

$f'''(x) = 8e^{2x} \qquad f'''(0) = 8$

$$P(x) = 1 + 2x + \frac{4x^2}{2!} + \frac{8x^3}{3!} = 1 + 2x + 2x^2 + \frac{4}{3}x^3$$

**(b)** $e^x = \sum_{n=0}^{\infty} \dfrac{x^n}{n!}, \; e^{2x} = \sum_{n=0}^{\infty} \dfrac{(2x)^n}{n!}$

$$P(x) = 1 + 2x + 2x^2 + \frac{4}{3}x^3$$

**(c)** $e^x \cdot e^x = \left(1 + x + \dfrac{x^2}{2!} + \cdots\right)\left(1 + x + \dfrac{x^2}{2!} + \cdots\right)$

$$P(x) = 1 + 2x + 2x^2 + \frac{4}{3}x^3$$

**105.** $e^x = \sum_{n=0}^{\infty} \frac{x^n}{n!} = 1 + x + \frac{x^2}{2!} + \frac{x^3}{3!} + \cdots$

$e^{6x} = \sum_{n=0}^{\infty} \frac{(6x)^n}{n!} = 1 + 6x + \frac{(6x)^2}{2!} + \frac{(6x)^3}{3!} + \cdots$

$= 1 + 6x + 18x^2 + 36x^3 + \cdots$

**107.** $\sin x = \sum_{n=0}^{\infty} \frac{(-1)^n x^{2n+1}}{(2n+1)!}$

$\sin 2x = \sum_{n=0}^{\infty} \frac{(-1)^n (2x)^{2n+1}}{(2n+1)!}$

$= 2x - \frac{4}{3}x^3 + \frac{4}{15}x^5 - \cdots$

**109.** $\arctan x = x - \frac{x^3}{3} + \frac{x^5}{5} - \frac{x^7}{7} + \frac{x^9}{9} - \cdots$

$\dfrac{\arctan x}{\sqrt{x}} = \sqrt{x} - \dfrac{x^{5/2}}{3} + \dfrac{x^{9/2}}{5} - \dfrac{x^{13/2}}{7} + \dfrac{x^{17/2}}{9} - \cdots$

$\lim_{x \to 0^+} \dfrac{\arctan x}{\sqrt{x}} = 0$

By L'Hôpital's Rule,

$\lim_{x \to 0^+} \dfrac{\arctan x}{\sqrt{x}} = \lim_{x \to 0^+} \dfrac{\left(\dfrac{1}{1+x^2}\right)}{\left(\dfrac{1}{2\sqrt{x}}\right)} = \lim_{x \to 0^+} \dfrac{2\sqrt{x}}{1+x^2} = 0.$

## Problem Solving for Chapter 9

**1. (a)** $1\left(\dfrac{1}{3}\right) + 2\left(\dfrac{1}{9}\right) + 4\left(\dfrac{1}{27}\right) + \cdots = \sum_{n=0}^{\infty} \dfrac{1}{3}\left(\dfrac{2}{3}\right)^n$

$= \dfrac{1/3}{1 - (2/3)}$

$= 1$

**(b)** $0, \dfrac{1}{3}, \dfrac{2}{3}, 1,$ etc.

**(c)** $\lim_{n \to \infty} C_n = 1 - \sum_{n=0}^{\infty} \dfrac{1}{3}\left(\dfrac{2}{3}\right)^n = 1 - 1 = 0$

**3.** Let $S = \sum_{n=1}^{\infty} \dfrac{1}{(2n-1)^2} = \dfrac{1}{1^2} + \dfrac{1}{3^2} + \dfrac{1}{5^2} + \cdots$.

Then

$\dfrac{\pi^2}{6} = \dfrac{1}{1^2} + \dfrac{1}{2^2} + \dfrac{1}{3^2} + \dfrac{1}{4^2} + \cdots$

$= S + \dfrac{1}{2^2} + \dfrac{1}{4^2} + \cdots$

$= S + \dfrac{1}{2^2}\left[1 + \dfrac{1}{2^2} + \dfrac{1}{3^2} + \cdots\right] = S + \dfrac{1}{2^2}\left(\dfrac{\pi^2}{6}\right).$

So, $S = \dfrac{\pi^2}{6} - \dfrac{1}{4}\dfrac{\pi^2}{6} = \dfrac{\pi^2}{6}\left(\dfrac{3}{4}\right) = \dfrac{\pi^2}{8}.$

**5. (a)** Position the three blocks as indicated in the figure. The bottom block extends 1/6 over the edge of the table, the middle block extends 1/4 over the edge of the bottom block, and the top block extends 1/2 over the edge of the middle block.

The centers of gravity are located at

bottom block: $\dfrac{1}{6} - \dfrac{1}{2} = -\dfrac{1}{3}$

middle block: $\dfrac{1}{6} + \dfrac{1}{4} - \dfrac{1}{2} = -\dfrac{1}{12}$

top block: $\dfrac{1}{6} + \dfrac{1}{4} + \dfrac{1}{2} - \dfrac{1}{2} = \dfrac{5}{12}$.

The center of gravity of the top 2 blocks is

$\left(-\dfrac{1}{12} + \dfrac{5}{12}\right)\bigg/2 = \dfrac{1}{6}$, which lies over the bottom block. The center of gravity of the 3 blocks is $\left(-\dfrac{1}{3} - \dfrac{1}{12} + \dfrac{5}{12}\right)\bigg/3 = 0$

which lies over the table. So, the far edge of the top block lies $\dfrac{1}{6} + \dfrac{1}{4} + \dfrac{1}{2} = \dfrac{11}{12}$ beyond the edge of the table.

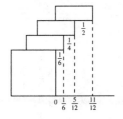

(b) Yes. If there are $n$ blocks, then the edge of the top block lies $\displaystyle\sum_{i=1}^{n} \frac{1}{2i}$ from the edge of the table. Using 4 blocks,

$$\sum_{i=1}^{4} \frac{1}{2i} = \frac{1}{2} + \frac{1}{4} + \frac{1}{6} + \frac{1}{8} = \frac{25}{24}$$

which shows that the top block extends beyond the table.

(c) The blocks can extend any distance beyond the table because the series diverges:

$$\sum_{i=1}^{\infty} \frac{1}{2i} = \frac{1}{2}\sum_{i=1}^{\infty} \frac{1}{i} = \infty.$$

**7.** (a)
$$e^x = 1 + x + \frac{x^2}{2!} + \cdots = \sum_{n=0}^{\infty} \frac{x^n}{n!}$$

$$xe^x = \sum_{n=0}^{\infty} \frac{x^{n+1}}{n!}$$

$$\int xe^x \, dx = xe^x - e^x + C = \sum_{n=0}^{\infty} \frac{x^{n+2}}{(n+2)n!}$$

Letting $x = 0$, you have $C = 1$. Letting $x = 1$,

$$e - e + 1 = \sum_{n=0}^{\infty} \frac{1}{(n+2)n!} = \frac{1}{2} + \sum_{n=1}^{\infty} \frac{1}{(n+2)n!}.$$

So, $\displaystyle\sum_{n=1}^{\infty} \frac{1}{(n+2)n!} = \frac{1}{2}.$

(b) Differentiating, $\displaystyle xe^x + e^x = \sum_{n=0}^{\infty} \frac{(n+1)x^n}{n!}.$

Letting $x = 1$, $\displaystyle 2e = \sum_{n=0}^{\infty} \frac{n+1}{n!} \approx 5.4366.$

**9.** $\displaystyle a - \frac{b}{2} + \frac{a}{3} - \frac{b}{4} + \cdots = \sum_{n=1}^{\infty} \frac{(-1)^{n+1}(a+b) + (a-b)}{2n}$

If $a = b$, $\displaystyle\sum_{n=1}^{\infty} \frac{(-1)^{n+1}(2a)}{2n} = a\sum_{n=1}^{\infty} \frac{(-1)^{n+1}}{n}$ converges conditionally.

If $a \ne b$, $\displaystyle\sum_{n=1}^{\infty} \frac{(-1)^{n+1}(a+b)}{2n} + \sum_{n=1}^{\infty} \frac{a-b}{2n}$ diverges.

No values of $a$ and $b$ give absolute convergence.
$a = b$ implies conditional convergence.

**11.** Let $b_n = a_n r^n$.

$$(b_n)^{1/n} = (a_n r^n)^{1/n} = a_n^{1/n} \cdot r \to Lr \text{ as } n \to \infty.$$

$$Lr < \frac{1}{r}r = 1.$$

By the Root Test, $\sum b_n$ converges $\Rightarrow$ $\sum a_n r^n$ converges.

**13.** (a)
$$\frac{1}{0.99} = \frac{1}{1 - 0.01} = \sum_{n=0}^{\infty} (0.01)^n$$
$$= 1 + 0.01 + (0.01)^2 + \cdots$$
$$= 1.010101\cdots$$

(b)
$$\frac{1}{0.98} = \frac{1}{1 - 0.02} = \sum_{n=0}^{\infty} (0.02)^n$$
$$= 1 + 0.02 + (0.02)^2 + \cdots$$
$$= 1 + 0.02 + 0.0004 + \cdots$$
$$= 1.0204081632\cdots$$

**15.** (a) Height $= 2\left[1 + \dfrac{1}{\sqrt{2}} + \dfrac{1}{\sqrt{3}} + \cdots\right]$

$$= 2\sum_{n=1}^{\infty} \frac{1}{n^{1/2}} = \infty \left(p\text{-series}, \ p = \frac{1}{2} < 1\right)$$

(b) $S = 4\pi\left[1 + \dfrac{1}{2} + \dfrac{1}{3} + \cdots\right] = 4\pi\displaystyle\sum_{n=1}^{\infty} \frac{1}{n} = \infty$

(c) $W = \dfrac{4}{3}\pi\left[1 + \dfrac{1}{2^{3/2}} + \dfrac{1}{3^{3/2}} + \cdots\right]$

$$= \frac{4}{3}\pi\sum_{n=1}^{\infty} \frac{1}{n^{3/2}} \text{ converges.}$$

# CHAPTER 10
# Conics, Parametric Equations, and Polar Coordinates

# C H A P T E R  1 0
## Conics, Parametric Equations, and Polar Coordinates

## Section 10.1   Conics and Calculus

**1.** $y^2 = 4x$   Parabola

Vertex: $(0, 0)$

$p = 1 > 0$

Opens to the right

Matches (a).

**2.** $(x + 4)^2 = -2(y - 2)$  Parabola

Vertex: $(-4, 2)$

Opens downward

Matches (e).

**3.** $\dfrac{y^2}{16} - \dfrac{x^2}{1} = 1$   Hyperbola

Vertices: $(0, \pm 4)$

Matches (c).

**4.** $\dfrac{(x - 2)^2}{16} + \dfrac{(y + 1)^2}{4} = 1$   Ellipse

Center: $(2, -1)$

Matches (b).

**5.** $\dfrac{x^2}{4} + \dfrac{y^2}{9} = 1$   Ellipse

Center: $(0, 0)$

Vertices: $(0, \pm 3)$

Matches (f).

**6.** $\dfrac{(x - 2)^2}{9} - \dfrac{y^2}{4} = 1$   Hyperbola

Vertices: $(5, 0), (-1, 0)$

Matches (d).

**7** $y^2 = -8x = 4(-2)x$

Vertex: $(0, 0)$

Focus: $(-2, 0)$

Directrix: $x = 2$

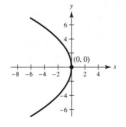

**9.** $(x + 5) + (y - 3)^2 = 0$

$\qquad (y - 3)^2 = -(x + 5) = 4\left(-\tfrac{1}{4}\right)(x + 5)$

Vertex: $(-5, 3)$

Focus: $\left(-\tfrac{21}{4}, 3\right)$

Directrix: $x = -\dfrac{19}{4}$

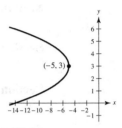

**11.** $y^2 - 4y - 4x = 0$

$\qquad y^2 - 4y + 4 = 4x + 4$

$\qquad (y - 2)^2 = 4(1)(x + 1)$

Vertex: $(-1, 2)$

Focus: $(0, 2)$

Directrix: $x = -2$

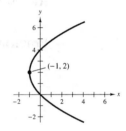

**13.** $x^2 + 4x + 4y - 4 = 0$

$\qquad x^2 + 4x + 4 = -4y + 4 + 4$

$\qquad (x + 2)^2 = 4(-1)(y - 2)$

Vertex: $(-2, 2)$

Focus: $(-2, 1)$

Directrix: $y = 3$

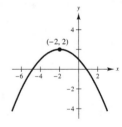

**15.** $\qquad (y - 4)^2 = 4(-2)(x - 5)$

$\qquad y^2 - 8y + 16 = -8x + 40$

$\qquad y^2 - 8y + 8x - 24 = 0$

**17.**
$$(x - 0)^2 = 4(8)(y - 5)$$
$$x^2 = 4(8)(y - 5)$$
$$x^2 - 32y + 160 = 0$$

**19.** Vertex: $(0, 4)$, vertical axis

$$(x - 0)^2 = 4p(y - 4)$$

$(-2, 0)$ on parabola: $(-2)^2 = 4p(-4)$
$$4 = -16p$$
$$p = -\tfrac{1}{4}$$

$$x^2 = 4\left(-\tfrac{1}{4}\right)(y - 4)$$
$$x^2 = -(y - 4)$$
$$x^2 + y - 4 = 0$$

**21.** Because the axis of the parabola is vertical, the form of the equation is $y = ax^2 + bx + c$. Now, substituting the values of the given coordinates into this equation, you obtain

$$3 = c, 4 = 9a + 3b + c, 11 = 16a + 4b + c.$$

Solving this system, you have $a = \tfrac{5}{3}, b = -\tfrac{14}{3}, c = 3$.

So,

$$y = \tfrac{5}{3}x^2 - \tfrac{14}{3}x + 3 \text{ or } 5x^2 - 14x - 3y + 9 = 0.$$

**23.** $16x^2 + y^2 = 16$

$$x^2 + \frac{y^2}{16} = 1$$

$$a^2 = 16, b^2 = 1, c^2 = 16 - 1 = 15$$

Center: $(0, 0)$

Foci: $\left(0, \pm\sqrt{15}\right)$

Vertices: $(0, \pm 4)$

$$e = \frac{c}{a} = \frac{\sqrt{15}}{4}$$

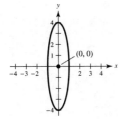

**25.** $\dfrac{(x - 3)^2}{16} + \dfrac{(y - 1)^2}{25} = 1$

$$a^2 = 25, b^2 = 16, c^2 = 25 - 16 = 9$$

Center: $(3, 1)$

Foci: $(3, 1 + 3) = (3, 4), (3, 1 - 3) = (3, -2)$

Vertices: $(3, 6), (3, -4)$

$$e = \frac{c}{a} = \frac{3}{5}$$

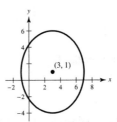

**27.**
$$9x^2 + 4y^2 + 36x - 24y + 36 = 0$$
$$9(x^2 + 4x + 4) + 4(y^2 - 6y + 9) = -36 + 36 + 36$$
$$= 36$$

$$\frac{(x + 2)^2}{4} + \frac{(y - 3)^2}{9} = 1$$

$$a^2 = 9, b^2 = 4, c^2 = 5$$

Center: $(-2, 3)$

Foci: $\left(-2, 3 \pm \sqrt{5}\right)$

Vertices: $(-2, 6), (-2, 0)$

$$e = \frac{\sqrt{5}}{3}$$

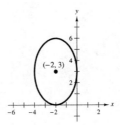

**29.** Center: $(0, 0)$

Focus: $(5, 0)$

Vertex: $(6, 0)$

Horizontal major axis

$$a = 6, c = 5 \Rightarrow b = \sqrt{a^2 - c^2} = \sqrt{11}$$

$$\frac{x^2}{36} + \frac{y^2}{11} = 1$$

**31.** Vertices: $(3, 1), (3, 9)$

Minor axis length: 6

Vertical major axis

Center: $(3, 5)$

$a = 4, b = 3$

$$\frac{(x - 3)^2}{9} + \frac{(y - 5)^2}{16} = 1$$

**33.** Center: $(0, 0)$

Horizontal major axis

Points on ellipse: $(3, 1), (4, 0)$

Because the major axis is horizontal,

$$\left(\frac{x^2}{a^2}\right) + \left(\frac{y^2}{b^2}\right) = 1.$$

Substituting the values of the coordinates of the given points into this equation, you have

$$\left(\frac{9}{a^2}\right) + \left(\frac{1}{b^2}\right) = 1, \text{ and } \frac{16}{a^2} = 1.$$

The solution to this system is $a^2 = 16, b^2 = \frac{16}{7}$.

So,

$$\frac{x^2}{16} + \frac{y^2}{16/7} = 1, \frac{x^2}{16} + \frac{7y^2}{16} = 1.$$

**35.** $\dfrac{x^2}{25} - \dfrac{y^2}{16} = 1$

$a = 5, b = 4, c = \sqrt{25 + 16} = \sqrt{41}$

Center: $(0, 0)$

Vertices: $(\pm 5, 0)$

Foci: $\left(\pm\sqrt{41}, 0\right)$

Asymptotes: $y = \pm\dfrac{b}{a}x = \pm\dfrac{4}{5}x$

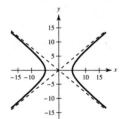

**37.** $9x^2 - y^2 - 36x - 6y + 18 = 0$

$9(x^2 - 4x + 4) - (y^2 + 6y + 9) = -18 + 36 - 9$

$$\frac{(x - 2)^2}{1} - \frac{(y + 3)^2}{9} = 1$$

$a = 1, b = 3, c = \sqrt{10}$

Center: $(2, -3)$

Vertices: $(1, -3), (3, -3)$

Foci: $\left(2 \pm \sqrt{10}, -3\right)$

Asymptotes: $y = -3 \pm 3(x - 2)$

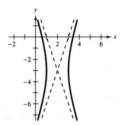

**39.** $x^2 - 9y^2 + 2x - 54y - 80 = 0$

$(x^2 + 2x + 1) - 9(y^2 + 6y + 9) = 80 + 1 - 81 = 0$

$(x + 1)^2 - 9(y + 3)^2 = 0$

$$y + 3 = \pm\frac{1}{3}(x + 1)$$

$$y = -3 \pm \frac{1}{3}(x + 1)$$

Degenerate hyperbola is two lines intersecting at $(-1, -3)$.

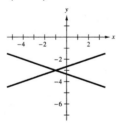

**41.** Vertices: $(\pm 1, 0)$

Asymptotes: $y = \pm 5x$

Horizontal transverse axis

Center: $(0, 0)$

$a = 1, \dfrac{b}{a} = 5 \Rightarrow b = 5$

$$\frac{x^2}{1} - \frac{y^2}{25} = 1$$

**43.** Vertices: $(2, \pm 3)$

Point on graph: $(0, 5)$

Vertical transverse axis

Center: $(2, 0)$

$a = 3$

So, the equation is of the form

$$\frac{y^2}{9} - \frac{(x-2)^2}{b^2} = 1.$$

Substituting the coordinates of the point $(0, 5)$, you have

$$\frac{25}{9} - \frac{4}{b^2} = 1 \quad \text{or} \quad b^2 = \frac{9}{4}.$$

So, the equation is $\dfrac{y^2}{9} - \dfrac{(x-2)^2}{9/4} = 1.$

**45.** Center: $(0, 0)$

Vertex: $(0, 2)$

Focus: $(0, 4)$

Vertical transverse axis

$a = 2, c = 4, b^2 = c^2 - a^2 = 12$

So, $\dfrac{y^2}{4} - \dfrac{x^2}{12} = 1.$

**47.** Vertices: $(0, 2), (6, 2)$

Asymptotes: $y = \dfrac{2}{3}x, \; y = 4 - \dfrac{2}{3}x$

Horizontal transverse axis

Center: $(3, 2)$

$a = 3$

Slopes of asymptotes: $\pm\dfrac{b}{a} = \pm\dfrac{2}{3}$

So, $b = 2.$ Therefore,

$$\frac{(x-3)^2}{9} - \frac{(y-2)^2}{4} = 1.$$

**49. (a)** $\dfrac{x^2}{9} - y^2 = 1, \dfrac{2x}{9} - 2yy' = 0, \dfrac{x}{9y} = y'$

At $x = 6$: $y = \pm\sqrt{3}, \; y' = \dfrac{\pm 6}{9\sqrt{3}} = \dfrac{\pm 2\sqrt{3}}{9}$

At $(6, \sqrt{3})$: $y - \sqrt{3} = \dfrac{2\sqrt{3}}{9}(x - 6)$

or $2x - 3\sqrt{3}y - 3 = 0$

At $(6, -\sqrt{3})$: $y + \sqrt{3} = \dfrac{-2\sqrt{3}}{9}(x - 6)$

or $2x + 3\sqrt{3}y - 3 = 0$

**(b)** From part (a) you know that the slopes of the normal lines must be $\mp 9/(2\sqrt{3})$.

At $(6, \sqrt{3})$: $y - \sqrt{3} = -\dfrac{9}{2\sqrt{3}}(x - 6)$

or $9x + 2\sqrt{3}y - 60 = 0$

At $(6, -\sqrt{3})$: $y + \sqrt{3} = \dfrac{9}{2\sqrt{3}}(x - 6)$

or $9x - 2\sqrt{3}y - 60 = 0$

**51.**  $x^2 + 4y^2 - 6x + 16y + 21 = 0$

$(x^2 - 6x + 9) + 4(y^2 + 4y + 4) = -21 + 9 + 16$

$(x - 3)^2 + 4(y + 2)^2 = 4$

Ellipse

**53.**  $25x^2 - 10x - 200y - 119 = 0$

$25\left(x^2 - \frac{2}{5}x + \frac{1}{25}\right) = 200y + 119 + 1$

$25\left(x - \frac{1}{5}\right)^2 = 200(y + 1)$

Parabola

**55.**  $9x^2 + 9y^2 - 36x + 6y + 34 = 0$

$9(x^2 - 4x + 4) + 9\left(y^2 + \frac{2}{3}y + \frac{1}{9}\right) = -34 + 36 + 1$

$9(x - 2)^2 + 9\left(y + \frac{1}{3}\right)^2 = 3$

Circle (Ellipse)

**57.**  $3(x - 1)^2 = 6 + 2(y + 1)^2$

$3(x - 1)^2 - 2(y + 1)^2 = 6$

$\dfrac{(x - 1)^2}{2} - \dfrac{(y + 1)^2}{3} = 1$

Hyperbola

**59.** (a) A parabola is the set of all points $(x, y)$ that are equidistant from a fixed line (directrix) and a fixed point (focus) not on the line.

(b) For directrix $y = k - p$: $(x - h)^2 = 4p(y - k)$

For directrix $x = h - p$: $(y - k)^2 = 4p(x - h)$

(c) If $P$ is a point on a parabola, then the tangent line to the parabola at $P$ makes equal angles with the line passing through $P$ and the focus, and with the line passing through $P$ parallel to the axis of the parabola.

**61.** (a) A hyperbola is the set of all points $(x, y)$ for which the absolute value of the difference between the distances from two distinct fixed points (foci) is constant.

(b) Transverse axis is horizontal:

$$\frac{(x - h)^2}{a^2} - \frac{(y - k)^2}{b^2} = 1$$

Transverse axis is vertical:

$$\frac{(y - k)^2}{a^2} - \frac{(x - h)^2}{b^2} = 1$$

(c) Transverse axis is horizontal:

$y = k + (b/a)(x - h)$ and $y = k - (b/a)(x - h)$

Transverse axis is vertical:

$y = k + (a/b)(x - h)$ and $y = k - (a/b)(x - h)$

**63.** $9x^2 + 4y^2 - 36x - 24y - 36 = 0$

(a) $9(x^2 - 4x + 4) + 4(y^2 - 6y + 9) = 36 + 36 + 36$

$9(x - 2)^2 + 4(y - 3)^2 = 108$

$$\frac{(x - 2)^2}{12} + \frac{(y - 3)^2}{27} = 1$$

Ellipse

(b) $9x^2 - 4y^2 - 36x - 24y - 36 = 0$

$9(x^2 - 4x + 4) - 4(y^2 + 6y + 9) = 36 + 36 - 36$

$$\frac{(x - 2)^2}{4} - \frac{(y + 3)^2}{9} = 1$$

Hyperbola

(c) $4x^2 + 4y^2 - 36x - 24y - 36 = 0$

$4\left(x^2 - 9x + \frac{81}{4}\right) + 4(y^2 - 6y + 9) = 36 + 81 + 36$

$$\left(x - \frac{9}{2}\right)^2 + (y - 3)^2 = \frac{153}{4}$$

Circle

(d) *Sample answer:* Eliminate the $y^2$-term

**65.** Assume that the vertex is at the origin.

$x^2 = 4py$

$(3)^2 = 4p(1)$

$\frac{9}{4} = p$

The pipe is located $\frac{9}{4}$ meters from the vertex.

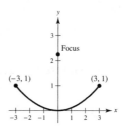

**67.** (a) Without loss of generality, place the coordinate system so that the equation of the parabola is $x^2 = 4py$ and, so,

$$y' = \left(\frac{1}{2p}\right)x.$$

So, for distinct tangent lines, the slopes are unequal and the lines intersect.

(b) $x^2 - 4x - 4y = 0$

$2x - 4 - 4\dfrac{dy}{dx} = 0$

$$\frac{dy}{dx} = \frac{1}{2}x - 1$$

At $(0, 0)$, the slope is $-1$: $y = -x$. At $(6, 3)$, the slope is $2$: $y = 2x - 9$. Solving for $x$,

$-x = 2x - 9$

$-3x = -9$

$x = 3$

$y = -3$.

Point of intersection: $(3, -3)$

**69.** $x^2 = 4py$, $p = \frac{1}{4}, \frac{1}{2}, 1, \frac{3}{2}, 2$

As $p$ increases, the graph becomes wider.

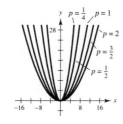

**71.** Parabola

Vertex: $(0, 4)$

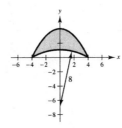

$x^2 = 4p(y - 4)$

$4^2 = 4p(0 - 4)$

$p = -1$

$x^2 = -4(y - 4)$

$y = 4 - \dfrac{x^2}{4}$

Circle

Center: $(0, k)$

Radius: 8

$x^2 + (y - k)^2 = 64$

$4^2 + (0 - k)^2 = 64$

$k^2 = 48$

$k = -4\sqrt{3}$   (Center is on the negative $y$-axis.)

$x^2 + \left(y + 4\sqrt{3}\right)^2 = 64$

$y = -4\sqrt{3} \pm \sqrt{64 - x^2}$

Because the $y$-value is positive when $x = 0$, we have $y = -4\sqrt{3} + \sqrt{64 - x^2}$.

$A = 2\displaystyle\int_0^4 \left[\left(4 - \frac{x^2}{4}\right) - \left(-4\sqrt{3} + \sqrt{64 - x^2}\right)\right] dx$

$\phantom{A} = 2\left[4x - \dfrac{x^3}{12} + 4\sqrt{3}x - \dfrac{1}{2}\left(x\sqrt{64 - x^2} + 64\arcsin\dfrac{x}{8}\right)\right]_0^4$

$\phantom{A} = 2\left(16 - \dfrac{64}{12} + 16\sqrt{3} - 2\sqrt{48} - 32\arcsin\dfrac{1}{2}\right) = \dfrac{16\left(4 + 3\sqrt{3} - 2\pi\right)}{3} \approx 15.536$ square feet

**73.** $e = \dfrac{c}{a}$

$0.0167 = \dfrac{c}{149{,}598{,}000}$

$c \approx 2{,}498{,}286.6$

Least distance: $a - c = 147{,}099{,}713.4$ km

Greatest distance: $a + c = 152{,}096{,}286.6$ km

**75.** $e = \dfrac{A - P}{A + P}$

$\phantom{e} = \dfrac{(123{,}000 + 4000) - (119 + 4000)}{(123{,}000 + 4000) + (119 + 4000)}$

$\phantom{e} = \dfrac{122{,}881}{131{,}119} \approx 0.9372$

**77.** $e = \dfrac{A - P}{A + P} = \dfrac{35.29 - 0.59}{35.29 + 0.59} = 0.9671$

**79. (a)** $A = 4\int_0^2 \frac{1}{2}\sqrt{4 - x^2}\, dx = \left[x\sqrt{4 - x^2} + 4\arcsin\left(\frac{x}{2}\right)\right]_0^2 = 2\pi \quad \left[\text{or, } A = \pi ab = \pi(2)(1) = 2\pi\right]$

**(b) Disk:** $V = 2\pi\int_0^2 \frac{1}{4}\left(4 - x^2\right) dx = \frac{1}{2}\pi\left[4x - \frac{1}{3}x^3\right]_0^2 = \frac{8\pi}{3}$

$$y = \frac{1}{2}\sqrt{4 - x^2}$$

$$y' = \frac{-x}{2\sqrt{4 - x^2}}$$

$$\sqrt{1 + (y')^2} = \sqrt{1 + \frac{x^2}{16 - 4x^2}} = \sqrt{\frac{16 - 3x^2}{4y}}$$

$$S = 2(2\pi)\int_0^2 y\left(\frac{\sqrt{16 - 3x^2}}{4y}\right) dx = \pi\int_0^2 \sqrt{16 - 3x^2}\, dx$$

$$= \frac{\pi}{2\sqrt{3}}\left[\sqrt{3}x\sqrt{16 - 3x^2} + 16\arcsin\left(\frac{\sqrt{3}x}{4}\right)\right]_0^2 = \frac{2\pi}{9}\left(9 + 4\sqrt{3}\pi\right) \approx 21.48$$

**(c) Shell:** $V = 2\pi\int_0^2 x\sqrt{4 - x^2}\, dx = -\pi\int_0^2 -2x\left(4 - x^2\right)^{1/2} dx = -\frac{2\pi}{3}\left[\left(4 - x^2\right)^{3/2}\right]_0^2 = \frac{16\pi}{3}$

$$x = 2\sqrt{1 - y^2}$$

$$x' = \frac{-2y}{\sqrt{1 - y^2}}$$

$$\sqrt{1 + (x')^2} = \sqrt{1 + \frac{4y^2}{1 - y^2}} = \frac{\sqrt{1 + 3y^2}}{\sqrt{1 - y^2}}$$

$$S = 2(2\pi)\int_0^1 2\sqrt{1 - y^2}\,\frac{\sqrt{1 + 3y^2}}{\sqrt{1 - y^2}}\, dy = 8\pi\int_0^1 \sqrt{1 + 3y^2}\, dy$$

$$= \frac{8\pi}{2\sqrt{3}}\left[\sqrt{3}y\sqrt{1 + 3y^2} + \ln\left|\sqrt{3}y + \sqrt{1 + 3y^2}\right|\right]_0^1$$

$$= \frac{4\pi}{3}\left|6 + \sqrt{3}\ln\left(2 + \sqrt{3}\right)\right| \approx 34.69$$

**81.** From Example 5,

$$C = 4a\int_0^{\pi/2} \sqrt{1 - e^2\sin^2\theta}\, d\theta$$

For $\dfrac{x^2}{25} + \dfrac{y^2}{49} = 1$, you have

$$a = 7, b = 5, c = \sqrt{49 - 25} = 2\sqrt{6}, e = \frac{c}{a} = \frac{2\sqrt{6}}{7}.$$

$$C = 4(7)\int_0^{\pi/2} \sqrt{1 - \frac{24}{49}\sin^2\theta}\, d\theta \approx 28(1.3558) \approx 37.96$$

**83.** Area circle $= \pi r^2 = 100\pi$

Area ellipse $= \pi ab = \pi a(10)$

$$2(100\pi) = 10\pi a \Rightarrow a = 20$$

So, the length of the major axis is $2a = 40$.

**85.** The transverse axis is horizontal since $(2, 2)$ and $(10, 2)$ are the foci (see definition of hyperbola).

Center: $(6, 2)$

$$c = 4, 2a = 6, b^2 = c^2 - a^2 = 7$$

So, the equation is $\dfrac{(x - 6)^2}{9} - \dfrac{(y - 2)^2}{7} = 1.$

**87.** $c = 150, 2a = 0.001(186{,}000), a = 93,$

$$b = \sqrt{150^2 - 93^2} = \sqrt{13{,}851}$$

$$\frac{x^2}{93^2} - \frac{y^2}{13{,}851} = 1$$

When $y = 75$, you have

$$x^2 = 93^2\left(1 + \frac{75^2}{13{,}851}\right)$$

$$x \approx 110.3 \text{ mi.}$$

**89.**

$$\frac{x^2}{a^2} - \frac{y^2}{b^2} = 1$$

$$\frac{2x}{a^2} - \frac{2yy'}{b^2} = 0 \text{ or } y' = \frac{b^2 x}{a^2 y}$$

$$y - y_0 = \frac{b^2 x_0}{a^2 y_0}(x - x_0)$$

$$a^2 y_0 y - a^2 y_0^2 = b^2 x_0 x - b^2 x_0^2$$

$$b^2 x_0^2 - a^2 y_0^2 = b^2 x_0 x - a^2 y_0 y$$

$$a^2 b^2 = b^2 x_0 x - a^2 y_0 y$$

$$\frac{x_0 x}{a^2} - \frac{y_0 y}{b^2} = 1$$

**91.** False. The parabola is equidistant from the directrix and focus and therefore cannot intersect the directrix.

**93.** True

**95.** True

**97.** Let $\dfrac{x^2}{a^2} + \dfrac{y^2}{b^2} = 1$ be the equation of the ellipse with $a > b > 0$. Let $(\pm c, 0)$ be the foci,

$c^2 = a^2 - b^2$. Let $(u, v)$ be a point on the tangent line at $P(x, y)$, as indicated in the figure.

$x^2b^2 + y^2a^2 = a^2b^2$

$2xb^2 + 2yy'a^2 = 0$

$$y' = -\frac{b^2x}{a^2y} \quad \text{Slope at } P(x, y)$$

Now, $\qquad \dfrac{y - v}{x - u} = -\dfrac{b^2x}{a^2y}$

$\qquad\qquad y^2a^2 - a^2vy = -b^2x^2 + b^2xu$

$\qquad\qquad y^2a^2 + x^2b^2 = a^2vy + b^2ux$

$\qquad\qquad\quad a^2b^2 = a^2vy + b^2ux$

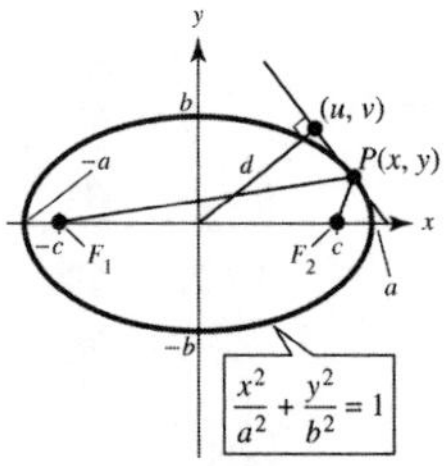

Because there is a right angle at $(u, v)$,

$$\frac{v}{u} = \frac{a^2y}{b^2x}$$

$vb^2x = a^2uy.$

You have two equations:

$a^2vy + b^2ux = a^2b^2$

$a^2uy - b^2vx = 0.$

Multiplying the first by $v$ and the second by $u$, and adding,

$a^2v^2y + a^2u^2y = a^2b^2v$

$\qquad y\left[u^2 + v^2\right] = b^2v$

$\qquad\qquad yd^2 = b^2v$

$\qquad\qquad\quad v = \dfrac{yd^2}{b^2}.$

Similarly, $u = \dfrac{xd^2}{a^2}.$

From the figure, $u = d \cos \theta$ and $v = d \sin \theta$. So, $\cos \theta = \dfrac{xd}{a^2}$ and $\sin \theta = \dfrac{yd}{b^2}.$

$$\cos^2 \theta + \sin^2 \theta = \frac{x^2d^2}{a^4} + \frac{y^2d^2}{b^4} = 1$$

$$x^2b^4d^2 + y^2a^4d^2 = a^4b^4$$

$$d^2 = \frac{a^4b^4}{x^2b^4 + y^2a^4}$$

Let $r_1 = PF_1$ and $r_2 = PF_2, \quad r_1 + r_2 = 2a.$

$$r_1r_2 = \frac{1}{2}\left[(r_1 + r_2)^2 - r_1^2 - r_2^2\right] = \frac{1}{2}\left[4a^2 - (x + c)^2 - y^2 - (x - c)^2 - y^2\right] = 2a^2 - x^2 - y^2 - c^2 = a^2 + b^2 - x^2 - y^2$$

Finally, $d^2r_1r_2 = \dfrac{a^4b^4}{x^2b^4 + y^2a^4} \cdot \left[a^2 + b^2 - x^2 - y^2\right]$

$$= \frac{a^4b^4}{b^2(b^2x^2) + a^2(a^2y^2)} \cdot \left[a^2 + b^2 - x^2 - y^2\right]$$

$$= \frac{a^4b^4}{b^2(a^2b^2 - a^2y^2) + a^2(a^2b^2 - b^2x^2)} \cdot \left[a^2 + b^2 - x^2 - y^2\right]$$

$$= \frac{a^4b^4}{a^2b^2\left[a^2 + b^2 - x^2 - y^2\right]} \cdot \left[a^2 + b^2 - x^2 - y^2\right] = a^2b^2, \quad \text{a constant!}$$

# Section 10.2   Plane Curves and Parametric Equations

**1.** $x = 2t - 3$

$y = 3t + 1$

$t = \dfrac{x + 3}{2}$

$y = 3\left(\dfrac{x + 3}{2}\right) + 1 = \dfrac{3}{2}x + \dfrac{11}{2}$

$3x - 2y + 11 = 0$

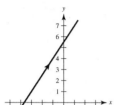

**3.** $x = t + 1$

$y = t^2$

$y = (x - 1)^2$

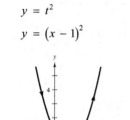

**5.** $x = t^3$

$y = \dfrac{1}{2}t^2$

$y = t^3$ implies $t = x^{1/3}$

$y = \dfrac{1}{2}x^{2/3}$

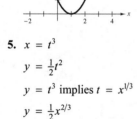

**7.** $x = \sqrt{t}$

$y = t - 5$

$x^2 = t$

$y = x^2 - 5, x \geq 0$

**9.** $x = t - 3$

$y = \dfrac{t}{t - 3}$

$t = x + 3$

$y = \dfrac{x + 3}{(x + 3) - 3} = 1 + \dfrac{3}{x} = \dfrac{x + 3}{x}$

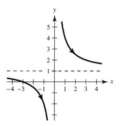

**11.** $x = 2t$

$y = |t - 2|$

$y = \left|\dfrac{x}{2} - 2\right| = \dfrac{|x - 4|}{2}$

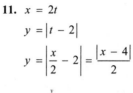

**13.** $x = e^t, x > 0$

$y = e^{3t} + 1$

$y = x^3 + 1, x > 0$

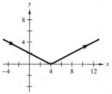

**15.** $x = \sec \theta$

$y = \cos \theta$

$0 \leq \theta < \dfrac{\pi}{2}, \dfrac{\pi}{2} < \theta \leq \pi$

$xy = 1$

$y = \dfrac{1}{x}$

$|x| \geq 1, |y| \leq 1$

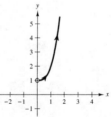

**17.** $x = 8 \cos \theta$

$y = 8 \sin \theta$

$x^2 + y^2 = 64 \cos^2 \theta + 64 \sin^2 \theta = 64(1) = 64$

Circle

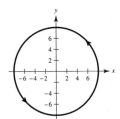

**19.** $x = 6 \sin 2\theta$

$y = 4 \cos 2\theta$

$\left(\dfrac{x}{6}\right)^2 + \left(\dfrac{y}{4}\right)^2 = \sin^2 2\theta + \cos^2 2\theta = 1$

$\dfrac{x^2}{36} + \dfrac{y^2}{16} = 1$

Ellipse

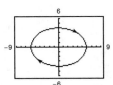

**21.**

$x = 4 + 2 \cos \theta$

$y = -1 + \sin \theta$

$\dfrac{(x - 4)^2}{4} = \cos^2 \theta$

$\dfrac{(y + 1)^2}{1} = \sin^2 \theta$

$\dfrac{(x - 4)^2}{4} + \dfrac{(y + 1)^2}{1} = 1$

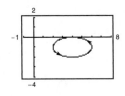

**23.** $x = -3 + 4 \cos \theta$

$y = 2 + 5 \sin \theta$

$x + 3 = 4 \cos \theta$

$y - 2 = 5 \sin \theta$

$\left(\dfrac{x + 3}{4}\right)^2 + \left(\dfrac{y - 2}{5}\right)^2 = \cos^2 \theta + \sin^2 \theta = 1$

$\dfrac{(x + 3)^2}{16} + \dfrac{(y - 2)^2}{25} = 1$

Ellipse

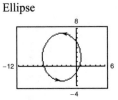

**25.**      $x = 4 \sec \theta$

$y = 3 \tan \theta$

$\dfrac{x^2}{16} = \sec^2 \theta$

$\dfrac{y^2}{9} = \tan^2 \theta$

$\dfrac{x^2}{16} - \dfrac{y^2}{9} = 1$

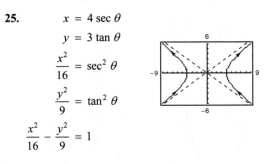

**27.** $x = t^3$

$y = 3 \ln t$

$y = 3 \ln \sqrt[3]{x} = \ln x$

**29.** $x = e^{-t}$

$y = e^{3t}$

$e^t = \dfrac{1}{x}$

$e^t = \sqrt[3]{y}$

$\sqrt[3]{y} = \dfrac{1}{x}$

$y = \dfrac{1}{x^3}$

$x > 0$

$y > 0$

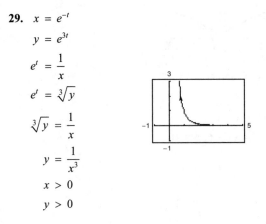

**31.** By eliminating the parameters in (a) – (d), you get $y = 2x + 1$. They differ from each other in orientation and in restricted domains. These curves are all smooth except for (b).

(a) $x = t$, $y = 2t + 1$

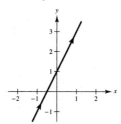

(b) $x = \cos\theta$     $y = 2\cos\theta + 1$

$-1 \le x \le 1$     $-1 \le y \le 3$

$\dfrac{dx}{d\theta} = \dfrac{dy}{d\theta} = 0$ when $\theta = 0, \pm\pi, \pm2\pi, \ldots$.

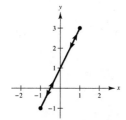

(c) $x = e^{-t}$     $y = 2e^{-t} + 1$

$x > 0$     $y > 1$

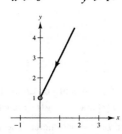

(d) $x = e^{t}$     $y = 2e^{t} + 1$

$x > 0$     $y > 1$

**33.** The curves are identical on $0 < \theta < \pi$. They are both smooth. They represent $y = 2(1 - x^2)$ for $-1 \le x \le 1$. The orientation is from right to left in part (a) and in part (b).

**35.** (a)

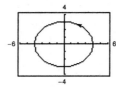

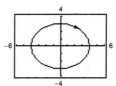

(b) The orientation of the second curve is reversed.

(c) The orientation will be reversed.

(d) Answers will vary. For example,

$x = 2\sec t$     $x = 2\sec(-t)$

$y = 5\sin t$     $y = 5\sin(-t)$

have the same graphs, but their orientations are reversed.

**37.**
$$x = x_1 + t(x_2 - x_1)$$
$$y = y_1 + t(y_2 - y_1)$$
$$\frac{x - x_1}{x_2 - x_1} = t$$
$$y = y_1 + \left(\frac{x - x_1}{x_2 - x_1}\right)(y_2 - y_1)$$
$$y - y_1 = \frac{y_2 - y_1}{x_2 - x_1}(x - x_1)$$
$$y - y_1 = m(x - x_1)$$

**39.**
$$x = h + a\cos\theta$$
$$y = k + b\sin\theta$$
$$\frac{x - h}{a} = \cos\theta$$
$$\frac{y - k}{b} = \sin\theta$$
$$\frac{(x - h)^2}{a^2} + \frac{(y - k)^2}{b^2} = 1$$

**41.** From Exercise 37 you have

$x = 4t$

$y = -7t$

Solution not unique

**43.** From Exercise 38 you have

$x = 3 + 2\cos\theta$

$y = 1 + 2\sin\theta$

Solution not unique

**45.** From Exercise 39 you have

$a = 10, c = 8 \Rightarrow b = 6$

$x = 10 \cos \theta$

$y = 6 \sin \theta$

Center: $(0, 0)$

Solution not unique

**47.** From Exercise 40 you have

$a = 4, c = 5 \Rightarrow b = 3$

$x = 4 \sec \theta$

$y = 3 \tan \theta.$

Center: $(0, 0)$

Solution not unique

**49.** $y = 6x - 5$

Examples:

$x = t, y = 6t - 5$

$x = t + 1, y = 6t + 1$

**51.** $y = x^3$

Example

$x = t, \qquad y = t^3$

$x = \sqrt[3]{t}, \qquad y = t$

$x = \tan t, \qquad y = \tan^3 t$

**53.** $y = 2x - 5$

At $(3, 1), t = 0: \quad x = 3 - t$

$\qquad\qquad\qquad y = 2(3 - t) - 5 = -2t + 1$

or, $x = t + 3$

$\qquad y = 2t + 1$

**55.** $y = x^2$

$t = 4$ at $(4, 16): \quad x = t$

$\qquad\qquad\qquad\qquad y = t^2$

**57.** $x = 2(\theta - \sin \theta)$

$y = 2(1 - \cos \theta)$

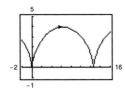

Not smooth at $\theta = 2n\pi$

**59.** $x = \theta - \frac{3}{2} \sin \theta$

$y = 1 - \frac{3}{2} \cos \theta$

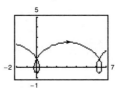

Smooth everywhere

**61.** $x = 3 \cos^3 \theta$

$y = 3 \sin^3 \theta$

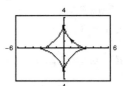

Not smooth at $(x, y) = (\pm 3, 0)$ and $(0, \pm 3)$, or

$\theta = \frac{1}{2} n\pi.$

**63.** $x = 2 \cot \theta$

$y = 2 \sin^2 \theta$

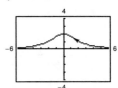

Smooth everywhere

**65.** If $f$ and $g$ are continuous functions of $t$ on an interval $I$, then the equations $x = f(t)$ and $y = g(t)$ are called parametric equations and $t$ is the parameter. The set of points $(x, y)$ obtained as $t$ varies over $I$ is the graph. Taken together, the parametric equations and the graph are called a plane curve $C$.

**67.** A curve $C$ represented by $x = f(t)$ and $y = g(t)$ on an interval $I$ is called smooth when $f'$ and $g'$ are continuous on $I$ and not simultaneously 0, except possibly at the endpoints of $I$.

**69.** Matches (d) because $(4, 0)$ is on the graph.

**70.** Matches (a) because $(0, 2)$ is on the graph.

**71.** Matches (b) because $(1, 0)$ is on the graph.

**72.** Matches (c) because the graph is undefined when $\theta = 0.$

**73.** When the circle has rolled $\theta$ radians, you know that the center is at $(a\theta, a)$.

$$\sin\theta = \sin(180° - \theta) = \frac{|AC|}{b} = \frac{|BD|}{b} \text{ or } |BD| = b\sin\theta$$

$$\cos\theta = -\cos(180° - \theta) = \frac{|AP|}{-b} \text{ or } |AP| = -b\cos\theta$$

So, $x = a\theta - b\sin\theta$ and $y = a - b\cos\theta$.

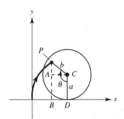

**79.** (a)  $100 \text{ mi/hr} = \frac{(100)(5280)}{3600} = \frac{440}{3} \text{ ft/sec}$

$$x = (v_0\cos\theta)t = \left(\frac{440}{3}\cos\theta\right)t$$

$$y = h + (v_0\sin\theta)t - 16t^2 = 3 + \left(\frac{440}{3}\sin\theta\right)t - 16t^2$$

(b)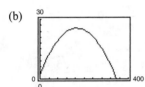

It is not a home run when $x = 400$, $y < 10$.

(c)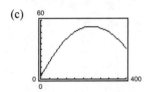

Yes, it's a home run when $x = 400$, $y > 10$.

(d) You need to find the angle $\theta$ (and time $t$) such that

$$x = \left(\frac{440}{3}\cos\theta\right)t = 400$$

$$y = 3 + \left(\frac{440}{3}\sin\theta\right)t - 16t^2 = 10.$$

From the first equation $t = 1200/440\cos\theta$. Substituting into the second equation,

$$10 = 3 + \left(\frac{440}{3}\sin\theta\right)\left(\frac{1200}{440\cos\theta}\right) - 16\left(\frac{1200}{440\cos\theta}\right)^2$$

$$7 = 400\tan\theta - 16\left(\frac{120}{44}\right)^2\sec^2\theta = 400\tan\theta - 16\left(\frac{120}{44}\right)^2(\tan^2\theta + 1).$$

You now solve the quadratic for $\tan\theta$:

$$16\left(\frac{120}{44}\right)^2\tan^2\theta - 400\tan\theta + 7 + 16\left(\frac{120}{44}\right)^2 = 0.$$

$\tan\theta \approx 0.35185 \Rightarrow \theta \approx 19.4°$

**75.** False

$x = t^2 \Rightarrow x \geq 0$

$y = t^2 \Rightarrow y \geq 0$

The graph of the parametric equations is only a portion of the line $y = x$ when $x \geq 0$.

**77.** True. $y = \cos x$

## Section 10.3 Parametric Equations and Calculus

**1.** $\dfrac{dy}{dx} = \dfrac{dy/dt}{dx/dt} = \dfrac{-6}{2t} = -\dfrac{3}{t}$

**3.** $\dfrac{dy}{dx} = \dfrac{dy/d\theta}{dx/d\theta} = \dfrac{-2\cos\theta\sin\theta}{2\sin\theta\cos\theta} = -1$

$\left[\text{Note: } x + y = 1 \Rightarrow y = 1 - x \text{ and } \dfrac{dy}{d\theta} = -1\right]$

**5.** $x = 4t,\ y = 3t - 2$

$\dfrac{dy}{dx} = \dfrac{dy/dt}{dx/dt} = \dfrac{3}{4}$

$\dfrac{d^2y}{dx^2} = 0$

At $t = 3$, slope is $\dfrac{3}{4}$. (Line)

Neither concave upward nor downward

**7.** $x = t + 1,\ y = t^2 + 3t$

$\dfrac{dy}{dx} = \dfrac{2t + 3}{1} = 1$ when $t = -1$.

$\dfrac{d^2y}{dx^2} = 2$

Concave upward

**9.** $x = 4\cos\theta,\ y = 4\sin\theta$

$\dfrac{dy}{dx} = \dfrac{dy/d\theta}{dx/d\theta} = \dfrac{4\cos\theta}{-4\sin\theta} = \dfrac{-\cos\theta}{\sin\theta} = -\cot\theta$

$\dfrac{d^2y}{dx^2} = \dfrac{\dfrac{d}{d\theta}[-\cot\theta]}{dx/d\theta} = \dfrac{\csc^2\theta}{-4\sin\theta} = \dfrac{-1}{4\sin^3\theta} = -\dfrac{1}{4}\csc^3\theta$

At $\theta = \dfrac{\pi}{4}$, $\dfrac{dy}{dx} = -1$.

$\dfrac{d^2y}{dx^2} = \dfrac{-1}{4(\sqrt{2}/2)^3} = \dfrac{-\sqrt{2}}{2}$

Concave downward

**11.** $x = 2 + \sec\theta,\ y = 1 + 2\tan\theta$

$\dfrac{dy}{dx} = \dfrac{2\sec^2\theta}{\sec\theta\tan\theta}$

$= \dfrac{2\sec\theta}{\tan\theta} = 2\csc\theta = 4$ when $\theta = \dfrac{\pi}{6}$.

$\dfrac{d^2y}{dx^2} = \dfrac{\dfrac{d}{d\theta}\left[\dfrac{dy}{dx}\right]}{\dfrac{dx}{d\theta}} = \dfrac{-2\csc\theta\cot\theta}{\sec\theta\tan\theta}$

$= -2\cot^3\theta = -6\sqrt{3}$ when $\theta = \dfrac{\pi}{6}$.

Concave downward

**13.** $x = \cos^3\theta,\ y = \sin^3\theta$

$\dfrac{dy}{dx} = \dfrac{3\sin^2\theta\cos\theta}{-3\cos^2\theta\sin\theta} = -\tan\theta = -1$ when $\theta = \dfrac{\pi}{4}$.

$\dfrac{d^2y}{dx^2} = \dfrac{-\sec^2\theta}{-3\cos^2\theta\sin\theta} = \dfrac{1}{3\cos^4\theta\sin\theta}$

$= \dfrac{\sec^4\theta\csc\theta}{3} = \dfrac{4\sqrt{2}}{3}$ when $\theta = \dfrac{\pi}{4}$.

Concave upward

**15.** $x = 2\cot\theta,\ y = 2\sin^2\theta$

$\dfrac{dy}{dx} = \dfrac{4\sin\theta\cos\theta}{-2\csc^2\theta} = -2\sin^3\theta\cos\theta$

At $\left(-\dfrac{2}{\sqrt{3}}, \dfrac{3}{2}\right)$, $\theta = \dfrac{2\pi}{3}$, and $\dfrac{dy}{dx} = \dfrac{3\sqrt{3}}{8}$.

Tangent line: $\quad y - \dfrac{3}{2} = \dfrac{3\sqrt{3}}{8}\left(x + \dfrac{2}{\sqrt{3}}\right)$

$3\sqrt{3}x - 8y + 18 = 0$

At $(0, 2)$, $\theta = \dfrac{\pi}{2}$, and $\dfrac{dy}{dx} = 0$.

Tangent line: $y - 2 = 0$

At $\left(2\sqrt{3}, \dfrac{1}{2}\right)$, $\theta = \dfrac{\pi}{6}$, and $\dfrac{dy}{dx} = -\dfrac{\sqrt{3}}{8}$.

Tangent line: $\quad y - \dfrac{1}{2} = -\dfrac{\sqrt{3}}{8}\left(x - 2\sqrt{3}\right)$

$\sqrt{3}x + 8y - 10 = 0$

**17.** $x = t^2 - 4$

$y = t^2 - 2t$

$\dfrac{dy}{dx} = \dfrac{dy/dt}{dx/dt} = \dfrac{2t - 2}{2t}$

At $(0, 0)$, $t = 2$, $\dfrac{dy}{dx} = \dfrac{1}{2}$.

Tangent line: $\quad y = \dfrac{1}{2}x$

$2y - x = 0$

At $(-3, -1)$, $t = 1$, $\dfrac{dy}{dx} = 0$.

Tangent line: $\quad y = -1$

$y + 1 = 0$

At $(-3, 3)$, $t = -1$, $\dfrac{dy}{dx} = 2$.

Tangent line: $\quad y - 3 = 2(x + 3)$

$2x - y + 9 = 0$

**19.** $x = 6t$, $y = t^2 + 4$, $t = 1$

(a), (d)

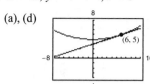

(b) At $t = 1$, $(x, y) = (6, 5)$, and

$$\frac{dx}{dt} = 6, \frac{dy}{dt} = 2, \frac{dy}{dx} = \frac{1}{3}.$$

(c) $y - 5 = \frac{1}{3}(x - 6)$

$$y = \frac{1}{3}x + 3$$

**21.** $x = t^2 - t + 2$, $y = t^3 - 3t$, $t = -1$

(a), (d)

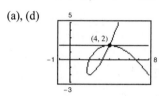

(b) At $t = -1$, $(x, y) = (4, 2)$, and

$$\frac{dx}{dt} = -3, \frac{dy}{dt} = 0, \frac{dy}{dx} = 0.$$

(c) $\frac{dy}{dx} = 0$. At $(4, 2)$, $y - 2 = 0(x - 4)$

$$y = 2.$$

**23.** $x = 2 \sin 2t$, $y = 3 \sin t$ crosses itself at the origin, $(x, y) = (0, 0)$.

At this point, $t = 0$ or $t = \pi$.

$$\frac{dy}{dx} = \frac{3 \cos t}{4 \cos 2t}$$

At $t = 0$: $\frac{dy}{dx} = \frac{3}{4}$ and $y = \frac{3}{4}x$.   Tangent Line

At $t = \pi$, $\frac{dy}{dx} = -\frac{3}{4}$ and $y = -\frac{3}{4}x$.   Tangent Line

**25.** $x = t^2 - t$, $y = t^3 - 3t - 1$ crosses itself at the point $(x, y) = (2, 1)$.

At this point, $t = -1$ or $t = 2$.

$$\frac{dy}{dx} = \frac{3t^2 - 3}{2t - 1}$$

At $t = -1$, $\frac{dy}{dx} = 0$ and $y = 1$.   Tangent Line

At $t = 2$, $\frac{dy}{dt} = \frac{9}{3} = 3$ and $y - 1 = 3(x - 2)$ or

$$y = 3x - 5.$$

Tangent Line

**27.** $x = \cos \theta + \theta \sin \theta$, $y = \sin \theta - \theta \cos \theta$

Horizontal tangents: $\frac{dy}{d\theta} = \theta \sin \theta = 0$ when

$$\theta = \pm\pi, \pm 2\pi, \pm 3\pi, \ldots$$

Points: $(-1, [2n - 1]\pi)$, $(1, 2n\pi)$ where $n$ is an integer.

Points shown: $(1, 0), (-1, \pi), (1, -2\pi)$

Vertical tangents: $\frac{dx}{d\theta} = \theta \cos \theta = 0$ when

$$\theta = \pm\frac{\pi}{2}, \pm\frac{3\pi}{2}, \pm\frac{5\pi}{2}, \ldots$$

**Note:** $\theta = 0$ corresponds to the cusp at $(x, y) = (1, 0)$.

$$\frac{dy}{dx} = \frac{\theta \sin \theta}{\theta \cos \theta} = \tan \theta = 0 \text{ at } \theta = 0$$

Points: $\left( \dfrac{(-1)^{n+1}(2n - 1)\pi}{2}, (-1)^{n+1} \right)$

Points shown: $\left( \dfrac{\pi}{2}, 1 \right), \left( -\dfrac{3\pi}{2}, -1 \right), \left( \dfrac{5\pi}{2}, 1 \right)$

**29.** $x = 4 - t$, $y = t^2$

Horizontal tangents: $\frac{dy}{dt} = 2t = 0$ when $t = 0$.

Point: $(4, 0)$

Vertical tangents: $\frac{dx}{dt} = -1 \neq 0$   None

**31.** $x = t + 4$, $y = t^3 - 3t$

Horizontal tangents:

$$\frac{dy}{dt} = 3t^2 - 3 = 3(t - 1)(t + 1) = 0 \Rightarrow t = \pm 1$$

Points: $(5, -2), (3, 2)$

Vertical tangents: $\frac{dx}{dt} = 1 \neq 0$   None

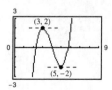

**33.** $x = 3\cos\theta,\ y = 3\sin\theta$

Horizontal tangents: $\dfrac{dy}{d\theta} = 3\cos\theta = 0$ when

$\theta = \dfrac{\pi}{2}, \dfrac{3\pi}{2}.$

Points: $(0, 3), (0, -3)$

Vertical tangents: $\dfrac{dx}{d\theta} = -3\sin\theta = 0$ when $\theta = 0, \pi.$

Points: $(3, 0), (-3, 0)$

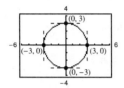

**35.** $x = 5 + 3\cos\theta,\ y = -2 + \sin\theta$

Horizontal tangents: $\dfrac{dy}{dt} = \cos\theta = 0 \Rightarrow \theta = \dfrac{\pi}{2}, \dfrac{3\pi}{2}$

Points: $(5, -1), (5, -3)$

Vertical tangents: $\dfrac{dx}{dt} = -3\sin\theta = 0 \Rightarrow \theta = 0, \pi$

Points: $(8, -2), (2, -2)$

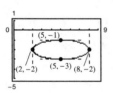

**37.** $x = \sec\theta,\ y = \tan\theta$

Horizontal tangents: $\dfrac{dy}{d\theta} = \sec^2\theta \neq 0;$ None

Vertical tangents: $\dfrac{dx}{d\theta} = \sec\theta\tan\theta = 0$ when

$x = 0, \pi.$

Points: $(1, 0), (-1, 0)$

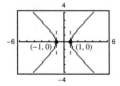

**39.** $x = 3t^2,\ y = t^3 - t$

$\dfrac{dy}{dx} = \dfrac{dy/dt}{dx/dt} = \dfrac{3t^2 - 1}{6t} = \dfrac{t}{2} - \dfrac{1}{6t}$

$\dfrac{d^2y}{dx^2} = \dfrac{\dfrac{d}{dt}\left[\dfrac{t}{2} - \dfrac{1}{6t}\right]}{dx/dt} = \dfrac{\dfrac{1}{2} + \dfrac{1}{6t^2}}{6t} = \dfrac{6t^2 + 2}{36t^3}$

Concave upward for $t > 0$

Concave downward for $t < 0$

**41.** $x = 2t + \ln t,\ y = 2t - \ln t,\ t > 0$

$\dfrac{dy}{dx} = \dfrac{2 - (1/t)}{2 + (1/t)} = \dfrac{2t - 1}{2t + 1}$

$\dfrac{d^2y}{dx^2} = \left[\dfrac{(2t + 1)2 - (2t - 1)2}{(2t + 1)^2}\right] \Big/ \left(2 + \dfrac{1}{t}\right)$

$= \dfrac{4}{(2t + 1)^2} \cdot \dfrac{t}{2t + 1} = \dfrac{4t}{(2t + 1)^3}$

Because $t > 0, \dfrac{d^2y}{dx^2} > 0$

Concave upward for $t > 0$

**43.** $x = \sin t,\ y = \cos t,\ 0 < t < \pi$

$\dfrac{dy}{dx} = -\dfrac{\sin t}{\cos t} = -\tan t$

$\dfrac{d^2y}{dx^2} = -\dfrac{\sec^2 t}{\cos t} = -\dfrac{1}{\cos^3 t}$

Concave upward on $\pi/2 < t < \pi$

Concave downward on $0 < t < \pi/2$

**45.** $x = 3t + 5,\ y = 7 - 2t,\ -1 \le t \le 3$

$\dfrac{dx}{dt} = 3, \dfrac{dy}{dt} = -2$

$s = \int_a^b \sqrt{\left(\dfrac{dx}{dt}\right)^2 + \left(\dfrac{dy}{dt}\right)^2}\ dt$

$= \int_{-1}^{3} \sqrt{9 + 4}\ dt$

$\left[\sqrt{13}\ t\right]_{-1}^{3} = 4\sqrt{13} \approx 14.422$

**47.** $x = e^{-t} \cos t, \ y = e^{-t} \sin t, \ 0 \le t \le \dfrac{\pi}{2}$

$$\frac{dx}{dt} = -e^{-t}(\sin t + \cos t), \frac{dy}{dt} = e^{-t}(\cos t - \sin t)$$

$$s = \int_0^{\pi/2} \sqrt{\left(\frac{dx}{dt}\right)^2 + \left(\frac{dy}{dt}\right)^2} \, dt$$

$$= \int_0^{\pi/2} \sqrt{2e^{-2t}} \, dt = -\sqrt{2} \int_0^{\pi/2} e^{-t}(-1) \, dt$$

$$= \left[ -\sqrt{2} e^{-t} \right]_0^{\pi/2}$$

$$= \sqrt{2}\left(1 - e^{-\pi/2}\right) \approx 1.12$$

**49.** $x = \sqrt{t}, \ y = 3t - 1, \ \dfrac{dx}{dt} = \dfrac{1}{2\sqrt{t}}, \ \dfrac{dy}{dt} = 3$

$$s = \int_0^1 \sqrt{\frac{1}{4t} + 9} \, dt = \frac{1}{2} \int_0^1 \frac{\sqrt{1 + 36t}}{\sqrt{t}} \, dt$$

$$= \frac{1}{6} \int_0^6 \sqrt{1 + u^2} \, du$$

$$= \frac{1}{12} \left[ \ln\left(\sqrt{1 + u^2} + u\right) + u\sqrt{1 + u^2} \right]_0^6$$

$$= \frac{1}{12} \left[ \ln\left(\sqrt{37} + 6\right) + 6\sqrt{37} \right] \approx 3.249$$

$$u = 6\sqrt{t}, \ du = \frac{3}{\sqrt{t}} \, dt$$

**51.** $x = a \cos^3 \theta, \ y = a \sin^3 \theta, \ \dfrac{dx}{d\theta} = -3a \cos^2 \theta \sin \theta,$

$$\frac{dy}{d\theta} = 3a \sin^2 \theta \cos \theta$$

$$s = 4 \int_0^{\pi/2} \sqrt{9a^2 \cos^4 \theta \sin^2 \theta + 9a^2 \sin^4 \theta \cos^2 \theta} \, d\theta$$

$$= 12a \int_0^{\pi/2} \sin \theta \cos \theta \sqrt{\cos^2\theta + \sin^2 \theta} \, d\theta$$

$$= 6a \int_0^{\pi/2} \sin 2\theta \, d\theta = \left[ -3a \cos 2\theta \right]_0^{\pi/2} = 6a$$

**53.** $x = a(\theta - \sin \theta), \ y = a(1 - \cos \theta),$

$$\frac{dx}{d\theta} = a(1 - \cos \theta), \frac{dy}{d\theta} = a \sin \theta$$

$$s = 2 \int_0^{\pi} \sqrt{a^2(1 - \cos \theta)^2 + a^2 \sin^2 \theta} \, d\theta$$

$$= 2\sqrt{2}a \int_0^{\pi} \sqrt{1 - \cos \theta} \, d\theta$$

$$= 2\sqrt{2}a \int_0^{\pi} \frac{\sin \theta}{\sqrt{1 + \cos \theta}} \, d\theta$$

$$= \left[ -4\sqrt{2}a\sqrt{1 + \cos \theta} \right]_0^{\pi} = 8a$$

**55.** $x = (90 \cos 30°)t, \ y = (90 \sin 30°)t - 16t^2$

(a)
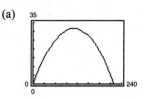

(b) Range: 219.2 ft, $\left( t = \dfrac{45}{16} \right)$

(c) $\dfrac{dx}{dt} = 90 \cos 30°, \dfrac{dy}{dt} = 90 \sin 30° - 32t$

$y = 0$ for $t = \dfrac{45}{16}$.

$$s = \int_0^{45/16} \sqrt{(90 \cos 30°)^2 + (90 \sin 30° - 32t)^2} \, dt$$

$$\approx 230.8 \text{ ft}$$

**57.** $x = \dfrac{4t}{1 + t^3}, \ y = \dfrac{4t^2}{1 + t^3}$

(a) $x^3 + y^3 = 4xy$

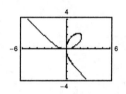

(b) $\dfrac{dy}{dt} = \dfrac{(1 + t^3)(8t) - 4t^2(3t^2)}{(1 + t^3)^2}$

$$= \frac{4t(2 - t^3)}{(1 + t^3)^2} = 0 \text{ when } t = 0 \text{ or } t = \sqrt[3]{2}.$$

Points: $(0, 0), \left( \dfrac{4\sqrt[3]{2}}{3}, \dfrac{4\sqrt[3]{4}}{3} \right) \approx (1.6799, 2.1165)$

(c) $s = 2 \int_0^1 \sqrt{\left[ \dfrac{4(1 - 2t^3)}{(1 + t^3)^2} \right]^2 + \left[ \dfrac{4t(2 - t^3)}{(1 + t^3)^2} \right]^2} \, dt$

$$= 2 \int_0^1 \sqrt{\frac{16}{(1 + t^3)^4} \left[ t^8 + 4t^6 - 4t^5 - 4t^3 + 4t^2 + 1 \right]} \, dt$$

$$= 8 \int_0^1 \frac{\sqrt{t^8 + 4t^6 - 4t^5 - 4t^3 + 4t^2 + 1}}{(1 + t^3)^2} \, dt \approx 6.557$$

**59.** (a) $x = t - \sin t$
$y = 1 - \cos t$
$0 \le t \le 2\pi$

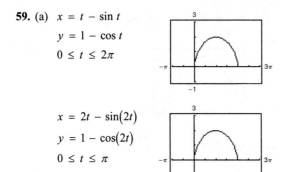

$x = 2t - \sin(2t)$
$y = 1 - \cos(2t)$
$0 \le t \le \pi$

(b) The average speed of the particle on the second path is twice the average speed of a particle on the first path.

(c) $x = \tfrac{1}{2}t - \sin\left(\tfrac{1}{2}t\right)$
$y = 1 - \cos\left(\tfrac{1}{2}t\right)$

The time required for the particle to traverse the same path is $t = 4\pi$.

**61.** $x = 3t, \dfrac{dx}{dt} = 3$

$y = t + 2, \dfrac{dy}{dt} = 1$

$S = 2\pi \displaystyle\int_0^4 (t + 2)\sqrt{3^2 + 1^2}\, dt$

$= 2\pi\sqrt{10}\left[\dfrac{t^2}{2} + 2t\right]_0^4$

$= 2\pi\sqrt{10}[8 + 8] = 32\sqrt{10}\,\pi \approx 317.9068$

**63.** $x = \cos^2\theta, \dfrac{dx}{d\theta} = -2\cos\theta\sin\theta$

$y = \cos\theta, \dfrac{dy}{d\theta} = -\sin\theta$

$S = 2\pi \displaystyle\int_0^{\pi/2} \cos\theta\sqrt{4\cos^2\theta\sin^2\theta + \sin^2\theta}\, d\theta$

$= 2\pi \displaystyle\int_0^{\pi/2} \cos\theta\sin\theta\sqrt{4\cos^2\theta + 1}\, d\theta$

$= \dfrac{\left(5\sqrt{5} - 1\right)\pi}{6}$

$\approx 5.3304$

**69.** $x = a\cos^3\theta,\ y = a\sin^3\theta, \dfrac{dx}{d\theta} = -3a\cos^2\theta\sin\theta, \dfrac{dy}{d\theta} = 3a\sin^2\theta\cos\theta$

$S = 4\pi \displaystyle\int_0^{\pi/2} a\sin^3\theta\sqrt{9a^2\cos^4\theta\sin^2\theta + 9a^2\sin^4\theta\cos^2\theta}\, d\theta$

$= 12a^2\pi \displaystyle\int_0^{\pi/2} \sin^4\theta\cos\theta\, d\theta = \dfrac{12\pi a^2}{5}\left[\sin^5\theta\right]_0^{\pi/2} = \dfrac{12}{5}\pi a^2$

**71.** $\dfrac{dy}{dx} = \dfrac{dy/dt}{dx/dt}$

See Theorem 10.7.

**65.** $x = 2t, \dfrac{dx}{dt} = 2$

$y = 3t, \dfrac{dy}{dt} = 3$

(a) $S = 2\pi \displaystyle\int_0^3 3t\sqrt{4 + 9}\, dt$

$= 6\sqrt{13}\pi\left[\dfrac{t^2}{2}\right]_0^3 = 6\sqrt{13}\pi\left(\dfrac{9}{2}\right) = 27\sqrt{13}\pi$

(b) $S = 2\pi \displaystyle\int_0^3 2t\sqrt{4 + 9}\, dt$

$= 4\sqrt{13}\pi\left[\dfrac{t^2}{2}\right]_0^3 = 4\sqrt{13}\pi\left(\dfrac{9}{2}\right) = 18\sqrt{13}\pi$

**67.** $x = 5\cos\theta\ \dfrac{dx}{d\theta} = -5\sin\theta$

$y = 5\sin\theta\ \dfrac{dy}{d\theta} = 5\cos\theta$

$S = 2\pi \displaystyle\int_0^{\pi/2} 5\cos\theta\sqrt{25\sin^2\theta + 25\cos^2\theta}\, d\theta$

$= 10\pi \displaystyle\int_0^{\pi/2} 5\cos\theta\, d\theta$

$= 50\pi\left[\sin\theta\right]_0^{\pi/2} = 50\pi$

[**Note:** This is the surface area of a hemisphere of radius 5]

**73.** $x = t,\ y = 6t - 5 \Rightarrow \dfrac{dy}{dx} = \dfrac{6}{1} = 6$

**75.** (a) $S = 2\pi \int_a^b g(t) \sqrt{\left(\dfrac{dx}{dt}\right)^2 + \left(\dfrac{dy}{dt}\right)^2}\, dt$

   (b) $S = 2\pi \int_a^b f(t) \sqrt{\left(\dfrac{dx}{dt}\right)^2 + \left(\dfrac{dy}{dt}\right)^2}\, dt$

**77.** Let $y$ be a continuous function of $x$ on $a \le x \le b$. Suppose that $x = f(t)$, $y = g(t)$, and $f(t_1) = a$, $f(t_2) = b$. Then using integration by substitution, $dx = f'(t)\, dt$ and

$$\int_a^b y\, dx = \int_{t_1}^{t_2} g(t) f'(t)\, dt.$$

**79.**   $x = 2\sin^2\theta$

   $y = 2\sin^2\theta \tan\theta$

   $\dfrac{dx}{d\theta} = 4\sin\theta\cos\theta$

   $A = \int_0^{\pi/2} 2\sin^2\theta \tan\theta (4\sin\theta\cos\theta)\, d\theta$

   $= 8 \int_0^{\pi/2} \sin^4\theta\, d\theta$

   $= 8\left[ \dfrac{-\sin^3\theta\cos\theta}{4} - \dfrac{3}{8}\sin\theta\cos\theta + \dfrac{3}{8}\theta \right]_0^{\pi/2} = \dfrac{3\pi}{2}$

$0 \le \theta < \dfrac{\pi}{2}$

**81.** $\pi ab$ is area of ellipse (d).

**83.** $6\pi a^2$ is area of cardioid (f).

**85.** $\dfrac{8}{3} ab$ is area of hourglass (a).

**87.** $x = \sqrt{t},\ y = 4 - t,\ 0 < t < 4$

$$A = \int_0^2 y\, dx = \int_0^4 (4 - t)\frac{1}{2\sqrt{t}}\, dt = \frac{1}{2}\int_0^4 \left(4t^{-1/2} - t^{1/2}\right) dt = \left[\frac{1}{2}\left(8\sqrt{t} - \frac{2}{3}t\sqrt{t}\right)\right]_0^4 = \frac{16}{3}$$

$$\bar{x} = \frac{1}{A}\int_0^2 yx\, dx = \frac{3}{16}\int_0^4 (4 - t)\sqrt{t}\left(\frac{1}{2\sqrt{t}}\right) dt = \frac{3}{32}\int_0^4 (4 - t)\, dt = \left[\frac{3}{32}\left(4t - \frac{t^2}{2}\right)\right]_0^4 = \frac{3}{4}$$

$$\bar{y} = \frac{1}{A}\int_0^2 \frac{y^2}{2}\, dx = \frac{3}{32}\int_0^4 (4 - t)^2 \frac{1}{2\sqrt{t}}\, dt = \frac{3}{64}\int_0^4 \left(16t^{-1/2} - 8t^{1/2} + t^{3/2}\right) dt = \frac{3}{64}\left[32\sqrt{t} - \frac{16}{3}t\sqrt{t} + \frac{2}{5}t^2\sqrt{t}\right]_0^4 = \frac{8}{5}$$

$(\bar{x}, \bar{y}) = \left(\dfrac{3}{4}, \dfrac{8}{5}\right)$

**89.** $x = 6 \cos \theta, \; y = 6 \sin \theta, \dfrac{dx}{d\theta} = -6 \sin \theta \, d\theta$

$$V = 2\pi \int_{\pi/2}^{0} (6 \sin \theta)^2 (-6 \sin \theta) \, d\theta$$

$$= -432\pi \int_{\pi/2}^{0} \sin^3 \theta \, d\theta$$

$$= -432\pi \int_{\pi/2}^{0} (1 - \cos^2 \theta) \sin \theta \, d\theta$$

$$= -432\pi \left[ -\cos \theta + \frac{\cos^3 \theta}{3} \right]_{\pi/2}^{0}$$

$$= -432\pi \left( -1 + \frac{1}{3} \right) = 288\pi$$

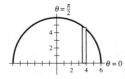

**Note:** Volume of sphere is $\dfrac{4}{3}\pi (6^3) = 288\pi$.

**91.** $x = a(\theta - \sin \theta), \; y = a(1 - \cos \theta)$

(a) $\quad \dfrac{dy}{d\theta} = a \sin \theta, \dfrac{dx}{d\theta} = a(1 - \cos \theta)$

$\quad \dfrac{dy}{dx} = \dfrac{a \sin \theta}{a(1 - \cos \theta)} = \dfrac{\sin \theta}{1 - \cos \theta}$

$\quad \dfrac{d^2 y}{dx^2} = \left[ \dfrac{(1 - \cos \theta) \cos \theta - \sin \theta (\sin \theta)}{(1 - \cos \theta)^2} \right] \Big/ \left[ a(1 - \cos \theta) \right] = \dfrac{\cos \theta - 1}{a(1 - \cos \theta)^3} = \dfrac{-1}{a(\cos \theta - 1)^2}$

(b) At $\theta = \dfrac{\pi}{6}, \; x = a\left( \dfrac{\pi}{6} - \dfrac{1}{2} \right), \; y = a\left( 1 - \dfrac{\sqrt{3}}{2} \right), \dfrac{dy}{dx} = \dfrac{1/2}{1 - \sqrt{3}/2} = 2 + \sqrt{3}.$

$\quad$ Tangent line: $y - a\left( 1 - \dfrac{\sqrt{3}}{2} \right) = (2 + \sqrt{3})\left( x - a\left( \dfrac{\pi}{6} - \dfrac{1}{2} \right) \right)$

(c) $\dfrac{dy}{dx} = \dfrac{\sin \theta}{1 - \cos \theta} = 0 \Rightarrow \sin \theta = 0, 1 - \cos \theta \neq 0$

$\quad$ Points of horizontal tangency: $(x, y) = (a(2n + 1)\pi, 2a)$

(d) Concave downward on all open $\theta$-intervals:

$\quad \ldots, (-2\pi, 0), (0, 2\pi), (2\pi, 4\pi), \ldots$

(e) $s = \displaystyle\int_{0}^{2\pi} \sqrt{a^2 \sin^2 \theta + a^2 (1 - \cos \theta)^2} \, d\theta$

$\quad = a\displaystyle\int_{0}^{2\pi} \sqrt{2 - 2\cos \theta} \, d\theta = a\displaystyle\int_{0}^{2\pi} \sqrt{4 \sin^2 \dfrac{\theta}{2}} \, d\theta = 2a\displaystyle\int_{0}^{2\pi} \sin \dfrac{\theta}{2} \, d\theta = \left[ -4a \cos\left( \dfrac{\theta}{2} \right) \right]_{0}^{2\pi} = 8a$

**93.** $x = t + u = r \cos \theta + r\theta \sin \theta$

$\quad = r(\cos \theta + \theta \sin \theta)$

$y = v - w = r \sin \theta - r\theta \cos \theta$

$\quad = r(\sin \theta - \theta \cos \theta)$

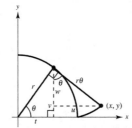

**95.** (a)

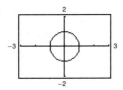

(b) $x = \dfrac{1 - t^2}{1 + t^2}, \; y = \dfrac{2t}{1 + t^2}, \; -20 \le t \le 20$

The graph (for $-\infty < t < \infty$) is the circle $x^2 + y^2 = 1$, except the point $(-1, 0)$.

Verify:

$$x^2 + y^2 = \left(\dfrac{1 - t^2}{1 + t^2}\right)^2 + \left(\dfrac{2t}{1 + t^2}\right)^2$$

$$= \dfrac{1 - 2t^2 + t^4 + 4t^2}{\left(1 + t^2\right)^2} = \dfrac{\left(1 + t^2\right)^2}{\left(1 + t^2\right)^2} = 1$$

(c) As $t$ increases from $-20$ to $0$, the speed increases, and as $t$ increases from $0$ to $20$, the speed decreases.

**97.** False. $\dfrac{d^2y}{dx^2} = \dfrac{\dfrac{d}{dt}\left[\dfrac{g'(t)}{f'(t)}\right]}{f'(t)} = \dfrac{f'(t)g''(t) - g'(t)f''(t)}{\left[f'(t)\right]^3}$

# Section 10.4   Polar Coordinates and Polar Graphs

**1.** $\left(8, \dfrac{\pi}{2}\right)$

$x = 8 \cos \dfrac{\pi}{2} = 0$

$y = 8 \sin \dfrac{\pi}{2} = 8$

$(x, y) = (0, 8)$

**3.** $\left(-4, -\dfrac{3\pi}{4}\right)$

$x = -4 \cos\left(\dfrac{-3\pi}{4}\right) = -4\left(-\dfrac{\sqrt{2}}{2}\right) = 2\sqrt{2}$

$y = -4 \sin\left(\dfrac{-3\pi}{4}\right) = -4\left(-\dfrac{\sqrt{2}}{2}\right) = 2\sqrt{2}$

$(x, y) = \left(2\sqrt{2}, 2\sqrt{2}\right)$

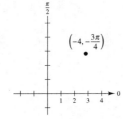

**5.** $(r, \theta) = \left(7, \dfrac{5\pi}{4}\right)$

$x = 7 \cos \dfrac{5\pi}{4} = 7\left(\dfrac{-\sqrt{2}}{2}\right) = -\dfrac{7\sqrt{2}}{2}$

$y = 7 \sin \dfrac{5\pi}{4} = 7\left(-\dfrac{\sqrt{2}}{2}\right) = -\dfrac{7\sqrt{2}}{2}$

$(x, y) = \left(-\dfrac{7\sqrt{2}}{2}, -\dfrac{7\sqrt{2}}{2}\right)$

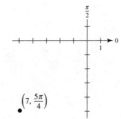

**7.** $\left(\sqrt{2}, 2.36\right)$

$x = \sqrt{2} \cos(2.36) \approx -1.004$

$y = \sqrt{2} \sin(2.36) \approx 0.996$

$(x, y) = (-1.004, 0.996)$

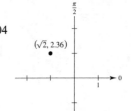

**9.** $(r, \theta) = (-4.5, 3.5)$

$\quad x = -4.5 \cos 3.5 \approx 4.2141$

$\quad y = -4.5 \sin 3.5 \approx 1.5785$

$(x, y) = (4.2141, 1.5785)$

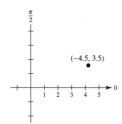

**11.** $(x, y) = (2, 2)$

$\quad r = \sqrt{2^2 + 2^2} = 2\sqrt{2}$

$\quad \tan \theta = \dfrac{2}{2} = 1$

$\quad \theta = \dfrac{\pi}{4}, \dfrac{5\pi}{4}$

$\quad \left(2\sqrt{2}, \dfrac{\pi}{4}\right), \left(-2\sqrt{2}, \dfrac{5\pi}{4}\right)$

**13.** $(x, y) = (-3, 4)$

$\quad r = \pm\sqrt{9 + 16} = \pm 5$

$\quad \tan \theta = -\dfrac{4}{3}$

$\quad \theta \approx 2.214, 5.356, (5, 2.214),$

$\quad\quad (-5, 5.356)$

**15.** $(x, y) = \left(-1, -\sqrt{3}\right)$

$\quad r = \sqrt{4} = 2$

$\quad \tan \theta = \dfrac{-\sqrt{3}}{-1} = \sqrt{3}$

$\quad \theta = \dfrac{\pi}{3}, \dfrac{4\pi}{3}$

$\quad \left(2, \dfrac{4\pi}{3}\right), \left(-2, \dfrac{\pi}{3}\right)$

**17.** $(x, y) = (3, -2)$

$\quad r = \sqrt{3^2 + (-2)^2} = \sqrt{13} \approx 3.6056$

$\quad \tan \theta = -\dfrac{2}{3} \Rightarrow \theta \approx 5.6952$

$(r, \theta) \approx (3.6056, 5.6952) = (-3.6056, 2.5536)$

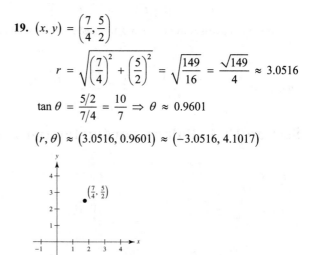

**19.** $(x, y) = \left(\dfrac{7}{4}, \dfrac{5}{2}\right)$

$\quad r = \sqrt{\left(\dfrac{7}{4}\right)^2 + \left(\dfrac{5}{2}\right)^2} = \sqrt{\dfrac{149}{16}} = \dfrac{\sqrt{149}}{4} \approx 3.0516$

$\quad \tan \theta = \dfrac{5/2}{7/4} = \dfrac{10}{7} \Rightarrow \theta \approx 0.9601$

$(r, \theta) \approx (3.0516, 0.9601) \approx (-3.0516, 4.1017)$

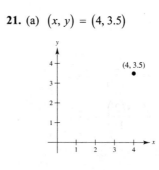

**21.** (a) $(x, y) = (4, 3.5)$

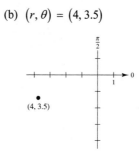

(b) $(r, \theta) = (4, 3.5)$

**23.** $x^2 + y^2 = 9$

$r^2 = 9$

$r = 3$

Circle

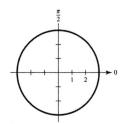

**25.** $x^2 + y^2 = a^2$

$r = a$

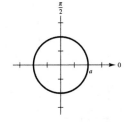

**27.** $y = 8$

$r \sin \theta = 8$

$r = 8 \csc \theta$

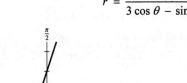

**29.** $3x - y + 2 = 0$

$3r \cos \theta - r \sin \theta + 2 = 0$

$r(3 \cos \theta - \sin \theta) = -2$

$$r = \frac{-2}{3 \cos \theta - \sin \theta}$$

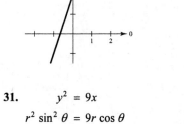

**31.** $y^2 = 9x$

$r^2 \sin^2 \theta = 9r \cos \theta$

$$r = \frac{9 \cos \theta}{\sin^2 \theta}$$

$r = 9 \csc^2 \theta \cos \theta$

**33.** $r = 4$

$r^2 = 16$

$x^2 + y^2 = 16$

Circle

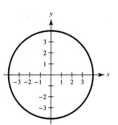

**35.** $r = 3 \sin \theta$

$r^2 = 3r \sin \theta$

$x^2 + y^2 = 3y$

$x^2 + \left(y^2 - 3y + \frac{9}{4}\right) = \frac{9}{4}$

$x^2 + \left(y - \frac{3}{2}\right)^2 = \frac{9}{4}$

Circle

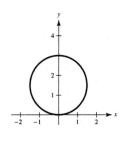

**37.** $r = \theta$

$\tan r = \tan \theta$

$\tan \sqrt{x^2 + y^2} = \dfrac{y}{x}$

$\sqrt{x^2 + y^2} = \arctan \dfrac{y}{x}$

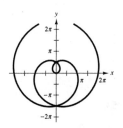

**39.** $r = 3 \sec \theta$

$r \cos \theta = 3$

$x = 3$

$x - 3 = 0$

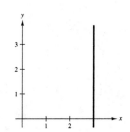

**41.** $r = \sec \theta \tan \theta$

$r \cos \theta = \tan \theta$

$x = \dfrac{y}{x}$

$y = x^2$

Parabola

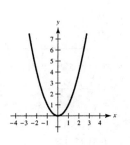

**43.** $r = 2 - 5 \cos \theta$

$0 \le \theta < 2\pi$

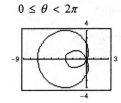

**45.** $r = 2 + \sin \theta$

$0 \le \theta < 2\pi$

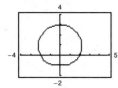

**47.** $r = \dfrac{2}{1 + \cos \theta}$

Traced out once on $-\pi < \theta < \pi$

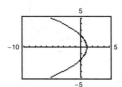

**49.** $r = 2 \cos\left(\dfrac{3\theta}{2}\right)$

$0 \le \theta < 4\pi$

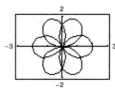

**51.** $r^2 = 4 \sin 2\theta$

$r_1 = 2\sqrt{\sin 2\theta}$

$r_2 = -2\sqrt{\sin 2\theta}$

$0 \le \theta < \dfrac{\pi}{2}$

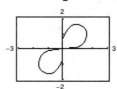

**53.**

$$r = 2(h \cos \theta + k \sin \theta)$$
$$r^2 = 2r(h \cos \theta + k \sin \theta)$$
$$r^2 = 2\big[h(r \cos \theta) + k(r \sin \theta)\big]$$
$$x^2 + y^2 = 2(hx + ky)$$
$$x^2 + y^2 - 2hx - 2ky = 0$$
$$\left(x^2 - 2hx + h^2\right) + \left(y^2 - 2ky + k^2\right) = 0 + h^2 + k^2 \qquad \text{Radius: } \sqrt{h^2 + k^2}$$
$$(x - h)^2 + (y - k)^2 = h^2 + k^2 \qquad \text{Center: } (h, k)$$

**55.** $\left(1, \dfrac{5\pi}{6}\right), \left(4, \dfrac{\pi}{3}\right)$

$$d = \sqrt{1^2 + 4^2 - 2(1)(4) \cos\left(\dfrac{5\pi}{6} - \dfrac{\pi}{3}\right)}$$
$$= \sqrt{17 - 8 \cos \dfrac{\pi}{2}} = \sqrt{17}$$

**57.** $(2, 0.5), (7, 1.2)$

$$d = \sqrt{2^2 + 7^2 - 2(2)(7) \cos(0.5 - 1.2)}$$
$$= \sqrt{53 - 28 \cos(-0.7)} \approx 5.6$$

**59.** $r = 2 + 3 \sin \theta$

$$\dfrac{dy}{dx} = \dfrac{3 \cos \theta \sin \theta + \cos \theta(2 + 3 \sin \theta)}{3 \cos \theta \cos \theta - \sin \theta(2 + 3 \sin \theta)}$$
$$= \dfrac{2 \cos \theta(3 \sin \theta + 1)}{3 \cos 2\theta - 2 \sin \theta} = \dfrac{2 \cos \theta(3 \sin \theta + 1)}{6 \cos^2 \theta - 2 \sin \theta - 3}$$

At $\left(5, \dfrac{\pi}{2}\right), \dfrac{dy}{dx} = 0$.

At $(2, \pi), \dfrac{dy}{dx} = -\dfrac{2}{3}$.

At $\left(-1, \dfrac{3\pi}{2}\right), \dfrac{dy}{dx} = 0$.

**61.** (a), (b)  $r = 3(1 - \cos\theta)$

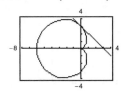

$(r, \theta) = \left(3, \dfrac{\pi}{2}\right) \Rightarrow (x, y) = (0, 3)$

Tangent line: $y - 3 = -1(x - 0)$

$y = -x + 3$

(c) At $\theta = \dfrac{\pi}{2}, \dfrac{dy}{dx} = -1.0.$

**63.** (a), (b)  $r = 3\sin\theta$

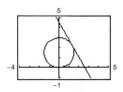

$(r, \theta) = \left(\dfrac{3\sqrt{3}}{2}, \dfrac{\pi}{3}\right) \Rightarrow (x, y) = \left(\dfrac{3\sqrt{3}}{4}, \dfrac{9}{4}\right)$

Tangent line: $y - \dfrac{9}{4} = -\sqrt{3}\left(x - \dfrac{3\sqrt{3}}{4}\right)$

$y = -\sqrt{3}x + \dfrac{9}{2}$

(c) At $\theta = \dfrac{\pi}{3}, \dfrac{dy}{dx} = -\sqrt{3} \approx -1.732.$

**65.**  $r = 1 - \sin\theta$

$\dfrac{dy}{d\theta} = (1 - \sin\theta)\cos\theta - \cos\theta\sin\theta$

$= \cos\theta(1 - 2\sin\theta) = 0$

$\cos\theta = 0$ or $\sin\theta = \dfrac{1}{2} \Rightarrow \theta = \dfrac{\pi}{2}, \dfrac{3\pi}{2}, \dfrac{\pi}{6}, \dfrac{5\pi}{6}$

Horizontal tangents: $\left(2, \dfrac{3\pi}{2}\right), \left(\dfrac{1}{2}, \dfrac{\pi}{6}\right), \left(\dfrac{1}{2}, \dfrac{5\pi}{6}\right)$

$\dfrac{dx}{d\theta} = (-1 + \sin\theta)\sin\theta - \cos\theta\cos\theta$

$= -\sin\theta + \sin^2\theta + \sin^2\theta - 1$

$= 2\sin^2\theta - \sin\theta - 1$

$= (2\sin\theta + 1)(\sin\theta - 1) = 0$

$\sin\theta = 1$ or $\sin\theta = -\dfrac{1}{2} \Rightarrow \theta = \dfrac{\pi}{2}, \dfrac{7\pi}{6}, \dfrac{11\pi}{6}$

Vertical tangents: $\left(\dfrac{3}{2}, \dfrac{7\pi}{6}\right), \left(\dfrac{3}{2}, \dfrac{11\pi}{6}\right)$

**67.**  $r = 2\csc\theta + 3$

$\dfrac{dy}{d\theta} = (2\csc\theta + 3)\cos\theta + (-2\csc\theta\cot\theta)\sin\theta$

$= 3\cos\theta = 0$

$\theta = \dfrac{\pi}{2}, \dfrac{3\pi}{2}$

Horizontal tangents: $\left(5, \dfrac{\pi}{2}\right), \left(1, \dfrac{3\pi}{2}\right)$

**69.**  $r = 5\sin\theta$

$r^2 = 5r\sin\theta$

$x^2 + y^2 = 5y$

$x^2 + \left(y^2 - 5y + \dfrac{25}{4}\right) = \dfrac{25}{4}$

$x^2 + \left(y - \dfrac{5}{2}\right)^2 = \dfrac{25}{4}$

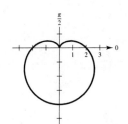

Circle: center: $\left(0, \dfrac{5}{2}\right)$, radius: $\dfrac{5}{2}$

Tangent at pole: $\theta = 0$

Note: $f(\theta) = r = 5\sin\theta$

$f(0) = 0, f'(0) \neq 0$

**71.**  $r = 2(1 - \sin\theta)$

Cardioid

Symmetric to $y$-axis, $\theta = \dfrac{\pi}{2}$

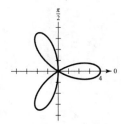

**73.**  $r = 4\cos 3\theta$

Rose curve with three petals.

Tangents at pole: $(r = 0, r' \neq 0)$:

$\theta = \dfrac{\pi}{6}, \dfrac{\pi}{2}, \dfrac{5\pi}{6}$

**75.** $r = 3 \sin 2\theta$

Rose curve with four petals

Symmetric to the polar axis, $\theta = \dfrac{\pi}{2}$, and pole

Relative extrema: $\left(\pm 3, \dfrac{\pi}{4}\right), \left(\pm 3, \dfrac{5\pi}{4}\right)$

Tangents at the pole: $\theta = 0, \dfrac{\pi}{2}$

$\left(\theta = \pi, \dfrac{3\pi}{2} \text{ give the same tangents.}\right)$

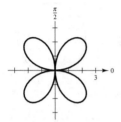

**77.** $r = 8$

Circle radius 8

$x^2 + y^2 = 64$

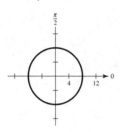

**79.** $r = 4(1 + \cos \theta)$

Cardioid

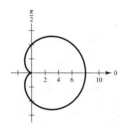

**81.** $r = 3 - 2 \cos \theta$

Limaçon

Symmetric to polar axis

| $\theta$ | 0 | $\dfrac{\pi}{3}$ | $\dfrac{\pi}{2}$ | $\dfrac{2\pi}{3}$ | $\pi$ |
|---|---|---|---|---|---|
| $r$ | 1 | 2 | 3 | 4 | 5 |

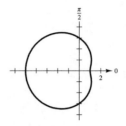

**83.**  $r = 3 \csc \theta$

$r \sin \theta = 3$

$y = 3$

Horizontal line

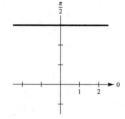

**85.** $r = 2\theta$

Spiral of Archimedes

Symmetric to $\theta = \dfrac{\pi}{2}$

| $\theta$ | 0 | $\dfrac{\pi}{4}$ | $\dfrac{\pi}{2}$ | $\dfrac{3\pi}{4}$ | $\pi$ | $\dfrac{5\pi}{4}$ | $\dfrac{3\pi}{2}$ |
|---|---|---|---|---|---|---|---|
| $r$ | 0 | $\dfrac{\pi}{2}$ | $\pi$ | $\dfrac{3\pi}{2}$ | $2\pi$ | $\dfrac{5\pi}{2}$ | $3\pi$ |

Tangent at the pole: $\theta = 0$

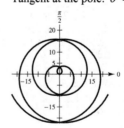

**87.** $r^2 = 4\cos(2\theta)$

$r = 2\sqrt{\cos 2\theta}, \quad 0 \le \theta \le 2\pi$

Lemniscate

Symmetric to the polar axis, $\theta = \dfrac{\pi}{2}$, and pole

Relative extrema: $(\pm 2, 0)$

| $\theta$ | $0$ | $\dfrac{\pi}{6}$ | $\dfrac{\pi}{4}$ |
|---|---|---|---|
| $r$ | $\pm 2$ | $\pm\sqrt{2}$ | $0$ |

Tangents at the pole: $\theta = \dfrac{\pi}{4}, \dfrac{3\pi}{4}$

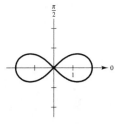

**89.** Because

$$r = 2 - \sec\theta = 2 - \frac{1}{\cos\theta},$$

the graph has polar axis symmetry and the tangents at the pole are

$$\theta = \frac{\pi}{3}, -\frac{\pi}{3}.$$

Furthermore,

$$r \Rightarrow -\infty \text{ as } \theta \Rightarrow \frac{\pi}{2^-}$$

$$r \Rightarrow \infty \text{ as } \theta \Rightarrow -\frac{\pi}{2^+}.$$

Also,

$$r = 2 - \frac{1}{\cos\theta}$$

$$= 2 - \frac{r}{r\cos\theta} = 2 - \frac{r}{x}$$

$$rx = 2x - r$$

$$r = \frac{2x}{1+x}.$$

So, $r \Rightarrow \pm\infty$ as $x \Rightarrow -1$.

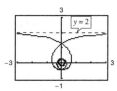

**91.** $r = \dfrac{2}{\theta}$

Hyperbolic spiral

$r \Rightarrow \infty$ as $\theta \Rightarrow 0$

$$r = \frac{2}{\theta} \Rightarrow \theta = \frac{2}{r} = \frac{2\sin\theta}{r\sin\theta} = \frac{2\sin\theta}{y}$$

$$y = \frac{2\sin\theta}{\theta}$$

$$\lim_{\theta\to 0}\frac{2\sin\theta}{\theta} = \lim_{\theta\to 0}\frac{2\cos\theta}{1}$$

$$= 2$$

**93.** The rectangular coordinate system consists of all points of the form $(x, y)$ where $x$ is the directed distance from the $y$-axis to the point, and $y$ is the directed distance from the $x$-axis to the point.

Every point has a unique representation.

The polar coordinate system uses $(r, \theta)$ to designate the location of a point.

$r$ is the directed distance to the origin and $\theta$ is the angle the point makes with the positive $x$-axis, measured counterclockwise.

Points do not have a unique polar representation.

**95.** Slope of tangent line to graph of $r = f(\theta)$ at $(r, \theta)$ is

$$\frac{dy}{dx} = \frac{f(\theta)\cos\theta + f'(\theta)\sin\theta}{-f(\theta)\sin\theta + f'(\theta)\cos\theta}.$$

If $f(\alpha) = 0$ and $f'(\alpha) \ne 0$, then $\theta = \alpha$ is tangent at the pole.

**97.** $r = 4\sin\theta$

(a) $0 \le \theta \le \dfrac{\pi}{2}$

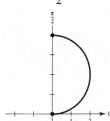

(b) $\dfrac{\pi}{2} \le \theta \le \pi$

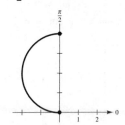

(c) $-\dfrac{\pi}{2} \le \theta \le \dfrac{\pi}{2}$

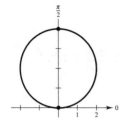

**99.** Let the curve $r = f(\theta)$ be rotated by $\phi$ to form the curve $r = g(\theta)$. If $(r_1, \theta_1)$ is a point on $r = f(\theta)$, then $(r_1, \theta_1 + \phi)$ is on $r = g(\theta)$. That is,

$$g(\theta_1 + \phi) = r_1 = f(\theta_1).$$

Letting $\theta = \theta_1 + \phi$, or $\theta_1 = \theta - \phi$, you see that

$$g(\theta) = g(\theta_1 + \phi) = f(\theta_1) = f(\theta - \phi).$$

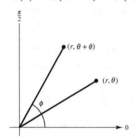

**101.** $r = 2 - \sin \theta$

(a) $r = 2 - \sin\left(\theta - \dfrac{\pi}{4}\right) = 2 - \dfrac{\sqrt{2}}{2}(\sin \theta - \cos \theta)$

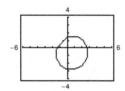

(b) $r = 2 - \sin\left(\theta - \dfrac{\pi}{2}\right) = 2 - (-\cos \theta) = 2 + \cos \theta$

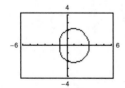

(c) $r = 2 - \sin(\theta - \pi) = 2 - (-\sin \theta) = 2 + \sin \theta$

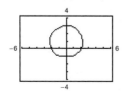

(d) $r = 2 - \sin\left(\theta - \dfrac{3\pi}{2}\right) = 2 - \cos \theta$

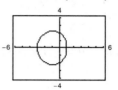

**103.** (a) $r = 1 - \sin \theta$

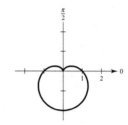

(b) $r = 1 - \sin\left(\theta - \dfrac{\pi}{4}\right)$

Rotate the graph of $r = 1 - \sin \theta$ through the angle $\pi/4$.

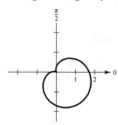

**105.** $\tan \psi = \dfrac{r}{dr/d\theta} = \dfrac{2(1 - \cos \theta)}{2 \sin \theta}$

At $\theta = \pi$, $\tan \psi$ is undefined $\Rightarrow \psi = \dfrac{\pi}{2}$.

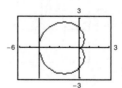

**107.** $r = 2 \cos 3\theta$

$$\tan \psi = \frac{r}{dr/d\theta} = \frac{2 \cos 3\theta}{-6 \sin 3\theta} = -\frac{1}{3} \cot 3\theta$$

At $\theta = \dfrac{\pi}{4}$, $\tan \psi = -\dfrac{1}{3} \cot\left(\dfrac{3\pi}{4}\right) = \dfrac{1}{3}$.

$$\psi = \arctan\left(\frac{1}{3}\right) \approx 18.4°$$

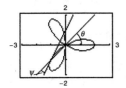

**109.** $r = \dfrac{6}{1 - \cos \theta} = 6(1 - \cos \theta)^{-1} \Rightarrow \dfrac{dr}{d\theta} = \dfrac{6 \sin \theta}{(1 - \cos \theta)^2}$

$$\tan \psi = \frac{r}{\dfrac{dr}{d\theta}} = \frac{\dfrac{6}{(1 - \cos \theta)}}{\dfrac{-6 \sin \theta}{(1 - \cos \theta)^2}} = \frac{1 - \cos \theta}{-\sin \theta}$$

At $\theta = \dfrac{2\pi}{3}$, $\tan \psi = \dfrac{1 - \left(-\dfrac{1}{2}\right)}{-\dfrac{\sqrt{3}}{2}} = -\sqrt{3}$.

$$\psi = \frac{\pi}{3}, (60°)$$

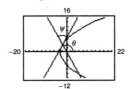

**111.** True

**113.** True

# Section 10.5   Area and Arc Length in Polar Coordinates

**1.** $A = \frac{1}{2} \int_\alpha^\beta \left[f(\theta)\right]^2 d\theta$

$= \frac{1}{2} \int_0^{\pi/2} \left[4 \sin \theta\right]^2 d\theta = 8 \int_0^{\pi/2} \sin^2 \theta \, d\theta$

**3.** $A = \frac{1}{2} \int_\alpha^\beta \left[f(\theta)\right]^2 d\theta = \frac{1}{2} \int_{\pi/2}^{3\pi/2} \left[3 - 2 \sin \theta\right]^2 d\theta$

**5.** $A = \frac{1}{2} \int_0^\pi \left[6 \sin \theta\right]^2 d\theta$

$= 18 \int_0^\pi \dfrac{1 - \cos 2\theta}{2} d\theta = 9\left[\theta - \dfrac{\sin 2\theta}{2}\right]_0^\pi = 9\pi$

Note: $r = 6 \sin \theta$ is circle of radius 3, $0 \le \theta \le \pi$.

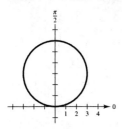

**7.** $A = 2\left[\dfrac{1}{2} \int_0^{\pi/6} (2 \cos 3\theta)^2 d\theta\right] = 2\left[\theta + \dfrac{1}{6} \sin 6\theta\right]_0^{\pi/6} = \dfrac{\pi}{3}$

**9.** $A = \dfrac{1}{2} \int_0^{\pi/2} \left[\sin 2\theta\right]^2 d\theta$

$= \dfrac{1}{2} \int_0^{\pi/2} \dfrac{1 - \cos 4\theta}{2} d\theta$

$= \dfrac{1}{4}\left[\theta - \dfrac{\sin 4\theta}{4}\right]_0^{\pi/2}$

$= \dfrac{1}{4}\left[\dfrac{\pi}{2}\right] = \dfrac{\pi}{8}$

**11.** $A = 2\left[\dfrac{1}{2} \int_{-\pi/2}^{\pi/2} (1 - \sin \theta)^2 d\theta\right]$

$= \left[\dfrac{3}{2}\theta + 2 \cos \theta - \dfrac{1}{4} \sin 2\theta\right]_{-\pi/2}^{\pi/2} = \dfrac{3\pi}{2}$

**13.** $A = \dfrac{1}{2} \int_0^{2\pi} \left[5 + 2 \sin \theta\right]^2 d\theta$

$= \dfrac{1}{2} \int_0^{2\pi} \left[25 + 20 \sin \theta + 4 \sin^2 \theta\right] d\theta$

$= \dfrac{1}{2} \int_0^{2\pi} \left[25 + 20 \sin \theta + 2(1 - \cos 2\theta)\right] d\theta$

$= \dfrac{1}{2}\left[27\theta - 20 \cos \theta - \sin 2\theta\right]_0^{2\pi}$

$= \dfrac{1}{2}\left[27(2\pi)\right] = 27\pi$

**15.** On the interval $-\dfrac{\pi}{4} \le \theta \le 0, r = 2\sqrt{\cos 2\theta}$ traces out one-half of one leaf of the lemniscate. So,

$$A = 4\frac{1}{2}\int_{-\pi/4}^{0} 4 \cos 2\theta \, d\theta$$

$$= 8\left[\frac{\sin 2\theta}{2}\right]_{-\pi/4}^{0} = 8\left[\frac{1}{2}\right] = 4.$$

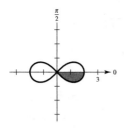

**17.** $A = \left[2\dfrac{1}{2}\displaystyle\int_{2\pi/3}^{\pi} \left(1 + 2 \cos \theta\right)^2 \, d\theta\right]$

$$= \left[3\theta + 4 \sin \theta + \sin 2\theta\right]_{2\pi/3}^{\pi} = \frac{2\pi - 3\sqrt{3}}{2}$$

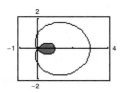

**21.** The area inside the outer loop is

$$2\left[\frac{1}{2}\int_{0}^{2\pi/3} \left(1 + 2 \cos \theta\right)^2 \, d\theta\right] = \left[3\theta + 4 \sin \theta + \sin 2\theta\right]_{0}^{2\pi/3}$$

$$= \frac{4\pi + 3\sqrt{3}}{2}.$$

From the result of Exercise 17, the area between the loops is

$$A = \left(\frac{4\pi + 3\sqrt{3}}{2}\right) - \left(\frac{2\pi - 3\sqrt{3}}{2}\right) = \pi + 3\sqrt{3}.$$

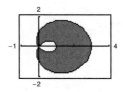

**19.** The inner loop of $r = 1 + 2 \sin \theta$ is traced out on the interval $\dfrac{7\pi}{6} \le \theta \le \dfrac{11\pi}{6}$. So,

$$A = \frac{1}{2}\int_{7\pi/6}^{11\pi/6} \left[1 + 2 \sin \theta\right]^2 \, d\theta$$

$$= \frac{1}{2}\int_{7\pi/6}^{11\pi/6} \left[1 + 4 \sin \theta + 4 \sin^2 \theta\right] d\theta$$

$$= \frac{1}{2}\int_{7\pi/6}^{11\pi/6} \left[1 + 4 \sin \theta + 2\left(1 - \cos 2\theta\right)\right] d\theta$$

$$= \frac{1}{2}\left[3\theta - 4 \cos \theta - \sin 2\theta\right]_{7\pi/6}^{11\pi/6}$$

$$= \frac{1}{2}\left[\left(\frac{11\pi}{2} - 2\sqrt{3} + \frac{\sqrt{3}}{2}\right) - \left(\frac{7\pi}{2} + 2\sqrt{3} - \frac{\sqrt{3}}{2}\right)\right]$$

$$= \frac{1}{2}\left[2\pi - 3\sqrt{3}\right].$$

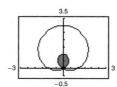

**23.** The area inside the outer loop is

$$A = 2 \cdot \frac{1}{2} \int_{5\pi/6}^{3\pi/2} [3 - 6\sin\theta]^2 \, d\theta$$

$$= \int_{5\pi/6}^{3\pi/2} \left[ 9 - 36\sin\theta + 36\sin^2\theta \right] d\theta$$

$$= \int_{5\pi/6}^{3\pi/2} \left[ 9 - 36\sin\theta + 18(1 - \cos2\theta) \right] d\theta$$

$$= \left[ 27\theta + 36\cos\theta - 9\sin2\theta \right]_{5\pi/6}^{3\pi/2} = \left[ \frac{81\pi}{2} - \left( \frac{45\pi}{2} - 18\sqrt{3} + \frac{9\sqrt{3}}{2} \right) \right] = 18\pi + \frac{27\sqrt{3}}{2}.$$

The area inside the inner loop is

$$A = 2 \cdot \frac{1}{2} \int_{\pi/6}^{\pi/2} [3 - 6\sin\theta]^2 \, d\theta$$

$$= \left[ 27\theta + 36\cos\theta - 9\sin2\theta \right]_{\pi/6}^{\pi/2} = \left[ \frac{27\pi}{2} - \left( \frac{9\pi}{2} + 18\sqrt{3} - \frac{9\sqrt{3}}{2} \right) \right] = 9\pi - \frac{27\sqrt{3}}{2}.$$

Finally, the area between the loops is

$$\left[ 18\pi + \frac{27\sqrt{3}}{2} \right] - \left[ 9\pi - \frac{27\sqrt{3}}{2} \right] = 9\pi + 27\sqrt{3}.$$

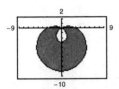

**25.** $r = 1 + \cos\theta$

$r = 1 - \cos\theta$

Solving simultaneously,

$1 + \cos\theta = 1 - \cos\theta$

$2\cos\theta = 0$

$$\theta = \frac{\pi}{2}, \frac{3\pi}{2}.$$

Replacing $r$ by $-r$ and $\theta$ by $\theta + \pi$ in the first equation and solving, $-1 + \cos\theta = 1 - \cos\theta$, $\cos\theta = 1$,

$\theta = 0$. Both curves pass through the pole, $(0, \pi)$, and

$(0, 0)$, respectively.

Points of intersection: $\left( 1, \dfrac{\pi}{2} \right), \left( 1, \dfrac{3\pi}{2} \right), (0, 0)$

**27.** $r = 1 + \cos\theta$

$r = 1 - \sin\theta$

Solving simultaneously,

$1 + \cos\theta = 1 - \sin\theta$

$\cos\theta = -\sin\theta$

$\tan\theta = -1$

$$\theta = \frac{3\pi}{4}, \frac{7\pi}{4}.$$

Replacing $r$ by $-r$ and $\theta$ by $\theta + \pi$ in the first equation and solving, $-1 + \cos\theta = 1 - \sin\theta$,

$\sin\theta + \cos\theta = 2$, which has no solution. Both curves pass through the pole, $(0, \pi)$, and $(0, \pi/2)$, respectively.

Points of intersection:

$\left( \dfrac{2 - \sqrt{2}}{2}, \dfrac{3\pi}{4} \right), \left( \dfrac{2 + \sqrt{2}}{2}, \dfrac{7\pi}{4} \right), (0, 0)$

**29.** $r = 4 - 5 \sin \theta$

$r = 3 \sin \theta$

Solving simultaneously,

$4 - 5 \sin \theta = 3 \sin \theta$

$\sin \theta = \dfrac{1}{2}$

$\theta = \dfrac{\pi}{6}, \dfrac{5\pi}{6}.$

Both curves pass through the pole, $\left(0, \arcsin 4/5\right)$, and $\left(0, 0\right)$, respectively.

Points of intersection: $\left(\dfrac{3}{2}, \dfrac{\pi}{6}\right), \left(\dfrac{3}{2}, \dfrac{5\pi}{6}\right), \left(0, 0\right)$

**31.** $r = \dfrac{\theta}{2}$

$r = 2$

Solving simultaneously, you have

$\theta/2 = 2, \theta = 4.$

Points of intersection:

$\left(2, 4\right), \left(-2, -4\right)$

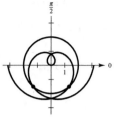

**33.** $r = \cos \theta$

$r = 2 - 3 \sin \theta$

Points of intersection:

$\left(0, 0\right), \left(0.935, 0.363\right), \left(0.535, -1.006\right)$

The graphs reach the pole at different times ($\theta$ values).

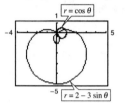

**35.** The points of intersection for one petal are $\left(2, \pi/12\right)$ and $\left(2, 5\pi/12\right)$. The area within one petal is

$$A = \frac{1}{2} \int_0^{\pi/12} \left(4 \sin 2\theta\right)^2 d\theta + \frac{1}{2} \int_{\pi/12}^{5\pi/12} \left(2\right)^2 d\theta + \frac{1}{2} \int_{5\pi/12}^{\pi/2} \left(4 \sin 2\theta\right)^2 d\theta$$

$$= 16 \int_0^{\pi/12} \sin^2\left(2\theta\right) d\theta + 2 \int_{\pi/12}^{5\pi/12} d\theta \text{ (by symmetry of the petal)}$$

$$= 8\left[\theta - \frac{1}{4} \sin 4\theta\right]_0^{\pi/12} + \left[2\theta\right]_{\pi/12}^{5\pi/12} = \frac{4\pi}{3} - \sqrt{3}.$$

$$\text{Total area } = 4\left(\frac{4\pi}{3} - \sqrt{3}\right) = \frac{16\pi}{3} - 4\sqrt{3} = \frac{4}{3}\left(4\pi - 3\sqrt{3}\right)$$

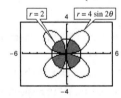

**37.** $A = 4\left[\dfrac{1}{2} \displaystyle\int_0^{\pi/2} \left(3 - 2 \sin \theta\right)^2 d\theta\right]$

$= 2\left[11\theta + 12 \cos \theta - \sin\left(2\theta\right)\right]_0^{\pi/2} = 11\pi - 24$

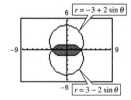

**39.** $A = 2\left[\dfrac{1}{2} \displaystyle\int_0^{\pi/6} \left(4 \sin \theta\right)^2 d\theta + \dfrac{1}{2} \displaystyle\int_{\pi/6}^{\pi/2} \left(2\right)^2 d\theta\right]$

$= 16\left[\dfrac{1}{2}\theta - \dfrac{1}{4} \sin\left(2\theta\right)\right]_0^{\pi/6} + \left[4\theta\right]_{\pi/6}^{\pi/2}$

$= \dfrac{8\pi}{3} - 2\sqrt{3} = \dfrac{2}{3}\left(4\pi - 3\sqrt{3}\right)$

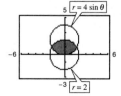

**41.** $r = 2 \cos \theta = 1 \Rightarrow \theta = \pi/3$

$$A = 2 \cdot \frac{1}{2} \int_0^{\pi/3} \left( [2 \cos \theta]^2 - 1 \right) d\theta = \int_0^{\pi/3} \left[ 2(1 + \cos 2\theta) - 1 \right] d\theta = \left[ \theta + \sin 2\theta \right]_0^{\pi/3} = \frac{\pi}{3} + \frac{\sqrt{3}}{2}$$

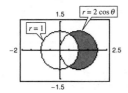

**43.** $A = 2 \left[ \frac{1}{2} \int_0^{\pi} \left[ a(1 + \cos \theta) \right]^2 d\theta \right] - \frac{a^2 \pi}{4} = a^2 \left[ \frac{3}{2} \theta + 2 \sin \theta + \frac{\sin 2\theta}{4} \right]_0^{\pi} - \frac{a^2 \pi}{4} = \frac{3 a^2 \pi}{2} - \frac{a^2 \pi}{4} = \frac{5 a^2 \pi}{4}$

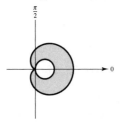

**45.** $A = \dfrac{\pi a^2}{8} + \dfrac{1}{2} \int_{\pi/2}^{\pi} \left[ a(1 + \cos \theta) \right]^2 d\theta$

$\quad\quad = \dfrac{\pi a^2}{8} + \dfrac{a^2}{2} \int_{\pi/2}^{\pi} \left( \dfrac{3}{2} + 2 \cos \theta + \dfrac{\cos 2\theta}{2} \right) d\theta$

$\quad\quad = \dfrac{\pi a^2}{8} + \dfrac{a^2}{2} \left[ \dfrac{3}{2} \theta + 2 \sin \theta + \dfrac{\sin 2\theta}{4} \right]_{\pi/2}^{\pi}$

$\quad\quad = \dfrac{\pi a^2}{8} + \dfrac{a^2}{2} \left[ \dfrac{3\pi}{2} - \dfrac{3\pi}{4} - 2 \right] = \dfrac{a^2}{2} [\pi - 2]$

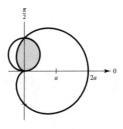

**47.** (a) $r = a \cos^2 \theta$

$\quad\quad r^3 = a r^2 \cos^2 \theta$

$\quad\quad \left( x^2 + y^2 \right)^{3/2} = a x^2$

(b)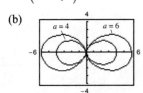

(c) $A = 4 \left( \dfrac{1}{2} \right) \int_0^{\pi/2} \left[ \left( 6 \cos^2 \theta \right)^2 - \left( 4 \cos^2 \theta \right)^2 \right] d\theta$

$\quad\quad = 40 \int_0^{\pi/2} \cos^4 \theta \, d\theta$

$\quad\quad = 10 \int_0^{\pi/2} \left( 1 + \cos 2\theta \right)^2 d\theta$

$\quad\quad = 10 \int_0^{\pi/2} \left( 1 + 2 \cos 2\theta + \dfrac{1 - \cos 4\theta}{2} \right) d\theta$

$\quad\quad = 10 \left[ \dfrac{3}{2} \theta + \sin 2\theta + \dfrac{1}{8} \sin 4\theta \right]_0^{\pi/2} = \dfrac{15\pi}{2}$

**49.** $r = a \cos(n\theta)$

For $n = 1$:

$r = a \cos \theta$

$$A = \pi \left( \frac{a}{2} \right)^2 = \frac{\pi a^2}{4}$$

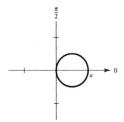

For $n = 2$:

$r = a \cos 2\theta$

$$A = 8 \left( \frac{1}{2} \right) \int_0^{\pi/4} (a \cos 2\theta)^2 \, d\theta = \frac{\pi a^2}{2}$$

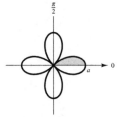

For $n = 3$:

$r = a \cos 3\theta$

$$A = 6 \left( \frac{1}{2} \right) \int_0^{\pi/6} (a \cos 3\theta)^2 \, d\theta = \frac{\pi a^2}{4}$$

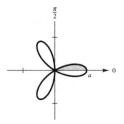

For $n = 4$:

$r = a \cos 4\theta$

$$A = 16 \left( \frac{1}{2} \right) \int_0^{\pi/8} (a \cos 4\theta)^2 \, d\theta = \frac{\pi a^2}{2}$$

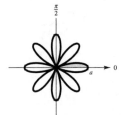

In general, the area of the region enclosed by
$r = a \cos(n\theta)$ for $n = 1, 2, 3, \ldots$ is $(\pi a^2)/4$ if $n$ is odd
and is $(\pi a^2)/2$ if $n$ is even.

**51.** $r = 8, r^1 = 0$

$$s = \int_0^{2\pi} \sqrt{8^2 + 0^2} \, d\theta = 8\theta \Big]_0^{2\pi} = 16\pi$$

(circumference of circle of radius 8)

**53.** $r = 4 \sin \theta$

$r' = 4 \cos \theta$

$$s = \int_0^{\pi} \sqrt{(4 \sin \theta)^2 + (4 \cos \theta)^2} \, d\theta$$

$$= \int_0^{\pi} 4 \, d\theta = [4\theta]_0^{\pi} = 4\pi$$

(circumference of circle of radius 2)

**55.** $r = 1 + \sin \theta$

$r' = \cos \theta$

$$s = 2 \int_{\pi/2}^{3\pi/2} \sqrt{(1 + \sin \theta)^2 + (\cos \theta)^2} \, d\theta$$

$$= 2\sqrt{2} \int_{\pi/2}^{3\pi/2} \sqrt{1 + \sin \theta} \, d\theta$$

$$= 2\sqrt{2} \int_{\pi/2}^{3\pi/2} \frac{-\cos \theta}{\sqrt{1 - \sin \theta}} \, d\theta$$

$$= \left[ 4\sqrt{2} \sqrt{1 - \sin \theta} \right]_{\pi/2}^{3\pi/2}$$

$$= 4\sqrt{2} \left( \sqrt{2} - 0 \right) = 8$$

**57.** $r = 2\theta, 0 \le \theta \le \dfrac{\pi}{2}$

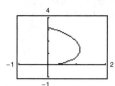

Length $\approx 4.16$

**59.** $r = \dfrac{1}{\theta}, \pi \le \theta \le 2\pi$

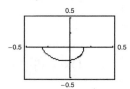

Length $\approx 0.71$

**61.** $r = \sin(3 \cos \theta), 0 \le \theta \le \pi$

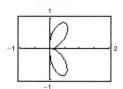

Length $\approx 4.39$

**63.** $r = 6 \cos \theta$

$r' = -6 \sin \theta$

$S = 2\pi \int_0^{\pi/2} 6 \cos \theta \sin \theta \sqrt{36 \cos^2 \theta + 36 \sin^2 \theta} \, d\theta$

$\qquad = 72\pi \int_0^{\pi/2} \sin \theta \cos \theta \, d\theta$

$\qquad = \left[ 36\pi \sin^2 \theta \right]_0^{\pi/2}$

$\qquad = 36\pi$

**65.** $r = e^{a\theta}$

$r' = ae^{a\theta}$

$S = 2\pi \int_0^{\pi/2} e^{a\theta} \cos \theta \sqrt{\left(e^{a\theta}\right)^2 + \left(ae^{a\theta}\right)^2} \, d\theta$

$\qquad = 2\pi \sqrt{1 + a^2} \int_0^{\pi/2} e^{2a\theta} \cos \theta \, d\theta$

$\qquad = 2\pi \sqrt{1 + a^2} \left[ \dfrac{e^{2a\theta}}{4a^2 + 1} (2a \cos \theta + \sin \theta) \right]_0^{\pi/2}$

$\qquad = \dfrac{2\pi \sqrt{1 + a^2}}{4a^2 + 1} \left( e^{\pi a} - 2a \right)$

**67.** $r = 4 \cos 2\theta$

$r' = -8 \sin 2\theta$

$S = 2\pi \int_0^{\pi/4} 4 \cos 2\theta \sin \theta \sqrt{16 \cos^2 2\theta + 64 \sin^2 \theta \, 2\theta} \, d\theta = 32\pi \int_0^{\pi/4} \cos 2\theta \sin \theta \sqrt{\cos^2 2\theta + 4 \sin^2 2\theta} \, d\theta \approx 21.87$

**69.** You will only find simultaneous points of intersection. There may be intersection points that do not occur with the same coordinates in the two graphs.

**71.** (a)  $r = 10 \cos \theta, 0 \le \theta < \pi$

Circle of radius 5

Area $= 25\pi$

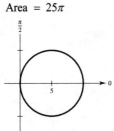

(b)  $r = 5 \sin \theta, 0 \le \theta < \pi$

Circle radius 5/2

Area $= \dfrac{25}{4}\pi$

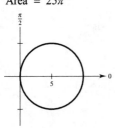

**73.** Revolve $r = 2$ about the line $r = 5 \sec \theta$.

$f(\theta) = 2, f'(\theta) = 0$

$S = 2\pi \int_0^{2\pi} (5 - 2 \cos \theta) \sqrt{2^2 + 0^2} \, d\theta$

$\qquad = 4\pi \int_0^{2\pi} (5 - 2 \cos \theta) \, d\theta$

$\qquad = 4\pi \left[ 5\theta - 2 \sin \theta \right]_0^{2\pi}$

$\qquad = 40\pi^2$

**75.** $r = 8 \cos \theta, 0 \le \theta \le \pi$

(a) $A = \dfrac{1}{2} \int_0^\pi r^2 \, d\theta = \dfrac{1}{2} \int_0^\pi 64 \cos^2 \theta \, d\theta = 32 \int_0^\pi \dfrac{1 + \cos 2\theta}{2} \, d\theta = 16 \left[ \theta + \dfrac{\sin 2\theta}{2} \right]_0^\pi = 16\pi$

$\left( \text{Area circle} = \pi r^2 = \pi 4^2 = 16\pi \right)$

(b)

| $\theta$ | 0.2 | 0.4 | 0.6 | 0.8 | 1.0 | 1.2 | 1.4 |
|---|---|---|---|---|---|---|---|
| $A$ | 6.32 | 12.14 | 17.06 | 20.80 | 23.27 | 24.60 | 25.08 |

(c), (d) For $\dfrac{1}{4}$ of area $(4\pi \approx 12.57)$: 0.42

For $\dfrac{1}{2}$ of area $(8\pi \approx 25.13)$: $1.57 \left( \dfrac{\pi}{2} \right)$

For $\dfrac{3}{4}$ of area $(12\pi \approx 37.70)$: 2.73

(e) No, it does not depend on the radius.

**77.**
$$r = a \sin \theta + b \cos \theta$$
$$r^2 = ar \sin \theta + br \cos \theta$$
$$x^2 + y^2 = ay + bx$$
$$x^2 + y^2 - bx - ay = 0 \text{ represents a circle.}$$

**79.** (a) $r = \theta, \theta \ge 0$

As $a$ increases, the spiral opens more rapidly. If $\theta < 0$, the spiral is reflected about the $y$-axis.

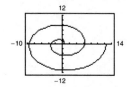

(b) $r = a\theta, \theta \ge 0$, crosses the polar axis for $\theta = n\pi$, $n$ and integer. To see this

$r = a\theta \Rightarrow r \sin \theta = y = a\theta \sin \theta = 0$

for $\theta = n\pi$. The points are $(r, \theta) = (an\pi, n\pi), n = 1, 2, 3, \ldots$

(c) $f(\theta) = \theta, f'(\theta) = 1$

$s = \displaystyle\int_0^{2\pi} \sqrt{\theta^2 + 1} \, d\theta$

$= \dfrac{1}{2} \left[ \ln \left( \sqrt{x^2 + 1} + x \right) + x\sqrt{x^2 + 1} \right]_0^{2\pi}$

$= \dfrac{1}{2} \ln \left( \sqrt{4\pi^2 + 1} + 2\pi \right) + \pi\sqrt{4\pi^2 + 1} \approx 21.2563$

(d) $A = \dfrac{1}{2} \displaystyle\int_\alpha^\beta r^2 \, dr = \dfrac{1}{2} \int_0^{2\pi} \theta^2 \, d\theta = \left[ \dfrac{\theta^3}{6} \right]_0^{2\pi} = \dfrac{4}{3}\pi^3$

**81.** The smaller circle has equation $r = a \cos \theta$. The area of the shaded lune is:

$$A = 2\left(\frac{1}{2}\right)\int_0^{\pi/4}\left[(a\cos\theta)^2 - 1\right]d\theta$$

$$= \int_0^{\pi/4}\left[\frac{a^2}{2}(1 + \cos 2\theta) - 1\right]d\theta$$

$$= \left[\frac{a^2}{2}\left(\theta + \frac{\sin 2\theta}{2}\right) - \theta\right]_0^{\pi/4}$$

$$= \frac{a^2}{2}\left(\frac{\pi}{4} + \frac{1}{2}\right) - \frac{\pi}{4}$$

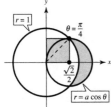

This equals the area of the square, $\left(\frac{\sqrt{2}}{2}\right)^2 = \frac{1}{2}$.

$$\frac{a^2}{2}\left(\frac{\pi}{4} + \frac{1}{2}\right) - \frac{\pi}{4} = \frac{1}{2}$$

$$\pi a^2 + 2a^2 - 2\pi - 4 = 0$$

$$a^2 = \frac{4 + 2\pi}{2 + \pi} = 2$$

$$a = \sqrt{2}$$

Smaller circle: $r = \sqrt{2}\cos\theta$

**83.** False. $f(\theta) = 1$ and $g(\theta) = -1$ have the same graphs.

**85.** In parametric form,

$$s = \int_a^b \sqrt{\left(\frac{dx}{dt}\right)^2 + \left(\frac{dy}{dt}\right)^2}\, dt.$$

Using $\theta$ instead of $t$, you have
$x = r\cos\theta = f(\theta)\cos\theta$ and
$y = r\sin\theta = f(\theta)\sin\theta.$ So,

$$\frac{dx}{d\theta} = f'(\theta)\cos\theta - f(\theta)\sin\theta \text{ and}$$

$$\frac{dy}{d\theta} = f'(\theta)\sin\theta + f(\theta)\cos\theta.$$

It follows that

$$\left(\frac{dx}{d\theta}\right)^2 + \left(\frac{dy}{d\theta}\right)^2 = [f(\theta)]^2 + [f'(\theta)]^2.$$

So, $s = \int_\alpha^\beta \sqrt{[f(\theta)]^2 + [f'(\theta)]^2}\, d\theta.$

# Section 10.6   Polar Equations of Conics and Kepler's Laws

**1.** $r = \dfrac{2e}{1 + e\cos\theta}$

(a) $e = 1, r = \dfrac{2}{1 + \cos\theta}$,   parabola

(b) $e = 0.5,$

$r = \dfrac{1}{1 + 0.5\cos\theta} = \dfrac{2}{2 + \cos\theta}$, ellipse

(c) $e = 1.5,$

$r = \dfrac{3}{1 + 1.5\cos\theta} = \dfrac{6}{2 + 3\cos\theta}$, hyperbola

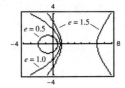

**3.** $r = \dfrac{2e}{1 - e\sin\theta}$

(a) $e = 1, r = \dfrac{2}{1 - \sin\theta}$, parabola

(b) $e = 0.5,$

$r = \dfrac{1}{1 - 0.5\sin\theta} = \dfrac{2}{2 - \sin\theta}$, ellipse

(c) $e = 1.5,$

$r = \dfrac{3}{1 - 1.5\sin\theta} = \dfrac{6}{2 - 3\sin\theta}$, hyperbola

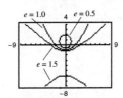

**5.** $r = \dfrac{4}{1 + e \sin \theta}$

(a) The conic is an ellipse. As $e \to 1^-$, the ellipse becomes more elliptical, and as $e \to 0^+$, it becomes more circular.

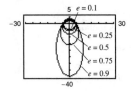

(b) The conic is a parabola.

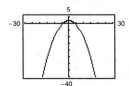

(c) The conic is a hyperbola. As $e \to 1^+$, the hyperbola opens more slowly, and as $e \to \infty$, it opens more rapidly.

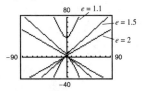

**7.** Parabola; Matches (c)

**8.** Ellipse; Matches (f )

**9.** Hyperbola; Matches (a)

**10.** Parabola; Matches (e)

**11.** Ellipse; Matches (b)

**12.** Hyperbola; Matches (d)

**13.** $r = \dfrac{1}{1 - \cos \theta}$

Parabola because $e = 1, d = 1$

Distance from pole to directrix: $|d| = 1$

Directrix: $x = -d = -1$

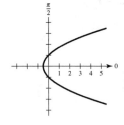

**15.** $r = \dfrac{3}{2 + 6 \sin \theta} = \dfrac{3/2}{1 + 3 \sin \theta}$

Hyperbola because $e = 3 > 0; d = 1/2$

Directrix: $y = 1/2$

Distance from pole to directrix: $|d| = 1/2$

Vertices: $(r, \theta) = (3/8, \pi/2), (-3/4, 3\pi/2)$

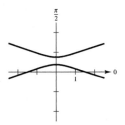

**17.** $r = \dfrac{5}{-1 + 2 \cos \theta} = \dfrac{-5}{1 - 2 \cos \theta}$

Hyperbola because $e = 2 > 1; d = -5/2$

Directrix: $x = 5/2$

Distance from pole to directrix: $|d| = 5/2$

Vertices: $(r, \theta) = (5, 0), (-5/3, \pi)$

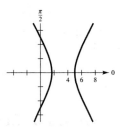

**19.** $r = \dfrac{6}{2 + \cos \theta} = \dfrac{3}{1 + (1/2) \cos \theta}$

Ellipse because $e = \dfrac{1}{2}; d = 6$

Directrix: $x = 6$

Distance from pole to directrix: $|d| = 6$

Vertices: $(r, \theta) = (2, 0), (6, \pi)$

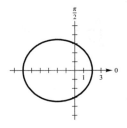

**21.** $r = \dfrac{300}{-12 + 6\sin\theta} = \dfrac{-25}{1 - \frac{1}{2}\sin\theta} = \dfrac{\frac{1}{2}(-50)}{1 - \frac{1}{2}\sin\theta}$

Ellipse because $e = \dfrac{1}{2}$, $d = -50$

Distance from pole to directrix: $|d| = 50$

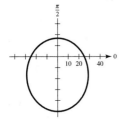

**23.** $r = \dfrac{3}{-4 + 2\sin\theta} = \dfrac{-\frac{3}{4}}{1 - \frac{1}{2}\sin\theta}$

$e = \dfrac{1}{2}$, Ellipse

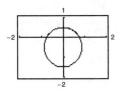

**25.** $r = \dfrac{-10}{1 - \cos\theta}$

$e = 1$,  Parabola

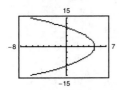

**27.** $r = \dfrac{4}{1 + \cos\left(\theta - \dfrac{\pi}{3}\right)}$

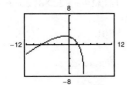

Rotate the graph of $r = \dfrac{4}{1 + \cos\theta}$

$\dfrac{\pi}{3}$ radian counterclockwise.

**29.** $r = \dfrac{6}{2 + \cos\left(\theta + \dfrac{\pi}{6}\right)}$

Rotate the graph of $r = \dfrac{6}{2 + \cos\theta}$

$\dfrac{\pi}{6}$ radian clockwise.

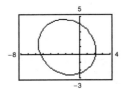

**31.** Change $\theta$ to $\theta + \dfrac{\pi}{6}$

$r = \dfrac{8}{8 + 5\cos\left(\theta + \dfrac{\pi}{6}\right)}$

**33.** Parabola

$e = 1$

$x = -3 \Rightarrow d = 3$

$r = \dfrac{ed}{1 - e\cos\theta} = \dfrac{3}{1 - \cos\theta}$

**35.** Ellipse

$e = \dfrac{1}{2}$, $y = 1$, $d = 1$

$r = \dfrac{ed}{1 + e\sin\theta} = \dfrac{1/2}{1 + (1/2)\sin\theta} = \dfrac{1}{2 + \sin\theta}$

**37.** Hyperbola

$e = 2$, $x = 1$, $d = 1$

$r = \dfrac{ed}{1 + e\cos\theta} = \dfrac{2}{1 + 2\cos\theta}$

**39.** Parabola

Vertex: $\left(1, -\dfrac{\pi}{2}\right)$

$e = 1$, $d = 2$, $r = \dfrac{2}{1 - \sin\theta}$

**41.** Ellipse

Vertices: $(2, 0), (8, \pi)$

$e = \dfrac{3}{5}$, $d = \dfrac{16}{3}$

$r = \dfrac{ed}{1 + e\cos\theta} = \dfrac{16/5}{1 + (3/5)\cos\theta} = \dfrac{16}{5 + 3\cos\theta}$

**43.** Hyperbola

Vertices: $\left(1, \frac{3\pi}{2}\right), \left(9, \frac{3\pi}{2}\right)$

$e = \frac{5}{4}, d = \frac{9}{5}$

$r = \dfrac{ed}{1 - e \sin \theta} = \dfrac{9/4}{1 - (5/4) \sin \theta} = \dfrac{9}{4 - 5 \sin \theta}$

**45.** Ellipse, $e = \dfrac{1}{2}$,

Directrix: $r = 4 \sec \theta \Rightarrow x = r \cos \theta = 4$

$r = \dfrac{ed}{1 + e \cos \theta} = \dfrac{\left(\frac{1}{2}\right)4}{1 + \frac{1}{2} \cos \theta} = \dfrac{4}{2 + \cos \theta}$

**47.** Ellipse if $0 < e < 1$, parabola if $e = 1$, hyperbola if $e > 1$.

**49.** If the foci are fixed and $e \to 0$, then $d \to \infty$. To see this, compare the ellipses

$r = \dfrac{1/2}{1 + (1/2) \cos \theta}, e = 1/2, d = 1$

$r = \dfrac{5/16}{1 + (1/4) \cos \theta}, e = 1/4, d = 5/4.$

**51.**
$$\frac{x^2}{a^2} + \frac{y^2}{b^2} = 1$$
$$x^2 b^2 + y^2 a^2 = a^2 b^2$$
$$b^2 r^2 \cos^2 \theta + a^2 r^2 \sin^2 \theta = a^2 b^2$$
$$r^2 \left[ b^2 \cos^2 \theta + a^2 \left(1 - \cos^2 \theta\right) \right] = a^2 b^2$$
$$r^2 \left[ a^2 + \cos^2 \theta \left(b^2 - a^2\right) \right] = a^2 b^2$$
$$r^2 = \frac{a^2 b^2}{a^2 + \left(b^2 - a^2\right) \cos^2 \theta} = \frac{a^2 b^2}{a^2 - c^2 \cos^2 \theta}$$
$$= \frac{b^2}{1 - (c/a)^2 \cos^2 \theta} = \frac{b^2}{1 - e^2 \cos^2 \theta}$$

**53.** $a = 5, c = 4, e = \dfrac{4}{5}, b = 3$

$r^2 = \dfrac{9}{1 - (16/25) \cos^2 \theta}$

**55.** $a = 3, b = 4, c = 5, e = \dfrac{5}{3}$

$r^2 = \dfrac{-16}{1 - (25/9) \cos^2 \theta}$

**57.** $A = 2 \left[ \dfrac{1}{2} \displaystyle\int_0^{\pi} \left( \dfrac{3}{2 - \cos \theta} \right)^2 d\theta \right]$

$= 9 \displaystyle\int_0^{\pi} \dfrac{1}{(2 - \cos \theta)^2} d\theta \approx 10.88$

**59.** $A = 2 \left[ \dfrac{1}{2} \displaystyle\int_{-\pi/2}^{\pi/2} \left( \dfrac{2}{3 - 2 \sin \theta} \right)^2 d\theta \right]$

$= 4 \displaystyle\int_{-\pi/2}^{\pi/2} \dfrac{1}{(3 - 2 \sin \theta)^2} d\theta \approx 3.37$

**61.** Vertices: $(123{,}000 + 4000, 0) = (127{,}000, 0)$

$(119 + 4000, \pi) = (4119, \pi)$

$a = \dfrac{127{,}000 + 4119}{2} = 65{,}559.5$

$c = 65{,}559.5 - 4119 = 61{,}440.5$

$e = \dfrac{c}{a} = \dfrac{122{,}881}{131{,}119} \approx 0.93717$

$r = \dfrac{ed}{1 - e \cos \theta}$

$\theta = 0: r = \dfrac{ed}{1 - e}, \theta = \pi: r = \dfrac{ed}{1 + e}$

$2a = 2(65{,}559.5) = \dfrac{ed}{1 - e} + \dfrac{ed}{1 + e}$

$131{,}119 = d\left( \dfrac{e}{1 - e} + \dfrac{e}{1 + e} \right) = d\left( \dfrac{2e}{1 - e^2} \right)$

$d = \dfrac{131{,}119\left(1 - e^2\right)}{2e} \approx 8514.1397$

$r = \dfrac{7979.21}{1 - 0.93717 \cos \theta} = \dfrac{1{,}046{,}226{,}000}{131{,}119 - 122{,}881 \cos \theta}$

When $\theta = 60° = \dfrac{\pi}{3}, r \approx 15{,}015.$

Distance between earth and the satellite is $r - 4000 \approx 11{,}015$ miles.

**63.** $a = 1.496 \times 10^8, e = 0.0167$

$$r = \frac{(1 - e^2)a}{1 - e \cos \theta} = \frac{149,558,278.1}{1 - 0.0167 \cos \theta}$$

Perihelion distance: $a(1 - e) \approx 147,101,680$ km

Aphelion distance: $a(1 + e) \approx 152,098,320$ km

**65.** $a = 4.498 \times 10^9, e = 0.0086$

$$r = \frac{(1 - e^2)a}{1 - e \cos \theta} = \frac{4,497,667,328}{1 - 0.0086 \cos \theta}$$

Perihelion distance: $a(1 - e) \approx 4,459,317,200$ km

Aphelion distance: $a(1 + e) \approx 4,536,682,800$ km

**67.** $r = \dfrac{4.498 \times 10^9}{1 - 0.0086 \cos \theta}$

(a) $A = \dfrac{1}{2} \displaystyle\int_0^{\pi/9} r^2 \, d\theta \approx 3.591 \times 10^{18}$ km$^2$

$$165 \left[ \frac{\dfrac{1}{2} \displaystyle\int_0^{\pi/2} r^2 \, d\theta}{\dfrac{1}{2} \displaystyle\int_0^{2\pi} r^2 \, d\theta} \right] \approx 9.322 \text{ yrs}$$

(b) $\dfrac{1}{2} \displaystyle\int_\pi^\alpha r^2 \, d\theta = 3.591 \times 10^{18}$

By trial and error, $\alpha \approx \pi + 0.361$

$0.361 > \pi/9 \approx 0.349$ because the rays in part (a) are longer than those in part (b)

(c) For part (a),

$$s = \int_0^{\pi/9} \sqrt{r^2 + (dr/d\theta)^2} \approx 1.583 \times 10^9 \text{ km}$$

Average per year $= \dfrac{1.583 \times 10^9}{9.322} \approx 1.698 \times 10^8$ km/yr

For part (b),

$$s = \int_\pi^{\pi + 0.361} \sqrt{r^2 + (dr/d\theta)^2} \, d\theta \approx 1.610 \times 10^9 \text{ km}$$

Average per year $= \dfrac{1.610 \times 10^9}{9.322} \approx 1.727 \times 10^8$ km/yr

**69.** $r_1 = a + c, r_0 = a - c, r_1 - r_0 = 2c, r_1 + r_0 = 2a$

$$e = \frac{c}{a} = \frac{r_1 - r_0}{r_1 + r_0}$$

$$\frac{1 + e}{1 - e} = \frac{1 + \dfrac{c}{a}}{1 - \dfrac{c}{a}} = \frac{a + c}{a - c} = \frac{r_1}{r_0}$$

# Review Exercises for Chapter 10

**1.** $4x^2 + y^2 = 4$

Ellipse

Vertex: $(1, 0)$.

Matches (e)

**2.** $4x^2 - y^2 = 4$

Hyperbola

Vertex: $(1, 0)$

Matches (c)

**3.** $y^2 = -4x$

Parabola opening to left.

Matches (b)

**4.** $y^2 - 4x^2 = 4$

Hyperbola

Vertex: $(0, 2)$

Matches (d)

**5.** $x^2 + 4y^2 = 4$

Ellipse

Vertex: $(0, 1)$

Matches (a)

**6.** $x^2 = 4y$

Parabola opening upward.

Matches (f)

**7.** $16x^2 + 16y^2 - 16x + 24y - 3 = 0$

$$\left(x^2 - x + \tfrac{1}{4}\right) + \left(y^2 + \tfrac{3}{2}y + \tfrac{9}{16}\right) = \tfrac{3}{16} + \tfrac{1}{4} + \tfrac{9}{16}$$

$$\left(x - \tfrac{1}{2}\right)^2 + \left(y + \tfrac{3}{4}\right)^2 = 1$$

Circle

Center: $\left(\tfrac{1}{2}, -\tfrac{3}{4}\right)$

Radius: 1

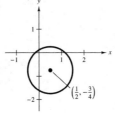

**9.** $3x^2 - 2y^2 + 24x + 12y + 24 = 0$

$$3\left(x^2 + 8x + 16\right) - 2\left(y^2 - 6y + 9\right) = -24 + 48 - 18$$

$$\frac{(x + 4)^2}{2} - \frac{(y - 3)^2}{3} = 1$$

Hyperbola

Center: $(-4, 3)$

Vertices: $\left(-4 \pm \sqrt{2}, 3\right)$

Foci: $\left(-4 \pm \sqrt{5}, 3\right)$

Eccentricity: $\dfrac{\sqrt{10}}{2}$

Asymptotes:

$$y = 3 \pm \sqrt{\tfrac{3}{2}}(x + 4)$$

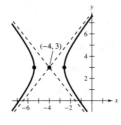

**11.** $3x^2 + 2y^2 - 12x + 12y + 29 = 0$

$$3\left(x^2 - 4x + 4\right) + 2\left(y^2 + 6y + 9\right) = -29 + 12 + 18$$

$$\frac{(x - 2)^2}{1/3} + \frac{(y + 3)^2}{1/2} = 1$$

Ellipse

Center: $(2, -3)$

Vertices: $\left(2, -3 \pm \dfrac{\sqrt{2}}{2}\right)$

Foci: $\left(2, -\dfrac{17}{6}\right), \left(2, -\dfrac{19}{6}\right)$

Eccentricity: $\dfrac{\sqrt{3}}{3}$

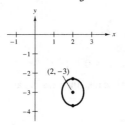

**13.** $x^2 - 6x - 8y + 1 = 0$

$$x^2 - 6x + 9 = 8y - 1 + 9$$

$$(x - 3)^2 = 8y + 8$$

$$(x - 3)^2 = 4(2)(y + 1)$$

Parabola

Vertex: $(3, -1)$

Directrix: $y = -2 - 1 = -3$

Focus: $(3, 1)$

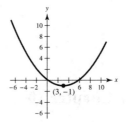

**15.** Vertex: $(0, 2)$

Directrix: $x = -3$

Parabola opens to the right.

$p = 3$

$(y - 2)^2 = 4(3)(x - 0)$

$y^2 - 4y - 12x + 4 = 0$

**17.** Center: $(0, 0)$

Vertices: $(7, 0), (-7, 0)$

Foci: $(5, 0), (-5, 0)$

Horizontal major axis

$a = 7, c = 5, b = \sqrt{49 - 25} = \sqrt{24} = 2\sqrt{6}$

$\dfrac{x^2}{49} + \dfrac{y^2}{24} = 1$

**19.** Vertices: $(3, 1), (3, 7)$

Center: $(3, 4)$

Eccentricity $= \dfrac{2}{3} = \dfrac{c}{a} \Rightarrow a = 3, c = 2$

Vertical major axis

$b = \sqrt{9 - 4} = \sqrt{5}$

$\dfrac{(x - 3)^2}{5} + \dfrac{(y - 4)^2}{9} = 1$

**21.** Vertices: $(0, \pm 8) \Rightarrow a = 8$

Center: $(0, 0)$

Vertical transverse axis

Asymptotes:

$y = \pm 2x \Rightarrow \dfrac{a}{b} = 2 \Rightarrow \dfrac{8}{b} = 2 \Rightarrow b = 4$

$\dfrac{y^2}{64} - \dfrac{x^2}{16} = 1$

**23.** Vertices: $(\pm 7, -1)$

Center: $(0, -1)$

Horizontal transverse axis

Foci: $(\pm 9, -1)$

$a = 7, c = 9, b = \sqrt{81 - 49} = \sqrt{32} = 4\sqrt{2}$

$\dfrac{x^2}{49} - \dfrac{(y + 1)^2}{32} = 1$

**25.** $y = \dfrac{1}{200}x^2$

(a) $x^2 = 200y$

$x^2 = 4(50)y$

Focus: $(0, 50)$

(b)
$y = \dfrac{1}{200}x^2$

$y' = \dfrac{1}{100}x$

$\sqrt{1 + (y')^2} = \sqrt{1 + \dfrac{x^2}{10{,}000}}$

$S = 2\pi \displaystyle\int_0^{100} x\sqrt{1 + \dfrac{x^2}{10{,}000}}\, dx \approx 38{,}294.49$

**27.** $x = 1 + 8t, \ y = 3 - 4t$

$t = \dfrac{x - 1}{8} \Rightarrow y = 3 - 4\left(\dfrac{x - 1}{8}\right) = \dfrac{7}{2} - \dfrac{x}{2}$

$x + 2y - 7 = 0, \ \text{Line}$

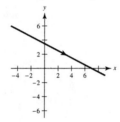

**29.** $x = e^t - 1, \ y = e^{3t}$

$e^t = x + 1 \Rightarrow y = (x + 1)^3, \ x > -1$

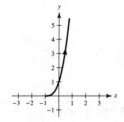

**31.** $x = 6\cos\theta, \ y = 6\sin\theta$

$\left(\dfrac{x}{6}\right)^2 + \left(\dfrac{y}{6}\right)^2 = 1$

$x^2 + y^2 = 36$

Circle

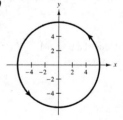

**33.** $x = 2 + \sec\theta, \ y = 3 + \tan\theta$

$$(x - 2)^2 = \sec^2\theta = 1 + \tan^2\theta = 1 + (y - 3)^2$$

$$(x - 2)^2 - (y - 3)^2 = 1$$

Hyperbola

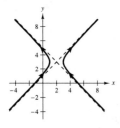

**35.** $y = 4x + 3$

Examples: $x = t, \ y = 4t + 3$

$\qquad\qquad x = t + 1, \ y = 4(t + 1) + 3 = 4t + 7$

**37.** $x = \cos 3\theta + 5\cos\theta$

$\quad\ y = \sin 3\theta + 5\sin\theta$

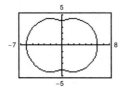

**39.** $x = 2 + 5t, \ y = 1 - 4t$

$$\frac{dy}{dx} = \frac{dy/dt}{dx/dt} = \frac{-4}{5}$$

$$\frac{d^2y}{dx^2} = 0$$

At $t = 3$, the slope is $-\dfrac{4}{5}$. (Line)

Neither concave upward nor downward

**41.** $x = \dfrac{1}{t}, \ y = 2t + 3$

$$\frac{dy}{dx} = \frac{dy/dt}{dx/dt} = \frac{2}{\left(-1/t^2\right)} = -2t^2$$

$$\frac{d^2y}{dx^2} = \frac{\dfrac{d}{dt}\left[-2t^2\right]}{dx/dt} = \frac{-4t}{\left(-1/t^2\right)} = 4t^3$$

At $t = -1$, the slope is $\dfrac{dy}{dx} = -2$ and $\dfrac{d^2y}{dx^2} = -4$.

Concave downward

**43.** $x = 5 + \cos\theta, \ y = 3 + 4\sin\theta$

$$\frac{dy}{dx} = \frac{dy/d\theta}{dx/d\theta} = \frac{4\cos\theta}{-\sin\theta} = -4\cot\theta$$

$$\frac{d^2y}{dx^2} = \frac{\dfrac{d}{d\theta}[-4\cot\theta]}{dx/d\theta} = \frac{4\csc^2\theta}{-\sin\theta} = -4\csc^3\theta$$

At $\theta = \dfrac{\pi}{6}$, the slope is $\dfrac{dy}{dx} = -4\sqrt{3}$ and $\dfrac{d^2y}{dx^2} = -32$.

Concave downward

**45.** $x = \cos^3\theta, \ y = 4\sin^3\theta$

$$\frac{dy}{dx} = \frac{dy/d\theta}{dx/d\theta} = \frac{12\sin^2\theta\cos\theta}{3\cos^2\theta(-\sin\theta)} = -\frac{4\sin\theta}{\cos\theta} = -4\tan\theta$$

$$\frac{d^2y}{dx^2} = \frac{\dfrac{d}{d\theta}[-4\tan\theta]}{dx/d\theta} = \frac{-4\sec^2\theta}{3\cos^2\theta(-\sin\theta)} = \frac{4}{3\cos^4\theta\sin\theta} = \frac{4}{3}\sec^4\theta\csc\theta$$

At $\theta = \dfrac{\pi}{3}$, the slope is $\dfrac{dy}{dx} = -4\sqrt{3}$ and $\dfrac{d^2y}{dx^2} = \dfrac{4}{3\left(\dfrac{1}{16}\right)\left(\dfrac{\sqrt{3}}{2}\right)} = \dfrac{128}{3\sqrt{3}} = \dfrac{128\sqrt{3}}{9}$.

Concave upward

**47.** $x = \cot \theta,\ y = \sin 2\theta,\ \theta = \dfrac{\pi}{6}$

(a), (d)

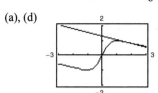

(b) At $\theta = \dfrac{\pi}{6},\ \dfrac{dx}{d\theta} = -4,\ \dfrac{dy}{d\theta} = 1,$ and $\dfrac{dy}{dx} = -\dfrac{1}{4}.$

(c) At $\theta = \dfrac{\pi}{6},\ (x, y) = \left(\sqrt{3}, \dfrac{\sqrt{3}}{2}\right).$

$$y - \frac{\sqrt{3}}{2} = -\frac{1}{4}\left(x - \sqrt{3}\right)$$

$$y = -\frac{1}{4}x + \frac{3\sqrt{3}}{4}$$

**49.** $x = 5 - t,\ y = 2t^2$

$$\frac{dx}{dt} = -1,\ \frac{dy}{dt} = 4t$$

Horizontal tangent at $t = 0:\ (5, 0)$

No vertical tangents

**51.** $x = 2 + 2 \sin \theta,\ y = 1 + \cos \theta$

$$\frac{dx}{d\theta} = 2 \cos \theta,\ \frac{dy}{d\theta} = -\sin \theta$$

$$\frac{dy}{d\theta} = 0 \text{ for } \theta = 0, \pi, 2\pi, \dots$$

Horizontal tangents: $(x, y) = (2, 2), (2, 0)$

$$\frac{dx}{d\theta} = 0 \text{ for } \theta = \frac{\pi}{2}, \frac{3\pi}{2}, \dots$$

Vertical tangents: $(x, y) = (4, 1), (0, 1)$

**53.** $x = t^2 + 1,\ y = 4t^3 + 3,\ 0 \le t \le 2$

$$\frac{dx}{dt} = 2t,\ \frac{dy}{dt} = 12t^2$$

$$s = \int_0^2 \sqrt{(2t)^2 + (12t^2)^2}\, dt$$

$$= \int_0^2 \sqrt{4t^2 + 144t^4}\, dt$$

$$= \int_0^2 2t\sqrt{1 + 36t^2}\, dt$$

$$= \frac{1}{36}\left[\frac{2}{3}\left(1 + 36t^2\right)^{3/2}\right]_0^2$$

$$= \frac{1}{54}\left[145^{3/2} - 1\right] \approx 32.3154$$

**55.** $x = t,\ y = 3t,\ 0 \le t \le 2$

$$\frac{dx}{dt} = 1,\ \frac{dy}{dt} = 3,\ \sqrt{\left(\frac{dx}{dt}\right)^2 + \left(\frac{dy}{dt}\right)^2} = \sqrt{1 + 9} = \sqrt{10}$$

(a) $S = 2\pi \displaystyle\int_0^2 3t\sqrt{10}\, dt = 6\sqrt{10}\ \pi \left[\dfrac{t^2}{2}\right]_0^2 = 12\sqrt{10}\ \pi \approx 119.215$

(b) $S = 2\pi \displaystyle\int_0^2 \sqrt{10}\, dt = 2\pi \left[\sqrt{10}t\right]_0^2 = 4\pi\sqrt{10} \approx 39.738$

**57.** $x = 3 \sin \theta,\ y = 2 \cos \theta$

$$A = \int_a^b y\, dx = \int_{-\pi/2}^{\pi/2} 2 \cos \theta (3 \cos \theta)\, d\theta$$

$$= 6 \int_{-\pi/2}^{\pi/2} \frac{1 + \cos 2\theta}{2}\, d\theta$$

$$= 3\left[\theta + \frac{\sin 2\theta}{2}\right]_{-\pi/2}^{\pi/2}$$

$$= 3\left[\frac{\pi}{2} + \frac{\pi}{2}\right] = 3\pi$$

**59.** $(r, \theta) = \left(5, \dfrac{3\pi}{2}\right)$

$$x = r \cos \theta = 5 \cos \frac{3\pi}{2} = 0$$

$$y = r \sin \theta = 5 \sin \frac{3\pi}{2} = -5$$

$$(x, y) = (0, -5)$$

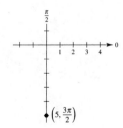

**61.** $(r, \theta) = \left( \sqrt{3}, 1.56 \right)$

$(x, y) = \left( \sqrt{3} \cos(1.56), \sqrt{3} \sin(1.56) \right)$

$\approx (0.0187, 1.7319)$

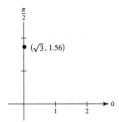

**63.** $(x, y) = (4, -4)$

$r = \sqrt{4^2 + (-4)^2} = 4\sqrt{2}$

$\theta = \dfrac{7\pi}{4}$

$(r, \theta) = \left( 4\sqrt{2}, \dfrac{7\pi}{4} \right), \left( -4\sqrt{2}, \dfrac{3\pi}{4} \right)$

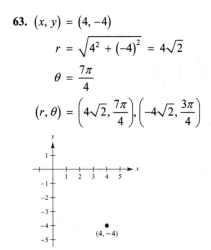

**65.** $(x, y) = (-1, 3)$

$r = \sqrt{(-1)^2 + 3^2} = \sqrt{10}$

$\theta = \arctan(-3) \approx 1.89 (108.43°)$

$(r, \theta) = \left( \sqrt{10}, 1.89 \right), \left( -\sqrt{10}, 5.03 \right)$

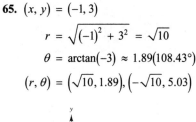

**67.** $x^2 + y^2 = 25$

$r^2 = 25$

$r = 5$

Circle

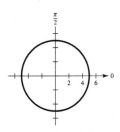

**69.**

$y = 9$

$r \sin \theta = 9$

$r = \dfrac{9}{\sin \theta} = 9 \csc \theta$

Horizontal line

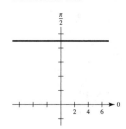

**71.**

$x^2 = 4y$

$r^2 \cos^2 \theta = 4r \sin \theta$

$r = \dfrac{4 \sin \theta}{\cos^2 \theta} = 4 \tan \theta \sec \theta$

Parabola

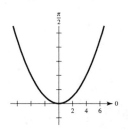

**73.**

$r = 3 \cos \theta$

$r^2 = 3r \cos \theta$

$x^2 + y^2 = 3x$

$x^2 - 3x + \dfrac{9}{4} + y^2 = \dfrac{9}{4}$

$\left( x - \dfrac{3}{2} \right)^2 + y^2 = \dfrac{9}{4}$

Circle

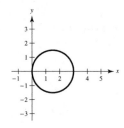

**75.**
$$r = 6 \sin \theta$$
$$r^2 = 6r \sin \theta$$
$$x^2 + y^2 = 6y$$
$$x^2 + y^2 - 6y + 9 = 9$$
$$x^2 + (y - 3)^2 = 9$$

Circle

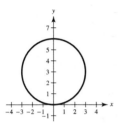

**77.**
$$r = -2 \sec \theta \tan \theta$$
$$r \cos \theta = -2 \tan \theta$$
$$x = -2\left(\frac{y}{x}\right)$$
$$x^2 = -2y$$
$$y = -\frac{1}{2}x^2$$

Parabola

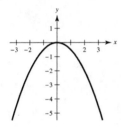

**79.** $r = \dfrac{3}{\cos \theta - (\pi/4)}$

Graph of $r = 3 \sec \theta$ rotated through an angle of $\pi/4$

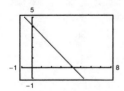

**81.** $r = 4 \cos 2\theta \sec \theta$

Strophoid

Symmetric to the polar axis

$$r \Rightarrow \infty \text{ as } \theta \Rightarrow \frac{\pi^-}{2}$$

$$r \Rightarrow \infty \text{ as } \theta \Rightarrow \frac{-\pi^+}{2}$$

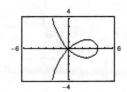

**83.** $r = 1 - \cos \theta$, Cardioid

$$\frac{dy}{dx} = \frac{(1 - \cos \theta)\cos \theta + (\sin \theta)\sin \theta}{-(1 - \cos \theta)\sin \theta + (\sin \theta)\cos \theta}$$

Horizontal tangents:
$$\cos \theta - \cos^2 \theta + \sin^2 \theta = 0$$
$$\cos \theta - \cos^2 \theta + (1 - \cos^2 \theta) = 0$$
$$2 \cos^2 \theta - \cos \theta - 1 = 0$$
$$(2 \cos \theta + 1)(\cos \theta - 1) = 0$$
$$\cos \theta = -\frac{1}{2} \Rightarrow \theta = \frac{2\pi}{3}, \frac{4\pi}{3}$$
$$\cos \theta = 1 \Rightarrow \theta = 0$$

Vertical tangents:
$$-\sin \theta + 2 \cos \theta \sin \theta = 0$$
$$\sin \theta(2 \cos \theta - 1) = 0$$
$$\sin \theta = 0 \Rightarrow \theta = 0, \pi$$
$$\cos \theta = \frac{1}{2} \Rightarrow \theta = \frac{\pi}{3}, \frac{5\pi}{3}$$

Horizontal tangents: $\left(\dfrac{3}{2}, \dfrac{2\pi}{3}\right), \left(\dfrac{3}{2}, \dfrac{4\pi}{3}\right)$

Vertical tangents: $\left(\dfrac{1}{2}, \dfrac{\pi}{3}\right), \left(\dfrac{1}{2}, \dfrac{5\pi}{3}\right), (2\pi)$

(There is a cusp at the pole.)

**85.** $r = 4 \sin 3\theta$, Rose curve with three petals

Tangents at the pole: $\sin 3\theta = 0$

$$\theta = 0, \frac{\pi}{3}, \frac{2\pi}{3}$$

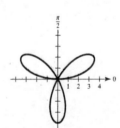

**87.** $r = 6$, Circle radius 6

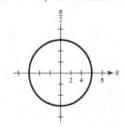

**89.** $r = -\sec \theta = \dfrac{-1}{\cos \theta}$

$r \cos \theta = -1, \ x = -1$

Vertical line

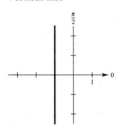

**91.** $r^2 = 4 \sin^2 2\theta$

$r = \pm 2 \sin(2\theta)$

Rose curve with four petals

Symmetric to the polar axis, $\theta = \dfrac{\pi}{2}$, and pole

Relative extrema: $\left( \pm 2, \dfrac{\pi}{4} \right), \left( \pm 2, \dfrac{3\pi}{4} \right)$

Tangents at the pole: $\theta = 0, \dfrac{\pi}{2}$

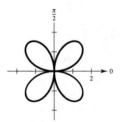

**93.** $r = 4 - 3 \cos \theta$

Limaçon

Symmetric to polar axis

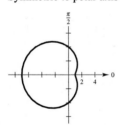

| $\theta$ | 0 | $\dfrac{\pi}{3}$ | $\dfrac{\pi}{2}$ | $\dfrac{2\pi}{3}$ | $\pi$ |
|---|---|---|---|---|---|
| $r$ | 1 | $\dfrac{5}{2}$ | 4 | $\dfrac{11}{2}$ | 7 |

**95.** $r = -3 \cos 2\theta$

Rose curve with four petals

Symmetric to polar axis, $\theta = \dfrac{\pi}{2}$, and pole

Relative extrema: $(-3, 0), \left( 3, \dfrac{\pi}{2} \right), (-3, \pi), \left( 3, \dfrac{3\pi}{2} \right)$

Tangents at the pole: $\theta = \dfrac{\pi}{4}, \dfrac{3\pi}{4}$

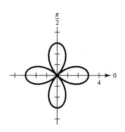

**97.** $A = 2 \cdot \dfrac{1}{2} \displaystyle\int_0^{\pi/10} \left[ 3 \cos 5\theta \right]^2 d\theta$

$= \displaystyle\int_0^{\pi/10} 9 \left( \dfrac{1 + \cos(10\theta)}{2} \right) d\theta$

$= \dfrac{9}{2} \left[ \theta + \dfrac{\sin(10\theta)}{2} \right]_0^{\pi/10} = \dfrac{9}{2} \left[ \dfrac{\pi}{10} \right] = \dfrac{9\pi}{20}$

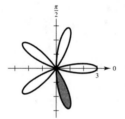

**99.** $r = 2 + \cos \theta$

$A = 2 \left[ \dfrac{1}{2} \displaystyle\int_0^{\pi} (2 + \cos \theta)^2 d\theta \right] \approx 14.14, \left( \dfrac{9\pi}{2} \right)$

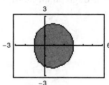

**101.** $r^2 = 4 \sin 2\theta$

$A = 2 \left[ \dfrac{1}{2} \displaystyle\int_0^{\pi/2} 4 \sin 2\theta \, d\theta \right] = 4$

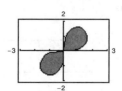

**103.** $r = 3 - 6 \cos \theta$

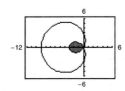

$$A = 2\left[\frac{1}{2}\int_0^{\pi/3} (3 - 6\cos\theta)^2\, d\theta\right]$$

$$= \int_0^{\pi/3} \left[9 - 36\cos\theta + 36\cos^2\theta\right]d\theta$$

$$= 9\int_0^{\pi/3}\left[1 - 4\cos\theta + 2(1 + \cos 2\theta)\right]d\theta$$

$$= 9\left[3\theta - 4\sin\theta + \sin 2\theta\right]_0^{\pi/3}$$

$$= 9\left[\pi - 2\sqrt{3} + \frac{\sqrt{3}}{2}\right] = \frac{18\pi - 27\sqrt{3}}{2}$$

**105.** $r = 3 - 6 \cos \theta$

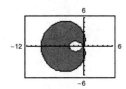

$$A = 2\left[\frac{1}{2}\int_{\pi/3}^{\pi} (3 - 6\cos\theta)^2\, d\theta - \frac{1}{2}\int_0^{\pi/3} (3 - 6\cos\theta)^2\, d\theta\right]$$

From Exercise 103 you have:

$$A = 9\left[3\theta - 4\sin\theta + \sin 2\theta\right]_{\pi/3}^{\pi} - 9\left[3\theta - 4\sin\theta + \sin 2\theta\right]_0^{\pi/3}$$

$$= 9\left[3\pi - \left(\pi - 2\sqrt{3} + \frac{\sqrt{3}}{2}\right)\right] - 9\left[\pi - 2\sqrt{3} + \frac{\sqrt{3}}{2}\right]$$

$$= 9\pi + 27\sqrt{3}$$

**107.** $r = 1 - \cos\theta, r = 1 + \sin\theta$

$$1 - \cos\theta = 1 + \sin\theta$$

$$\tan\theta = -1 \Rightarrow \theta = \frac{3\pi}{4}, \frac{7\pi}{4}$$

The graphs also intersect at the pole.

Points of intersection:

$$\left(1 + \frac{\sqrt{2}}{2}, \frac{3\pi}{4}\right), \left(1 - \frac{\sqrt{2}}{2}, \frac{7\pi}{4}\right), (0, 0)$$

**109.** $r = 5\cos\theta, \dfrac{\pi}{2} \le \theta \le \pi$

$$\frac{dr}{d\theta} = -5\sin\theta$$

$$s = \int_{\pi/2}^{\pi} \sqrt{(25\cos^2\theta) + (25\sin^2\theta)}\, d\theta$$

$$= \int_{\pi/2}^{\pi} 5\, d\theta = \left[5\theta\right]_{\pi/2}^{\pi} = \frac{5\pi}{2} \quad \text{(Semicircle)}$$

**111.** $f(\theta) = 1 + 4\cos\theta$

$$f'(\theta) = -4\sin\theta$$

$$\sqrt{f(\theta)^2 + f'(\theta)^2} = \sqrt{(1 + 4\cos\theta)^2 + (-4\sin\theta)^2}$$

$$= \sqrt{17 + 8\cos\theta}$$

$$S = 2\pi\int_0^{\pi/2} (1 + 4\cos\theta)\sin\theta\sqrt{17 + 8\cos\theta}\, d\theta$$

$$= \frac{34\pi\sqrt{17}}{5} \approx 88.08$$

**113.** $r = \dfrac{6}{1 - \sin\theta}$

$e = 1,$

Parabola

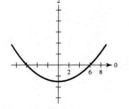

**115.** $r = \dfrac{6}{3 + 2 \cos \theta} = \dfrac{2}{1 + (2/3) \cos \theta}, e = \dfrac{2}{3}$

Ellipse

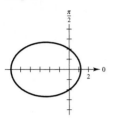

**117.** $r = \dfrac{4}{2 - 3 \sin \theta} = \dfrac{2}{1 - (3/2) \sin \theta}, e = \dfrac{3}{2}$

Hyperbola

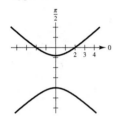

**119.** Parabola

$e = 1$

$x = 4 \Rightarrow d = 4$

$r = \dfrac{ed}{1 + e \cos \theta} = \dfrac{4}{1 + \cos \theta}$

**121.** Hyperbola, $e = 3$, $y = 3$

$d = 3$

$r = \dfrac{ed}{1 + e \sin \theta} = \dfrac{3(3)}{1 + 3 \sin \theta} = \dfrac{9}{1 + 3 \sin \theta}$

**123.** Ellipse

Vertices: $(5, 0), (1, \pi)$

Focus: $(0, 0)$

$$a = 3, c = 2, e = \frac{2}{3}, d = \frac{5}{2}$$

$$r = \dfrac{\left(\dfrac{2}{3}\right)\left(\dfrac{5}{2}\right)}{1 - \left(\dfrac{2}{3}\right) \cos \theta} = \dfrac{5}{3 - 2 \cos \theta}$$

# Problem Solving for Chapter 10

**1. (a)**

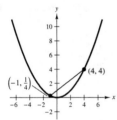

**(b)** $x^2 = 4y$

$2x = 4y'$

$y' = \dfrac{1}{2}x$

$y - 4 = 2(x - 4) \Rightarrow y = 2x - 4$  Tangent line at $(4, 4)$

$y - \frac{1}{4} = -\frac{1}{2}(x + 1) \Rightarrow y = -\frac{1}{2}x - \frac{1}{4}$  Tangent line at $\left(-1, \frac{1}{4}\right)$

Tangent lines have slopes of 2 and $-\frac{1}{2} \Rightarrow$ perpendicular.

**(c)** Intersection:

$2x - 4 = -\frac{1}{2}x - \frac{1}{4}$

$8x - 16 = -2x - 1$

$10x = 15$

$x = \frac{3}{2} \Rightarrow \left(\frac{3}{2}, -1\right)$

Point of intersection, $\left(\frac{3}{2}, -1\right)$, is on directrix $y = -1$.

**3.** Consider $x^2 = 4py$ with focus $F = (0, p)$.

Let $P(a, b)$ be point on parabola.

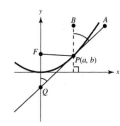

$2x = 4py' \Rightarrow y' = \dfrac{x}{2p}$

$y - b = \dfrac{a}{2p}(x - a)$    Tangent line at $P$

For $x = 0$, $y = b + \dfrac{a}{2p}(-a) = b - \dfrac{a^2}{2p} = b - \dfrac{4pb}{2p} = -b.$

So, $Q = (0, -b)$.

$\Delta FQP$ is isosceles because

$|FQ| = p + b$

$|FP| = \sqrt{(a - 0)^2 + (b - p)^2} = \sqrt{a^2 + b^2 - 2bp + p^2} = \sqrt{4pb + b^2 - 2bp + p^2} = \sqrt{(b + p)^2} = b + p.$

So, $\angle FQP = \angle BPA = \angle FPQ$.

**5. (a)** $y^2 = \dfrac{t^2(1 - t^2)^2}{(1 + t^2)^2}$, $x^2 = \dfrac{(1 - t^2)^2}{(1 + t^2)^2}$

$\dfrac{1 - x}{1 + x} = \dfrac{1 - \left(\dfrac{1 - t^2}{1 + t^2}\right)}{1 + \left(\dfrac{1 - t^2}{1 + t^2}\right)} = \dfrac{2t^2}{2} = t^2$

So, $y^2 = x^2\left(\dfrac{1 - x}{1 + x}\right)$.

**(b)**

$$r^2 \sin^2 \theta = r^2 \cos^2 \theta\left(\dfrac{1 - r \cos \theta}{1 + r \cos \theta}\right)$$

$$\sin^2 \theta(1 + r \cos \theta) = \cos^2 \theta(1 - r \cos \theta)$$

$$r \cos \theta \sin^2 \theta + \sin^2 \theta = \cos^2 \theta - r \cos^3 \theta$$

$$r \cos \theta(\sin^2 \theta + \cos^2 \theta) = \cos^2 \theta - \sin^2 \theta$$

$$r \cos \theta = \cos 2\theta$$

$$r = \cos 2\theta \cdot \sec \theta$$

**(c)**

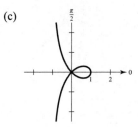

**(d)** $r(\theta) = 0$ for $\theta = \dfrac{\pi}{4}, \dfrac{3\pi}{4}$.

So, $y = x$ and $y = -x$ are tangent lines to curve at the origin.

(e)  $y'(t) = \dfrac{\left(1 + t^2\right)\left(1 - 3t^2\right) - \left(t - t^3\right)(2t)}{\left(1 + t^2\right)^2} = \dfrac{1 - 4t^2 - t^4}{\left(1 + t^2\right)^2} = 0$

$t^4 + 4t^2 - 1 = 0 \Rightarrow t^2 = -2 \pm \sqrt{5} \Rightarrow x = \dfrac{1 - \left(-2 \pm \sqrt{5}\right)}{1 + \left(-2 \pm \sqrt{5}\right)} = \dfrac{3 \mp \sqrt{5}}{-1 \pm \sqrt{5}} = \dfrac{3 - \sqrt{5}}{-1 + \sqrt{5}} = \dfrac{\sqrt{5} - 1}{2}$

$\left(\dfrac{\sqrt{5} - 1}{2}, \pm \dfrac{\sqrt{5} - 1}{2}\sqrt{-2 + \sqrt{5}}\right)$

**7. (a)**

*Generated by Mathematica*

(b)  $(-x, -y) = \left(-\displaystyle\int_0^t \cos\dfrac{\pi u^2}{2}\, du, -\int_0^t \sin\dfrac{\pi u^2}{2}\, du\right)$ is

on the curve whenever $(x, y)$ is on the curve.

(c)  $x'(t) = \cos\dfrac{\pi t^2}{2},\; y'(t) = \sin\dfrac{\pi t^2}{2},$

$x'(t)^2 + y'(t)^2 = 1$

So, $s = \displaystyle\int_0^a dt = a.$

On $[-\pi, \pi]$, $s = 2\pi$.

**9.**  $r = \dfrac{ab}{a\sin\theta + b\cos\theta},\quad 0 \le \theta \le \dfrac{\pi}{2}$

$r(a\sin\theta + b\cos\theta) = ab$

$ay + bx = ab$

$\dfrac{y}{b} + \dfrac{x}{a} = 1$

Line segment

Area $= \dfrac{1}{2}ab$

**11.**  Let $(r, \theta)$ be on the graph.

$\sqrt{r^2 + 1 + 2r\cos\theta}\,\sqrt{r^2 + 1 - 2r\cos\theta} = 1$

$\left(r^2 + 1\right)^2 - 4r^2\cos^2\theta = 1$

$r^4 + 2r^2 + 1 - 4r^2\cos^2\theta = 1$

$r^2\left(r^2 - 4\cos^2\theta + 2\right) = 0$

$r^2 = 4\cos^2\theta - 2$

$r^2 = 2\left(2\cos^2\theta - 1\right)$

$r^2 = 2\cos 2\theta$

**13.**  If a dog is located at $(r, \theta)$ in the first quadrant, then its

neighbor is at $\left(r, \theta + \dfrac{\pi}{2}\right)$:

$(x_1, y_1) = (r\cos\theta, r\sin\theta)$ and

$(x_2, y_2) = (-r\sin\theta, r\cos\theta).$

The slope joining these points is

$\dfrac{r\cos\theta - r\sin\theta}{-r\sin\theta - r\cos\theta} = \dfrac{\sin\theta - \cos\theta}{\sin\theta + \cos\theta}$

$= $ slope of tangent line at $(r, \theta).$

$\dfrac{dy}{dx} = \dfrac{\dfrac{dy}{dr}}{\dfrac{dx}{dr}} = \dfrac{\dfrac{dr}{d\theta}\sin\theta + r\cos\theta}{\dfrac{dr}{d\theta}\cos\theta - r\sin\theta} = \dfrac{\sin\theta - \cos\theta}{\sin\theta + \cos\theta}$

$\Rightarrow \dfrac{dr}{d\theta} = -r$

$\dfrac{dr}{r} = -d\theta$

$\ln r = -\theta + C_1$

$r = e^{-\theta + C_1}$

$r = Ce^{-\theta}$

$r\left(\dfrac{\pi}{4}\right) = \dfrac{d}{\sqrt{2}} \Rightarrow r = Ce^{-\pi/4} = \dfrac{d}{\sqrt{2}} \Rightarrow C = \dfrac{d}{\sqrt{2}}e^{\pi/4}$

Finally, $r = \dfrac{d}{\sqrt{2}}e^{\left((\pi/4) - \theta\right)},\; \theta \ge \dfrac{\pi}{4}.$

**15.** (a) In $\triangle OCB$, $\cos \theta = \dfrac{2a}{OB} \Rightarrow OB = 2a \cdot \sec \theta$.

In $\triangle OAC$, $\cos \theta = \dfrac{OA}{2a} \Rightarrow OA = 2a \cdot \cos \theta$.

$$r = OP = AB = OB - OA = 2a(\sec \theta - \cos \theta)$$

$$= 2a\left(\frac{1}{\cos \theta} - \cos \theta\right)$$

$$= 2a \cdot \frac{\sin^2 \theta}{\cos \theta}$$

$$= 2a \cdot \tan \theta \sin \theta$$

(b) $x = r \cos \theta = (2a \tan \theta \sin \theta) \cos \theta = 2a \sin^2 \theta$

$y = r \sin \theta = (2a \tan \theta \sin \theta) \sin \theta = 2a \tan \theta \cdot \sin^2 \theta, -\dfrac{\pi}{2} < \theta < \dfrac{\pi}{2}$

Let $t = \tan \theta, -\infty < t < \infty$.

Then $\sin^2 \theta = \dfrac{t^2}{1 + t^2}$ and $x = 2a\dfrac{t^2}{1 + t^2}, y = 2a\dfrac{t^3}{1 + t^2}$.

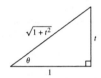

(c) $\quad\quad r = 2a \tan \theta \sin \theta$

$\quad\quad r \cos \theta = 2a \sin^2 \theta$

$\quad\quad r^3 \cos \theta = 2a\, r^2 \sin^2 \theta$

$\quad (x^2 + y^2)x = 2ay^2$

$$y^2 = \frac{x^3}{(2a - x)}$$

**17.**

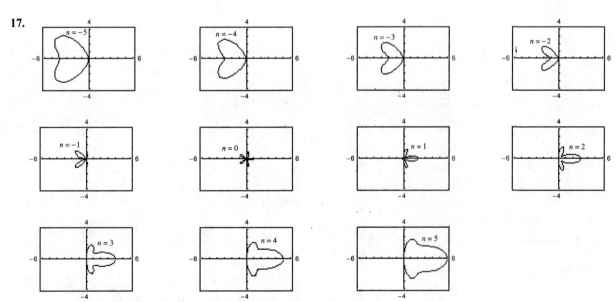

$n = 1, 2, 3, 4, 5$ produce "bells"; $n = -1, -2, -3, -4, -5$ produce "hearts".

# Appendix C.1

**1.** $0.7 = \frac{7}{10}$

Rational

**3.** $\frac{3\pi}{2}$

Irrational (because $\pi$ is irrational)

**5.** $4.345\overline{1451}$

Rational

**7.** $\sqrt[3]{64} = 4$

Rational

**9.** $4\frac{5}{8} = \frac{37}{8}$

Rational

**11.** Let $x = 0.36\overline{36}$.

$100x = 36.36\overline{36}$

$\underline{-x = -0.36\overline{36}}$

$99x = 36$

$x = \frac{36}{99} = \frac{4}{11}$

**13.** Let $x = 0.297\overline{297}$.

$1000x = \ \ \ 0.297\overline{297}$

$\underline{-x = -297.297\overline{297}}$

$999x = \ \ \ 297$

$x = \frac{297}{999} = \frac{11}{37}$

**15.** Given $a < b$:

(a) $a + 2 < b + 2$; True

(b) $5b < 5a$; False

(c) $5 - a > 5 - b$; True

(d) $\dfrac{1}{a} < \dfrac{1}{b}$; False

(e) $(a - b)(b - a) > 0$; False

(f) $a^2 < b^2$; False

**17.** $x$ is greater than $-3$ and less than 3.

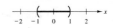

The interval is bounded.

**19.** $x$ is less than, or equal to, 5.

The interval is unbounded.

**21.** $y \geq 4$, $[4, \infty)$

**23.** $0.03 < r \leq 0.07$, $(0.03, 0.07]$

**25.** $2x - 1 \geq 0$

$2x \geq 1$

$x \geq \frac{1}{2}$

**27.** $-4 < 2x - 3 < 4$

$-1 < \ \ \ 2x \ \ \ < 7$

$-\frac{1}{2} < \ \ \ x \ \ \ < \frac{7}{2}$

**29.** $\dfrac{x}{2} + \dfrac{x}{3} > 5$

$3x + 2x > 30$

$5x > 30$

$x > 6$

**31.** $|x| < 1 \Rightarrow -1 < x < 1$

**33.** $\left|\dfrac{x-3}{2}\right| \geq 5$

$x - 3 \geq 10 \ \ \text{or} \ \ x - 3 \leq -10$

$x \geq 13 \ \ \ \ \ \ \ \ \ \ \ \ \ x \leq -7$

**35.** $|x - a| < b$

$-b \ \ < x - a < \ \ b$

$a - b < \ \ \ x \ \ \ < a + b$

**37.** $|2x + 1| < 5$

$-5 < 2x + 1 < 5$

$-6 < \ \ \ 2x \ \ \ < 4$

$-3 < \ \ \ x \ \ \ < 2$

**39.** $\left|1 - \dfrac{2x}{3}\right| < 1$

$-1 < 1 - \dfrac{2x}{3} < 1$

$-2 < \ \ -\dfrac{2x}{3} \ \ < 0$

$3 > \ \ \ x \ \ \ > 0$

**41.** $x^2 \le 3 - 2x$

$x^2 + 2x - 3 \le 0$

$(x + 3)(x - 1) \le 0$

Test intervals: $(-\infty, -3), (-3, 1), (1, \infty)$

Solution: $-3 \le x \le 1$

**43.** $x^2 + x - 1 \le 5$

$x^2 + x - 6 \le 0$

$(x + 3)(x - 2) \le 0$

$x = -3$

$x = 2$

Test intervals: $(-\infty, -3), (-3, 2), (2, \infty)$

Solution: $-3 \le x \le 2$

**45.** $a = -1, b = 3$

Directed distance from $a$ to $b$: 4

Directed distance from $b$ to $a$: $-4$

Distance between $a$ and $b$: 4

**47.** (a) $a = 126, b = 75$

Directed distance from $a$ to $b$: $-51$

Directed distance from $b$ to $a$: 51

Distance between $a$ and $b$: 51

(b) $a = -126, b = -75$

Directed distance from $a$ to $b$: 51

Directed distance from $b$ to $a$: $-51$

Distance between $a$ and $b$: 51

**49.** $a = -2, b = 2$

Midpoint: 0

Distance between midpoint and each endpoint: 2

$|x - 0| \le 2$

$|x| \le 2$

**51.** $a = 0, b = 4$

Midpoint: 2

Distance between midpoint and each endpoint: 2

$|x - 2| > 2$

**53.** (a) All numbers that are at most 10 units from 12

$|x - 12| \le 10$

(b) All numbers that are at least 10 units from 12

$|x - 12| \ge 10$

**55.** $a = -1, b = 3$

Midpoint: $\dfrac{-1 + 3}{2} = 1$

**57.** (a) $[7, 21]$

Midpoint: 14

(b) $[8.6, 11.4]$

Midpoint: 10

**59.** $R = 115.95x, C = 95x + 750, R > C$

$115.95x > 95x + 750$

$20.95x > 750$

$x > 35.7995$

$x \ge 36$ units

**61.** $\left|\dfrac{x - 50}{5}\right| \ge 1.645$

$\dfrac{x - 50}{5} \le -1.645$ or $\dfrac{x - 50}{5} \ge 1.645$

$x - 50 \le -8.225 \qquad x - 50 \ge 8.225$

$x \le 41.775 \qquad\quad x \ge 58.225$

$x \le 41 \qquad\qquad\quad x \ge 59$

**63.** (a) $\pi \approx 3.1415926535$

$\dfrac{355}{113} \approx 3.141592920$

$\dfrac{355}{113} > \pi$

(b) $\pi \approx 3.1415926535$

$\dfrac{22}{7} \approx 3.142857143$

$\dfrac{22}{7} > \pi$

**65.** Speed of light: $2.998 \times 10^8$ meters per second

Distance traveled in one year $=$ rate $\times$ time

$d = (2.998 \times 10^8) \times (365 \times 24 \times 60 \times 60)$

days $\times$ hours $\times$ minutes $\times$ seconds

$= (2.998 \times 10^8) \times (3.1536 \times 10^7) \approx 9.45 \times 10^{15}$

This is best estimated by (b).

**67.** False; 2 is a nonzero integer and the reciprocal of 2 is $\frac{1}{2}$.

**69.** True

**71.** True; if $x < 0$, then $|x| = -x = \sqrt{x^2}$.

**73.** If $a \geq 0$ and $b \geq 0$, then $|ab| = ab = |a||b|$.

If $a < 0$ and $b < 0$, then $|ab| = ab = (-a)(-b) = |a||b|$.

If $a \geq 0$ and $b < 0$, then $|ab| = -ab = a(-b) = |a||b|$.

If $a < 0$ and $b \geq 0$, then $|ab| = -ab = (-a)b = |a||b|$.

**75.** $\left|\dfrac{a}{b}\right| = \left|a\left(\dfrac{1}{b}\right)\right| = |a|\left|\dfrac{1}{b}\right| = |a| \cdot \dfrac{1}{|b|} = \dfrac{|a|}{|b|}, \ b \neq 0$

**77.** $n = 1,$ $\quad |a| = |a|$

$n = 2,$ $\quad \left|a^2\right| = |a \cdot a| = |a||a| = |a|^2$

$n = 3,$ $\quad \left|a^3\right| = \left|a^2 \cdot a\right| = \left|a^2\right||a| = |a|^2|a| = |a|^3$

$\quad\quad \vdots$

$\left|a^n\right| = \left|a^{n-1}a\right| = \left|a^{n-1}\right||a| = |a|^{n-1}|a| = |a|^n$

**79.** $|a| \leq k \Leftrightarrow \sqrt{a^2} \leq k \Leftrightarrow a^2 \leq k^2 \Leftrightarrow a^2 - k^2 \leq 0 \Leftrightarrow (a + k)(a - k) \leq 0 \Leftrightarrow -k \leq a \leq k, \ k > 0$

**81.** $\left.\begin{array}{l} |7 - 12| = |-5| = 5 \\ |7| - |12| = 7 - 12 = -5 \end{array}\right\} \ |7 - 12| > |7| - |12|$

$\left.\begin{array}{l} |12 - 7| = |5| = 5 \\ |12| - |7| = 12 - 7 = 5 \end{array}\right\} \ |12 - 7| = |12| - |7|$

You know that $|a||b| \geq ab$. So, $-2|a||b| \leq -2ab$. Because $a^2 = |a|^2$ and $b^2 = |b|^2$, you have

$|a|^2 + |b|^2 - 2|a||b| \leq a^2 + b^2 - 2ab$

$0 \leq \left(|a| - |b|\right)^2 \leq (a - b)^2$

$\sqrt{\left(|a| - |b|\right)^2} \leq \sqrt{(a - b)^2}$

$\left||a| - |b|\right| \leq |a - b|.$

Because $|a| - |b| \leq \left||a| - |b|\right|$, you have $|a| - |b| \leq |a - b|$. So, $|a - b| \geq |a| - |b|$.

# Appendix C.2

**1. (a)**

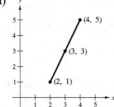

**(b)** $d = \sqrt{(4 - 2)^2 + (5 - 1)^2}$

$\quad\quad = \sqrt{4 + 16} = \sqrt{20} = 2\sqrt{5}$

**(c)** Midpoint: $\left(\dfrac{4 + 2}{2}, \dfrac{5 + 1}{2}\right) = (3, 3)$

**3.** (a)

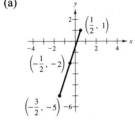

(b) $d = \sqrt{\left(\dfrac{1}{2} + \dfrac{3}{2}\right)^2 + \left(1 + 5\right)^2}$

$\phantom{d} = \sqrt{4 + 36} = \sqrt{40} = 2\sqrt{10}$

(c) Midpoint: $\left(\dfrac{(-3/2) + (1/2)}{2}, \dfrac{-5 + 1}{2}\right) = \left(-\dfrac{1}{2}, -2\right)$

**5.** (a)

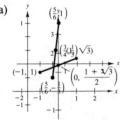

(b) $d = \sqrt{\left(-1 - 1\right)^2 + \left(1 - \sqrt{3}\right)^2}$

$\phantom{d} = \sqrt{4 + 1 - 2\sqrt{3} + 3} = \sqrt{8 - 2\sqrt{3}}$

(c) Midpoint: $\left(\dfrac{-1 + 1}{2}, \dfrac{1 + \sqrt{3}}{2}\right) = \left(0, \dfrac{1 + \sqrt{3}}{2}\right)$

**7.** $x = -2 \Rightarrow$ quadrants II, III

$\phantom{7.}$ $y > 0 \Rightarrow$ quadrants I, II

$\phantom{7.}$ Therefore, quadrant II

**9.** $xy > 0 \Rightarrow$ quadrants I or III

**11.** $d_1 = \sqrt{9 + 36} = \sqrt{45}$

$\phantom{11.}$ $d_2 = \sqrt{4 + 1} = \sqrt{5}$

$\phantom{11.}$ $d_3 = \sqrt{25 + 25} = \sqrt{50}$

$\phantom{11.}$ $\left(d_1\right)^2 + \left(d_2\right)^2 = \left(d_3\right)^2$

$\phantom{11.}$ Right triangle

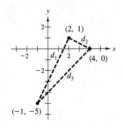

**13.** $d_1 = d_2 = d_3 = d_4 = \sqrt{5}$

$\phantom{13.}$ Rhombus

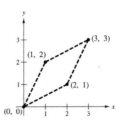

**15.**

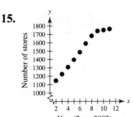

**17.** $d_1 = \sqrt{4 + 16} = \sqrt{20} = 2\sqrt{5}$

$\phantom{17.}$ $d_2 = \sqrt{1 + 4} = \sqrt{5}$

$\phantom{17.}$ $d_3 = \sqrt{9 + 36} = 3\sqrt{5}$

$\phantom{17.}$ $d_1 + d_2 = d_3$

$\phantom{17.}$ Collinear

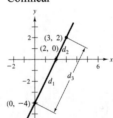

**19.** $d_1 = \sqrt{1 + 1} = \sqrt{2}$

$\phantom{19.}$ $d_2 = \sqrt{9 + 4} = \sqrt{13}$

$\phantom{19.}$ $d_3 = \sqrt{16 + 9} = 5$

$\phantom{19.}$ $d_1 + d_2 \neq d_3$

$\phantom{19.}$ Not collinear

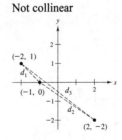

**21.** $5 = \sqrt{(x - 0)^2 + (-4 - 0)^2}$

$\quad\quad 5 = \sqrt{x^2 + 16}$

$\quad\quad 25 = x^2 + 16$

$\quad\quad 9 = x^2$

$\quad\quad x = \pm 3$

**23.** $8 = \sqrt{(3 - 0)^2 + (y - 0)^2}$

$\quad\quad 8 = \sqrt{9 + y^2}$

$\quad\quad 64 = 9 + y^2$

$\quad\quad 55 = y^2$

$\quad\quad y = \pm\sqrt{55}$

**25.** The midpoint of the given line segment is $\left(\dfrac{x_1 + x_2}{2}, \dfrac{y_1 + y_2}{2}\right)$.

The midpoint between $(x_1, y_1)$ and $\left(\dfrac{x_1 + x_2}{2}, \dfrac{y_1 + y_2}{2}\right)$ is $\left(\dfrac{x_1 + (x_1 + x_2)/2}{2}, \dfrac{y_1 + (y_1 + y_2)/2}{2}\right) = \left(\dfrac{3x_1 + x_2}{4}, \dfrac{3y_1 + y_2}{4}\right)$.

The midpoint between $\left(\dfrac{x_1 + x_2}{2}, \dfrac{y_1 + y_2}{2}\right)$ and $(x_2, y_2)$ is $\left(\dfrac{(x_1 + x_2)/2 + x_2}{2}, \dfrac{(y_1 + y_2)/2 + y_2}{2}\right) = \left(\dfrac{x_1 + 3x_2}{4}, \dfrac{y_1 + 3y_2}{4}\right)$.

Thus, the three points are $\left(\dfrac{3x_1 + x_2}{4}, \dfrac{3y_1 + y_2}{4}\right), \left(\dfrac{x_1 + x_2}{2}, \dfrac{y_1 + y_2}{2}\right), \left(\dfrac{x_1 + 3x_2}{4}, \dfrac{y_1 + 3y_2}{4}\right)$.

**27.** Center: $(0, 0)$

Radius: 1

Matches graph (c)

**29.** Center: $(1, 0)$

Radius: 0

Matches graph (a)

**31.** $(x - 0)^2 + (y - 0)^2 = (3)^2$

$\quad\quad x^2 + y^2 - 9 = 0$

**33.** $\quad\quad (x - 2)^2 + (y + 1)^2 = (4)^2$

$\quad x^2 + y^2 - 4x + 2y - 11 = 0$

**35.** Radius $= \sqrt{(-1 - 0)^2 + (2 - 0)^2} = \sqrt{5}$

$\quad\quad (x + 1)^2 + (y - 2)^2 = 5$

$\quad x^2 + 2x + 1 + y^2 - 4y + 4 = 5$

$\quad\quad x^2 + y^2 + 2x - 4y = 0$

**37.** Center $=$ Midpoint $= (3, 2)$

Radius $= \sqrt{10}$

$\quad\quad (x - 3)^2 + (y - 2)^2 = \left(\sqrt{10}\right)^2$

$\quad x^2 - 6x + 9 + y^2 - 4y + 4 = 10$

$\quad\quad x^2 + y^2 - 6x - 4y + 3 = 0$

**39.** Place the center of Earth at the origin. Then you have

$x^2 + y^2 = (22{,}000 + 4000)^2$

$x^2 + y^2 = 26{,}000^2$.

**41.** $\quad\quad x^2 + y^2 - 2x + 6y + 6 = 0$

$(x^2 - 2x + 1) + (y^2 + 6y + 9) = -6 + 1 + 9$

$\quad\quad (x - 1)^2 + (y + 3)^2 = 4$

Center: $(1, -3)$

Radius: 2

**43.** $\quad\quad x^2 + y^2 - 2x + 6y + 10 = 0$

$(x^2 - 2x + 1) + (y^2 + 6y + 9) = -10 + 1 + 9$

$\quad\quad (x - 1)^2 + (y + 3)^2 = 0$

Only a point $(1, -3)$

**45.**
$$2x^2 + 2y^2 - 2x - 2y - 3 = 0$$
$$2\left(x^2 - x + \tfrac{1}{4}\right) + 2\left(y^2 - y + \tfrac{1}{4}\right) = 3 + \tfrac{1}{2} + \tfrac{1}{2}$$
$$\left(x - \tfrac{1}{2}\right)^2 + \left(y - \tfrac{1}{2}\right)^2 = 2$$

Center: $\left(\tfrac{1}{2}, \tfrac{1}{2}\right)$

Radius: $\sqrt{2}$

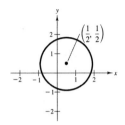

**47.**
$$16x^2 + 16y^2 + 16x + 40y - 7 = 0$$
$$16\left(x^2 + x + \tfrac{1}{4}\right) + 16\left(y^2 + \tfrac{5y}{2} + \tfrac{25}{16}\right) = 7 + 4 + 25$$
$$16\left(x + \tfrac{1}{2}\right)^2 + 16\left(y + \tfrac{5}{4}\right)^2 = 36$$
$$\left(x + \tfrac{1}{2}\right)^2 + \left(y + \tfrac{5}{4}\right)^2 = \tfrac{9}{4}$$

Center: $\left(-\tfrac{1}{2}, -\tfrac{5}{4}\right)$

Radius: $\tfrac{3}{2}$

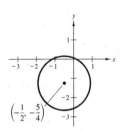

**49.**
$$4x^2 + 4y^2 - 4x + 24y - 63 = 0$$
$$x^2 + y^2 - x + 6y = \tfrac{63}{4}$$
$$\left(x^2 - x + \tfrac{1}{4}\right) + \left(y^2 + 6y + 9\right) = \tfrac{63}{4} + \tfrac{1}{4} + 9$$
$$\left(x - \tfrac{1}{2}\right)^2 + \left(y + 3\right)^2 = 25$$
$$\left(y + 3\right)^2 = 25 - \left(x - \tfrac{1}{2}\right)^2$$
$$y + 3 = \pm\sqrt{25 - \left(x - \tfrac{1}{2}\right)^2}$$
$$y = -3 \pm \sqrt{25 - \left(x - \tfrac{1}{2}\right)^2}$$
$$= \frac{-6 \pm \sqrt{99 + 4x - 4x^2}}{2}$$

**51.**
$$x^2 + y^2 - 4x + 2y + 1 \le 0$$
$$\left(x^2 - 4x + 4\right) + \left(y^2 + 2y + 1\right) \le -1 + 4 + 1$$
$$\left(x - 2\right)^2 + \left(y + 1\right)^2 \le 4$$

Center: $(2, -1)$

Radius: $2$

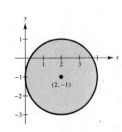

**53.** The distance between $(x_1, y_1)$ and $\left(\dfrac{2x_1 + x_2}{3}, \dfrac{2y_1 + y_2}{3}\right)$ is

$$d = \sqrt{\left(x_1 - \frac{2x_1 + x_2}{3}\right)^2 + \left(y_1 - \frac{2y_1 + y_2}{3}\right)^2}$$

$$= \sqrt{\left(\frac{x_1 - x_2}{3}\right)^2 + \left(\frac{y_1 - y_2}{3}\right)^2}$$

$$= \sqrt{\frac{1}{9}\left[(x_1 - x_2)^2 + (y_1 - y_2)^2\right]} = \frac{1}{3}\sqrt{(x_1 - x_2)^2 + (y_1 - y_2)^2}$$

which is $\frac{1}{3}$ of the distance between $(x_1, y_1)$ and $(x_2, y_2)$.

$$\left(\frac{\left(\dfrac{2x_1 + x_2}{3}\right) + x_2}{2}, \frac{\left(\dfrac{2y_1 + y_2}{3}\right) + y_2}{2}\right) = \left(\frac{x_1 + 2x_2}{3}, \frac{y_1 + 2y_2}{3}\right)$$

is the second point of the trisection.

**55.** True; if $ab < 0$ then either $a$ is positive and $b$ is negative (Quadrant IV) or $a$ is negative and $b$ is positive (Quadrant II).

**57.** True

**59.** Let one vertex be at $(0, 0)$ and another at $(a, 0)$.

Midpoint of $(0, 0)$ and $(d, e)$ is $\left(\dfrac{d}{2}, \dfrac{e}{2}\right)$.

Midpoint of $(b, c)$ and $(a, 0)$ is $\left(\dfrac{a + b}{2}, \dfrac{c}{2}\right)$.

Midpoint of $(0, 0)$ and $(a, 0)$ is $\left(\dfrac{a}{2}, 0\right)$.

Midpoint of $(b, c)$ and $(d, e)$ is $\left(\dfrac{b + d}{2}, \dfrac{c + e}{2}\right)$.

Midpoint of line segment joining $\left(\dfrac{d}{2}, \dfrac{e}{2}\right)$ and $\left(\dfrac{a + b}{2}, \dfrac{c}{2}\right)$ is $\left(\dfrac{a + b + d}{4}, \dfrac{c + e}{4}\right)$.

Midpoint of line segment joining $\left(\dfrac{a}{2}, 0\right)$ and $\left(\dfrac{b + d}{2}, \dfrac{c + e}{2}\right)$ is $\left(\dfrac{a + b + d}{4}, \dfrac{c + e}{4}\right)$.

Therefore the line segments intersect at their midpoints.

**61.** Let $(a, b)$ be a point on the semicircle of radius $r$, centered at the origin. We will show that the angle at $(a, b)$ is a right angle by verifying that $d_1^2 + d_2^2 = d_3^2$.

$$d_1^2 = (a + r)^2 + (b - 0)^2$$

$$d_2^2 = (a - r)^2 + (b - 0)^2$$

$$d_1^2 + d_2^2 = \left(a^2 + 2ar + r^2 + b^2\right) + \left(a^2 - 2ar + r^2 + b^2\right)$$

$$= 2a^2 + 2b^2 + 2r^2$$

$$= 2\left(a^2 + b^2\right) + 2r^2$$

$$= 2r^2 + 2r^2$$

$$= 4r^2 = (2r)^2 = d_3^2$$

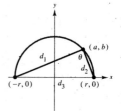

# Appendix C.3

**1.** (a) 396°, −324°

   (b) 240°, −480°

**3.** (a) $\dfrac{19\pi}{9}, -\dfrac{17\pi}{9}$

   (b) $\dfrac{10\pi}{3}, -\dfrac{2\pi}{3}$

**5.** (a) $30\left(\dfrac{\pi}{180}\right) = \dfrac{\pi}{6} \approx 0.524$

   (b) $150\left(\dfrac{\pi}{180}\right) = \dfrac{5\pi}{6} \approx 2.618$

   (c) $315\left(\dfrac{\pi}{180}\right) = \dfrac{7\pi}{4} \approx 5.498$

   (d) $120\left(\dfrac{\pi}{180}\right) = \dfrac{2\pi}{3} \approx 2.094$

**7.** (a) $\dfrac{3\pi}{2}\left(\dfrac{180}{\pi}\right) = 270°$

   (b) $\dfrac{7\pi}{6}\left(\dfrac{180}{\pi}\right) = 210°$

   (c) $-\dfrac{7\pi}{12}\left(\dfrac{180}{\pi}\right) = -105°$

   (d) $-2.637\left(\dfrac{180}{\pi}\right) \approx -151.1°$

**9.**

| $r$ | 8 ft | 15 in. | 85 cm | 24 in. | $\dfrac{12{,}963}{\pi}$ mi |
|---|---|---|---|---|---|
| $s$ | 12 ft | 24 in. | $63.75\pi$ cm | 96 in. | 8642 mi |
| $\theta$ | 1.5 | 1.6 | $\dfrac{3\pi}{4}$ | 4 | $\dfrac{2\pi}{3}$ |

**11.** (a) $x = 3, y = 4, r = 5$

   $\sin \theta = \frac{4}{5}$   $\csc \theta = \frac{5}{4}$

   $\cos \theta = \frac{3}{5}$   $\sec \theta = \frac{5}{3}$

   $\tan \theta = \frac{4}{3}$   $\cot \theta = \frac{3}{4}$

   (b) $x = -12, y = -5, r = 13$

   $\sin \theta = -\frac{5}{13}$   $\csc \theta = -\frac{13}{5}$

   $\cos \theta = -\frac{12}{13}$   $\sec \theta = -\frac{13}{12}$

   $\tan \theta = \frac{5}{12}$   $\cot \theta = \frac{12}{5}$

**13.** (a) $\sin \theta < 0 \Rightarrow \theta$ is in Quadrant III or IV.

   $\cos \theta < 0 \Rightarrow \theta$ is in Quadrant II or III.

   $\sin \theta < 0$ **and** $\cos \theta < 0 \Rightarrow \theta$ is in Quadrant III.

   (b) $\sec \theta > 0 \Rightarrow \theta$ is in Quadrant I or IV.

   $\cot \theta < 0 \Rightarrow \theta$ is in Quadrant II or IV.

   $\sec \theta > 0$ **and** $\cot \theta < 0 \Rightarrow \theta$ is in Quadrant IV.

**15.** $x^2 + 1^2 = 2^2 \Rightarrow x = \sqrt{3}$

   $\cos \theta = \dfrac{x}{2} = \dfrac{\sqrt{3}}{2}$

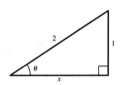

**17.** $4^2 + y^2 = 5^2 \Rightarrow y = 3$

   $\cot \theta = \dfrac{4}{y} = \dfrac{4}{3}$

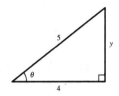

**19.** (a) $\sin 60° = \dfrac{\sqrt{3}}{2}$

   $\cos 60° = \dfrac{1}{2}$

   $\tan 60° = \sqrt{3}$

   (b) $\sin 120° = \sin 60° = \dfrac{\sqrt{3}}{2}$

   $\cos 120° = -\cos 60° = -\dfrac{1}{2}$

   $\tan 120° = -\tan 60° = -\sqrt{3}$

   (c) $\sin \dfrac{\pi}{4} = \dfrac{\sqrt{2}}{2}$

   $\cos \dfrac{\pi}{4} = \dfrac{\sqrt{2}}{2}$

   $\tan \dfrac{\pi}{4} = 1$

   (d) $\sin \dfrac{5\pi}{4} - \sin \dfrac{\pi}{4} = -\dfrac{\sqrt{2}}{2}$

   $\cos \dfrac{5\pi}{4} = \cos \dfrac{\pi}{4} = -\dfrac{\sqrt{2}}{2}$

   $\tan \dfrac{5\pi}{4} = \tan \dfrac{\pi}{4} = 1$

**21.** (a) $\sin 225° = -\sin 45° = -\dfrac{\sqrt{2}}{2}$

$\cos 225° = -\cos 45° = -\dfrac{\sqrt{2}}{2}$

$\tan 225° = \tan 45° = 1$

(b) $\sin(-225°) = \sin 45° = \dfrac{\sqrt{2}}{2}$

$\cos(-225°) = -\cos 45° = -\dfrac{\sqrt{2}}{2}$

$\tan(-225°) = -\tan 45° = -1$

(c) $\sin\dfrac{5\pi}{3} = -\sin\dfrac{\pi}{3} = -\dfrac{\sqrt{3}}{2}$

$\cos\dfrac{5\pi}{3} = \cos\dfrac{\pi}{3} = \dfrac{1}{2}$

$\tan\dfrac{5\pi}{3} = -\tan\dfrac{\pi}{3} = -\sqrt{3}$

(d) $\sin\dfrac{11\pi}{6} = -\sin\dfrac{\pi}{6} = -\dfrac{1}{2}$

$\cos\dfrac{11\pi}{6} = \cos\dfrac{\pi}{6} = \dfrac{\sqrt{3}}{2}$

$\tan\dfrac{11\pi}{6} = -\tan\dfrac{\pi}{6} = -\dfrac{\sqrt{3}}{3}$

**23.** (a) $\sin 10° \approx 0.1736$

(b) $\csc 10° \approx 5.759$

**25.** (a) $\tan\dfrac{\pi}{9} \approx 0.3640$

(b) $\tan\dfrac{10\pi}{9} \approx 0.3640$

**27.** (a) $\cos\theta = \dfrac{\sqrt{2}}{2}$

$\theta = \dfrac{\pi}{4}, \dfrac{7\pi}{4}$

(b) $\cos\theta = -\dfrac{\sqrt{2}}{2}$

$\theta = \dfrac{3\pi}{4}, \dfrac{5\pi}{4}$

**29.** (a) $\tan\theta = 1$

$\theta = \dfrac{\pi}{4}, \dfrac{5\pi}{4}$

(b) $\cot\theta = -\sqrt{3}$

$\theta = \dfrac{5\pi}{6}, \dfrac{11\pi}{6}$

**31.** $2\sin^2\theta = 1$

$\sin\theta = \pm\dfrac{\sqrt{2}}{2}$

$\theta = \dfrac{\pi}{4}, \dfrac{3\pi}{4}, \dfrac{5\pi}{4}, \dfrac{7\pi}{4}$

**33.** $\tan^2\theta = \tan\theta = 0$

$\tan\theta(\tan\theta - 1) = 0$

$\tan\theta = 0 \qquad\qquad \tan\theta = 1$

$\theta = 0, \pi \qquad\qquad \theta = \dfrac{\pi}{4}, \dfrac{5\pi}{4}$

**35.** $\sec\theta\csc\theta - 2\csc\theta = 0$

$\csc\theta(\sec\theta - 2) = 0$

$\left(\csc\theta \neq 0 \text{ for any value of } \theta\right)$

$\sec\theta = 2$

$\theta = \dfrac{\pi}{3}, \dfrac{5\pi}{3}$

**37.** $\cos^2\theta + \sin\theta = 1$

$1 - \sin^2\theta + \sin\theta = 1$

$\sin^2\theta - \sin\theta = 0$

$\sin\theta(\sin\theta - 1) = 0$

$\sin\theta = 0 \qquad\qquad \sin\theta = 1$

$\theta = 0, \pi \qquad\qquad \theta = \dfrac{\pi}{2}$

**39.** $(275 \text{ ft/sec})(60 \text{ sec}) = 16{,}500 \text{ feet}$

$\sin 18° = \dfrac{a}{16{,}500}$

$a = 16{,}500\sin 18° \approx 5099 \text{ feet}$

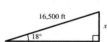

**41.** (a) Period: $\pi$

Amplitude: 2

(b) Period: 2

Amplitude: $\frac{1}{2}$

**43.** Period: $\frac{1}{2}$

Amplitude: 3

**45.** Period: $\dfrac{\pi}{2}$

**47.** Period: $\dfrac{2\pi}{5}$

**49.** (a)  $f(x) = c \sin x$; changing $c$ changes the amplitude.

When $c = -2$: $f(x) = -2 \sin x$.

When $c = -1$: $f(x) = -\sin x$.

When $c = 1$: $f(x) = \sin x$.

When $c = 2$: $f(x) = 2 \sin x$.

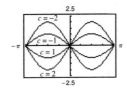

(b)  $f(x) = \cos(cx)$; changing $c$ changes the period.

When $c = -2$: $f(x) = \cos(-2x) = \cos 2x$.

When $c = -1$: $f(x) = \cos(-x) = \cos x$.

When $c = 1$: $f(x) = \cos x$.

When $c = 2$: $f(x) = \cos 2x$.

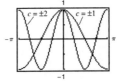

(c)  $f(x) = \cos(\pi x - c)$; changing $c$ causes a horizontal shift.

When $c = -2$: $f(x) = \cos(\pi x + 2)$.

When $c = -1$: $f(x) = \cos(\pi x + 1)$.

When $c = 1$: $f(x) = \cos(\pi x - 1)$.

When $c = 2$: $f(x) = \cos(\pi x - 2)$.

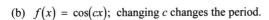

**51.** $y = \sin \dfrac{x}{2}$

Period: $4\pi$

Amplitude: 1

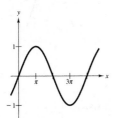

**53.** $y = -\sin \dfrac{2\pi x}{3}$

Period: 3

Amplitude: 1

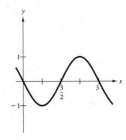

**55.** $y = \csc \dfrac{x}{2}$

Period: $4\pi$

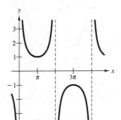

**57.** $y = 2 \sec 2x$

Period: $\pi$

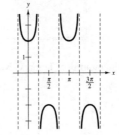

**59.** $y = \sin(x + \pi)$

Period: $2\pi$

Amplitude: 1

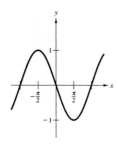

**61.** $y = 1 + \cos\left(x - \dfrac{\pi}{2}\right)$

Period: $2\pi$

Amplitude: 1

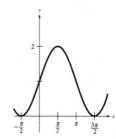

**63.** $y = a\cos(bx - c)$

From the graph, we see that the amplitude is 3, the period is $4\pi$, and the horizontal shift is $\pi$. Thus,

$$a = 3$$

$$\frac{2\pi}{b} = 4\pi \Rightarrow b = \frac{1}{2}$$

$$\frac{c}{d} = \pi \Rightarrow c = \frac{\pi}{2}.$$

Therefore, $y = 3\cos\left[(1/2)x - (\pi/2)\right]$.

**65.** $f(x) = \sin x$

$g(x) = |\sin x|$

$h(x) = \sin|x|$

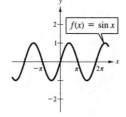 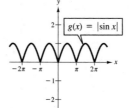 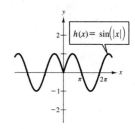

The graph of $\left|f(x)\right|$ will reflect any parts of the graph of $f(x)$ below the $x$-axis about the $y$-axis.

The graph of $f(|x|)$ will reflect the part of the graph of $f(x)$ to the right of the $y$-axis about the $y$-axis.

**67.** $S = 58.3 + 32.5\cos\dfrac{\pi t}{6}$

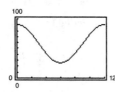

Sales exceed 75,000 during the months of January, November, and December.

**69.** $f(x) = \dfrac{4}{\pi}\left(\sin \pi x + \dfrac{1}{3}\sin 3\pi x\right)$

$g(x) = \dfrac{4}{\pi}\left(\sin \pi x + \dfrac{1}{3}\sin 3\pi x + \dfrac{1}{5}\sin 5\pi x\right)$

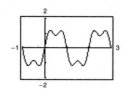

Pattern: $f(x) = \dfrac{4}{\pi}\left(\sin \pi x + \dfrac{1}{3}\sin 3\pi x + \dfrac{1}{5}\sin 5\pi x + \cdots + \dfrac{1}{2n-1}\sin(2n-1)\pi x\right)$, $n = 1, 2, 3\ldots$